STUDENT SOLUTIONS MANUAL

DANIEL S. MILLER
Niagara County Community College

INTRODUCTORY ALGEBRA
FOR COLLEGE STUDENTS
SIXTH EDITION

Robert Blitzer
Miami Dade College

PEARSON

Boston Columbus Indianapolis New York San Francisco Upper Saddle River
Amsterdam Cape Town Dubai London Madrid Milan Munich Paris Montreal Toronto
Delhi Mexico City Sao Paulo Sydney Hong Kong Seoul Singapore Taipei Tokyo

ISBN-13: 978-0-321-75980-1
ISBN-10: 0-321-75980-X

3 4 5 6 V031 17 16 15 14

www.pearsonhighered.com

PEARSON

TABLE OF CONTENTS for STUDENT SOLUTIONS

INTRODUCTORY ALGEBRA FOR COLLEGE STUDENTS 6E

Chapter 1
Variables, Real Numbers, and Mathematical Models

1.1 Check Points

1. **a.** $6 + 2x = 6 + 2(10) = 26$

   b. $2(x + 6) = 2(10 + 6) = 32$

2. **a.** $7x + 2y = 7 \cdot 3 + 2 \cdot 8 = 21 + 16 = 37$

   b. $\dfrac{6x - y}{2y - x - 8} = \dfrac{6 \cdot 3 - 8}{2 \cdot 8 - 3 - 8} = \dfrac{10}{5} = 2$

3. **a.** $6x$

   b. $4 + x$

   c. $3x + 5$

   d. $12 - 2x$

   e. $\dfrac{15}{x}$

4. **a.** $9x - 3 = 42$
      $9(6) - 3 = 42$
      $54 - 3 = 42$
      $\qquad 51 = 42, \text{ false}$
      6 is not a solution.

   b. $2(y + 3) = 5y - 3$
      $2(3 + 3) = 5(3) - 3$
      $\quad 2(6) = 15 - 3$
      $\qquad 12 = 12, \text{ true}$
      3 is a solution.

5. **a.** $\dfrac{x}{6} = 5$

   b. $7 - 2x = 1$

6. **a.** $d = 4n + 5$
      $d = 4(15) + 5 = 65$
      65% of marriages end in divorce after 15 years when the wife is under 18 at the time of marriage.

   b. According to the line graph, 60% of marriages end in divorce after 15 years when the wife is under 18 at the time of marriage.

   c. The mathematical model overestimates the actual percentage shown in the graph by 5%.

1.1 Concept and Vocabulary Check

1. variable

2. expression

3. substituting; evaluating

4. equation; solution

5. formula; modeling; models

1.1 Exercise Set

1. $x + 8 = 4 + 8 = 12$

3. $12 - x = 12 - 4 = 8$

5. $5x = 5 \cdot 4 = 20$

7. $\dfrac{28}{x} = \dfrac{28}{4} = 7$

9. $5 + 3x = 5 + 3 \cdot 4 = 5 + 12 = 17$

11. $2(x + 5) = 2(4 + 5) = 2(9) = 18$

13. $\dfrac{12x - 8}{2x} = \dfrac{12 \cdot 4 - 8}{2 \cdot 4} = \dfrac{48 - 8}{8} = \dfrac{40}{8} = 5$

15. $2x + y = 2 \cdot 7 + 5 = 14 + 5 = 19$

17. $2(x + y) = 2(7 + 5) = 2(12) = 24$

19. $4x - 3y = 4 \cdot 7 - 3 \cdot 5 = 28 - 15 = 13$

21. $\dfrac{21}{x} + \dfrac{35}{y} = \dfrac{21}{7} + \dfrac{35}{5} = 3 + 7 = 10$

23. $\dfrac{2x - y + 6}{2y - x} = \dfrac{2 \cdot 7 - 5 + 6}{2 \cdot 5 - 7} = \dfrac{14 - 5 + 6}{10 - 7} = \dfrac{15}{3} = 5$

25. $x + 4$

27. $x - 4$

29. $x + 4$

31. $x - 9$

33. $9 - x$

35. $3x - 5$

37. $12x - 1$

39. $\dfrac{10}{x} + \dfrac{x}{10}$

41. $\dfrac{x}{30} + 6$

43. $x + 14 = 20$
$6 + 14 = 20$
$\quad 20 = 20, \text{ true}$
The number is a solution.

45. $\quad 30 - y = 10$
$30 - 20 = 10$
$\quad\quad 10 = 10, \text{ true}$
The number is a solution.

47. $\quad 4z = 20$
$4(10) = 20$
$\quad\quad 40 = 20, \text{ false}$
The number is not a solution.

49. $\dfrac{r}{6} = 8$
$\dfrac{48}{6} = 8$
$\quad 8 = 8, \text{ true}$
The number is a solution.

51. $\quad 4m + 3 = 23$
$4(6) + 3 = 23$
$\quad 24 + 3 = 23$
$\quad\quad 27 = 23, \text{ false}$
The number is not a solution.

53. $\quad 5a - 4 = 2a + 5$
$5(3) - 4 = 2(3) + 5$
$\quad 15 - 4 = 6 + 5$
$\quad\quad 11 = 11, \text{ true}$
The number is a solution.

55. $6(p - 4) = 3p$
$6(8 - 4) = 3(8)$
$\quad 6(4) = 24$
$\quad\quad 24 = 24, \text{ true}$
The number is a solution.

57. $2(w + 1) = 3(w - 1)$
$2(7 + 1) = 3(7 - 1)$
$\quad 2(8) = 3(6)$
$\quad\quad 16 = 18, \text{ false}$
The number is not a solution.

59. $4x = 28$

61. $\dfrac{14}{x} = \dfrac{1}{2}$

63. $20 - x = 5$

65. $2x + 6 = 16$

67. $3x - 5 = 7$

69. $4x + 5 = 33$

71. $4(x + 5) = 33$

73. $5x = 24 - x$

75. First find x.
$x = 7y + 2$
$x = 7(5) + 2 = 37$
Evaluate the expression.
$\dfrac{x - y}{4} = \dfrac{37 - 5}{4} = \dfrac{32}{4} = 8$

77. First find x.
$x = \dfrac{y}{4} - 1$
$x = \dfrac{12}{4} - 1 = 3 - 1 = 2$
Evaluate the expression.
$4x + 3(y + 5) = 4(2) + 3(12 + 5)$
$\quad\quad\quad\quad\quad = 8 + 3(17)$
$\quad\quad\quad\quad\quad = 8 + 51$
$\quad\quad\quad\quad\quad = 59$

79. a. $2(x + 3y) = 2(4 + 3 \cdot 1) = 2(7) = 14$

b. $\quad 5z - 30 = 40$
$5(14) - 30 = 40$
$\quad 70 - 30 = 40$
$\quad\quad\quad 40 = 40, \text{ true}$
Yes, it is a solution.

81. a. $6x - 2y = 6 \cdot 3 - 2 \cdot 6 = 18 - 12 = 6$

b. $7w = 45 - 2w$
$7(6) = 45 - 2(6)$
$42 = 45 - 12$
$42 = 33$, false
No, it is not a solution.

83. a. $T = 0.15n + 2.72$
$T = 0.15(10) + 2.72$
$= 4.22$
According to the formula, the average price in 1990 was $4.22.
The model underestimates the actual amount shown in the bar graph by $0.01.

b. $T = 0.15n + 2.72$
$T = 0.15(30) + 2.72$
$= 7.22$
According to the formula, the average price in 2010 was $7.22.
The bar graph shows a price in 2010 of $7.85.
$7.85 - $7.22 = $0.63 $
The model underestimates the actual amount shown in the bar graph by $0.63.

85. $p = 52 - 0.3a$
$p = 52 - 0.3(24)$
$= 44.8$
According to the formula, 44.8% of 24-year-olds say that marriage is obsolete.
The model overestimates the actual percentage shown in the bar graph by 0.8%.

87. a. $H = 0.8(200 - A)$
$H = 0.8(200 - 145)$
$= 44$
A bowler with an average score of 145 will have a handicap of 44.

b. The bowler's final score will be $120 + 44$, or 164.

89. – 97. Answers will vary.

99. makes sense

101. makes sense

103. true

105. true

107. Choices of variables may vary.
Let $h =$ hours worked.
Let $s =$ salary.
$s = 20h$

109. $\dfrac{3}{7} \cdot \dfrac{2}{5} = \dfrac{3 \cdot 2}{7 \cdot 5} = \dfrac{6}{35}$

110. $\dfrac{2}{3} \div \dfrac{7}{5} = \dfrac{2}{3} \cdot \dfrac{5}{7} = \dfrac{2 \cdot 5}{3 \cdot 7} = \dfrac{10}{21}$

111. $\dfrac{9}{17} - \dfrac{5}{17} = \dfrac{9 - 5}{17} = \dfrac{4}{17}$

1.2 Check Points

1. $2\dfrac{5}{8} = \dfrac{2 \cdot 8 + 5}{8} = \dfrac{16 + 5}{8} = \dfrac{21}{8}$

2. 5 divided by 3 is 1 with a remainder of 2, so $\dfrac{5}{3} = 1\dfrac{2}{3}$.

3. Begin by selecting any two numbers whose product is 36.
Here is one possibility: $36 = 4 \cdot 9$
Because the factors 4 and 9 are not prime, factor each of these composite numbers.
$36 = 4 \cdot 9$
$\quad = 2 \cdot 2 \cdot 3 \cdot 3$
Notice that 2 and 3 are both prime. The prime factorization of 36 is $2 \cdot 2 \cdot 3 \cdot 3$.

4. a. $\dfrac{10}{15} = \dfrac{2 \cdot \cancel{5}}{3 \cdot \cancel{5}} = \dfrac{2}{3}$

b. $\dfrac{42}{24} = \dfrac{\cancel{2} \cdot \cancel{3} \cdot 7}{2 \cdot 2 \cdot \cancel{2} \cdot \cancel{3}} = \dfrac{7}{4}$

When reducing fractions, it may not be necessary to write prime factorizations. We can use the greatest common factor to reduce this fraction.
$\dfrac{42}{24} = \dfrac{7 \cdot \cancel{6}}{4 \cdot \cancel{6}} = \dfrac{7}{4}$

c. $\dfrac{13}{15}$; Because 13 and 15 share no common factors (other than 1), $\dfrac{13}{15}$ is already reduced to its lowest terms.

d. $\dfrac{9}{45} = \dfrac{1 \cdot \cancel{9}}{5 \cdot \cancel{9}} = \dfrac{1}{5}$

5. a. $\dfrac{4}{11} \cdot \dfrac{2}{3} = \dfrac{4 \cdot 2}{11 \cdot 3} = \dfrac{8}{33}$

b. $6 \cdot \dfrac{3}{5} = \dfrac{6}{1} \cdot \dfrac{3}{5} = \dfrac{18}{5} = 3\dfrac{3}{5}$

c. $\dfrac{3}{7} \cdot \dfrac{2}{3} = \dfrac{3 \cdot 2}{7 \cdot 3} = \dfrac{6}{21} = \dfrac{2 \cdot \cancel{3}}{7 \cdot \cancel{3}} = \dfrac{2}{7}$

Remember that you can divide numerators and denominators by common factors before performing multiplication.

$\dfrac{3}{7} \cdot \dfrac{2}{3} = \dfrac{\cancel{3}}{7} \cdot \dfrac{2}{\cancel{3}} = \dfrac{2}{7}$

d. $\left(3\dfrac{2}{5}\right)\left(1\dfrac{1}{2}\right) = \dfrac{17}{5} \cdot \dfrac{3}{2} = \dfrac{51}{10} = 5\dfrac{1}{10}$

6. a. $\dfrac{5}{4} \div \dfrac{3}{8} = \dfrac{5}{4} \cdot \dfrac{8}{3} = \dfrac{5}{\cancel{4}} \cdot \dfrac{\cancel{4} \cdot 2}{3} = \dfrac{10}{3} = 3\dfrac{1}{3}$

b. $\dfrac{2}{3} \div 3 = \dfrac{2}{3} \div \dfrac{3}{1} = \dfrac{2}{3} \cdot \dfrac{1}{3} = \dfrac{2}{9}$

c.

$3\dfrac{3}{8} \div 2\dfrac{1}{4} = \dfrac{27}{8} \div \dfrac{9}{4} = \dfrac{27}{8} \cdot \dfrac{4}{9} = \dfrac{\cancel{9} \cdot 3}{\cancel{4} \cdot 2} \cdot \dfrac{\cancel{4}}{\cancel{9}} = \dfrac{3}{2} = 1\dfrac{1}{2}$

7. a. $\dfrac{2}{11} + \dfrac{3}{11} = \dfrac{5}{11}$

b. $\dfrac{5}{6} - \dfrac{1}{6} = \dfrac{4}{6} = \dfrac{2}{3}$

c. $3\dfrac{3}{8} - 1\dfrac{1}{8} = \dfrac{27}{8} - \dfrac{9}{8} = \dfrac{18}{8} = \dfrac{9}{4} = 2\dfrac{1}{4}$

8. $\dfrac{2}{3} = \dfrac{2 \cdot 7}{3 \cdot 7} = \dfrac{14}{21}$

9. a. $\dfrac{1}{2} + \dfrac{3}{5} = \dfrac{1 \cdot 5}{2 \cdot 5} + \dfrac{3 \cdot 2}{5 \cdot 2} = \dfrac{5}{10} + \dfrac{6}{10} = \dfrac{11}{10}$

b. $\dfrac{4}{3} - \dfrac{3}{4} = \dfrac{4 \cdot 4}{3 \cdot 4} - \dfrac{3 \cdot 3}{4 \cdot 3} = \dfrac{16}{12} - \dfrac{9}{12} = \dfrac{7}{12}$

c. $3\dfrac{1}{6} - 1\dfrac{11}{12} = \dfrac{19}{6} - \dfrac{23}{12} = \dfrac{19 \cdot 2}{6 \cdot 2} - \dfrac{23}{12}$

$= \dfrac{38}{12} - \dfrac{23}{12} = \dfrac{15}{12}$

$= \dfrac{5}{4} = 1\dfrac{1}{4}$

10. $10 = 2 \cdot 5$
$12 = 2 \cdot 2 \cdot 3$
$\text{LCD} = 2 \cdot 2 \cdot 3 \cdot 5 = 60$
$\dfrac{3}{10} + \dfrac{7}{12} = \dfrac{3 \cdot 6}{10 \cdot 6} + \dfrac{7 \cdot 5}{12 \cdot 5} = \dfrac{18}{60} + \dfrac{35}{60} = \dfrac{53}{60}$

11. a. $x - \dfrac{2}{9}x = 1$

$1\dfrac{2}{7} - \dfrac{2}{9}\left(1\dfrac{2}{7}\right) = 1$

$\dfrac{9}{7} - \dfrac{2}{9}\left(\dfrac{9}{7}\right) = 1$

$\dfrac{9}{7} - \dfrac{2}{7} = 1$

$\dfrac{7}{7} = 1$

$1 = 1, \text{ true}$

The given fraction is a solution.

b. $\dfrac{1}{5} - w = \dfrac{1}{3}w$

$\dfrac{1}{5} - \dfrac{3}{20} = \dfrac{1}{3}\left(\dfrac{3}{20}\right)$

$\dfrac{4}{20} - \dfrac{3}{20} = \dfrac{1}{20}$

$\dfrac{1}{20} = \dfrac{1}{20}, \text{ true}$

The given fraction is a solution.

12. a. $\dfrac{2}{3}(x - 6)$

b. $\dfrac{3}{4}x - 2 = \dfrac{1}{5}x$

13. $C = \dfrac{5}{9}(F - 32)$

$C = \dfrac{5}{9}(77 - 32) = \dfrac{5}{9}(45) = 25$

$77°F$ is equivalent to $25°C$.

1.2 Concept and Vocabulary Check

1. numerator; denominator

2. mixed; improper

3. 5; 3; 2; 5

4. natural

5. prime

6. factors; product

7. $\dfrac{a}{b}$

8. $\dfrac{a \cdot c}{b \cdot d}$

9. reciprocals

10. $\dfrac{d}{c}$

11. $\dfrac{a + c}{b}$

12. least common denominator

1.2 Exercise Set

1. $2\dfrac{3}{8} = \dfrac{2 \cdot 8 + 3}{8} = \dfrac{16 + 3}{8} = \dfrac{19}{8}$

3. $7\dfrac{3}{5} = \dfrac{7 \cdot 5 + 3}{5} = \dfrac{35 + 3}{5} = \dfrac{38}{5}$

5. $8\dfrac{7}{16} = \dfrac{8 \cdot 16 + 7}{16} = \dfrac{128 + 7}{16} = \dfrac{135}{16}$

7. 23 divided by 5 is 4 with a remainder of 3, so $\dfrac{23}{5} = 4\dfrac{3}{5}$.

9. 76 divided by 9 is 8 with a remainder of 4, so $\dfrac{76}{9} = 8\dfrac{4}{9}$.

11. 711 divided by 20 is 35 with a remainder of 11, so $\dfrac{711}{20} = 35\dfrac{11}{20}$.

13. composite; $22 = 2 \cdot 11$

15. composite; $20 = 4 \cdot 5 = 2 \cdot 2 \cdot 5$

17. 37 has no factors other than 1 and 37, so 37 is prime.

19. composite; $36 = 4 \cdot 9 = 2 \cdot 2 \cdot 3 \cdot 3$

21. composite; $140 = 10 \cdot 14 = 2 \cdot 5 \cdot 2 \cdot 7$
$= 2 \cdot 2 \cdot 5 \cdot 7$

23. 79 has no factors other than 1 and 79, so 79 is prime.

25. composite; $81 = 9 \cdot 9 = 3 \cdot 3 \cdot 3 \cdot 3$

27. composite; $240 = 10 \cdot 24$
$= 2 \cdot 5 \cdot 2 \cdot 12$
$= 2 \cdot 5 \cdot 2 \cdot 3 \cdot 4$
$= 2 \cdot 5 \cdot 2 \cdot 3 \cdot 2 \cdot 2$
$= 2 \cdot 2 \cdot 2 \cdot 2 \cdot 3 \cdot 5$

29. $\dfrac{10}{16} = \dfrac{\cancel{2} \cdot 5}{\cancel{2} \cdot 8} = \dfrac{5}{8}$

31. $\dfrac{15}{18} = \dfrac{\cancel{3} \cdot 5}{\cancel{3} \cdot 6} = \dfrac{5}{6}$

33. $\dfrac{35}{50} = \dfrac{\cancel{5} \cdot 7}{\cancel{5} \cdot 10} = \dfrac{7}{10}$

35. $\dfrac{32}{80} = \dfrac{\cancel{16} \cdot 2}{\cancel{16} \cdot 5} = \dfrac{2}{5}$

37. $\dfrac{44}{50} = \dfrac{\cancel{2} \cdot 22}{\cancel{2} \cdot 25} = \dfrac{22}{25}$

39. $\dfrac{120}{86} = \dfrac{\cancel{2} \cdot 60}{\cancel{2} \cdot 43} = \dfrac{60}{43}$

41. $\dfrac{2}{5} \cdot \dfrac{1}{3} = \dfrac{2 \cdot 1}{5 \cdot 3} = \dfrac{2}{15}$

43. $\dfrac{3}{8} \cdot \dfrac{7}{11} = \dfrac{3 \cdot 7}{8 \cdot 11} = \dfrac{21}{88}$

45. $9 \cdot \dfrac{4}{7} = \dfrac{9}{1} \cdot \dfrac{4}{7} = \dfrac{9 \cdot 4}{1 \cdot 7} = \dfrac{36}{7}$ or $5\dfrac{1}{7}$

47. $\dfrac{1}{10} \cdot \dfrac{5}{6} = \dfrac{1 \cdot 5}{10 \cdot 6} = \dfrac{5}{60} = \dfrac{5 \cdot 1}{5 \cdot 12} = \dfrac{1}{12}$

49. $\dfrac{5}{4} \cdot \dfrac{6}{7} = \dfrac{5 \cdot 6}{4 \cdot 7} = \dfrac{30}{28} = \dfrac{2 \cdot 15}{2 \cdot 14} = \dfrac{15}{14}$ or $1\dfrac{1}{14}$

51. $\left(3\dfrac{3}{4}\right)\left(1\dfrac{3}{5}\right) = \dfrac{15}{4} \cdot \dfrac{8}{5} = \dfrac{120}{20} = \dfrac{20 \cdot 6}{20 \cdot 1} = 6$

53. $\dfrac{5}{4} \div \dfrac{4}{3} = \dfrac{5}{4} \cdot \dfrac{3}{4} = \dfrac{5 \cdot 3}{4 \cdot 4} = \dfrac{15}{16}$

55. $\dfrac{18}{5} \div 2 = \dfrac{18}{5} \cdot \dfrac{1}{2}$

$= \dfrac{18 \cdot 1}{5 \cdot 2} = \dfrac{18}{10} = \dfrac{2 \cdot 9}{2 \cdot 5} = \dfrac{9}{5}$ or $1\dfrac{4}{5}$

57. $2 \div \dfrac{18}{5} = \dfrac{2}{1} \cdot \dfrac{5}{18} = \dfrac{10}{18} = \dfrac{2 \cdot 5}{2 \cdot 9} = \dfrac{5}{9}$

59. $\dfrac{3}{4} \div \dfrac{1}{4} = \dfrac{3}{4} \cdot \dfrac{4}{1} = \dfrac{3 \cdot 4}{4 \cdot 1} = \dfrac{12}{4} = 3$

61. $\dfrac{7}{6} \div \dfrac{5}{3} = \dfrac{7}{6} \cdot \dfrac{3}{5} = \dfrac{7 \cdot 3}{6 \cdot 5} = \dfrac{21}{30} = \dfrac{3 \cdot 7}{3 \cdot 10} = \dfrac{7}{10}$

63. $\dfrac{1}{14} \div \dfrac{1}{7} = \dfrac{1}{14} \cdot \dfrac{7}{1} = \dfrac{7}{14} = \dfrac{7 \cdot 1}{7 \cdot 2} = \dfrac{1}{2}$

65. $6\dfrac{3}{5} \div 1\dfrac{1}{10} = \dfrac{33}{5} \div \dfrac{11}{10}$

$= \dfrac{33}{5} \cdot \dfrac{10}{11} = \dfrac{11 \cdot 3}{5} \cdot \dfrac{5 \cdot 2}{11} = \dfrac{6}{1} = 6$

67. $\dfrac{2}{11} + \dfrac{4}{11} = \dfrac{2+4}{11} = \dfrac{6}{11}$

69. $\dfrac{7}{12} + \dfrac{1}{12} = \dfrac{8}{12} = \dfrac{4 \cdot 2}{4 \cdot 3} = \dfrac{2}{3}$

71. $\dfrac{5}{8} + \dfrac{5}{8} = \dfrac{10}{8} = \dfrac{2 \cdot 5}{2 \cdot 4} = \dfrac{5}{4}$ or $1\dfrac{1}{4}$

73. $\dfrac{7}{12} - \dfrac{5}{12} = \dfrac{2}{12} = \dfrac{2 \cdot 1}{2 \cdot 6} = \dfrac{1}{6}$

75. $\dfrac{16}{7} - \dfrac{2}{7} = \dfrac{14}{7} = \dfrac{7 \cdot 2}{7 \cdot 1} = 2$

77. $\dfrac{1}{2} + \dfrac{1}{5} = \dfrac{1}{2} \cdot \dfrac{5}{5} + \dfrac{1}{5} \cdot \dfrac{2}{2}$

$= \dfrac{5}{10} + \dfrac{2}{10} = \dfrac{5+2}{10} = \dfrac{7}{10}$

79. $\dfrac{3}{4} + \dfrac{3}{20} = \dfrac{3}{4} \cdot \dfrac{5}{5} + \dfrac{3}{20}$

$= \dfrac{15}{20} + \dfrac{3}{20}$

$= \dfrac{18}{20} = \dfrac{2 \cdot 9}{2 \cdot 10} = \dfrac{9}{10}$

81. $\dfrac{3}{8} + \dfrac{5}{12} = \dfrac{3}{8} \cdot \dfrac{3}{3} + \dfrac{5}{12} \cdot \dfrac{2}{2}$

$= \dfrac{9}{24} + \dfrac{10}{24} = \dfrac{19}{24}$

83. $\dfrac{11}{18} - \dfrac{2}{9} = \dfrac{11}{18} - \dfrac{2}{9} \cdot \dfrac{2}{2} = \dfrac{11}{18} - \dfrac{4}{18} = \dfrac{7}{18}$

85. $\dfrac{4}{3} - \dfrac{3}{4} = \dfrac{4}{3} \cdot \dfrac{4}{4} - \dfrac{3}{4} \cdot \dfrac{3}{3}$

$= \dfrac{16}{12} - \dfrac{9}{12} = \dfrac{7}{12}$

87. $\dfrac{7}{10} - \dfrac{3}{16} = \dfrac{7}{10} \cdot \dfrac{8}{8} - \dfrac{3}{16} \cdot \dfrac{5}{5}$

$= \dfrac{56}{80} - \dfrac{15}{80} = \dfrac{41}{80}$

89. $3\dfrac{3}{4} - 2\dfrac{1}{3} = \dfrac{15}{4} - \dfrac{7}{3}$

$= \dfrac{15}{4} \cdot \dfrac{3}{3} - \dfrac{7}{3} \cdot \dfrac{4}{4}$

$= \dfrac{45}{12} - \dfrac{28}{12} = \dfrac{17}{12}$ or $1\dfrac{5}{12}$

91.
$$\frac{7}{2}x = 28$$
$$\frac{7}{2}\cdot 8 = 28$$
$$\frac{7}{2}\cdot\frac{8}{1} = 28$$
$$\frac{7}{\cancel{2}}\cdot\frac{4\cdot\cancel{2}}{1} = 28$$
$$28 = 28,\ \text{true}$$
The given number is a solution.

93.
$$w - \frac{2}{3} = \frac{3}{4}$$
$$1\frac{5}{12} - \frac{2}{3} = \frac{3}{4}$$
$$\frac{17}{12} - \frac{2}{3} = \frac{3}{4}$$
$$\frac{17}{12} - \frac{8}{12} = \frac{3}{4}$$
$$\frac{9}{12} = \frac{3}{4}$$
$$\frac{3}{4} = \frac{3}{4},\ \text{true}$$
The given number is a solution.

95.
$$20 - \frac{1}{3}z = \frac{1}{2}z$$
$$20 - \frac{1}{3}\cdot 12 = \frac{1}{2}\cdot 12$$
$$20 - \frac{1}{3}\cdot\frac{12}{1} = \frac{1}{2}\cdot\frac{12}{1}$$
$$20 - 4 = 12$$
$$16 = 12,\ \text{false}$$
The given number is not a solution.

97.
$$\frac{2}{9}y + \frac{1}{3}y = \frac{3}{7}$$
$$\frac{2}{9}\cdot\frac{27}{35} + \frac{1}{3}\cdot\frac{27}{35} = \frac{3}{7}$$
$$\frac{2}{\cancel{9}}\cdot\frac{\cancel{9}\cdot 3}{35} + \frac{1}{\cancel{3}}\cdot\frac{\cancel{3}\cdot 9}{35} = \frac{3}{7}$$
$$\frac{6}{35} + \frac{9}{35} = \frac{3}{7}$$
$$\frac{15}{35} = \frac{3}{7}$$
$$\frac{3\cdot\cancel{5}}{7\cdot\cancel{5}} = \frac{3}{7}$$
$$\frac{3}{7} = \frac{3}{7},\ \text{true}$$
The given number is a solution.

99.
$$\frac{1}{3}(x-2) = \frac{1}{5}(x+4)$$
$$\frac{1}{3}(26-2) = \frac{1}{5}(26+4)$$
$$\frac{1}{3}(24) = \frac{1}{5}(30)$$
$$8 = 6,\ \text{false}$$
The given number is not a solution.

101.
$$(y \div 6) + \frac{2}{3} = (y \div 2) - \frac{7}{9}$$
$$\left(4\frac{1}{3} \div 6\right) + \frac{2}{3} = \left(4\frac{1}{3} \div 2\right) - \frac{7}{9}$$
$$\left(\frac{13}{3} \div \frac{6}{1}\right) + \frac{2}{3} = \left(\frac{13}{3} \div \frac{2}{1}\right) - \frac{7}{9}$$
$$\left(\frac{13}{3} \cdot \frac{1}{6}\right) + \frac{2}{3} = \left(\frac{13}{3} \cdot \frac{1}{2}\right) - \frac{7}{9}$$
$$\frac{13}{18} + \frac{2}{3} = \frac{13}{6} - \frac{7}{9}$$
$$\frac{13}{18} + \frac{12}{18} = \frac{39}{18} - \frac{14}{18}$$
$$\frac{25}{18} = \frac{25}{18},\ \text{true}$$
The given number is a solution.

103. $\dfrac{1}{5}x$

105. $x - \dfrac{1}{4}x$

107. $x - \dfrac{1}{4} = \dfrac{1}{2}x$

109. $\dfrac{1}{7}x + \dfrac{1}{8}x = 12$

111. $\dfrac{2}{3}(x+6)$

113. $\dfrac{2}{3}x + 6 = x - 3$

115. $\dfrac{3}{4} \cdot \dfrac{a}{5} = \dfrac{3 \cdot a}{4 \cdot 5} = \dfrac{3a}{20}$

117. $\dfrac{11}{x} + \dfrac{9}{x} = \dfrac{11+9}{x} = \dfrac{20}{x}$

119. $\left(\dfrac{1}{2} - \dfrac{1}{3}\right) \div \dfrac{5}{8} = \left(\dfrac{3}{6} - \dfrac{2}{6}\right) \div \dfrac{5}{8}$

$\qquad\qquad = \dfrac{1}{6} \div \dfrac{5}{8}$

$\qquad\qquad = \dfrac{1}{6} \cdot \dfrac{8}{5} = \dfrac{8}{30} = \dfrac{\cancel{2} \cdot 4}{\cancel{2} \cdot 15} = \dfrac{4}{15}$

121. $\dfrac{1}{5}(x+2) = \dfrac{1}{2}\left(x - \dfrac{1}{5}\right)$

$\qquad \dfrac{1}{5}\left(\dfrac{5}{8} + 2\right) = \dfrac{1}{2}\left(\dfrac{5}{8} - \dfrac{1}{5}\right)$

$\qquad \dfrac{1}{5}\left(\dfrac{5}{8} + \dfrac{2}{1}\right) = \dfrac{1}{2}\left(\dfrac{5}{8} - \dfrac{1}{5}\right)$

$\qquad \dfrac{1}{5}\left(\dfrac{5}{8} + \dfrac{16}{8}\right) = \dfrac{1}{2}\left(\dfrac{25}{40} - \dfrac{8}{40}\right)$

$\qquad\quad \dfrac{1}{5}\left(\dfrac{21}{8}\right) = \dfrac{1}{2}\left(\dfrac{17}{40}\right)$

$\qquad\qquad \dfrac{21}{40} = \dfrac{17}{80}$

$\qquad\qquad \dfrac{42}{80} = \dfrac{17}{80},$ false

The given number is not a solution.

123. $C = \dfrac{5}{9}(F - 32)$

$\quad C = \dfrac{5}{9}(68 - 32) = \dfrac{5}{9}(36) = 20$

68°F is equivalent to 20°C.

125. a. $H = \dfrac{7}{10}(220 - a)$

$\qquad H = \dfrac{7}{10}(220 - 20)$

$\qquad\quad = \dfrac{7}{10}(200)$

$\qquad\quad = 140$

The lower limit of the heart rate for a 20-year-old with this exercise goal is 140 beats per minute.

b. $H = \dfrac{4}{5}(220 - a)$

$\qquad H = \dfrac{4}{5}(220 - 20)$

$\qquad\quad = \dfrac{4}{5}(200)$

$\qquad\quad = 160$

The upper limit of the heart rate for a 20-year-old with this exercise goal is 160 beats per minute.

127. a. $H = \dfrac{9}{10}(220 - a)$

b. $H = \dfrac{9}{10}(220 - a)$

$\qquad H = \dfrac{9}{10}(220 - 40)$

$\qquad\quad = \dfrac{9}{10}(180)$

$\qquad\quad = 162$

The heart rate for a 40-year-old with this exercise goal is 162 beats per minute.

129. 2008 is 1 year after 2007.

$\quad P = 23\dfrac{1}{5} - 2\dfrac{1}{5}n$

$\quad P = 23\dfrac{1}{5} - 2\dfrac{1}{5}(1)$

$\qquad = 21$

The model estimates that the average number of presents in 2008 was 21.
This underestimates the actual total shown in the bar graph by $\dfrac{1}{2}$ present.

131. – 139. Answers will vary.

141. makes sense

143. makes sense

145. false; Changes to make the statement true will vary.

A sample change is: $\frac{1}{2} + \frac{1}{5} = \frac{5}{10} + \frac{2}{10} = \frac{7}{10}$.

147. true

149.

150. 5

151. $2\frac{1}{2}$ or $\frac{5}{2}$

152. −4

1.3 Check Points

1. a. −500

  b. −282

2. a.

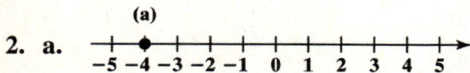

  b.

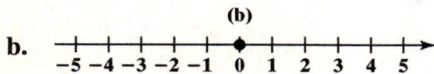

  c.

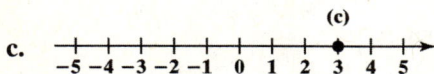

3. a.

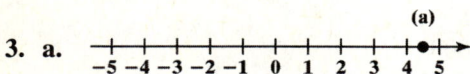

  b.

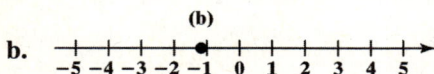

4. a.
$$8\overline{)3.000} = 0.375$$
$$\begin{array}{r} 0.375 \\ 8\overline{)3.000} \\ \underline{24} \\ 60 \\ \underline{56} \\ 40 \\ \underline{40} \\ 0 \end{array}$$

$\frac{3}{8} = 0.375$

b.
$$\begin{array}{r} 0.454... \\ 11\overline{)5.000...} \\ \underline{44} \\ 60 \\ \underline{55} \\ 50 \\ \underline{44} \\ 60 \end{array}$$

$\frac{5}{11} = 0.\overline{45}$

5. a. $\sqrt{9}$

  b. $0, \sqrt{9}$

  c. $-9, 0, \sqrt{9}$

  d. $-9, -1.3, 0, 0.\overline{3}, \sqrt{9}$

  e. $\frac{\pi}{2}, \sqrt{10}$

  f. $-9, -1.3, 0, 0.\overline{3}, \frac{\pi}{2}, \sqrt{9}, \sqrt{10}$

6. a. $14 > 5$ since 14 is to the right of 5 on the number line.

  b. $-5.4 < 2.3$ since -5.4 is to the left of 2.3 on the number line.

  c. $-19 < -6$ since -19 is to the left of -6 on the number line.

  d. $\frac{1}{4} < \frac{1}{2}$ since $\frac{1}{4}$ is to the left of $\frac{1}{2}$ on the number line.

7. a. $-2 \le 3$ is true because $-2 < 3$ is true.

  b. $-2 \ge -2$ is true because $-2 = -2$ is true.

  c. $-4 \ge 1$ is false because neither $-4 > 1$ nor $-4 = 1$ is true.

8. a. $|-4| = 4$

  b. $|6| = 6$

  c. $|-\sqrt{2}| = \sqrt{2}$

1.3 Concept and Vocabulary Check

1. natural

2. whole

3. integers

4. rational

5. irrational

6. rational; irrational

7. left

8. absolute value; $|a|$

1.3 Exercise Set

1. -20

3. 8

5. -3000

7. -4 billion

9. 2 is shown as a dot on the number line.

11. -5 is shown as a dot on the number line.

13. $3\frac{1}{2}$ is shown as a dot on the number line.

15. $\frac{11}{3}$ is shown as a dot on the number line.

17. -1.8 is shown as a dot on the number line.

19. $-\frac{16}{5}$ is shown as a dot on the number line.

21.
$$\begin{array}{r} 0.75 \\ 4\overline{)3.00} \\ \underline{28} \\ 20 \\ \underline{20} \\ 0 \end{array}$$

$\frac{3}{4}=0.75$

23.
$$\begin{array}{r} 0.35 \\ 20\overline{)7.00} \\ \underline{60} \\ 100 \\ \underline{100} \\ 0 \end{array}$$

$\frac{7}{20}=0.35$

25.
$$\begin{array}{r} 0.875 \\ 8\overline{)7.000} \\ \underline{64} \\ 60 \\ \underline{56} \\ 40 \\ \underline{40} \\ 0 \end{array}$$

$\frac{7}{8}=0.875$

27.
$$\begin{array}{r} 0.818... \\ 11\overline{)9.000...} \\ \underline{88} \\ 20 \\ \underline{11} \\ 90 \\ \underline{88} \\ 20 \end{array}$$

$\frac{9}{11}=0.\overline{81}$

29.

$$2\overline{)1.0}$$
$$\underline{1.0}$$
$$0$$

$$-\frac{1}{2} = -0.5$$

31.

$$6\overline{)5.000...}$$
$$\underline{48}$$
$$20$$
$$\underline{18}$$
$$20$$
$$\underline{18}$$
$$20$$

$$\frac{5}{6} = 0.8\overline{3}$$

33.
  a. $\sqrt{100}$ $(=10)$
  b. $0, \sqrt{100}$
  c. $-9, 0, \sqrt{100}$
  d. $-9, -\frac{4}{5}, 0, 0.25, 9.2, \sqrt{100}$
  e. $\sqrt{3}$
  f. $-9, -\frac{4}{5}, 0, 0.25, \sqrt{3}, 9.2, \sqrt{100}$

35.
  a. $\sqrt{64}$ $(=8)$
  b. $0, \sqrt{64}$
  c. $-11, 0, \sqrt{64}$
  d. $-11, -\frac{5}{6}, 0, 0.75, \sqrt{64}$
  e. $\sqrt{5}, \pi$
  f. $-11, -\frac{5}{6}, 0, 0.75, \sqrt{5}, \pi, \sqrt{64}$

37. The only whole number that is not a natural number is 0.

39. Answers will vary. As an example, one rational number that is not an integer is $\frac{1}{2}$.

41. Answers will vary. As an example, 6 is a number that is an integer, a whole number, and a natural number.

43. Answers will vary. As an example, one number that is an irrational number and a real number is π.

45. $\frac{1}{2} < 2$ since $\frac{1}{2}$ is to the left of 2 on the number line.

47. $3 > -\frac{5}{2}$ since 3 is to the right of $-\frac{5}{2} = -2\frac{1}{2}$.

49. $-4 > -6$ since -4 is to the right of -6.

51. $-2.5 < 1.5$ since -2.5 is to the left of 1.5.

53. $-\frac{3}{4} > -\frac{5}{4}$ since $-\frac{3}{4}$ is to the right of $-\frac{5}{4}$.

55. $-4.5 < 3$ since -4.5 is to the left of 3.

57. $\sqrt{2} < 1.5$ since $\sqrt{2} \approx 1.414$ is to the left of 1.5.

59. $0.\overline{3} > 0.3$ since $0.\overline{3} = 0.333...$ is to the right of 0.3.

61. $-\pi > -3.5$ since $-\pi \approx -3.14$ is to the right of -3.5.

63. $-5 \geq -13$ is true because $-5 > -13$ is true.

65. $-9 \geq -9$ is true because $-9 = -9$ is true.

67. $0 \geq -6$ is true because $0 > -6$ is true.

69. $-17 \geq 6$ is false because neither $-17 > 6$ nor $-17 = 6$ is true.

71. $|6| = 6$ because the distance between 6 and 0 on the number line is 6 units.

73. $|-7| = 7$ because the distance between -7 and 0 on the number line is 7 units.

75. $\left|\frac{5}{6}\right| = \frac{5}{6}$ because the distance between $\frac{5}{6}$ and 0 on the number line is $\frac{5}{6}$ units.

77. $\left|-\sqrt{11}\right| = \sqrt{11}$ because the distance between $-\sqrt{11}$ and 0 on the number line is $\sqrt{11}$ units.

79. $|-6| \ \square \ |-3|$

$\quad\quad 6 \ \square \ 3$

$\quad\quad 6 > 3$

$\quad$ Since $6 > 3$, $|-6| > |-3|$.

81. $\left|\dfrac{3}{5}\right| \ \square \ |-0.6|$

$\quad |0.6| \ \square \ |-0.6|$

$\quad\quad 0.6 \ \square \ 0.6$

$\quad\quad 0.6 = 0.6$

$\quad$ Since $0.6 = 0.6$, $\left|\dfrac{3}{5}\right| = |-0.6|$.

83. $\dfrac{30}{40} - \dfrac{3}{4} \ \square \ \dfrac{14}{15} \cdot \dfrac{15}{14}$

$\quad \dfrac{30}{40} - \dfrac{30}{40} \ \square \ \dfrac{14}{15} \cdot \dfrac{15}{14}$

$\quad\quad\quad 0 \ \square \ 1$

$\quad\quad\quad 0 < 1$

$\quad$ Since $0 < 1$, $\dfrac{30}{40} - \dfrac{3}{4} < \dfrac{14}{15} \cdot \dfrac{15}{14}$.

85. $\dfrac{8}{13} \div \dfrac{8}{13} \ \square \ |-1|$

$\quad \dfrac{8}{13} \cdot \dfrac{13}{8} \ \square \ 1$

$\quad\quad\quad 1 \ \square \ 1$

$\quad\quad\quad 1 = 1$

$\quad$ Since $1 = 1$, $\dfrac{8}{13} \div \dfrac{8}{13} = |-1|$.

87. rational numbers

89. integers

91. all real numbers

93. whole numbers

95. a.

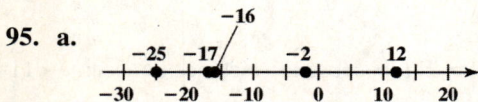

b. Rhode Island, Georgia, Louisiana, Florida, Hawaii

97. – 107. Answers will vary

109. does not make sense; Explanations will vary. Sample explanation: The Bismarck's resting place is lower because it is further below sea level.

111. makes sense

113. false; Changes to make the statement true will vary. A sample change is: All whole numbers are integers.

115. false; Changes to make the statement true will vary. A sample change is: Irrational numbers can be negative.

117. false; Changes to make the statement true will vary. A sample change is: All integers are rational numbers.

119. $-\dfrac{1}{2}d$

121. $-\sqrt{12} \approx -3.464$ and should be graphed between -4 and -3.

123. $2 - \sqrt{5} \approx -0.236$ and should be graphed between -1 and 0.

124. $3(x+5) = 3(4+5) = 3(9) = 27$

$\quad$ and

$\quad 3x + 15 = 3(4) + 15 = 12 + 15 = 27$

$\quad$ Both expressions have the same value.

125. $3x + 5x = 3(4) + 5(4) = 12 + 20 = 32$

$\quad$ and

$\quad 8x = 8(4) = 32$

$\quad$ Both expressions have the same value.

126. $9x - 2x = 9(4) - 2(4) = 36 - 8 = 28$

$\quad$ and

$\quad 7x = 7(4) = 28$

$\quad$ Both expressions have the same value.

1.4 Check Points

1. **a.** 3 terms

    b. 6

    c. 11

    d. $6x$ and $2x$

2. **a.** $x + 14 = 14 + x$

    b. $7y = y7$

3. **a.** $5x + 17 = 17 + 5x$

    b. $5x + 17 = x5 + 17$

4. **a.** $8 + (12 + x) = (8 + 12) + x$
    $$= 20 + x \text{ or } x + 20$$

    b. $6(5x) = (6 \cdot 5)x = 30x$

5. $8 + (x + 4) = 8 + (4 + x)$
    $$= (8 + 4) + x$$
    $$= 12 + x \text{ or } x + 12$$

6. $5(x + 3) = 5 \cdot x + 5 \cdot 3$
    $$= 5x + 15$$

7. $6(4y + 7) = 6 \cdot 4y + 6 \cdot 7$
    $$= 24y + 42$$

8. **a.** $7x + 3x = (7 + 3)x = 10x$

    b. $9a - 4a = (9 - 4)a = 5a$

9. **a.** $8x + 7 + 10x + 3 = (8x + 10x) + (7 + 3)$
    $$= 18x + 10$$

    b. $9x + 6y + 5x + 2y = (9x + 5x) + (6y + 2y)$
    $$= 14x + 8y$$

10. $7(2x + 3) + 11x = 7 \cdot 2x + 7 \cdot 3 + 11x$
    $$= 14x + 21 + 11x$$
    $$= (14x + 11x) + 21$$
    $$= 25x + 21$$

11. $7(4x + 3y) + 2(5x + y) = 7 \cdot 4x + 7 \cdot 3y + 2 \cdot 5x + 2 \cdot y$
    $$= 28x + 21y + 10x + 2y$$
    $$= (28x + 10x) + (21y + 2y)$$
    $$= 38x + 23y$$

1.4 Concept and Vocabulary Check

1. like

2. $b + a$

3. ab

4. $a + (b + c)$

5. $(ab)c$

6. $ab + ac$

7. simplified

1.4 Exercise Set

1. $3x + 5$

    a. 2 terms
    b. 3
    c. 5
    d. no like terms

3. $x + 2 + 5x$

    a. 3 terms
    b. 1
    c. 2
    d. x and $5x$ are like terms.

5. $4y + 1 + 3$

    a. 3 terms
    b. 4
    c. 1
    d. no like terms

7. $y + 4 = 4 + y$

9. $5 + 3x = 3x + 5$

11. $4x + 5y = 5y + 4x$

13. $5(x + 3) = 5(3 + x)$

15. $9x = x \cdot 9$ or $x9$

17. $x + y6 = x + 6y$

19. $7x + 23 = x7 + 23$

21. $5(x+3) = (x+3)5$

23. $7 + (5 + x) = (7 + 5) + x = 12 + x$

25. $7(4x) = (7 \cdot 4)x = 28x$

27. $3(x+5) = 3(x) + 3(5) = 3x + 15$

29. $8(2x+3) = 8(2x) + 8(3) = 16x + 24$

31. $\dfrac{1}{3}(12 + 6r) = \dfrac{1}{3}(12) + \dfrac{1}{3}(12) + \dfrac{1}{3}(6r)$
$\qquad = 4 + 2r$

33. $5(x + y) = 5x + 5y$

35. $3(x - 2) = 3(x) - 3(2) = 3x - 6$

37. $2(4x - 5) = 2(4x) - 2(5) = 8x - 10$

39. $\dfrac{1}{2}(5x - 12) = \dfrac{1}{2}(5x) + \dfrac{1}{2}(-12)$
$\qquad = \dfrac{5}{2}x - 6$

41. $(2x+7)4 = 2x(4) + 7(4) = 8x + 28$

43. $6(x + 3 + 2y) = 6(x) + 6(3) + 6(2y)$
$\qquad = 6x + 18 + 12y$

45. $5(3x - 2 + 4y) = 5(3x) - 5(2) + 5(4y)$
$\qquad = 15x - 10 + 20y$

47. $7x + 10x = (7 + 10)x = 17x$

49. $11a - 3a = (11 - 3)a = 8a$

51. $3 + (x + 11) = (3 + 11) + x = 14 + x$

53. $5y + 3 + 6y = (5y + 6y) + 3 = 11y + 3$

55. $2x + 5 + 7x - 4 = (2x + 7x) + (5 - 4)$
$\qquad = 9x + 1$

57. $11a + 12 + 3a + 2 = (11a + 3a) + (12 + 2)$
$\qquad = 14a + 14$

59. $5(3x + 2) - 4 = 15x + 10 - 4 = 15x + 6$

61. $12 + 5(3x - 2) = 12 + 15x - 10$
$\qquad = 15x + 12 - 10 = 15x + 2$

63. $7(3a + 2b) + 5(4a + 2b)$
$\qquad = 21a + 14b + 20a + 10b$
$\qquad = 21a + 20a + 14b + 10b$
$\qquad = 41a + 24b$

65. $7 + 2(x + 9)$
$\qquad = 7 + (2x + 18)$ Distributive Property
$\qquad = 7 + (18 + 2x)$ Commutative Property of Addition
$\qquad = (7 + 18) + 2x$ Associative Property of Addition
$\qquad = 25 + 2x$
$\qquad = 2x + 25$ Commutative Property of Addition

67. $7x + 2x$
$\qquad 7x + 2x = 9x$

69. $12x - 3x$
$\qquad 12x - 3x = 9x$

71. $6(4x)$
$\qquad 6(4x) = 24x$

73. $6(4 + x)$
$\qquad 6(4 + x) = 24 + 6x$

75. $8 + 5(x - 1)$
$\qquad 8 + 5(x - 1) = 8 + 5x - 5 = 5x + 3$

77. a. $U = 4(2n + 32) + 5(n + 1)$
$\qquad = 8n + 128 + 5n + 5$
$\qquad = 13n + 133$

b. $U = 13n + 133$
$\qquad = 13(6) + 133$
$\qquad = 211$
According to the formula, there were 211 million U.S. internet users in 2006. This overestimates the actual amount shown in the bar graph by 1 million.

79. – 87. Answers will vary.

89. does not make sense; Explanations will vary. Sample explanation: Subtraction does not have a commutative property.

91. makes sense

93. false; Changes to make the statement true will vary. A sample change is: $(24 \div 6) \div 2 \neq 24 \div (6 \div 2)$.

95. false; Changes to make the statement true will vary. A sample change is: Addition cannot be distributed over multiplication.

97. 60 because $150 - 90 = 60$

98. -60 because $-50 - 10 = -60$

99. -5 because $30 - 35 = -5$

Mid-Chapter Check Point – Chapter 1

1. $2 + 10x = 2 + 10(6)$
$= 2 + 60$
$= 62$

2. $10x - 4 = 10\left(\dfrac{3}{5}\right) - 4$
$= 6 - 4$
$= 2$

3. $\dfrac{xy}{2} + 4(y - x) = \dfrac{3 \cdot 10}{2} + 4(10 - 3)$
$= \dfrac{30}{2} + 4(7)$
$= 15 + 28$
$= 43$

4. $\dfrac{1}{4}x - 2$

5. $\dfrac{x}{6} + 5 = 19$

6. $3(x + 2) = 4x - 1$
$3(6 + 2) = 4 \cdot 6 - 1$
$3(8) = 24 - 1$
$24 = 23$, false
The number is not a solution.

7. $8y = 12\left(y - \dfrac{1}{2}\right)$
$8 \cdot \dfrac{3}{4} = 12\left(\dfrac{3}{4} - \dfrac{1}{2}\right)$
$6 = 12\left(\dfrac{3}{4} - \dfrac{2}{4}\right)$
$6 = 12\left(\dfrac{1}{4}\right)$
$6 = 3$, false
The number is not a solution.

8. a. $M = 12,624 - 235n$
$= 12,624 - 235(3)$
$= 11,919$
The average number of miles per passenger car in 2008 was 11,919 miles.
This overestimates the actual number shown in the bar graph by 131 miles.

b. $M = 12,624 - 235n$
$= 12,624 - 235(10)$
$= 10,274$
If trends continue, the number of miles per passenger car in 2015 will be 10,274 miles.

9. $\dfrac{7}{10} - \dfrac{8}{15} = \dfrac{7}{10} \cdot \dfrac{3}{3} - \dfrac{8}{15} \cdot \dfrac{2}{2}$
$= \dfrac{21}{30} - \dfrac{16}{30} = \dfrac{5}{30} = \dfrac{1}{6}$

10. $\dfrac{2}{3} \cdot \dfrac{3}{4} = \dfrac{2}{\cancel{3}} \cdot \dfrac{\cancel{3}}{4} = \dfrac{2}{4} = \dfrac{1}{2}$

11. $\dfrac{5}{22} + \dfrac{5}{33} = \dfrac{5}{22} \cdot \dfrac{3}{3} + \dfrac{5}{33} \cdot \dfrac{2}{2} = \dfrac{15}{66} + \dfrac{10}{66} = \dfrac{25}{66}$

12. $\dfrac{3}{5} \div \dfrac{9}{10} = \dfrac{3}{5} \cdot \dfrac{10}{9} = \dfrac{3}{5} \cdot \dfrac{2 \cdot 5}{3 \cdot 3} = \dfrac{\cancel{3}}{\cancel{5}} \cdot \dfrac{2 \cdot \cancel{5}}{\cancel{3} \cdot 3} = \dfrac{2}{3}$

13. $\dfrac{23}{105} - \dfrac{2}{105} = \dfrac{21}{105} = \dfrac{\cancel{3} \cdot \cancel{7}}{\cancel{3} \cdot 5 \cdot \cancel{7}} = \dfrac{1}{5}$

14. $2\dfrac{7}{9} \div 3 = \dfrac{25}{9} \div \dfrac{3}{1} = \dfrac{25}{9} \cdot \dfrac{1}{3} = \dfrac{25}{27}$

15. $5\dfrac{2}{9} - 3\dfrac{1}{6} = \dfrac{47}{9} - \dfrac{19}{6}$

$\qquad\qquad = \dfrac{47}{9} \cdot \dfrac{2}{2} - \dfrac{19}{6} \cdot \dfrac{3}{3}$

$\qquad\qquad = \dfrac{94}{18} - \dfrac{57}{18} = \dfrac{37}{18} \text{ or } 2\dfrac{1}{18}$

16. $C = \dfrac{5}{9}(F - 32)$

$C = \dfrac{5}{9}(50 - 32) = \dfrac{5}{9}(18) = 10$

50°F is equivalent to 10°C.

17. $-8000 < -8\dfrac{1}{4}$

18. $\dfrac{1}{11} = 0.\overline{09}$

19. $\left|-19.3\right| = 19.3$

20. $-11, \ -\dfrac{3}{7}, \ 0, \ 0.45, \text{ and } \sqrt{25}$ are rational numbers.

21. $5(x + 3) = (x + 3)5$

22. $5(x + 3) = 5(3 + x)$

23. $5(x + 3) = 5x + 15$

24. $7(9x + 3) + \dfrac{1}{3}(6x) = 63x + 21 + 2x$

$\qquad\qquad\qquad\qquad = 65x + 21$

25. $2(3x + 5y) + 4(x + 6y) = 6x + 10y + 4x + 24y$

$\qquad\qquad\qquad\qquad\qquad = 10x + 34y$

1.5 Check Points

1. $4 + (-7) = -3$

Start at 4 and move 7 units to the left.

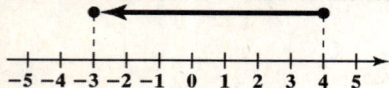

2. a. $-1 + (-3) = -4$

Start at -1 and move 3 units to the left.

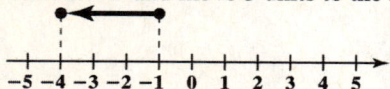

b. $-5 + 3 = -2$

Start at -5 and move 3 units to the right.

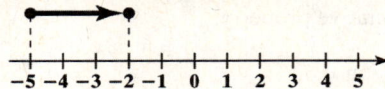

3. a. $-10 + (-25) = -35$

b. $-0.3 + (-1.2) = -1.5$

c. $-\dfrac{2}{3} + \left(-\dfrac{1}{6}\right) = -\dfrac{4}{6} + \left(-\dfrac{1}{6}\right) = -\dfrac{5}{6}$

4. a. $-15 + 2 = -13$

b. $-0.4 + 1.6 = 1.2$

c. $-\dfrac{2}{3} + \dfrac{1}{6} = -\dfrac{4}{6} + \dfrac{1}{6} = -\dfrac{3}{6} = -\dfrac{1}{2}$

5. a. $-20x + 3x = (-20 + 3)x = -17x$

b. $3y + (-10z) + (-10y) + 16z$

$\qquad = 3y + (-10y) + (-10z) + 16z$

$\qquad = \left[3 + (-10)\right]y + \left[(-10) + 16\right]z$

$\qquad = -7y + 6z$

c. $5(2x + 3) + (-30x) = 10x + 15 + (-30x)$

$\qquad\qquad\qquad\qquad = 10x + (-30x) + 15$

$\qquad\qquad\qquad\qquad = \left[10 + (-30)\right]x + 15$

$\qquad\qquad\qquad\qquad = -20x + 15$

6. $2 + (-4) + 1 + (-5) + 3 = (2 + 1 + 3) + \left[(-4) + (-5)\right]$

$\qquad\qquad\qquad\qquad\qquad = 6 + (-9)$

$\qquad\qquad\qquad\qquad\qquad = -3$

At the end of 5 months the water level was down 3 feet.

1.5 Concept and Vocabulary Check

1. additive inverses

2. zero

3. negative number

4. positive number

5. 0

6. negative number

7. positive number

8. 0

1.5 Exercise Set

1. $7+(-3)=4$

3. $-2+(-5)=-7$

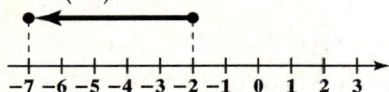

5. $-6+2=-4$

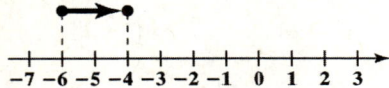

7. $3+(-3)=0$

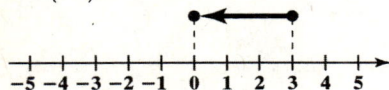

9. $-7+0=-7$

11. $30+(-30)=0$

13. $-30+(-30)=-60$

15. $-8+(-10)=-18$

17. $-0.4+(-0.9)=-1.3$

19. $-\dfrac{7}{10}+\left(-\dfrac{3}{10}\right)=-\dfrac{10}{10}=-1$

21. $-9+4=-5$

23. $12+(-8)=4$

25. $6+(-9)=-3$

27. $-3.6+2.1=-1.5$

29. $-3.6+(-2.1)=-5.7$

31. $\dfrac{9}{10}+\left(-\dfrac{3}{5}\right)=\dfrac{9}{10}+\left(-\dfrac{6}{10}\right)=\dfrac{3}{10}$

33. $-\dfrac{5}{8}+\dfrac{3}{4}=-\dfrac{5}{8}+\dfrac{6}{8}=\dfrac{1}{8}$

35. $-\dfrac{3}{7}+\left(-\dfrac{4}{5}\right)=-\dfrac{15}{35}+\left(-\dfrac{28}{35}\right)=-\dfrac{43}{35}$

37. $4+(-7)+(-5)=\left[4+(-7)\right]+(-5)$
$$=-3+(-5)$$
$$=-8$$

39. $85+(-15)+(-20)+12$
$$=\left[85+(-15)\right]+(-20)+12$$
$$=70+(-20)+12$$
$$=\left[70+(-20)\right]+12$$
$$=50+12$$
$$=62$$

41. $17+(-4)+2+3+(-10)$
$$=13+2+3+(-10)$$
$$=15+3+(-10)$$
$$=18+(-10)$$
$$=8$$

43. $-45+\left(-\dfrac{3}{7}\right)+25+\left(-\dfrac{4}{7}\right)$
$$=(-45+25)+\left[-\dfrac{3}{7}+\left(-\dfrac{4}{7}\right)\right]$$
$$=-20+\left(-\dfrac{7}{7}\right)$$
$$=-20+(-1)$$
$$=-21$$

45. $3.5+(-45)+(-8.4)+72$
$$=\left[3.5+(-8.4)\right]+(-45+72)$$
$$=-4.9+27$$
$$=22.1$$

47. $-10x+2x=(-10+2)x=-8x$

49. $25y+(-12y)=\left[25+(-12)\right]y=13y$

51. $-8a + (-15a) = [-8 + (-15)]a$
$$= -23a$$

53. $4y + (-13z) + (-10y) + 17z$
$= 4y + (-10y) + (-13z) + 17z$
$= -6y + 4z$

55. $-7b + 10 + (-b) + (-6)$
$= -7b + (-b) + 10 + (-6)$
$= -8b + 4$

57. $7x + (-5y) + (-9x) + 19y$
$= 7x + (-9x) + (-5y) + 19y$
$= -2x + 14y$

59. $8(4y + 3) + (-35y)$
$= 32y + 24 + (-35y)$
$= 32y + (-35y) + 24$
$= -3y + 24$

61. $\left|-3 + (-5)\right| + \left|2 + (-6)\right| = \left|-8\right| + \left|-4\right|$
$$= 8 + 4$$
$$= 12$$

63. $-20 + \left[-\left|15 + (-25)\right|\right]$
$= -20 + \left[-\left|-10\right|\right]$
$= -20 + [-10]$
$= -30$

65. $6 + \left[2 + (-13)\right]\square -3 + \left[4 + (-8)\right]$
$6 + [-11]\square -3 + [-4]$
$-5 \square -7$
$-5 > -7$

67. $-6x + (-13x)$
$-6x + (-13x) = -19x$

69. $\dfrac{-20}{x} + \dfrac{3}{x}$
$\dfrac{-20}{x} + \dfrac{3}{x} = \dfrac{-17}{x}$

71. $-56 + 100 = 44$
The high temperature was 44°F.

73. $-1312 + 712 = -600$
The elevation of the person is 600 feet below sea level.

75. $-7 + 15 - 5 = 3$
The temperature at 4:00 P.M. was 3°F.

77. $27 + 4 - 2 + 8 - 12$
$= (27 + 4 + 8) + (-2 - 12)$
$= 34 - 14$
$= 25$
The location of the football at the end of the fourth play is at the 25-yard line.

79. a. $2521 + (-2931) = -410$
The deficit in 2008 was $-\$410$ billion.

b. $2700 + (-3107) = -407$
The deficit in 2009 was $-\$407$ billion.

c. $-410 + (-407) = -817$
The combined deficit in 2008 and 2009 was $-\$817$ billion.

81. – 87. Answers will vary.

89. makes sense

91. makes sense

93. true

95. false; Changes to make the statement true will vary. A sample change is: The sum of a positive number and a negative number is sometimes negative.

97. The sum is negative. When finding the sum of numbers with different signs, use the sign of the number with the greater absolute value as the sign of the sum. Since a is further from 0 than c, we use a negative sign.

99. Though the sum inside the absolute value is negative, the absolute value of this sum is positive.

101. The calculator verifies your results.

102. a. $\sqrt{4}\ (= 2)$
b. $0, \sqrt{4}$
c. $-6, 0, \sqrt{4}$
d. $-6, 0, 0.\overline{7}, \sqrt{4}$
e. $-\pi, \sqrt{3}$
f. $-6, -\pi, 0, 0.\overline{7}, \sqrt{3}, \sqrt{4}$

103. $19 \geq -18$ is true because 19 is to the right of -18 on the number line.

104. $16 = 2(x-1)-x$

$16 = 2(18-1)-18$

$16 = 2(17)-18$

$16 = 34-18$

$16 = 16, \text{ true}$

3 is a solution.

105. $7-10 = 7+(-10) = -3$

106. $-8-13 = -8+(-13) = -21$

107. $-8-(-13) = -8+13 = 5$

1.6 Check Points

1. a. $3-11 = 3+(-11) = -8$

   b. $4-(-5) = 4+5 = 9$

   c. $-7-(-2) = -7+2 = -5$

2. a. $-3.4-(-12.6) = -3.4+12.6 = 9.2$

   b. $-\dfrac{3}{5}-\dfrac{1}{3} = -\dfrac{3}{5}+\left(-\dfrac{1}{3}\right) = -\dfrac{9}{15}+\left(-\dfrac{5}{15}\right) = -\dfrac{14}{15}$

   c. $5\pi-(-2\pi) = 5\pi+2\pi = 7\pi$

3. $10-(-12)-4-(-3)-6$

$= 10+12+(-4)+3+(-6)$

$= (10+12+3)+\left[(-4)+(-6)\right]$

$= 25+(-10)$

$= 15$

4. $-6+4a-7ab$ has terms of -6, $4a$, and $-7ab$.

5. a. $4+2x-9x = 4+(2-9)x$

$\qquad\qquad = 4+\left[2+(-9)\right]x$

$\qquad\qquad = 4-7x$

   b. $-3x-10y-6x+14y = -3x-6x-10y+14y$

$\qquad\qquad\qquad\qquad\qquad = (-3-6)x+(-10+14)y$

$\qquad\qquad\qquad\qquad\qquad = -9x+4y$

6. $8848-(-10,915) = 8848+10,915 = 19,763$

The difference in elevation between the peak of Mount Everest and the Marianas Trench is 19,763 meters.

1.6 Concept and Vocabulary Check

1. (-14)

2. 14

3. 14

4. -8; (-14)

5. 3; (-12); (-23)

6. $(-4y)$; 6

7. three; addition

1.6 Exercise Set

1. a. -12

   b. $5-12 = 5+(-12)$

3. a. 7

   b. $5-(-7) = 5+7$

5. $14-8 = 14+(-8) = 6$

7. $8-14 = 8+(-14) = -6$

9. $3-(-20) = 3+20 = 23$

11. $-7-(-18) = -7+18 = 11$

13. $-13-(-2) = -13+2 = -11$

15. $-21-17 = -21+(-17) = -38$

17. $-45-(-45) = -45+45 = 0$

19. $23-23 = 23+(-23) = 0$

21. $13-(-13) = 13+13 = 26$

23. $0-13 = 0+(-13) = -13$

25. $0-(-13)=0+13=13$

27. $\dfrac{3}{7}-\dfrac{5}{7}=\dfrac{3}{7}+\left(-\dfrac{5}{7}\right)=-\dfrac{2}{7}$

29. $\dfrac{1}{5}-\left(-\dfrac{3}{5}\right)=\dfrac{1}{5}+\dfrac{3}{5}=\dfrac{4}{5}$

31. $-\dfrac{4}{5}-\dfrac{1}{5}=-\dfrac{4}{5}+\left(-\dfrac{1}{5}\right)=-\dfrac{5}{5}=-1$

33. $-\dfrac{4}{5}-\left(-\dfrac{1}{5}\right)=-\dfrac{4}{5}+\dfrac{1}{5}=-\dfrac{3}{5}$

35. $\dfrac{1}{2}-\left(-\dfrac{1}{4}\right)=\dfrac{1}{2}+\dfrac{1}{4}=\dfrac{2}{4}+\dfrac{1}{4}=\dfrac{3}{4}$

37. $\dfrac{1}{2}-\dfrac{1}{4}=\dfrac{1}{2}+\left(-\dfrac{1}{4}\right)=\dfrac{2}{4}+\left(-\dfrac{1}{4}\right)=\dfrac{1}{4}$

39. $9.8-2.2=9.8+(-2.2)=7.6$

41. $-3.1-(-1.1)=-3.1+1.1=-2$

43. $1.3-(-1.3)=1.3+1.3=2.6$

45. $-2.06-(-2.06)=-2.06+2.06=0$

47. $5\pi-2\pi=5\pi+(-2\pi)=3\pi$

49. $3\pi-(-10\pi)=3\pi+10\pi=13\pi$

51. $\begin{aligned}13-2-(-8)&=13+(-2)+8\\&=(13+8)+(-2)\\&=21+(-2)\\&=19\end{aligned}$

53. $\begin{aligned}9-8+3-7&=9+(-8)+3+(-7)\\&=(9+3)+\left[(-8)+(-7)\right]\\&=12+(-15)\\&=-3\end{aligned}$

55. $\begin{aligned}-6-2+3-10&\\=-6+(-2)&+3+(-10)\\=\left[(-6)+(-2)+(-10)\right]&+3\\=-18+3&\\=-15&\end{aligned}$

57. $\begin{aligned}-10-(-5)+7-2&\\=-10+5+7+(-2)&\\=\left[(-10)+(-2)\right]+(5+7)&\\=-12+12&\\=0&\end{aligned}$

59. $\begin{aligned}-23-11-(-7)+(-25)&\\=(-23)+(-11)+7+(-25)&\\=\left[(-23)+(-11)+(-25)\right]+7&\\=-59+7&\\=-52&\end{aligned}$

61. $\begin{aligned}-823-146-50-(-832)&\\=-823+(-146)+(-50)+832&\\=\left[(-823)+(-146)+(-50)\right]+832&\\=-1019+832&\\=-187&\end{aligned}$

63. $\begin{aligned}1-\dfrac{2}{3}-\left(-\dfrac{5}{6}\right)&=1+\left(-\dfrac{2}{3}\right)+\dfrac{5}{6}\\&=\left(1+\dfrac{5}{6}\right)+\left(-\dfrac{2}{3}\right)\\&=\left(\dfrac{6}{6}+\dfrac{5}{6}\right)+\left(-\dfrac{2}{3}\right)\\&=\dfrac{11}{6}+\left(-\dfrac{2}{3}\cdot\dfrac{2}{2}\right)\\&=\dfrac{11}{6}+\left(-\dfrac{4}{6}\right)\\&=\dfrac{7}{6}\ \text{ or }\ 1\dfrac{1}{6}\end{aligned}$

65. $\begin{aligned}-0.16-5.2-(-0.87)&\\=-0.16+(-5.2)+0.87&\\=\left[(-0.16)+(-5.2)\right]+0.87&\\=-5.36+0.87&\\=-4.49&\end{aligned}$

67. $-\dfrac{3}{4}-\dfrac{1}{4}-\left(-\dfrac{5}{8}\right)=-\dfrac{3}{4}+\left(-\dfrac{1}{4}\right)+\dfrac{5}{8}$

$\qquad\qquad\qquad = -\dfrac{4}{4}+\dfrac{5}{8}$

$\qquad\qquad\qquad = -\dfrac{8}{8}+\dfrac{5}{8}=-\dfrac{3}{8}$

69. $-3x-8y=-3x+(-8y)$

The terms are $-3x$ and $-8y$.

71. $12x-5xy-4=12x+(-5xy)+(-4)$

The terms are $12x$, $-5xy$, and -4.

73. $3x-9x=3x+(-9x)$

$\qquad\quad =\left[3+(-9)\right]x=-6x$

75. $4+7y-17y=4+7y+(-17y)$

$\qquad\qquad\quad =4+\left[7+(-17)\right]y$

$\qquad\qquad\quad =4-10y$

77. $2a+5-9a=2a+5+(-9a)$

$\qquad\qquad\;\; =2a+(-9a)+5$

$\qquad\qquad\;\; =\left[2+(-9)\right]a+5$

$\qquad\qquad\;\; =-7a+5 \text{ or } 5-7a$

79. $4-6b-8-3b$

$\quad =4+(-6b)+(-8)+(-3b)$

$\quad =4+(-8)+(-6b)+(-3b)$

$\quad =4+(-8)+\left[-6+(-3)\right]b$

$\quad =-4-9b$

81. $13-(-7x)+4x-(-11)$

$\quad =13+7x+4x+11$

$\quad =13+11+7x+4x$

$\quad =24+11x$

83. $-5x-10y-3x+13y$

$\quad =-5x+(-10y)+(-3x)+13y$

$\quad =-5x+(-3x)+(-10y)+13y$

$\quad =\left[-5+(-3)\right]x+(-10+13)\,y$

$\quad =-8x+3y \text{ or } 3y-8x$

85. $-\left|-9-(-6)\right|-(-12)=-\left|-9+6\right|+12$

$\qquad\qquad\qquad\quad = -\left|-3\right|+12$

$\qquad\qquad\qquad\quad = -3+12$

$\qquad\qquad\qquad\quad = 9$

87. $\dfrac{5}{8}-\left(\dfrac{1}{2}\cdot\dfrac{3}{4}\right)=\dfrac{5}{8}-\left(\dfrac{1}{2}\cdot\dfrac{2}{2}-\dfrac{3}{4}\right)$

$\qquad\qquad\quad =\dfrac{5}{8}-\left(\dfrac{2}{4}+\left(-\dfrac{3}{4}\right)\right)$

$\qquad\qquad\quad =\dfrac{5}{8}-\left(-\dfrac{1}{4}\right)$

$\qquad\qquad\quad =\dfrac{5}{8}+\dfrac{1}{4}$

$\qquad\qquad\quad =\dfrac{5}{8}+\dfrac{1}{4}\cdot\dfrac{2}{2}$

$\qquad\qquad\quad =\dfrac{5}{8}+\dfrac{2}{8}$

$\qquad\qquad\quad =\dfrac{7}{8}$

89. $\left|-9-(-3+7)\right|-\left|-17-(-2)\right|$

$\quad =\left|-9-4\right|-\left|-17+2\right|$

$\quad =\left|-9+(-4)\right|-\left|-17+2\right|$

$\quad =\left|-13\right|-\left|-15\right|$

$\quad =13+(-15)$

$\quad =-2$

91. $6x-(-5x)$

$\quad 6x-(-5x)=6x+5x=11x$

93. $\dfrac{-5}{x}-\left(\dfrac{-2}{x}\right)$

$\quad \dfrac{-5}{x}-\left(\dfrac{-2}{x}\right)=\dfrac{-5}{x}+\dfrac{2}{x}=\dfrac{-3}{x}$

95. Elevation of Mount Kilimanjaro – elevation of Qattara Depression

$\quad =19,321-(-436)=19,757$

The difference in elevation between the two geographic locations is 19,757 feet.

97. $2-(-19)=2+19=21$

The difference between the average daily low temperature for March and February is 21°F.

99. $-19-(-22)=-19+22=3$

February's average low temperature is 3°F warmer than January's.

101. $10+(-15)=-5$

shrink by 5 years

103. $-5+(-6)=-11$

shrink by 11 years

105. $5-(-5)=5+5=10$ years

107. $5+(-5)=0$

no change

109. $-6-(-15)=-6+15=9$ years

111. – 115. Answers will vary.

117. makes sense

119. makes sense

121. false; Changes to make the statement true will vary. A sample change is: $7-(-2)=7+2=9$

123. true

125. $a-b=\overbrace{(\text{negative number})}^{a}-\overbrace{(\text{negative number})}^{b}$

$\underbrace{}_{\substack{a \text{ has the greater} \\ \text{absolute value}}}$

$=(\text{negative number})+(\text{positive number})$

$=\text{negative number}$

127. $0-b=0-\overbrace{(\text{negative number})}^{b}$

$=0+(\text{positive number})$

$=\text{positive number}$

129. The calculator verifies your results.

131.
$$13x+3=3(5x-1)$$
$$13\cdot2+3=3(5\cdot2-1)$$
$$26+3=3(10-1)$$
$$29=3(9)$$
$$29=27, \text{ false}$$

The number is not a solution.

132. $5(3x+2y)+6(5y)=15x+10y+30y$
$$=15x+40y$$

133. Answers will vary. -17 is an example of an integer that is not a natural number.

134. $4(-3)=(-3)+(-3)+(-3)+(-3)=-12$

135. $3(-3)=(-3)+(-3)+(-3)=-9$

136.
$$2(-3)=-6$$
$$1(-3)=-3$$
$$0(-3)=\ 0$$
$$-1(-3)=\ 3$$
$$-2(-3)=\ 6$$
$$-3(-3)=\ 9$$
$$-4(-3)=\boxed{12}$$

1.7 Check Points

1. **a.** $8(-5)=-40$

b. $-\dfrac{1}{3}\cdot\dfrac{4}{7}=-\dfrac{4}{21}$

c. $(-12)(-3)=36$

d. $(-1.1)(-5)=5.5$

e. $(-543)(0)=0$

2. **a.** $(-2)(3)(-1)(4)=24$

b. $(-1)(-3)(2)(-1)(5)=-30$

3. **a.** The multiplicative inverse of 7 is $\dfrac{1}{7}$ because

$7\cdot\dfrac{1}{7}=1.$

b. The multiplicative inverse of $\dfrac{1}{8}$ is 8 because

$\dfrac{1}{8}\cdot8=1.$

c. The multiplicative inverse of -6 is $-\dfrac{1}{6}$ because

$(-6)\left(-\dfrac{1}{6}\right)=1.$

d. The multiplicative inverse of $-\dfrac{7}{13}$ is $-\dfrac{13}{7}$

because $\left(-\dfrac{7}{13}\right)\left(-\dfrac{13}{7}\right)=1.$

4. a. $-28 \div 7 = -28 \cdot \dfrac{1}{7} = -4$

b. $\dfrac{-16}{-2} = -16 \cdot \left(-\dfrac{1}{2}\right) = 8$

5. a. $\dfrac{-32}{-4} = 8$

b. $-\dfrac{2}{3} \div \dfrac{5}{4} = -\dfrac{2}{3} \cdot \dfrac{4}{5} = -\dfrac{8}{15}$

c. $\dfrac{21.9}{-3} = -7.3$

d. $\dfrac{0}{-5} = 0$

6. a. $-4(5x) = (-4 \cdot 5)x = -20x$

b. $9x + x = 9x + 1x = (9+1)x = 10x$

c. $13b - 14b = (13-14)b = -1b = -b$

d. $-7(3x - 4) = -7(3x) - 7(-4) = -21x + 28$

e. $-(7y - 6) = -(7y) - (-6) = -7y + 6$

7. $4(3y - 7) - (13y - 2) = 12y - 28 - 13y + 2$
$$= 12y - 13y - 28 + 2$$
$$= -1y - 26$$
$$= -y - 26$$

8. $2x - 5 = 8x + 7$
$$2(-3) - 5 = 8(-3) + 7$$
$$-6 - 5 = -24 + 7$$
$$-11 = -17, \text{ false}$$
The number is not a solution.

9. $M = -0.6n + 64.4$
$$M = -0.6(25) + 64.4$$
$$= -15 + 64.4$$
$$= 49.4$$
According to this model, 49.4% of doctorate degrees will be awarded to men in 2014. This overestimates the actual value shown in the bar graph by 0.4%.

1.7 Concept and Vocabulary Check

1. positive

2. negative

3. negative

4. positive

5. 0

6. negative

7. negative

8. positive

9. 0

10. undefined

11. positive

1.7 Exercise Set

1. $5(-9) = -(5 \cdot 9) = -45$

3. $(-8)(-3) = +(8 \cdot 3) = 24$

5. $(-3)(7) = -21$

7. $(-19)(-1) = 19$

9. $0(-19) = 0$

11. $\dfrac{1}{2}(-24) = -12$

13. $\left(-\dfrac{3}{4}\right)(-12) = \dfrac{3 \cdot 12}{4 \cdot 1} = 9$

15. $-\dfrac{3}{5} \cdot \left(-\dfrac{4}{7}\right) = \dfrac{3 \cdot 4}{5 \cdot 7} = \dfrac{12}{35}$

17. $-\dfrac{7}{9} \cdot \dfrac{2}{3} = -\dfrac{7 \cdot 2}{9 \cdot 3} = -\dfrac{14}{27}$

19. $3(-1.2) = -3.6$

21. $-0.2(-0.6) = 0.12$

23. $(-5)(-2)(3) = 30$

25. $(-4)(-3)(-1)(6) = -72$

27. $-2(-3)(-4)(-1) = 24$

29. $(-3)(-3)(-3) = 9(-3) = -27$

31. $5(-3)(-1)(2)(3) = 90$

33. $(-8)(-4)(0)(-17)(-6) = 0$

35. The multiplicative inverse of 4 is $\frac{1}{4}$.

37. The multiplicative inverse of $\frac{1}{5}$ is 5.

39. The multiplicative inverse of -10 is $-\frac{1}{10}$.

41. The multiplicative inverse of $-\frac{2}{5}$ is $-\frac{5}{2}$.

43. a. $-32 \div 4 = -32 \cdot \frac{1}{4}$

b. $-32 \cdot \frac{1}{4} = -8$

45. a. $\frac{-60}{-5} = -60 \cdot \left(-\frac{1}{5}\right)$

b. $-60 \cdot \left(-\frac{1}{5}\right) = 12$

47. $\frac{12}{-4} = 12 \cdot \left(-\frac{1}{4}\right) = -3$

49. $\frac{-21}{3} = -21 \cdot \frac{1}{3} = -7$

51. $\frac{-90}{-3} = -90 \cdot \left(-\frac{1}{3}\right) = 30$

53. $\frac{0}{-7} = 0$

55. $\frac{7}{0}$ is undefined.

57. $-15 \div 3 = -15 \cdot \frac{1}{3} = -5$

59. $12 \div (-10) = 120 \cdot \left(-\frac{1}{10}\right) = -12$

61. $(-180) \div (-30) = -180 \cdot \left(-\frac{1}{30}\right) = 6$

63. $0 \div (-4) = 0$

65. $-4 \div 0$ is undefined.

67. $\frac{-12.9}{3} = -12.9 \cdot \frac{1}{3} = -4.3$

69. $-\frac{1}{2} \div \left(-\frac{3}{5}\right) = -\frac{1}{2} \cdot \left(-\frac{5}{3}\right) = \frac{5}{6}$

71. $-\frac{14}{9} \div \frac{7}{8} = -\frac{14}{9} \cdot \frac{8}{7}$

$= -\frac{112}{63} = \frac{\cancel{7} \cdot 16}{\cancel{7} \cdot 9} = -\frac{16}{9}$

73. $\frac{1}{3} \div \left(-\frac{1}{3}\right) = \frac{1}{3} \cdot (-3) = -1$

75. $6 \div \left(-\frac{2}{5}\right) = 6 \cdot \left(-\frac{5}{2}\right) = -\frac{30}{2} = -15$

77. $-5(2x) = (-5 \cdot 2)x = -10x$

79. $-4\left(-\frac{3}{4}y\right) = \left[-4 \cdot \left(-\frac{3}{4}\right)\right]y = 3y$

81. $8x + x = 8x + 1x = (8+1)x = 9x$

83. $-5x + x = -5x + 1x = (-5+1)x = -4x$

85. $6b - 7b = (6-7)b = -1b = -b$

87. $-y + 4y = -1y + 4y = (-1+4)y = 3y$

89. $-4(2x-3) = -4(2x) - 4(-3) = -8x + 12$

91. $-3(-2x+4) = -3(-2x) - 3(4) = 6x - 12$

93. $-(2y-5) = -2y + 5$

95. $4(2y-3)-(7y+2)$

$\quad = 4(2y)+4(-3)-7y-2$

$\quad = 8y-12+7y-2$

$\quad = 8y-7y-12-2$

$\quad = y-14$

97. $\quad 4x=2x-10$

$\quad 4(-5)=2(-5)-10$

$\quad -20=-10-10$

$\quad -20=-20,\ \text{true}$

The number is a solution.

99. $\quad -7y+18=-10y+6$

$\quad -7(-4)+18=-10(-4)+6$

$\quad 28+18=40+6$

$\quad 46=46,\ \text{true}$

The number is a solution.

101. $\quad 5(w+3)=2w-21$

$\quad 5(-10+3)=2(-10)-21$

$\quad 5(-7)=-20-21$

$\quad -35=-41,\ \text{false}$

The number is not a solution.

103. $\quad 4(6-z)+7z=0$

$\quad 4(6-(-8))+7(-8)=0$

$\quad 4(6+8)-56=0$

$\quad 4(14)-56=0$

$\quad 56-56=0$

$\quad 0=0,\ \text{true}$

The number is a solution.

105. $\quad 14-2x=-4x+7$

$\quad 14-2\left(-2\dfrac{1}{2}\right)=-4\left(-2\dfrac{1}{2}\right)+7$

$\quad 14-2\left(-\dfrac{5}{2}\right)=-4\left(-\dfrac{5}{2}\right)+7$

$\quad 14+5=10+7$

$\quad 19=17,\ \text{false}$

The number is not a solution.

107. $\quad \dfrac{5m-1}{6}=\dfrac{3m-2}{4}$

$\quad \dfrac{5(-4)-1}{6}=\dfrac{3(-4)-2}{4}$

$\quad \dfrac{-20-1}{6}=\dfrac{-12-2}{4}$

$\quad \dfrac{-21}{6}=\dfrac{-14}{4}$

$\quad \dfrac{-7}{2}=\dfrac{-7}{2},\ \text{true}$

The number is a solution.

109. $4(-10)+8=-40+8=-32$

111. $(-9)(-3)-(-2)=27+2=29$

113. $\dfrac{-18}{-15+12}=\dfrac{-18}{-3}=6$

115. $-6-\left(\dfrac{12}{-4}\right)=-6-(-3)=-6+3=-3$

117. $C=\dfrac{5}{9}(F-32)$

$\quad C=\dfrac{5}{9}(-22-32)=\dfrac{5}{9}(-54)=-30$

$-22°\text{F}$ is equivalent to $-30°\text{C}$.

119. a. 2009 is 44 years after 1965.

$\quad C=-0.5x+41$

$\quad C=-0.5(44)+41$

$\quad =19$

According to the formula, about 19% of American adults smoked in 2009. This overestimates the actual amount shown in the bar graph by 1%.

b. 2015 is 50 years after 1965.

$\quad C=-0.5x+41$

$\quad C=-0.5(50)+41$

$\quad =16$

According to the formula, about 16% of American adults will smoke in 2015.

121. a. According to the bar graph, wives devoted about 25 hours per week to household chores and child care in 2005.

b. $W = -0.4n + 39$

$W = -0.4(40) + 39$

$\quad = 23$

According to the formula, wives devoted 23 hours per week to household chores and child care in 2005.
This is less than the estimate in part a.

c. $D = \overset{W}{\overbrace{(-0.4n+39)}} - \overset{H}{\overbrace{(0.2n+12)}}$

$D = -0.4n + 39 - 0.2n - 12$

$D = -0.6n + 27$

d. $D = -0.6n + 27$

$D = -0.6(40) + 27$

$\quad = 3$

According to the formula, the difference in hours per week between wives and husbands will be 3 hours in 2005.
This underestimates the difference shown in the bar graph.

123. – 129. Answers will vary.

131. does not make sense; Explanations will vary.
Sample explanation: The sign rules for dividing real numbers and multiplying real numbers are the same.

133. makes sense

135. false; Changes to make the statement true will vary.
A sample change is: The sum of two negative numbers is a negative number.

137. false; Changes to make the statement true will vary.
A sample change is: $0 \div \left(-\sqrt{2}\right) = \dfrac{0}{-\sqrt{2}} = 0$

139. $5x$

141. $\dfrac{x}{12}$

143. The calculator verifies your results.

145. $0.3(4.7x - 5.9) - 0.07(3.8x - 61)$

$= 0.3(4.7x) + 0.3(-5.9) - (0.07)(3.8x) - (0.07)(-61)$

$= 1.41x - 1.77 - 0.266x + 4.27$

$= \left[1.41x + (-0.266x)\right] + (-1.77 + 4.27)$

$= 1.144x + 2.5$

147. $-6 + (-3) = -9$

148. $-6 - (-3) = -6 + 3 = -3$

149. $-6 \div (-3) = -6\left(-\dfrac{1}{3}\right) = 2$

150. $(-6)^2 = (-6)(-6) = 36$

151. $(-5)^3 = (-5)(-5)(-5) = -125$

152. $(-2)^4 = (-2)(-2)(-2)(-2) = 16$

1.8 Check Points

1. a. $6^2 = 6 \cdot 6 = 36$

b. $(-4)^3 = (-4)(-4)(-4) = -64$

c. $(-1)^4 = (-1)(-1)(-1)(-1) = 1$

d. $-1^4 = -(1 \cdot 1 \cdot 1 \cdot 1) = -1$

2. a. $16x^2 + 5x^2 = (16 + 5)x^2 = 21x^2$

b. $7x^3 + x^3 = 7x^3 + 1x^3 = (7+1)x^3 = 8x^3$

c. $10x^2 + 8x^3$ cannot be simplified.

3. $20 + 4 \cdot 3 - 17 = 20 + 12 - 17$

$\qquad\qquad\qquad\quad = 20 + 12 - 17$

$\qquad\qquad\qquad\quad = 15$

4. $7^2 - 48 \div 4^2 \cdot 5 - 2 = 49 - 48 \div 16 \cdot 5 - 2$

$\qquad\qquad\qquad\qquad\quad = 49 - 3 \cdot 5 - 2$

$\qquad\qquad\qquad\qquad\quad = 49 - 15 - 2$

$\qquad\qquad\qquad\qquad\quad = 34 - 2$

$\qquad\qquad\qquad\qquad\quad = 32$

5. a. $(3 \cdot 2)^2 = 6^2 = 36$

b. $3 \cdot 2^2 = 3 \cdot 4 = 12$

6. $\left(-\dfrac{1}{2}\right)^2 - \left(\dfrac{7}{10} - \dfrac{8}{15}\right)^2 (-18) = \left(-\dfrac{1}{2}\right)^2 - \left(\dfrac{1}{6}\right)^2 (-18)$

$$= \dfrac{1}{4} - \dfrac{1}{36} \cdot \dfrac{-18}{1}$$

$$= \dfrac{1}{4} + \dfrac{1}{2}$$

$$= \dfrac{3}{4}$$

7. $4[3(6-11)+5] = 4[3(-5)+5]$

$$= 4[-15+5]$$

$$= 4[-10]$$

$$= -40$$

8. $25 \div 5 + 3[4 + 2(7-9)^3] = 25 \div 5 + 3[4 + 2(-2)^3]$

$$= 25 \div 5 + 3[4 + 2(-8)]$$

$$= 25 \div 5 + 3[4 - 16]$$

$$= 25 \div 5 + 3[-12]$$

$$= 5 + (-36)$$

$$= -31$$

9. $\dfrac{5(4-9)+10 \cdot 3}{2^3 - 1} = \dfrac{5(-5)+10 \cdot 3}{8-1}$

$$= \dfrac{-25+30}{7}$$

$$= \dfrac{5}{7}$$

10. $-x^2 - 4x = -(-5)^2 - 4(-5)$

$$= -25 + 20$$

$$= -5$$

11. $14x^2 + 5 - [7(x^2 - 2) + 4]$

$$= 14x^2 + 5 - [7x^2 - 14 + 4]$$

$$= 14x^2 + 5 - [7x^2 - 10]$$

$$= 14x^2 + 5 - 7x^2 + 10$$

$$= 14x^2 - 7x^2 + 5 + 10$$

$$= 7x^2 + 15$$

12. $M = -120x^2 + 998x + 590$

$$M = -120(4)^2 + 998(4) + 590$$

$$= 2662$$

According to the model, males between the ages of 19 and 30 with this lifestyle need 2662 calories per day. This underestimates the actual value shown in the bar graph by 38 calories.

13. a. $\bar{C} = \dfrac{30x + 300,000}{x}$

$$\bar{C} = \dfrac{30(1000) + 300,000}{1000} = \dfrac{30,000 + 300,000}{1000}$$

$$= \dfrac{330,000}{1000}$$

$$= 330$$

The average cost is \$330.

b. $\bar{C} = \dfrac{30x + 300,000}{x}$

$$\bar{C} = \dfrac{30(10,000) + 300,000}{10,000} = \dfrac{300,000 + 300,000}{10,000}$$

$$= \dfrac{600,000}{10,000}$$

$$= 60$$

The average cost is \$60.

c. $\bar{C} = \dfrac{30x + 300,000}{x}$

$$\bar{C} = \dfrac{30(100,000) + 300,000}{100,000}$$

$$= \dfrac{3,000,000 + 300,000}{100,000}$$

$$= \dfrac{3,300,000}{100,000}$$

$$= 33$$

The average cost is \$33.

1.8 Concept and Vocabulary Check

1. base; exponent

2. *b* to the *n*th power

3. multiply

4. add

5. divide

6. subtract

7. multiply

1.8 Exercise Set

1. $9^2 = 9 \cdot 9 = 81$

3. $4^3 = 4 \cdot 4 \cdot 4 = 64$

5. $(-4)^2 = (-4)(-4) = 16$

7. $(-4)^3 = (-4)(-4)(-4) = -64$

9. $(-5)^4 = (-5)(-5)(-5)(-5) = 625$

11. $-5^4 = -5 \cdot 5 \cdot 5 \cdot 5 = -625$

13. $-10^2 = -10 \cdot 10 = -100$

15. $7x^2 + 12x^2 = (7+12)x^2 = 19x^2$

17. $10x^3 + 5x^3 = (10+5)x^3 = 15x^3$

19. $8x^4 + x^4 = 8x^4 + 1x^4 = (8+1)x^4 = 9x^4$

21. $26x^2 - 27x^2 = 26x^2 + (-27x^2)$
$$= \left[26 + (-27)\right]x^2$$
$$= -1x^2 = -x^2$$

23. $27x^3 - 26x^3 = 27x^3 + (-26x^2)$
$$= 1x^3 = x^3$$

25. $5x^2 + 5x^3$ cannot be simplified. The terms $5x^2$ and $5x^3$ are not like terms because they have different variable factors, namely, x^2 and x^3.

27. $16x^2 - 16x^2 = 16x^2 + (-16x^2)$
$$= \left[16 + (-16)\right]x^2$$
$$= 0x^2 = 0$$

29. $7 + 6 \cdot 3 = 7 + 18 = 25$

31. $45 \div 5 \cdot 3 = 9 + 18 = 27$

33. $6 \cdot 8 \div 4 = 48 \div 4 = 12$

35. $14 - 2 \cdot 6 + 3 = 14 - 12 + 3 = 2 + 3 = 5$

37. $8^2 - 16 \div 2^2 \cdot 4 - 3 = 64 - 16 \div 4 \cdot 4 - 3$
$$= 64 - 4 \cdot 4 - 3$$
$$= 64 - 16 - 3$$
$$= 48 - 3$$
$$= 45$$

39. $3(-2)^2 - 4(-3)^2 = 3 \cdot 4 - 4 \cdot 9$
$$= 12 - 36$$
$$= 12 + (-36)$$
$$= -24$$

41. $(4 \cdot 5)^2 - 4 \cdot 5^2 = 20^2 - 4 \cdot 25$
$$= 400 - 100$$
$$= 300$$

43. $(2-6)^2 - (3-7)^2 = (-4)^2 - (-4)^2$
$$= 16 - 16$$
$$= 0$$

45. $6(3-5)^3 - 2(1-3)^3$
$$= 6(-2)^3 - 2(-2)^3$$
$$= 6(-8) - 2(-8)$$
$$= -48 + 16$$
$$= -32$$

47. $\left[2(6-2)\right]^2 = (2 \cdot 4)^2 = 8^2 = 64$

49. $2\left[5 + 2(9-4)\right] = 2\left[5 + 2(5)\right]$
$$= 2(5 + 10)$$
$$= 2 \cdot 15 = 30$$

51. $\left[7 + 3\left(2^3 - 1\right)\right] \div 21 = \left[7 + 3(8-1)\right] \div 21$
$$= (7 + 3 \cdot 7) \div 21$$
$$= (7 + 21) \div 21$$
$$= 28 \div 21$$
$$= \frac{28}{21} = \frac{\cancel{7} \cdot 4}{\cancel{7} \cdot 3}$$
$$= \frac{4}{3}$$

53. $\dfrac{10+8}{5^2 - 4^2} = \dfrac{18}{25-16} = \dfrac{18}{9} = 2$

55. $\dfrac{37+15\div(-3)}{2^4}=\dfrac{37+(-5)}{16}=\dfrac{32}{16}=2$

57. $\dfrac{(-11)(-4)+2(-7)}{7-(-3)}=\dfrac{44+(-14)}{7+3}$

$\qquad\qquad\qquad =\dfrac{30}{10}=3$

59. $4\left|10-(8-20)\right|=4\left|10-(-12)\right|=4\left|10+12\right|$

$\qquad\qquad\qquad =4\left|22\right|=4\cdot 22$

$\qquad\qquad\qquad =88$

61. $8(-10)+\left|4(-5)\right|=-80+\left|-20\right|$

$\qquad\qquad\qquad =-80+20=-60$

63. $-2^2+4\left[16+(3-5)\right]$

$\quad =-4+4\left[16+(-2)\right]$

$\quad =-4+4(-8)=-4-32=-36$

65. $24\div\dfrac{3^2}{8-5}-(-6)=24\div\dfrac{9}{3}-(-6)$

$\qquad\qquad\qquad =24\div 3-(-6)$

$\qquad\qquad\qquad =8+6=14$

67. $\dfrac{\frac{1}{4}-\frac{1}{2}}{\frac{1}{3}}=\dfrac{\frac{1}{4}-\frac{2}{4}}{\frac{1}{3}}=\dfrac{-\frac{1}{4}}{\frac{1}{3}}=-\dfrac{1}{4}\cdot\dfrac{3}{1}=-\dfrac{3}{4}$

69. $-\dfrac{9}{4}\left(\dfrac{1}{2}\right)+\dfrac{3}{4}\div\dfrac{5}{6}=-\dfrac{9}{4}\left(\dfrac{1}{2}\right)+\dfrac{3}{4}\cdot\dfrac{6}{5}$

$\qquad\qquad\qquad =-\dfrac{9}{8}+\dfrac{18}{20}$

$\qquad\qquad\qquad =-\dfrac{45}{40}+\dfrac{36}{40}=-\dfrac{9}{40}$

71. $\dfrac{\frac{7}{9}-3}{\frac{5}{6}}\div\dfrac{3}{2}+\dfrac{3}{4}=\dfrac{\frac{7}{9}-\frac{27}{9}}{\frac{5}{6}}\cdot\dfrac{3}{2}+\dfrac{3}{4}$

$\qquad =\dfrac{-\frac{20}{9}}{\frac{5}{6}}\cdot\dfrac{3}{2}+\dfrac{3}{4}$

$\qquad =-\dfrac{20}{9}\cdot\dfrac{6}{5}\cdot\dfrac{2}{3}+\dfrac{3}{4}$

$\qquad =-\dfrac{240}{135}+\dfrac{3}{4}$

$\qquad =-\dfrac{15\cdot 16}{15\cdot 9}+\dfrac{3}{4}$

$\qquad =-\dfrac{16}{9}+\dfrac{3}{4}$

$\qquad =-\dfrac{64}{36}+\dfrac{27}{36}$

$\qquad =-\dfrac{37}{36}\ \text{ or }\ -1\dfrac{1}{36}$

73. $x^2+5x;\ x=3$

$\quad x^2+5x=3^2+5\cdot 3$

$\qquad\qquad =9+5\cdot 3=9+15=24$

75. $3x^2-8x;\ x=-2$

$\quad 3x^2-8x=3(-2)^2-9(-2)$

$\qquad\qquad =3\cdot 4-8(-2)=12+16=28$

77. $-x^2-10x;\ x=-1$

$\quad -x^2-10x=-(-1)^2-10(-1)$

$\qquad\qquad =-1+10=9$

79. $\dfrac{6y-4y^2}{y^2-15};\ y=5$

$\quad \dfrac{6y-4y^2}{y^2-15}=\dfrac{6(5)-4\left(5^2\right)}{5^2-15}$

$\qquad\qquad =\dfrac{6(5)-4(25)}{25-15}$

$\qquad\qquad =\dfrac{30-100}{25-15}=\dfrac{-70}{10}=-7$

81. $3[5(x-2)+1] = 3(5x-10+1)$
$$= 3(5x-9)$$
$$= 15x-27$$

83. $3[6-(y+1)] = 3(6-y-1)$
$$= 3(5-y)$$
$$= 15-3y$$

85. $7-4[3-(4y-5)]$
$$= 7-4(3-4y+5)$$
$$= 7-12+16y-20$$
$$= -25+16y \text{ or } 16y-25$$

87. $2(3x^2-5)-[4(2x^2-1)+3]$
$$= 6x^2-10-(8x^2-4+3)$$
$$= 6x^2-10-(8x^2-1)$$
$$= 6x^2-10-8x^2-1$$
$$= -2x^2-9$$

89. $-10-(-2)^3 = -10-(-8) = -10+8 = -2$

91. $[2(7-10)]^2 = [2(-3)]^2 = [-6]^2 = 36$

93. $x-(5x+8) = x-5x-8 = -4x-8$

95. $5(x^3-4) = 5x^3-20$

97. $F = -82x^2+654x+620$

$F = -82(4)^2+654(4)+620 = 1924$

1924 calories per day are needed.
This underestimates the value given in the bar graph by 76 calories.

99. a. According to the line graph, about 22% of students anticipated a starting salary of $30 thousand.

b. $p = -0.01s^2+0.8s+3.7$

$p = -0.01(30)^2+0.8(30)+3.7 = 18.7$

According to the formula, about 18.7% of students anticipated a starting salary of $30 thousand.
This is less than the value obtained from the line graph.

101. a. $C = \dfrac{200x}{100-x}$

$C = \dfrac{200(50)}{100-50} = 200$

It will cost $200 tens of thousands, or $2,000,000 to remove 50% of the contamination.

b. $C = \dfrac{200x}{100-x}$

$C = \dfrac{200(80)}{100-80} = 800$

It will cost $800 tens of thousands, or $8,000,000 to remove 80% of the contamination.

c. The cost of cleanup increases as the percentage of contaminant removed increases.

103. – 105. Answers will vary.

107. does not make sense; Explanations will vary.
Sample explanation: $10^4 = 10,000$

109. makes sense

111. false; Changes to make the statement true will vary.
A sample change is:
$$\dfrac{6(-3)+6}{-3+1} = \dfrac{-18+6}{-2} = \dfrac{-12}{-2} = 6 \text{ and}$$
$$\dfrac{6(2)+6}{2+1} = \dfrac{12+6}{3} = \dfrac{18}{3} = 6$$

113. false; Changes to make the statement true will vary.
A sample change is: $-2(6-4^2)^3 = -2(6-16)^3$
$$= -2(-10)^3$$
$$= -2(-1000)$$
$$= 2000$$

115. $(2 \cdot 3+3) \cdot 5 = (6+3) \cdot 5 = 9 \cdot 5 = 45$

117. $-8-2-(-5)+11$
$$= -8+(-2)+5+11$$
$$= [(-8)+(-2)]+(5+11)$$
$$= -10+16 = 6$$

118. $-4(-1)(-3)(2) = -24$

119. Answers will vary. One example is 5.

120. $-\dfrac{1}{2} = x - \dfrac{2}{3}$

$-\dfrac{1}{2} = \dfrac{1}{6} - \dfrac{2}{3}$

$-\dfrac{1}{2} = \dfrac{1}{6} - \dfrac{4}{6}$

$-\dfrac{1}{2} = -\dfrac{3}{6}$

$-\dfrac{1}{2} = -\dfrac{1}{2}$, true

The number is a solution.

121. $5y + 3 - 4y - 8 = 15$

$5(20) + 3 - 4(20) - 8 = 15$

$100 + 3 - 80 - 8 = 15$

$15 = 15$, true

The number is a solution.

122. $4x + 2 = 3(x - 6) + 8$

$4(-11) + 2 = 3(-11 - 6) + 8$

$-44 + 2 = 3(-17) + 8$

$-42 = -51 + 8$

$-42 = -43$, false

The number is not a solution.

Chapter 1 Review Exercises

1. $10 + 5x = 10 + 5(6) = 10 + 30 = 40$

2. $8(x - 2) + 3x = 8(6 - 2) + 3(6)$

$= 8(4) + 18$

$= 32 + 18$

$= 50$

3. $\dfrac{40}{x} - \dfrac{y}{5} = \dfrac{40}{8} - \dfrac{10}{5}$

$= \dfrac{200}{40} - \dfrac{80}{40}$

$= \dfrac{120}{40}$

$= 3$

4. $3(2y + x) = 3(2(10) + 8)$

$= 3(20 + 8)$

$= 3(28)$

$= 84$

5. $7x - 6$

6. $\dfrac{x}{5} - 2 = 18$

7. $9 - 2x = 14$

8. $3(x + 7)$

9. $4x + 5 = 13$

$4(3) + 5 = 13$

$12 + 5 = 13$

$17 = 13$, false

The number is not a solution.

10. $2y + 7 = 4y - 5$

$2(6) + 7 = 4(6) - 5$

$12 + 7 = 24 - 5$

$19 = 19$, true

The number is a solution.

11. $3(w + 1) + 11 = 2(w + 8)$

$3(2 + 1) + 11 = 2(2 + 8)$

$3(3) + 11 = 2(10)$

$9 + 11 = 20$

$20 = 20$, true

The number is a solution.

12. According to the line graph, the average number of Latin words that the class remembered after 5 days was about 11.

13. $L = \dfrac{5n + 30}{n}$

$L = \dfrac{5(5) + 30}{5} = \dfrac{25 + 30}{5} = \dfrac{55}{5} = 11$

According to the mathematical model, the average number of Latin words that the class remembered after 5 days was 11. This is the same value given in the line graph.

14. a. $C = 783n + 6522$

$C = 783(9) + 6522 = 13{,}569$

According to the formula, the average cost of a family health insurance plan was $13,569 in 2009.
This overestimates the actual value shown in the bar graph by $194.

b. $C = 783n + 6522$

$C = 783(20) + 6522 = 22{,}182$

The average cost of a family health insurance plan is expected to be \$22,182 in 2020.

15. $3\dfrac{2}{7} = \dfrac{3 \cdot 7 + 2}{7} = \dfrac{21 + 2}{7} = \dfrac{23}{7}$

16. $5\dfrac{9}{11} = \dfrac{5 \cdot 11 + 9}{11} = \dfrac{55 + 9}{11} = \dfrac{64}{11}$

17. 17 divided by 9 is 1 with a remainder of 8, so $\dfrac{17}{9} = 1\dfrac{8}{9}$.

18. 27 divided by 5 is 5 with a remainder of 2, so $\dfrac{27}{5} = 5\dfrac{2}{5}$.

19. Composite

$60 = 6 \cdot 10 = 2 \cdot 3 \cdot 2 \cdot 5 = 2 \cdot 2 \cdot 3 \cdot 5$

20. Composite

$63 = 7 \cdot 9 = 7 \cdot 3 \cdot 3 = 3 \cdot 3 \cdot 7$

21. 67 is a prime number.

22. $\dfrac{15}{33} = \dfrac{\cancel{3} \cdot 5}{\cancel{3} \cdot 11} = \dfrac{5}{11}$

23. $\dfrac{40}{75} = \dfrac{\cancel{5} \cdot 8}{\cancel{5} \cdot 15} = \dfrac{8}{15}$

24. $\dfrac{3}{5} \cdot \dfrac{7}{10} = \dfrac{3 \cdot 7}{5 \cdot 10} = \dfrac{21}{50}$

25. $\dfrac{4}{5} \div \dfrac{3}{10} = \dfrac{4}{5} \cdot \dfrac{10}{3} = \dfrac{40}{15} = \dfrac{\cancel{5} \cdot 8}{\cancel{5} \cdot 3} = \dfrac{8}{3}$

26. $1\dfrac{2}{3} \div 6\dfrac{2}{3} = \dfrac{5}{3} \div \dfrac{20}{3}$

$\qquad = \dfrac{5}{\cancel{3}} \cdot \dfrac{\cancel{3}}{20} = \dfrac{5}{20} = \dfrac{1 \cdot \cancel{5}}{4 \cdot \cancel{5}} = \dfrac{1}{4}$

27. $\dfrac{2}{9} + \dfrac{4}{9} = \dfrac{2 + 4}{9} = \dfrac{6}{9} = \dfrac{2 \cdot \cancel{3}}{3 \cdot \cancel{3}} = \dfrac{2}{3}$

28. $\dfrac{5}{6} + \dfrac{7}{9} = \dfrac{5}{6} \cdot \dfrac{3}{3} + \dfrac{7}{9} \cdot \dfrac{2}{2}$

$\qquad = \dfrac{15}{18} + \dfrac{14}{18} = \dfrac{29}{18}$ or $1\dfrac{11}{18}$

29. $\dfrac{3}{4} - \dfrac{2}{15} = \dfrac{3}{4} \cdot \dfrac{15}{15} - \dfrac{2}{15} \cdot \dfrac{4}{4} = \dfrac{45}{60} - \dfrac{8}{60} = \dfrac{37}{60}$

30. $x - \dfrac{3}{4} = \dfrac{7}{4}$

$2\dfrac{1}{2} - \dfrac{3}{4} = \dfrac{7}{4}$

$\dfrac{5}{2} - \dfrac{3}{4} = \dfrac{7}{4}$

$\dfrac{10}{4} - \dfrac{3}{4} = \dfrac{7}{4}$

$\dfrac{7}{4} = \dfrac{7}{4}$, true

The number is a solution.

31. $\dfrac{2}{3}w = \dfrac{1}{15}w + \dfrac{3}{5}$

$\dfrac{2}{3} \cdot 2 = \dfrac{1}{15} \cdot 2 + \dfrac{3}{5}$

$\dfrac{4}{3} = \dfrac{2}{15} + \dfrac{3}{5}$

$\dfrac{4}{3} = \dfrac{2}{15} + \dfrac{9}{15}$

$\dfrac{4}{3} = \dfrac{11}{15}$, false

The number is not a solution.

32. $2 - \dfrac{1}{2}x = \dfrac{1}{4}x$

33. $\dfrac{3}{5}(x + 6)$

34. $H = \dfrac{4}{5}(220 - a)$

$H = \dfrac{4}{5}(220 - 30) = \dfrac{4}{5}(190) = 152$

The target heart rate of a 30-year-old is 152 beats per minute.

35. -2.5

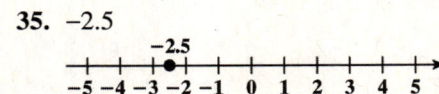

36. $4\dfrac{3}{4}$

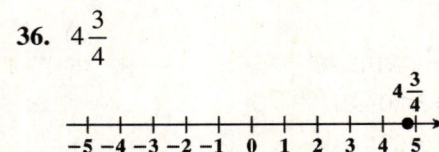

37.

$$8\overline{)5.000} \quad \begin{array}{r} 0.625 \\ \end{array}$$

$$\begin{array}{r} 48 \\ \hline 20 \\ 16 \\ \hline 40 \\ 40 \\ \hline 0 \end{array}$$

$$\frac{5}{8} = 0.625$$

38.

$$11\overline{)3.0000...} \quad \begin{array}{r} 0.2727... \\ \end{array}$$

$$\begin{array}{r} 22 \\ \hline 80 \\ 77 \\ \hline 30 \\ 27 \\ \hline 30 \\ 22 \\ \hline 8 \\ \vdots \end{array}$$

$$\frac{3}{11} = 0.\overline{27}$$

39.
  a. $\sqrt{81} \ (=9)$

  b. $0, \sqrt{81}$

  c. $-17, 0, \sqrt{81}$

  d. $-17, -\dfrac{9}{13}, 0, 0.75, \sqrt{81}$

  e. $\sqrt{2}, \pi$

  f. $-17, -\dfrac{9}{13}, 0, 0.75, \sqrt{2}, \pi, \sqrt{81}$

40. Answers will vary. One example of an integer that is not a natural number is -7.

41. Answers will vary. One example of a rational number that is not an integer is $\dfrac{3}{4}$.

42. Answers will vary. One example of a real number that is not a rational number is π.

43. $-93 < 17$; -93 is to the left of 17, so $-93 < 17$.

44. $-2 > -200$; -2 is to the right of -200, so $-2 > -200$.

45. $0 > -\dfrac{1}{3}$; 0 is to the right of $-\dfrac{1}{3}$, so $0 > -\dfrac{1}{3}$.

46.

$$-\frac{1}{4} < -\frac{1}{5}; \quad -\frac{1}{4} = -0.25 \text{ is to the left of}$$

$$-\frac{1}{5} = -0.2, \text{ so } -\frac{1}{4} < -\frac{1}{5}.$$

47. $-13 \geq -11$ is false because neither $-13 > -11$ nor $-13 = -11$ is true.

48. $-126 \leq -126$ is true because $-126 = -126$.

49. $|-58| = 58$ because the distance between -58 and 0 on the number line is 58.

50. $|2.75| = 2.75$ because the distance between 2.75 and 0 on the number line is 2.75.

51. $7 + 13y = 13y + 7$

52. $9(x+7) = (x+7)9$

53. $6 + (4+y) = (6+4) + y = 10 + y$

54. $7(10x) = (7 \cdot 10)x = 70x$

55. $6(4x - 2 + 5y) = 6(4x) + 6(-2) + 6(5y)$
$$= 24x - 12 + 30y$$

56. $4a + 9 + 3a - 7 = 4a + 3a + 9 - 7$
$$= (4+3)a + (9-7)$$
$$= 7a + 2$$

57. $6(3x+4) + 5(2x-1)$
$$= 6(3x) + 6(4) + 5(2x) + 5(-1)$$
$$= 18x + 24 + 10x - 5$$
$$= 18x + 10x + 24 - 5$$
$$= (18+10)x + \left[24 + (-5)\right]$$
$$= 28x + 19$$

58. $-6 + 8 = +2$ or 2.
Start at -6. Move 8 units to the right because 8 is positive.

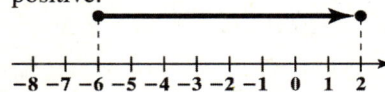

59. $8 + (-11) = -3$

60.
$$-\frac{3}{4} + \frac{1}{5} = -\frac{3}{4} \cdot \frac{5}{5} + \frac{1}{5} \cdot \frac{4}{4}$$
$$= -\frac{15}{20} + \frac{4}{20} = -\frac{11}{20}$$

61. $7 + (-5) + (-13) + 4$
$$= [7 + (-5)] + (-13) + 4$$
$$= 2 + (-13) + 4$$
$$= [2 + (-13)] + 4 = -11 + 4 = -7$$

62. $8x + (-6y) + (-12x) + 11y$
$$= 8x + (-12x) + (-6y) + 11y$$
$$= [8 + (-12)]x + (-6 + 11)y$$
$$= -4x + 5y \text{ or } 5y - 4x$$

63. $10(3y + 4) + (-40y) = 30y + 40 + (-40y)$
$$= 30y + (-40y) + 40$$
$$= -10y + 40$$

64. $-1312 + 512 = -800$
The person's elevation is 800 feet below sea level.

65. $25 - 3 + 2 + 1 - 4 + 2$
$$= 25 + (-3) + 2 + 1 + (-4) + 2$$
$$= 23$$
The reservoir's water level at the end of five months is 23 feet.

66. $9 - 13 = 9 + (-13)$

67. $-9 - (-13) = -9 + 13 = 4$

68.
$$-\frac{7}{10} - \frac{1}{2} = -\frac{7}{10} - \frac{1}{2} \cdot \frac{5}{5}$$
$$= -\frac{7}{10} - \frac{5}{10} = -\frac{12}{10} = -\frac{6}{5}$$

69. $-3.6 - (-2.1) = -3.6 + 2.1 = -1.5$

70. $-7 - (-5) + 11 - 16$
$$= -7 + 5 + 11 + (-16)$$
$$= [(-7) + (-16)] + (5 + 11)$$
$$= -23 + 16$$
$$= -7$$

71. $-25 - 4 - (-10) + 16$
$$= -25 + (-4) + 10 + 16$$
$$= [(-25) + (-4)] + (10 + 16)$$
$$= -29 + 26$$
$$= -3$$

72. $3 - 6a - 8 - 2a = 3 - 8 - 6a - 2a$
$$= [3 + (-8)] + [-6a - 2a]$$
$$= -5 + (-6 - 2)a$$
$$= -5 - 8a$$

73. $26,000 - (-650) = 26,500 + 650$
$$= 27,150$$
The difference in elevation is 27,150 feet.

74. $(-7)(-12) = 84$

75.
$$\frac{3}{5}\left(-\frac{5}{11}\right) = -\frac{3 \cdot \cancel{5}}{\cancel{5} \cdot 11} = -\frac{3}{11}$$

76. $5(-3)(-2)(-4) = -120$

77. $\frac{45}{-5} = 45\left(-\frac{1}{5}\right) = -9$

78. $-17 \div 0$ is undefined.

79.
$$-\frac{4}{5} \div \left(-\frac{2}{5}\right) = -\frac{4}{5}\left(-\frac{5}{2}\right) = \frac{20}{10} = 2$$

80. $-4\left(-\frac{3}{4}x\right) = \left[-4\left(-\frac{3}{4}\right)\right]x = 3x$

81. $-3(2x - 1) - (4 - 5x)$
$$= -3(2x) + (-3)(-1) - 4 + 5x$$
$$= -6x + 3 - 4 + 5x$$
$$= -6x + 5x + 3 - 4$$
$$= (-6 + 5)x + [3 + (-4)]$$
$$= -1x - 1$$
$$= -x - 1$$

82. $5x + 16 = -8 - x$
$5(-6) + 16 = -8 - (-6)$
$-30 + 16 = -8 + 6$
$\quad\quad -14 = -2, \text{ false}$
The number is not a solution.

83. $2(x + 3) - 18 = 5x$
$2(-4 + 3) - 18 = 5(-4)$
$2(-1) - 18 = -20$
$-2 - 18 = -20$
$\quad\quad\quad -20 = -20, \text{ true}$
The number is a solution.

84. $p = -0.6a + 93$
$p = -0.6(20) + 93 = 81$
According to the formula, 81% of 20-year-old taxpayers expect a refund.
This overestimates the actual value shown in the bar graph by 4%.

85. $(-6)^2 = (-6)(-6) = 36$

86. $-6^2 = -6 \cdot 6 = -36$

87. $(-2)^5 = (-2)(-2)(-2)(-2)(-2) = -32$

88. $4x^3 + 2x^3 = (4 + 2)x^3 = 6x^3$

89. $4x^3 + 4x^2$ cannot be simplified. The terms $4x^3$ and $4x^2$ are not like terms because they have different variable factors.

90. $-40 \div 5 \cdot 2 = -8 \cdot 2 = -16$

91. $-6 + (-2) \cdot 5 = -6 + (-10) = -16$

92. $6 - 5(-3 + 2) = 6 - 4(-1) = 6 + 4 = 10$

93. $28 \div (2 - 4^2) = 28 \div (2 - 16)$
$= 28 \div [2 + (-16)]$
$= 28 \div (-14)$
$= -2$

94. $36 - 24 \div 4 \cdot 3 - 1 = 36 - 6 \cdot 3 - 1$
$= 36 - 18 - 1$
$= 18 - 1$
$= 17$

95. $-8[-4 - 5(-3)] = -8(-4 + 15)$
$= -8(11) = -88$

96. $\dfrac{6(-10 + 3)}{2(-15) - 9(-3)} = \dfrac{6(-7)}{-30 + 27}$
$= \dfrac{-42}{-3} = 14$

97. $\left(\dfrac{1}{2} + \dfrac{1}{3}\right) \div \left(\dfrac{1}{4} - \dfrac{3}{8}\right)$
$= \left(\dfrac{3}{6} + \dfrac{2}{6}\right) \div \left(\dfrac{2}{8} - \dfrac{3}{8}\right)$
$= \dfrac{5}{6} \div \left(-\dfrac{1}{8}\right) = \dfrac{5}{6} \cdot \left(-\dfrac{8}{1}\right) = -\dfrac{40}{6} = -\dfrac{20}{3}$

98. $\dfrac{1}{2} - \dfrac{2}{3} \div \dfrac{5}{9} + \dfrac{3}{10}$
$= \dfrac{1}{2} - \dfrac{2}{\cancel{3}_1} \cdot \dfrac{\cancel{9}^3}{5} + \dfrac{3}{10}$
$= \dfrac{1}{2} - \dfrac{6}{5} + \dfrac{3}{10}$
$= \dfrac{5}{10} - \dfrac{12}{10} + \dfrac{3}{10} = -\dfrac{4}{10} = -\dfrac{2}{5}$

99. $x^2 - 2x + 3; \; x = -1$
$x^2 - 2x + 3 = (-1)^2 - 2(-1) + 3$
$= 1 + 2 + 3$
$= 6$

100. $-x^2 - 7x; \; x = -2$
$-x^2 - 7x = -(-2)^2 - 7(-2)$
$= -4 + 14$
$= 10$

101. $4[7(a - 1) + 2] = 4(7a - 7 + 2)$
$= 4(7a - 5)$
$= 4(7a) + 4(-5)$
$= 28a - 20$

102. $-6[4 - (y + 2)] = -6(4 - y - 2)$
$= -6(2 - y)$
$= -6(2) + (-6)(-y)$
$= -12 + 6y \text{ or } 6y - 12$

103. a. $p = 0.002n^2 + 0.3n + 5$

$p = 0.002(50)^2 + 0.3(50) + 5 = 25$

According to the formula, 25% of people 25 years of age and older were college graduates in 2000.
This underestimates the actual value shown in the bar graph by 1%.

b. $p = 0.002n^2 + 0.3n + 5$

$p = 0.002(70)^2 + 0.3(70) + 5 = 35.8$

35.8% of people 25 years of age and older are expected to be college graduates in 2020.

Chapter 1 Test

1. $1.4 - (-2.6) = 1.4 + 2.6 = 4$

$-9 + 3 + (-11) + 6$

2. $= [-9 + (-11)] + (3 + 6)$

$= -20 + 9 = -11$

3. $3(-17) = -51$

4. $\left(-\dfrac{3}{7}\right) \div \left(-\dfrac{15}{7}\right) = \left(-\dfrac{3}{7}\right)\left(-\dfrac{7}{15}\right)$

$= \dfrac{21}{105} = \dfrac{\cancel{21} \cdot 1}{\cancel{21} \cdot 5} = \dfrac{1}{5}$

5. $\left(3\dfrac{1}{3}\right)\left(-1\dfrac{3}{4}\right) = \left(\dfrac{10}{3}\right)\left(-\dfrac{7}{4}\right)$

$= -\dfrac{10 \cdot 7}{3 \cdot 4} = -\dfrac{70}{12}$

$= -\dfrac{\cancel{2} \cdot 35}{\cancel{2} \cdot 6}$

$= -\dfrac{35}{6}$ or $-5\dfrac{5}{6}$

6. $-50 \div 10 = -50\left(\dfrac{1}{10}\right) = -5$

7. $-6 - (5 - 12) = -6 - (-7) = -6 + 7 = 1$

8. $(-3)(-4) \div (7 - 10)$

$= (-3)(-4) \div [7 + (-10)]$

$= (-3)(-4) \div (-3)$

$= 12 \div (-3)$

$= -4$

9. $(6-8)^2 (5-7)^3 = (-2)^2 (-2)^3$

$= 4(-8) = -32$

10. $\dfrac{3(-2) - 2(2)}{-2(8-3)} = \dfrac{-6 - 4}{-2(5)}$

$= \dfrac{-6 + (-4)}{-2(5)} = \dfrac{-10}{-10} = 1$

11. $11x - (7x - 4) = 11x - 7x + 4$

$= 11x + (-7x) + 4$

$= [11 + (-7)]x + 4$

$= 4x + 4$

12. $5(3x - 4y) - (2x - y)$

$= 5(3x) - 5(4y) - 2x + y$

$= 15x - 20y - 2x + y$

$= 15x - 2x - 20y + y$

$= 13x - 19y$

13. $6 - 2[3(x+1) - 5] = 6 - 2[3x + 3 - 5]$

$= 6 - 2(3x - 2)$

$= 6 - 6x + 4$

$= 10 - 6x$

14. Rational numbers can be written as the quotient of two integers.

$-7 = -\dfrac{7}{1}, -\dfrac{4}{5} = \dfrac{-4}{5}, 0 = \dfrac{0}{1}, 0.25 = \dfrac{1}{4},$

$\sqrt{4} = 2 = \dfrac{2}{1},$ and $\dfrac{22}{7} = \dfrac{22}{7}.$

Thus, $-7,$ $-\dfrac{4}{5},$ $0,$ $0.25,$ $\sqrt{4},$ and $\dfrac{22}{7}$ are the rational numbers of the set.

15. $-1 > -100;$ -1 is to the right of -100 on the number line, so -1 is greater than -100.

16. $|-12.8| = 12.8$ because the distance between 12.8 and 0 on the number line is 12.8

17. $5(x-7)$; $x=4$

$5(x-7) = 5(4-7) = 5(-3) = -15$

18. $x^2 - 5x$; $x = -10$

$x^2 - 5x = (-10)^2 - 5(-10)$

$= 100 + 50 = 150$

19. $2(x+3) = 2(3+x)$

20. $-6(4x) = (-6 \cdot 4)x = -24x$

21. $7(5x - 1 + 2y) = 7(5x) - 7(1) + 7(2y)$

$= 35x - 7 + 14y$

22. $16,200 - (-830) = 17,030$

The difference in elevations is 17,030 feet.

23. $\dfrac{1}{5}(x+2) = \dfrac{1}{10}x + \dfrac{3}{5}$

$\dfrac{1}{5}(3+2) = \dfrac{1}{10} \cdot 3 + \dfrac{3}{5}$

$\dfrac{1}{5}(5) = \dfrac{3}{10} + \dfrac{3}{5}$

$1 = \dfrac{3}{10} + \dfrac{6}{10}$

$1 = \dfrac{9}{10}$, false

The number is not a solution.

24. $3(x+2) - 15 = 4x$

$3(-9 + 2) - 15 = 4(-9)$

$3(-7) - 15 = -36$

$-21 - 15 = -36$

$-36 = -36$, true

The number is a solution.

25. $\dfrac{1}{4}x - 5 = 32$

26. $5(x+4) - 7$

27. $H = 0.01n^2 - 0.5n + 31$

$H = 0.01(20)^2 - 0.5(20) + 31 = 25$

According to the formula, there were 25 million multigenerational households in 1970.
This underestimates the actual number shown in the bar graph by 1 million.

28. According to the line graph, the target heart rate for a 40-year-old taking a stress test is about 144 beats per minute.

29. $H = \dfrac{4}{5}(220 - a)$

$H = \dfrac{4}{5}(220 - 40) = \dfrac{4}{5}(180) = 144$

According to the formula, the target heart rate for a 40-year-old taking a stress test is about 144 beats per minute.
This value is the same as the value estimated from the line graph.

Chapter 2
Linear Equations and Inequalities in One Variable

2.1 Check Points

1.
$$x - 5 = 12$$
$$x - 5 + 5 = 12 + 5$$
$$x + 0 = 17$$
$$x = 17$$
Check:
$$x - 5 = 12$$
$$17 - 5 = 12$$
$$12 = 12$$
The solution set is $\{17\}$.

2.
$$z + 2.8 = 5.09$$
$$z + 2.8 - 2.8 = 5.09 - 2.8$$
$$z + 0 = 2.29$$
$$z = 2.29$$
Check:
$$z + 2.8 = 5.09$$
$$2.29 + 2.8 = 5.09$$
$$5.09 = 5.09$$
The solution set is $\{2.29\}$.

3.
$$-\frac{1}{2} = x - \frac{3}{4}$$
$$-\frac{1}{2} + \frac{3}{4} = x - \frac{3}{4} + \frac{3}{4}$$
$$-\frac{2}{4} + \frac{3}{4} = x$$
$$\frac{1}{4} = x$$
Check:
$$-\frac{1}{2} = x - \frac{3}{4}$$
$$-\frac{1}{2} = \frac{1}{4} - \frac{3}{4}$$
$$-\frac{1}{2} = -\frac{2}{4}$$
$$-\frac{1}{2} = -\frac{1}{2}$$
The solution set is $\left\{\frac{1}{4}\right\}$.

4.
$$8y + 7 - 7y - 10 = 6 + 4$$
$$y - 3 = 10$$
$$y - 3 + 3 = 10 + 3$$
$$y = 13$$
Check:
$$8y + 7 - 7y - 10 = 6 + 4$$
$$8(13) + 7 - 7(13) - 10 = 6 + 4$$
$$104 + 7 - 91 - 10 = 10$$
$$111 - 101 = 10$$
$$10 = 10$$
The solution set is $\{13\}$.

5.
$$7x = 12 + 6x$$
$$7x - 6x = 12 + 6x - 6x$$
$$x = 12$$
Check:
$$7(12) = 12 + 6(12)$$
$$84 = 12 + 72$$
$$84 = 84$$
The solution set is $\{12\}$.

6.
$$3x - 6 = 2x + 5$$
$$3x - 2x - 6 = 2x - 2x + 5$$
$$x - 6 = 5$$
$$x - 6 + 6 = 5 + 6$$
$$x = 11$$
Check:
$$3x - 6 = 2x + 5$$
$$3(11) - 6 = 2(11) + 5$$
$$33 - 6 = 22 + 5$$
$$27 = 27$$
The solution set is $\{11\}$.

7.
$$V + 900 = 60A$$
$$V + 900 = 60(50)$$
$$V + 900 = 3000$$
$$V + 900 - 900 = 3000 - 900$$
$$V = 2100$$
At 50 months, a child will have a vocabulary of 2100 words.

2.1 Concept and Vocabulary Check

1. solving

2. linear

3. equivalent

4. $b + c$

5. subtract; solution

6. adding 7

7. subtracting $6x$

2.1 Exercise Set

1. linear

3. not linear

5. not linear

7. linear

9. not linear

11.
$$x - 4 = 19$$
$$x - 4 + 4 = 19 + 4$$
$$x + 0 = 23$$
$$x = 23$$
Check:
$$x - 4 = 19$$
$$23 - 4 = 19$$
$$19 = 19$$
The solution set is $\{23\}$.

13.
$$z + 8 = -12$$
$$z + 8 - 8 = -12 - 8$$
$$z + 0 = -20$$
$$z = -20$$
Check:
$$z + 8 = -12$$
$$-20 + 8 = -12$$
$$-12 = -12$$
The solution set is $\{-20\}$.

15.
$$-2 = x + 14$$
$$-2 - 14 = x + 14 - 14$$
$$-16 = x$$
Check:
$$-2 = -16 + 14$$
$$-2 = -2$$
The solution set is $\{-16\}$.

17.
$$-17 = y - 5$$
$$-17 + 5 = y - 5 + 5$$
$$-12 = y$$
Check:
$$-17 = -12 - 5$$
$$-17 = -17$$
The solution set is $\{-12\}$.

19.
$$7 + z = 11$$
$$z = 11 - 7$$
$$z = 4$$
Check:
$$7 + 4 = 11$$
$$11 = 11$$
The solution set is $\{4\}$.

21.
$$-6 + y = -17$$
$$y = -17 + 6$$
$$y = -11$$
Check:
$$-6 - 11 = -17$$
$$-17 = -17$$
The solution set is $\{-11\}$.

23.
$$x + \frac{1}{3} = \frac{7}{3}$$
$$x = \frac{7}{3} - \frac{1}{3}$$
$$x = 2$$
Check:
$$2 + \frac{1}{3} = \frac{7}{3}$$
$$\frac{6}{3} + \frac{1}{3} = \frac{7}{3}$$
$$\frac{7}{3} = \frac{7}{3}$$
The solution set is $\{2\}$.

25.
$$t + \frac{5}{6} = -\frac{7}{12}$$
$$t = -\frac{7}{12} - \frac{5}{6}$$
$$t = -\frac{7}{12} - \frac{10}{12} = -\frac{17}{12}$$

Check:
$$-\frac{17}{15} + \frac{5}{6} = -\frac{7}{12}$$
$$-\frac{17}{12} + \frac{10}{12} = -\frac{7}{12}$$
$$-\frac{7}{12} = -\frac{7}{12}$$

The solution set is $\left\{-\frac{17}{12}\right\}$.

27.
$$x - \frac{3}{4} = \frac{9}{2}$$
$$x - \frac{3}{4} + \frac{3}{4} = \frac{9}{2} + \frac{3}{4}$$
$$x = \frac{21}{4}$$

Check:
$$\frac{21}{4} - \frac{3}{4} = \frac{9}{2}$$
$$\frac{18}{4} = \frac{9}{2}$$
$$\frac{9}{2} = \frac{9}{2}$$

The solution set is $\left\{\frac{21}{4}\right\}$.

29.
$$-\frac{1}{5} + y = -\frac{3}{4}$$
$$y = -\frac{3}{4} + \frac{1}{5}$$
$$y = -\frac{15}{20} + \frac{4}{20} = -\frac{11}{20}$$

Check:
$$-\frac{1}{5} + \left(-\frac{11}{20}\right) = -\frac{3}{4}$$
$$-\frac{4}{20} - \frac{11}{20} = -\frac{3}{4}$$
$$-\frac{15}{20} = -\frac{3}{4}$$
$$-\frac{3}{4} = -\frac{3}{4}$$

The solution set is $\left\{-\frac{11}{20}\right\}$.

31.
$$3.2 + x = 7.5$$
$$3.2 + x - 3.2 = 7.5 - 3.2$$
$$x = 4.3$$

Check:
$$3.2 + 4.3 = 7.5$$
$$7.5 = 7.5$$

The solution set is $\{4.3\}$.

33.
$$x + \frac{3}{4} = -\frac{9}{2}$$
$$x + \frac{3}{4} - \frac{3}{4} = -\frac{9}{2} - \frac{3}{4}$$
$$x = -\frac{21}{4}$$

Check:
$$-\frac{21}{4} + \frac{3}{4} = -\frac{9}{2}$$
$$-\frac{18}{4} = -\frac{9}{2}$$
$$-\frac{9}{2} = -\frac{9}{2}$$

The solution set is $\left\{-\frac{21}{4}\right\}$.

35.
$$5 = -13 + y$$
$$5 + 13 = y$$
$$18 = y$$
Check:
$$5 = -13 + 18$$
$$5 = 5$$
The solution set is $\{18\}$.

37.
$$-\frac{3}{5} = -\frac{3}{2} + s$$
$$-\frac{3}{5} + \frac{3}{2} = s$$
$$-\frac{6}{10} + \frac{15}{10} = s$$
$$\frac{9}{10} = s$$
Check:
$$-\frac{3}{5} = -\frac{3}{2} + \frac{9}{10}$$
$$-\frac{6}{10} = -\frac{15}{10} + \frac{9}{10}$$
$$-\frac{6}{10} = -\frac{6}{10}$$
The solution set is $\left\{\frac{9}{10}\right\}$.

39.
$$830 + y = 520$$
$$y = 520 - 830$$
$$y = -310$$
Check:
$$830 - 310 = 520$$
$$520 = 520$$
The solution set is $\{-310\}$.

41. $r + 3.7 = 8$
$$r = 8 - 3.7$$
$$r = 4.3$$
Check:
$$4.3 + 3.7 = 8$$
$$8 = 8$$
The solution set is $\{4.3\}$.

43. $-3.7 + m = -3.7$
$$m = -3.7 + 3.7$$
$$m = 0$$
Check:
$$-3.7 + 0 = -3.7$$
$$-3.7 = -3.7$$
The solution set is $\{0\}$.

45. $6y + 3 - 5y = 14$
$$y + 3 = 14$$
$$y = 14 - 3$$
$$y = 11$$
Check:
$$6(11) + 3 - 5(11) = 14$$
$$66 + 3 - 55 = 14$$
$$14 = 14$$
The solution set is $\{11\}$.

47.
$$7 - 5x + 8 + 2x + 4x - 3 = 2 + 3 \cdot 5$$
$$(-5x + 2x + 4x) + (7 + 8 - 3) = 2 + 15$$
$$x + 12 = 17$$
$$x + 12 - 12 = 17 - 12$$
$$x = 5$$
Check:
$$7 - 5(5) + 8 + 2(5) + 4(5) - 3 = 2 + 3 \cdot 5$$
$$7 - 25 + 8 + 10 + 20 - 3 = 2 + 15$$
$$17 = 17$$
The solution set is $\{5\}$.

49.
$$7y + 4 = 6y - 9$$
$$7y - 6y + 4 = -9$$
$$y = -9 - 4$$
$$y = -13$$
Check:
$$7(-13) + 4 = 6(-13) - 9$$
$$-91 + 4 = -78 - 9$$
$$-87 = -87$$
The solution set is $\{-13\}$.

51. $12 - 6x = 18 - 7x$

$12 + x = 18$

$x = 6$

Check:

$12 - 6(6) = 18 - 7(6)$

$12 - 36 = 18 - 42$

$-24 = -24$

The solution set is $\{6\}$.

53. $4x + 2 = 3(x - 6) + 8$

$4x + 2 = 3x - 18 + 8$

$4x + 2 = 3x - 10$

$4x - 3x + 2 = -10$

$x + 2 = -10$

$x = -10 - 2$

$x = -12$

Check:

$4(-12) + 2 = 3(-12 - 6) + 8$

$-48 + 2 = 3(-18) + 8$

$-46 = -54 + 8$

$-46 = -46$

The solution set is $\{-12\}$.

55. $x - \square = \triangle$

$x - \square + \square = \triangle + \square$

$x = \triangle + \square$

57. $2x + \triangle = 3x + \square$

$\triangle = 3x - 2x + \square$

$\triangle = x + \square$

$\triangle - \square = x + \square - \square$

$\triangle - \square = x$

59. $x - 12 = -2$

$x = -2 + 12$

$x = 10$

The number is 10.

61. $\dfrac{2}{5}x - 8 = \dfrac{7}{5}x$

$-8 = \dfrac{7}{5}x - \dfrac{2}{5}x$

$-8 = \dfrac{5}{5}x$

$-8 = x$

The number is -8.

63. $S = 1850, \ M = 150$

$C + M = S$

$C + 150 = 1850$

$C = 1850 - 150$

$C = 1700$

The cost of the computer is $1700.

65. $d - 257x = 8328$

$d - 257(8) = 8328$

$d - 2056 = 8328$

$d - 2056 + 2056 = 8328 + 2056$

$d = 10,384$

According to the formula, the average credit-card debt per U.S. household was $10,384 in 2008. This underestimates the value given in the bar graph by $307.

67. a. According to the line graph, the U.S. diversity index was about 52 in 2009.

b. 2009 is 29 years after 1980.

$I - 0.6x = 34$

$I - 0.6(29) = 34$

$I - 17.4 = 34$

$I - 17.4 + 17.4 = 34 + 17.4$

$I = 51.4$

According to the formula, the U.S. diversity index was 51.4 in 2009. This matches the line graph very well.

69. – 71. Answers will vary.

73. does not make sense; Explanations will vary. Sample explanation: It does not matter whether the number is added beside or below, as long as it is added to both sides of the equation.

75. makes sense

77. false; Changes to make the statement true will vary. A sample change is: If $y - a = -b$, then $y = a - b$.

79. true

81. Answers will vary. An example is: $x - 100 = -101$

83. $6.9825 = 4.2296 + y$

$6.9825 - 4.2296 = y$

$2.7529 = y$

The solution set is $\{2.7529\}$.

84. $\dfrac{9}{x} - 4x$

85. $-16 - 8 \div 4 \cdot (-2) = -16 - 2 \cdot (-2)$
$$= -16 + (-2)(-2)$$
$$= -16 + 4$$
$$= -12$$

86. $3\left[7x - 2(5x - 1)\right] = 3\left[7x - 10x + 2\right]$
$$= 3\left[-3x + 2\right]$$
$$= -9x + 6 \text{ or } 6 - 9x$$

87. $5 \cdot \dfrac{x}{5} = \dfrac{5}{1} \cdot \dfrac{x}{5} = x$

88. $\dfrac{-7y}{-7} = y$

89. $\quad 3x - 14 = -2x + 6$
$$3(4) - 14 = -2(4) + 6$$
$$12 - 14 = -8 + 6$$
$$-2 = -2, \text{ true}$$
Yes, 4 is a solution of the equation.

2.2 Check Points

1. $\quad \dfrac{x}{3} = 12$
$$3 \cdot \dfrac{x}{3} = 12 \cdot 3$$
$$1x = 36$$
$$x = 36$$
Check:
$$\dfrac{x}{3} = 12$$
$$\dfrac{36}{3} = 12$$
$$12 = 12$$
The solution set is $\{36\}$.

2. a. $\quad 4x = 84$
$$\dfrac{4x}{4} = \dfrac{84}{4}$$
$$1x = 21$$
$$x = 21$$
The solution set is $\{21\}$.

b. $\quad -11y = 44$
$$\dfrac{-11y}{-11} = \dfrac{44}{-11}$$
$$1x = -4$$
$$x = -4$$
The solution set is $\{-4\}$.

c. $\quad -15.5 = 5z$
$$\dfrac{-15.5}{5} = \dfrac{5z}{5}$$
$$-3.1 = 1z$$
$$-3.1 = z$$
The solution set is $\{-3.1\}$.

3. a. $\quad \dfrac{2}{3}y = 16$
$$\dfrac{3}{2}\left(\dfrac{2}{3}y\right) = \dfrac{3}{2} \cdot 16$$
$$1y = 24$$
$$y = 24$$
The solution set is $\{24\}$.

b. $\quad 28 = -\dfrac{7}{4}x$
$$-\dfrac{4}{7} \cdot 28 = -\dfrac{4}{7}\left(-\dfrac{7}{4}x\right)$$
$$-16 = 1x$$
$$-16 = x$$
The solution set is $\{-16\}$.

4. a. $\quad -x = 5$
$$-1x = 5$$
$$(-1)(-1x) = (-1)5$$
$$1x = -5$$
$$x = -5$$
The solution set is $\{-5\}$.

b. $\quad -x = -3$
$$-1x = -3$$
$$(-1)(-1x) = (-1)(-3)$$
$$1x = 3$$
$$x = 3$$
The solution set is $\{3\}$.

5.
$$4x + 3 = 27$$
$$4x + 3 - 3 = 27 - 3$$
$$4x = 24$$
$$\frac{4x}{4} = \frac{24}{4}$$
$$x = 6$$
The solution set is $\{6\}$.

6.
$$-4y - 15 = 25$$
$$-4y - 15 + 15 = 25 + 15$$
$$-4y = 40$$
$$\frac{-4y}{-4} = \frac{40}{-4}$$
$$y = -10$$
The solution set is $\{-10\}$.

7.
$$2x - 15 = -4x + 21$$
$$2x + 4x - 15 = -4x + 4x + 21$$
$$6x - 15 = 21$$
$$6x - 15 + 15 = 21 + 15$$
$$6x = 36$$
$$\frac{6x}{6} = \frac{36}{6}$$
$$x = 6$$
The solution set is $\{6\}$.

8. a. The bar graph indicates that the price of a Westie puppy was $2000 in 2009. Since 2009 is 69 years after 1940, substitute 69 into the formula for n.
$$P = 18n + 765$$
$$P = 18(69) + 765$$
$$P = 1242 + 765$$
$$P = 2007$$
The formula indicates that the price of a Westie puppy was $2007 in 2009.
The formula overestimates by $7.

b.
$$P = 18n + 765$$
$$2151 = 18n + 765$$
$$2151 - 765 = 18n + 765 - 765$$
$$1386 = 18n$$
$$\frac{1386}{18} = \frac{18n}{18}$$
$$77 = n$$
The formula estimates that the price will be $2151 for a Westie puppy 77 years after 1940, or in 2017.

2.2 Concept and Vocabulary Check

1. bc

2. divide

3. multiplying; 7

4. dividing; -8
Alternatively, multiplying; $-\frac{1}{8}$

5. multiplying; $\frac{5}{3}$

6. multiplying/dividing; -1

7. subtracting 2; dividing; 5

2.2 Exercise Set

1.
$$\frac{x}{6} = 5$$
$$6 \cdot \frac{x}{6} = 6 \cdot 5$$
$$1x = 30$$
$$x = 30$$
Check:
$$\frac{30}{6} = 5$$
$$5 = 5$$
The solution set is $\{30\}$.

3.
$$\frac{x}{-3} = 11$$
$$-3 \cdot \frac{x}{-3} = -3(11)$$
$$1x = -33$$
$$x = -33$$
Check:
$$\frac{-33}{-3} = 11$$
$$11 = 11$$
The solution set is $\{-33\}$.

5. $5y = 35$

$\dfrac{5y}{5} = \dfrac{35}{5}$

$y = 7$

Check:

$5(7) = 35$

$35 = 35$

The solution set is $\{7\}$.

7. $-7y = 63$

$\dfrac{-7y}{-7} = \dfrac{63}{-7}$

$y = -9$

Check:

$-7(-9) = 63$

$63 = 63$

The solution set is $\{-9\}$.

9. $-28 = 8z$

$\dfrac{-28}{8} = \dfrac{8z}{8}$

$-\dfrac{7}{2} = z$

Check:

$-28 = 8\left(-\dfrac{7}{2}\right)$

$-28 = -\dfrac{56}{2}$

$-28 = -28$

The solution set is $\left\{-\dfrac{7}{2}\right\}$. or $\left\{-3\dfrac{1}{2}\right\}$.

11. $-18 = -3z$

$\dfrac{-18}{-3} = \dfrac{-3z}{-3}$

$6 = z$

Check:

$-18 = -3(6)$

$-18 = -18$

The solution set is $\{6\}$.

13. $-8x = 6$

$\dfrac{-8x}{-8} = \dfrac{6}{-8}$

$x = -\dfrac{6}{8} = -\dfrac{3}{4}$

Check:

$-8\left(-\dfrac{3}{4}\right) = 6$

$\dfrac{24}{4} = 6$

$6 = 6$

The solution set is $\left\{-\dfrac{3}{4}\right\}$.

15. $17y = 0$

$\dfrac{17y}{17} = \dfrac{0}{17}$

$y = 0$

Check:

$17(0) = 0$

$0 = 0$

The solution set is $\{0\}$.

17. $\dfrac{2}{3}y = 12$

$\dfrac{3}{2}\left(\dfrac{2}{3}y\right) = \dfrac{3}{2}(12)$

$1y = \dfrac{3}{2} \cdot \dfrac{12}{1} = \dfrac{36}{2}$

$y = 18$

Check:

$\dfrac{2}{3}(18) = 12$

$\dfrac{36}{3} = 12$

$12 = 12$

The solution set is $\{18\}$.

19.
$$28 = -\frac{7}{2}x$$
$$-\frac{2}{7}(28) = -\frac{2}{7}\left(-\frac{7}{2}x\right)$$
$$-\frac{56}{7} = 1x$$
$$-8 = x$$

Check:
$$28 = -\frac{7}{2}(-8)$$
$$28 = \frac{56}{2}$$
$$28 = 28$$

The solution set is $\{-8\}$.

21.
$$-x = 17$$
$$-1x = 17$$
$$-1(-1x) = -1(17)$$
$$x = -17$$

Check:
$$-(-17) = 17$$
$$17 = 17$$

The solution set is $\{-17\}$.

23.
$$-47 = -y$$
$$-47 = -1(-y)$$
$$-1(-47) = -1(-1)(-y)$$
$$47 = y$$

Check:
$$-47 = -y$$
$$-47 = -(47)$$
$$-47 = -47$$

The solution set is $\{47\}$.

25.
$$-\frac{x}{5} = -9$$
$$5\left(-\frac{x}{5}\right) = 5(-9)$$
$$-x = -45$$
$$x = 45$$

Check:
$$-\frac{45}{5} = -9$$
$$-9 = -9$$

The solution set is $\{45\}$.

27.
$$2x - 12x = 50$$
$$(2-12)x = 50$$
$$-10x = 50$$
$$\frac{-10x}{-10} = \frac{50}{-10}$$
$$x = -5$$

Check:
$$2(-5) - 12(-5) = 50$$
$$-10 + 60 = 50$$
$$50 = 50$$

The solution set is $\{-5\}$.

29.
$$2x + 1 = 11$$
$$2x + 1 - 1 = 11 - 1$$
$$2x = 10$$
$$\frac{2x}{2} = \frac{10}{2}$$
$$x = 5$$

Check:
$$2(5) + 1 = 11$$
$$10 + 1 = 11$$
$$11 = 11$$

The solution set is $\{5\}$.

31.
$$2x - 3 = 9$$
$$2x - 3 + 3 = 9 + 3$$
$$2x = 12$$
$$\frac{2x}{2} = \frac{12}{2}$$
$$x = 6$$

Check:
$$2(6) - 3 = 9$$
$$12 - 3 = 9$$
$$9 = 9$$

The solution set is $\{6\}$.

33.
$$-2y + 5 = 7$$
$$-2y + 5 - 5 = 7 - 5$$
$$-2y = 2$$
$$\frac{-2y}{2} = \frac{2}{-2}$$
$$y = -1$$
Check:
$$-2(-1) + 5 = 7$$
$$2 + 5 = 7$$
$$7 = 7$$
The solution set is $\{-1\}$.

35.
$$-3y - 7 = -1$$
$$-3y - 7 + 7 = -1 + 7$$
$$-3y = 6$$
$$\frac{-3y}{-3} = \frac{6}{-3}$$
$$y = -2$$
Check:
$$-3(-2) - 7 = -1$$
$$6 - 7 = -1$$
$$-1 = -1$$
The solution set is $\{-2\}$.

37.
$$12 = 4z + 3$$
$$12 - 3 = 4z + 3 - 3$$
$$9 = 4z$$
$$\frac{9}{4} = \frac{4z}{4}$$
$$\frac{9}{4} = z$$
Check:
$$12 = 4\left(\frac{9}{4}\right) + 3$$
$$12 = 9 + 3$$
$$12 = 12$$
The solution set is $\left\{\frac{9}{4}\right\}$.

39.
$$-x - 3 = 3$$
$$-x - 3 + 3 = 3 + 3$$
$$-x = 6$$
$$x = -6$$
Check:
$$-(-6) - 3 = 3$$
$$6 - 3 = 3$$
$$3 = 3$$
The solution set is $\{-6\}$.

41.
$$6y = 2y - 12$$
$$6y + 12 = 2y - 12 + 12$$
$$6y + 12 = 2y$$
$$6y + 12 - 6y = 2y - 6y$$
$$12 = -4y$$
$$\frac{12}{-4} = \frac{-4y}{-4}$$
$$-3 = y$$
Check:
$$6(-3) = 2(-3) - 12$$
$$-18 = -6 - 12$$
$$-18 = -18$$
The solution set is $\{-3\}$.

43.
$$3z = -2z - 15$$
$$3z + 2z = -2z - 15 + 2z$$
$$5z = -15$$
$$\frac{5z}{5} = \frac{-15}{3}$$
$$z = -3$$
Check:
$$3(-3) = -2(-3) - 15$$
$$-9 = 6 - 15$$
$$-9 = -9$$
The solution set is $\{-3\}$.

45.
$$-5x = -2x - 12$$
$$-5x + 2x = -2x - 12 + 2x$$
$$-3x = -12$$
$$\frac{-3x}{3} = \frac{-12}{-3}$$
$$x = 4$$
Check:
$$-5(4) = 2(4) - 12$$
$$-20 = -8 - 12$$
$$-20 = -20$$
The solution set is $\{4\}$.

47.
$$8y + 4 = 2y - 5$$
$$8y + 4 - 2y = 2y - 5 - 2y$$
$$6y + 4 = -5$$
$$6y + 4 - 4 = -5 - 4$$
$$6y = -9$$
$$\frac{6y}{6} = \frac{-9}{6}$$
$$y = -\frac{3}{2}$$
Check:
$$8\left(-\frac{3}{2}\right) + 4 = 2\left(-\frac{3}{2}\right) - 5$$
$$-12 + 4 = -3 - 5$$
$$-8 = -8$$
The solution set is $\left\{-\frac{3}{2}\right\}$.

49.
$$6z - 5 = z + 5$$
$$6z - 5 - z = z + 5 - z$$
$$5z - 5 = 5$$
$$5z - 5 + 5 = 5 + 5$$
$$5z = 10$$
$$\frac{5z}{5} = \frac{10}{5}$$
$$z = 2$$
Check:
$$6(2) - 5 = 2 + 5$$
$$12 - 5 = 2 + 5$$
$$7 = 7$$
The solution set is $\{2\}$.

51.
$$6x + 14 = 2x - 2$$
$$6x - 2x + 14 = -2$$
$$4x = -2 - 14$$
$$4x = -16$$
$$x = -4$$
Check:
$$6(-4) + 14 = 2(-4) - 2$$
$$-24 + 14 = -8 - 2$$
$$-10 = -10$$
The solution set is $\{-4\}$.

53.
$$-3y - 1 = 5 - 2y$$
$$-3y + 2y - 1 = 5$$
$$-y = 5 + 1$$
$$-y = 6$$
$$y = -6$$
Check:
$$-3(-6) - 1 = 5 - 2(-6)$$
$$18 - 1 = 5 + 12$$
$$17 = 17$$
The solution set is $\{-6\}$.

55.
$$\frac{x}{\square} = \triangle$$
$$\square \cdot \frac{x}{\square} = \triangle \cdot \square$$
$$x = \triangle \square$$

57.
$$\triangle = -x$$
$$\triangle(-1) = -x(-1)$$
$$-\triangle = x$$

59.
$$6x = 10$$
$$\frac{6x}{6} = \frac{10}{6}$$
$$x = \frac{10}{6} = \frac{5}{3}$$
The number is $\frac{5}{3}$.

61.
$$\frac{x}{-9} = 5$$
$$\frac{x}{-9}(-9) = 5(-9)$$
$$x = -45$$
The number is -45.

63. $4x - 8 = 56$

$4x - 8 + 8 = 56 + 8$

$4x = 64$

$\dfrac{4x}{4} = \dfrac{64}{4}$

$x = 16$

The number is 16.

65. $-3x + 15 = -6$

$-3x + 15 - 15 = -6 - 15$

$-3x = -21$

$\dfrac{-3x}{-3} = \dfrac{-21}{-3}$

$x = 7$

The number is 7.

67. $M = \dfrac{n}{5}$

$2 = \dfrac{n}{5}$

$5(2) = 5\left(\dfrac{n}{5}\right)$

$10 = n$

If you are 2 miles away from the lightning flash, it will take 10 seconds for the sound of thunder to reach you.

69. $M = \dfrac{A}{740}$

$2.03 = \dfrac{A}{740}$

$740(2.03) = 740 \cdot \dfrac{A}{740}$

$1502.2 = A$

The speed of the Concorde is 1502.2 miles per hour.

71. a. The bar graph indicates that the least expensive Ford automobile was $16,000.
Since 2009 is 105 years after 1904, substitute 105 into the formula for n.
$F = -48n + 21,000$
$F = -48(105) + 21,000$
$F = -5040 + 21,000$
$F = 15,960$
The formula indicates that the least expensive Ford automobile was $15,960 in 2009. The formula underestimates by $40.

b. $F = -48n + 21,000$

$15,000 = -48n + 21,000$

$15,000 - 21,000 = -48n + 21,000 - 21,000$

$-6000 = -48n$

$\dfrac{-6000}{-48} = \dfrac{-48n}{-48}$

$125 = n$

The formula estimates that $15,000 will be the cost of the least expensive Ford automobile 125 years after 1904, or 2029.

73. – 75. Answers will vary.

77. does not make sense; Explanations will vary. Sample explanation: When you subtract 12 from $12 - 3x$, you should obtain $-3x$, not positive $3x$.

79. does not make sense; Explanations will vary. Sample explanation: To determine the price in 2009, substitute 69 in for n and simplify.

81. false; Changes to make the statement true will vary. A sample change is: If $3x - 4 = 16$, then $3x = 20$.

83. true

85. Answers will vary. As an example, start with an integer solution, such as 10, and set it equal to x. That is, we have $x = 10$. The solution was obtained by multiplying both sides by $\dfrac{4}{5}$. To undo this, we multiply both sides of our equation by the reciprocal, $\dfrac{5}{4}$. This gives, $\dfrac{5}{4}x = \dfrac{5}{4}(10)$

$\dfrac{5}{4}x = \dfrac{25}{2}$

Therefore, an example equation would be $\dfrac{5}{4}x = \dfrac{25}{2}$.

87. $-72.8y - 14.6 = -455.43 - 4.98y$

$-72.8y - 14.6 + 4.98y =$
$\qquad\qquad -455.43 - 4.98y + 4.98y$

$-67.82y - 14.6 = -455.43$

$-67.82y - 14.6 + 14.6 = -455.43 + 14.6$

$-67.82y = -440.83$

$\dfrac{-67.82y}{-67.82} = \dfrac{-440.83}{-67.82}$

$y = 6.5$

The solution set is $\{6.5\}$.

88. $(-10)^2 = (-10)(-10) = 100$

89. $-10^2 = -1 \cdot 10^2 = -1(10)(10) = -100$

90. $x^3 - 4x = (-1)^3 - 4(-1)$
$\qquad\qquad = -1 + 4$
$\qquad\qquad = 3$

91. $13 - 3(x+2) = 13 - 3x - 6$
$\qquad\qquad\qquad = -3x + 7$

92. $2(x-3) - 17 = 13 - 3(x+2)$
$\quad 2(6-3) - 17 = 13 - 3(6+2)$
$\qquad 2(3) - 17 = 13 - 3(8)$
$\qquad\quad 6 - 17 = 13 - 24$
$\qquad\qquad -11 = -11, \text{ true}$
Yes, 6 is a solution of the equation.

93. $10\left(\dfrac{x}{5} - \dfrac{39}{5}\right) = 10 \cdot \dfrac{x}{5} - 10 \cdot \dfrac{39}{5}$
$\qquad\qquad\qquad\quad = 2x - 78$

2.3 Check Points

1. Simplify the algebraic expression on each side.
$-7x + 25 + 3x = 16 - 2x - 3$
$\qquad -4x + 25 = 13 - 2x$
Collect variable terms on one side and constant terms on the other side.
$\qquad\quad -4x + 25 = 13 - 2x$
$-4x + 25 + 2x = 13 - 2x + 2x$
$\qquad\quad -2x + 25 = 13$
$-2x + 25 - 25 = 13 - 25$
$\qquad\qquad -2x = -12$
Isolate the variable and solve.
$\qquad \dfrac{-2x}{-2} = \dfrac{-12}{-2}$
$\qquad\qquad x = 6$
The solution set is $\{6\}$.

2. Simplify the algebraic expression on each side.
$\qquad 8x = 2(x+6)$
$\qquad 8x = 2x + 12$
Collect variable terms on one side and constant terms on the other side.
$8x - 2x = 2x - 2x + 12$
$\qquad 6x = 12$
Isolate the variable and solve.
$\qquad \dfrac{6x}{6} = \dfrac{12}{6}$
$\qquad\quad x = 2$
The solution set is $\{2\}$.

3. Simplify the algebraic expression on each side.
$4(2x+1) - 29 = 3(2x-5)$
$\quad 8x + 4 - 29 = 6x - 15$
$\qquad\quad 8x - 25 = 6x - 15$
Collect variable terms on one side and constant terms on the other side.
$8x - 6x - 25 = 6x - 6x - 15$
$\qquad 2x - 25 = -15$
$2x - 25 + 25 = -15 + 25$
$\qquad\qquad 2x = 10$
Isolate the variable and solve.
$\qquad \dfrac{2x}{2} = \dfrac{10}{2}$
$\qquad\quad x = 5$
The solution set is $\{5\}$.

4. Begin by multiplying both sides of the equation by 12, the least common denominator.
$$\frac{x}{4} = \frac{2x}{3} + \frac{5}{6}$$
$$12 \cdot \frac{x}{4} = 12\left(\frac{2x}{3} + \frac{5}{6}\right)$$
$$12 \cdot \frac{x}{4} = 12 \cdot \frac{2x}{3} + 12 \cdot \frac{5}{6}$$
$$3x = 8x + 10$$
$$3x - 8x = 8x - 8x + 10$$
$$-5x = 10$$
$$\frac{-5x}{-5} = \frac{10}{-5}$$
$$x = -2$$
The solution set is $\{-2\}$.

5. First apply the distributive property to remove the parentheses, and then multiply both sides by 100 to clear the decimals.

$$0.48x + 3 = 0.2(x - 6)$$
$$0.48x + 3 = 0.2x - 1.2$$
$$100(0.48x + 3) = 100(0.2x - 1.2)$$
$$48x + 300 = 20x - 120$$
$$48x + 300 - 300 = 20x - 120 - 300$$
$$48x = 20x - 420$$
$$48x - 20x = 20x - 20x - 420$$
$$28x = -420$$
$$\frac{28x}{28} = \frac{-420}{28}$$
$$x = -15$$

The solution set is $\{-15\}$.

6.
$$3x + 7 = 3(x + 1)$$
$$3x + 7 = 3x + 3$$
$$3x - 3x + 7 = 3x - 3x + 3$$
$$7 = 3$$

The original equation is equivalent to the false statement $7 = 3$.
The equation has no solution. The solution set is $\{\ \}$.

7.
$$3(x - 1) + 9 = 8x + 6 - 5x$$
$$3x - 3 + 9 = 3x + 6$$
$$3x + 6 = 3x + 6$$
$$3x - 3x + 6 = 3x - 3x + 6$$
$$6 = 6$$

The original equation is equivalent to $6 = 6$, which is true for every value of x.
The equation's solution is all real numbers or $\{x | x \text{ is a real number}\}$.

8.
$$D = \frac{10}{9}x + \frac{53}{9}$$
$$10 = \frac{10}{9}x + \frac{53}{9}$$
$$9 \cdot 10 = 9\left(\frac{10}{9}x + \frac{53}{9}\right)$$
$$90 = 10x + 53$$
$$90 - 53 = 10x + 53 - 53$$
$$37 = 10x$$
$$\frac{37}{10} = \frac{10x}{10}$$
$$3.7 = x$$
$$x = 3.7$$

The formula indicates that if the low-humor group averages a level of depression of 10 in response to a negative life event, the intensity of that event is 3.7. This is shown as the point whose corresponding value on the vertical axis is 10 and whose value on the horizontal axis is 3.7.

2.3 Concept and Vocabulary Check

1. simplify each side; combine like terms

2. 30

3. 100

4. inconsistent

5. identity

6. inconsistent

7. identity

2.3 Exercise Set

1.
$$5x + 3x - 4x = 10 + 2$$
$$8x - 4x = 12$$
$$4x = 12$$
$$\frac{4x}{4} = \frac{12}{4}$$
$$x = 3$$

The solution set is $\{3\}$.

3.
$$4x - 9x + 22 = 3x + 30$$
$$-5x + 22 = 3x + 30$$
$$-5x - 3x + 22 = 30$$
$$-8x + 22 = 30$$
$$-8x = 30 - 22$$
$$-8x = 8$$
$$\frac{-8x}{-8} = \frac{8}{-8}$$
$$x = -1$$
The solution set is $\{-1\}$.

5.
$$3x + 6 - x = 8 + 3x - 6$$
$$2x + 6 = 2 + 3x$$
$$2x + 6 - 2 = 2 + 3x - 2$$
$$2x + 4 = 3x$$
$$2x + 4 - 2x = 3x - 2x$$
$$4 = x$$
The solution set is $\{4\}$.

7.
$$4(x + 1) = 20$$
$$4x + 4 = 20$$
$$4x = 20 - 4$$
$$4x = 16$$
$$\frac{4x}{4} = \frac{16}{4}$$
$$x = 4$$
The solution set is $\{4\}$.

9.
$$4(x + 1) = 20$$
$$4x + 4 = 20$$
$$4x = 20 - 4$$
$$4x = 16$$
$$\frac{4x}{4} = \frac{16}{4}$$
$$x = 4$$
The solution set is $\{4\}$.

11.
$$38 = 30 - 2(x - 1)$$
$$38 = 30 - 2x + 2$$
$$38 = 32 - 2x$$
$$38 - 32 = -2x$$
$$6 = -2x$$
$$\frac{6}{-2} = \frac{-2x}{-2}$$
$$-3 = x$$
The solution set is $\{-3\}$.

13.
$$2(4z + 3) - 8 = 46$$
$$8z + 6 - 8 = 46$$
$$8z - 2 = 46$$
$$8z - 2 + 2 = 46 + 2$$
$$8z = 48$$
$$\frac{8z}{3} = \frac{48}{8}$$
$$z = 6$$
The solution set is $\{6\}$.

15.
$$6x - (3x + 10) = 14$$
$$6x - 3x - 10 = 14$$
$$3x - 10 = 14$$
$$3x - 10 + 10 = 14 + 10$$
$$3x = 24$$
$$\frac{3x}{3} = \frac{24}{3}$$
$$x = 8$$
The solution set is $\{8\}$.

17.
$$5(2x + 1) = 12x - 3$$
$$10x + 5 = 12x - 3$$
$$10x - 10x + 5 = 12x - 10x - 3$$
$$5 = 2x - 3$$
$$5 + 3 = 2x - 3 + 3$$
$$8 = 2x$$
$$\frac{8}{2} = \frac{2x}{2}$$
$$x = 4$$
The solution set is $\{4\}$.

19.
$$3(5-x) = 4(2x+1)$$
$$15-3x = 8x+4$$
$$15-3x-8x = 8x+4-8x$$
$$15-11x = 4$$
$$15-11x-15 = 4-15$$
$$-11x = -11$$
$$\frac{-11x}{-11} = \frac{-11}{-11}$$
$$x = 1$$
The solution set is $\{1\}$.

21.
$$8(y+2) = 2(3y+4)$$
$$8y+16 = 6y+8$$
$$8y+16-16 = 6y+8-16$$
$$8y = 6y-8$$
$$8y-6y = 6y-8-6y$$
$$2y = -8$$
$$y = -4$$
The solution set is $\{-4\}$.

23.
$$3x+3 = 7x-14-3$$
$$3x+3 = 7x-17$$
$$3x+3-3 = 7x-17-3$$
$$3x = 7x-20$$
$$3x-7x = 7x-20-7x$$
$$-4x = -20$$
$$\frac{-4x}{-4} = \frac{-20}{-4}$$
$$x = 5$$
The solution set is $\{5\}$.

25.
$$5(2x-8)-2 = 5(x-3)+3$$
$$10x-40-2 = 5x-15+3$$
$$10x-42 = 5x-12$$
$$10x-42+42 = 5x-12+42$$
$$10x = 5x+30$$
$$10x = 5x+30-5x$$
$$5x = 30$$
$$\frac{5x}{5} = \frac{30}{5}$$
$$x = 6$$
The solution set is $\{6\}$.

27.
$$6 = -4(1-x)+3(x+1)$$
$$6 = -4+4x+3x+3$$
$$6 = -1+7x$$
$$6+1 = -1+7x+1$$
$$7 = 7x$$
$$\frac{7}{7} = \frac{7x}{7}$$
$$1 = x$$
The solution set is $\{1\}$.

29.
$$10(z+4)-4(z-2) = 3(z-1)+2(z-3)$$
$$10z+40-4z+8 = 3z-3+2z-6$$
$$6z+48 = 5z-9$$
$$6z+48-48 = 5z-9-48$$
$$6z-5z = 5z-57-5z$$
$$z = -57$$
The solution set is $\{-57\}$.

31. $\frac{x}{5}-4 = -6$

To clear the equation of fractions, multiply both sides by the least common denominator (LCD), which is 5.
$$5\left(\frac{x}{5}-4\right) = 5(-6)$$
$$5 \cdot \frac{x}{5}-5 \cdot 4 = -30$$
$$x-20 = -30$$
$$x-20+20 = -30+20$$
$$x = -10$$
The solution set is $\{-10\}$.

33. $\dfrac{2x}{3} - 5 = 7$

To clear the equation of fractions, multiply both sides by the least common denominator (LCD), which is 3.

$$3\left(\dfrac{2}{3}x - 5\right) = 3(7)$$

$$3 \cdot \dfrac{2}{3}x - 3 \cdot 5 = 21$$

$$2x - 15 = 21$$

$$2x - 15 + 15 = 21 + 15$$

$$2x = 36$$

$$\dfrac{2x}{2} = \dfrac{36}{2}$$

$$x = 18$$

The solution set is $\{18\}$.

35. $\dfrac{2y}{3} - \dfrac{3}{4} = \dfrac{5}{12}$

To clear the equation of fractions, multiply both sides by the least common denominator (LCD), which is 12.

$$12\left(\dfrac{2y}{3} - \dfrac{3}{4}\right) = 12\left(\dfrac{5}{12}\right)$$

$$12\left(\dfrac{2y}{3}\right) - 12\left(\dfrac{3}{4}\right) = 5$$

$$8y - 9 = 5$$

$$8y - 9 + 9 = 5 + 9$$

$$8y = 14$$

$$\dfrac{8y}{8} = \dfrac{14}{8}$$

$$y = \dfrac{14}{8} = \dfrac{7}{4}$$

The solution set is $\left\{\dfrac{7}{4}\right\}$.

37. $\dfrac{x}{3} + \dfrac{x}{2} = \dfrac{5}{6}$

To clear the equation of fractions, multiply both sides by the least common denominator (LCD), which is 6.

$$6\left(\dfrac{x}{3} + \dfrac{x}{2}\right) = 6\left(\dfrac{5}{6}\right)$$

$$2x + 3x = 5$$

$$5x = 5$$

$$\dfrac{5x}{5} = \dfrac{5}{5}$$

$$x = 1$$

The solution set is $\{1\}$.

39. $20 - \dfrac{z}{3} = \dfrac{z}{2}$

To clear the equation of fractions, multiply both sides by the least common denominator (LCD), which is 6.

$$6\left(20 - \dfrac{z}{3}\right) = 6\left(\dfrac{z}{2}\right)$$

$$120 - 2z = 3z$$

$$120 - 2z + 2z = 3z + 2z$$

$$120 = 5z$$

$$\dfrac{120}{5} = \dfrac{5z}{5}$$

$$24 = z$$

The solution set is $\{24\}$.

41. $\dfrac{y}{3} + \dfrac{2}{5} = \dfrac{y}{5} - \dfrac{2}{5}$

To clear the equation of fractions, multiply both sides by the least common denominator (LCD), which is 15.

$$15\left(\dfrac{y}{3} + \dfrac{2}{5}\right) = 15\left(\dfrac{y}{5} + \dfrac{2}{5}\right)$$

$$15\left(\dfrac{y}{3}\right) + 15\left(\dfrac{2}{5}\right) = 15\left(\dfrac{y}{5}\right) + 15\left(-\dfrac{2}{5}\right)$$

$$5y + 6 = 3y - 6$$
$$5y + 6 - 3y = 3y - 6 - 3y$$
$$2y + 6 = -6$$
$$2y + 6 - 6 = -6 - 6$$
$$2y = -12$$
$$\dfrac{2y}{2} = \dfrac{-12}{2}$$
$$y = -6$$

The solution set is $\{-6\}$.

43. $\dfrac{3x}{4} - 3 = \dfrac{x}{2} + 2$

To clear the equation of fractions, multiply both sides by the least common denominator (LCD), which is 8.

$$8\left(\dfrac{3x}{4} - 3\right) = 8\left(\dfrac{x}{2} + 2\right)$$

$$8\left(\dfrac{3x}{4}\right) - 8\cdot 3 = 8\left(\dfrac{x}{2}\right) + 8\cdot 2$$

$$6x - 24 = 4x + 16$$
$$6x - 24 - 4x = 4x + 16 - 4x$$
$$2x - 24 = 16$$
$$2x - 24 + 24 = 16 + 24$$
$$2x = 40$$
$$\dfrac{2x}{2} = \dfrac{40}{2}$$
$$x = 20$$

The solution set is $\{20\}$.

45. $\dfrac{x-3}{5} - 1 = \dfrac{x-5}{4}$

To clear the equation of fractions, multiply both sides by the least common denominator (LCD), which is 20.

$$20\left(\dfrac{x-3}{5} - 1\right) = 20\left(\dfrac{x-5}{4}\right)$$

$$4(x-3) - 20 = 5(x-5)$$
$$4x - 12 - 20 = 5x - 25$$
$$4x - 5x - 32 = 5x - 5x - 25$$
$$-x - 32 = -25$$
$$-x - 32 + 32 = -25 + 32$$
$$-x = 7$$
$$-1(-x) = -1(7)$$
$$x = -7$$

The solution set is $\{-7\}$.

47. $3.6x = 2.9x + 6.3$

To clear the equation of decimals, multiply both sides by 10.

$$10(3.6x) = 10(2.9x + 6.3)$$
$$36x = 29x + 63$$
$$7x = 63$$
$$x = 9$$

The solution set is $\{9\}$.

49. $0.92y + 2 = y - 0.4$

To clear the equation of decimals, multiply both sides by 100.

$$100(0.92y + 2) = 100(y - 0.4)$$
$$92y + 200 = 100y - 40$$
$$92y = 100y - 240$$
$$-8y = -240$$
$$y = 30$$

The solution set is $\{30\}$.

51. $0.3x - 4 = 0.1(x + 10)$

$$0.3x - 4 = 0.1x + 1$$

To clear the equation of decimals, multiply both sides by 10.

$$10(0.3x - 4) = 10(0.1x + 1)$$
$$3x - 40 = x + 10$$
$$3x = x + 50$$
$$2x = 50$$
$$x = 25$$

The solution set is $\{25\}$.

53. $0.4(2z+6)+0.1=0.5(2z-3)$

$0.8z+2.4+0.1=z-1.5$

$0.8z+2.5=z-1.5$

To clear the equation of decimals, multiply both sides by 10.

$10(0.8z+2.5)=10(z-1.5)$

$8z+25=10z-15$

$8z=10z-40$

$-2z=-40$

$z=20$

The solution set is $\{20\}$.

55. $0.01(x+4)-0.04=0.01(5x+4)$

$0.01x+0.4-0.04=0.05x+0.4$

$0.01x+0.36=0.05x+0.4$

To clear the equation of decimals, multiply both sides by 100.

$100(0.01x+0.36)=100(0.05x+0.4)$

$x+36=5x+40$

$x=5x+4$

$-4x=4$

$x=-1$

The solution set is $\{-1\}$.

57. $0.6(x+300)=0.65x-205$

$0.6x+180=0.65x-205$

To clear the equation of decimals, multiply both sides by 100.

$100(0.6x+180)=100(0.65x-205)$

$60x+18,000=65x-20,500$

$60x=65x-38,500$

$-5x=-38,500$

$x=7700$

The solution set is $\{7700\}$.

59. $3x-7=3(x+1)$

$3x-7=3x+3$

$3x-7-3x=3x+3-3x$

$-7=3$

The original equation is equivalent to the false statement $-7=3$, so the equation is inconsistent and has no solution. The solution set is $\{\ \}$.

61. $2(x+4)=4x+5-2x+3$

$2x+8=2x+8$

$2x-8-2x=2x+8-2x$

$8=8$

The original equation is equivalent to the true statement $8=8$, so the equation is an identity and the solution set is all real numbers $\{x|x\text{ is a real number}\}$.

63. $7+2(3x-5)=8-3(2x+1)$

$7+6x-10=8-6x-3$

$6x-3=5-6x$

$6x+6x-3=5-6x+6x$

$12x-3=5$

$12x-3+3=5+3$

$12x=8$

$\dfrac{12x}{12}=\dfrac{8}{12}$

$x=\dfrac{2}{3}$

The solution set is $\left\{\dfrac{2}{3}\right\}$.

65. $4x+1-5x=5-(x+4)$

$-x+1=5-x-4$

$-x+1=1-x$

$-x+1+x=1-x+x$

$1=1$

The original equation is equivalent to the true statement $1=1$, so the equation is an identity and the solution set is all real numbers $\{x|x\text{ is a real number}\}$.

67. $4(x+2)+1=7x-3(x-2)$

$4x+8+1=7x-3x+6$

$4x+9=4x+6$

$4x-4x+9=4x-4x+6$

$9=6$

Since $9=6$ is a false statement, the original equation is inconsistent and has no solution. The solution set is $\{\ \}$.

69.
$$3 - x = 2x + 3$$
$$3 - x + x = 2x + x + 3$$
$$3 = 3x + 3$$
$$3 - 3 = 3x + 3 - 3$$
$$0 = 3x$$
$$\frac{0}{3} = \frac{3x}{3}$$
$$0 = x$$

The solution set is $\{0\}$.

71. $\dfrac{x}{3} + 2 = \dfrac{x}{3}$

Multiply by the LCD, which is 3.
$$3\left(\frac{x}{3} + 2\right) = 3\left(\frac{x}{3}\right)$$
$$x + 6 = x$$
$$x - x + 6 = x - x$$
$$6 = 0$$

Since $6 = 0$ is a false statement, the original equation has no solution. The solution set is $\{\ \}$.

73. $\dfrac{x}{2} - \dfrac{x}{4} + 4 = x + 4$

Multiply by the LCD, which is 4.
$$4\left(\frac{x}{2} - \frac{x}{4} + 4\right) = 4(x + 4)$$
$$4\left(\frac{x}{2}\right) - 4\left(\frac{x}{4}\right) + 16 = 4x + 16$$
$$2x - x + 16 = 4x + 16$$
$$x + 16 = 4x + 16$$
$$x - x + 16 = 4x - x + 16$$
$$16 = 3x + 16$$
$$16 - 16 = 3x + 16 - 16$$
$$0 = 3x$$
$$\frac{0}{3} = \frac{3x}{3}$$
$$0 = x$$

The solution set is $\{0\}$.

75. $\dfrac{2}{3}x = 2 - \dfrac{5}{6}x$

Multiply both sides by the LCD which is 6.
$$6\left(\frac{2}{3}x\right) = 6(2) - 6\left(\frac{5}{6}x\right)$$
$$2(2x) = 12 - 5x$$
$$4x = 12 - 5x$$
$$4x + 5x = 12 - 5x + 5x$$
$$9x = 12$$
$$\frac{9x}{9} = \frac{12}{9}$$
$$x = \frac{12}{9} = \frac{4}{3}$$

The solution set is $\left\{\dfrac{4}{3}\right\}$.

77.
$$0.06(x + 5) = 0.03(2x + 7) + 0.09$$
$$0.06x + 0.3 = 0.06x + 0.21 + 0.09$$
$$0.06x + 0.3 = 0.06x + 0.3$$

To clear the equation of decimals, multiply both sides by 100.
$$100(0.06x + 0.3) = 100(0.06x + 0.3)$$
$$6x + 30 = 6x + 30$$
$$30 = 30$$

The original equation is equivalent to the true statement $30 = 30$, so the equation is an identity and the solution set is all real numbers $\{x | x \text{ is a real number}\}$.

79.
$$\frac{x}{\square} + \triangle = \$$$
$$\frac{x}{\square} + \triangle - \triangle = \$ - \triangle$$
$$\frac{x}{\square} = \$ - \triangle$$
$$\square\left(\frac{x}{\square}\right) = \square(\$ - \triangle)$$
$$x = \square\$ - \square\triangle$$

81. First solve the equation for x.

$$\frac{x}{5} - 2 = \frac{x}{3}$$

$$\frac{x}{5} - \frac{x}{5} - 2 = \frac{x}{3} - \frac{x}{5}$$

$$-2 = \frac{5x}{15} - \frac{3x}{15}$$

$$-2 = \frac{2x}{15}$$

$$15(-2) = 15\left(\frac{2x}{15}\right)$$

$$-30 = 2x$$

$$\frac{-30}{2} = \frac{2x}{2}$$

$$-15 = x$$

Now evaluate the expression $x^2 - x$ for $x = -15$.

$$x^2 - x = (-15)^2 - (-15)$$
$$= 225 + 15$$
$$= 240$$

83.

$$\frac{1}{3}x + \frac{1}{5}x = 16$$

$$\text{LCD} = 15$$

$$15\left(\frac{1}{3}x\right) + 15\left(\frac{1}{5}x\right) = 15(16)$$

$$5x + 3x = 240$$

$$8x = 240$$

$$\frac{8x}{8} = \frac{240}{8}$$

$$x = 30$$

The number is 30.

85.

$$\frac{3}{4}x - 3 = \frac{1}{2}x$$

$$4\left(\frac{3}{4}x\right) - 4(3) = 4\left(\frac{1}{2}x\right)$$

$$3x - 12 = 2x$$

$$3x - 2x - 12 = 2x - 2x$$

$$x - 12 = 0$$

$$x - 12 + 12 = 0 + 12$$

$$x = 12$$

The number is 12.

87.

$$F = 10(x - 65) + 50$$

$$250 = 10(x - 65) + 50$$

$$250 - 50 = 10(x - 65) + 50 - 50$$

$$200 = 10x - 650$$

$$200 + 650 = 10x - 650 + 650$$

$$850 = 10x$$

$$\frac{850}{10} = \frac{10x}{10}$$

$$85 = x$$

A person receiving a $250 fine was driving 85 miles per hour.

89.

$$\frac{W}{2} - 3H = 53$$

$$\frac{W}{2} - 3(6) = 53$$

$$\frac{W}{2} - 18 = 53$$

$$\frac{W}{2} - 18 + 18 = 53 + 18$$

$$\frac{W}{2} = 71$$

$$2 \cdot \frac{W}{2} = 2 \cdot 71$$

$$W = 142$$

According to the formula, the healthy weight of a person of height 5'6" is 142 pounds. This is 13 pounds below the upper end of the range shown in the bar graph.

91.

$$p = 15 + \frac{5d}{11}$$

$$201 = 15 + \frac{5d}{11}$$

$$201 - 15 = 15 + \frac{5d}{11} - 15$$

$$186 = \frac{5d}{11}$$

$$11(186) = 11\left(\frac{5d}{11}\right)$$

$$2046 = 5d$$

$$\frac{2046}{5} = d$$

$$409.2 = d$$

He descended to a depth of 409.2 feet below the surface.

93. – 97. Answers will vary.

99. makes sense

101. does not make sense; Explanations will vary. Sample explanation: For this equation it would have been sufficient to multiply by 10.

103. false; Changes to make the statement true will vary. A sample change is: The equation $2y + 5 = 0$ is equivalent to $2y = -5$.

105. false; Changes to make the statement true will vary. A sample change is: The equation $x + \dfrac{1}{3} = \dfrac{1}{2}$ is

equivalent to $6 \cdot x + 6 \cdot \dfrac{1}{3} = 6 \cdot \dfrac{1}{2}$ or $6x + 2 = 3$.

107.
$$\frac{2x-3}{9} + \frac{x-3}{2} = \frac{x+5}{6} - 1$$
$$18\left(\frac{2x-3}{9} + \frac{x-3}{2}\right) = 18\left(\frac{x+5}{6} - 1\right)$$
$$18\left(\frac{2x-3}{9}\right) + 18\left(\frac{x-3}{2}\right) = 18\left(\frac{x+5}{6}\right) - 18 \cdot 1$$
$$2(2x-3) + 9(x-3) = 3(x+5) - 18$$
$$4x - 6 + 9x - 27 = 3x + 15 - 18$$
$$13x - 33 = 3x - 3$$
$$13x - 33 - 3x = 3x - 3 - 3x$$
$$10x - 33 = -3$$
$$10x - 33 + 33 = -3 + 33$$
$$10x = 30$$
$$\frac{10x}{10} = \frac{30}{10}$$
$$x = 3$$

The solution set is $\{3\}$.

109. $-24 < -20$ because -24 lies further to the left on a number line.

110. $-\dfrac{1}{3} < -\dfrac{1}{5}$ because $-\dfrac{1}{3}$ lies further to the left on a number line.

111. $-9 - 11 + 7 - (-3) = -9 - 11 + 7 + 3$
$$= -20 + 10$$
$$= -10$$

112. a. $T = D + pm$
$$T - D = pm$$

b. $T - D = pm$
$$\frac{T - D}{p} = \frac{pm}{p}$$
$$\frac{T - D}{p} = m$$

113. $4 = 0.25B$
$$\frac{4}{0.25} = \frac{0.25B}{0.25}$$
$$16 = B$$
The solution set is $\{16\}$.

114. $1.3 = P \cdot 26$
$$\frac{1.3}{26} = \frac{P \cdot 26}{26}$$
$$0.05 = P$$
The solution set is $\{0.05\}$.

2.4 Check Points

1. $A = lw$
$$\frac{A}{w} = \frac{lw}{w}$$
$$\frac{A}{w} = l$$

2. $2l + 2w = P$
$$2l + 2w - 2w = P - 2w$$
$$2l = P - 2w$$
$$\frac{2l}{2} = \frac{P - 2w}{2}$$
$$l = \frac{P - 2w}{2}$$

3. $T = D + pm$
$$T - D = pm$$
$$\frac{T - D}{p} = \frac{pm}{p}$$
$$\frac{T - D}{p} = m$$
$$m = \frac{T - D}{p}$$

4. $\dfrac{x}{3} - 4y = 5$

$$3\left(\dfrac{x}{3} - 4y\right) = 3 \cdot 5$$

$$3 \cdot \dfrac{x}{3} - 3 \cdot 4y = 3 \cdot 5$$

$$x - 12y = 15$$

$$x - 12y + 12y = 15 + 12y$$

$$x = 15 + 12y$$

5. Use the formula $A = PB$: A is P percent of B.

$$\boxed{\text{What}} \ \boxed{\text{is}} \ \boxed{9\%} \ \boxed{\text{of}} \ \boxed{50?}$$

$$A \quad = 0.09 \ \cdot \ 50$$

$$A = 4.5$$

6. Use the formula $A = PB$: A is P percent of B.

$$\boxed{9} \ \boxed{\text{is}} \ \boxed{60\%} \ \boxed{\text{of}} \ \boxed{\text{what?}}$$

$$9 = 0.60 \ \cdot \ B$$

$$\dfrac{9}{0.60} = \dfrac{0.60B}{0.60}$$

$$15 = B$$

7. Use the formula $A = PB$: A is P percent of B.

$$\boxed{18} \ \boxed{\text{is}} \ \boxed{\text{what percent}} \ \boxed{\text{of}} \ \boxed{50?}$$

$$18 = \quad P \quad \cdot \ 50$$

$$18 = P \cdot 50$$

$$\dfrac{18}{50} = \dfrac{50P}{50}$$

$$0.36 = P$$

To change 0.36 to a percent, move the decimal point two places to the right and add a percent sign.
$0.36 = 36\%$

8. Use the formula $A = PB$: A is P percent of B.

Find the price decrease: $\$940 - \$611 = \$329$

$$\boxed{\text{The price decrease}} \ \boxed{\text{is}} \ \boxed{\text{what percent}} \ \boxed{\text{of}} \ \boxed{\text{the original price?}}$$

$$329 \quad = \quad P \quad \cdot \quad 940$$

$$329 = P \cdot 940$$

$$\dfrac{329}{940} = \dfrac{940P}{940}$$

$$0.35 = P$$

To change 0.35 to a percent, move the decimal point two places to the right and add a percent sign.
$0.35 = 35\%$

9. a.

Year	Tax Paid the Year Before	increase/decrease	Taxes Paid This Year
1	$1200	20% decrease : $0.20 \cdot \$1200 = \$240$	$\$1200 - \$240 = \$960$
2	$960	20% increase : $0.20 \cdot \$960 = \$192$	$\$960 + \$192 = \$1152$

The taxes for year 2 will be $1152.

b. The taxes for year 2 are less than those originally paid.
Find the tax decrease: $\$1200 - \$1152 = \$48$

$$\underset{\substack{\text{The tax}\\\text{decrease}}}{48} \underset{\text{is}}{=} \underset{\substack{\text{what}\\\text{percent}}}{P} \underset{\text{of}}{\cdot} \underset{\substack{\text{the original}\\\text{tax?}}}{1200}$$

$$48 = P \cdot 1200$$

$$\frac{48}{1200} = \frac{1200P}{1200}$$

$$0.04 = P$$

To change 0.04 to a percent, move the decimal point two places to the right and add a percent sign.
$0.04 = 4\%$
The overall tax decrease is 4%.

2.4 Concept and Vocabulary Check

1. isolated on one side

2. $A = lw$

3. $P = 2l \times 2w$

4. $A = PB$

5. subtract b; divide by m

2.4 Exercise Set

1. $d = rt$ for r

$$\frac{d}{t} = \frac{rt}{t}$$

$$\frac{d}{t} = r \text{ or } r = \frac{d}{t}$$

This is the distance traveled formula:
distance = rate · time.

3. $I = Prt$ for P

$$\frac{I}{rt} = \frac{Prt}{rt}$$

$$\frac{I}{rt} = P \text{ or } P = \frac{I}{rt}$$

This is the formula for simple interest:
interest = principal · rate · time.

5. $C = 2\pi r$ for r

$$\frac{C}{2\pi} = \frac{2\pi r}{2\pi}$$

$$\frac{C}{2\pi} = r \text{ or } r = \frac{C}{2\pi}$$

This is the formula for finding the circumference of a circle if you know its radius.

7. $E = mc^2$

$$\frac{E}{c^2} = \frac{mc^2}{c^2}$$

$$\frac{E}{c^2} = m \text{ or } m = \frac{E}{c^2}$$

This is Einstein's formula relating energy, mass, and the speed of light.

9. $y = mx + b$ for m

$$y - b = mx$$

$$\frac{y-b}{x} = \frac{mx}{x}$$

$$\frac{y-b}{x} = m \text{ or } m = \frac{y-b}{x}$$

This is the slope-intercept formula for the equation of a line.

11. $T = D + pm$ for D

$$T - pm = D + pm - pm$$

$$T - pm = D$$

$$D = T - pm$$

13. $A = \frac{1}{2}bh$ for b

$$2A = 2\left(\frac{1}{2}bh\right)$$

$$2A = bh$$

$$\frac{2A}{h} = \frac{bh}{h}$$

$$\frac{2A}{h} = b \text{ or } b = \frac{2A}{h}$$

This is the formula for the area of a triangle: area = $\frac{1}{2}$ · base · height.

15. $M = \frac{n}{5}$ for n

$$5M = 5\left(\frac{n}{5}\right)$$

$$5M = n \text{ or } n = 5M$$

17. $\frac{c}{2} + 80 = 2F$ for c

$$\frac{c}{2} + 80 - 80 = 2F - 80$$

$$\frac{c}{2} = 2F - 80$$

$$2\left(\frac{c}{2}\right) = 2(2F - 80)$$

$$c = 4F - 160$$

19. $A = \frac{1}{2}(a + b)$ for a

$$2A = 2\left[\frac{1}{2}(a + b)\right]$$

$$2A = a + b$$

$$2A - b = a + b - b$$

$$2A - b = a \text{ or } a = 2A - b$$

This is the formula for finding the average of two numbers.

21. $S = P + Prt$ for r

$$S - P = P + Prt - P$$

$$S - P = Prt$$

$$\frac{S - P}{Pt} = \frac{Prt}{Pt}$$

$$\frac{S - P}{Pt} = r \text{ or } r = \frac{S - P}{Pt}$$

This is the formula for finding the sum of principle and interest for simple interest problems.

23. $A = \dfrac{1}{2}h(a+b)$ for b

$$2A = 2\left[\dfrac{1}{2}h(a+b)\right]$$
$$2A = h(a+b)$$
$$2A = ha + hb$$
$$2A - ha = ha + hb - ha$$
$$2A - ha = hb$$
$$\dfrac{2A - ha}{h} = \dfrac{hb}{h}$$
$$\dfrac{2A - ha}{h} = b \quad \text{or} \quad b = \dfrac{2A}{h} - a$$

This is the formula for the area of a trapezoid.

25. $Ax + By = C$ for x

$$Ax + By - By = C - By$$
$$Ax = C - By$$
$$\dfrac{Ax}{A} = \dfrac{C - By}{A}$$
$$x = \dfrac{C - By}{A}$$

This is the standard form of the equation of a line.

27. $A = PB$; $P = 3\% = 0.03, B = 200$

$$A = PB$$
$$A = 0.03 \cdot 200$$
$$A = 6$$

3% of 200 is 6.

29. $A = PB$; $P = 18\% = 0.18, \ B = 40$

$$A = PB$$
$$A = 0.18 \cdot 40$$
$$A = 7.2$$

18% of 40 is 7.2.

31. $A = PB$; $A = 3, P = 60\% = 0.6$

$$A = PB$$
$$3 = 0.6 \cdot B$$
$$\dfrac{3}{0.6} = \dfrac{0.6B}{0.6}$$
$$5 = B$$

3 is 60% of 5.

33. $A = PB$; $A = 40.8, P = 24\% = 0.24$

$$A = PB$$
$$40.8 = 0.24 \cdot B$$
$$\dfrac{40.8}{0.24} = \dfrac{0.24B}{0.24}$$
$$170 = B$$

24% of 170 is 40.8.

35. $A = PB$; $A = 3, \ B = 15$

$$A = PB$$
$$3 = P \cdot 15$$
$$\dfrac{3}{15} = \dfrac{P \cdot 15}{15}$$
$$0.2 = P$$
$$0.2 = 20\%$$

3 is 20% of 15.

37. $A = PB$; $A = 0.3, \ B = 2.5$

$$A = PB$$
$$0.3 = P \cdot 2.5$$
$$\dfrac{0.3}{2.5} = \dfrac{P - 2.5}{2.5}$$
$$0.12 = P$$
$$0.12 = 12\%$$

0.3 is 12% of 2.5.

39. The increase is $8 - 5 = 3$.

$$A = PB$$
$$3 = P \cdot 5$$
$$\dfrac{3}{5} = \dfrac{P \cdot 5}{5}$$
$$0.60 = P$$

This is a 60% increase.

41. The decrease is $4 - 1 = 3$.

$$A = PB$$
$$3 = P \cdot 4$$
$$\dfrac{3}{4} = \dfrac{4P}{4}$$
$$0.75 = P$$

This is a 75% decrease.

43.
$$y = (a+b)x$$
$$\dfrac{y}{(a+b)} = \dfrac{(a+b)x}{(a+b)}$$
$$\dfrac{y}{a+b} = x \quad \text{or} \quad x = \dfrac{y}{a+b}$$

45.

$$y = (a-b)x + 5$$
$$y - 5 = (a-b)x + 5 - 5$$
$$y - 5 = (a-b)x$$
$$\frac{y-5}{a-b} = \frac{(a-b)x}{a-b}$$
$$\frac{y-5}{a-b} = x \quad \text{or} \quad x = \frac{y-5}{a-b}$$

47.

$$y = cx + dx$$
$$y = (c+d)x$$
$$\frac{y}{c+d} = \frac{(c+d)x}{c+d}$$
$$\frac{y}{c+d} = x \quad \text{or} \quad x = \frac{y}{c+d}$$

49.

$$y = Ax - Bx - C$$
$$y = (A-B)x - C$$
$$y + C = (A-B)x - C + C$$
$$y + C = (A-B)x$$
$$\frac{y+C}{A-B} = \frac{(A-B)x}{A-B}$$
$$\frac{y+C}{A-B} = x \quad \text{or} \quad x = \frac{y+C}{A-B}$$

51. a. $A = \dfrac{x+y+z}{3}$ for z

$$3A = 3\left(\frac{x+y+z}{3}\right)$$
$$3A = x + y + z$$
$$3A - x - y = x + y + z - x - y$$
$$3A - x - y = z$$

b. $A = 90, x = 86, y = 88$

$$z = 3A - x - y$$
$$z = 3(90) - 86 - 88 = 96$$

You need to get 96% on the third exam to have an average of 90%

53. a. $d = rt$ for t

$$\frac{d}{r} = \frac{rt}{r}$$
$$\frac{d}{r} = t$$

b. $t = \dfrac{d}{r}; \; d = 100, r = 40$

$$t = \frac{100}{40} = 2.5$$

You would travel for 2.5 $\left(\text{or } 2\dfrac{1}{2}\right)$ hours.

55. $0.29 \cdot 1800 = 522$

522 workers stated that religion is the most taboo topic to discuss at work.

57. This is the equivalent of asking: 175 is 35% of what?

$$A = P \cdot B$$
$$175 = 0.35 \cdot B$$
$$\frac{175}{0.35} = \frac{0.35B}{0.35}$$
$$500 = B$$

Americans throw away 500 billion pounds of trash each year.

59. a. The total number of countries in 1974 was $41 + 48 + 63 = 152$.

$$A = P \cdot B$$
$$41 = P \cdot 152$$
$$\frac{41}{152} = \frac{152B}{152}$$
$$0.27 \approx B$$

About 27% of countries were free in 1974.

b. The total number of countries in 2009 was $89 + 62 + 42 = 193$.

$$A = P \cdot B$$
$$89 = P \cdot 193$$
$$\frac{89}{193} = \frac{193B}{193}$$
$$0.46 \approx B$$

About 46% of countries were free in 2009.

c. The increase is $89 - 41 = 48$.

$$A = P \cdot B$$
$$48 = P \cdot 41$$
$$\frac{48}{41} = \frac{41B}{41}$$
$$1.17 \approx B$$

There was approximately a 117% increase in the number of free countries from 1974 to 2009.

61. $A = PB$; $A = 7500, B = 60,000$

$$A = PB$$
$$7500 = P \cdot 60,000$$
$$\frac{7500}{60,000} = \frac{P \cdot 60,000}{60,000}$$
$$0.125 = P$$

The charity has raised $0.125 = 12.5\%$ of its goal.

63. $A = PB$; $p = 15\% = 0.15, B = 60$

$$A = 0.15 \cdot 60 = 09$$

The tip was $9.

65. a. The sales tax is 6% of $16,800.

$$0.06(16,800) = 1008$$

The sales tax due on the car is $1008.

b. The total cost is the sum of the price of the car and the sales tax.
$16,800 + $1008 = $17,808$
The car's total cost is $17,808.

67. a. The discount is 12% of $860.

$$0.12(860) = 103.20$$

The discount amount is $103.20.

b. The sale price is the regular price minus the discount amount:
$860 - $103.20 = 756.80

69. The decrease is $840 - $714 = $126.

$$A = P \cdot B$$
$$126 = P \cdot 840$$
$$\frac{126}{840} = \frac{P \cdot 840}{840}$$
$$0.15 = P$$

This is a $0.15 = 15\%$ decrease.

71. Investment dollars decreased in year 1 are $0.30 \cdot \$10,000 = \$3000$. This means that $10,000 − $3000 = $7000 remains. Investment dollars increased in year 2 are $0.40 \cdot \$7000 = \$2800$. This means that $7000 + $2800 = $9800 of the original investment remains. This is an overall loss of $200 over the two years.

$$A = P \cdot B$$
$$200 = P \cdot 10,000$$
$$\frac{200}{10,000} = \frac{P \cdot 10,000}{10,000}$$
$$0.02 = P$$

The financial advisor is not using percentages properly. Instead of a 10% gain, this is a $0.02 = 2\%$ loss.

73. Answers will vary.

75. makes sense

77. does not make sense; Explanations will vary. Sample explanation: $100 is more than enough because 20% of $80 is $0.20 \cdot \$80 = \$16$.

79. false; Changes to make the statement true will vary. A sample change is: If $ax + b = 0$, then $ax = -b$ and $x = \dfrac{-b}{a}$.

81. false; Changes to make the statement true will vary. A sample change is: If $A = \dfrac{1}{2}bh$, then $\dfrac{2A}{h} = b$.

83. $Q = \dfrac{100M}{C}$ for C

$$CQ = C\left(\frac{100M}{C}\right)$$
$$CQ = 100M$$
$$\frac{CQ}{Q} = \frac{100M}{Q}$$
$$C = \frac{100M}{Q}$$

84.
$$5x + 20 = 8x - 16$$
$$5x + 20 - 8x = 8x - 16 - 8x$$
$$-3x + 20 = -16$$
$$-3x + 20 - 20 = -16 - 20$$
$$-3x = -36$$
$$\frac{-3x}{-3} = \frac{-36}{-3}$$
$$x = 12$$

Check:
$$5(12) + 20 = 8(12) - 16$$
$$60 + 20 = 96 - 16$$
$$80 = 80$$

The solution set is $\{12\}$.

85.
$$5(2y-3)-1=4(6+2y)$$
$$10y-15-1=24+8y$$
$$10y-16=24+8y$$
$$10y-16-8y=24+8y-8y$$
$$2y-16=24$$
$$2y-16+16=24+16$$
$$2y=40$$
$$\frac{2y}{2}=\frac{40}{2}$$
$$y=20$$
Check:
$$5(2\cdot20-3)-1=4(6+2\cdot20)$$
$$5(40-3)-1=4(6+40)$$
$$5(37)-1=4(46)$$
$$185-1=184$$
$$184=184$$
The solution set is $\{20\}$.

86. $x-0.3x=1x-0.3x=(1-0.3)x=0.7x$

87. $\dfrac{13}{x}-7x$

88. $8(x+14)$

89. $9(x-5)$

Mid-Chapter Check Point

1. Begin by multiplying both sides of the equation by 4, the least common denominator.
$$\frac{x}{2}=12-\frac{x}{4}$$
$$4\left(\frac{x}{2}\right)=4(12)-4\left(\frac{x}{4}\right)$$
$$2x=48-x$$
$$2x+x=48-x+x$$
$$3x=48$$
$$\frac{3x}{3}=\frac{48}{3}$$
$$x=16$$
The solution set is $\{16\}$.

2.
$$5x-42=-57$$
$$5x-42+42=-57+42$$
$$5x=-15$$
$$\frac{5x}{5}=\frac{-15}{5}$$
$$x=-3$$
The solution set is $\{-3\}$.

3.
$$H=\frac{EC}{825}$$
$$H\cdot825=\frac{EC}{825}\cdot825$$
$$825H=EC$$
$$\frac{825H}{E}=\frac{EC}{E}$$
$$\frac{825H}{E}=C$$

4. $A=P\cdot B$
$$A=0.06\cdot140$$
$$A=8.4$$
8.4 is 6% of 140.

5.
$$\frac{-x}{10}=-3$$
$$10\left(\frac{-x}{10}\right)=10(-3)$$
$$-x=-30$$
$$-1(-x)=-1(-30)$$
$$x=30$$
The solution set is $\{30\}$.

6.
$$1-3(y-5)=4(2-3y)$$
$$1-3y+15=8-12y$$
$$-3y+16=8-12y$$
$$-3y+12y+16=8-12y+12y$$
$$9y+16=8$$
$$9y+16-16=8-16$$
$$9y=-8$$
$$\frac{9y}{9}=\frac{-8}{9}$$
$$y=-\frac{8}{9}$$
The solution set is $\left\{-\dfrac{8}{9}\right\}$.

7.

$$S = 2\pi rh$$

$$\frac{S}{2\pi h} = \frac{2\pi rh}{2\pi h}$$

$$\frac{S}{2\pi h} = r$$

8.

$$A = P \cdot B$$

$$12 = 0.30 \cdot B$$

$$\frac{12}{0.30} = \frac{0.30 \cdot B}{0.30}$$

$$40 = B$$

12 is 30% of 40.

9. $\dfrac{3y}{5} + \dfrac{y}{2} = \dfrac{5y}{4} - 3$

To clear fractions, multiply both sides by the LCD, 20.

$$20\left(\frac{3y}{5}\right) + 20\left(\frac{y}{2}\right) = 20\left(\frac{5y}{4}\right) - 20(3)$$

$$4(3y) + 10y = 5(5y) - 60$$

$$12y + 10y = 25y - 60$$

$$22y = 25y - 60$$

$$22y - 25y = 25y - 25y - 60$$

$$-3y = -60$$

$$\frac{-3y}{-3} = \frac{-60}{-3}$$

$$y = 20$$

The solution set is $\{20\}$.

10.

$$2.4x + 6 = 1.4x + 0.5(6x - 9)$$

$$2.4x + 6 = 1.4x + 3x - 4.5$$

$$2.4x + 6 = 4.4x - 4.5$$

To clear decimals, multiply both sides by 10.

$$10(2.4x + 6) = 10(4.4x - 4.5)$$

$$24x + 60 = 44x - 45$$

$$24x = 44x - 105$$

$$-20x = -105$$

$$\frac{-20x}{-20} = \frac{-105}{-20}$$

$$x = 5.25$$

The solution set is $\{5.25\}$.

11.

$$5z + 7 = 6(z - 2) - 4(2z - 3)$$

$$5z + 7 = 6z - 12 - 8z + 12$$

$$5z + 7 = -2z$$

$$5z - 5z + 7 = -2z - 5z$$

$$7 = -7z$$

$$\frac{7}{-7} = \frac{-7z}{-7}$$

$$-1 = z$$

The solution set is $\{-1\}$.

12.

$$Ax - By = C$$

$$Ax - By + By = C + By$$

$$Ax = C + By$$

$$\frac{Ax}{A} = \frac{C + By}{A}$$

$$x = \frac{C + By}{A} \text{ or } \frac{By + C}{A}$$

13. $6y + 7 + 3y = 3(3y - 1)$

$$9y + 7 = 9y - 3$$

$$9y - 9y + 7 = 9y - 9y - 3$$

$$7 = -3$$

Since this is a false statement, there is no solution or $\{\ \}$.

14.

$$10\left(\frac{1}{2}x + 3\right) = 10\left(\frac{3}{5}x - 1\right)$$

$$10\left(\frac{1}{2}x\right) + 10(3) = 10\left(\frac{3}{5}x\right) - 10(1)$$

$$5x + 30 = 6x - 10$$

$$5x - 5x + 30 = 6x - 5x - 10$$

$$30 = x - 10$$

$$30 + 10 = x - 10 + 10$$

$$40 = x$$

The solution set is $\{40\}$.

15.

$$A = P \cdot B$$

$$50 = P \cdot 400$$

$$\frac{50}{400} = \frac{P \cdot 400}{400}$$

$$0.125 = P$$

50 is $0.125 = 12.5\%$ of 400.

16.

$$\frac{3(m+2)}{4} = 2m+3$$

$$4 \cdot \frac{3(m+2)}{4} = 4(2m+3)$$

$$3(m+2) = 4(2m+3)$$

$$3m+6 = 8m+12$$

$$3m-3m+6 = 8m-3m+12$$

$$6 = 5m+12$$

$$6-12 = 5m+12-12$$

$$-6 = 5m$$

$$\frac{-6}{5} = \frac{5m}{5}$$

$$-\frac{6}{5} = m$$

The solution set is $\left\{ -\frac{6}{5} \right\}$.

17. The increase is $50 - 40 = 10$.

$$A = P \cdot B$$

$$10 = P \cdot 40$$

$$\frac{10}{40} = \frac{P \cdot 40}{40}$$

$$0.25 = P$$

This is a $0.25 = 25\%$ increase.

18.

$$12w-4+8w-4 = 4(5w-2)$$

$$20w-8 = 20w-8$$

$$20w-20w-8 = 20w-20w-8$$

$$-8 = -8$$

Since $-8 = -8$ is a true statement, the solution is all real numbers or $\{x \mid x \text{ is a real number}\}$.

19. a.

$$B = -\frac{5}{2}a + 82$$

$$B = -\frac{5}{2}(14) + 82$$

$$= -35 + 82$$

$$= 47$$

According to the formula, 47% of 14-year-olds believe that reading books is important.
This underestimates the actual percentage shown in the bar graph by 2%

b.

$$B = -\frac{5}{2}a + 82$$

$$22 = -\frac{5}{2}a + 82$$

$$2(22) = 2\left(-\frac{5}{2}a + 82 \right)$$

$$44 = -5a + 164$$

$$-120 = -5a$$

$$24 = a$$

According to the formula, 22% of 24-year-olds will believe that reading books is important.

2.5 Check Points

1. Let x = the number.

$$6x - 4 = 68$$

$$6x - 4 + 4 = 68 + 4$$

$$6x = 72$$

$$x = 12$$

The number is 12.

2. Let x = the median starting salary, in thousands of dollars, for English majors.
Let $x + 18$ = the median starting salary, in thousands of dollars, for computer science majors.

$$x + (x+18) = 94$$

$$x + x + 18 = 94$$

$$2x + 18 = 94$$

$$2x = 76$$

$$x = 38$$

$$x + 18 = 56$$

The average salary for English majors is $18 thousand and the average salary for computer science majors is $38 + $18 = $56.

3. Let x = the page number of the first facing page.
Let $x + 1$ = the page number of the second facing page.

$$x + (x+1) = 145$$

$$x + x + 1 = 145$$

$$2x + 1 = 145$$

$$2x + 1 - 1 = 145 - 1$$

$$2x = 144$$

$$x = 72$$

$$x + 1 = 73$$

The page numbers are 72 and 73.

4. Let x = the number of eighths of a mile traveled.
$$2 + 0.25x = 10$$
$$2 - 2 + 0.25x = 10 - 2$$
$$0.25x = 8$$
$$\frac{0.25x}{0.25} = \frac{8}{0.25}$$
$$x = 32$$
You can go 32 eighths of a mile. That is equivalent to $\frac{32}{8} = 4$ miles.

5. Let x = the width of the swimming pool.
Let $3x$ = the length of the swimming pool.
$$P = 2l + 2w$$
$$320 = 2 \cdot 3x + 2 \cdot x$$
$$320 = 6x + 2x$$
$$320 = 8x$$
$$\frac{320}{8} = \frac{8x}{8}$$
$$40 = x$$
$$x = 40$$
$$3x = 120$$
The pool is 40 feet wide and 120 feet long.

6. Let x = the original price.

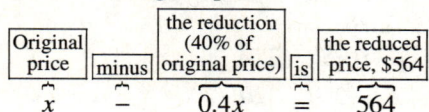

Original price	minus	the reduction (40% of original price)	is	the reduced price, $564
x	$-$	$0.4x$	$=$	564

$$x - 0.4x = 564$$
$$0.6x = 564$$
$$\frac{0.6x}{0.6} = \frac{564}{0.6}$$
$$x = 940$$
The original price was $940.

2.5 Concept and Vocabulary Check

1. $4x - 6$

2. $x + 215$

3. $x + 1$

4. $125 + 0.15x$

5. $2 \cdot 4x + 2x$ or $2x + 2 \cdot 4x$

6. $x - 0.35x$ or $0.65x$

2.5 Exercise Set

1.
$$x + 60 = 410$$
$$x + 60 - 60 = 410 - 60$$
$$x = 350$$
The number is 350.

3.
$$x - 23 = 214$$
$$x - 23 + 23 = 214 + 23$$
$$x = 237$$
The number is 237.

5.
$$7x = 126$$
$$\frac{7x}{7} = \frac{126}{7}$$
$$x = 18$$
The number is 18.

7.
$$\frac{x}{19} = 5$$
$$19\left(\frac{x}{19}\right) = 19(5)$$
$$x = 95$$
The number is 95.

9.
$$4 + 2x = 56$$
$$4 - 4 + 2x = 56 - 4$$
$$2x = 52$$
$$\frac{2x}{2} = \frac{52}{2}$$
$$x = 26$$
The number is 26.

11.
$$5x - 7 = 178$$
$$5x - 7 + 7 = 178 + 7$$
$$5x = 185$$
$$\frac{5x}{5} = \frac{185}{5}$$
$$x = 37$$
The number is 37.

13.
$$x + 5 = 2x$$
$$x + 5 - x = 2x - x$$
$$5 = x$$
The number is 5.

15. $2(x+4)=36$
$2x+8=36$
$2x=28$
$x=14$
The number is 14.

17. $9x=30+3x$
$6x=30$
$x=5$
The number is 5.

19. $\dfrac{3x}{5}+4=34$
$\dfrac{3x}{5}=30$
$3x=150$
$x=50$
The number is 50.

21. Let $x=$ the number of years spent watching TV.
Let $x+19=$ the number of years spent sleeping.
$x+(x+19)=37$
$x+x+19=37$
$2x+19=37$
$2x=18$
$x=9$
$x+19=28$
Americans will spend 9 years watching TV and 28 years sleeping.

23. Let $x=$ the average salary, in thousands, for an American whose final degree is a bachelor's.
Let $2x-49=$ the average salary, in thousands, for an American whose final degree is a master's.
$x+(2x-49)=116$
$x+2x-49=116$
$3x-49=116$
$3x=165$
$x=55$
$2x-49=61$
The average salary for an American whose final degree is a bachelor's is $55 thousand and for an American whose final degree is a master's is $61 thousand.

25. Let $x=$ the number of the left-hand page.
Let $x+1=$ the number of the right-hand page.
$x+(x+1)=629$
$x+x+1=629$
$2x+1=629$
$2x+1-1=629-1$
$2x=628$
$\dfrac{2x}{2}=\dfrac{628}{2}$
$x=314$
The pages are 314 and 315.

27. Let $x=$ the first consecutive odd integer (Babe Ruth).
Let $x+2=$ the second consecutive odd integer (Roger Maris).
$x+(x+2)=120$
$x+x+2=120$
$2x+2=120$
$2x=118$
$x=59$
$x+2=61$
Babe Ruth had 59 home runs and Roger Maris had 61.

29. Let $x=$ the number of miles you can travel in one week for $320.
$200+0.15x=320$
$200+0.15x-200=320-200$
$0.15x=120$
$\dfrac{0.15x}{0.15}=\dfrac{120}{0.15}$
$x=800$
You can travel 800 miles in one week for $320. This checks because $200 + 0.15($800) = $320.

31. Let $x=$ the number of years after 2008.
$1514+20x=1714$
$20x=200$
$\dfrac{20x}{20}=\dfrac{200}{20}$
$x=10$
Mortgage payments will average $1714 ten years after 2008, or 2018.

33. Let $x =$ the width of the field.
Let $4x =$ the length of the field.

$$P = 2l + 2w$$
$$500 = 2 \cdot 4x + 2 \cdot x$$
$$500 = 8x + 2x$$
$$500 = 10x$$
$$\frac{500}{10} = \frac{10x}{10}$$
$$50 = x$$
$$x = 50$$
$$4x = 200$$

The field is 50 yards wide and 200 yards long.

35. Let $x =$ the width of a football field.
Let $x + 200 =$ the length of a football field.

$$P = 2l + 2w$$
$$1040 = 2(x + 200) + 2 \cdot x$$
$$1040 = 2x + 400 + 2x$$
$$1040 = 4x + 400$$
$$640 = 4x$$
$$160 = x$$
$$x = 160$$
$$x + 200 = 360$$

A football field is 160 feet wide and 360 feet long.

37. As shown in the diagram,
let $x =$ the height and $3x =$ the length.
To construct the bookcase, 3 heights and 4 lengths are needed.
Since 60 feet of lumber is available,

$$3x + 4(3x) = 60$$
$$3x + 12x = 60$$
$$15x = 60$$
$$x = 4$$
$$3x = 12$$

The bookcase is 12 feet long and 4 feet high.

39. Let $x =$ the price before the reduction.

$$x - 0.20x = 320$$
$$0.80x = 320$$
$$\frac{0.80x}{0.80} = \frac{320}{0.80}$$
$$x = 400$$

The price before the reduction was $400.

41. Let $x =$ the last year's salary.

$$x + 0.08x = 50,220$$
$$1.08x = 50,220$$
$$\frac{1.08x}{1.08} = \frac{50,220}{1.08}$$
$$x = 46,500$$

Last year's salary was $46,500.

43. Let $x =$ the price of the car without tax.

$$x + 0.06x = 23,850$$
$$1.06x = 23,850$$
$$\frac{1.06x}{1.06} = \frac{23,850}{1.06}$$
$$x = 22,500$$

The price of the car without sales tax was $14,500.

45. Let $x =$ the number of hours of labor.

$$63 + 35x = 448$$
$$63 + 35x - 63 = 448 - 63$$
$$35x = 385$$
$$\frac{35x}{35} = \frac{385}{35}$$
$$x = 11$$

It took 11 hours of labor to repair the car.

47. – 49. Answers will vary.

51. does not make sense; Explanations will vary.

53. makes sense

55. false; Changes to make the statement true will vary. A sample change is: This should be modeled by $x - 10 = 160$.

57. true

59. Let $x =$ the number of inches over 5 feet.

$$W = 100 + 5x$$
$$135 = 100 + 5x$$
$$135 - 100 = 100 - 100 + 5x$$
$$35 = 5x$$
$$\frac{35}{5} = \frac{5x}{5}$$
$$7 = x$$

The height 5' 7" corresponds to 135 pounds.

61. Let x = the woman's age.
Let $3x$ = the "uncle's" age.

$$3x + 20 = 2(x + 20)$$
$$3x + 20 = 2x + 40$$
$$3x - 2x + 20 = 2x - 2x + 40$$
$$x + 20 = 40$$
$$x + 20 - 20 = 40 - 20$$
$$x = 20$$

The woman is 20 years old and the "uncle" is $3x = 3(20) = 60$ years old.

63. $$\frac{4}{5}x = -16$$
$$\frac{5}{4}\left(\frac{4}{5}x\right) = \frac{5}{4}(-16)$$
$$x = -20$$

Check:
$$\frac{4}{5}(-20) = -16$$
$$\frac{4}{5} \cdot \frac{-20}{1} = -16$$
$$\frac{-80}{5} = -16$$
$$-16 = -16$$

The solution set is $\{-20\}$.

64. $$6(y-1) + 7 = 9y - y + 1$$
$$6y - 6 + 7 = 9y - y + 1$$
$$6y + 1 = 8y + 1$$
$$6y + 1 - 1 = 8y + 1 - 1$$
$$6y = 8y$$
$$6y - 8y = 8y - 8y$$
$$-2y = 0$$
$$y = 0$$

Check:
$$6(0-1) + 7 = 9(0) - 0 + 1$$
$$6 - 10 + 7 = 0 - 0 + 1$$
$$1 = 1$$

The solution set is $\{0\}$.

65. $V = \frac{1}{3}lwh$ for w

$$V = \frac{1}{3}lwh$$
$$3V = 3\left(\frac{1}{3}lwh\right)$$
$$3V = lwh$$
$$\frac{3V}{lh} = \frac{lwh}{lh}$$
$$\frac{3V}{lh} = w \quad \text{or} \quad w = \frac{3V}{lh}$$

66. $A = \frac{1}{2}bh$

$$30 = \frac{1}{2} \cdot 12h$$
$$30 = 6h$$
$$\frac{30}{6} = \frac{6h}{6}$$
$$5 = h$$

67. $A = \frac{1}{2}h(a+b)$

$$A = \frac{1}{2}(7)(10+16)$$
$$A = \frac{1}{2}(7)(26)$$
$$A = 91$$

68. $$x = 4(90 - x) - 40$$
$$x = 360 - 4x - 40$$
$$x = 320 - 4x$$
$$5x = 320$$
$$x = 64$$

The solution set is $\{64\}$.

2.6 Check Points

1. $A = 24, b = 4$

$$A = \frac{1}{2}bh$$

$$24 = \frac{1}{2} \cdot 4 \cdot h$$

$$24 = 2h$$

$$12 = h$$

The height of the sail is 12 ft.

2. Use the formulas for the area and circumference of a circle. The radius is 20 ft.

$$A = \pi r^2$$

$$A = \pi(20)^2$$

$$= 400\pi$$

$$\approx 1256 \text{ or } 1257$$

The area is 400π ft^2 or approximately 1256 ft^2 or 1257 ft^2.

$$C = 2\pi r$$

$$C = 2\pi(20)$$

$$= 40\pi$$

$$\approx 126$$

The circumference is 40π ft or approximately 126 ft.

3. The radius of the large pizza is 9 inches, and the radius of the medium pizza is 7 inches.
large pizza:

$$A = \pi r^2 = \pi(9 \text{ in.})^2 = 81\pi \text{ in.}^2 \approx 254 \text{ in.}^2$$

medium pizza:

$$A = \pi r^2 = \pi(7 \text{ in.})^2 = 49\pi \text{ in.}^2 \approx 154 \text{ in.}^2$$

For each pizza, find the price per inch by dividing the price by the area.
Price per square inch for the large pizza

$$= \frac{\$20.00}{81\pi \text{ in.}^2} \approx \frac{\$20.00}{254 \text{ in.}^2} \approx \frac{\$0.08}{\text{in.}^2}$$

Price per square inch for the medium pizza

$$= \frac{\$14.00}{49\pi \text{ in.}^2} \approx \frac{\$14.00}{154 \text{ in.}^2} \approx \frac{\$0.09}{\text{in.}^2}.$$

The large pizza is the better buy.

4. Smaller cylinder: $r = 3$ in., $h = 5$ in.

$$V = \pi r^2 h$$

$$V = \pi(3)^2 \cdot 5$$

$$= 45\pi$$

The volume of the smaller cylinder is 45π in.^{3}.
Larger cylinder: $r = 3$ in., $h = 10$ in.

$$V = \pi r^2 h$$

$$V = \pi(3)^2 \cdot 10$$

$$= 90\pi$$

The volume of the smaller cylinder is 90π in.^{3}.
The ratio of the volumes of the two cylinders is

$$\frac{V_{\text{larger}}}{V_{\text{smaller}}} = \frac{90\pi \text{ in.}^3}{45\pi \text{ in.}^3} = \frac{2}{1}.$$

So, the volume of the larger cylinder is 2 times the volume of the smaller cylinder.

5. Use the formula for the volume of a sphere. The radius is 4.5 in.

$$V = \frac{4}{3}\pi r^3$$

$$V = \frac{4}{3}\pi(4.5)^3$$

$$= 121.5\pi$$

$$\approx 382$$

The volume is approximately 382 in.^{3}. Thus the 350 cubic inches will not be enough to fill the ball. About 32 more cubic inches are needed.

6. Let $3x$ = the measure of the first angle.
Let x = the measure of the second angle.
Let $x - 20$ = the measure of the third angle.

$$3x + x + (x - 20) = 180$$

$$5x - 20 = 180$$

$$5x = 200$$

$$x = 40$$

$$3x = 120$$

$$x - 20 = 20$$

The three angle measures are 120°, 40°, and 20°.

7. *Step 1* Let x = the measure of the angle.

Step 2 Let $90 - x$ = the measure of its complement.

Step 3 The angle's measure is twice that of its complement, so the equation is
$x = 2 \cdot (90 - x)$.

Step 4 Solve this equation
$$x = 2 \cdot (90 - x)$$
$$x = 180 - 2x$$
$$x + 2x = 180 - 2x + 2x$$
$$3x = 180$$
$$x = 60$$
The measure of the angle is 60°.

Step 5 The complement of the angle is
$90° - 60° = 30°$, and 60° is indeed twice 30°.

2.6 Concept and Vocabulary Check

1. $A = \dfrac{1}{2} bh$

2. $A = \pi r^2$

3. $C = 2\pi r$

4. radius; diameter

5. $V - lwh$

6. $V = \pi r^2 h$

7. 180°

8. complementary

9. supplementary

10. $90 - x$; $180 - x$

2.6 Exercise Set

1. Use the formulas for the perimeter and area of a rectangle. The length is 6 m and the width is 3 m.
$$P = 2l + 2w$$
$$= 2(6) + 2(3) = 12 + 6 = 18$$
$$A = lw = 6 \cdot 3 = 18$$
The perimeter is 18 meters, and the area is 18 square meters.

3. Use the formula for the area of a triangle. The base is 14 in and the height is 8 in.
$$A = \frac{1}{2} bh = \frac{1}{2} (14)(8) = 56$$
The area is 56 square inches.

5. Use the formula for the area of a trapezoid. The bases are 16 m and 10 m and the height is 7 m.
$$A = \frac{1}{2} h(a + b)$$
$$= \frac{1}{2} (7)(16 + 10) = \frac{1}{2} \cdot 7 \cdot 26 = 91$$
The area is 91 square meters.

7. $A = 1250$, $w = 25$
$$A = lw$$
$$1250 = l \cdot 25$$
$$50 = l$$
The length of the swimming pool is 50 feet.

9. $A = 20, b = 5$
$$A = \frac{1}{2} bh$$
$$20 = \frac{1}{2} \cdot 5 \cdot h$$
$$20 = \frac{5}{2} h$$
$$\frac{2}{5} (20) = \frac{2}{5} \left(\frac{5}{2} h \right)$$
$$8 = h$$
The height of the triangle is 8 feet.

11. $P = 188$, $w = 44$
$$188 = 2l + 2(44)$$
$$188 = 2l + 88$$
$$100 = 2l$$
$$50 = l$$
The length of the rectangle is 50 cm.

13. Use the formulas for the area and circumference of a circle. The radius is 4 cm.

$$A = \pi r^2$$

$$A = \pi(4)^2$$

$$= 16\pi$$

$$\approx 50$$

The area is 16π cm^2 or approximately 50 cm^2.

$$C = 2\pi r$$

$$C = 2\pi(4)$$

$$= 8\pi$$

$$\approx 25$$

The circumference is 8π cm or approximately 25 cm.

15. Since the diameter is 12 yd, the radius is $\dfrac{12}{2} = 6$ yd.

$$A = \pi r^2$$

$$A = \pi(6)^2$$

$$= 36\pi$$

$$\approx 113$$

The area is 36π yd^2 or approximately 113 yd^2.

$$C = 2\pi r$$

$$C = 2\pi \cdot 6$$

$$= 12\pi$$

$$\approx 38$$

The circumference is 12π yd or approximately 38 yd.

17.

$$C = 2\pi r$$

$$14\pi = 2\pi r$$

$$\frac{14\pi}{2\pi} = \frac{2\pi r}{2\pi}$$

$$7 = r$$

The radius is 7 in. and the diameter is 2(7 in) = 14 in.

19. Use the formula for the volume of a rectangular solid. The length and width are each 3 inches and the height is 4 inches.

$$V = lwh$$

$$V = 3 \cdot 3 \cdot 4$$

$$= 36$$

The volume is 36 in.^{3}.

21. Use the formula for the volume of a cylinder. The radius is 5 cm and the height is 6 cm.

$$V = \pi r^2 h$$

$$V = \pi(5)^2 \, 6$$

$$= \pi(25)6$$

$$= 150\pi$$

$$\approx 471$$

The volume of the cylinder is 150π cm^3 or approximately 471 cm^3.

23. Use the formula for the volume of a sphere. The diameter is 18 cm, so the radius is 9 cm.

$$V = \frac{4}{3}\pi r^3$$

$$V = \frac{4}{3}\pi(9)^3$$

$$= 972\pi$$

$$\approx 3052$$

The volume is 972π cm^3 or approximately 3052 cm^3.

25. Use the formula for the volume of a cone. The radius is 4 m and the height is 9 m.

$$V = \frac{1}{3}\pi r^2 h$$

$$V = \frac{1}{3}\pi(4)^2 \cdot 9$$

$$= 48\pi$$

$$\approx 151$$

The volume is 48π m^3 or approximately 151 m^3.

27.

$$\frac{V}{\pi r^2} = \frac{\pi r^2 h}{\pi r^2}$$

$$\frac{V}{\pi r^2} = h$$

29. Smaller cylinder: $r = 3$ in, $h = 4$ in.

$$V = \pi r^2 h = \pi(3)^2 \cdot 4 = 36\pi$$

The volume of the smaller cylinder is $36\pi \, in^3$

Larger cylinder: $r = 3(3 \text{ in}) = 9$ in, $h = 4$ in.

$$V = \pi r^2 h = \pi(9)^2 \cdot 4 = 324\pi$$

The volume of the larger cylinder is $324\pi \, in.^3$

The ratio of the volumes of the two cylinders is

$$\frac{V_{larger}}{V_{smaller}} = \frac{324\pi}{36\pi} = \frac{9}{1}.$$

So, the volume of the larger cylinder is 9 times the volume of the smaller cylinder.

31. The sum of the measures of the three angles of any triangle is $180°$.

$$x + x + (x + 30) = 180$$
$$3x + 30 = 180$$
$$3x = 150$$
$$x = 50$$
$$x + 30 = 80$$

The three angle measures are $50°, 50°,$ and $80°$.

33.
$$4x + (3x + 4) + (2x + 5) = 180$$
$$9x + 9 = 180$$
$$9x = 171$$
$$x = 19$$
$$3x + 4 = 61$$
$$2x + 5 = 43$$

The three angle measures are $76°,$ $61°,$ and $43°$.

35. Let x = the measure of the smallest angle.
Let $2x$ = the measure of the second angle.
Let $x + 20$ = the measure of the third angle.
$$x + 2x + (x + 20) = 180$$
$$4x + 20 = 180$$
$$4x = 160$$
$$x = 40$$
$$2x = 80$$
$$x + 20 = 60$$

The three angle measures are $40°, 80°,$ and $60°$.

37. If the measure of an angle is $58°$, the measure of its complement is $90° - 58° = 32°$.

39. If the measure of an angle is $88°$, the measure of its complement is $2°$.

41. If the measure of an angle is $132°$, the measure of its supplement is $180° - 132° = 48°$.

43. If the measure of an angle is $90°$, the measure of its supplement is $180° - 90° = 90°$.

45. *Step 1* Let x = the measure of the angle.

Step 2 Let $90 - x$ = the measure of its complement.

Step 3 The angle's measure is $60°$ more than that of its complement, so the equation is
$$x = (90 - x) + 60.$$

Step 4 Solve this equation
$$x = 90 - x + 60$$
$$x = 150 - x$$
$$2x = 150$$
$$x = 75$$
The measure of the angle is $75°$.

Step 5 The complement of the angle is $90° - 75° = 15°$, and $75°$ is $60°$ more than $15°$.

47. *Step 1* Let x = the measure of the angle.

Step 2 Then $180 - x$ = the measure of its supplement.

Step 3 The angle's measure is three times that of its supplement, so the equation is
$$x = 3(180 - x).$$

Step 4 Solve this equation
$$x = 3(180 - x)$$
$$x = 540 - 3x$$
$$4x = 540$$
$$x = 135$$
The measure of the angle is $135°$.

Step 5 The measure of its supplement is $180° - 135° = 45°$, and $135° = 3(45°)$, so the proposed solution checks.

49. *Step 1* Let x = the measure of the angle.

Step 2 Let $180 - x$ = the measure of its supplement, and, $90 - x$ = the measure of its complement.

Step 3 The measure of the angle's supplement is $10°$ more than three times that of its complement, so the equation is
$$180 - x = 3(90 - x) + 10.$$

Step 4 Solve this equation
$$180 - x = 3(90 - x) + 10$$
$$180 - x = 270 - 3x + 10$$
$$180 - x = 280 - 3x$$
$$2x = 100$$
$$x = 50$$
The measure of the angle is $50°$.

Step 5 The measure of its supplement is $130°$ and the measure of its complement is $40°$. Since $130° = 3(40°) + 10°$, the proposed solution checks.

51. Divide the shape into two rectangles.

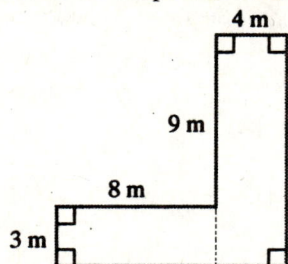

$$A_{\text{entire figure}} = A_{\text{bottom rectangle}} + A_{\text{side rectangle}}$$

$$A_{\text{entire figure}} = 3 \cdot 8 + 4(9 + 3)$$

$$= 24 + 4(12)$$

$$= 24 + 48$$

$$= 72$$

The area of the figure is 72 square meters.

53. Divide the shape into a rectangle and a triangle.

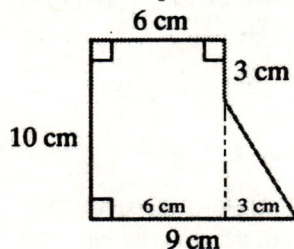

$$A_{\text{entire figure}} = A_{\text{rectangle}} + A_{\text{triangle}}$$

$$A_{\text{entire figure}} = lw + \frac{1}{2}bh$$

$$= 10(6) + \frac{1}{2}(3)(10 - 3)$$

$$= 60 + \frac{1}{2}(3)(7)$$

$$= 60 + 10.5 = 70.5$$

The area of the figure is 70.5 cm^2.

55. Subtract the volume of the three hollow portions from the volume of the whole rectangular solid.

$$V_{\text{cement block}} = V_{\text{rectangular solid}} - 3 \cdot V_{\text{hollow}}$$

$$= LWH - 3 \cdot lwh$$

$$= (8)(8)(16) - 3 \cdot (4)(6)(8)$$

$$= 1024 - 576$$

$$= 448$$

The volume of the cement block is 448 cubic inches.

57. The area of the office is $(20 \text{ ft})(16 \text{ ft}) = 320 \text{ ft}^2$.

Use a proportion to determine how much of the yearly electric bill is deductible.
Let $x =$ the amount of the electric bill that is deductible.

$$\frac{320}{2200} = \frac{x}{4800}$$

$$2200x = (320)(4800)$$

$$2200x = 1,536,000$$

$$\frac{2200x}{2200} = \frac{1,546,000}{2200}$$

$$x \approx 698.18$$

$698.18 of the yearly electric bill is deductible.

59. The radius of the large pizza is $\frac{1}{2} \cdot 14 = 7$ inches, and the radius of the medium pizza is
$\frac{1}{2} \cdot 7$ inches $= 3.5$ inches.

large pizza:
$$A = \pi r^2 = \pi(7 \text{ in.})^2$$
$$= 49\pi \text{ in.}^2 \approx 154 \text{ in.}^2$$

medium pizza:
$$A = \pi r^2 = \pi(3.5 \text{ in.})^2$$
$$= 12.25 \text{ in.}^2 \approx 38.465 \text{ in.}^2$$

For each pizza, find the price per inch by dividing the price by the area.
Price per square inch for the large pizza
$$= \frac{\$12.00}{154 \text{ in.}^2} \approx \frac{\$0.08}{\text{in.}^2}$$ and the price per square inch

for the medium pizza $= \dfrac{\$5.00}{28.465 \text{ in.}^2} \approx \dfrac{\$0.13}{\text{in.}^2}$.

The large pizza is the better buy.

61. The area of the larger circle is
$$A = \pi r^2 = \pi \cdot 50^2 = 2500\pi \text{ ft}^2.$$
The area of the smaller circle is
$$A = \pi r^2 = \pi \cdot 40^2 = 1600\pi \text{ ft}^2.$$
The area of the circular road is the difference between the area of the larger circle and the area of the smaller circle.
$$A = 2500\pi \text{ ft}^2 - 1600\pi \text{ ft}^2 = 900\pi \text{ ft}^2$$
The cost to pave the circular road is
$\$0.80(900\pi) \approx \$2262.$

63. To find the perimeter of the entire window, first find the perimeter of the lower rectangular portion. This is the bottom and two sides of the window, which is 3 ft + 6 ft + 6 ft = 15 ft. Next, find the perimeter or circumference of the semicircular portion of the window. The radius of the semicircle is $\frac{1}{2}\cdot 3\,\text{ft} = 1.5\,\text{ft}$, so the circumference is

$$\frac{1}{2}\cdot 2\pi r \approx 3.14(1.5) = 4.7\,\text{ft.}$$

So, approximately 15 ft + 4.7 ft = 19.7 ft of stripping would be needed to frame the window.

65. First, find the volume of water when the reservoir was full.
$V = lwh = 50\cdot 0\cdot 20 = 30,000$
The volume was 30,000 yd^3.
Next, find the volume when the height of the water was 6 yards.
$V = 50\cdot 30\cdot 6 = 9000$
The volume was 9000 yd^3. The amount of water used in the three-month period was 30,000 yd^3 – 9000 yd^3 = 21,000 yd^3.

67. For the first can, the diameter is 6 in. so the radius is 3 in. and $V = \pi r^2 h = \pi(3)^2\cdot 5 = 45\pi \approx 141.3$. The volume of the first can is 141.3 in^3. For the second can, the diameter is 5 in., so the radius is 2.5 in. and $V = \pi r^2 h = \pi(2.5)^2\cdot 6 = 37.5\pi \approx 117.75$. The volume of the second can is 117.75 in^2. Since the cans are the same price, the can with the greater volume is the better buy. Choose the can with the diameter of 6 inches and height of 5 inches.

69. Find the volume of a cylinder with radius 3 feet and height 2 feet 4 inches.
$2\,\text{ft}\,4\,\text{in} = 2\frac{1}{3}\ \text{feet} = \frac{7}{3}\ \text{feet}$
$V = \pi r^2 h$
$= \pi(3)^2\left(\frac{7}{3}\right) = \pi\cdot 9\cdot\frac{7}{3} = 21\pi \approx 65.94$
The volume of the tank is approximately 65.94 ft^3. This is a little over 1 ft^3 smaller than 67 ft^3 so it is too small to hold 500 gallons of water. Yes, you should be able to win your case.

71. – 77. Answers will vary.

79. does not make sense; Explanations will vary. Sample explanation: Though the heights of the books are proportional to the data, the widths are also changing. This cause the larger values to be visually exaggerated.

81. does not make sense; Explanations will vary. Sample explanation: If the radius is doubled, the area is multiplied by 4.
$$A_{\text{radius }x} = \pi r^2 = \pi(x)^2 = \pi x^2$$
$$A_{\text{radius }2x} = \pi r^2 = \pi(2x)^2 = 4\pi x^2$$

83. true

85. false; Changes to make the statement true will vary. A sample change is: $90°$ does not have a complement.

87. Area of smaller deck = $(8\,\text{ft})(10) = 80\,\text{ft}^2$.
Area of larger deck = $(12\,\text{ft})(15) = 180\,\text{ft}^2$.
Find the ratio of the areas.
$$\frac{A_{\text{larger}}}{A_{\text{smaller}}} = \frac{180\,ft^2}{80\,ft^2} = \frac{2.25}{1}\ or\ 2.25:1$$
The cost will increase 2.25 times.

89. Let x = the radius of the original sphere.
Let $2x$ = the radius of the larger sphere.
Find the ratio of the volumes of the two spheres.
$$\frac{A_{\text{larger}}}{A_{\text{original}}} = \frac{\frac{4}{3}\pi(2x)^3}{\frac{4}{3}\pi x^3} = \frac{8x^3}{x^3} = \frac{8}{1}\ or\ 8:1$$
If the radius of a sphere is doubled, the volume increases 8 times.

91. The angles marked $2x$ and $2x+40$ in the figure are supplementary, so their sum is 180°.
$2x + (2x+40) = 180$
$2x + 2x + 40 = 180$
$4x + 40 = 180$
$4x = 10$
$x = 35$
The angle of inclination is $35°$.

92. $P = 2s + b$ for s
$P - b = 2s$
$\dfrac{P-b}{2} = \dfrac{2s}{2}$
$\dfrac{P-b}{2} = s\ $ or $\ s = \dfrac{P-b}{2}$

93. $\dfrac{x}{2} + 7 = 13 - \dfrac{x}{4}$

Multiply both sides by the LCD, 4.

$$4\left(\dfrac{x}{2} + 7\right) = 4\left(13 - \dfrac{x}{4}\right)$$

$$2x + 28 = 52 - x$$

$$2x + 28 + x = 52 - x + x$$

$$3x + 28 = 52$$

$$3x + 28 - 28 = 52 - 28$$

$$3x = 24$$

$$\dfrac{3x}{3} = \dfrac{24}{3}$$

$$x = 8$$

The solution set is $\{8\}$.

94. $\left[3\left(12 \div 2^2 - 3\right)^2\right]^2$

$$= \left[3\left(12 \div 4 - 3\right)^2\right]^2$$

$$= \left[3\left(3 - 3\right)^2\right]^2 = \left(3 \cdot 0^2\right)^2 = 0^2 = 0$$

95. $x + 3 < 8$

$2 + 3 < 8$

$5 < 8,$ true

2 is a solution to the inequality.

96. $4y - 7 \geq 5$

$4(6) - 7 \geq 5$

$24 - 7 \geq 5$

$17 \geq 5,$ true

6 is a solution to the inequality.

97. $2(x - 3) + 5x = 8(x - 1)$

$$2x - 6 + 5x = 8x - 8$$

$$7x - 6 = 8x - 8$$

$$7x - 8x - 6 = 8x - 8x - 8$$

$$-x - 6 + 6 = -8 + 6$$

$$-x = -2$$

$$x = 2$$

The solution set is $\{2\}$.

2.7 Check Points

1. a.

b.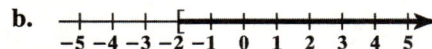

c.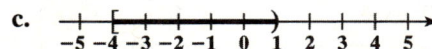

2. a. $[0, \infty)$

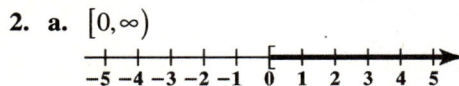

b. $(-\infty, 5)$

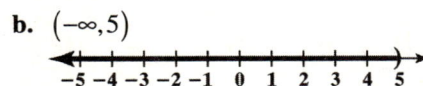

3. $x + 6 < 9$

$x + 6 - 6 < 9 - 6$

$x < 3$

The solution set is $(-\infty, 3)$ or $\{x \mid x < 3\}$.

4. $8x - 2 \geq 7x - 4$

$8x - 7x - 2 \geq 7x - 7x - 4$

$x - 2 \geq -4$

$x - 2 + 2 \geq -4 + 2$

$x \geq -2$

The solution set is $[-2, \infty)$ or $\{x \mid x \geq -2\}$.

5. a. $\dfrac{1}{4}x < 2$

$4 \cdot \dfrac{1}{4}x < 4 \cdot 2$

$x < 8$

The solution set is $(-\infty, 8)$ or $\{x \mid x < 8\}$.

b. $-6x < 18$

$\dfrac{-6x}{-6} > \dfrac{18}{-6}$

$x > -3$

The solution set is $(-3, \infty)$ or $\{x \mid x > -3\}$.

6.
$$5y - 3 \geq 17$$
$$5y - 3 + 3 \geq 17 + 3$$
$$5y \geq 20$$
$$\frac{5y}{5} \geq \frac{20}{5}$$
$$y \geq 4$$

The solution set is $[4, \infty)$ or $\{y \mid y \geq 4\}$.

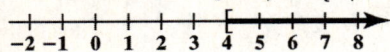

7.
$$6 - 3x \leq 5x - 2$$
$$6 - 3x - 5x \leq 5x - 5x - 2$$
$$6 - 8x \leq -2$$
$$6 - 6 - 8x \leq -2 - 6$$
$$-8x \leq -8$$
$$\frac{-8x}{-8} \geq \frac{-8}{-8}$$
$$x \geq 1$$

The solution set is $[1, \infty)$ or $\{x \mid x \geq 1\}$.

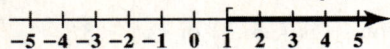

8.
$$2(x - 3) - 1 \leq 3(x + 2) - 14$$
$$2x - 6 - 1 \leq 3x + 6 - 14$$
$$2x - 7 \leq 3x - 8$$
$$2x - 3x - 7 \leq 3x - 3x - 8$$
$$-x - 7 \leq -8$$
$$-x - 7 + 7 \leq -8 + 7$$
$$-x \leq -1$$
$$\frac{-x}{-1} \geq \frac{-1}{-1}$$
$$x \geq 1$$

The solution set is $[1, \infty)$ or $\{x \mid x \geq 1\}$.

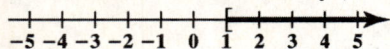

9.
$$4(x + 2) > 4x + 15$$
$$4x + 8 > 4x + 15$$
$$4x - 4x + 8 > 4x - 4x + 15$$
$$8 > 15, \text{ false}$$

There is no solution or $\{\ \}$.

10.
$$3(x + 1) \geq 2x + 1 + x$$
$$3x + 3 \geq 3x + 1$$
$$3x - 3x + 3 \geq 3x - 3x + 1$$
$$3 \geq 1, \text{ true}$$

The solution is $(-\infty, \infty)$ or $\{x \mid x \text{ is a real number}\}$.

11. Let x = your grade on the final examination.
$$\frac{82 + 74 + 78 + x + x}{5} \geq 80$$
$$\frac{234 + 2x}{5} \geq 80$$
$$5\left(\frac{234 + 2x}{5}\right) \geq 5 \cdot 80$$
$$234 + 2x \geq 400$$
$$234 - 234 + 2x \geq 400 - 234$$
$$2x \geq 166$$
$$x \geq 83$$

To earn a B you must get at least an 83% on the final examination.

12. Let x = the number of people you invite to the picnic.
$$95 + 35x \leq 1600$$
$$35x \leq 1505$$
$$\frac{35x}{35} \leq \frac{1505}{35}$$
$$x \leq 43$$

To can invite at most 43 people to the picnic.

2.7 Concept and Vocabulary Check

1. $(-\infty, 5)$

2. $(2, \infty)$

3. $< b + c$

4. $< bc$

5. $> bc$

6. subtracting 4; dividing; -3; direction; $>$; $<$

7. $\varnothing$ or the empty set

8. $(-\infty, \infty)$

2.7 Exercise Set

1. $x > 5$

3. $x < -2$

5. $x \geq -4$

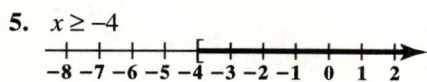

7. $x \leq 4.5$

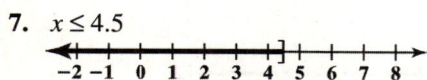

9. $-2 < x \leq 6$

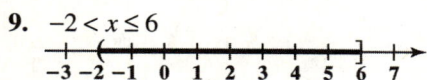

11. $-1 < x < 3$

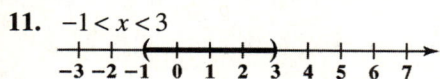

13. $(-\infty, 3]$

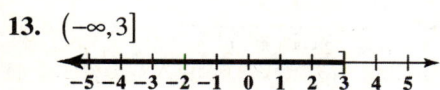

15. $\left(\dfrac{5}{2}, \infty\right)$

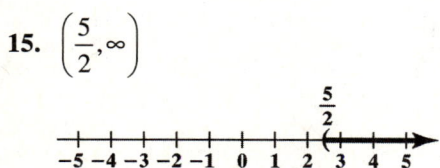

17. $(-\infty, 0]$

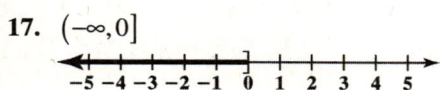

19. $(-\infty, 4)$

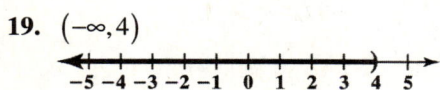

21. $\quad x - 3 > 4$

$x - 3 + 3 > 4 + 3$

$\qquad x > 7$

$(7, \infty)$

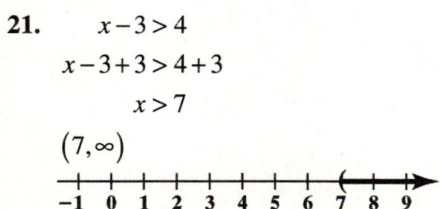

23. $\quad x + 4 \leq 10$

$x + 4 - 4 \leq 10 - 4$

$\qquad x \leq 6$

$(-\infty, 6]$

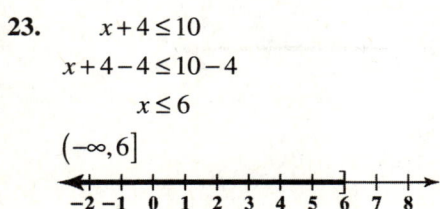

25. $\quad y - 2 < 0$

$y - 2 + 2 < 0 + 2$

$\qquad y < 2$

$(-\infty, 2)$

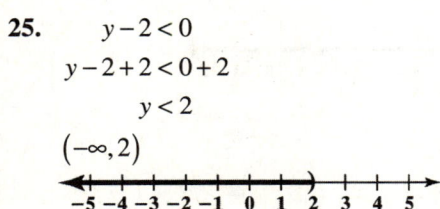

27. $\quad 3x + 4 \leq 2x + 7$

$3x - 2x \leq 7 - 4$

$\qquad x \leq 3$

$(-\infty, 3]$

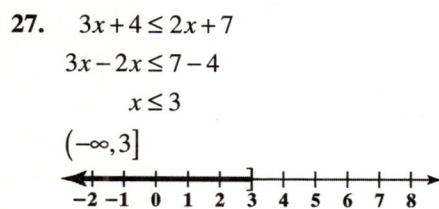

29. $\quad 5x - 9 < 4x + 7$

$5x - 4x < 7 + 9$

$\qquad x < 16$

$(-\infty, 16)$

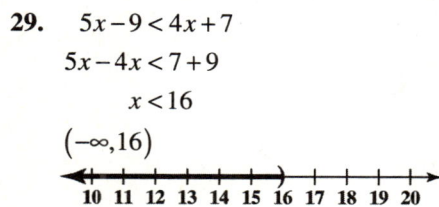

31. $\quad 7x - 7 > 6x - 3$

$7x - 6x > -3 + 7$

$\qquad x > 4$

$(4, \infty)$

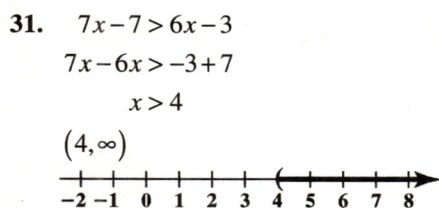

33. $\qquad x - \dfrac{2}{3} > \dfrac{1}{2}$

$x - \dfrac{2}{3} + \dfrac{2}{3} > \dfrac{1}{2} + \dfrac{2}{3}$

$\qquad x > \dfrac{3}{6} + \dfrac{4}{6}$

$\qquad x > \dfrac{7}{6}$

$\left(\dfrac{7}{6}, \infty\right)$

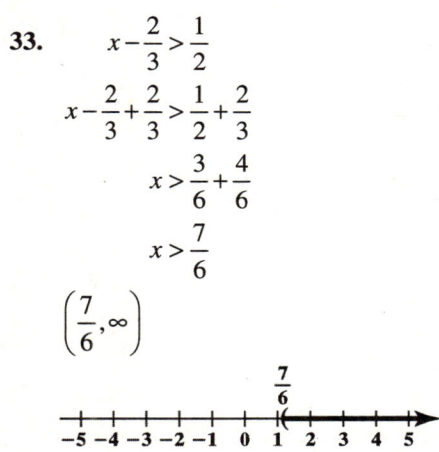

35. $\qquad y + \dfrac{7}{8} \leq \dfrac{1}{2}$

$y + \dfrac{7}{8} - \dfrac{7}{8} \leq \dfrac{1}{2} - \dfrac{7}{8}$

$\qquad y \leq \dfrac{4}{8} - \dfrac{7}{8}$

$\qquad y \leq -\dfrac{3}{8}$

$\left(-\infty, -\dfrac{3}{8}\right]$

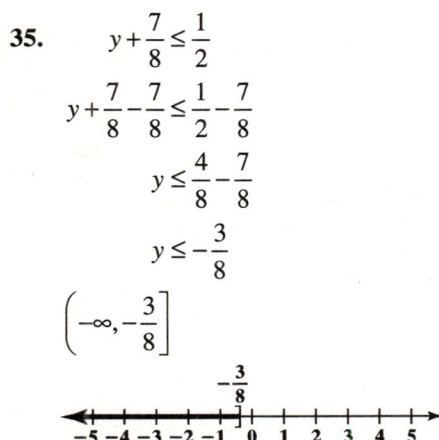

37.
$$-15y+13 > 13-16y$$
$$-15y+13+16y > 13-16y+16y$$
$$y+13 > 13$$
$$y+13-13 > 13-13$$
$$y > 0$$
$$(0, \infty)$$

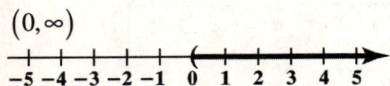

39.
$$\frac{1}{2}x < 4$$
$$2\left(\frac{1}{2}x\right) < 2(4)$$
$$1x < 8$$
$$x < 8$$
$$(-\infty, 8)$$

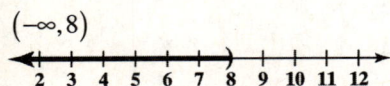

41.
$$\frac{x}{3} > -2$$
$$3\left(\frac{x}{3}\right) > 3(-2)$$
$$x > -6$$
$$(-6, \infty)$$

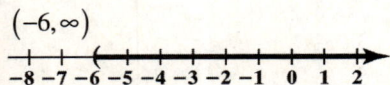

43.
$$4x < 20$$
$$\frac{4x}{4} < 20$$
$$x < 5$$
$$(-\infty, 5)$$

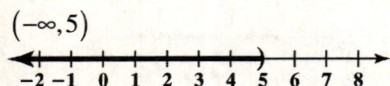

45.
$$3x \geq -21$$
$$\frac{3x}{3} \geq \frac{-21}{3}$$
$$x \geq -7$$
$$[-7, \infty)$$

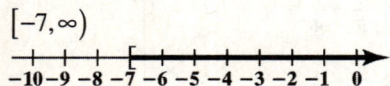

47.
$$-3x < 15$$
$$\frac{-3x}{-3} > \frac{15}{-3}$$
$$x > -5$$
$$(-5, \infty)$$

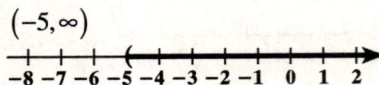

49.
$$-3x \geq 15$$
$$\frac{-3x}{-3} \leq \frac{15}{-3}$$
$$x \leq -5$$
$$(-\infty, -5]$$

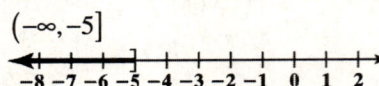

51.
$$-16x > -48$$
$$\frac{-16x}{-16} < \frac{-48}{-16}$$
$$x < 3$$
$$(-\infty, 3)$$

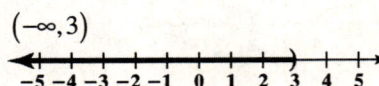

53.
$$-4y \leq \frac{1}{2}$$
$$2(-4y) \leq 2\left(\frac{1}{2}\right)$$
$$-8y \leq 1$$
$$\frac{-8y}{-8} \geq \frac{1}{-8}$$
$$y \geq -\frac{1}{8}$$
$$\left[-\frac{1}{8}, \infty\right)$$

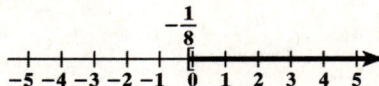

55.
$$-x < 4$$
$$-1(-x) > -1(4)$$
$$x > -4$$
$$(-4, \infty)$$

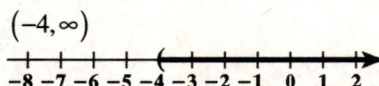

57. $2x - 3 > 7$

$2x - 3 + 3 > 7 + 3$

$2x > 10$

$\dfrac{2x}{2} > \dfrac{10}{2}$

$x > 5$

$(5, \infty)$

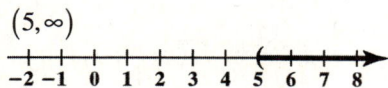

59. $3x + 3 < 18$

$3x + 3 - 3 < 18 - 3$

$3x < 15$

$\dfrac{3x}{3} < \dfrac{15}{3}$

$x < 5$

$(-\infty, 5)$

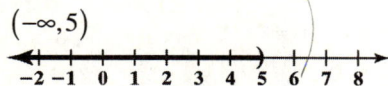

61. $3 - 7x \le 17$

$3 - 7x - 3 \le 17 - 3$

$-7x \le 14$

$\dfrac{-7x}{-7} \ge \dfrac{14}{-7}$

$x \ge -2$

$[-2, \infty)$

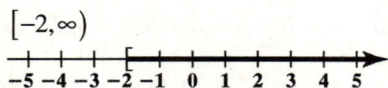

63. $-2x - 3 < 3$

$-2x - 3 + 3 < 3 + 3$

$-2x < 6$

$\dfrac{-2x}{-2} > \dfrac{6}{-2}$

$x > -3$

$(-3, \infty)$

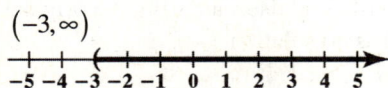

65. $5 - x \le 1$

$5 - x - 5 \le 1 - 5$

$-x \le -4$

$-1(-x) \ge -1(-4)$

$x \ge 4$

$[4, \infty)$

67. $2x - 5 > -x + 6$

$2x - 5 + x > -x + 6 + x$

$3x - 5 > 6$

$3x - 5 + 5 > 6 + 5$

$3x > 11$

$\dfrac{3x}{3} > \dfrac{11}{3}$

$x > \dfrac{11}{3}$

$\left(\dfrac{11}{3}, \infty \right)$

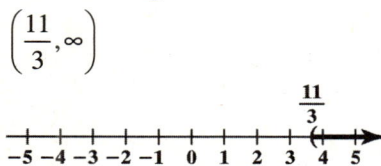

69. $2y - 5 < 5y - 11$

$2y - 5 - 5y < 5y - 11 - 5y$

$-3y - 5 < -11$

$-3y - 5 + 5 < -11 + 5$

$-3y < -6$

$\dfrac{-3y}{-3} > \dfrac{-6}{-3}$

$y > 2$

$(2, \infty)$

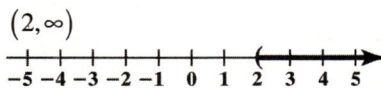

71. $3(2y - 1) < 9$

$6y - 3 < 9$

$6y - 3 + 3 < 9 + 3$

$6y < 12$

$\dfrac{6y}{6} < \dfrac{12}{6}$

$y < 2$

$(-\infty, 2)$

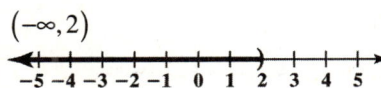

73. $3(x+1)-5 < 2x+1$

$\qquad 3x+3-5 < 2x+1$

$\qquad 3x-2 < 2x+1$

$\quad 3x-2-2x < 2x+1-2x$

$\qquad\quad x-2 < 1$

$\qquad x-2+2 < 1+2$

$\qquad\qquad x < 3$

$(-\infty, 3)$

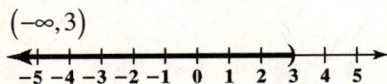

75. $\qquad 8x+3 > 3(2x+1)-x+5$

$\qquad 8x+3 > 6x+3-x+5$

$\qquad 8x+3 > 5x+8$

$\quad 8x+3-5x > 5x+8-5x$

$\qquad 3x+3 > 8$

$\qquad 3x+3-3 > 8-3$

$\qquad\qquad 3x > 5$

$\qquad\qquad x > \dfrac{5}{3}$

$\left(\dfrac{5}{3}, \infty\right)$

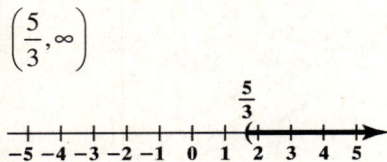

77. $\qquad \dfrac{x}{3}-2 \ge 1$

$\qquad \dfrac{x}{3}-2+2 \ge 1+2$

$\qquad\qquad \dfrac{x}{3} \ge 3$

$\qquad 3\left(\dfrac{x}{3}\right) \ge 3(3)$

$\qquad\qquad x \ge 9$

$[9, \infty)$

79. $1-\dfrac{x}{2}-1 > 4-1$

$\qquad -\dfrac{x}{2} > 3$

$\quad 2\left(-\dfrac{x}{2}\right) > 2(3)$

$\qquad -x > 6$

$\quad -1(-x) < -1(6)$

$\qquad x < -6$

$(-\infty, -6)$

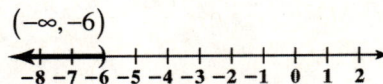

81. $\qquad 4x-4 < 4(x-5)$

$\qquad 4x-4 < 4x-20$

$\quad 4x-4+4 < 4x-20+4$

$\qquad\quad 4x < 4x-16$

$\quad 4x-4x < 4x-16-4x$

$\qquad\quad 0 < -16$

The original inequality is equivalent to the false statement $0 < -16$, so the inequality has no solution. The solution set is $\{\ \}$.

83. $\qquad x+3 < x+7$

$\quad x+3-x < x+7-x$

$\qquad\quad 3 < 7$

The original inequality is equivalent to the true statement $3 < 7$.

The solution is the set of all real numbers, written $\{x \mid x \text{ is a real number}\}$ or $(-\infty, \infty)$.

85. $\qquad 7x \le 7(x-2)$

$\qquad 7x \le 7x-14$

$\quad 7x-7x \le 7x-14-7x$

$\qquad\quad 0 \le -14$

Since $0 \le -14$ is a false statement, the original inequality has no solution.

The solution set is $\{\ \}$.

87. $\qquad 2(x+3) > 2x+1$

$\qquad 2x+6 > 2x+1$

$\quad 2x+6-2x > 2x+1-2x$

$\qquad\quad 6 > 1$

Since $6 > 1$ is a true statement, the original inequality is true for all real numbers the solution set is $\{x \mid x \text{ is a real number}\}$ or $(-\infty, \infty)$.

89.
$$5x - 4 \leq 4(x-1)$$
$$5x - 4 \leq 4x - 4$$
$$5x - 4 + 4 \leq 4x - 4 + 4$$
$$5x \leq 4x$$
$$5x - 4x \leq 4x - 4x$$
$$x \leq 0$$
$$(-\infty, 0]$$

91.
$$3x + a > b$$
$$3x > b - a$$
$$\frac{3x}{3} > \frac{b-a}{3}$$
$$x > \frac{b-a}{3}$$

93.
$$y \leq mx + b$$
$$y - b \leq mx$$
$$\frac{y-b}{m} \geq \frac{mx}{m}$$
$$\frac{y-b}{m} \geq x \quad \text{or} \quad x \leq \frac{y-b}{m}$$

95. x is between -2 and 2, so $|x| < 2$.

97. x is less than -2 or greater than 2, so $|x| > 2$.

99. weird, cemetery, accommodation

101. supersede, inoculate

103. harass

105.
$$S = 55 - 2.5x$$
$$\overbrace{S \leq 15}^{S \text{ is no more than 15 billion}}$$
$$55 - 2.5x \leq 15$$
$$-2.5x \leq -40$$
$$\frac{-2.5x}{-2.5} \geq \frac{-40}{-2.5}$$
$$x \geq 16$$
According to the model, there will be no more than 15 billion stamped letters mailed in the U.S. 16 years after 2000. In other words from 2016 onward.

107. a. Let x = your grade on the final exam.
$$\frac{86 + 88 + x}{3} \geq 90$$
$$3\left(\frac{86 + 88 + x}{3}\right) \geq 3(90)$$
$$86 + 88 + x \geq 270$$
$$174 + x \geq 270$$
$$174 + x - 174 \geq 270 - 174$$
$$x \geq 96$$
You must get at least a 96% on the final exam to earn an A in the course.

b.
$$\frac{86 + 88 + x}{3} < 80$$
$$3\left(\frac{86 + 88 + x}{3}\right) < 3(80)$$
$$86 + 88 + x < 240$$
$$174 + x < 240$$
$$174 + x - 174 < 240 - 174$$
$$x < 66$$
If you get less than a 66 on the final exam, your grade will be below a B.

109. Let x = number of miles driven.
$$80 + 0.25x \leq 400$$
$$80 + 0.25x - 80 \leq 400 - 80$$
$$0.25x \leq 320$$
$$\frac{0.25x}{0.25} \leq \frac{320}{0.25}$$
$$x \leq 1280$$
You can drive up to 1280 miles.

111. Let x = number of cement bags.
$$245 + 95x \leq 3000$$
$$245 + 95x - 245 \leq 3000 - 245$$
$$95x \leq 2755$$
$$\frac{95x}{95} \leq \frac{2755}{95}$$
$$x \leq 29$$
Up to 29 bags of cement can safely be listed on the elevator in one trip.

113. – 115. Answers will vary.

117. makes sense

119. makes sense

121. false; Changes to make the statement true will vary. A sample change is: The inequality $x - 3 > 0$ is equivalent to $x > 3$.

123. false; Changes to make the statement true will vary. A sample change is: The inequality $-4x < -20$ is equivalent to $x > 5$.

125. Let x = number of miles driven.
Weekly cost for Basic Rental: $260.
Weekly cost for Continental: $80 + 0.25x$
The cost for Basic Rental is a better deal if
$80 + 0.25x > 260$.
Solve this inequality.
$80 + 0.25x - 80 > 260 - 80$
$$0.25x > 180$$
$$\frac{0.25x}{0.25} > \frac{180}{0.25}$$
$$x > 720$$

Basic Car Rental is a better deal if you drive more than 720 miles in a week.

127.
$$1.45 - 7.23x > -1.442$$
$$1.45 - 7.23x - 1.45 > -1.442 - 1.45$$
$$-7.23x > -2.892$$
$$\frac{-7.23x}{-7.23} < \frac{-2.892}{-7.23}$$
$$x < 0.4$$

$(-\infty, 0.4)$

129. $A = PB$, $A = 8$, $P = 40\% = 0.4$
$$A = PB$$
$$8 = 0.4B$$
$$\frac{8}{0.4} = \frac{0.4B}{0.4}$$
$$20 = B$$
8 is 40% of 20.

130. Let x = the width of the rectangle.
Let $x + 5$ = the length of the rectangle.

$$P = 2l + 2w$$
$$34 = 2(x + 5) + 2 \cdot x$$
$$34 = 2x + 10 + 2x$$
$$34 = 4x + 10$$
$$34 - 10 = 4x + 10 - 10$$
$$24 = 4x$$
$$6 = x$$
$$x = 6$$
$$x + 5 = 11$$
The width is 6 inches and the length is 11 inches.

131.
$$5x + 16 = 3(x + 8)$$
$$5x + 16 = 3x + 24$$
$$5x + 16 - 3x = 3x + 24 - 3x$$
$$2x + 16 = 24$$
$$2x + 16 - 16 = 24 - 16$$
$$2x = 8$$
$$\frac{2x}{2} = \frac{8}{2}$$
$$x = 4$$
Check: $5(4) + 16 = 3(4 + 8)$
$$20 + 16 = 3(12)$$
$$36 = 36, \text{ true}$$
The solution is set is $\{4\}$.

132.
$$x - 4y = 14$$
$$2 - 4(-3) = 14$$
$$2 + 12 = 14$$
$$14 = 14, \text{ true}$$
Yes, the values make it a true statement.

133.
$$x - 4y = 14$$
$$12 - 4(1) = 14$$
$$12 - 4 = 14$$
$$8 = 14, \text{ false}$$
No, the values make it a false statement.

134. $y = \frac{2}{3}x + 1$
$$y = \frac{2}{3}(-6) + 1$$
$$y = -4 + 1$$
$$y = -3$$

Chapter 2 Review Exercises

1.　　$x - 10 = 22$

　　　$x - 10 + 10 = 22 + 10$

　　　　　　　$x = 32$

The solution is set is $\{32\}$.

2.　　$-14 = y + 8$

　　$-14 - 8 = y + 8 - 8$

　　　　$-22 = y$

The solution is set is $\{-22\}$.

3.　　　$7z - 3 = 6z + 9$

　　$7z - 3 - 6z = 6z + 9 - 6z$

　　　　　　$z - 3 = 9$

　　　$z - 3 + 3 = 9 + 3$

　　　　　　　$z = 12$

The solution is set is $\{12\}$.

4.　　$4(x + 3) = 3x - 10$

　　　$4x + 12 = 3x - 10$

　$4x + 12 - 3x = 3x - 10 - 3x$

　　　　$x + 12 = -10$

　　$x + 12 - 12 = -10 - 12$

　　　　　　$x = -22$

The solution is set is $\{-22\}$.

5.　$6x - 3x - 9 + 1 = -5x + 7x - 3$

　　　　$3x - 8 = 2x - 3$

　　$3x - 8 - 2x = 2x - 3 - 2x$

　　　　　$x - 8 = -3$

　　　$x - 8 + 8 = -3 + 8$

　　　　　　　$x = 5$

The solution is set is $\{5\}$.

6.　　$\dfrac{x}{8} = 10$

　　$8\left(\dfrac{x}{8}\right) = 8(10)$

　　　　$x = 80$

The solution is set is $\{80\}$.

7.　　　$\dfrac{y}{-8} = 7$

　　$-8\left(\dfrac{y}{-8}\right) = -8(7)$

　　　　　$y = -56$

The solution is set is $\{-56\}$.

8.　$7z = 77$

　　$\dfrac{7z}{7} = \dfrac{77}{7}$

　　　$z = 11$

The solution is set is $\{11\}$.

9.　$-36 = -9y$

　　$\dfrac{-36}{-9} = \dfrac{-9y}{-9}$

　　　$4 = y$

The solution is set is $\{4\}$.

10.　　$\dfrac{3}{5}x = -9$

　　$\dfrac{5}{3}\left(\dfrac{3}{5}x\right) = \dfrac{5}{3}(-9)$

　　　　$1x = -15$

　　　　$x = -15$

The solution is set is $\{-15\}$.

11.　　$30 = -\dfrac{5}{2}y$

　　$-\dfrac{2}{5}(30) = -\dfrac{2}{5}\left(-\dfrac{5}{2}y\right)$

　　　　$-12 = y$

The solution is set is $\{-12\}$.

12.　　$-x = 25$

　　$-1(-x) = -1(25)$

　　　　$x = -25$

The solution is set is $\{-25\}$.

13.
$$\frac{-x}{10} = -1$$
$$10\left(\frac{-x}{10}\right) = 10(-1)$$
$$-x = -10$$
$$-1(-x) = -1(-10)$$
$$x = 10$$
The solution is set is $\{10\}$.

14.
$$4x + 9 = 33$$
$$4x + 9 - 9 = 33 - 9$$
$$4x = 24$$
$$\frac{4x}{4} = \frac{24}{4}$$
$$x = 6$$
The solution is set is $\{6\}$.

15.
$$-3y - 2 = 13$$
$$-3y - 2 + 2 = 13 + 2$$
$$-3y = 15$$
$$\frac{-3y}{-3} = \frac{15}{-3}$$
$$y = -5$$
The solution is set is $\{-5\}$.

16.
$$5z + 20 = 3z$$
$$5z + 20 - 3z = 3z - 3z$$
$$2z + 20 = 0$$
$$2z + 20 - 20 = 0 - 20$$
$$2z = -20$$
$$\frac{2z}{2} = \frac{-20}{2}$$
$$z = -10$$
The solution is set is $\{-10\}$.

17.
$$5x - 3 = x + 5$$
$$5x - 3 - x = x + 5 - x$$
$$4x - 3 = 5$$
$$4x - 3 + 3 = 5 + 3$$
$$4x = 8$$
$$\frac{4x}{4} = \frac{8}{4}$$
$$x = 2$$
The solution is set is $\{2\}$.

18.
$$3 - 2x = 9 - 8x$$
$$3 - 2x + 8x = 9 - 8x + 8x$$
$$3 + 6x = 9$$
$$3 + 6x - 3 = 9 - 3$$
$$6x = 6$$
$$\frac{6x}{6} = \frac{6}{6}$$
$$x = 1$$
The solution is set is $\{1\}$.

19. a. 2009 is 4 years after 2005.
$$P = 3.5n + 51$$
$$P = 3.5(4) + 51 = 65$$
According to the formula, 65% of returns are filed electronically.
The formula underestimates the actual value given in the bar graph by 1%.

b.
$$P = 3.5n + 51$$
$$93 = 3.5n + 51$$
$$42 = 3.5n$$
$$\frac{42}{3.5} = \frac{3.5n}{3.5}$$
$$12 = n$$
If trends continue, 93% of returns will be filed electronically 12 years after 2005, or 2017.

20.
$$5x + 9 - 7x + 6 = x + 18$$
$$-2x + 15 = x + 18$$
$$-2x + 15 - x = x + 18 - x$$
$$-3x + 15 = 18$$
$$-3x + 15 - 15 = 18 - 15$$
$$-3x = 3$$
$$\frac{-3x}{-3} = \frac{3}{-3}$$
$$x = -1$$
The solution is set is $\{-1\}$.

21.
$$3(x+4) = 5x-12$$
$$3x+12 = 5x-12$$
$$3x+12-5x = 5x-12-5x$$
$$-2x+12 = -12$$
$$-2x+12-12 = -12-12$$
$$-2x = -24$$
$$\frac{-2x}{-2} = \frac{-24}{-2}$$
$$x = 12$$

The solution is set is $\{12\}$.

22.
$$1-2(6-y) = 3y+2$$
$$1-12+2y = 3y+2$$
$$2y-11 = 3y+2$$
$$2y-11-3y = 3y+2-3y$$
$$-y-11 = 2$$
$$-y-11+11 = 2+11$$
$$-y = 13$$
$$y = -13$$

The solution is set is $\{-13\}$.

23.
$$2x-8+3x+15 = 2x-2$$
$$5x+7 = 2x-2$$
$$5x+7-2x = 2x-2-2x$$
$$3x+7 = -2$$
$$3x+7-7 = -2-7$$
$$3x = -9$$
$$\frac{3x}{3} = \frac{-9}{3}$$
$$x = -3$$

The solution is set is $\{-3\}$.

24.
$$-2(y-4)-(3y-2) = -2-(6y-2)$$
$$-2y+8-3y+2 = -2-6y+2$$
$$-5y+10 = -6y$$
$$-5y+10+6y = -6y+6y$$
$$10+y = 0$$
$$10+y-10 = 0-10$$
$$y = -10$$

The solution is set is $\{-10\}$.

25. $\dfrac{2x}{3} = \dfrac{x}{6}+1$

To clear fractions, multiply both sides by the LCD, which is 6.

$$6\left(\frac{2x}{3}\right) = 6\left(\frac{x}{6}+1\right)$$
$$6\left(\frac{2x}{3}\right) = 6\left(\frac{x}{6}\right)+6(1)$$
$$4x = x+6$$
$$4x-x = x+6-x$$
$$3x = 6$$
$$\frac{3x}{3} = \frac{6}{3}$$
$$x = 2$$

The solution is set is $\{2\}$.

26. $\dfrac{x}{2}-\dfrac{1}{10} = \dfrac{x}{5}+\dfrac{1}{2}$

Multiply both sides by the LCD, which is 10.

$$10\left(\frac{x}{2}-\frac{1}{10}\right) = 10\left(\frac{x}{5}+\frac{1}{2}\right)$$
$$10\left(\frac{x}{2}\right)-10\left(\frac{1}{10}\right) = 10\left(\frac{x}{5}\right)+10\left(\frac{1}{2}\right)$$
$$5x-1 = 2x+5$$
$$5x-1-2x = 2x+5-2x$$
$$3x-1 = 5$$
$$3x-1+1 = 5+1$$
$$3x = 6$$
$$\frac{3x}{3} = \frac{6}{3}$$
$$x = 2$$

The solution is set is $\{2\}$.

27. Multiply both sides by 100 to clear the decimals.
$$0.5x+8.75 = 13.25$$
$$100(0.5x+8.75) = 100(13.25)$$
$$50x+875 = 1325$$
$$50x = 450$$
$$x = 9$$

The solution set is $\{9\}$.

28. First apply the distributive property to remove the parentheses, and then multiply both sides by 100 to clear the decimals.

$$0.1(x-3) = 1.1 - 0.25x$$
$$0.1x - 0.3 = 1.1 - 0.25x$$
$$100(0.1x - 0.3) = 100(1.1 - 0.25x)$$
$$10x - 30 = 110 - 25x$$
$$10x = 140 - 25x$$
$$35x = 140$$
$$\frac{35x}{35} = \frac{140}{35}$$
$$x = 4$$

The solution set is $\{4\}$.

29.
$$3(8x-1) = 6(5+4x)$$
$$24x - 3 = 30 + 24x$$
$$24x - 3 - 24x = 30 + 24x - 24x$$
$$-3 = 30$$

Since $-3 = 30$ is a false statement, the original equation is inconsistent and has no solution or $\{\ \}$.

30.
$$4(2x-3) + 4 = 8x - 8$$
$$8x - 12 + 4 = 8x - 8$$
$$8x - 8 = 8x - 8$$
$$8x - 8 - 8x = 8x - 8 - 8x$$
$$-8 = -8$$

Since $-8 = -8$ is a true statement, so the solution is the set of all real numbers, written $\{x | x \text{ is a real number}\}$.

31.
$$H = 0.7(220 - a)$$
$$133 = 0.7(220 - a)$$
$$133 - 154 = 154 - 154 - 0.7a$$
$$-21 = -0.7a$$
$$\frac{-21}{-0.7} = \frac{-0.7a}{-0.7}$$
$$30 = a$$

If the optimal heart rate is 133 beats per minute, the person is 30 years old.

32. $I = Pr$ for r
$$\frac{I}{P} = \frac{Pr}{P}$$
$$\frac{I}{P} = r \ \text{ or } \ r = \frac{I}{P}$$

33. $V = \frac{1}{3}Bh$ for h
$$3V = 3\left(\frac{1}{3}Bh\right)$$
$$3V = Bh$$
$$\frac{3V}{B} = \frac{Bh}{B}$$
$$\frac{3V}{B} = h \ \text{ or } \ h = \frac{3V}{B}$$

34. $P = 2l + 2w$ for w
$$P - 2l = 2l + 2w - 2l$$
$$P - 2l = 2w$$
$$\frac{P - 2l}{2} = \frac{2w}{2}$$
$$\frac{P - 2l}{2} = w \ \text{ or } \ w = \frac{P - 2l}{2}$$

35. $A = \frac{B+C}{2}$ for B
$$2A = 2\left(\frac{B+C}{2}\right)$$
$$2A = B + C$$
$$2A - C = B + C - C$$
$$2A - C = B \ \text{ or } \ B = 2A - C$$

36. $T = D + pm$ for m
$$T - D = D + pm - D$$
$$T - D = pm$$
$$\frac{T - D}{p} = \frac{pm}{p}$$
$$\frac{T - D}{p} = m \ \text{ or } \ m = \frac{T - D}{p}$$

37. $A = PB$; $P = 8\% = 0.08$, $B = 120$
$$A = 0.08 \cdot 120$$
$$A = 9.6$$
8% of 120 is 9.6

38. $A = PB$; $A = 90$, $P = 45\% = 0.45$
$$90 = 0.45B$$
$$\frac{90}{0.45} = \frac{0.45B}{0.45}$$
$$200 = B$$
90 is 45% of 200.

39. $A = PB;\ A = 36,\ B = 75$

$$36 = P \cdot 75$$

$$\frac{36}{75} = \frac{P \cdot 75}{75}$$

$$0.48 = P$$

36 is 48% of 75.

40. Increase = Percent · Original

First, find the increase: $12 - 6 = 6$

$$6 = P \cdot 6$$

$$\frac{6}{6} = \frac{P \cdot 6}{6}$$

$$1 = P$$

The percent increase is 100%.

41. Decrease = Percent · Original

First, find the decrease: $5 - 3 = 2$

$$2 = P \cdot 5$$

$$\frac{2}{5} = \frac{P \cdot 5}{5}$$

$$0.4 = P$$

The percent decrease is 40%.

42. Increase = Percent · Original

First, find the increase: $45 - 40 = 5$

$$5 = P \cdot 40$$

$$\frac{5}{40} = \frac{P \cdot 40}{40}$$

$$0.125 = P$$

The percent increase is 12.5%.

43. Investment dollars lost last year were $0.10 \cdot \$10,000 = \$1000$. This means that $\$10,000 - \$1000 = \$9000$ remains. Investment dollars gained this year are $0.10 \cdot \$9000 = \$900$. This means that $\$9000 + \$900 = \$9900$ of the original investment remains. This is an overall loss of $100.

decrease = percent · original

$$100 = P \cdot 10,000$$

$$\frac{100}{10,000} = \frac{P \cdot 10,000}{10,000}$$

$$0.01 = P$$

The statement is not true. Instead of recouping losses, there is an overall 1% decrease in the portfolio.

44. a. $r = \dfrac{h}{7}$

$$7r = 7\left(\frac{h}{7}\right)$$

$$7r = h \text{ or } h = 7r$$

b. $h = 7r;\ r = 9$

$$h = 7(9) = 63$$

The woman's height is 63 inches or 5 feet, 3 inches.

45. $A = P \cdot B$

$$91 = 0.26 \cdot B$$

$$\frac{91}{0.26} = \frac{0.26 \cdot B}{0.26}$$

$$350 = B$$

The average U.S. household uses 350 gallons of water per day.

46. Let $x =$ the unknown number.

$$6x - 20 = 4x$$

$$6x - 20 - 4x = 4x - 4x$$

$$2x - 20 = 0$$

$$2x - 20 + 20 = 0 + 20$$

$$2x = 20$$

$$x = 10$$

The number is 10.

47. Let $x =$ Buffett's net worth.

Let $x + 9 =$ Gate's net worth.

$$x + (x + 9) = 99$$

$$x + x + 9 = 99$$

$$2x + 9 = 99$$

$$2x = 90$$

$$x = 45$$

$$x + 9 = 54$$

In 2010 Buffett's net worth was $45 billion and Gate's net worth was $54 billion.

48. Let $x =$ the smaller page number.

Let $x + 1 =$ the larger page number.

$$x + (x + 1) = 93$$

$$2x + 1 = 93$$

$$2x = 92$$

$$x = 46$$

The page numbers are 46 and 47.

49. Let $x =$ the percentage of females.
Let $x + 2 =$ the percentage of males.
$$x + (x + 2) = 100$$
$$x + x + 2 = 100$$
$$2x + 2 = 100$$
$$2x + 2 - 2 = 100 - 2$$
$$2x = 98$$
$$x = 49$$
$$x + 2 = 51$$
For Americans under 20, 49% are female and 51% are male.

50. Let $x =$ number of years after 2001.
$$316 + 42x = 904$$
$$42x = 588$$
$$\frac{42x}{42} = \frac{588}{42}$$
$$x = 14$$
According to this model, the U.S. defense budget will reach $904 in 14 years after 2001, or 2015.

51. Let $x =$ the number of checks written.

$$6 + 0.05x = 6.90$$
$$6 + 0.05x - 6 = 6.90 - 6$$
$$0.05x = 0.90$$
$$\frac{0.05x}{0.05} = \frac{0.90}{0.05}$$
$$x = 18$$

You wrote 18 checks that month.

52. Let $x =$ the width of the field.
Let $3x =$ the length of the field.
$$P = 2l + 2w$$
$$400 = 2 \cdot 3x + 2 \cdot x$$
$$400 = 6x + 2x$$
$$400 = 8x$$
$$\frac{400}{8} = \frac{8x}{8}$$
$$50 = x$$
$$x = 50$$
$$3x = 150$$
The field is 50 yards wide and 150 yards long.

53. Let $x =$ the original price of the table.
$$x - 0.25x = 180$$
$$0.75x = 180$$
$$\frac{0.75x}{0.75} = \frac{180}{0.75}$$
$$x = 240$$
The table's price before the reduction was $240.

54. Find the area of a rectangle with length 6.5 ft and width 5 ft.
$$A = lw = (6.5)(5) = 32.5$$
The area is 32.5 ft^2.

55. Find the area of a triangle with base 20 cm and height 5 cm.
$$A = \frac{1}{2}bh = \frac{1}{2}(20)(5) = 50$$
The area is 50 cm^2.

56. Find the area of a trapezoid with bases 22 yd and 5 yd and height 10 yd.
$$A = \frac{1}{2}h(a + b)$$
$$= \frac{1}{2}(10)(22 + 5)$$
$$= \frac{1}{2} \cdot 10 \cdot 27 = 135$$
The area is 135 yd^2.

57. Notice that the height of the middle rectangle is $64 - 12 - 12 = 40$ m.

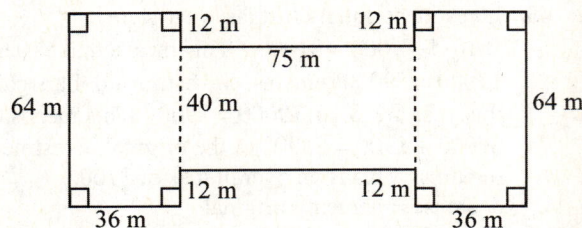

Using $A = lw$ we must find the sum of areas of the middle rectangle and the two side rectangles.
$$A = (40)(75) + 2 \cdot (64)(36)$$
$$= 3000 + 2 \cdot 2304$$
$$= 3000 + 4608$$
$$= 7608$$
The area is 7608 m^2.

58. Since the diameter is 20 m, the radius is $\frac{20}{2} = 10$ m.

$C = 2\pi = 2\pi(10) = 20\pi \approx 63$

$A = \pi r^2 = \pi(10)^2 = 100\pi \approx 314$

The circumference is 20π m or approximately 63 m; the area is 100π m^2 or approximately 314 m^2.

59. $A = 42, b = 14$

$A = \frac{1}{2}bh$

$42 = \frac{1}{2} \cdot 14 \cdot h$

$42 = 7h$

$6 = h$

The height of the sail is 6 ft.

60. Area of floor:

$A = bh = (12\,\text{ft})(15\,\text{ft}) = 180\,\text{ft}^2$

Area of base of stove:

$A = bh = (3\,\text{ft})(4\,\text{ft}) = 12\,\text{ft}^2$

Area of bottom of refrigerator:

$A = bh = (3\,\text{ft})(4\,\text{ft}) = 12\,\text{ft}^2$

The area to be covered with floor tile is

$180\,\text{ft}^2 - 12\,\text{ft}^2 - 12\,\text{ft}^2 = 156\,\text{ft}^2$.

61. First, find the area of a trapezoid with bases 80 ft and 100 ft and height 60 ft.

$A = \frac{1}{2}h(a+b)$

$= \frac{1}{2}(60)(80+100) = 5400$

The area of the yard is 5400 ft^2. The cost is $0.35(5400) = \$1890$.

62. The radius of the medium pizza is

$\frac{1}{2} \cdot 14$ inches $= 7$ inches, and the radius of each

small pizza is $\frac{1}{2} \cdot 8$ inches $= 4$ inches.

Medium pizza:

$A = \pi r^2 = \pi(7\,\text{in.})^2$

$= 49\pi\,\text{in.}^2 \approx 154\,\text{in.}^2$

Small pizza:

$A = \pi r^2 = \pi(4\,\text{in.})^2$

$= 16\pi\,\text{in.}^2 \approx 50\,\text{in.}^2$

The area of one medium pizza is approximately 154 in.^{2} and the area of two small pizzas is approximately $2(50) = 100$ in.^{2}. Since the price of one medium pizza is the same as the price of two small pizzas and the medium pizza has the greater area, the medium pizza is the better buy. (Because the prices are the same, it is not necessary to find price per square inch in this case.)

63. Find the volume of a rectangular solid with length 5 cm, width 3 cm, and height 4 cm.

$A = lwh = 5 \cdot 3 \cdot 4 = 60$

The volume is 60 cm^3.

64. Find the volume of a cylinder with radius 4 yd and height 8 yd.

$V = \pi r^2 h$

$= \pi(4)^2 \cdot 8 = 128\pi \approx 402$

The volume is 128π yd$^3 \approx 402$ yd^3.

65. Find the volume of a sphere with radius 6 m.

$V = \frac{4}{3}\pi r^3$

$= \frac{4}{3}\pi(6)^3 = \frac{4}{3} \cdot \pi \cdot 216$

$= 288\pi \approx 905$

The volume is 288π m$^3 \approx 905$ m^3.

66. Find the volume of each box.

$V = lwh = (8\text{m})(4\text{m})(3\text{m}) = 96\text{m}^3$

The space required for 50 containers is

$50(96\,\text{m}^3) = 4800$ m^3.

67. Since the diameter of the fish tank 6 ft, the radius is 3 ft.

$$V = \pi r^2 h = \pi(3)^2 \cdot 3 = 27\pi \approx 84.82$$

The volume of the tank is approximately 85 ft³. Divide by 5 to determine how many fish can be put in the tank.

$$\frac{84.82}{5} \approx 16.96$$

There is enough water in the tank for 16 fish. Round down to 16, since 0.96 of a fish cannot be purchased.

68. The sum of the measures of the angles of any triangle is $180°$, so $x + 3x + 2x = 180$.

$$x + 3x + 2x = 180$$
$$6x = 180$$
$$x = 30$$

If $x = 30$, then $3x = 90$ and $2x = 60$, so the angles measure $30°$, $60°$, and $90°$.

69. Let x = the measure of the second angle.
Let $2x + 15$ = the measure of the first angle.
Let $x + 25$ = the measure of the third angle.

$$x + (2x + 15) + (x + 25) = 180$$
$$4x + 40 = 180$$
$$4x = 140$$
$$x = 35$$

If $x = 35$, then $2x + 15 = 2(35) + 15 = 85$ and $x + 25 = 35 + 25 = 60$. The angles measure $85°$, $35°$, and $60°$.

70. If the measure of an angle is $57°$, the measure of its complement is $90° - 57° = 33°$

71. If the measure of an angle is $75°$, the measure of its supplement is $180° - 75° = 105°$.

72. Let x = the measure of the angle.
Let $90 - x$ = the measure of its complement.

$$x = (90 - x) + 25$$
$$x = 115 - x$$
$$2x = 115$$
$$x = 57.5$$

The measure of the angle is $57.5°$.

73. Let x = the measure of the angle.
Let $180 - x$ = the measure of its supplement.

$$180 - x = 4x - 45$$
$$180 - 5x = -45$$
$$-5x = -225$$
$$x = 45$$

If $x = 45$, then $180 - x = 135$. The measure of the angle is $45°$ and the measure of its supplement is $135°$.

74. $x < -1$

75. $-2 < x \le 4$

76. $\left[\dfrac{3}{2}, \infty \right)$

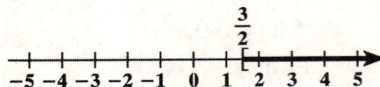

77. $(-\infty, 0)$

78.
$$2x - 5 < 3$$
$$2x - 5 + 5 < 3 + 5$$
$$2x < 8$$
$$\frac{2x}{2} < \frac{8}{2}$$
$$x < 4$$

$(-\infty, 4)$

79.
$$\frac{x}{2} > -4$$
$$2\left(\frac{x}{2} \right) > 2(-4)$$
$$x > -8$$

$(-8, \infty)$

80.
$$3 - 5x \le 18$$
$$3 - 5x - 3 \le 18 - 3$$
$$-5x \le 15$$
$$\frac{-5x}{-5} \ge \frac{15}{-5}$$
$$x \ge -3$$
$$[-3, \infty)$$

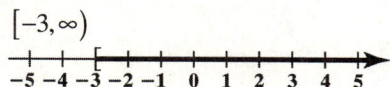

81.
$$4x + 6 < 5x$$
$$4x + 6 - 5x < 5x - 5x$$
$$-x + 6 < 0$$
$$-x + 6 - 6 < 0 - 6$$
$$-x < -6$$
$$-1(-x) > -1(-6)$$
$$x > 6$$
$$(6, \infty)$$

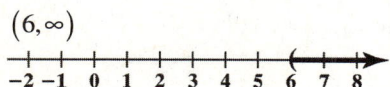

82.
$$6x - 10 \ge 2(x + 3)$$
$$6x - 10 \ge 2x + 6$$
$$6x - 10 - 2x \ge 2x + 6 - 2x$$
$$4x - 10 \ge 6$$
$$4x - 10 + 10 \ge 6 + 10$$
$$4x \ge 16$$
$$\frac{4x}{4} \ge \frac{16}{4}$$
$$x \ge 4$$
$$[4, \infty)$$

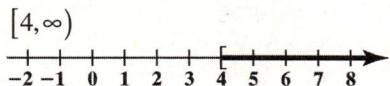

83.
$$4x + 3(2x - 7) \le x - 3$$
$$4x + 6x - 21 \le x - 3$$
$$10x - 21 \le x - 3$$
$$10x - 21 - x \le x - 3 - x$$
$$9x - 21 \le -3$$
$$9x - 21 + 21 \le -3 + 21$$
$$9x \le 18$$
$$\frac{9x}{9} \le \frac{18}{9}$$
$$x \le 2$$
$$(-\infty, 2]$$

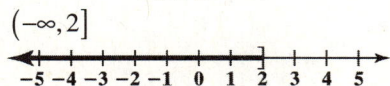

84.
$$2(2x + 4) > 4(x + 2) - 6$$
$$4x + 8 > 4x + 8 - 6$$
$$4x + 8 > 4x + 2$$
$$4x + 8 - 4x > 4x + 2 - 4x$$
$$8 > 2$$
Since $8 > 2$ is a true statement, the original inequality is true for all real numbers, and the solution set is $\{x \mid x \text{ is a real number}\}$.

85.
$$-2(x - 4) \le 3x + 1 - 5x$$
$$-2x + 8 \le -2x + 1$$
$$-2x + 8 + 2x \le -2x + 1 + 2x$$
$$8 \le 1$$
Since $8 \le 1$ is a false statement, the original inequality has no solution. The solution set is $\{\ \}$.

86. Let x = the student's score on the third test.
$$\frac{42 + 74 + x}{3} \ge 60$$
$$3\left(\frac{42 + 74 + x}{3}\right) \ge 3(60)$$
$$42 + 74 + x \ge 180$$
$$116 + x \ge 180$$
$$116 + x - 116 \ge 180 - 116$$
$$x \ge 64$$
The student must score at least 64 on the third test to pass the course.

87. Let x = the number of people you invite to the picnic.
$$350 + 55x \le 2000$$
$$55x \le 1650$$
$$\frac{55x}{55} \le \frac{1650}{55}$$
$$x \le 30$$
To can invite at most 30 people to the party.

Chapter 2 Test

1. $$4x - 5 = 13$$
    $$4x + 5 + 5 = 13 + 5$$
    $$4x = 18$$
    $$\frac{4x}{4} = \frac{18}{4} = \frac{9}{2}$$
    $$x = \frac{9}{2}$$
    The solution set is $\left\{ \frac{9}{2} \right\}$.

2. $$12x + 4 = 7x - 21$$
    $$12x + 4 - 7x = 7x - 21 - 7x$$
    $$5x + 4 = -21$$
    $$5x + 4 - 4 = -21 - 4$$
    $$5x = -25$$
    $$\frac{5x}{5} = \frac{-25}{5}$$
    $$x = -5$$
    The solution set is $\{-5\}$.

3. $$8 - 5(x - 2) = x + 26$$
    $$8 - 5x + 10 = x + 26$$
    $$18 - 5x = x + 26$$
    $$18 - 5x - x = x + 26 - x$$
    $$18 - 6x = 26$$
    $$18 - 6x - 18 = 26 - 18$$
    $$-6x = 8$$
    $$\frac{-6x}{-6} = \frac{8}{-6}$$
    $$x = -\frac{8}{6} = -\frac{4}{3}$$
    The solution set is $\left\{ -\frac{4}{3} \right\}$.

4. $$3(2y - 4) = 9 - 3(y + 1)$$
    $$6y - 12 = 9 - 3y - 3$$
    $$6y - 12 = 6 - 3y$$
    $$6y - 12 + 3y = 6 - 3y + 3y$$
    $$9y - 12 = 6$$
    $$9y - 12 + 12 = 6 + 12$$
    $$9y = 18$$
    $$\frac{9y}{9} = \frac{18}{9}$$
    $$y = 2$$
    The solution set is $\{2\}$.

5. $$\frac{3}{4}x = -15$$
    $$\frac{4}{3}\left(\frac{3}{4}x \right) = \frac{4}{3}(-15)$$
    $$x = -20$$
    The solution set is $\{-20\}$.

6. $$\frac{x}{10} + \frac{1}{3} = \frac{x}{5} + \frac{1}{2}$$
    Multiply both sides by the LCD, 30.
    $$30\left(\frac{x}{10} + \frac{1}{3} \right) = 30\left(\frac{x}{5} + \frac{1}{2} \right)$$
    $$30\left(\frac{x}{10} \right) + 30\left(\frac{1}{3} \right) = 30\left(\frac{x}{5} \right) + 30\left(\frac{1}{2} \right)$$
    $$3x + 10 = 6x + 15$$
    $$3x + 10 - 6x = 6x + 15 - 6x$$
    $$-3x + 10 = 15$$
    $$-3x + 10 - 10 = 15 - 10$$
    $$-3x = 5$$
    $$\frac{-3x}{-3} = \frac{5}{-3}$$
    $$x = -\frac{5}{3}$$
    The solution set is $\left\{ -\frac{5}{3} \right\}$.

7.　$9.2x - 80.1 = 21.3x - 19.6$
To clear the equation of decimals, multiply both sides by 100.
$$10(9.2x - 80.1) = 10(21.3x - 19.6)$$
$$92x - 801 = 213x - 196$$
$$92x = 213x + 605$$
$$-121x = 605$$
$$\frac{-121x}{-121} = \frac{605}{-121}$$
$$x = -5$$
The solution set is $\{-5\}$.

8.　$N = 2.4x + 180; \; N = 324$
$$2.4x + 180 = 324$$
$$2.4x + 180 - 180 = 324 - 180$$
$$2.4x = 144$$
$$\frac{2.4x}{2.4} = \frac{144}{2.4}$$
$$x = 60$$
The US population is expected to reach 324 million 60 years after 1960, in the year 2020.

9.　$V = \pi r^2 h$ for h
$$\frac{V}{\pi r^2} = \frac{\pi r^2 h}{\pi r^2}$$
$$\frac{V}{\pi r^2} = h \text{ or } h = \frac{V}{\pi r^2}$$

10.　$l = \dfrac{P - 2w}{2}$ for w
$$2l = 2\left(\frac{P - 2w}{2}\right)$$
$$2l = P - 2w$$
$$2l - P = P - 2w - P$$
$$2l - P = -2w$$
$$\frac{2l - P}{-2} = \frac{-2w}{-2}$$
$$\frac{2l - P}{-2} = w \text{ or } w = \frac{P - 2l}{2}$$

11.　$A = PB; \; P = 6\% = 0.06, \; B = 140$
$$A = 0.06(140)$$
$$A = 8.4$$
6% of 140 is 8.4.

12.　$A = PB; \; A = 120, \; P = 80\% = 0.80$
$$120 = 0.80B$$
$$\frac{120}{0.80} = \frac{0.80B}{0.80}$$
$$150 = B$$
120 is 80% of 150.

13.　$A = PB; \; A = 12, \; B = 240$
$$12 = P \cdot 240$$
$$\frac{12}{240} = \frac{P \cdot 240}{240}$$
$$0.05 = P$$
12 is 5% of 240.

14.　Let x = the unknown number.
$$5x - 9 = 306$$
$$5x - 9 + 9 = 306 + 9$$
$$5x = 315$$
$$\frac{5x}{5} = \frac{315}{5}$$
$$x = 63$$
The number is 63.

15.　Let x = the average number of vacation days for Americans.
Let $x + 29$ = the average number of vacation days for Italians.
$$x + (x + 29) = 55$$
$$x + x + 29 = 55$$
$$2x + 29 = 55$$
$$2x = 26$$
$$x = 13$$
$$x + 29 = 42$$
Americans average 13 vacation days and Italians average 42 vacation days.

16.　Let x = number of monthly calling minutes.
$$15 + 0.05x = 45$$
$$0.05x = 30$$
$$x = \frac{30}{0.05}$$
$$x = 600$$
You can talk for 600 minutes.

17. Let $x =$ the width of the field.
Let $2x =$ the length of the field.

$$P = 2l + 2w$$
$$450 = 2 \cdot 2x + 2 \cdot x$$
$$450 = 4x + 2x$$
$$450 = 6x$$
$$\frac{450}{6} = \frac{6x}{6}$$
$$75 = x$$
$$x = 75$$
$$2x = 150$$

The field is 75 yards wide and 150 yards long.

18. Let $x =$ the book's original price.

$$x - 0.20x = 28$$
$$0.80x = 28$$
$$x = \frac{28}{0.80}$$
$$x = 35$$

The price of the book before the reduction was \$35.

19. Find the area of a triangle with base 47 meters and height 22 meters.

$$A = \frac{1}{2}bh = \frac{1}{2}(47)(22) = 517$$

The area of the triangle is 517 m^2.

20. Find the area of a trapezoid with height 15 in, lower base 40 in and upper base 30 in.

$$A = \frac{1}{2}h(a+b)$$
$$= \frac{1}{2}(15)(40+30)$$
$$= \frac{1}{2} \cdot 15 \cdot 70 = 525$$

The area is 525 in^2.

21. Notice that the height of the side rectangle is $6 + 3 = 9$ ft.

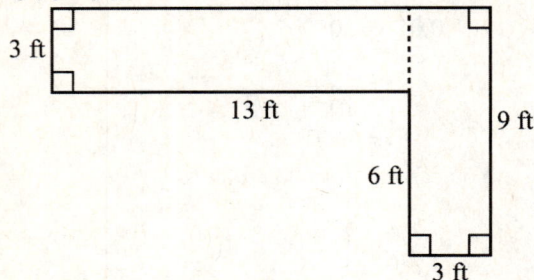

Using $A = lw$ we must find the sum of areas of the upper rectangle and the side rectangle.

$$A = (3)(13) + (3)(9)$$
$$= 39 + 27$$
$$= 66$$

The area is 66 ft^2.

22. Find the volume of a rectangular solid with length 3 in, width 2 in, and height 3 in.

$$V = lwh = 3 \cdot 2 \cdot 3 = 18$$

The volume is 18 in^3.

23. Find the volume of a cylinder with radius 5 cm and height 7 cm.

$$V = \pi r^2 h$$
$$= \pi(5)^2 \cdot 7 = \pi \cdot 25 \cdot 7$$
$$= 175\pi \approx 550$$

The volume is 175π cm^3 or approximately 550 cm^3.

24. The area of the floor is $A = (40\,\text{ft})(50\,\text{ft}) = 2000\,\text{ft}^2$.

The area of each tile is $A = (2\,\text{ft})(2\,\text{ft}) = 4\,\text{ft}^2$.

The number of tiles needed is $\frac{2000\,\text{ft}^2}{4\,\text{ft}^2} = 500$.

Since there are 10 tiles in a package, the number of packages needed is $\frac{500}{10} = 50$.

Since each package costs \$13, the cost for enough tiles to cover the floor is $50(\$13) = \$650$.

25. $A = 56, b = 8$

$$A = \frac{1}{2}bh$$
$$56 = \frac{1}{2} \cdot 8 \cdot h$$
$$56 = 4h$$
$$14 = h$$

The height of the sail is 14 feet.

26. Let $x =$ the measure of the second angle.
Let $3x =$ the measure of the first angle.
Let $x - 30 =$ the measure of the third angle.

$$x + 3x + (x - 30) = 180$$
$$5x - 30 = 180$$
$$5x = 210$$
$$x = 42$$

The measure of the first angle: $3x = 3(42°) = 126°$.
The measure of the second angle: $x = 42°$.
The measure of the third angle: $x - 30 = 42° - 30° = 12°$.

27. Let x = the measure of the angle.

Let $90 - x$ = the measure of its complement.

$$x = (90 - x) + 16$$
$$x = 106 - x$$
$$2x = 106$$
$$x = 53$$

The measure of the angle is $53\,^\circ$.

28. $(-2, \infty)$

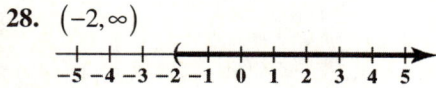

29. $(-\infty, 3]$

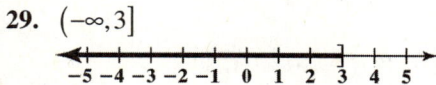

30. $\dfrac{x}{2} < -3$

$$2\left(\dfrac{x}{2}\right) < 2(-3)$$
$$x < -6$$
$$(-\infty, -6)$$

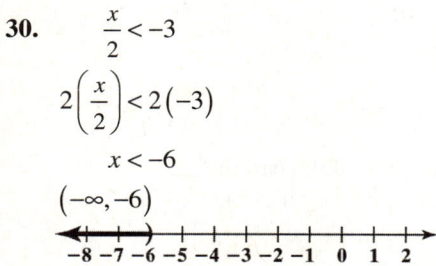

31. $6 - 9x \geq 33$

$$6 - 9x - 6 \geq 33 - 6$$
$$-9x \geq 27$$
$$\dfrac{-9x}{-9} \leq \dfrac{27}{-9}$$
$$x \leq -3$$
$$(-\infty, -3]$$

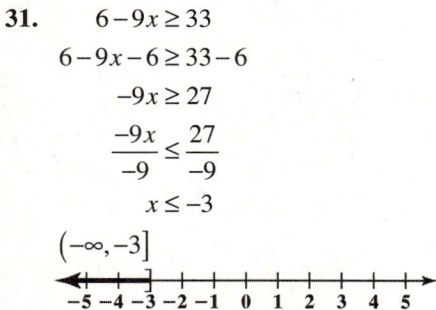

32. $4x - 2 > 2(x + 6)$

$$4x - 2 > 2x + 12$$
$$4x - 2 - 2x > 2x + 12 - 2x$$
$$2x - 2 > 12$$
$$2x > 14$$
$$x > 7$$
$$(7, \infty)$$

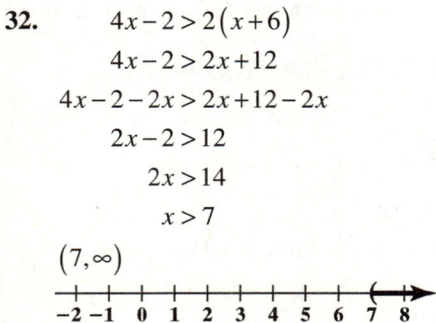

33. Let x = the student's score on the fourth exam.

$$\dfrac{76 + 80 + 72 + x}{4} \geq 80$$
$$4\left(\dfrac{76 + 80 + 72 + x}{4}\right) \geq 4(80)$$
$$76 + 80 + 72 + x \geq 320$$
$$228 + x \geq 320$$
$$x \geq 92$$

The student must score at least 92 on the fourth exam to have an average of at least 80.

34. Let x = the width of the rectangle.

$$2(20) + 2x > 56$$
$$40 + 2x > 56$$
$$40 - 40 + 2x > 56 - 40$$
$$2x > 16$$
$$x > 8$$

The perimeter is greater than 56 inches when the width is greater than 8 inches.

Cumulative Review Exercises (Chapters 1-2)

1. $-8 - (12 - 16) = -8 - (-4) = -8 + 4 = -4$

2. $(-3)(-2) + (-2)(4) = 6 + (-8) = -2$

3. $(8 - 10)^3 (7 - 11)^2 = (-2)^3 (-4)^2$
$$= -8(16) = -128$$

4. $2 - 5\left[x + 3(x + 7)\right]$
$$= 2 - 5(x + 3x + 21)$$
$$= 2 - 5(4x + 21)$$
$$= 2 - 20x - 105$$
$$= -103 - 20x$$

5. The rational numbers are
$$-4, -\dfrac{1}{3}, 0, \sqrt{4}\,(= 2), \text{ and } 1063.$$

6. $\dfrac{5}{x} - (x + 2)$

7. $-10{,}000 < -2$ since $-10{,}000$ is to the left of -2 on the number line.

8. $6(4x - 1 - 5y) = 6(4x) - 6(1) - 6(5y)$
$$= 24x - 6 - 30y$$

9. $A = -0.9n + 69$

    $A = -0.9(20) + 69$

    $A = -18 + 69$

    $A = 51$

    According to the formula, 51% of seniors used alcohol in 2000.

    This overestimates the actual value shown in the bar graph by 1%.

10. $\qquad A = -0.9n + 69$

    $\qquad 33 = -0.9n + 69$

    $\quad -36 = -0.9n$

    $\dfrac{-36}{-0.9} = \dfrac{-0.9n}{-0.9}$

    $\qquad 40 = n$

    If trends continue, 33% of seniors will seniors will use alcohol 40 years after 1980, or 2020.

11. $5 - 6(x + 2) = x - 14$

    $5 - 6x - 12 = x - 14$

    $-7 - 6x = x - 14$

    $-7 - 6x - x = x - 14 - x$

    $-7 - 7x = -14$

    $-7 - 7x + 7 = -14 + 7$

    $-7x = -7$

    $\dfrac{-7x}{-7} = \dfrac{-7}{-7}$

    $x = 1$

    The solution set is $\{1\}$.

12. $\dfrac{x}{5} - 2 = \dfrac{x}{3}$

    Multiply both sides by the LCD, 15.

    $15\left(\dfrac{x}{5} - 2\right) = 15\left(\dfrac{x}{3}\right)$

    $15\left(\dfrac{x}{5}\right) - 15(2) = 15\left(\dfrac{x}{3}\right)$

    $3x - 30 = 5x$

    $3x - 30 - 3x = 5x - 3x$

    $-30 = 2x$

    $\dfrac{-30}{2} = \dfrac{2x}{2}$

    $-15 = x$

    The solution set is $\{-15\}$.

13. $V = \dfrac{1}{3}Ah$ for A

    $V = \dfrac{1}{3}Ah$

    $3V = 3\left(\dfrac{1}{3}Ah\right)$

    $3V = Ah$

    $\dfrac{3V}{h} = \dfrac{Ah}{h}$

    $\dfrac{3V}{h} = A$ or $A = \dfrac{3V}{h}$

14. $A = PB;\ A = 48,\ P = 30\% = 0.30$

    $\qquad 48 = 0.30B$

    $\dfrac{48}{0.30} = \dfrac{0.30B}{0.30}$

    $\quad 160 = B$

    48 is 30% of 160.

15. Let x = the width of the parking lot.

    Let $2x - 10$ = the length of the parking lot.

    $\qquad P = 2l + 2w$

    $\qquad 400 = 2(2x - 10) + 2 \cdot x$

    $\qquad 400 = 4x - 20 + 2x$

    $\qquad 400 = 6x - 20$

    $400 + 20 = 6x - 20 + 20$

    $\qquad 420 = 6x$

    $\qquad \dfrac{420}{6} = \dfrac{6x}{6}$

    $\qquad 70 = x$

    $\qquad x = 70$

    $2x - 10 = 130$

    The parking lot is 70 yards wide and 130 yards long.

16. Let x = number of gallons of gasoline.

    $0.40x = 30,000$

    $\dfrac{0.40x}{0.40} = \dfrac{30,000}{0.40}$

    $x = 75,000$

    75,000 gallons of gasoline must be sold

17. $\left(-\infty, \dfrac{1}{2}\right]$

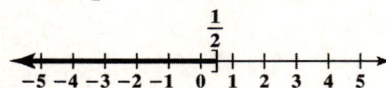

18.
$$3 - 3x > 12$$
$$3 - 3x - 3 > 12 - 3$$
$$-3x > 9$$
$$\frac{-3x}{-3} < \frac{9}{-3}$$
$$x < -3$$
$$(-\infty, -3)$$

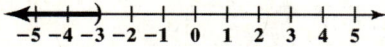

19.
$$5 - 2(3 - x) \le 2(2x + 5) + 1$$
$$5 - 6 + 2x \le 4x + 10 + 1$$
$$2x - 1 \le 4x + 11$$
$$2x - 1 - 4x \le 4x + 11 - 4x$$
$$-2x - 1 \le 11$$
$$-2x - 1 + 1 \le 11 + 1$$
$$-2x \le 12$$
$$\frac{-2x}{-2} \ge \frac{12}{-2}$$
$$x \ge -6$$
$$[-6, \infty)$$

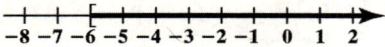

20. Let x = value of medical supplies sold.
$$600 + 0.04x > 2500$$
$$600 + 0.04x - 600 > 2500 - 600$$
$$0.04x > 1900$$
$$\frac{0.04x}{0.04} > \frac{1900}{0.04}$$
$$x > 47,500$$

You must sell more than $47,500 worth of medical supplies.

Chapter 3
Linear Equations and Inequalities in Two Variables

3.1 Check Points

1.

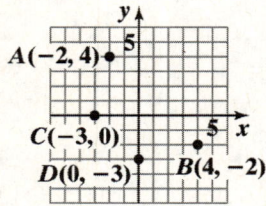

2. $E(-4,-2)$ $F(-2,0)$ $G(6,0)$

3. a. $x - 3y = 9$
$$3 - 3(-2) = 9$$
$$3 + 6 = 9$$
$$9 = 9, \ \text{true}$$
$(3, -2)$ is a solution.

b. $x - 3y = 9$
$$-2 - 3(3) = 9$$
$$-2 - 9 = 9$$
$$-11 = 9, \ \text{false}$$
$(-2, 3)$ is not a solution.

4.

x	$y = 3x + 2$	(x, y)
-2	$y = 3(-2) + 2 = -4$	$(-2, -4)$
-1	$y = 3(-1) + 2 = -1$	$(-1, -1)$
0	$y = 3(0) + 2 = 2$	$(0, 2)$
1	$y = 3(1) + 2 = 5$	$(1, 5)$
2	$y = 3(2) + 2 = 8$	$(2, 8)$

5.

x	$y = 2x$	(x, y)
-2	$y = 2(-2) = -4$	$(-2, -4)$
-1	$y = 2(-1) = -2$	$(-1, -2)$
0	$y = 2(0) = 0$	$(0, 0)$
1	$y = 2(1) = 2$	$(1, 2)$
2	$y = 2(2) = 4$	$(2, 4)$

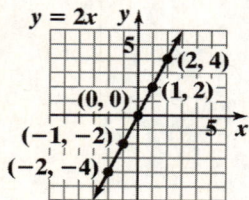

6.

x	$y = 2x - 2$	(x, y)
-2	$y = 2(-2) - 2 = -6$	$(-2, -6)$
-1	$y = 2(-1) - 2 = -4$	$(-1, -4)$
0	$y = 2(0) - 2 = -2$	$(0, -2)$
1	$y = 2(1) - 2 = 0$	$(1, 0)$
2	$y = 2(2) - 2 = 2$	$(2, 2)$

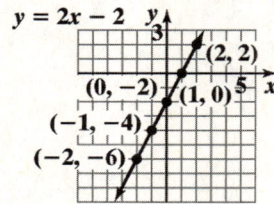

7.

x	$y = \frac{1}{2}x + 2$	(x, y)
-4	$y = \frac{1}{2}(-4) + 2 = 0$	$(-4, 0)$
-2	$y = \frac{1}{2}(-2) + 2 = 1$	$(-2, 1)$
0	$y = \frac{1}{2}(0) + 2 = 2$	$(0, 2)$
2	$y = \frac{1}{2}(2) + 2 = 3$	$(2, 3)$
4	$y = \frac{1}{2}(4) + 2 = 4$	$(4, 4)$

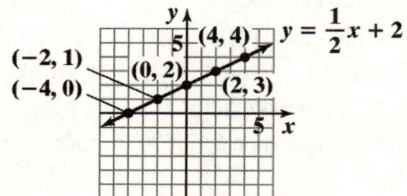

8. a.

n	$D = 1.4n + 1$	(n, D)
0	$D = 1.4(0) + 1 = 1$	$(0, 1)$
5	$D = 1.4(5) + 1 = 8$	$(5, 8)$
10	$D = 1.4(10) + 1 = 15$	$(10, 15)$
15	$D = 1.4(15) + 1 = 22$	$(15, 22)$

b. Graph formula:

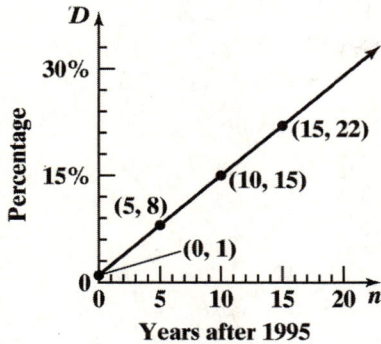

c. According to the graph, about 29% of consumers will pay primarily with debit cards in 2015.

d. $D = 1.4n + 1$

$D = 1.4(20) + 1$

$= 29$

According to the formula, about 29% of consumers will pay primarily with debit cards in 2015.

3.1 Concept and Vocabulary Check

1. x-axis

2. y-axis

3. origin

4. quadrants; four

5. x-coordinate; y-coordinate

6. solution; satisfies

7. a/one

8. $mx + b$

3.1 Exercise Set

1. Quadrant I

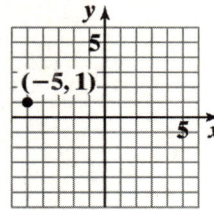

3. Quadrant II

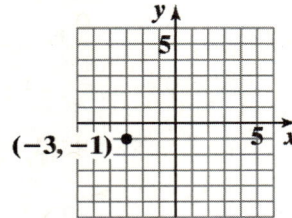

5. Quadrant III

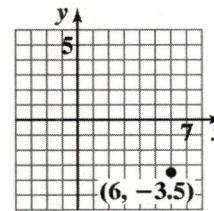

7. Quadrant IV

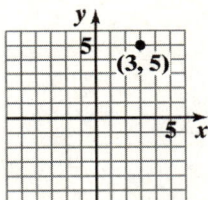

9. – 23.

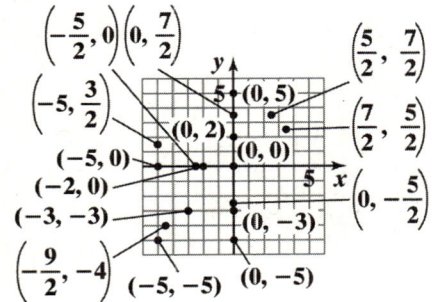

25. A (5,2)

27. C (−6,5)

29. $E(-2,-3)$

31. $G(5,-3)$

33. The y-coordinates are positive in Quadrants I and II.

35. The x- and y-coordinates have the same sign in Quadrants I and III.

37. $y = 3x$

$y = 3x$

$3 = 3(2)$

$3 = 6,$ false

$(2,3)$ is not a solution.

$y = 3x$

$2 = 3(3)$

$2 = 9,$ false

$(3,2)$ is not a solution.

$y = 3x$

$-12 = 3(-4)$

$-12 = -12,$ true

$(-4,-12)$ is a solution.

39. $y = -4x$

$-20 = -4(-5)$

$-20 = 20,$ false

$(-5,-20)$ is not a solution.

$y = -4x$

$0 = -4(0)$

$0 = 0,$ true

$(0,0)$ is a solution.

$y = -4x$

$-36 = -4(9)$

$-36 = -36,$ true

$(9,-36)$ is a solution.

41. $y = 2x + 6$

$6 = 2(0) + 6$

$6 = 6,$ true

$(0,6)$ is a solution.

$y = 2x + 6$

$0 = 2(-3) + 6$

$0 = 0,$ true

$(-3,0)$ is a solution.

$y = 2x + 6$

$-2 = 2(2) + 6$

$-2 = 10,$ false

$(2,-2)$ is not a solution.

43. $3x + 5y = 15$

$3(-5) + 5(6) = 15$

$-15 + 30 = 15$

$15 = 15,$ true

$(-5,6)$ is a solution.

$3x + 5y = 15$

$3(0) + 5(5) = 15$

$0 + 25 = 15$

$25 = 15,$ false

$(0,5)$ is not a solution.

$3x + 5y = 15$

$3(10) + 5(-3) = 15$

$30 - 15 = 15$

$15 = 15,$ true

$(10,-3)$ is a solution.

45. $x + 3y = 0$

$0 + 3(0) = 0$

$0 = 0,$ true

$(0,0)$ is a solution.

$x + 3y = 0$

$1 + 3\left(\dfrac{1}{3}\right) = 0$

$1 + 1 = 0$

$2 = 0,$ false

$\left(1,\dfrac{1}{3}\right)$ is not a solution.

$x + 3y = 0$

$2 + 3\left(-\dfrac{2}{3}\right) = 0$

$2 - 2 = 0$

$0 = 0,$ true

$\left(2,-\dfrac{2}{3}\right)$ is a solution.

47. $x - 4 = 0$

$4 - 4 = 0$

$0 = 0$, true

$(4, 7)$ is a solution.

$x - 4 = 0$

$3 - 4 = 0$

$-1 = 0$, false

$(3, 4)$ is not a solution.

$x - 4 = 0$

$0 - 4 = 0$

$-4 = 0$, false

$(0, -4)$ is not a solution.

49.

x	$y = 12x$	(x, y)
-2	$y = 12(-2) = -24$	$(-2, -24)$
-1	$y = 12(-1) = -12$	$(-1, -12)$
0	$y = 12(0) = 0$	$(0, 0)$
1	$y = 12(1) = 12$	$(1, 12)$
2	$y = 12(2) = 24$	$(2, 24)$

51.

x	$y = -10x$	(x, y)
-2	$y = -10(-2) = 20$	$(-2, 20)$
-1	$y = -10(-1) = 10$	$(-1, 10)$
0	$y = -10(0) = 0$	$(0, 0)$
1	$y = -10(1) = -10$	$(1, -10)$
2	$y = -10(2) = -20$	$(2, -20)$

53.

x	$y = 8x - 5$	(x, y)
-2	$y = 8(-2) - 5 = -21$	$(-2, -21)$
-1	$y = 8(-1) - 5 = -13$	$(-1, -13)$
0	$y = 8(0) - 5 = -5$	$(0, -5)$
1	$y = 8(1) - 5 = 3$	$(1, 3)$
2	$y = 8(2) - 5 = 11$	$(2, 11)$

55.

x	$y = -3x + 7$	(x, y)
-2	$y = -3(-2) + 7 = 13$	$(-2, 13)$
-1	$y = -3(-1) + 7 = 10$	$(-1, 10)$
0	$y = -3(0) + 7 = 7$	$(0, 7)$
1	$y = -3(1) + 7 = 4$	$(1, 4)$
2	$y = -3(2) + 7 = 1$	$(2, 1)$

57.

x	$y = x$	(x, y)
-2	$y = -2$	$(-2, -2)$
-1	$y = -1$	$(-1, -1)$
0	$y = 0$	$(0, 0)$
1	$y = 1$	$(1, 1)$
2	$y = 2$	$(2, 2)$

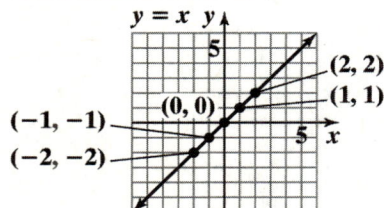

59.

x	$y = x - 1$	(x, y)
-2	$y = -2 - 1 = -3$	$(-2, -3)$
-1	$y = -1 - 1 = -2$	$(-1, -2)$
0	$y = 0 - 1 = -1$	$(0, -1)$
1	$y = 1 - 1 = 0$	$(1, 0)$
2	$y = 2 - 1 = 1$	$(2, 1)$

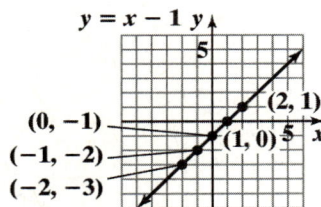

61.

x	$y = 2x + 1$	(x, y)
-2	$y = 2(-2) + 1 = -3$	$(-2, -3)$
-1	$y = 2(-1) + 1 = -1$	$(-1, -1)$
0	$y = 2(0) + 1 = 1$	$(0, 1)$
1	$y = 2(1) + 1 = 3$	$(1, 3)$
2	$y = 2(2) + 1 = 5$	$(2, 5)$

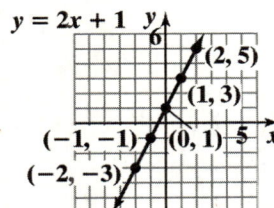

63.

x	$y = -x + 2$	(x, y)
-2	$y = -(-2) + 2 = 4$	$(-2, 4)$
-1	$y = -(-1) + 2 = 3$	$(-1, 3)$
0	$y = -0 + 2 = 2$	$(0, 2)$
1	$y = -1 + 2 = 1$	$(1, 1)$
2	$y = -2 + 2 = 0$	$(2, 0)$

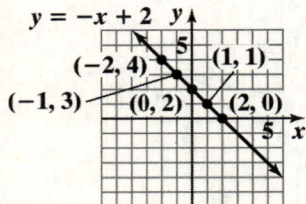

65.

x	$y = -3x - 1$	(x, y)
-2	$y = -3(-2) - 1 = 5$	$(-2, 5)$
-1	$y = -3(-1) - 1 = 2$	$(-1, 2)$
0	$y = -3(0) - 1 = -1$	$(0, -1)$
1	$y = -3(1) - 1 = -4$	$(1, -4)$
2	$y = -3(2) - 1 = -7$	$(2, -7)$

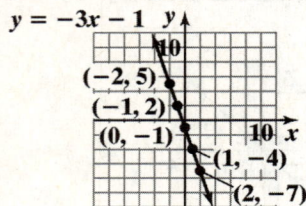

67.

x	$y = \frac{1}{2}x$	(x, y)
-4	$y = \frac{1}{2}(-4) = -2$	$(-4, -2)$
-2	$y = \frac{1}{2}(-2) = -1$	$(-2, -1)$
0	$y = \frac{1}{2}(0) = 0$	$(0, 0)$
2	$y = \frac{1}{2}(2) = 1$	$(2, 1)$
4	$y = \frac{1}{2}(4) = 2$	$(4, 2)$

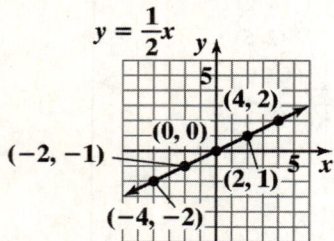

69.

x	$y = -\frac{1}{4}x$	(x, y)
-8	$y = -\frac{1}{4}(-8) = 2$	$(-8, 2)$
-4	$y = -\frac{1}{4}(-4) = 1$	$(-4, 1)$
0	$y = -\frac{1}{4}(0) = 0$	$(0, 0)$
4	$y = -\frac{1}{4}(4) = -1$	$(4, -1)$
8	$y = -\frac{1}{4}(8) = -2$	$(8, -2)$

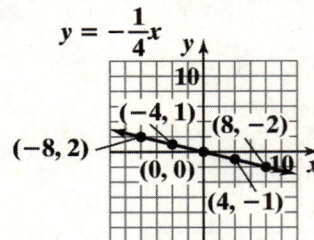

71.

x	$y = \frac{1}{3}x + 1$	(x, y)
-6	$y = \frac{1}{3}(-6) + 1 = -1$	$(-6, -1)$
-3	$y = \frac{1}{3}(-3) + 1 = 0$	$(-3, 0)$
0	$y = \frac{1}{3}(0) + 1 = 1$	$(0, -1)$
3	$y = \frac{1}{3}(3) + 1 = 2$	$(3, 2)$
6	$y = \frac{1}{3}(6) + 1 = 3$	$(6, 3)$

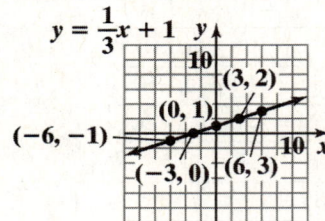

73.

x	$y = -\frac{3}{2}x + 1$	(x, y)
-4	$y = -\frac{3}{2}(-4) + 1 = 7$	$(-4, 7)$
-2	$y = -\frac{3}{2}(-2) + 1 = 4$	$(-2, 4)$
0	$y = -\frac{3}{2}(0) + 1 = 1$	$(0, 1)$
2	$y = -\frac{3}{2}(2) + 1 = -2$	$(2, -2)$
4	$y = -\frac{3}{2}(4) + 1 = -5$	$(4, -5)$

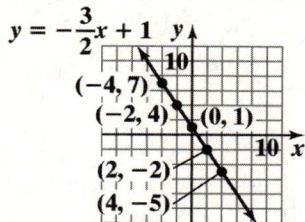

75.

x	$y = -\frac{5}{2}x - 1$	(x, y)
-4	$y = -\frac{5}{2}(-4) - 1 = 9$	$(-4, 9)$
-2	$y = -\frac{5}{2}(-2) - 1 = 4$	$(-2, 4)$
0	$y = -\frac{5}{2}(0) - 1 = -1$	$(0, -1)$
2	$y = -\frac{5}{2}(2) - 1 = -6$	$(2, -6)$
4	$y = -\frac{5}{2}(4) - 1 = -11$	$(4, -11)$

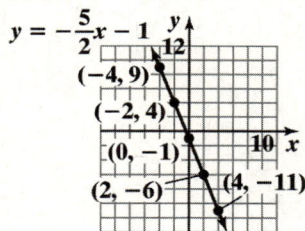

77.

x	$y = x + \frac{1}{2}$	(x, y)
-4	$y = -4 + \frac{1}{2} = -3.5$	$(-4, -3.5)$
-2	$y = -2 + \frac{1}{2} = -1.5$	$(-2, -1.5)$
0	$y = 0 + \frac{1}{2} = 0.5$	$(0, 0.5)$
2	$y = 2 + \frac{1}{2} = 2.5$	$(2, 2.5)$
4	$y = 4 + \frac{1}{2} = 4.5$	$(4, 4.5)$

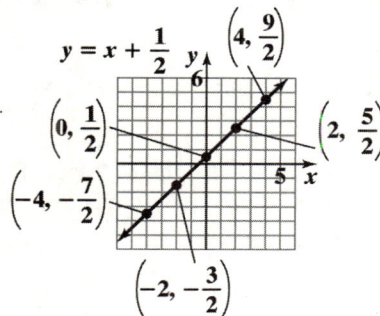

79.

x	$y = 0x + 4$	(x, y)
-6	$y = 0(-6) + 4 = 4$	$(-6, 4)$
-3	$y = 0(-3) + 4 = 4$	$(-3, 4)$
0	$y = 0(0) + 4 = 4$	$(0, 4)$
3	$y = 0(3) + 4 = 4$	$(3, 4)$
6	$y = 0(6) + 4 = 4$	$(6, 4)$

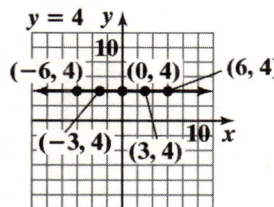

81. $y = x + 3$

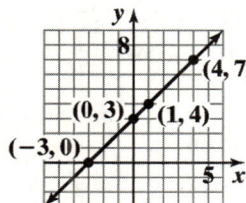

83. $y = 2x + 5$

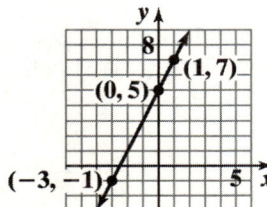

85. a. $8x + 6y = 14.50$

b. $8x + 6(0.75) = 14.50$

$8x + 4.50 = 14.50$

$8x = 10.00$

$x = 1.25$

One pen costs $1.25.

87. The coordinates of point *A* are (2,7). When the football is 2 yards from the quarterback, its height is 7 feet.

89. The coordinates of point *C* are approximately (6, 9.25).

91. The football's maximum height is 12 feet. It reaches this height when it is 15 yards from the quarterback.

93. a.

n	$S = 2.4n + 31$	(n, S)
0	$S = 2.4(0) + 31 = 31$	$(0, 31)$
5	$S = 2.4(5) + 31 = 43$	$(5, 43)$
10	$S = 2.4(10) + 31 = 55$	$(10, 55)$
15	$S = 2.4(15) + 31 = 67$	$(15, 67)$
20	$S = 2.4(20) + 31 = 79$	$(20, 79)$

b. Graph of formula:

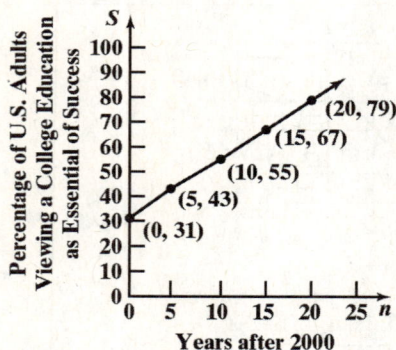

c. According to the graph in part (b), the percentage is approximately 74%.

d. $S = 2.4n + 31$

$S = 2.4(18) + 31$

$= 74.2$

According to the formula, about 74.2% of U.S. adults will believe that college is essential in 2018.

95. – 101. Answers will vary.

103. makes sense

105. does not make sense; Explanations will vary. Sample explanation: These points do not lie along a line.

107. false; Changes to make the statement true will vary. A sample change is: The lines are not parallel as they both contain the point $(-1, -2)$.

109. false; Changes to make the statement true will vary. A sample change is: The point $(4, 3)$ satisfies the equation.

111. a. $\left(1, \dfrac{1}{2}\right)$, $(2, 1)$, $\left(3, \dfrac{3}{2}\right)$, $(4, 2)$

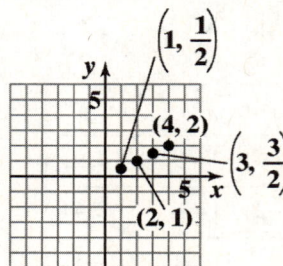

b. In order for the resulting graph to be a mirror-image reflection about the *y*-axis of the graph in part (a), the sign of each *x*-coordinate should be changed: $\left(-1, \dfrac{1}{2}\right)$, $(-2, 1)$, $\left(-3, \dfrac{3}{2}\right)$, $(-4, 2)$

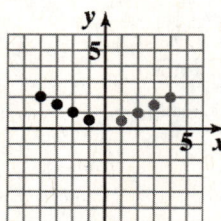

c. In order for the resulting graph to be a mirror-image reflection about the *x*-axis of the graph in part (a), the sign of each *y*-coordinate should be changed: $\left(1, -\dfrac{1}{2}\right)$, $(2, -1)$, $\left(3, -\dfrac{3}{2}\right)$, $(4, -2)$

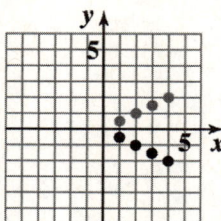

d. In order for the resulting graph to be a straight-line extension of the graph in part (a), the signs of both coordinates of each ordered pair should be changed:

$$\left(-1, -\frac{1}{2}\right), \ (-2, -1), \ \left(-3, -\frac{3}{2}\right), \ (-4, -2)$$

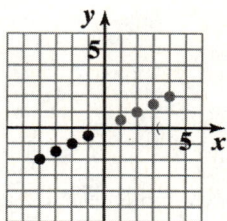

113. Answers will vary depending upon the points chosen. One example is shown here.

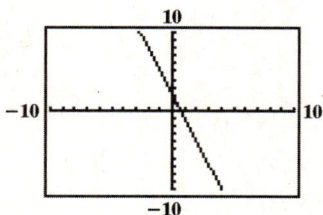

115. Answers will vary depending upon the points chosen. One example is shown here.

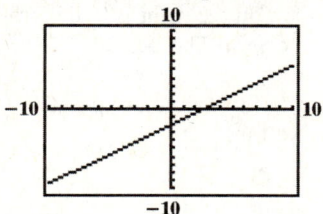

117.
$$3x + 5 = 4(2x - 3) + 7$$
$$3x + 5 = 8x - 12 + 7$$
$$3x + 5 = 8x - 5$$
$$3x + 5 - 8x = 8x - 5 - 8x$$
$$-5x + 5 = -5$$
$$-5x + 5 - 5 = -5 - 5$$
$$-5x = -10$$
$$\frac{-5x}{-5} = \frac{-10}{-5}$$
$$x = 2$$
The solution set is $\{2\}$.

118. $3(1 - 2 \cdot 5) - (-28) = 3(1 - 10) + 28$
$$= 3(-9) + 28$$
$$= -27 + 28 = 1$$

119. $V = \dfrac{1}{3}Ah$ for h

$$V = \frac{1}{3}Ah$$
$$3V = 3\left(\frac{1}{3}Ah\right)$$
$$3V = Ah$$
$$\frac{3V}{A} = \frac{Ah}{A}$$
$$\frac{3V}{A} = h \text{ or } h = \frac{3V}{A}$$

120. $3x - 4y = 24$
$$3x - 4(0) = 24$$
$$3x = 24$$
$$x = 8$$
The equation is satisfied by the ordered pair $(8, 0)$.

121. $3x - 4y = 24$
$$3(0) - 4y = 24$$
$$-4y = 24$$
$$y = -6$$
The equation is satisfied by the ordered pair $(0, -6)$.

122. $x + 2y = 0$
$$0 + 2y = 0$$
$$2y = 0$$
$$y = 0$$
The equation is satisfied by the ordered pair $(0, 0)$.

3.2 Check Points

1. a. The graph crosses the x-axis at $(-3, 0)$. Thus, the x-intercept is -3.
The graph crosses the y-axis at $(0, 5)$. Thus, the y-intercept is 5.

   b. The graph does not cross the x-axis. Thus, there is no x-intercept.
The graph crosses the y-axis at $(0, 4)$. Thus, the y-intercept is 4.

   c. The graph crosses the x-axis at $(0, 0)$. Thus, the x-intercept is 0.
The graph crosses the y-axis at $(0, 0)$. Thus, the y-intercept is 0.

2. To find the *x*-intercept, let *y* = 0 and solve for *x*.

$$4x - 3y = 12$$
$$4x - 3(0) = 12$$
$$4x = 12$$
$$x = 3$$

The *x*-intercept is 3.

3. To find the *y*-intercept, let *x* = 0 and solve for *y*.

$$4x - 3y = 12$$
$$4(0) - 3y = 12$$
$$-3y = 12$$
$$y = -4$$

The *y*-intercept is –4.

4. Find the *x*-intercept. Let *y* = 0 and solve for *x*.

$$2x + 3y = 6$$
$$2x + 3(0) = 6$$
$$2x = 6$$
$$x = 3$$

The *x*-intercept is 3.
Find the *y*- intercept. Let *x* = 0 and solve for *y*.

$$2x + 3y = 6$$
$$2(0) + 3y = 6$$
$$3y = 6$$
$$y = 2$$

The *y*-intercept is 2.
Find a checkpoint. For example, let *x* = 1 and solve for *y*.

$$2x + 3y = 6$$
$$2(1) + 3y = 6$$
$$2 + 3y = 6$$
$$3y = 4$$
$$y = \frac{4}{3} \text{ or } 1\frac{1}{3}$$

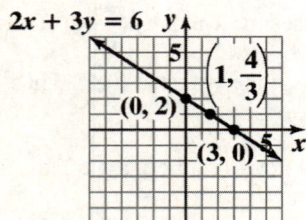

5. Find the *x*-intercept. Let *y* = 0 and solve for *x*.

$$x - 2y = 4$$
$$x - 2(0) = 4$$
$$x = 4$$

The *x*-intercept is 4.
Find the *y*- intercept. Let *x* = 0 and solve for *y*.

$$x - 2y = 4$$
$$0 - 2y = 4$$
$$-2y = 4$$
$$y = -2$$

The *y*-intercept is –2.
Find a checkpoint. For example, let *x* = 2 and solve for *y*.

$$x - 2y = 4$$
$$2 - 2y = 4$$
$$-2y = 2$$
$$y = -1$$

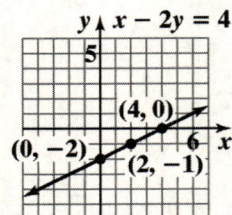

6. Because the constant on the right is 0, the graph passes through the origin. The *x*- and *y*-intercepts are both 0.
Thus we will need to find two more points.
Let *y* = –1 and solve for *x*.

$$x + 3y = 0$$
$$x + 3(-1) = 0$$
$$x - 3 = 0$$
$$x = 3$$

Let *y* = 1 and solve for *x*.

$$x + 3y = 0$$
$$x + 3(1) = 0$$
$$x + 3 = 0$$
$$x = -3$$

Use these three solutions of (0,0), (3,–1), and (–3,1).

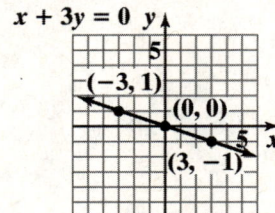

7. As demonstrated in the table below, all ordered pairs that are solutions of $y = 3$ have a value of y that is always 3.

x	$y = 3$	(x, y)
-2	3	$(-2, 3)$
0	3	$(0, 3)$
1	3	$(1, 3)$

Thus the line is horizontal.

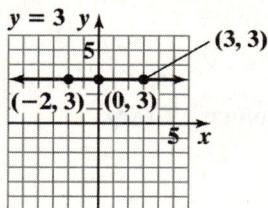

8. As demonstrated in the table below, all ordered pairs that are solutions of $x = -2$ have a value of x that is always -2.

$x = -2$	y	(x, y)
-2	-3	$(-2, -3)$
-2	0	$(-2, 0)$
-2	2	$(-2, 2)$

Thus the line is vertical.

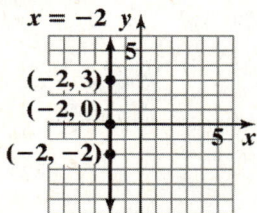

3.2 Concept and Vocabulary Check

1. x-intercept

2. y-intercept

3. x-intercept

4. y-intercept

5. standard

6. y; x

7. horizontal

8. vertical

3.2 Exercise Set

1. a. The graph crosses the x-axis at $(3, 0)$. Thus, the x-intercept is 3.

    b. The graph crosses the y-axis at $(0, 4)$. Thus, the y-intercept is 4.

3. a. The graph crosses the x-axis at $(-4, 0)$. Thus, the x-intercept is -4.

    b. The graph crosses the y-axis at $(0, -2)$. Thus, the y-intercept is -2.

5. a. The graph crosses the x-axis at $(0, 0)$ (the origin). Thus, the x-intercept is 0.

    b. The graph also crosses the y-axis at $(0, 0)$. Thus, the y-intercept is 0.

7. a. The graph does not cross the x-axis. Thus, there is no x-intercept.

    b. The graph crosses the y-axis at $(0, -2)$. Thus the y-intercept is -2.

9. To find the x-intercept, let $y = 0$ and solve for x.
$$2x + 5y = 20$$
$$2x + 5(0) = 20$$
$$2x = 20$$
$$x = 10$$
The x-intercept is 10.
To find the y-intercept, let $x = 0$ and solve for y.
$$2x + 5y = 20$$
$$2(0) + 5y = 20$$
$$5y = 20$$
$$y = 4$$
The y-intercept is 4.

11. To find the x-intercept, let $y = 0$ and solve for x.
$$2x - 3y = 15$$
$$2x - 3(0) = 15$$
$$2x = 15$$
$$x = \frac{15}{2}$$
The x-intercept is $\frac{15}{2}$.
To find the y-intercept, let $x = 0$ and solve for y.
$$2x - 3y = 15$$
$$2(0) - 3y = 15$$
$$-3y = 15$$
$$y = -5$$
The y-intercept is -5.

13. To find the *x*-intercept, let $y = 0$ and solve for *x*.

$$-x + 3y = -8$$
$$-x + 3(0) = -8$$
$$-x = -8$$
$$x = 8$$

The *x*-intercept is 8.

To find the *y*-intercept, let $x = 0$ and solve for *y*.

$$-x + 3y = -8$$
$$-0 + 3y = -8$$
$$3y = -8$$
$$y = -\frac{8}{3}$$

The *y*-intercept is $-\frac{8}{3}$.

15. To find the *x*-intercept, let $y = 0$ and solve for *x*.

$$7x - 9y = 0$$
$$7x - 9(0) = 0$$
$$7x = 0$$
$$x = 0$$

The *x*-intercept is 0.

To find the *y*-intercept, let $x = 0$ and solve for *y*.

$$7x - 9y = 0$$
$$7(0) - 9y = 0$$
$$-9y = 0$$
$$y = 0$$

The *y*-intercept is 0.

17. To find the *x*-intercept, let $y = 0$ and solve for *x*.

$$2x = 3y - 11$$
$$2x = 3(0) - 11$$
$$2x = -11$$
$$x = -\frac{11}{2}$$

The *x*-intercept is $-\frac{11}{2}$.

To find the *y*-intercept, let $x = 0$ and solve for *y*.

$$2x = 3y - 11$$
$$2(0) = 3y - 11$$
$$0 = 3y - 11$$
$$11 = 3y$$
$$\frac{11}{3} = y$$

The *y*-intercept is $\frac{11}{3}$.

19. $x + y = 5$

x-intercept: 5
y-intercept: 5
checkpoint: (2,3)
Draw a line through (5,0), (0,5), and (2,3).

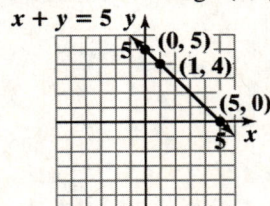

In Exercises 21-39, checkpoints will vary.

21. $x + 3y = 6$

x-intercept: 6
y-intercept: 2
checkpoint: (3,1)
Draw a line through (6,0), (0,2), and (3,1).

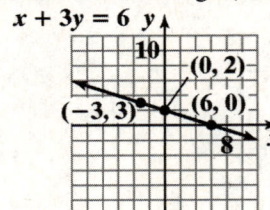

23. $6x - 9y = 18$

x-intercept: 3
y-intercept: −2

checkpoint: $\left(1, -\frac{4}{3}\right)$

Draw a line through (3,0), (0,−2), and $\left(1, -\frac{4}{3}\right)$.

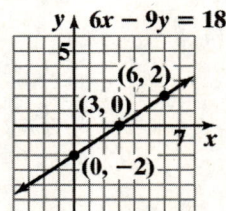

25. $-x + 4y = 6$

x-intercept: -6

y-intercept: $\dfrac{3}{2}$

checkpoint: $(2,2)$

Draw a line through $(-6,0)$, $\left(0, \dfrac{3}{2}\right)$, and $(2,2)$.

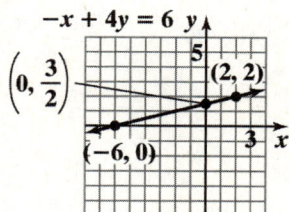

27. $2x - y = 7$

x-intercept: $\dfrac{7}{2}$

y-intercept: -7

checkpoint: $(1,-5)$

Draw a line through $\left(\dfrac{7}{2}, 0\right)$, $(0,7)$, and $(1,-5)$.

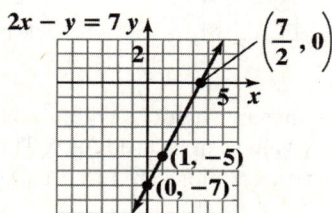

29. $3x = 5y - 15$

x-intercept: -5

y-intercept: 3

checkpoint: $\left(-\dfrac{10}{3}, 1\right)$

Draw a line through $(-5,0)$, $(0,3)$, and $\left(-\dfrac{10}{3}, 1\right)$.

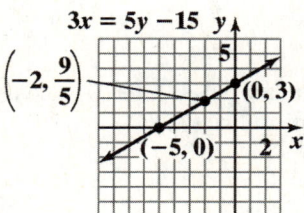

31. $25y = 100 - 50x$

x-intercept: 2

y-intercept: 4

checkpoint: $(1, 2)$

Draw a line through $(2,0)$, $(0,4)$, and $(1, 2)$.

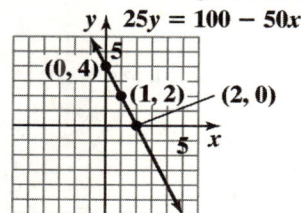

33. $2x - 8y = 12$

x-intercept: 6

y-intercept: $-\dfrac{3}{2}$

checkpoint: $(2, -1)$

Draw a line through $(6,0)$, $\left(0, -\dfrac{3}{2}\right)$, and $(2, -1)$.

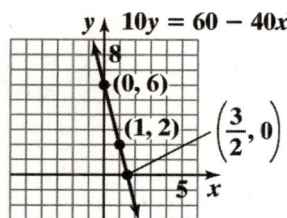

35. $x + 2y = 0$

x-intercept: 0

y-intercept: 0

Since the line goes through the origin, find two additional points.

checkpoint: $(2, -1)$

checkpoint: $(4, -2)$

Draw a line through $(0,0)$, $(2, -1)$, and $(4, -2)$.

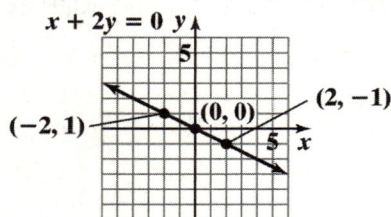

37. $y - 3x = 0$

x-intercept: 0
y-intercept: 0
Since the line goes through the origin, find two additional points.
checkpoint: (1, 3)
checkpoint: (2, 6)
Draw a line through (0,0), (1, 3), and (2, 6).

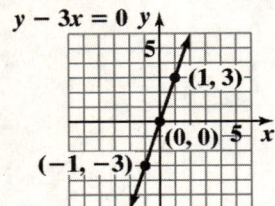

39. $2x - 3y = -11$

x-intercept: $-\dfrac{11}{2}$

y-intercept: $\dfrac{11}{3}$

checkpoint: (−1, 3)

Draw a line through $\left(-\dfrac{11}{2}, 0\right)$, $\left(0, \dfrac{11}{3}\right)$ and (−1, 3).

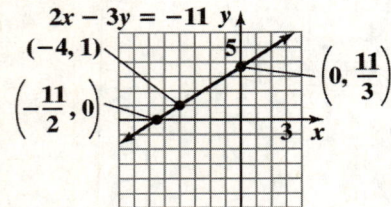

41. The equation for this horizontal line is $y = 3$.

43. The equation for this vertical line is $x = -3$.

45. The equation for this horizontal line, which is the *x*-axis is $y = 0$.

47. $y = 4$

All ordered pairs that are solutions will have a value of *y* that is 4. Any value can be used for *x*. Three ordered pairs that are solutions are (−2,4), (0,4), and (3,4).
Plot these points and draw the line through them.
The graph is a horizontal line.

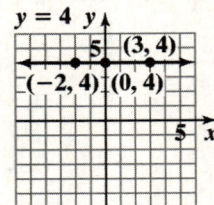

49. $y = -2$

Three ordered pairs are (−3,−2), (0,−2), and (4,−2).
The graph is a horizontal line.

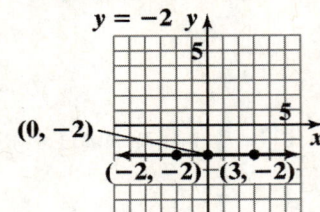

51. $x = 2$

All ordered pairs that are solutions will have a value of *x* that is 2. Any value can be used for *y*. Three ordered pairs that are solutions are (2, −3), (2,0), and (2,2).
The graph is a vertical line.

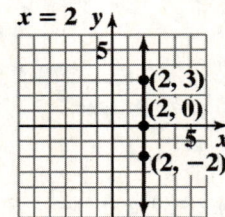

53. $x + 1 = 0$

$x = -1$

Three ordered pairs are (−1,−3), (−1,0), and (−1,3).
The graph is a vertical line.

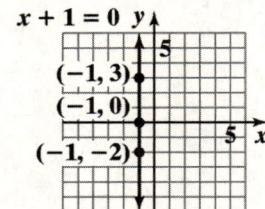

55. $y - 3.5 = 0$

$y = 3.5$

Three ordered pairs are $(-2, 3.5)$, $(0, 3.5)$, and $(3.5, 3.5)$. The graph is a horizontal line.

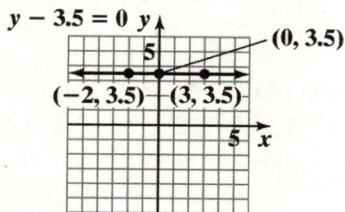

57. $x = 0$

Three ordered pairs are $(0, -2)$, $(0, 0)$, and $(0, 4)$. The graph is a vertical line, the y-axis.

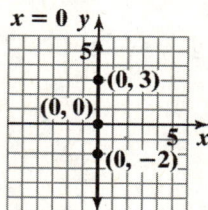

59. $3y = 9$

$y = 3$

Three ordered pairs are $(-3, 3)$, $(0, 3)$, and $(3, 3)$. The graph is a horizontal line.

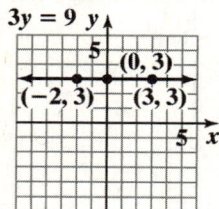

61. $12 - 3x = 0$

$-3x = -12$

$x = 4$

Three ordered pairs are $(4, -2)$, $(4, 1)$, and $(4, 3)$. The graph is a vertical line.

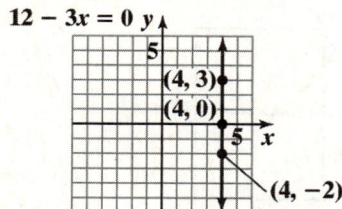

63. Using intercepts, we see that $3x + 2y = -6$ corresponds to Exercise 4.

x-intercept: $3x + 2y = -6$

$3x + 2(0) = -6$

$3x = -6$

$x = -2$

y-intercept: $3x + 2y = -6$

$3(0) + 2y = -6$

$2y = -6$

$y = -3$

65. Since $y = -2$ is a horizontal line at -2, it corresponds to Exercise 7.

67. Using intercepts, we see that $4x + 3y = 12$ corresponds to Exercise 1.

x-intercept: $4x + 3y = 12$

$4x + 3(0) = 12$

$4x = 12$

$x = 3$

y-intercept: $4x + 3y = 12$

$4(0) + 3y = 12$

$3y = 12$

$y = 4$

69. a. Let $x + 5 + 5 = x + 10$ = the length.
Let $y + 8$ = width.
Using the formula for the perimeter of a rectangle, we have

$2l + 2w = P$

$2(x + 10) + 2(y + 8) = 58$

$2x + 20 + 2y + 16 = 58$

$2x + 2y + 36 = 58$

$2x + 2y = 22$

$x + y = 11$

b. x and y must be non-negative because they are dimensions.

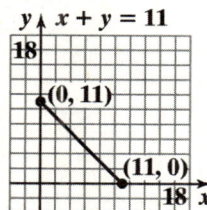

71. The eagle's height is decreasing from 3 seconds to 12 seconds.

73. The *y*-intercept is 45. This means that the eagle's height was 45 meters at the beginning of the observation.

75. Five *x*-intercepts of the graph are 12, 13, 14, 15, and 16. During these times (12-16 minutes), the eagle was on the ground.

77. a.
$$y = -5000x + 45,000$$
$$0 = -5000x + 45,000$$
$$5000x = 45,000$$
$$x = 9$$
After 9 years, the car is worth nothing.

b.
$$y = -5000x + 45,000$$
$$y = -5000(0) + 45,000$$
$$y = 45,000$$
The new car is worth $45,000.

c. *x* and *y* must be non-negative because they represent time and the car's value.

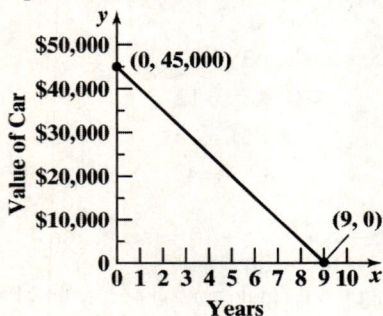

d. According to the graph, the car's value will be about $20,000 after five years. Estimates will vary.

79. – 85. Answers will vary.

87. makes sense

89. makes sense

91. Since the *x*-intercept is 5, *y* = 0 when *x* = 5.
$$\boxed{?}x + \boxed{}y = 10$$
$$\boxed{?}(5) + \boxed{}(0) = 10$$
$$\boxed{?}(5) = 10$$
$$\boxed{?} = 2$$
So, the coefficient of *x* is 2.
Similarly, since the *y*-intercept is 2, *x* = 0 when *y* = 2.
$$2x + \boxed{?}y = 10$$
$$2(0) + \boxed{?}(2) = 10$$
$$\boxed{?}(2) = 10$$
$$\boxed{?} = 5$$
So, the coefficient of *y* is 5.
The equation of the line is $2x + 5y = 10$.

93. Answers will vary.

95.
$$3x - y = 9$$
$$-y = -3x + 9$$
$$(-1)(-y) = -1(3x + 9)$$
$$y = 3x - 9$$

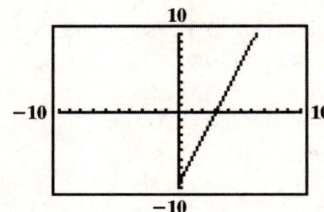

The *y*-intercept is −9.
The *x*-intercept is 3.

97.
$$4x - 2y = -40$$
$$-2y = -4x - 40$$
$$\frac{-2y}{-2} = \frac{-4x - 40}{-2}$$
$$y = 2x + 20$$

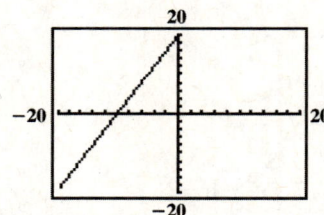

The *y*-intercept is 20.
The *x*-intercept is −10.

98. $\left| -13.4 \right| = 13.4$

99. $7x - (3x - 5) = 7x - 3x + 5 = 4x + 5$

100. $8(x - 2) - 2(x - 3) \le 8x$

$$8x - 16 - 2x + 6 \le 8x$$
$$6x - 10 \le 8x$$
$$6x - 8x - 10 \le 8x - 8x$$
$$-2x - 10 \le 0$$
$$-2x \le 10$$
$$\frac{-2x}{-2} \ge \frac{10}{-2}$$
$$x \ge -5$$

The solution set is $[-5, \infty)$.

101. $\dfrac{y_2 - y_1}{x_2 - x_1} = \dfrac{13 - 3}{6 - 1} = \dfrac{10}{5} = 2$

102. $\dfrac{y_2 - y_1}{x_2 - x_1} = \dfrac{-4 - (-2)}{6 - 4} = \dfrac{-2}{2} = -1$

103. $\dfrac{y_2 - y_1}{x_2 - x_1} = \dfrac{4 - 4}{5 - 3} = \dfrac{0}{2} = 0$

3.3 Check Points

1. a. Let $(x_1, y_1) = (-3, 4)$ and $(x_2, y_2) = (-4, -2)$.

$$m = \frac{\text{Change in } y}{\text{Change in } x} = \frac{y_2 - y_1}{x_2 - x_1} = \frac{-2 - 4}{-4 - (-3)} = \frac{-6}{-1} = 6$$

The slope is 6. Since the slope is positive, the line rises from left to right.

b. Let $(x_1, y_1) = (4, -2)$ and $(x_2, y_2) = (-1, 5)$.

$$m = \frac{\text{Change in } y}{\text{Change in } x} = \frac{y_2 - y_1}{x_2 - x_1} = \frac{5 - (-2)}{-1 - 4} = \frac{7}{-5} = -\frac{7}{5}$$

The slope is $-\dfrac{7}{5}$. Since the slope is negative, the line falls from left to right.

2. a. Let $(x_1, y_1) = (6, 5)$ and $(x_2, y_2) = (2, 5)$.

$$m = \frac{\text{Change in } y}{\text{Change in } x} = \frac{y_2 - y_1}{x_2 - x_1} = \frac{5 - 5}{2 - 6} = \frac{0}{-4} = 0$$

Since the slope is 0, the line is horizontal.

b. Let $(x_1, y_1) = (1, 6)$ and $(x_2, y_2) = (1, 4)$.

$$m = \frac{\text{Change in } y}{\text{Change in } x} = \frac{y_2 - y_1}{x_2 - x_1} = \frac{4 - 6}{1 - 1} = \frac{-2}{0}$$

Because division by 0 is undefined the slope is undefined. Since the slope is undefined, the line is vertical.

3. Line through $(4, 2)$ and $(6, 6)$:

$$m = \frac{\text{Change in } y}{\text{Change in } x} = \frac{6 - 2}{6 - 4} = \frac{4}{2} = 2$$

Line through $(0, -2)$ and $(1, 0)$:

$$m = \frac{\text{Change in } y}{\text{Change in } x} = \frac{0 - (-2)}{1 - 0} = \frac{2}{1} = 2$$

Since their slopes are equal, the lines are parallel.

4. Line through $(-1, 4)$ and $(3, 2)$:

$$m = \frac{\text{Change in } y}{\text{Change in } x} = \frac{2 - 4}{3 - (-1)} = \frac{-2}{4} = -\frac{1}{2}$$

Line through $(-2, -1)$ and $(2, 7)$:

$$m = \frac{\text{Change in } y}{\text{Change in } x} = \frac{7 - (-1)}{2 - (-2)} = \frac{8}{4} = 2$$

Since the product of their slopes is $-\dfrac{1}{2}(2) = -1$, the lines are perpendicular.

5. Let $(x_1, y_1) = (1990, 9.0)$ and $(x_2, y_2) = (2008, 14.7)$.

$$m = \frac{\text{Change in } y}{\text{Change in } x} = \frac{y_2 - y_1}{x_2 - x_1} = \frac{14.7 - 9.0}{2008 - 1990} = \frac{5.7}{18} \approx 0.32$$

The number of men living alone increased at a rate of 0.32 million per year. The rate of change is 0.32 million men per year.

3.3 Concept and Vocabulary Check

1. $\dfrac{y_2 - y_1}{x_2 - x_1}$

2. y; x

3. positive

4. negative

5. 0

6. undefined

7. parallel

8. perpendicular

3.3 Exercise Set

1. Let $(x_1, y_1) = (4, 7)$ and $(x_2, y_2) = (8, 10)$.
$$m = \frac{\text{Change in } y}{\text{Change in } x} = \frac{y_2 - y_1}{x_2 - x_1} = \frac{10 - 7}{8 - 4} = \frac{3}{4}$$
Since the slope is positive, the line rises from left to right.

3. Let $(x_1, y_1) = (-2, 1)$ and $(x_2, y_2) = (2, 2)$.
$$m = \frac{\text{Change in } y}{\text{Change in } x} = \frac{y_2 - y_1}{x_2 - x_1} = \frac{2 - 1}{2 - (-2)} = \frac{1}{4}$$
Since the slope is positive, the line rises from left to right.

5. Let $(x_1, y_1) = (4, -2)$ and $(x_2, y_2) = (3, -2)$.
$$m = \frac{\text{Change in } y}{\text{Change in } x} = \frac{y_2 - y_1}{x_2 - x_1} = \frac{-2 - (-2)}{3 - 4} = \frac{0}{-1} = 0$$
Since the slope is zero, the line is horizontal.

7. Let $(x_1, y_1) = (-2, 4)$ and $(x_2, y_2) = (-1, -1)$.
$$m = \frac{\text{Change in } y}{\text{Change in } x} = \frac{y_2 - y_1}{x_2 - x_1} = \frac{-1 - 4}{-1 - (-2)} = \frac{-5}{1} = -5$$
Since the slope is negative, the line falls from left to right.

9. Let $(x_1, y_1) = (5, 3)$ and $(x_2, y_2) = (5, -2)$.
$$m = \frac{\text{Change in } y}{\text{Change in } x} = \frac{y_2 - y_1}{x_2 - x_1} = \frac{-2 - 3}{5 - 5} = \frac{-5}{0}$$
Since the slope is undefined, the line is vertical.

11. Line through $(-2, 2)$ and $(2, 4)$:
$$m = \frac{4 - 2}{2 - (-2)} = \frac{2}{4} = \frac{1}{2}$$

13. Line through $(-3, 4)$ and $(3, 2)$:
$$m = \frac{2 - 4}{3 - (-3)} = \frac{-2}{6} = -\frac{1}{3}$$

15. Line through $(-2, 1)$, $(0, 0)$, and $(2, -1)$
Use any two of these points to find the slope.
$$m = \frac{0 - 1}{0 - (-2)} = \frac{-1}{2} = -\frac{1}{2}$$

17. Line through $(0, 4)$ and $(3, 0)$:
$$m = \frac{0 - 4}{3 - 0} = -\frac{4}{3}$$

19. Line through $(-2, 1)$ and $(4, 1)$:
$$m = \frac{1 - 1}{4 - (-2)} = \frac{0}{6} = 0$$
(Since the line is horizontal, it is not necessary to do this computation. The slope of every horizontal line is 0.)

21. Line through $(-3, 4)$ and $(-3, -2)$:
$$m = \frac{-2 - 4}{-3 - (-3)} = \frac{-6}{0}; \text{ undefined}$$
(Since the line is vertical, it is not necessary to do this computation. The slope of every vertical line is undefined.)

23. Line through $(-2, 0)$ and $(0, 6)$:
$$m = \frac{6 - 0}{0 - (-2)} = 3$$
Line through $(1, 8)$ and $(0, 5)$:
$$m = \frac{5 - 8}{0 - 1} = \frac{-3}{-1} = 3$$
Since their slopes are equal, the lines are parallel.

25. Line through $(0, 3)$ and $(1, 5)$:
$$m = \frac{5 - 3}{1 - 0} = \frac{2}{1} = 2$$
Line through $(-1, 7)$ and $(1, 10)$:
$$m = \frac{10 - 7}{1 - (-1)} = \frac{3}{2}$$
Since their slopes are not equal, the lines are not parallel.

27. Line through $(1, 5)$ and $(0, 3)$:
$$m = \frac{3 - 5}{0 - 1} = 2$$
Line through $(-2, 8)$ and $(2, 6)$:
$$m = \frac{6 - 8}{2 - (-2)} = -\frac{1}{2}$$
Since the product of their slopes is $2\left(-\frac{1}{2}\right) = -1$, the lines are perpendicular.

29. Line through $(-1, -6)$ and $(2, 9)$:
$$m = \frac{9 - (-6)}{2 - (-1)} = 5$$
Line through $(-15, -1)$ and $(5, 3)$:
$$m = \frac{3 - (-1)}{5 - (-15)} = \frac{1}{5}$$
Since the product of their slopes is $5\left(\frac{1}{5}\right) = 1 \neq -1$, the lines are not perpendicular.

31. Line through $(-2,-5)$ and $(3,10)$:

$$m = \frac{10-(-5)}{3-(-2)} = 3$$

Line through $(-1,-9)$ and $(4,6)$:

$$m = \frac{6-(-9)}{4-(-1)} = 3$$

Since their slopes are equal the lines are parallel.

33. Line through $(-4,-12)$ and $(0,-4)$:

$$m = \frac{-4-(-12)}{0-(-4)} = 2$$

Line through $(0,-5)$ and $(2,-4)$:

$$m = \frac{-4-(-5)}{2-0} = \frac{1}{2}$$

Since their slopes are not equal, nor are the slopes negative reciprocals, the lines are neither parallel nor perpendicular.

35. Line through $(-5,-1)$ and $(0,2)$:

$$m = \frac{2-(-1)}{0-(-5)} = \frac{3}{5}$$

Line through $(-6,9)$ and $(3,-6)$:

$$m = \frac{-6-9}{3-(-6)} = -\frac{5}{3}$$

Since the product of their slopes is $\frac{3}{5}\left(-\frac{5}{3}\right) = -1$,

the lines are perpendicular.

37.

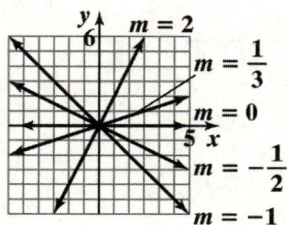

39. $m = \dfrac{y_2 - y_1}{x_2 - x_1} = \dfrac{-3-1}{-3-0} = \dfrac{-4}{-3} = \dfrac{4}{3}$

$m = \dfrac{y_2 - y_1}{x_2 - x_1} = \dfrac{-1-1}{5-0} = \dfrac{-2}{5} = -\dfrac{2}{5}$

$m = \dfrac{y_2 - y_1}{x_2 - x_1} = \dfrac{-5-(-1)}{2-5} = \dfrac{-4}{-3} = \dfrac{4}{3}$

$m = \dfrac{y_2 - y_1}{x_2 - x_1} = \dfrac{-3-(-5)}{-3-2} = \dfrac{2}{-5} = -\dfrac{2}{5}$

Slopes of opposite sides are equal, so the figure is a parallelogram.

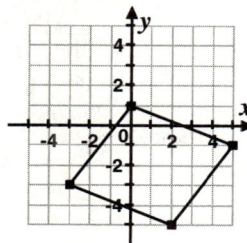

41. First find the slope of the line passing through $(2, 3)$ and $(-2, 1)$.

$$m = \frac{y_2 - y_1}{x_2 - x_1} = \frac{3-1}{2-(-2)} = \frac{2}{4} = \frac{1}{2}$$

Now, use the slope formula, the slope and the points $(5, y)$ and $(1, 0)$ to find y.

$$\frac{1}{2} = \frac{y-0}{5-1}$$

$$\frac{1}{2} = \frac{y}{4}$$

$$4\left(\frac{1}{2}\right) = 4\left(\frac{y}{4}\right)$$

$$2 = y$$

43. Find the slope of the line passing through $(-1, y)$ and $(1,0)$.

$$m = \frac{y_2 - y_1}{x_2 - x_1} = \frac{0-y}{1-(-1)} = \frac{-y}{2}$$

Find the slope of the line passing through $(2,3)$ and $(-2,1)$.

$$m = \frac{y_2 - y_1}{x_2 - x_1} = \frac{1-3}{-2-2} = \frac{1}{2}$$

Since the lines are perpendicular, the product of their slopes is -1.

$$\left(\frac{-y}{2}\right)\left(\frac{1}{2}\right) = -1$$

$$\frac{-y}{4} = -1$$

$$\frac{y}{4} = 1$$

$$y = 4$$

45. a. $m = \dfrac{\text{Change in } y}{\text{Change in } x} = \dfrac{42-26}{180-80} = \dfrac{16}{100} = 0.16$

b. For each minute of brisk walking, the percentage of patients with depression in remission increased by <u>0.16</u>%. The rate of change is <u>0.16</u>% per <u>minute of brisk walking</u>.

47. $m = \dfrac{\text{Change in } y}{\text{Change in } x} = \dfrac{6}{18} = \dfrac{1}{3}$

The pitch of the roof is $\dfrac{1}{3}$.

49. The grade of an access ramp is

$\dfrac{1 \text{ foot}}{12 \text{ feet}} = \dfrac{1}{12} \approx 0.083 = 8.3\%.$

51. – 55. Answers will vary.

57. does not make sense; Explanations will vary. Sample explanation: Either point can be considered (x_1, y_1) or (x_2, y_2).

59. makes sense

61. false; Changes to make the statement true will vary. A sample change is: Slope is rise divided by run.

63. false; Changes to make the statement true will vary. A sample change is: A line with slope 3 cannot be parallel to a line with slope -3.

65. The positive slopes are m_1 and m_2, and of these, the line with slope m_1 is the steeper one so has the larger slope. The negative slopes are m_2 and m_4, and of these, the line with slope m_4 is the steeper one, so has the slope with the largest absolute value, which is the smaller slope. Therefore, in decreasing size, the slopes are m_1, m_3, m_2 and m_4.

67. $y = 2x + 4$

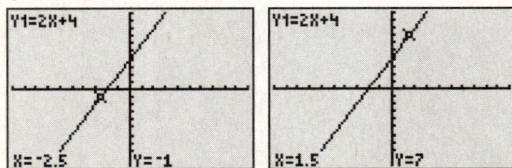

Two points on the graph are $(-2.5, -1)$, and $(1.5, 7)$.

$m = \dfrac{7 - (-1)}{1.5 - (-2.5)} = \dfrac{8}{4} = 2$

69. $y = -\dfrac{1}{2}x - 5$

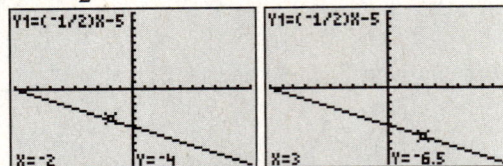

Two points on the graph are $(-2, -4)$ and $(3, -6.5)$.

$m = \dfrac{-6.5 - (-4)}{3 - (-2)} = \dfrac{-2.5}{5} = -0.5 \text{ or } -\dfrac{1}{2}$

71. The slope is always the coefficient of x.

72. Let x = length of shorter piece (in inches). Let $2x$ = length of longer piece.

$x + 2x = 36$

$3x = 36$

$x = 12$

The pieces are 12 inches and 24 inches.

73. $-10 + 16 \div 2(-4) = -10 + 8(-4)$

$= -10 - 32$

$= -10 + (-32) = -42$

74. $2x - 3 \le 5$

$2x \le 8$

$x \le 4$

$(-\infty, 4]$

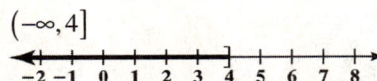

75. $(0 \overset{\boxed{\substack{1 \text{ unit} \\ \text{right}}}}{+ 1}, -3 \overset{\boxed{\substack{4 \text{ units} \\ \text{up}}}}{+ 4}) = (1, 1)$

76. $(0 \overset{\boxed{\substack{3 \text{ units} \\ \text{right}}}}{+ 3}, 1 \overset{\boxed{\substack{2 \text{ units} \\ \text{down}}}}{- 2}) = (3, -1)$

77. $2x + 5y = 0$

$2x - 2x + 5y = 0 - 2x$

$5y = -2x$

$y = \dfrac{-2x}{5}$

$y = -\dfrac{2}{5}x$

3.4 Check Points

1. **a.** $y = 5x - 3$

    The slope is the x-coefficient, which is $m = 5$.
    The y-intercept is the constant term, which is -3.

    b. $y = \dfrac{2}{3}x + 4$

    The slope is the x-coefficient, which is $m = \dfrac{2}{3}$.
    The y-intercept is the constant term, which is 4.

    c. $7x + y = 6 \rightarrow y = -7x + 6$

    The slope is the x-coefficient, which is $m = -7$.
    The y-intercept is the constant term, which is 6.

2. $y = 3x - 2$

    The y-intercept is -2, so plot the point $(0, -2)$.

    The slope is $m = 3$ or $m = \dfrac{3}{1}$. Find another point
    by going up 3 units and to the right 1 unit.
    Use a straightedge to draw a line through the two
    points.

    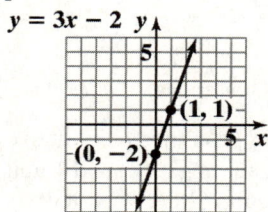

3. $y = \dfrac{3}{5}x + 1$

    The y-intercept is 1, so plot the point $(0, 1)$.

    The slope is $m = \dfrac{3}{5}$. Find another point by going up
    3 units and to the right 5 units.
    Use a straightedge to draw a line through the two
    points.

    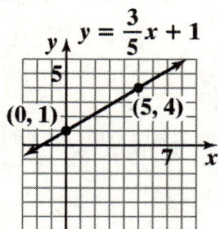

4. $3x + 4y = 0$

    $4y = -3x$

    $y = -\dfrac{3}{4}x$

    The y-intercept is 0, so plot the point $(0, 0)$.

    The slope is $m = \dfrac{-3}{4}$. Find another point by going

    down 3 units and to the right 4 units.
    Use a straightedge to draw a line through the two
    points.

    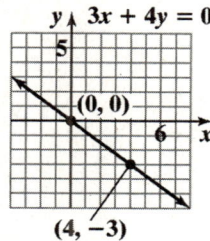

5. **a.** The y-intercept is 8 and the slope is

    $$m = \frac{\text{Change in } y}{\text{Change in } x} = \frac{24 - 8}{50 - 0} = \frac{16}{50} = 0.32$$

    The equation is $y = 0.32x + 8$.

    b. $y = 0.32x + 8 = 0.32(60) + 8 = 27.2$

    The model projects that 27.2% of the U.S.
    population will be college graduates in 2020.

3.4 Concept and Vocabulary Check

1. $y = mx + b$; slope; y-intercept

2. $(0, 3)$; 2; 5

3. y

3.4 Exercise Set

1. $y = 3x + 2$

    The slope is the x-coefficient, which is 3. The y-intercept is the constant term, which is 2.

3. $y = 3x - 5$

    $y = 3x + (-5)$

    $m = 3$; $y - \text{intercept} = -5$

5. $y = -\dfrac{1}{2}x + 5$

$m = -\dfrac{1}{2}$; y-intercept $= 5$

7. $y = 7x$

$y = 7x + 0$

$m = 7$; y-intercept $= 0$

9. $y = 10$

$y = 0x + 10$

$m = 0$; y-intercept $= 10$

11. $y = 4 - x$

$y = -x + 4 = -1x + 4$

$m = -1$; y-intercept $= 4$

13. $\quad -5x + y = 7$

$-5x + y + 5x = 5x + 7$

$\qquad\quad y = 5x + 7$

$m = 5$; y-intercept $= 7$

15. $x + y = 6$

$\quad y = -x + 6 = -1x + 6$

$m = -1$; y-intercept $= 6$

17. $6x + y = 0$

$\quad y = -6x = -6x + 0$

$m = -6$; y-intercept $= 0$

19. $3y = 6x$

$\quad y = 2x$

$m = 2$; y-intercept $= 0$

21. $2x + 7y = 0$

$\qquad 7y = -2x$

$\qquad\quad y = -\dfrac{2}{7}x$

$m = -\dfrac{2}{7}$; y-intercept $= 0$

23. $3x + 2y = 3$

$\qquad 2y = -3x + 3$

$\qquad\quad y = -\dfrac{3}{2}x + \dfrac{3}{2}$

$m = -\dfrac{3}{2}$; y-intercept $= \dfrac{3}{2}$

25. $3x - 4y = 12$

$\quad -4y = -3x + 12$

$\qquad\quad y = \dfrac{3}{4}x - 3$

$m = \dfrac{3}{4}$; y-intercept $= -3$

27. $y = 2x + 4$

Step 1. Plot $(0,4)$ on the y-axis.

Step 2. $m = \dfrac{2}{1} = \dfrac{\text{rise}}{\text{run}}$

Start at $(0,4)$. Using the slope, move 2 units *up* (the rise) and 1 unit to the *right* (the run) to reach the point $(1,6)$.

Step 3. Draw a line through $(0,4)$ and $(1,6)$.

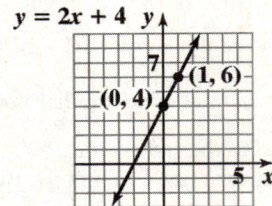

29. $y = -3x + 5$

Slope $= -3 = \dfrac{-3}{1}$; y-intercept $= 5$.

Plot $(0,5)$ on the y-axis. From this point, move 3 units *down* (because -3 is negative) and 1 unit to the *right* to reach the point $(1,2)$. Draw a line through $(0,5)$ and $(1,2)$.

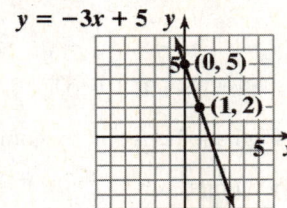

31. $y = \dfrac{1}{2}x + 1$

Slope $= \dfrac{1}{2}$; y-intercept $= 1$

Plot $(0,1)$. From this point, move 1 unit *up* and 2 units to the *right* to reach the point $(2,2)$. Draw a line through $(0,1)$ and $(2,2)$.

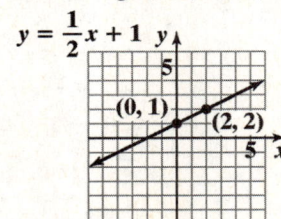

33. $y = \dfrac{2}{3}x - 5$

Slope = $\dfrac{2}{3}$; y-intercept = -5

Plot $(0,-5)$. From this point move 2 units *up* and 3 units to the *right* to reach the point $(3,-3)$. Draw a line through $(0,-5)$ and $(3,-3)$.

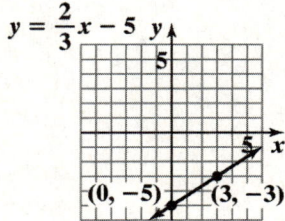

35. $y = -\dfrac{3}{4}x + 2$

Slope = $-\dfrac{3}{4} = \dfrac{-3}{4}$; y-intercept = 2

Plot $(0,2)$. From this point move 3 units *down* and 4 units to the *right* to reach the point $(4,-1)$. Draw a line through $(0,2)$ and $(4,-1)$

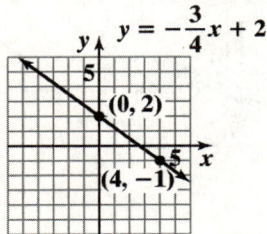

37. $y = -\dfrac{5}{3}x$

Slope = $-\dfrac{5}{3} = \dfrac{-5}{3}$; y-intercept = 0

Plot $(0,0)$. From this point, move 5 units *down* and 3 units to the *right* to reach the point $(3,-5)$. Draw a line through $(0,0)$ and $(3,-5)$.

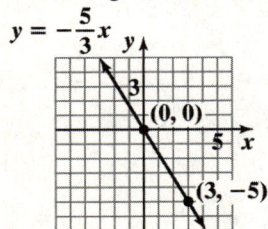

39. a. $3x + y = 0$
$$y = -3x$$

b. $m = -3$; y-intercept = 0

c. Plot $(0,0)$. Since $m = -3 = -\dfrac{3}{1}$, move 3 units *down* and 1 unit to the *right* to reach the point $(1,-3)$. Draw a line through $(0,0)$ and $(1,-3)$.

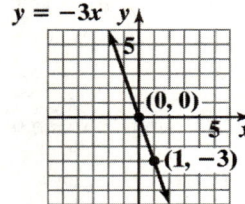

41. a. $3y = 4x$
$$y = \dfrac{4}{3}x$$

b. $m = \dfrac{4}{3}$; y-intercept = 0

c. Plot $(0,0)$. Move 4 units *up* and 3 units to the *right* to reach the point $(3,4)$.
Draw a line through $(0,0)$ and $(3,4)$.

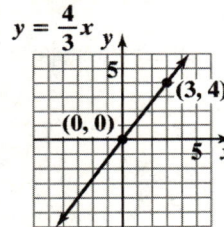

43. a. $2x + y = 3$
$$y = -2x + 3$$

b. $m = -2$; y-intercept = 3

c. Plot $(0,3)$. Since $m = -2 = -\dfrac{2}{1}$, move 2 units *down* and 1 unit to the *right* to reach the point $(1,1)$.
Draw a line through $(0,3)$ and $(1,1)$.

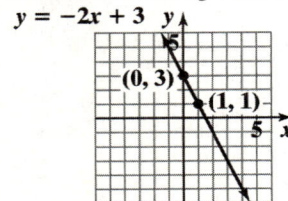

45. a. $7x + 2y = 14$

$$2y = -7x + 14$$

$$\frac{2y}{2} = \frac{-7x + 14}{2}$$

$$y = -\frac{7}{2}x + 7$$

b. $m = -\frac{7}{2}$; y-intercept $= 7$

c. Plot $(0,7)$. Since $m = -\frac{7}{2} = -\frac{7}{2}$, move 7 units *down* and 2 units to the *right* to reach the point $(2,0)$.
Draw a line through $(0,7)$ and $(2,0)$.

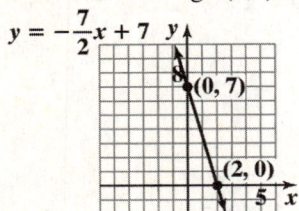

47. $y = 3x + 1$:

$m = 3$; y-intercept $= 1$

$y = 3x - 3$:

$m = 3$; y-intercept $= -3$

The lines are parallel because their slopes are equal.

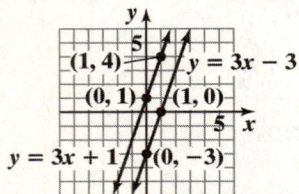

49. $y = -3x + 2$:

$m = -3$; y-intercept $= 2$

$y = 3x + 2$:

$m = 3$; y-intercept $= 2$

The lines are not parallel because their slopes are not equal.
The lines are not perpendicular because the product of their slopes is not -1.

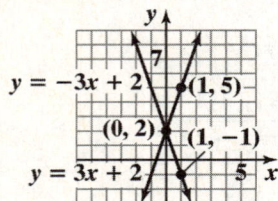

51. $y = x + 3$

$m = 1$; y-intercept $= 3$

$y = -x + 1$

$m = -1$; y-intercept $= 1$

The lines are perpendicular because the product of their slopes is -1.

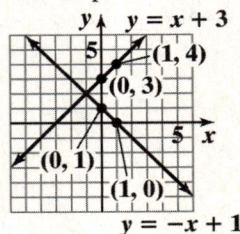

53. $x - 2y = 2 \rightarrow y = \frac{1}{2}x - 1$

$2x - 4y = 3 \rightarrow y = \frac{1}{2}x - \frac{3}{4}$

The lines are parallel because their slopes are equal.

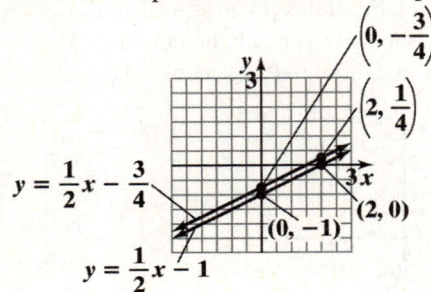

55. $2x - y = -1 \rightarrow y = 2x + 1$

$x + 2y = -6 \rightarrow y = -\frac{1}{2}x - 3$

The lines are perpendicular because the product of their slopes is $2\left(-\frac{1}{2}\right) = -1$.

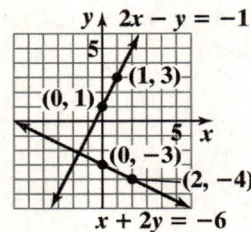

57. Find the slope of the parallel line. $3x + y = 6$

$$y = -3x + 6$$

The slope is -3. We are given that the y-intercept is 5, so using slope-intercept form, we have
$y = -3x + 5$.

59. The slope of the line $y = 5x - 1$ is 5. The negative

reciprocal of 5 is $-\dfrac{1}{5}$.

We are given that the y-intercept is 6, so using the

slope-intercept form, we have $y = -\dfrac{1}{5}x + 6$.

61. Find the y-intercept of the line.

$16y = 8x + 32$

$\dfrac{16}{16}y = \dfrac{8}{16}x + \dfrac{32}{16}$

$y = \dfrac{1}{2}x + 2$

The y-intercept is 2.
Find the slope of the parallel line.
$3x + 3y = 9$

$3y = -3x + 9$

$y = -x + 3$

The slope is -1. Using slope-intercept form, we
have $y = -x + 2$.

63. If the line rises from left to right, it has a positive
slope. It passes through the origin, (0, 0) and a
second point with equal x- and y-coordinates. The
point (2, 2) is one example.
Use the two points to find the slope.

$m = \dfrac{0 - 2}{0 - 2} = \dfrac{-2}{-2} = 1$

The slope is 1. The y-intercept is 0. Using slope-
intercept form, we have $y = 1x + 0$ or $y = x$.

65. a. The y-intercept is 68 and the slope is

$m = \dfrac{\text{Change in } y}{\text{Change in } x} = \dfrac{46 - 68}{42 - 0} = \dfrac{-22}{42} \approx -0.52$

The equation is $y = -0.52x + 68$.

b. $y = -0.52x + 68 = -0.52(100) + 68 = 16$

The model projects that 16% of the U.S.
population will be white non-Hispanic in 2108.

67. – 69. Answers will vary.

71. does not make sense; Explanations will vary.
Sample explanation: The slope can be determined if
the equation is solved for y.

73. does not make sense; Explanations will vary.
Sample explanation: Under these circumstances,
you can either find a different point to plot or
change the scale on your graph paper.

75. false; Changes to make the statement true will vary.
A sample change is: Vertical lines cannot be
expressed in slope-intercept form.

77. false; Changes to make the statement true will vary.
A sample change is: By solving the equation for y,

you can determine that the y-intercept is $\dfrac{7}{2}$.

79. $\dfrac{x}{2} + 7 = 13 - \dfrac{x}{4}$

Multiply by the LCD, which is 4.

$4\left(\dfrac{x}{2} + 7\right) = 4\left(13 - \dfrac{x}{4}\right)$

$2x + 28 = 52 - x$

$3x + 28 = 52$

$3x = 24$

$x = 8$

The solution is {8}.

81. $A = 14, P = 25\% = 0.25$

$A = PB$

$14 = 0.25 \cdot B$

$\dfrac{14}{0.25} = \dfrac{0.25B}{0.25}$

$56 = B$

14 is 25% of 56.

83. $y + 3 = -\dfrac{3}{2}(x - 4)$

$y + 3 = -\dfrac{3}{2}x + 6$

$y = -\dfrac{3}{2}x + 3$

Chapter 3 Mid-Chapter Check Point

1. a. The *x*-intercept is 4.

    b. The *y*-intercept is 2.

    c. The points (4, 0) and (0, 2) lie on the line.
    $$m = \frac{2-0}{0-4} = \frac{2}{-4} = -\frac{1}{2}$$

2. a. The *x*-intercept is −5.

    b. There is no *y*-intercept.

    c. It is a vertical line, so the slope is undefined.

3. a. The *x*-intercept is 0.

    b. The *y*-intercept is 0.

    c. The points (0, 0) and (5, 3) lie on the line.
    $$m = \frac{3-0}{5-0} = \frac{3}{5}$$

4. $y = -2x$

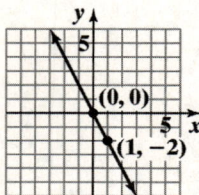

5. $y = -2$

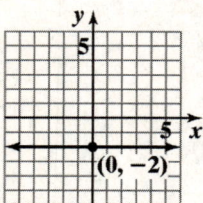

6. $x + y = -2$
    $$y = -x - 2$$

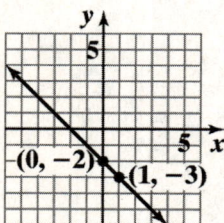

7. $y = \frac{1}{3}x - 2$

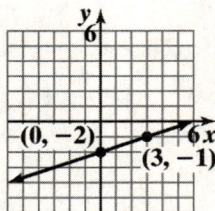

8. $x = 3.5$

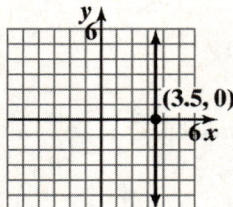

9. $4x - 2y = 8$
    $$-2y = -4x + 8$$
    $$\frac{-2y}{-2} = \frac{-4x}{-2} + \frac{8}{-2}$$
    $$y = 2x - 4$$

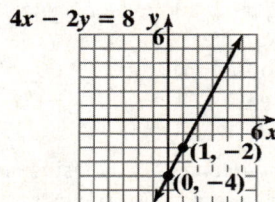

10. $y = 3x + 2$

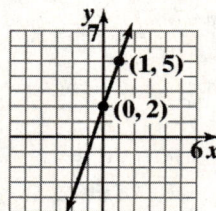

11. $3x + y = 0$
    $$y = -3x$$

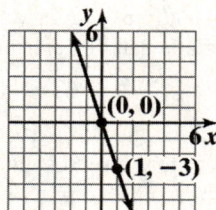

12. $y = -x + 4$

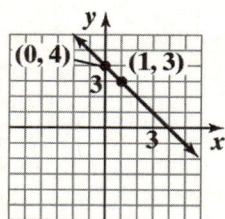

13. $y = x - 4$

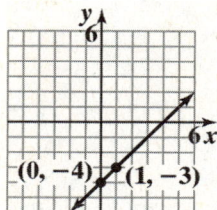

14. $5y = -3x$

$y = -\dfrac{3}{5}x$

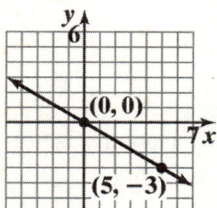

15. $5y = 20$

$y = 4$

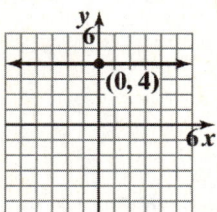

16. $5x - 2y = 10$

$-2y = -5x + 10$

$y = \dfrac{5}{2}x - 5$

The slope is $\dfrac{5}{2}$ and the y-intercept is -5.

17. Line through $(-5,-3)$ and $(0,-4)$:

$m = \dfrac{\text{Change in } y}{\text{Change in } x} = \dfrac{-4-(-3)}{0-(-5)} = \dfrac{-1}{5} = -\dfrac{1}{5}$

Line through $(-2,-8)$ and $(1,7)$:

$m = \dfrac{\text{Change in } y}{\text{Change in } x} = \dfrac{7-(-8)}{1-(-2)} = \dfrac{15}{3} = 5$

Since the product of their slopes is $-\dfrac{1}{5}(5) = -1$, the lines are perpendicular.

18. Line through $(-4,1)$ and $(2,7)$:

$m = \dfrac{\text{Change in } y}{\text{Change in } x} = \dfrac{7-1}{2-(-4)} = \dfrac{6}{6} = 1$

Line through $(-5,13)$ and $(4,-5)$:

$m = \dfrac{\text{Change in } y}{\text{Change in } x} = \dfrac{-5-13}{4-(-5)} = \dfrac{-18}{9} = -2$

Since their slopes are not equal, the lines are not parallel.

Since the product of their slopes is

$1(-2) = -2 \neq -1$, the lines are not perpendicular.

19. Line through $(2,-4)$ and $(7,0)$:

$m = \dfrac{\text{Change in } y}{\text{Change in } x} = \dfrac{0-(-4)}{7-2} = \dfrac{4}{5}$

Line through $(-4,2)$ and $(1,6)$:

$m = \dfrac{\text{Change in } y}{\text{Change in } x} = \dfrac{6-2}{1-(-4)} = \dfrac{4}{5}$

Since their slopes are equal, the lines are parallel.

20. a. The y-intercept is 98 and the slope is

$m = \dfrac{\text{Change in } y}{\text{Change in } x} = \dfrac{74-98}{10-0} = \dfrac{-24}{10} = -2.4$

The equation is $y = -2.4x + 98$.

b. 2014 is 15 years after 1999.

$y = -2.4x + 98 = -2.4(15) + 98 = 62$

The model projects that 62% of new military recruits will be high school graduates in 2014.

3.5 Check Points

1. Begin with the point-slope equation of a line.
$$y - y_1 = m(x - x_1)$$
$$y - (-5) = 6(x - 2)$$
$$y + 5 = 6(x - 2)$$
Now solve this equation for y to write the equation in slope-intercept form.
$$y + 5 = 6(x - 2)$$
$$y + 5 = 6x - 12$$
$$y = 6x - 17$$

2. **a.** Begin by finding the slope:
$$m = \frac{-6 - (-1)}{-1 - (-2)} = \frac{-5}{1} = -5$$
Using the slope and either point, find the point-slope equation of a line.

$$y - y_1 = m(x - x_1) \quad \text{or} \quad y - y_1 = m(x - x_1)$$
$$y - (-1) = -5(x - (-2)) \qquad y - (-6) = -5(x - (-1))$$
$$y + 1 = -5(x + 2) \qquad\qquad y + 6 = -5(x + 1)$$

   b. To obtain slope-intercept form, solve the above equation for y:

$$y + 1 = -5(x + 2) \quad \text{or} \quad y + 6 = -5(x + 1)$$
$$y + 1 = -5x - 10 \qquad\qquad y + 6 = -5x - 5$$
$$y = -5x - 11 \qquad\qquad\quad y = -5x - 11$$

3. Find slope: $m = \dfrac{32.8 - 30.0}{20 - 10} = \dfrac{2.8}{10} = 0.28$
Use the point-slope form to write the equation. Then solve for y to obtain slope-intercept form.
$$y - y_1 = m(x - x_1)$$
$$y - 30.0 = 0.28(x - 10)$$
$$y - 30.0 = 0.28x - 2.8$$
$$y = 0.28x + 27.2$$
Because 2020 is 50 years after 1970, substitute 50 for x and compute y.
$$y = 0.28x + 27.2 = 0.28(50) + 27.2 = 14 + 27.2 = 41.2$$

The model predicts that 41.2 will be the median age in 2020.

3.5 Concept and Vocabulary Check

1. $y - y_1 = m(x - x_2)$

2. standard

3. slope-intercept

4. point-slope

5. horizontal

6. vertical

3.5 Exercise Set

1. Begin with the point-slope equation of a line.
$$y - y_1 = m(x - x_1)$$
$$y - 5 = 3(x - 2)$$
Now solve this equation for y to write the equation in slope-intercept form.
$$y - 5 = 3x - 6$$
$$y = 3x - 1$$

3. Begin with the point-slope equation of a line.
$$y - y_1 = m(x - x_1)$$
$$y - 6 = 5(x - (-2))$$
$$y - 6 = 5(x + 2)$$
Now solve this equation for y to write the equation in slope-intercept form.
$$y - 6 = 5(x + 2)$$
$$y - 6 = 5x + 10$$
$$y = 5x + 16$$

5. Begin with the point-slope equation of a line.
$$y - y_1 = m(x - x_1)$$
$$y - (-2) = -8(x - (-3))$$
$$y + 2 = -8(x + 3)$$
Now solve this equation for y to write the equation in slope-intercept form.
$$y + 2 = -8(x + 3)$$
$$y + 2 = -8x - 24$$
$$y = -8x - 26$$

7. Begin with the point-slope equation of a line.

$$y - y_1 = m(x - x_1)$$
$$y - 0 = -12(x - (-8))$$
$$y = -12(x + 8)$$
$$y = -12x - 96$$

Now solve this equation for y to write the equation in slope-intercept form.

$$y = -12(x + 8)$$
$$y = -12x - 96$$

9. Begin with the point-slope equation of a line.

$$y - y_1 = m(x - x_1)$$
$$y - (-2) = -1\left(x - \left(-\frac{1}{2}\right)\right)$$
$$y + 2 = -1\left(x + \frac{1}{2}\right)$$

Now solve this equation for y to write the equation in slope-intercept form.

$$y + 2 = -1\left(x + \frac{1}{2}\right)$$
$$y + 2 = -x - \frac{1}{2}$$
$$y = -x - \frac{5}{2}$$

11. Begin with the point-slope equation of a line.

$$y - y_1 = m(x - x_1)$$
$$y - 0 = \frac{1}{2}(x - 0)$$

Now solve this equation for y to write the equation in slope-intercept form.

$$y - 0 = \frac{1}{2}(x - 0)$$
$$y = \frac{1}{2}x$$

13. Begin with the point-slope equation of a line.

$$y - y_1 = m(x - x_1)$$
$$y - (-2) = -\frac{2}{3}(x - 6)$$
$$y + 2 = -\frac{2}{3}(x - 6)$$

Now solve this equation for y to write the equation

in slope-intercept form.

$$y + 2 = -\frac{2}{3}(x - 6)$$
$$y + 2 = -\frac{2}{3}x + 4$$
$$y = -\frac{2}{3}x + 2$$

15. slope $= \dfrac{10 - 2}{5 - 1} = \dfrac{8}{4} = 2$

Using the slope and either point, find the point-slope equation of a line.

$$y - y_1 = m(x - x_1)$$
$$y - 2 = 2(x - 1) \text{ or } y - 10 = 2(x - 5)$$

Now solve this equation for y to write the equation in slope-intercept form.

$$y - 2 = 2(x - 1)$$
$$y - 2 = 2x - 2$$
$$y = 2x$$

17. slope $= \dfrac{3 - 0}{0 + 3} = \dfrac{3}{3} = 1$

Using the slope and either point, find the point-slope equation of a line.

$$y - y_1 = m(x - x_1)$$
$$y - 0 = 1(x + 3) \text{ or } y - 3 = 1(x - 0)$$

Now solve this equation for y to write the equation in slope-intercept form.

$$y - 0 = 1(x + 3)$$
$$y = x + 3$$

19. slope $= \dfrac{4 + 1}{2 + 3} = \dfrac{5}{5} = 1$

Using the slope and either point, find the point-slope equation of a line.

$$y - y_1 = m(x - x_1)$$
$$y - (-1) = 1(x - (-3))$$
$$y + 1 = 1(x + 3) \text{ or } y - 4 = 1(x - 2)$$

Now solve this equation for y to write the equation in slope-intercept form.

$$y + 1 = 1(x + 3)$$
$$y + 1 = x + 3$$
$$y = x + 2$$

21. slope $= \dfrac{4-(-1)}{3-(-4)} = \dfrac{5}{7}$

Using the slope and either point, find the point-slope equation of a line.

$$y - y_1 = m(x - x_1)$$

$$y - 4 = \frac{5}{7}(x - 3) \text{ or } y + 1 = \frac{5}{7}(x + 4)$$

Now solve this equation for y to write the equation in slope-intercept form.

$$y - 4 = \frac{5}{7}(x - 3)$$

$$y - 4 = \frac{5}{7}x - \frac{15}{7}$$

$$y = \frac{5}{7}x + \frac{13}{7}$$

23. slope $= \dfrac{-1+1}{4+3} = \dfrac{0}{7} = 0$

Using the slope and either point, find the point-slope equation of a line.

$$y - y_1 = m(x - x_1)$$

$$y - (-1) = 0(x - (-3))$$

$$y + 1 = 0(x + 3) \text{ or } y + 1 = 0(x - 4)$$

Now solve this equation for y to write the equation in slope-intercept form.

$$y + 1 = 0(x + 3)$$

$$y + 1 = 0$$

$$y = -1$$

25. Use the points $(2,4)$ and $(-2,0)$ to find the slope.

$$\text{slope} = \frac{0+4}{-2-2} = \frac{-4}{-4} = 1$$

Find the point-slope equation of a line.

$$y - y_1 = m(x - x_1)$$

$$y - 4 = 1(x - 2)$$

Now solve this equation for y to write the equation in slope-intercept form.

$$y - 4 = 1(x - 2)$$

$$y - 4 = x - 2$$

$$y = x + 2$$

27. Use the points $\left(-\dfrac{1}{2}, 0\right)$ and $(0,4)$ to find the slope.

$$\text{slope} = \frac{4-0}{0+\dfrac{1}{2}} = \frac{4}{\dfrac{1}{2}} = 8$$

Find the point-slope equation of a line.

$$y - y_1 = m(x - x_1)$$

$$y - 0 = 8\left(x - \left(-\frac{1}{2}\right)\right)$$

$$y - 0 = 8\left(x + \frac{1}{2}\right) \text{ or } y - 4 = 8(x - 0)$$

Now solve this equation for y to write the equation in slope-intercept form.

$$y - 4 = 8(x - 0)$$

$$y - 4 = 8x$$

$$y = 8x + 4$$

29. The slope of the line is 4 since the line is parallel to a line with a slope of 4.
Use point-slope form to find the equation of the line.

$$y - y_1 = m(x - x_1)$$

$$y - 2 = 4(x - (-3))$$

$$y - 2 = 4(x + 3)$$

$$y - 2 = 4x + 12$$

$$y = 4x + 14$$

31. Solve $3x + y = 6$ for y to obtain the slope.
$$3x + y = 6$$

$$y = -3x + 6$$

The slope of the line is -3 since the line is parallel to a line with a slope of -3.
Use point-slope form to find the equation of the line.

$$y - y_1 = m(x - x_1)$$

$$y - (-5) = -3(x - (-1))$$

$$y + 5 = -3(x + 1)$$

$$y + 5 = -3x - 3$$

$$y = -3x - 8$$

33. Solve $x - 2y = 3$ for y to obtain the slope.

$$x - 2y = 3$$
$$-2y = -x + 3$$
$$y = \frac{1}{2}x - \frac{3}{2}$$

The slope of the line is -2 since the line is parallel to a line with a slope of $\frac{1}{2}$.

Use point-slope form to find the equation of the line.

$$y - y_1 = m(x - x_1)$$
$$y - (-7) = -2(x - 4)$$
$$y + 7 = -2x + 8$$
$$y = -2x + 1$$

35. Through (2, 4) and same y-intercept as $x - 4y = 8$. Solve the equation to obtain the y-intercept.

$$x - 4y = 8$$
$$-4y = -x + 8$$
$$y = \frac{1}{4}x - 2$$

Now, use the two points to find the slope.

$$m = \frac{4 - (-2)}{2 - 0} = \frac{6}{2} = 3$$

Now use the slope and one of the points to find the equation of the line.

$$y - 4 = 3(x - 2)$$
$$y - 4 = 3x - 6$$
$$y = 3x - 2$$

37. x-intercept at -4 and parallel to the line containing (3, 1) and (2, 6)

First, find the slope of the line going through the points (3, 1) and (2, 6).

$$m = \frac{6 - 1}{2 - 3} = \frac{5}{-1} = -5$$

The slope of the line is -5. Since this line is parallel to the line we are writing the equation for, its slope is also -5. Since the x-intercept is -4, the line goes through the point $(-4, 0)$. Use the point and the slope to find the equation of the line.

$$y - 0 = -5(x - (-4))$$
$$y = -5(x + 4)$$
$$y = -5x - 20$$

39. a. Find slope: $m = \dfrac{5870 - 4571}{4 - 1} = \dfrac{1299}{3} = 433$

$$y - y_1 = m(x - x_1)$$
$$y - 4571 = 433(x - 1)$$
$$y - 4571 = 433x - 433$$
$$y = 433x + 4138$$

b. $y = 433x + 4138 = 433(10) + 4138 = 8468$

According to the model, 8468 fatalities will involve distracted driving in 2014.

41. Answers will vary.

43. does not make sense; Explanations will vary. Sample explanation: In this situation, a better choice would be to use the point-slope form.

45. makes sense

47. false; Changes to make the statement true will vary. A sample change is: If a line has undefined slope, then its equation is of the form $x = c$, where c is a constant.

49. false; Changes to make the statement true will vary. A sample change is: This line has an undefined slope.

51. Use the points $(M, E) = (25, 40)$ and $(M, E) = (125, 280)$ to find the equation.

$$m = \frac{280 - 40}{125 - 25} = \frac{240}{100} = 2.4$$
$$E = mM + b$$
$$40 = 2.4(25) + b$$
$$40 = 60 + b$$
$$-20 = b$$

Therefore, we get the equation $E = 2.4M - 20$.

53. a.

x	0	10	20	30	40	49
y	5	11	18	28	33	41

b. Scatter plot:

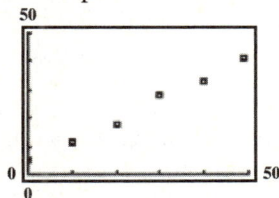

c. $a = 0.7417932386$

$b = 4.245467908$

$r = 0.9971076103$

d. Graph of regression equation:

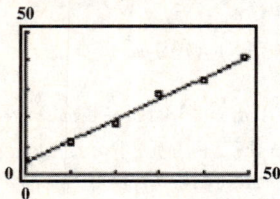

54. Let x = the number of sheets of paper.

$4 + 2x \le 29$

$2x \le 25$

$x \le \dfrac{25}{2}$ or $12\dfrac{1}{2}$

Since the number of sheets of paper must be a whole number, at most 12 sheets of paper can be put in the envelope.

55. The only natural numbers in the given set are 1 and $\sqrt{4}\,(=2)$.

56. $3x - 5y = 15$

x-intercept:

$3x - 5(0) = 15$

$3x = 15$

$x = 5$

y-intercept:

$3(0) - 5y = 15$

$-5y = 15$

$y = -3$

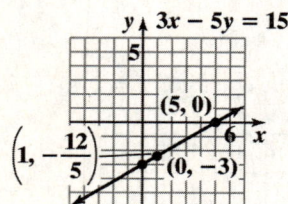

57. $2x - 3y \ge 6$

$2(3) - 3(-1) \ge 6$

$6 + 4 \ge 6$

$10 \ge 6$, true

Yes, the values make the inequality a true statement.

58. $2x - 3y \ge 6$

$2(0) - 3(0) \ge 6$

$0 \ge 6$, false

No, the values make the inequality a false statement.

59. $y \le \dfrac{2}{3}x$

$1 \le \dfrac{2}{3}(1)$

$1 \le \dfrac{2}{3}$, false

No, the values make the inequality a false statement.

3.6 Check Points

1. a. $5x + 4y \le 20$

$5(0) + 4(0) \le 20$

$0 \le 20$, true

$(0,0)$ is a solution.

b. $5x + 4y \le 20$

$5(6) + 4(2) \le 20$

$30 + 8 \le 20$

$38 \le 20$, false

$(6,2)$ is not a solution.

2. $2x - 4y < 8$

Step 1. Replace $<$ with $=$ and graph the linear equation $2x - 4y = 8$ The x-intercept is 4 and the y-intercept is -2, so the line passes through $(4,0)$ and $(0,-2)$. Draw a dashed line because the inequality contains a $<$ symbol.

Step 2. Use $(0,0)$ as a test point.

$2x - 4y < 8$

$2(0) - 4(0) < 8$

$0 < 8$, true

Step 3. The test point $(0,0)$ is part of the solution set, so shade the half-plane containing $(0,0)$.

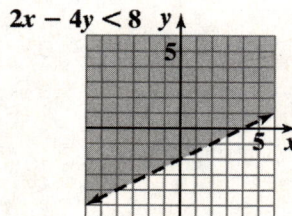

Copyright © 2013 Pearson Education, Inc

3. $y \geq \dfrac{1}{2}x$

Step 1. Replace $\geq$ with = and graph the linear

equation $y = \dfrac{1}{2}x$ The line passes through the

origin so we must find an additional point. For example, using the slope we get the point (2,1). Draw a solid line because the inequality contains a $\geq$ symbol.

Step 2. We cannot use (0,0) as a test point because the line passes through it.

Use (0,10) as a test point $y \geq \dfrac{1}{2}x$

$$10 \geq \dfrac{1}{2}(0)$$

$$10 \geq 0, \text{ true}$$

Step 3. The test point (0,10) is part of the solution set, so shade the half-plane containing (0,10).

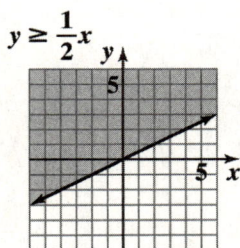

$y \geq \dfrac{1}{2}x$

4. a. $y > 1$

Graph the horizontal line, $y = 1$, with a dashed line and shade the half-plane above the line.

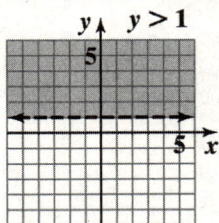

$y > 1$

b. $x \leq -2$

Graph the vertical line, $x = -2$, with a solid line and shade the half-plane to the left of the line.

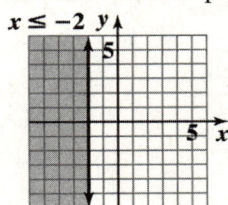

$x \leq -2$

5. a. The coordinates of B are (60, 20). A region that has an average annual temperature of $60°F$ and an average annual precipitation of 20 inches is a grassland.

b.
$$5T - 7P \geq 70$$
$$5(60) - 7(20) \geq 70$$
$$300 - 140 \geq 70$$
$$160 \geq 70, \text{ true}$$

$$3T - 35P \leq -140$$
$$3(60) - 35(20) \leq -140$$
$$180 - 700 \leq -140$$
$$-520 \leq -140, \text{ true}$$

3.6 Concept and Vocabulary Check

1. solution; x; y; $9 > 2$

2. graph

3. half-plane

4. false

5. true

6. false

3.6 Exercise Set

1. $x + y > 4$

$(2,2)$: $2 + 2 > 4$; $4 > 4$, false

$(2,2)$ is not a solution.

$(3,2)$: $3 + 2 > 4$; $5 > 4$, true

$(3,2)$ is a solution.

$(-3,8)$: $-3 + 8 > 4$; $5 > 4$, true

$(-3,8)$ is a solution.

3. $2x + y \geq 5$

$(4,0)$: $8 + 0 \geq 5$, true

$(4,0)$ is a solution.

$(1,3)$: $2 + 3 \geq 5$, true

$(1,3)$ is a solution.

$(0,0)$: $0 + 0 \geq 5$, false

$(0,0)$ is not a solution.

5. $y \geq -2x + 4$

$(4,0)$: $0 \geq -8 + 4$, true

$(4,0)$ is a solution.

$(1,3)$: $3 \geq -2 + 4 = 2$, true

$(1,3)$ is a solution.

$(-2,-4)$: $-4 \geq 4 + 4 = 8$, false

$(-2,-4)$ is not a solution.

7. $y > -2x + 1$

$(2,3)$: $3 > -4 + 1 = -3$, true

$(2,3)$ is a solution.

$(0,0)$: $0 > 0 + 1 = 1$, false

$(0,0)$ is not a solution.

$(0,5)$: $5 > 0 + 1 = 1$, true

$(0,5)$ is a solution.

9. $x + y \geq 3$

Step 1. Replace $\geq$ with $=$ and graph the linear equation $x + y = 3$. The x-intercept is 4 and the y-intercept is 3, so the line passes through $(3,0)$ and $(0,3)$. Draw a solid line because the inequality contains a $\geq$ symbol.

Step 2. Use $(0,0)$ as a test point. $x + y \geq 3$

$$0 + 0 \geq 3$$

$$0 \geq 3, \text{ false}$$

Step 3. The test point $(0,0)$ is not part of the solution set, so shade the half-plane *not* containing $(0,0)$.

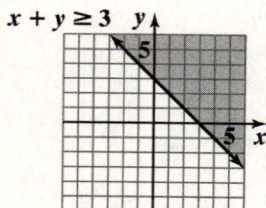

11. $x - y < 5$

Graph the equation $x - y = 5$, which passes through the points $(3,0)$ and $(0, -5)$. Draw a dashed line because the inequality contains a $<$ symbol. Use $(0,0)$ as a test point. Since $0 - 0 < 5$ is a true statement, shade the half-plane containing $(0,0)$.

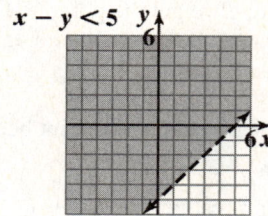

13. $x + 2y > 4$

Graph the equation $x + 2y = 4$ as a dashed line through $(4,0)$ and $(0,2)$. Use $(0,0)$ as a test point. Since $0 + 2(0) > 4$ is false, shade the half-plane *not* containing $(0,0)$.

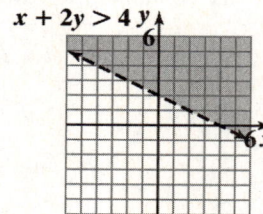

15. $3x - y \leq 6$

Graph the equation $3x - y = 6$ as a solid line through $(2,0)$ and $(0, -6)$. Use $(0,0)$ as a test point. Since $3(0) - \leq 6$ is true, shade the half-plane containing $(0,0)$.

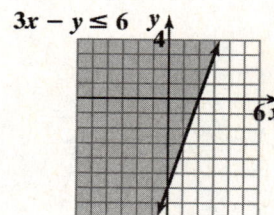

17. $3x - 2y \leq 8$

Graph the equation $3x - 2y = 8$ as a solid line

through $\left(\dfrac{8}{3}, 0\right)$ and $(0, -4)$. Use $(0,0)$ as a test

point. Since $3(0) - 2(0) \leq 8$ is true, shade the half-plane containing $(0,0)$.

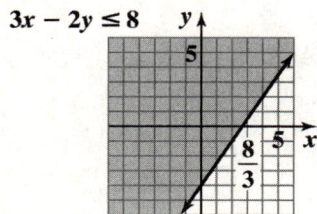

19. $4x + 3y > 15$

Graph $4x + 3y = 15$ as a dashed line through

$\left(\dfrac{15}{4}, 0\right)$ and $(0,5)$.

Use $(0,0)$ as a test point. Since $4(0) + 3(0) > 15$ is false, shade the half-plane not containing $(0,0)$.

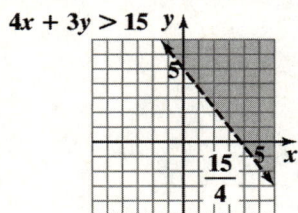

21. $5x - y < -7$

Graph the equation $5x - y = -7$ as a dashed line

through $\left(-\dfrac{7}{5}, 0\right)$ and $(0,7)$. Use $(0,0)$ as a test

point. Since $5(0) - 0 < -7$ is false, shade the half-plane *not* containing $(0,0)$.

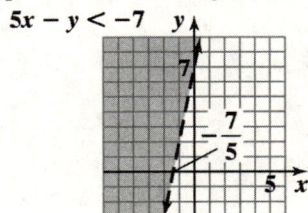

23. $y \leq \dfrac{1}{3}x$

Graph the equation $y = \dfrac{1}{3}x$ as a solid line through

$(0,0)$ and $(3,1)$. Because $(0,0)$ lies on the line, it cannot be used as a test point. Instead use a point

not on the line, such as $(0,6)$. Since $6 \leq \dfrac{1}{2}(0)$ is

false, shade the half-plane *not* containing $(0,6)$.

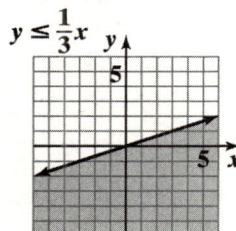

25. $y > 2x$

Graph the equation $y = 2x$ as a dashed line through the origin with slope 2. Use $(3,3)$ as a test point. Since $3 > 2(3)$ is false, shade the half-plane *not* containing $(3,3)$.

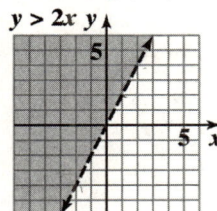

27. $y > 3x + 2$

Graph $y = 3x + 2$ as a dashed line using the slope and y-intercept. Use $(0,0)$ as a test point. Since $0 > 3(0) + 2$ is false, shade the half-plane *not* containing $(0,0)$.

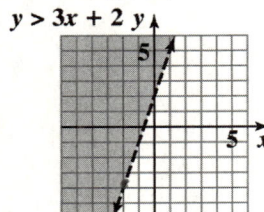

29. $y < \frac{3}{4}x - 3$

Graph $y = \frac{3}{4}x - 3$ as a dashed line using the slope and y-intercept. (Plot $(0, -3)$ and move 3 units up and 4 units to the right to the point $(4,0)$.) Use $(0,0)$ as a test point. Since $0 < \frac{3}{4}(0) - 3$ is false, shade the half-plane *not* containing $(0,0)$.

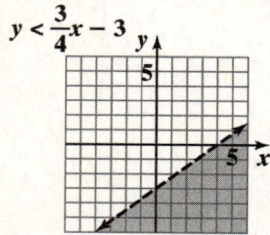

31. $x \le 1$

Graph the vertical line $x = 1$ as a solid line. Use $(0,0)$ as a test point. Since $0 \le 1$ is true, shade the half-plane containing $(0,0)$, which is the half-plane to the *left* of the line.

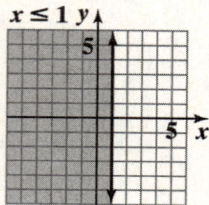

33. $y > 1$

Graph the horizontal line $y = 1$ as a dashed line. Use $(0,0)$ as a test point. Since $0 > 1$ is false, shade the half-plane *not* containing $(0,0)$, which is the half-plane *above* the line.

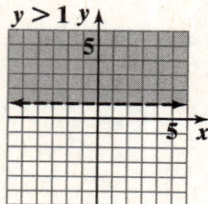

35. $x \ge 0$

Graph the vertical line $x = 0$ (the y-axis) as a solid line. Since $(0,0)$ is on the line, choose a different test point, such as $(3,3)$. Since $3 > 0$ is true, shade the half-plane containing $(3,3)$, which is the half-plane to the *right* of the y-axis.

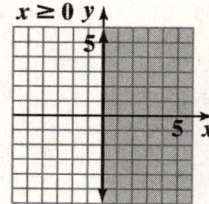

37. $x + y \ge 2$

Rewrite: $y \ge -x + 2$

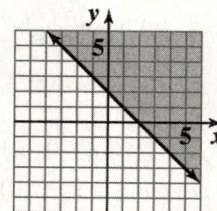

39. $5x - 2y \le 10$

Rewrite: $y \ge \frac{5}{2}x - 5$

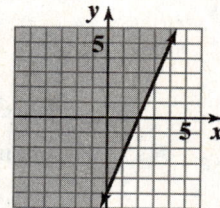

41. $y \ge \frac{1}{2}x$

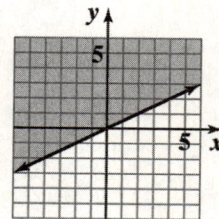

43. $y \le -1$

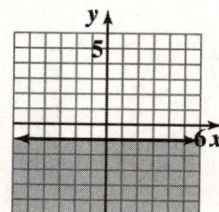

45. a. $20x + 10y \le 80,000$

Graph the line $20x + 10y = 80,000$ as a solid line, using the x-intercept 4000 and the y-intercept 8000. Use $(1000,1000)$ as a test point. Since $20(1000) + 10(1000) = 30,000 \le 80,000$ is true, shade the half-plane containing $(1000,1000)$. Draw the graph in quadrant I only.

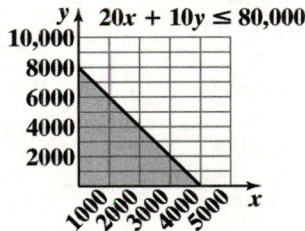

b. Answers will vary. One example is $(1000, 2000)$ since

$$20(1000) + 10(2000) = 20,000 + 20,000$$
$$= 40,000 < 80,000.$$

This indicates that the plane can carry 1000 bottles of water and 2000 medical kits.

47. a. $50x + 150y > 2000$

b. Graph $50x + 150y = 2000$ as a dashed line, using the x-intercept 40 and y-intercept $\dfrac{2000}{150} \approx 13.3$. Use $(10,20)$ as a test point. Since

$50(10) + 150(20) = 3500 > 2000$ is true, shade the half-plane containing $(10,20)$. Draw the graph in quadrant I only.

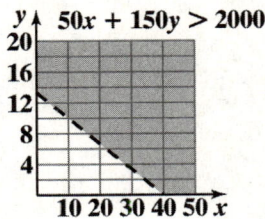

c. Answers will vary. One example is $(20,15)$ since $50(20) + 150(150) = 3200 > 2000$. This indicates that the elevator cannot carry 20 children and 15 adults.

49. a. $BMI = \dfrac{703W}{H^2}$; $W = 200, H = 72$

The man's $BMI = \dfrac{703(200)}{72^2} \approx 27.1$

b. Locate the point $(20, 27.1)$ on the graph for males. The point falls in the "Overweight" region, so the man is overweight.

51. – 57. Answers will vary.

59. does not make sense; Explanations will vary. Sample explanation: This inequality is satisfied by the points above the line

61. makes sense

63. false; Changes to make the statement true will vary. A sample change is: This inequality is satisfied by the points in the half-plane above the line.

65. true

67. The x-intercept of the line is 2 and the y-intercept is 4, so the equation of the line is $2x + y = 4$, which can be written in slope-intercept form as $y = -2x + 4$. The line is solid, so the inequality symbol must be either $\ge$ or $\le$. Choose a test point in the shaded region, for example $(0,0)$. Since $2(0) + 0 = 0 < 4$, the inequality symbol must be $\le$. Therefore, the inequality is $2x + y \le 4$ or $y \le -2x + 4$.

69. $y \ge x - 2$

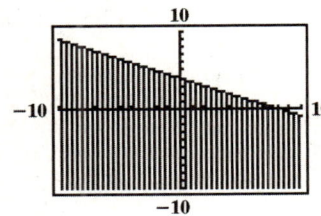

71. $y \le -\dfrac{1}{2}x + 4$

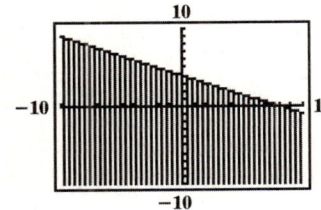

72. $V = lwh$

$$\frac{V}{lw} = \frac{lwh}{lw}$$

$$\frac{V}{lw} = h \text{ or } h = \frac{V}{lw}$$

73. $\dfrac{2}{3} \div \left(-\dfrac{5}{4}\right) = \dfrac{2}{3} \cdot \left(-\dfrac{4}{5}\right) = -\dfrac{8}{15}$

74. $x^2 - 4 = (-3)^2 - 4 = 9 - 4 = 5$

75.
$$x + 2y = 2$$
$$4 + 2(-1) = 2$$
$$4 - 2 = 2$$
$$2 = 2, \text{ true}$$

$$x - 2y = 6$$
$$4 - 2(-1) = 6$$
$$4 + 2 = 6$$
$$6 = 6, \text{ true}$$

Yes, the values are a solution of both equations.

76.
$$x + 2y = 2$$
$$-4 + 2(3) = 2$$
$$-4 + 6 = 2$$
$$2 = 2, \text{ true}$$

$$x - 2y = 6$$
$$-4 - 2(3) = 6$$
$$-4 - 6 = 6$$
$$-10 = 6, \text{ false}$$

No, the values are not a solution of both equations.

77. Graph $2x + 3y = 6$ by finding intercepts.
Find the *x*-intercept.
$$2x + 3y = 6$$
$$2x + 3(0) = 6$$
$$2x = 6$$
$$x = 3$$
Find the *y*-intercept.
$$2x + 3y = 6$$
$$2(0) + 3y = 6$$
$$3y = 6$$
$$y = 2$$

Graph $2x + y = -2$ by finding intercepts.
Find the *x*-intercept.
$$2x + y = -2$$
$$2x + 0 = -2$$
$$2x = -2$$
$$x = -1$$
Find the *y*-intercept.

$$2x + y = -2$$
$$2(0) + y = -2$$
$$y = -2$$

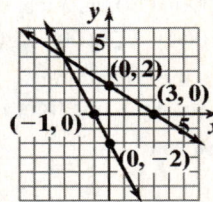

The point of intersection is $(-3,4)$.

Chapter 3 Review Exercises

1. Quadrant IV

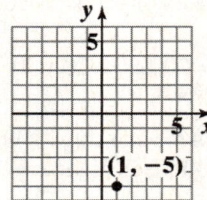

2. Quadrant IV

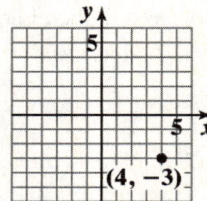

3. Quadrant I

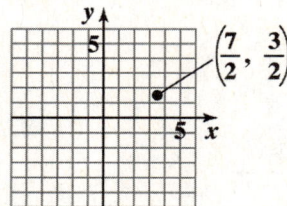

4. Quadrant II

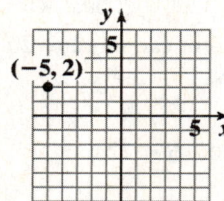

5. $A(5,6) \quad B(-3,0) \quad C(-5,2)$
$\quad D(-4,-2) \quad E(0,-5) \quad F(3,-1)$

6. $y = 3x + 6$

$3 = 3(-3) + 6$

$3 = -6 + 9$

$3 = -3$, false

$(-3, 3)$ is not a solution.

$y = 3x + 6$

$6 = 3(0) + 6$

$6 = 6$, true

$(0, 6)$ is a solution.

$y = 3x + 6$

$9 = 3(1) + 6$

$9 = 9$, true

$(1, 9)$ is a solution.

7. $3x - y = 12$

$3(0) - 4 = 12$

$-4 = 12$, false

$(0, 4)$ is not a solution.

$3x - y = 12$

$3(4) - 0 = 12$

$12 = 12$, true

$(4, 0)$ is a solution.

$3x - y = 12$

$3(-1) - 15 = 12$

$-3 - 15 = 12$

$-18 = 12$, false

$(-1, 15)$ is not a solution.

8. a.

x	$y = 2x - 3$	(x, y)
-2	$y = 2(-2) - 3 = -7$	$(-2, -7)$
-1	$y = 2(-1) - 3 = -5$	$(-1, -5)$
0	$y = 2(0) - 3 = -3$	$(0, -3)$
1	$y = 2(1) - 3 = -1$	$(1, -1)$
2	$y = 2(2) - 3 = 1$	$(2, 1)$

b. $y = 2x - 3$

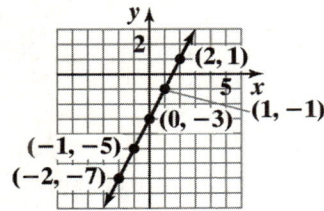

9. a.

x	$y = \frac{1}{2}x + 1$	(x, y)
-2	$y = \frac{1}{2}(-2) + 1 = 0$	$(-2, 0)$
-1	$y = \frac{1}{2}(-1) + 1 = \frac{1}{2}$	$\left(-1, \frac{1}{2}\right)$
0	$y = \frac{1}{2}(0) + 1 = 1$	$(0, 1)$
1	$y = \frac{1}{2}(1) + 1 = \frac{3}{2}$	$\left(1, \frac{3}{2}\right)$
2	$y = \frac{1}{2}(2) + 1 = 2$	$(2, 2)$

b. $y = \frac{1}{2}x + 1$

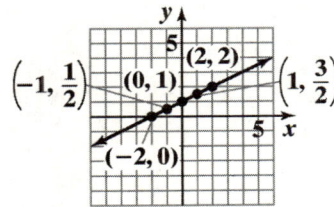

10. a. The graph crosses the *x*-axis at $(-2, 0)$, so the *x*-intercept is -2.

b. The graph crosses the *y*-axis at $(0, -4)$, so the *y*-intercept is -4.

11. a. The graph does not cross the *x*-axis, so there is no *x*-intercept.

b. The graph crosses the *y*-axis at $(0, 2)$, so the *y*-intercept is 2.

12. a. The graph crosses the *x*-axis at $(0, 0)$ (the origin), so the *x*-intercept is 0.

b. The graph also crosses the *y*-axis at $(0, 0)$, so the *y*-intercept is 0.

13. Find the *x*-intercept. Let $y = 0$ and solve for *x*.

$$2x + y = 4$$
$$2x + 0 = 4$$
$$2x = 4$$
$$x = 2$$

The *x*-intercept is 2.

Find the *y*- intercept. Let $x = 0$ and solve for *y*.

$$2x + y = 4$$
$$2(0) + y = 4$$
$$y = 4$$

The *y*-intercept is 4.

Find a checkpoint. For example, let $x = 1$ and solve for *y*.

$$2x + y = 4$$
$$2(1) + y = 4$$
$$2 + y = 4$$
$$y = 2$$

The checkpoint is (1,2).

Draw the line through these three points.

$2x + y = 4$

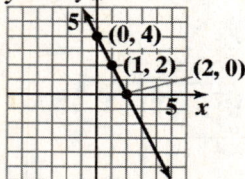

14. Find the *x*-intercept. Let $y = 0$ and solve for *x*.

$$3x - 2y = 12$$
$$3x - 2(0) = 12$$
$$3x = 12$$
$$x = 4$$

The *x*-intercept is 4.

Find the *y*- intercept. Let $x = 0$ and solve for *y*.

$$3x - 2y = 12$$
$$3(0) - 2y = 12$$
$$-2y = 12$$
$$y = -6$$

The *y*-intercept is –6.

Find a checkpoint. For example, let $x = 2$ and solve for *y*.

$$3x - 2y = 12$$
$$3(2) - 2y = 12$$
$$6 - 2y = 12$$
$$-2y = 6$$
$$y = -3$$

The checkpoint is (2,–3).

Draw the line through these three points.

$3x - 2y = 12$

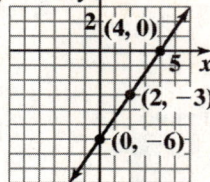

15. Find the *x*-intercept. Let $y = 0$ and solve for *x*.

$$3x = 6 - 2y$$
$$3x = 6 - 2(0)$$
$$3x = 6$$
$$x = 2$$

The *x*-intercept is 2.

Find the *y*- intercept. Let $x = 0$ and solve for *y*.

$$3x = 6 - 2y$$
$$3(0) = 6 - 2y$$
$$0 = 6 - 2y$$
$$2y = 6$$
$$y = 3$$

The *y*-intercept is 3.

Find a checkpoint. For example, let $x = 4$ and solve for *y*.

$$3x = 6 - 2y$$
$$3(4) = 6 - 2y$$
$$12 = 6 - 2y$$
$$6 = -2y$$
$$-3 = y$$

The checkpoint is (4,–3).

Draw the line through these three points.

$3x = 6 - 2y$

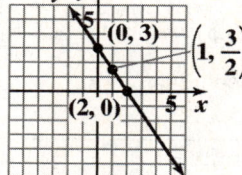

16. Because the constant on the right is 0, the graph passes through the origin. The *x*- and *y*-intercepts are both 0.

Thus we will need to find two more points.

Let $x = 1$ and solve for *y*.

$$3x - y = 0$$
$$3(1) - y = 0$$
$$3 - y = 0$$
$$3 = y$$

This gives the point (1,3).

Let $x = -1$ and solve for *y*.

$$3x - y = 0$$
$$3(-1) - y = 0$$
$$-3 - y = 0$$
$$-3 = y$$

This gives the point $(-1, -3)$.
Draw the line through these three points.

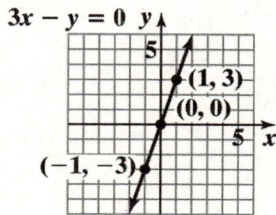

17. $x = 3$

Three ordered pairs are $(3, -2)$, $(3, 0)$, and $(3, 2)$. The graph is a vertical line.

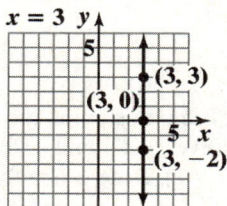

18. $y = -5$

Three ordered pairs are $(-2, -5)$, $(0, -5)$, and $(2, -5)$. The graph is a horizontal line.

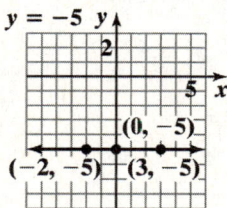

19. $y + 3 = 5$
$$y = 2$$

Three ordered pairs are $(-2, 2)$, $(0, 2)$, and $(2, 2)$. The graph is a horizontal line.

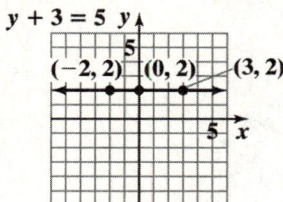

20. $2x = -8$
$$x = -4$$

Three ordered pairs $(-4, -2)$, $(-4, 0)$, and $(-4, 2)$. The graph is a vertical line.

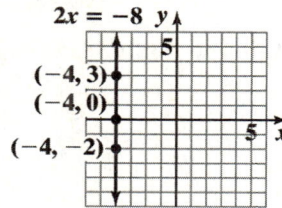

21. **a.** The minimum temperature occurred at 5 P.M. and was $-4°F$.

b. The maximum temperature occurred at 8 P.M. and was at $16°F$.

c. The *x*-intercepts are 4 and 6. This indicates that 4 P.M. and 6 P.M., the temperature was $0°F$.

d. The *y*-intercept is 12. This indicates that at noon the temperature was $12°F$.

e. This indicates that the temperature stayed the same, at $12°F$, from 9 P.M. until midnight.

22. $m = \dfrac{y_2 - y_1}{x_2 - x_1} = \dfrac{1 - 2}{5 - 3} = -\dfrac{1}{2}$

The slope is $-\dfrac{1}{2}$. Since the slope is negative, the line falls from left to right.

23. $m = \dfrac{-4 - 2}{-3 - (-1)} = \dfrac{-6}{-2} = 3$

Since the slope is positive, the line rises from left to right.

24. $m = \dfrac{4 - 4}{6 - (-3)} = \dfrac{0}{9} = 0$

Since the slope is 0, the line is horizontal.

25. $m = \dfrac{-3 - 3}{5 - 5} = \dfrac{-6}{0}$; undefined

Since the slope is undefined, the line is vertical.

26. $m = \dfrac{1 - (-2)}{2 - (-3)} = \dfrac{3}{5}$

27. The line is vertical, so its slope is undefined.

28. $m = \dfrac{-3-(-1)}{2-(-4)} = \dfrac{-2}{6} = -\dfrac{1}{3}$

29. The line is horizontal, so its slope is 0.

30. Line through $(-1, -3)$ and $(2, -8)$:
$m = \dfrac{-8-(-3)}{2-(-1)} = \dfrac{-5}{3} = -\dfrac{5}{3}$
Line through $(8, -7)$ and $(9,10)$:
$m = \dfrac{10-(-7)}{9-8} = \dfrac{17}{1} = 17$
Since their slopes are not equal, the lines are not parallel.
Since the product of their slopes is not -1, the lines are not perpendicular.

31. Line through $(0,-4)$ and $(5, -1)$:
$m = \dfrac{-1-(-4)}{5-0} = \dfrac{3}{5}$
Line through $(-6,8)$ and $(3,-7)$:
$m = \dfrac{-7-8}{3-(-6)} = \dfrac{-15}{9} = \dfrac{-5}{3}$
Since the product of their slopes is
$\dfrac{3}{5}\left(-\dfrac{5}{3}\right) = -1$, the lines are perpendicular.

32. Line through $(5,4)$ and $(9,7)$:
$m = \dfrac{7-4}{9-5} = \dfrac{3}{4}$
Line through $(-6,0)$ and $(-2,3)$:
$m = \dfrac{3-0}{-2-(-6)} = \dfrac{3}{4}$
Since their slopes are equal, the lines are parallel.

33. a. $m = \dfrac{52-64}{2010-1985} = \dfrac{-12}{25} = -0.48$

b. For each year from 1985 through 2010, the percentage of U.S. college freshmen rating their emotional health high or above average decreased by 0.48. The rate of change was -0.48% per year.

34. $y = 5x - 7$
$y = 5x + (-7)$
The slope is the x-coefficient, which is 5. The y-intercept is the constant term, which is -7.

35. $y = 6 - 4x$
$y = -4x + 6$
$m = -4$; y-intercept $= 6$

36. $y = 3$
$m = 0$; y-intercept $= 3$

37. $2x + 3y = 6$
$3y = -2x + 6$
$y = \dfrac{-2x+6}{3}$
$y = -\dfrac{2}{3}x + 2$
$m = -\dfrac{2}{3}$; y-intercept $= 2$

38. $y = 2x - 4$
slope $= 2 = \dfrac{2}{1}$; y-intercept $= -4$
Plot $(0, -4)$ on the y-axis. From this point, move 2 units *up* (because 2 is positive) and 1 unit to the *right* to reach the point $(1, -2)$. Draw a line through $(0, -4)$ and $(1, -2)$.

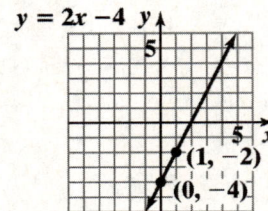

39. $y = \dfrac{1}{2}x - 1$
slope $= \dfrac{1}{2}$; y-intercept $= -1$
Plot $(0, -1)$. From the point, move 1 unit *up* and 2 units to the *right* to reach the point $(2,0)$. Draw a line through $(0, -1)$ and $(2,0)$.

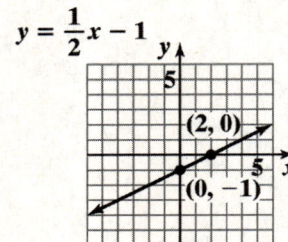

40. $y = -\dfrac{2}{3}x + 5$

slope $= -\dfrac{2}{3} = \dfrac{-2}{3}$; y-intercept $= 5$

Plot (0,5). Move 2 units *down* (because −2 is negative) and 3 units to the *right* to reach the point (3,3). Draw a line through (0,5) and (3,3).

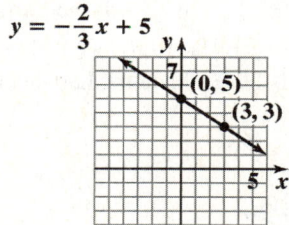

$y = -\dfrac{2}{3}x + 5$

41. $y - 2x = 0$

$\qquad y = 2x$

slope $= 2 = \dfrac{2}{1}$; y-intercept $= 0$

Plot (0,0) (the origin). Move 2 units *up* and 1 unit to the *right* to reach the point (1,2). Draw a line through (0,0) and (1,2).

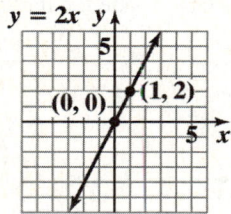

$y = 2x$

42. $\dfrac{1}{3}x + y = 2$

$\qquad y = -\dfrac{1}{3}x + 2$

slope $= -\dfrac{1}{3} = \dfrac{-1}{3}$; y-intercept $= 2$

Plot (0,2). Move 1 unit *down* and 3 units to the *right* to reach the point (3,1). Draw line through (0,2) and (3,1).

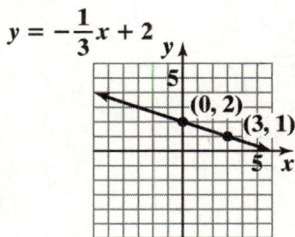

$y = -\dfrac{1}{3}x + 2$

43. $y = -\dfrac{1}{2}x + 4$

slope $= -\dfrac{1}{2} = \dfrac{-1}{2}$

y-intercept $= 4$

$y = -\dfrac{1}{2}x - 1$

slope $= -\dfrac{1}{2} = \dfrac{-1}{2}$

y-intercept $= -1$

Graph each line using its slope and y-intercept.

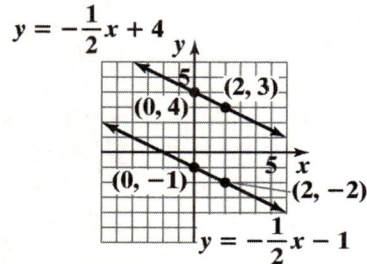

$y = -\dfrac{1}{2}x + 4$

$y = -\dfrac{1}{2}x - 1$

Yes, they are parallel since both lines have a slope of $-\dfrac{1}{2}$ and different y-intercepts.

44. a. $m = \dfrac{15 - 30}{10 - 0} = \dfrac{-15}{10} = -1.5$

The y-intercept is 30 as shown in the graph.

$y = mx + b$

$y = -1.5x + 30$

b. $y = -1.5x + 30$

$y = -1.5(5) + 30 = 22.5$

The model predicts that there were 22.5 million viewers in 1994.

This overestimates the actual value shown in the graph by 2.5 million.

45. Slope $= 6$, passing through $(-4, 7)$

point-slope form:

$y - y_1 = m(x - x_1)$

$y - 7 = 6\left[x - (-4)\right]$

$y - 7 = 6(x + 4)$

slope-intercept form:

$y - 7 = 6(x + 4)$

$y - 7 = 6x + 24$

$\qquad y = 6x + 31$

46. Passing through (3,4) and (2,1)
First, find the slope.

$$m = \frac{1-4}{2-3} = \frac{-3}{-1} = 3$$

Next, use the slope and one of the points to write the equation of the line in point-slope form.

$$y - y_1 = m(x - x_1)$$
$$y - 4 = 3(x - 3)$$

Solve for y to obtain slope-intercept form.

$$y - 4 = 3x - 9$$
$$y = 3x - 5$$

47. a. First, find the slope.

$$m = \frac{5.3 - 3.7}{40 - 20} = \frac{1.6}{20} = 0.08$$

Next, use the slope and one of the points to write the equation of the line in point-slope form.

$$y - y_1 = m(x - x_1)$$
$$y - 3.7 = 0.08(x - 20)$$

Solve for y to obtain slope-intercept form.

$$y - 3.7 = 0.08(x - 20)$$
$$y - 3.7 = 0.08x - 1.6$$
$$y = 0.08x + 2.1$$

b. $y = 0.08x + 2.1$

$$y = 0.08(75) + 2.1 = 8.1$$

The model projects a world population of 8.1 billion in 2025.

48.
$$3x - 4y > 7$$
$$3(0) - 4(0) > 7$$
$$0 - 0 > 7$$
$$0 > 7, \text{ false}$$
$(0,0)$ is not a solution.

$$3x - 4y > 7$$
$$3(3) - 4(-6) > 7$$
$$9 + 24 > 7$$
$$33 > 7, \text{ true}$$
$(3,-6)$ is a solution.

$$3x - 4y > 7$$
$$3(-2) - 4(-5) > 7$$
$$-6 + 20 > 7$$
$$14 > 7, \text{ true}$$
$(-2,-5)$ is a solution.

$$3x - 4y > 7$$
$$3(-3) - 4(4) > 7$$
$$-9 - 16 > 7$$
$$-25 > 7, \text{ false}$$
$(-3,4)$ is not a solution.

49. $x - 2y > 6$

Graph the equation $x - 2y = 6$ as a dashed line through (6,0) and (0, −3). Use (0,0) as a test point. Since $0 - 2(0) > 6$ is false, shade the half-plane *not* containing (0,0).

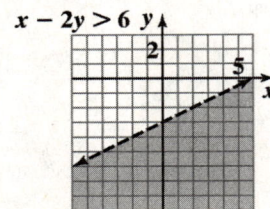

50. $4x - 6y \leq 12$

Graph the equation $4x - 6y = 12$ as a solid line through (3,0) and (0, −2). Use (0,0) as a test point. Since $4(0) - 6(0) \leq 12$ is true, shade the half-plane containing (0,0).

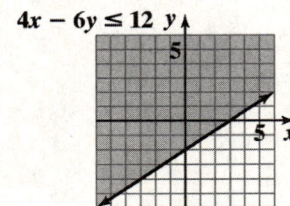

51. $y > 3x + 2$

Graph $y = 3x + 2$ as a dashed line using the slope and y-intercept. Use (0,0) as a test point. Since $0 > 3(0) + 2$ is false, shade the half-plane *not* containing (0,0).

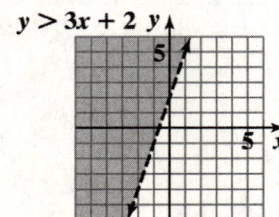

52. $y \le \frac{1}{3}x - 1$

Graph $y = \frac{1}{3}x - 1$ as a solid line using slope and y-intercept. Use $(0,0)$ as a test point. Since $0 \le \frac{1}{3}(0) - 1$ is false, shade the half-plane *not* containing $(0,0)$.

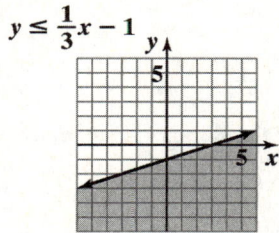

53. $y < -\frac{1}{2}x$

Graph $y = -\frac{1}{2}x$ as a solid line using the slope and y-intercept. Since the line passes through the origin, a point other than $(0,0)$ must be chosen as the test point, for example $(4,4)$. Since $4 \le -\frac{1}{2}(4)$ is false, shade the half-plane not containing $(4,4)$.

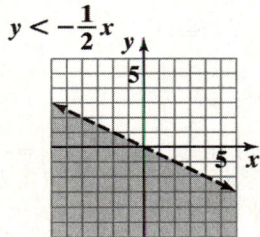

54. $x < 4$

Graph the vertical line $x = 4$ as a dashed line. Use $(0,0)$ as a test point. Since $0 < 4$ is true, shade the half-plane containing $(0,0)$, which is the half-plane to the *left* of the line.

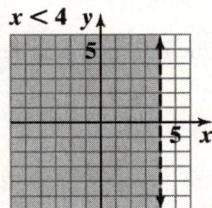

55. $y \ge -2$

Graph the horizontal line $y = -2$ as a solid line. Use $(0,0)$ as a test point. Since $0 \ge -2$ is true, shade the half-plane containing $(0,0)$, which is the half-plane *above* the line.

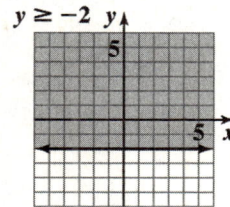

56. $x + 2y \le 0$

Graph $x + 2y = 0$ as a solid line through $(0,0)$ and $(2, -1)$. Since the line goes through the origin, choose another point as the test point, for example $(1,1)$.
Since $1 + 2(1) \le 0$ is false, shade the half-plane *not* containing $(1,1)$.

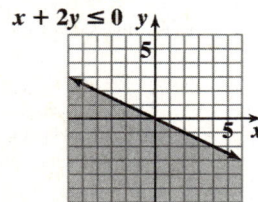

Chapter 3 Test

1.
$$4x - 2y = 10$$
$$4(0) - 2(-5) = 10$$
$$0 + 10 = 10$$
$$10 = 10, \text{ true}$$
$(0,-5)$ is a solution.

$$4x - 2y = 10$$
$$4(-2) - 2(1) = 10$$
$$-8 - 2 = 10$$
$$-10 = 10, \text{ false}$$
$(-2,1)$ is not a solution.

$$4x - 2y = 10$$
$$4(4) - 2(3) = 10$$
$$16 - 6 = 10$$
$$10 = 10, \text{ true}$$
$(4,3)$ is a solution.

2.

x	$y = 3x+1$	(x, y)
-2	$y = 3(-2)+1 = -5$	$(-2,-5)$
-1	$y = 3(-1)+1 = -2$	$(-1,-2)$
0	$y = 3(0)+1 = 1$	$(0,1)$
1	$y = 3(1)+1 = 4$	$(1,4)$
2	$y = 3(2)+1 = 7$	$(2,7)$

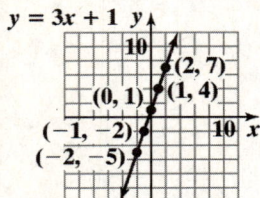

3. a. The graph crosses the x-axis at $(2,0)$, so the x-intercept is 2.

b. The graph crosses the y-axis at $(0, -3)$, so the y-intercept is -3.

4. Find the x-intercept. Let $y = 0$ and solve for x.
$$4x - 2y = -8$$
$$4x - 2(0) = -8$$
$$4x = -8$$
$$x = -2$$
The x-intercept is -2.
Find the y-intercept. Let $x = 0$ and solve for y.
$$4x - 2y = -8$$
$$4(0) - 2y = -8$$
$$-2y = -8$$
$$y = 4$$
The y-intercept is 4.
Find a checkpoint. For example, let $x = -1$ and solve for y.
$$4x - 2y = -8$$
$$4(-1) - 2y = -8$$
$$-4 - 2y = -8$$
$$-2y = -4$$
$$y = 2$$
The checkpoint is $(-1,2)$.
Draw the line through these three points.

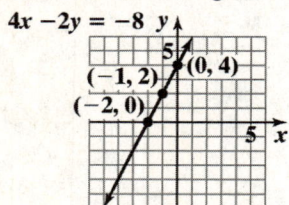

5. $y = 4$
The graph is a horizontal line.

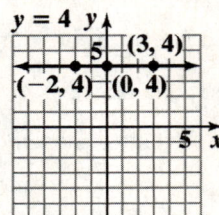

6. $m = \dfrac{-2-4}{-5-(-3)} = \dfrac{-6}{-2} = 3$

The slope is 3. Since the slope is positive, the line rises from left to right.

7. $m = \dfrac{3-(-1)}{6-6} = \dfrac{4}{0}$; undefined

Since the slope is undefined, the line is vertical.

8. Use the points $(-1, -2)$ and $(1,1)$.
$$m = \dfrac{1-(-2)}{1-(-1)} = \dfrac{3}{2}$$

9. Line through $(-2,10)$ and $(0,2)$:
$$m = \dfrac{2-10}{0-(-2)} = \dfrac{-8}{2} = -4$$
Line through $(-8,-7)$ and $(24,1)$:
$$m = \dfrac{1-(-7)}{24-(-8)} = \dfrac{8}{32} = \dfrac{1}{4}$$

Since the product of their slopes is $-4\left(\dfrac{1}{4}\right) = -1$, the lines are perpendicular.

10. Line through $(2,4)$ and $(6,1)$:
$$m = \dfrac{1-4}{6-2} = \dfrac{-3}{4} = -\dfrac{3}{4}$$
Line through $(-3,1)$ and $(1, -2)$:
$$m = \dfrac{-2-1}{1-(-3)} = \dfrac{-3}{4} = -\dfrac{3}{4}$$

Since the slopes are equal, the lines are parallel.

11. $y = -x + 10$
$$y = -1x + 10$$
The slope is the coefficient of x, which is -1. The y-intercept is the constant term, which is 10.

12. $2x + y = 6$
$$y = -2x + 6$$
$m = -2$; y-intercept $= 6$

13. $y = \dfrac{2}{3}x - 1$

slope $= \dfrac{2}{3}$; y-intercept $= -1$

Plot $(0, -1)$. From this point, move 2 units *up* and 3 units to the *right* to reach the point $(3,1)$. Draw a line through $(0, -1)$ and $(3,1)$.

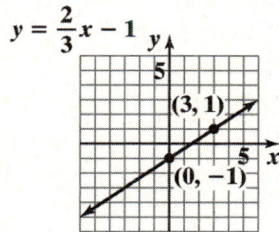

14. $y = -2x + 3$

slope $= -2 = \dfrac{-2}{1}$; y-intercept $= 3$

Plot $(0,3)$. Move 2 units *down* and 1 unit to the right to reach the point $(1,1)$. Draw a line through $(0,3)$ and $(1,1)$.

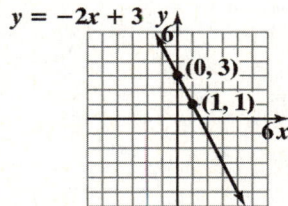

15. point-slope form:

$$y - y_1 = m(x - x_1)$$
$$y - 4 = -2\big[x - (-1)\big]$$
$$y - 4 = -2(x + 1)$$

slope-intercept form:

$$y - 4 = -2(x + 1)$$
$$y - 4 = -2x - 2$$
$$y = -2x + 2$$

16. Passing through $(2,1)$ and $(-1, -8)$
First, find the slope.

$$m = \frac{-8 - 1}{-1 - 2} = \frac{-9}{-3} = 3$$

Next, use the slope and one of the points to write the equation of the line in point-slope form.

$$y - y_1 = m\left(x - x_1\right)$$
$$y - 1 = 3(x - 2)$$

Solve for y to obtain slope-intercept form.

$$y - 1 = 3(x - 2)$$
$$y - 1 = 3x - 6$$
$$y = 3x - 5$$

17. $3x - 2y < 6$

Graph the line $3x - 2y = 6$ as a dashed line through $(2,0)$ and $(0, -3)$. Use $(0,0)$ as a test point. Since $3(0) - 2(0) < 6$ is true, shade the half-plane containing $(0,0)$.

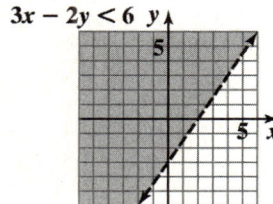

18. $y \geq 2x - 2$

Graph the line $y = 2x - 2$ as a solid line using the slope and y-intercept. Use $(0,0)$ as a test point. Since $0 \geq 2(0) - 2$ is true, shade the half-plane containing $(0,0)$.

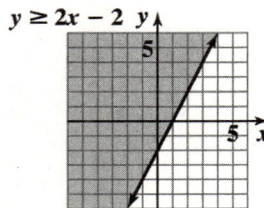

19. $x > -1$

Graph the vertical line $x = -1$ as a dashed line. Use $(0,0)$ as a test point. Since $0 > -1$ is true, shade the half-plane containing $(0,0)$, which is the half-plane to the *right* of the line.

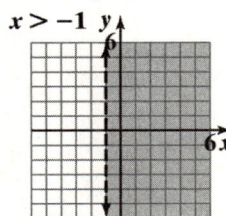

20. a. $m = \dfrac{2515 - 14{,}035}{2008 - 1993} = \dfrac{-11{,}520}{15} = -768$

b. For the period shown, the number of inmates who escaped from U.S. prisons decreased each year by approximately <u>768</u>. The rate of change was <u>−768 inmates</u> per <u>year</u>.

Cumulative Review Exercises (Chapters 1-3)

1. $\dfrac{10-(-6)}{3^2-(4-3)} = \dfrac{10+6}{9-1} = \dfrac{16}{8} = 2$

2. $6 - 2\big[3(x-1)+4\big]$
$= 6 - 2(3x - 3 + 4) = 6 - 2(3x + 1)$
$= 6 - 6x - 2 = 4 - 6x$

3. The only irrational number in the given set is $\sqrt{5}$.

4. $6(2x-1) - 6 = 11x + 7$
$12x - 6 - 6 = 11x + 7$
$12x - 12 = 11x + 7$
$x - 12 = 7$
$x = 19$
The solution set is $\{19\}$.

5. $x - \dfrac{3}{4} = \dfrac{1}{2}$
$x - \dfrac{3}{4} + \dfrac{3}{4} = \dfrac{1}{2} + \dfrac{3}{4}$
$x = \dfrac{2}{4} + \dfrac{3}{4} = \dfrac{5}{4}$
The solution set is $\left\{\dfrac{5}{4}\right\}$.

6. $y = mx + b$
$y - b = mx + b - b$
$y - b = mx$
$\dfrac{y-b}{m} = \dfrac{mx}{m}$
$\dfrac{y-b}{m} = x$ or $x = \dfrac{y-b}{m}$

7. $A = 120;\ P = 15\% = 0.15$
$A = PB$
$120 = 0.15 \cdot B$
$\dfrac{120}{0.15} = \dfrac{0.15B}{0.15}$
$800 = B$
120 is 15% of 800.

8. $y = 4.5x - 46.7$
$133.3 = 4.5x - 46.7$
$133.3 + 46.7 = 4.5x - 46.7 + 46.7$
$180 = 4.5x$
$\dfrac{180}{4.5} = \dfrac{4.5x}{4.5}$
$40 = x$
The car is traveling 40 miles per hour.

9. $2 - 6x \geq 2(5 - x)$
$2 - 6x \geq 10 - 2x$
$2 - 6x + 2x \geq 10 - 2x + 2x$
$2 - 4x \geq 10$
$2 - 4x - 2 \geq 10 - 2$
$-4x \geq 8$
$\dfrac{-4x}{-4} \leq \dfrac{8}{-4}$
$x \leq -2$
$(-\infty, -2]$

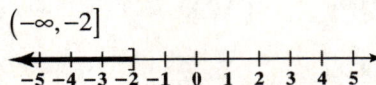

10. $6(2 - x) > 12$
$12 - 6x > 12$
$12 - 6x - 12 > 12 - 12$
$-6x > 0$
$\dfrac{-6x}{-6} < \dfrac{0}{-6}$
$x < 0$
$(-\infty, 0)$

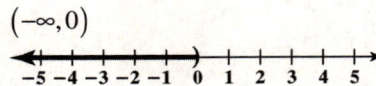

11. Let x = the number of hours the plumber worked.
$18 + 35x = 228$
$35x = 210$
$x = 6$
The plumber worked 6 hours.

12. Let $x =$ the width of a football field.

Let $2x + 14 =$ the length of a football field.

$$P = 2l + 2w$$

$$346 = 2(2x + 14) + 2(x)$$

$$346 = 4x + 28 + 2x$$

$$346 = 6x + 28$$

$$318 = 6x$$

$$53 = x$$

$$x = 53$$

$$2x + 14 = 120$$

The width is 53 meters and the length is 120 meters.

13. Let $x =$ the weight before the 10% loss.

$$x - 0.10x = 180$$

$$0.9x = 180$$

$$\frac{0.9x}{0.9} = \frac{180}{0.9}$$

$$x = 200$$

The weight was 200 pounds.

14. Let $x =$ the measure of the first angle.

Let $x + 20 =$ the measure of the second angle.

Let $2x =$ the measure of third angle.

$$x + (x + 20) + 2x = 180$$

$$4x + 20 = 180$$

$$4x = 160$$

$$x = 40$$

The angles measure $x = 40°$, $x + 20 = 60°$, and $2x = 80°$.

15.

$$x^2 - 10x = (-3)^2 - 10(-3)$$

$$= 9 - 10(-3)$$

$$= 9 + 30$$

$$= 39$$

16. x-intercept:

$$2x - y = 4$$

$$2x - 0 = 4$$

$$2x = 4$$

$$x = 2$$

y-intercept :

$$2x - y = 4$$

$$2(0) - y = 4$$

$$-y = 4$$

$$y = -4$$

checkpoint:

$$2x - y = 4$$

$$2(1) - y = 4$$

$$2 - y = 4$$

$$-y = 2$$

$$y = -2$$

Draw a line through $(2,0)$, $(0, -4)$, and $(1,-2)$.

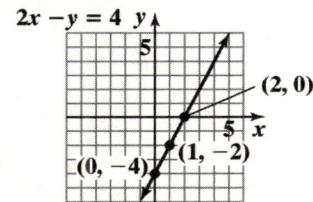

17. $x = -5$

The graph is a vertical line.

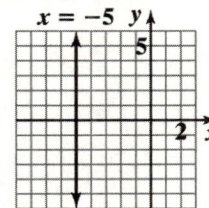

18. $y = -4x + 3$

slope $= -4 = \dfrac{-4}{1}$; y-intercept $= 3$

Plot $(0,3)$. Move 4 units *down* and 1 unit to the *right* to reach the point $(1, -1)$. Draw a line through $(0,3)$ and $(1, -1)$.

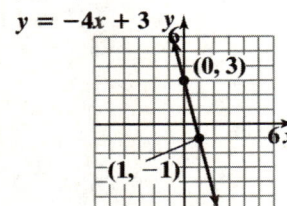

19. $3x - 2y < -6$

Graph the equation $3x - 2y = -6$ as a dashed line through $(-2,0)$ and $(0,3)$.

Use $(0,0)$ as a test point. Since $3(0) - 2(0) < -6$ is false, shade the half-plane not containing $(0,0)$.

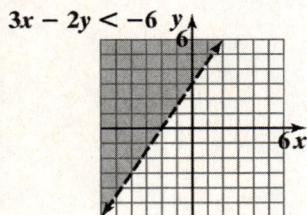

20. $y \geq -1$

Graph the horizontal line $y = -1$ as a solid line. Use $(0,0)$ as a test point. Since $0 \geq -1$ is true, shade the half-plane containing $(0,0)$.

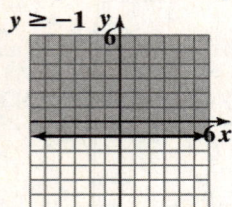

Chapter 4
Systems of Linear Equations and Inequalities

4.1 Check Points

1. a. To determine if $(1,2)$ is a solution to the system, replace x with 1 and y with 2 in both equations.

$$2x - 3y = -4$$
$$2(1) - 3(2) = -4$$
$$2 - 6 = -4$$
$$-4 = -4, \text{ true}$$

$$2x + y = 4$$
$$2(1) + 2 = 4$$
$$2 + 2 = 4$$
$$4 = 4, \text{ true}$$

The ordered pair satisfies both equations, so it is a solution to the system.

b. To determine if $(7,6)$ is a solution to the system, replace x with 7 and y with 6 in both equations.

$$2x - 3y = -4$$
$$2(7) - 3(6) = -4$$
$$14 - 18 = -4$$
$$-4 = -4, \text{ true}$$

$$2x + y = 4$$
$$2(7) + 6 = 4$$
$$14 + 6 = 4$$
$$20 = 4, \text{ false}$$

The ordered pair does not satisfy both equations, so it is not a solution to the system.

2. Graph $2x + y = 6$ by using intercepts.

x-intercept (Set $y = 0$.)
$$2x + y = 6$$
$$2x + 0 = 6$$
$$2x = 6$$
$$x = 3$$

y-intercept (Set $x = 0$.)
$$2x + y = 6$$
$$2(0) + y = 6$$
$$0 + y = 6$$
$$y = 6$$

Graph $2x - y = -2$ by using intercepts.

x-intercept (Set $y = 0$.)
$$2x - y = -2$$
$$2x - 0 = -2$$
$$2x = -2$$
$$x = -1$$

y-intercept (Set $x = 0$.)
$$2x - y = -2$$
$$2(0) - y = -2$$
$$-y = -2$$
$$y = 2$$

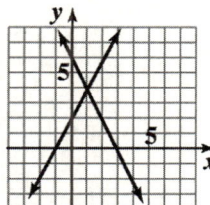

The lines intersect at $(1,4)$.
The solution set is $\{(1,4)\}$.

3. Graph $y = -x + 6$ by using the y-intercept of 6 and the slope of -1.
Graph $y = 3x - 6$ by using the y-intercept of -6 and the slope of 3.

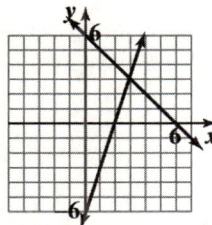

The lines intersect at $(3,3)$.
The solution set is $\{(3,3)\}$.

4. Graph $y = 3x - 2$ by using the y-intercept of -2 and the slope of 3.
Graph $y = 3x + 1$ by using the y-intercept of 1 and the slope of 3.

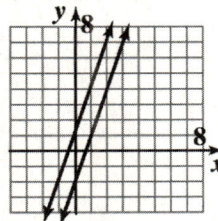

Because both equations have the same slope, 3, but different y-intercepts, the lines are parallel. Thus, the system is inconsistent and has no solution. The solution set is the empty set, $\{ \ \}$.

5. Graph $x + y = 3$ by using intercepts.

x-intercept (Set $y = 0$.)

$x + y = 3$

$x + 0 = 3$

$x = 3$

y-intercept (Set $x = 0$.)

$x + y = 3$

$0 + y = 3$

$y = 3$

Graph $2x + 2y = 6$ by using intercepts.

x-intercept (Set $y = 0$.)

$2x + 2y = 6$

$2x + 2(0) = 6$

$2x = 6$

$x = 3$

y-intercept (Set $x = 0$.)

$2x + 2y = 6$

$2(0) + 2y = 6$

$2y = 6$

$y = 3$

Both lines have the same x-intercept and the same y-intercept. Thus, the graphs of the two equations in the system are the same line.

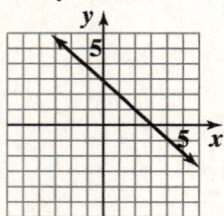

Any ordered pair that is a solution to one equation is a solution to the other, and, consequently, a solution of the system. The system has an infinite number of solutions, namely all points that are solutions of either line.

The solution set is $\{(x, y) | x + y = 3\}$.

6. a. Graph $y = 2x$ by using the y-intercept of 0 and the slope of 2.

Graph $y = x + 10$ by using the y-intercept of 10 and the slope of 1.

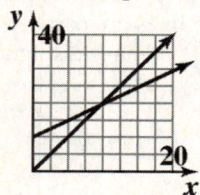

The solution is the ordered pair (10, 20).

b. If the bridge is used 10 times in a month, the total monthly cost without the discount pass is the same as the monthly cost with the discount pass, namely $20.

4.1 Concept and Vocabulary Check

1. satisfies both equations in the system

2. the intersection point

3. inconsistent; parallel

4. dependent; identical or coincide

4.1 Exercise Set

1. To determine if $(2, -3)$ is a solution to the system, replace x with 2 and y with -3 in both equations.

$2x + 3y = -5$

$2(2) + 3(-3) = -5$

$4 + (-9) = -5$

$-5 = -5$, true

$7x - 3y = 23$

$7(2) - 3(-3) = 23$

$14 + 9 = 23$

$23 = 23$, true

The ordered pair satisfies both equations, so it is a solution to the system.

3.

$x + 3y = 1$

$\dfrac{2}{3} + 3\left(\dfrac{1}{9}\right) = 1$

$\dfrac{2}{3} + \dfrac{1}{3} = 1$

$1 = 1$, true

$4x + 3y = 3$

$4\left(\dfrac{2}{3}\right) + 3\left(\dfrac{1}{9}\right) = 3$

$\dfrac{8}{3} + \dfrac{1}{3} = 3$

$\dfrac{9}{3} = 3$

$3 = 3$, true

The ordered pair satisfies both equations, so it is a solution to the system.

5.
$$5x + 3y = 2$$
$$5(-5) + 3(9) = 2$$
$$-25 + 27 = 2$$
$$2 = 2, \text{ true}$$

$$x + 4y = 14$$
$$-5 + 4(9) = 14$$
$$-5 + 36 = 14$$
$$31 = 14, \text{ false}$$

The ordered pair does not satisfy both equations, so it is not a solution to the system.

7.
$$x - 2y = 500$$
$$1400 - 2(450) = 500$$
$$1400 - 900 = 500$$
$$500 = 500, \text{ true}$$

$$0.03x + 0.02y = 51$$
$$0.03(1400) + 0.02(450) = 51$$
$$42 + 9 = 51$$
$$51 = 51, \text{ true}$$

The ordered pair satisfies both equations, so the ordered pair is a solution to the system.

9.
$$5x - 4y = 20$$
$$5(8) - 4(5) = 20$$
$$40 - 20 = 20$$
$$20 = 20, \text{ true}$$

$$3y = 2x + 1$$
$$3(5) = 2(8) + 1$$
$$15 = 16 + 1$$
$$15 = 17, \text{ false}$$

The ordered pair does not satisfy both equations, so it is not a solution to the system.

11. Graph both equations on the same axes.
$x + y = 6$:
x-intercept = 6; y-intercept = 6
$x - y = 2$:
x-intercept = 2; y-intercept = -2

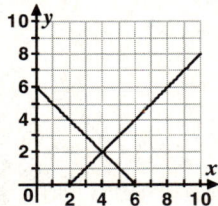

The solution set is $\{(4, 2)\}$.

13. Graph both equations on the same axes.
$x + y = 1$:
x-intercept = 1; y-intercept = 1
$y - x = 3$:
x-intercept = -3; y-intercept = 3

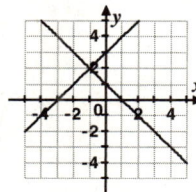

The solution set is $\{(-1, 2)\}$.

15. Graph both equations.
$2x - 3y = 6$:
x-intercept = 3: y-intercept = -2
$4x + 3y = 12$:
x-intercept = 3: y-intercept = 4

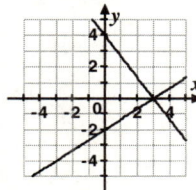

The solution set is $\{(3, 0)\}$.

17. Graph both equations.
$4x + y = 4$:
x-intercept = 1: y-intercept = 4
$3x - y = 3$:
x-intercept = 1: y-intercept = -3

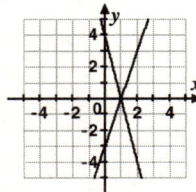

The solution set is $\{(1, 0)\}$.

19. Graph both equations.

$y = x + 5$:

Slope = 1; y-intercept = 5

$y = -x + 3$:

Slope = -1; y-intercept = 3

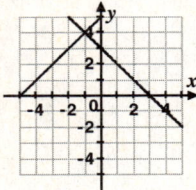

The solution set is $\{(-1, 4)\}$.

21. Graph both equations.

$y = 2x$:

slope = 2; y-intercept = 0

$y = -x + 6$:

slope = -1; y-intercept = 6

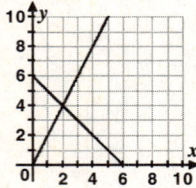

The solution set is $\{(2, 4)\}$.

23. Graph both equations.

$y = -2x + 3$:

slope = -2; y-intercept = 3

$y = -x + 1$:

slope = -1; y-intercept = 1

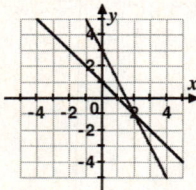

The solution set is $\{(2, -1)\}$.

25. Graph both equations.

$y = 2x - 1$:

Slope = 2; y-intercept = -1

$y = 2x + 1$:

Slope = 2; y-intercept = 1

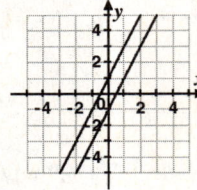

The lines are parallel, so the solution set is $\{\ \}$.

27. Graph each equation.

$x + y = 4$:

x-intercept = 4; y-intercept = 4

$x = -2$:

vertical line with x-intercept -2

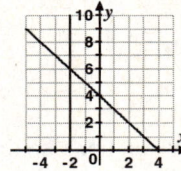

The solution set is $\{(-2, 6)\}$.

29. Graph each equation.

$x - 2y = 4$:

x-intercept = 4; y-intercept = -2

$2x - 4y = 8$:

x-intercept = 4; y-intercept = -2

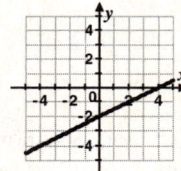

The graph of the two equations are the same line. (Note that they have the same slope and same y-intercept.) Because the lines coincide, the system has an infinite number of solutions. The solution set is $\{(x, y) | x - 2y = 4\}$ or $\{(x, y) | 2x - 4y = 8\}$.

31. Graph both lines.

$y = 2x - 1$:

slope $= 2$; y-intercept $= -1$

$x - 2y = -4$:

x-intercept $= -4$; y-intercept $= 2$

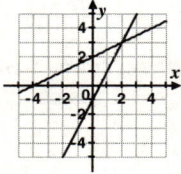

The solution set is $\{(2, 3)\}$.

33. Graph both lines.

$x + y = 5$:

x-intercept $= 5$; y-intercept $= 5$

$2x + 2y = 12$:

x-intercept $= 6$; y-intercept $= 6$

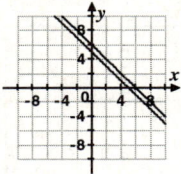

The lines are parallel, so the solution set is $\{\ \}$.

35. $x - y = 0$

$y = x$

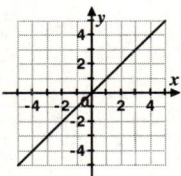

Because the lines coincide, the system has an infinite number of solutions. The solution set is $\{(x, y) | x - y = 0\}$ or $\{(x, y) | y = x\}$.

37. $x = 2$

$y = 4$

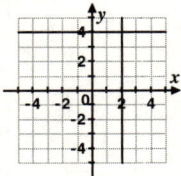

The solution set is $\{(2, 4)\}$.

39. $x = 2$

$x = -1$

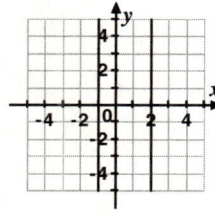

The two vertical lines are parallel, so the solution set is $\{\ \}$.

41. $y = 0$

$y = 4$

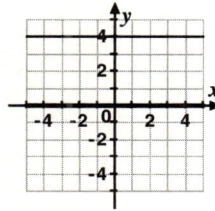

The two horizontal lines are parallel, so the solution set is $\{\ \}$.

43. $y = \dfrac{1}{2}x - 3$:

slope $= \dfrac{1}{2}$, y-intercept $= -3$

$y = \dfrac{1}{2}x - 5$:

slope $= \dfrac{1}{2}$, y-intercept $= -5$

Since the slopes are the same, but the y-intercepts are different, the lines are parallel and there is no solution.

45. $y = -\dfrac{1}{2}x + 4$

slope $= -\dfrac{1}{2}$, y-intercept $= 4$

$3x - y = -4$

$-y = -3x - 4$

$y = 3x + 4$

slope $= 3$, y-intercept $= 4$

Since the lines have different slopes, there will be one solution.

47. $3x - y = 6$

$-y = -3x + 6$

$y = 3x - 6$

slope = 3, y-intercept = -6

$x = \dfrac{y}{3} + 2$

$3x = y + 6$

$3x - 6 = y$

$y = 3x - 6$

slope = 3, y-intercept = -6
Since the lines have the same slopes and y-intercepts, the graphs will coincide and there are an infinite number of solutions.

49. $3x + y = 0$

$y = -3x$

slope = -3, y-intercept = 0
$y = -3x + 1$

slope = -3, y-intercept = 1
Since the slopes are the same, but the y-intercepts are different, the lines are parallel and there is no solution.

51. a. The x-coordinate of the intersection point is 40. Both companies charge the same for 40 miles driven.

b. The y-coordinate of the intersection point is about 55.

c. $y = 0.35x + 40$

$y = 0.35(40) + 40 = 54$

$y = 0.45x + 36$

$y = 0.45(40) + 36 = 54$
Both companies charge $54 for 40 miles driven.

53. a. The solution is the ordered pair (5,20).

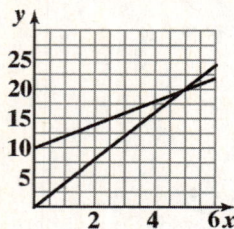

b. Nonmembers and members pay the same amount per month for taking 5 classes, namely $20.

55. – 61. Answers will vary.

63. does not make sense; Explanations will vary. Sample explanation: Some linear systems have no solutions or one solution.

65. makes sense

67. false; Changes to make the statement true will vary. A sample change is: The system's lines could have different slopes.

69. false; Changes to make the statement true will vary. A sample change is: The two lines of a linear system that has one solution must have different slopes.

71. Answers will vary.

73. $y = 2x + 2$

$y = -2x + 6$

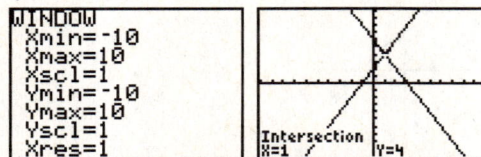

The solution set is $\{(1,4)\}$.

75. $x + 2y = 4$

$x - y = 4$

In order to enter the equations into a graphing calculator, each of them must be solved for y.

$x + 2y = 4$

$2y = -x + 4$

$\dfrac{2y}{2} = \dfrac{-x+4}{2}$

$y = -\dfrac{1}{2}x + 2$

$x - y = 4$

$-y = -x + 4$

$y = x - 4$

The solution set is $\{(4,0)\}$.

77. $3x - y = 5$
       $-5x + 2y = -10$
Solve each equation for y.
$3x - y = 5$
  $-y = -3x + 5$
   $y = 3x - 5$

$-5x + 2y = -10$
    $2y = 5x - 10$
     $y = \dfrac{5}{2}x - 5$

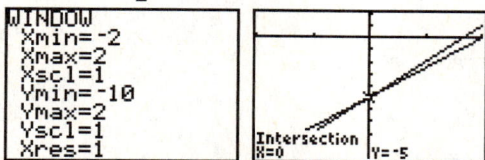

```
WINDOW
 Xmin=-2
 Xmax=2
 Xscl=1
 Ymin=-10
 Ymax=2
 Yscl=1
 Xres=1
```
Intersection
X=0   Y=-5

The solution set is $\{(0, -5)\}$.

79. $y = \dfrac{1}{3}x + \dfrac{2}{3}$

   $y = \dfrac{5}{7}x - 2$

```
WINDOW
 Xmin=-10
 Xmax=10
 Xscl=1
 Ymin=-10
 Ymax=10
 Yscl=1
 Xres=1
```
Intersection
X=7   Y=3

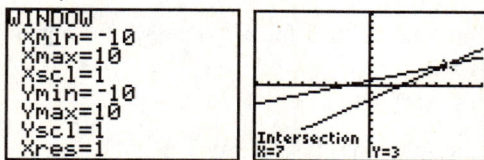

The solution set is $\{(7, 3)\}$.

81. $-3 + (-9) = -3 - 9 = -12$

82. $-3 - (-9) = -3 + 9 = 6$

83. $-3(-9) = 27$

84. $4x - 3(-x - 1) = 24$
     $4x + 3x + 3 = 24$
       $7x + 3 = 24$
         $7x = 21$
          $x = 3$
The solution set is $\{3\}$.

85. $5(2y - 3) - 4y = 9$
     $10y - 15 - 4y = 9$
        $6y - 15 = 9$
           $6y = 24$
            $y = 4$
The solution set is $\{4\}$.

86. $(5x - 1) + 1 = 5x + 5$
     $5x - 1 + 1 = 5x + 5$
         $5x = 5x + 5$
          $0 = 5$
The solution set is $\{\ \}$.

4.2 Check Points

1. $y = 5x - 13$
      $2x + 3y = 12$
Substitute $5x - 13$ for y in the second equation.
$2x + 3(5x - 13) = 12$
  $2x + 15x - 39 = 12$
       $17x - 39 = 12$
            $17x = 51$
              $x = 3$
Back substitute 3 for x into the first equation.
$y = 5x - 13$
$y = 5(3) - 13 = 2$
The solution set is $\{(3, 2)\}$.

2. $3x + 2y = -1$
     $x - y = 3$
Solve the second equation for x.
$x - y = 3$
  $x = y + 3$
Substitute $y + 3$ for x in the first equation.
  $3x + 2y = -1$
$3(y + 3) + 2y = -1$
 $3y + 9 + 2y = -1$
     $5y + 9 = -1$
        $5y = -10$
         $y = -2$
Back substitute -2 for y into $x = y + 3$.
$x = y + 3$
$x = -2 + 3 = 1$
The solution set is $\{(1, -2)\}$.

3. $3x + y = -5$

$\qquad y = -3x + 3$

Substitute $-3x + 3$ for y in the first equation.

$\qquad 3x + y = -5$

$\qquad 3x + \overbrace{(-3x + 3)}^{y} = -5$

$\qquad 3x - 3x + 3 = -5$

$\qquad\qquad\qquad 3 = -5, \text{ false}$

The false statement indicates that the system is inconsistent and has no solution.

The solution set is $\{\ \}$.

4. $\qquad y = 3x - 4$

$9x - 3y = 12$

Substitute $3x - 4$ for y in the second equation.

$\qquad 9x - 3y = 12$

$\qquad 9x - 3\overbrace{(3x - 4)}^{y} = 12$

$\qquad 9x - 3(3x - 4) = 12$

$\qquad 9x - 9x + 12 = 12$

$\qquad\qquad\qquad 12 = 12, \text{ true}$

The true statement indicates that the system contains dependent equations and has infinitely many solutions.

The solution set is $\{(x, y) \mid y = 3x - 4\}$.

5. a. $p = -30x + 1800$

$\qquad p = 30x$

Substitute $30x$ for p in the first equation.

$\qquad p = -30x + 1800$

$\qquad \overbrace{30x}^{p} = -30x + 1800$

$\qquad 60x = 1800$

$\qquad\quad x = 30$

Back-substitute to find p.

$\qquad p = 30x$

$\qquad p = 30(30) = 900$

The solution set is $\{(30, 900)\}$.

Equilibrium quantity: 30,000

Equilibrium price: $900

b. When rents are <u>$900</u> per month, consumers will demand <u>30,000</u> apartments and suppliers will offer <u>30,000</u> apartments for rent.

4.2 Concept and Vocabulary Check

1. $\{(4, 1)\}$

2. $\{(-13)\}$

3. $\varnothing$

4. $\{(x, y) \mid 2x - 6y = 8\}$ or $\{(x, y) \mid x = 3y + 4\}$

5. equilibrium

4.2 Exercise Set

1. $x + y = 4$

$\quad y = 3x$

Substitute $3x$ for y in the first equation.

$\qquad x + y = 4$

$\qquad x + (3x) = 4$

Solve this equation for x.

$\qquad 4x = 4$

$\qquad\quad x = 1$

Back substitute 1 for x into the second equation.

$\qquad y = 3x$

$\qquad y = 3(1) = 3$

The solution set is $\{(1, 3)\}$.

3. $x + 3y = 8$

$\quad y = 2x - 9$

Substitute $2x - 9$ for y in the first equation and solve for x.

$\qquad\quad x + 3y = 8$

$\qquad x + 3(2x - 9) = 8$

$\qquad\quad x + 6x - 27 = 8$

$\qquad\qquad 7x - 27 = 8$

$\qquad\qquad\qquad 7x = 35$

$\qquad\qquad\qquad\quad x = 5$

Back-substitute 5 for x into the second equation and solve for y.

$\qquad y = 2x - 9$

$\qquad y = 2(5) - 9 = 1$

The solution set is $\{(5, 1)\}$.

5. $x + 3y = 5$

$4x + 5y = 13$

Solve the first equation for x.

$x + 3y = 5$

$\quad x = 5 - 3y$

Substitute $5 - 3y$ for x in the second equation and solve for y.

$\quad\quad 4x + 5y = 13$

$4(5 - 3y) + 5y = 13$

$20 - 12y + 5y = 13$

$\quad 20 - 7y = 13$

$\quad\quad -7y = -7$

$\quad\quad\quad y = 1$

Back-substitute 1 for y in the equation $x = 5 - 3y$ and solve for x.

$x = 5 - 3y$

$x = 5 - 3(1) = 2$

The solution set is $\{(2, 1)\}$.

7. $2x - y = -5$

$x + 5y = 14$

Solve the second equation for x.

$x + 5y = 14$

$\quad x = 14 - 5y$

Substitute $14 - 5y$ for x in the first equation.

$2(14 - 5y) - y = -5$

$28 - 10y - y = -5$

$\quad 28 - 11y = -5$

$\quad\quad -11y = -33$

$\quad\quad\quad y = 3$

Back-substitute.

$x = 14 - 5y$

$x = 14 - 5(3) = 14 - 15 = -1$

The solution set is $\{(-1, 3)\}$.

9. $2x - y = 3$

$5x - 2y = 10$

Solve the first equation for y.

$2x - y = 3$

$\quad -y = -2x + 3$

$\quad\quad y = 2x - 3$

Substitute $2x - 3$ for y in the second equation.

$5x - 2(2x - 3) = 10$

$\quad 5x - 4x + 6 = 10$

$\quad\quad\quad x + 6 = 10$

$\quad\quad\quad\quad\quad x = 4$

Back-substitute.

$y = 2x - 3$

$y = 2(4) - 3 = 8 - 3 = 5$

The solution set is $\{(4, 5)\}$.

11. $-3x + y = -1$

$x - 2y = 4$

Solve the second equation for x.

$x - 2y = 4$

$\quad x = 2y + 4$

Substitute $2y + 4$ for x in the first equation.

$\quad\quad -3x + y = -1$

$-3(2y + 4) + y = -1$

$\quad -6y - 12 + y = -1$

$\quad\quad -5y - 12 = -1$

$\quad\quad\quad -5y = 11$

$\quad\quad\quad\quad y = -\dfrac{11}{5}$

Back-substitute.

$x = 2y + 4$

$x = 2\left(-\dfrac{11}{5}\right) + 4$

$x = -\dfrac{22}{5} + 4$

$x = -\dfrac{22}{5} + \dfrac{20}{5}$

$x = -\dfrac{2}{5}$

The solution set is $\left\{\left(-\dfrac{2}{5}, -\dfrac{11}{5}\right)\right\}$.

13. $x = 9 - 2y$

$x + 2y = 13$

The first equation is already solved for x.

Substitute $9 - 2y$ for x in the second equation.

$$x + 2y = 13$$
$$(9 - 2y) + 2y = 13$$
$$9 = 13, \text{ false}$$

The false statement 9=13 indicates that the system is inconsistent and has no solution.

The solution set is $\{\ \}$.

15. $y = 3x - 5$

$21x - 35 = 7y$

Substitute $3x–5$ for y in the second equation.

$$21x - 35 = 7y$$
$$21x - 35 = 7(3x - 5)$$
$$21x - 35 = 21x - 35$$
$$-35 = -35, \text{ true}$$

The true statement $-35 = -35$ indicates that the system contains dependent equations and has an infinite number of solutions.

The solution set is $\{(x, y) | y = 3x - 5\}$.

17. $5x + 2y = 0$

$x - 3y = 0$

Solve the second equation for x.

$x - 3y = 0$

$x = 3y$

Substitute $3y$ for x in the first equation.

$$5x + 2y = 0$$
$$5(3y) + 2y = 0$$
$$15y + 2y = 0$$
$$17y = 0$$
$$y = 0$$

Back-substitute to find x.

$x = 3y$

$x = 3(0) = 0$

The solution set is $\{(0,0)\}$.

19. $2x - y = 6$

$3x + 2y = 5$

Solve the first equation for y.

$$2x - y = 6$$
$$-y = -2x + 6$$
$$y = 2x - 6$$

Substitute $2x - 6$ for y in the second equation.

$$3x + 2y = 5$$
$$3x + 2(2x - 6) = 5$$
$$3x + 4x - 12 = 5$$
$$7x - 12 = 5$$
$$7x = 17$$
$$x = \frac{17}{7}$$

Back-substitute to find y.

$$y = 2x - 6 = 2\left(\frac{17}{7}\right) - 6 = -\frac{8}{7}$$

The solution set is $\left\{\left(\frac{17}{7}, -\frac{8}{7}\right)\right\}$.

21. $2(x - 1) - y = -3$

$y = 2x + 3$

Substitute $2x + 3$ for y in the first equation.

$$2(x - 1) - (2x + 3) = -3$$
$$2x - 2 - 2x - 3 = -3$$
$$-5 = -3, \text{ false}$$

The false statement $-5 = -5$ indicates that the system has no solution.

The solution set is $\{\ \}$.

23. $x = 2y + 9$

$x = 7y + 10$

Substitute $7y + 10$ for x in the first equation.

$$x = 2y + 9$$
$$7y + 10 = 2y + 9$$
$$5y + 10 = 9$$
$$5y = -1$$
$$y = -\frac{1}{5}$$

Back-substitute to find x.

$$x = 2y + 9 = 2\left(-\frac{1}{5}\right) + 9$$
$$= -\frac{2}{5} + 9 = -\frac{2}{5} + \frac{45}{5} = \frac{43}{5}$$

The solution set is $\left\{\left(\frac{43}{5}, -\frac{1}{5}\right)\right\}$.

25. $4x - y = 100$

$0.05x - 0.06y = -32$

Solve the first equation for y.

$4x - y = 100$

$-y = -4x + 100$

$y = 4x - 100$

Substitute $4x - 100$ for y in the second equation.

$0.05x - 0.06y = -32$

$0.05x - 0.06(4x - 100) = -32$

$0.05x - 0.24x + 6 = -32$

$-0.19x + 6 = -32$

$-0.19x = -38$

$x = 200$

Back-substitute to find y.

$y = 4x - 100$

$= 4(200) - 100$

$= 800 - 100 = 700$

$\left(-\dfrac{44}{3}, -\dfrac{7}{3} \right)$

The solution set is $\{(200, 700)\}$.

27. $y = \dfrac{1}{3}x + \dfrac{2}{3}$

$y = \dfrac{5}{7}x - 2$

First, clear both equations of fractions. Multiply the first equation by the LCD, 3.

$3y = 3\left(\dfrac{1}{3}x + \dfrac{2}{3} \right)$

$3y = 3x + 2$

Multiply the second equation by the LCD, 7.

$7y = 7\left(\dfrac{5}{7}x - 2 \right)$

$7y = 5x - 14$

Now solve the new system

$3y = x + 2$

$7y = 5x - 14$

Solve the first of these equations for x.

$3y = x + 2$

$3y - 2 = x$

Substitute $3y - 2$ for x in the second equation of the new system.

$7y = 5x - 14$

$7y = 5(3y - 2) - 14$

$7y = 15y - 10 - 14$

$7y = 15y - 24$

$-8y = -24$

$y = 3$

Back-substitute to find x.

$x = 3y - 2$

$x = 3(3) - 2 = 9 - 2 = 7$

The solution set is $\{(7, 3)\}$.

29. $\dfrac{x}{6} - \dfrac{y}{2} = \dfrac{1}{3}$

$x + 2y = -3$

Clear the first equation of fractions by multiplying 6.

$6\left(\dfrac{x}{6} - \dfrac{y}{2} \right) = 6\left(\dfrac{1}{3} \right)$

$x - 3y = 2$

Solve this equation for x.

$x = 3y + 2$

Substitute $3y + 2$ for x in the second equation of the system.

$(3y + 2) + 2y = -3$

$5y + 2 = -3$

$5y = -5$

$y = -1$

Back-substitute to find x.

$x = 3y + 2 = 3(-1) + 2 = -1$

The solution set is $\{(-1, -1)\}$.

31. $2x - 3y = 8 - 2x$

$3x + 4y = x + 3y + 14$

Simplify the first equation.

$2x - 3y = 8 - 2x$

$2x - 3y + 2x = 8 - 2x + 2x$

$4x - 3y = 8$

Simplify the second equation.

$3x + 4y = x + 3y + 14$

$3x + 4y - x - 3y = x + 3y + 14 - x - 3y$

$2x + y = 14$

Solve the last equation for y.

$2x + y = 14$

$y = 14 - 2x$

Substitute $14 - 2x$ for y in the equation $4x - 3y = 8$.

$4x - 3y = 8$

$4x - 3(14 - 2x) = 8$

$4x - 42 + 6x = 8$

$10x - 42 = 8$

$10x = 50$

$x = 5$

Back-substitute to find y.

$y = 14 - 2x$

$y = 14 - 2(5) = 4$

The solution set is $\{(5, 4)\}$.

33. $x + y = 81$

$x = y + 41$

Substitute $y + 41$ for x in the first equation.

$x + y = 81$

$(y + 41) + y = 81$

$y + 41 + y = 81$

$2y + 41 = 81$

$2y = 40$

$y = 20$

Back-substitute.

$x = y + 41 = 20 + 41 = 61$

The numbers are 20 and 61.

35. $x - y = 5$

$4x = 6y$

Solve the first equation for x.

$x - y = 5$

$x = y + 5$

Substitute $y + 5$ for x in the second equation.

$4x = 6y$

$4(y + 5) = 6y$

$4y + 20 = 6y$

$20 = 2y$

$10 = y$

Back-substitute.

$x = y + 5 = 10 + 5 = 15$

The numbers are 10 and 15.

37. $x - y = 1$

$x + 2y = 7$

Solve the first equation for x.

$x - y = 1$

$x = y + 1$

Substitute $y + 1$ for x in the second equation.

$x + 2y = 7$

$(y + 1) + 2y = 7$

$y + 1 + 2y = 7$

$3y + 1 = 7$

$3y = 6$

$y = 2$

Back-substitute.

$x = y + 1 = 2 + 1 = 3$

The numbers are 2 and 3.

39. $0.7x - 0.1y = 0.6$

$0.8x - 0.3y = -0.8$

Multiply both sides of both equations by 10.

$7x - y = 6$

$8x - 3y = -8$

Solve the first equation for y.

$7x - y = 6$

$7x = 6 + y$

$7x - 6 = y$

Substitute $7x - 6$ for y in the second equation.

$8x - 3y = -8$

$8x - 3(7x - 6) = -8$

$8x - 21x + 18 = -8$

$-13x + 18 = -8$

$-13x = -26$

$x = 2$

Back-substitute.

$y = 7x - 6 = 7(2) - 6 = 14 - 6 = 8$

The solution set is $\{(2,8)\}$.

41. **a.** Substitute $0.375x + 3$ for p in the first equation.

$p = -0.325x + 5.8$

$\overbrace{0.375x + 3}^{p} = -0.325x + 5.8$

$0.375x + 3 = -0.325x + 5.8$

$0.375x + 0.325x + 3 = -0.325x + 0.325x + 5.8$

$0.7x + 3 = 5.8$

$0.7x + 3 - 3 = 5.8 - 3$

$0.7x = 2.8$

$\dfrac{0.7x}{0.7} = \dfrac{2.8}{0.7}$

$x = 4$

Back-substitute to find p.

$p = -0.325x + 5.8$

$p = -0.325(4) + 5.8 = 4.5$

The ordered pair is (4,4.5).
Equilibrium number of workers: 4 million
Equilibrium hourly wage: $4.50

b. If workers are paid $4.50 per hour, there will be 4 million available workers and 4 million workers will be hired. In this state of market equilibrium, there is no unemployment.

c. $p = -0.325x + 5.8$

$5.15 = -0.325x + 5.8$

$0.65 = -0.325x$

$\dfrac{-0.65}{-0.325} = \dfrac{-0.325x}{-0.325}$

$2 = x$

At $5.15 per hour, 2 million workers will be hired.

d. $p = 0.375x + 3$

$5.15 = 0.375x + 3$

$2.15 = 0.375x$

$\dfrac{2.15}{0.375} = \dfrac{0.375x}{0.375}$

$x \approx 5.7$

At $5.15 per hour, there will be about 5.7 million available workers.

e. $5.7 - 2 = 3.7$

At $5.15 per hour, there will be about 3.7 million more people looking for work than employers are willing to hire.

43. – 47. Answers will vary.

49. does not make sense; Explanations will vary. Sample explanation: Solving for x in the second equation will allow us to avoid fractions.

51. does not make sense; Explanations will vary. Sample explanation: Equilibrium is the point at which demand is equal to supply.

53. true

55. false; Changes to make the statement true will vary. A sample change is: Replace y in the second equation with $2x - 5$.

57. $y = mx + 3$

$5x - 2y = 7$

Start by writing the second equation in slope-intercept form.

$5x - 2y = 7$

$-2y = -5x + 7$

$y = \dfrac{5}{2}x - \dfrac{7}{2}$

The system will be inconsistent if the graphs of the two equations have the same slope and different y-intercepts. The y-intercepts are different. Therefore, the system will be inconsistent if $m = \dfrac{5}{2}$.

58. $4x + 6y = 12$

x-intercept:

$4x + 6y = 12$

$4x + 6(0) = 12$

$4x = 12$

$x = 3$

y-intercept:

$4x + 6y = 12$

$4(0) + 6y = 12$

$6y = 12$

$y = 2$

Checkpoint:

$4x + 6y = 12$

$4(-3) + 6y = 12$

$-12 + 6y = 12$

$6y = 24$

$y = 4$

Draw a line through (3,0), (0, 2), and (−3, 4).

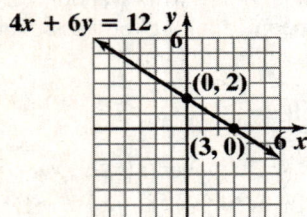

59. $4(x+1) = 25 + 3(x-3)$

$4x + 4 = 25 + 3x - 9$

$x + 4 = 16$

$x = 12$

The solution set is $\{12\}$.

60. The integers in the given set are −73, 0, and $\dfrac{3}{1} = 3$.

61. $3x + 2y = 48$

$3x + 2(12) = 48$

$3x + 24 = 48$

$3x = 24$

$x = 8$

$9x - 8y = -24$

$9x - 8(12) = -24$

$9x - 96 = -24$

$9x = 72$

$x = 8$

The same value of x is obtained in both equations, so (8,12) is the solution.

62. $-14y = -168$

$\dfrac{-14y}{-14} = \dfrac{-168}{-14}$

$y = 12$

The solution set is $\{12\}$.

63. $x - 5y = 3$

$-4(x - 5y) = -4(3)$

$-4x + 20y = -12$

4.3 Check Points

1. $x + y = 5$

$x - y = 9$

Add the equations to eliminate the y-terms.

$x + y = 5$

$\underline{x - y = 9}$

$2x \quad = 14$

Now solve for x.

$2x = 14$

$x = 7$

Back-substitute into either of the original equations to solve for y.

$x + y = 5$

$7 + y = 5$

$y = -2$

The solution set is $\{(7,-2)\}$.

2. $4x - y = 22$

$3x + 4y = 26$

Multiply each term of the first equation by 4 and add the equations to eliminate y.

$16x - 4y = 88$

$\underline{3x + 4y = 26}$

$19x \quad\quad = 114$

$x = 6$

Back-substitute into either of the original equations to solve for y.

$4x - y = 22$

$4(6) - y = 22$

$24 - y = 22$

$-y = -2$

$y = 2$

The solution set is $\{(6, 2)\}$.

3. $4x + 5y = 3$

$2x - 3y = 7$

Multiply each term of the second equation by -2 and add the equations to eliminate x.

$4x + 5y = 3$

$\underline{-4x + 6y = -14}$

$11y = -11$

$y = -1$

Back-substitute into either of the original equations to solve for x.

$2x - 3y = 7$

$2x - 3(-1) = 7$

$2x + 3 = 7$

$2x = 4$

$x = 2$

The solution set is $\{(2, -1)\}$.

4. $2x = 9 + 3y$

$4y = 8 - 3x$

Rewrite each equation in the form $Ax + By = C$.

$2x - 3y = 9$

$3x + 4y = 8$

Multiply the top equation by 4 and multiply the bottom equation by 3.

$8x - 12y = 36$

$\underline{9x + 12y = 24}$

$17x \quad\quad = 60$

$x = \dfrac{60}{17}$

Back-substitution of $\dfrac{60}{17}$ to find y would cause cumbersome arithmetic.

Instead, use the system that is in the form $Ax + By = C$ to eliminate x and find y.

$2x - 3y = 9$

$3x + 4y = 8$

Multiply the top equation by -3 and multiply the bottom equation by 2.

$-6x + 9y = -27$

$\underline{6x + 8y = 16}$

$17y = -11$

$y = \dfrac{-11}{17}$

The solution set is $\left\{ \left(\dfrac{60}{17}, -\dfrac{11}{17} \right) \right\}$.

5. $x + 2y = 4$

$3x + 6y = 13$

Multiply the first equation by -3.

$-3x - 6y = -12$

$\underline{3x + 6y = 13}$

$0 = 1, \text{ false}$

The false statement indicates that the system is inconsistent and has no solution.

The solution set is $\{ \ \}$.

6. $x - 5y = 7$

$3x - 15y = 21$

Multiply the first equation by -3.

$-3x + 15y = -21$

$\underline{3x - 15y = 21}$

$0 = 0, \text{ true}$

The true statement indicates that the system has infinitely many solutions.

The solution set is $\{(x, y) | x - 5y = 7\}$ or

$\{(x, y) | 3x - 15y = 21\}$.

4.3 Concept and Vocabulary Check

1. -3

2. -2

3. -2

4. 3

4.3 Exercise Set

1. $x + y = -3$
   $x - y = 11$
   Add the equations to eliminate the *y*-terms.
   $$x + y = -3$$
   $$\underline{x - y = 11}$$
   $$2x \quad = 8$$
   Now solve for *x*.
   $$2x = 8$$
   $$x = 4$$
   Back-substitute into either of the original equations to solve for *y*.
   $$x + y = -3$$
   $$4 + y = -3$$
   $$y = -7$$
   The solution set is $\{(4, -7)\}$.

3. $2x + 3y = 6$
   $$\underline{2x - 3y = 6}$$
   $$4x = 12$$
   $$x = 3$$
   Back-substitute into either of the original equations to solve for *y*.
   $$2x + 3y = 6$$
   $$2(3) + 3y = 6$$
   $$3y = 0$$
   $$y = 0$$
   The solution set is $\{(3, 0)\}$.

5. $x + 2y = 7$
   $$\underline{-x + 3y = 18}$$
   $$5y = 25$$
   $$y = 5$$
   Back-substitute into either of the original equations to solve for *x*.
   $$x + 2y = 7$$
   $$x + 2(5) = 7$$
   $$x + 10 = 7$$
   $$x = -3$$
   The solution set is $\{(-3, 5)\}$.

7. $5x - y = 14$
   $$\underline{-5x + 2y = -13}$$
   $$y = 1$$
   Back-substitute into either of the original equations to solve for *x*.
   $$5x - (1) = 14$$
   $$5x = 15$$
   $$x = 3$$
   The solution set is $\{(3, 1)\}$.

9. $3x + y = 7$
   $2x - 5y = -1$
   Multiply each term of the first equation by 5 and add the equations to eliminate *y*.
   $$15x + 5y = 35$$
   $$\underline{2x - 5y = -1}$$
   $$17x \quad = 34$$
   $$x \quad = 2$$
   Back-substitute into either of the original equations to solve for *y*.
   $$3x + y = 7$$
   $$3(2) + y = 7$$
   $$6 + y = 7$$
   $$y = 1$$
   The solution set is $\{(2, 1)\}$.

11. $x + 3y = 4$
    $4x + 5y = 2$
    Multiply each term of the first equation by -4 and add the equations to eliminate *x*.
    $$-4x - 12y = -16$$
    $$\underline{4x + 5y = 2}$$
    $$-7y = -14$$
    $$y = 2$$
    Back-substitute into either of the original equations to solve for *x*.
    $$x + 3y = 4$$
    $$x + 3(2) = 4$$
    $$x + 6 = 4$$
    $$x = -2$$
    The solution set is $\{(-2, 2)\}$.

13. $-3x + 7y = 14$
  $2x - y = -13$

Multiply each term of the second equation by 7 and add the equations to eliminate y.

$-3x + 7y = 14$

$\underline{14x - 7y = -91}$

$11x = -77$

$x = -7$

Back-substitute into either of the original equations to solve for y.

$2x - y = -13$

$2(-7) - y = -13$

$-14 - y = -13$

$-y = 1$

$y = -1$

The solution set is $\{(-7, -1)\}$.

15. $3x - 14y = 6$
  $5x + 7y = 10$

Multiply each term of the second equation by 2 and add the equations to eliminate y.

$3x - 14y = 6$

$\underline{10x + 14y = 20}$

$13x = 26$

$x = 2$

Back-substitute into either of the original equations to solve for y.

$5x + 7y = 10$

$5(2) + 7y = 10$

$10 + 7y = 10$

$7y = 0$

$y = 0$

The solution set is $\{(2, 0)\}$.

17. $3x - 4y = 11$
  $2x + 3y = -4$

Multiply the first equation by 3, and the second equation by 4.

$9x - 12y = 33$

$\underline{8x + 12y = -16}$

$17x = 17$

$x = 1$

Back-substitute into either of the original equations to solve for y.

$2x + 3y = -4$

$2(1) + 3y = -4$

$3y = -6$

$y = -2$

The solution set is $\{(1, -2)\}$.

19. $3x + 2y = -1$
  $-2x + 7y = 9$

Multiply the first equation by 2 and the second equation by 3.

$6x + 4y = -2$

$\underline{-6x + 21y = 27}$

$25y = 25$

$y = 1$

Back-substitute into either of the original equations to solve for x.

$3x + 2(1) = -1$

$3x = -3$

$x = -1$

The solution set is $\{(-1, 1)\}$.

21. $3x = 2y + 7$
  $5x = 2y + 13$

Rewrite each equation in the form $Ax + By = C$.

$3x - 2y = 7$

$5x - 2y = 13$

Multiply the first equation by -1 and add the equations to eliminate y.

$-3x + 2y = -7$

$\underline{5x - 2y = 13}$

$2x = 6$

$x = 3$

Back-substitute into either of the original equations to solve for y.

$3x = 2y + 7$

$3(3) = 2y + 7$

$2 = 2y$

$1 = y$

The solution set is $\{(3, 1)\}$.

23. $2x = 3y - 4$

$-6x + 12y = 6$

Rewrite the first equation in the form $Ax + By = C$.

$2x - 3y = -4$

Multiply the first equation by 3 and add to eliminate x.

$$\begin{array}{r} 6x - 9y = -12 \\ \underline{-6x + 12y = 6} \\ 3y = -6 \\ y = -2 \end{array}$$

Back-substitute into either of the original equations to solve for x.

$2x = 3(-2) - 4$

$2x = -6 - 4$

$2x = -10$

$x = -5$

The solution set is $\{(-5, -2)\}$.

25. $2x - y = 3$

$4x + 4y = -1$

Multiply the first equation by 4 and add to eliminate y.

$$\begin{array}{r} 8x - 4y = 12 \\ \underline{4x + 4y = -1} \\ 12x = 11 \\ x = \dfrac{11}{12} \end{array}$$

Instead of back-substituting $\dfrac{11}{12}$ and working with fractions, go back to the original system. Multiply the first equation by -2 and add the equations to eliminate x.

$$\begin{array}{r} -4x + 2y = -6 \\ \underline{4x + 4y = -1} \\ 6y = -7 \\ y = -\dfrac{7}{6} \end{array}$$

The solution set is $\left\{\left(\dfrac{11}{12}, -\dfrac{7}{6}\right)\right\}$.

27. $4x = 5 + 2y$

$2x + 3y = 4$

Rewrite the first equation in the form $Ax + By = C$, and multiply the second equation by -2.

$$\begin{array}{r} 4x - 2y = 5 \\ \underline{-4x - 6y = -8} \\ -8y = -3 \\ y = \dfrac{3}{8} \end{array}$$

Instead of back-substituting $\dfrac{3}{8}$ and working with fractions, go back to the original system. Use the rewritten form of the first equation, and multiply by -3. Solve by addition.

$$\begin{array}{r} -12x + 6y = -15 \\ \underline{-4x - 6y = -8} \\ -16x = -23 \\ x = \dfrac{23}{16} \end{array}$$

The solution set is $\left\{\left(\dfrac{23}{16}, \dfrac{3}{8}\right)\right\}$.

29. $3x - y = 1$

$3x - y = 2$

Multiply the first equation by -1.

$$\begin{array}{r} -3x + y = -1 \\ \underline{3x - y = 2} \\ 0 = 1, \text{ false} \end{array}$$

The false statement indicates that the system is inconsistent and has no solution.

The solution set is $\{\ \}$.

31. $x + 3y = 2$

$3x + 9y = 6$

Multiply the first equation by -3.

$$\begin{array}{r} -3x - 9y = -6 \\ \underline{3x + 9y = 6} \\ 0 = 0, \text{ true} \end{array}$$

The true statement indicates that the system has infinitely many solutions.

The solution set is $\{(x, y) \mid x + 3y = 2\}$ or

$\{(x, y) \mid 3x + 9y = 6\}$.

33. $7x - 3y = 4$
            $-14x + 6y = -7$
Multiply the first equation by 2.
            $14x - 6y = 8$
            $\underline{-14x + 6y = -7}$
                    $0 = 1,$ false
The false statement indicates that the system is inconsistent and has no solution.
The solution set is $\{\ \}$.

35. $5x + y = 2$
         $3x + y = 1$
Multiply the second equation by -1.
            $5x + y = 2$
            $\underline{-3x - y = -1}$
                 $2x = 1$
                 $x = \dfrac{1}{2}$

Back-substitute $\dfrac{1}{2}$ for x and solve for y.

$3\left(\dfrac{1}{2}\right) + y = 1$

                 $y = -\dfrac{1}{2}$

The solution set is $\left\{\left(\dfrac{1}{2}, -\dfrac{1}{2}\right)\right\}$.

37. $x = 5 - 3y$
         $2x + 6y = 10$
Rewrite the first equation in the form
$Ax + By = C$, and multiply the second equation by -2.
            $-2x - 6y = -10$
            $\underline{2x + 6y = 10}$
                 $0 = 0,$ true
The true statement indicates that the system has infinitely many solutions.
The solution set is $\{(x, y) \mid x = 5 - 3y\}$ or
$\{(x, y) \mid 2x + 6y = 10\}$.

39. $4(3x - y) = 0$
            $3(x + 3) = 10y$
Rewrite both equations.
         $12x - 4y = 0$
         $3x - 10y = -9$
Multiply the second equation by -4 and add the equations to eliminate x.
            $12x - 4y = 0$
            $\underline{-12x + 40y = 36}$
                  $36y = 36$
                    $y = 1$
Back-substitute 1 for y in one of the original equations and solve for x.
            $12x - 4y = 0$
            $12x - 4(1) = 0$
                 $12x = 4$
                    $x = \dfrac{1}{3}$

The solution set is $\left\{\left(\dfrac{1}{3}, 1\right)\right\}$.

41. $x + y = 11$
            $\dfrac{x}{5} + \dfrac{y}{7} = 1$
Multiply the second equation by the LCD, 35, to clear fractions.
$35\left(\dfrac{x}{5} + \dfrac{y}{7}\right) = 35(1)$
            $7x + 5y = 35$
Now solve the system.
            $x + y = 11$
            $7x + 5y = 35$
Multiply the top equation by -5 and add the result to the second equation.
            $-5x - 5y = -55$
            $\underline{7x + 5y = 35}$
                 $2x = -20$
                  $x = -10$
Back-substitute to find y.
            $-10 + y = 11$
                    $y = 21$
The solution set is $\{(-10, 21)\}$.

43. $\dfrac{4}{5}x - y = -1$

$\dfrac{2}{5}x + y = 1$

Multiply both equations by 5 to clear fractions.

$4x - 5y = -5$

$\underline{2x + 5y = 5}$

$6x = 0$

$x = 0$

Back-substitute 0 for x and solve for y.

$\dfrac{2}{5}(0) + y = 1$

$\phantom{\dfrac{2}{5}(0)+}y = 1$

The solution set is $\{(0,1)\}$.

45. $3x - 2y = 8$

$x = -2y$

The substitution method is a good choice because the second equation is already solved for x.

Substitute $-2y$ for x in the first equation

$3x - 2y = 8$

$3(-2y) - 2y = 8$

$-6y - 2y = 8$

$-8y = 8$

$y = -1$

Back-substitute -1 for y in the second equation.

$x = -2y = -2(-1) = 2$

The solution set is $\{(2,-1)\}$.

47. $3x + 2y = -3$

$2x - 5y = 17$

The addition method is a good choice because both equations are written in the form $Ax + By = C$.

Multiply the first equation by 2 and the second equation by -3.

$6x + 4y = -6$

$\underline{-6x + 15y = -51}$

$19y = -57$

$y = -3$

Back-substitute -3 for y and solve for x.

$3x + 2(-3) = -3$

$3x - 6 = -3$

$3x = 3$

$x = 1$

The solution set is $\{(1,-3)\}$.

49. $3x - 2y = 6$

$y = 3$

The substitution method is a good choice because the second equation is already solved for y.

Substitute 3 for y in the first equation.

$3x - 2y = 6$

$3x - 2(3) = 6$

$3x - 6 = 6$

$3x = 12$

$x = 4$

It is not necessary to back-substitute to find the value of y because $y = 3$ is one of the equations of the given system.

The solution set is $\{(4,3)\}$.

51. $y = 2x + 1$

$y = 2x - 3$

The substitution method is a good choice, because both equations are already solved for y. Substitute $2x + 1$ for y in the second equation.

$y = 2x - 3$

$2x + 1 = 2x - 3$

$2x + 1 - 2x = 2x - 3 - 2x$

$1 = -3,\ \text{ false}$

The false statement indicates that the system has no solution.

The solution set is $\{\ \}$.

53. $2(x + 2y) = 6$

$3(x + 2y - 3) = 0$

The addition method is a good choice since the equations can easily be simplified to give equations of the form $Ax + By = C$.

$2x + 4y = 6$ $\qquad\qquad 3x + 6y - 9 = 0$

$\ 3x + 6y = 9$

Solve the resulting system.

$2x + 4y = 6$

$3x + 6y = 9$

Multiply the first equation by -3 and the second by 2 and solve by addition.

$-6x - 12y = -18$

$\underline{6x + 12y = 18}$

$0 = 0,\ \text{true}$

The true statement indicates that the system has infinitely many solutions.

The solution set is $\{(x, y)\,|\,2(x + 2y) = 6\}$ or

$\{(x, y)\,|\,3(x + 2y - 3) = 0\}$.

55.
$$3y = 2x$$
$$2x + 9y = 24$$

The substitution method is a good choice because the first equation can easily be solved for one of the variables. Solve this equation for y.
$$3y = 2x$$
$$y = \frac{2}{3}x$$

Substitute $\frac{2}{3}x$ for y in the second equation.
$$2x + 9y = 24$$
$$2x + 9\left(\frac{2}{3}x\right) = 24$$
$$2x + 6x = 24$$
$$8x = 24$$
$$x = 3$$

Back-substitute 3 for x in the equation, $y = \frac{2}{3}x$.
$$y = \frac{2}{3}x = \frac{2}{3}(3) = 2$$

The solution set is $\{(3, 2)\}$.

57.
$$5x + y = 14$$
$$4x - y = 4$$

Add the equations.
$$5x + y = 14$$
$$\underline{4x - y = 4}$$
$$9x \quad = 18$$
$$x = 2$$

Back-substitute into the first equation of the original system.
$$5x + y = 14$$
$$5(2) + y = 14$$
$$10 + y = 14$$
$$y = 4$$

The numbers are 2 and 4.

59.
$$4x - 3y = 0$$
$$x + y = -7$$

Multiply the second equation by 3 and then add the equations.
$$4x - 3y = 0$$
$$\underline{3x + 3y = -21}$$
$$7x \quad = -21$$
$$x = -3$$

Back-substitute into the second equation of the original system.
$$x + y = -7$$
$$-3 + y = -7$$
$$y = -4$$

The numbers are -3 and -4.

61.
$$\frac{3x}{5} + \frac{4y}{5} = 1$$
$$\frac{x}{4} - \frac{3y}{8} = -1$$

Multiply the first equation by 5 and the second equation by 8 to clear fractions.
$$\frac{3x}{5} + \frac{4y}{5} = 1 \qquad \frac{x}{4} - \frac{3y}{8} = -1$$
$$3x + 4y = 5 \qquad 2x - 3y = -8$$

The addition method is a good choice since both equations are of the form $Ax + By = C$.
$$3x + 4y = 5$$
$$2x - 3y = -8$$

Multiply the first equation by 3 and the second equation by 4.
$$9x + 12y = \quad 15$$
$$\underline{8x - 12y = -32}$$
$$17x \quad = -17$$
$$x \quad = -1$$

Back-substitute -1 for x in the equation and solve for y.
$$3x + 4y = 5$$
$$3(-1) + 4y = 5$$
$$-3 + 4y = 5$$
$$4y = 8$$
$$y = 2$$

The solution set is $\{(-1, 2)\}$.

63. $5(x+1) = 7(y+1) - 7$

$6(x+1) + 5 = 5(y+1)$

Simplify both equations.

$5(x+1) = 7(y+1) - 7$

$5x + 5 = 7y + 7 - 7$

$5x + 5 = 7y$

$5x - 7y + 5 = 0$

$5x - 7y = -5$

$6(x+1) + 5 = 5(y+1)$

$6x + 6 + 5 = 5y + 5$

$6x + 11 = 5y + 5$

$6x - 5y + 11 = 5$

$6x - 5y = -6$

Solve the rewritten system.

$5x - 7y = -5$

$6x - 5y = -6$

Multiply the first equation by -6 and the second equation by 5, and solve by addition.

$-30x + 42y = 30$

$\underline{30x - 25y = -30}$

$17y = 0$

$y = 0$

Back-substitute 0 for y to find x.

$5x - 7y = -5$

$5x - 7(0) = -5$

$5x - 0 = -5$

$5x = -5$

$x = -1$

The solution set is $\{(-1, 0)\}$.

65. $0.4x + y = 2.2$

$0.5x - 1.2y = 0.3$

Multiply the first equation by 1.2 and solve by addition.

$0.48x + 1.2y = 2.64$

$\underline{0.50x - 1.2y = 0.30}$

$0.98x = 2.94$

$x = 3$

Back-substitute 3 for x to find y.

$0.4x + y = 2.2$

$0.4(3) + y = 2.2$

$1.2 + y = 2.2$

$y = 1$

The solution set is $\{(3, 1)\}$.

67. $\dfrac{x}{2} = \dfrac{y+8}{3}$

$\dfrac{x+2}{2} = \dfrac{y+11}{3}$

Simplify the first equation.

$\dfrac{x}{2} = \dfrac{y+8}{3}$

$3x = 2(y+8)$

$3x = 2y + 16$

$3x - 2y = 16$

Simplify the second equation.

$\dfrac{x+2}{2} = \dfrac{y+11}{3}$

$3(x+2) = 2(y+11)$

$3x + 6 = 2y + 22$

$3x - 2y + 6 = 22$

$3x - 2y = 16$

When simplified, the equations are the same. This means that the system is dependent and there are an infinite number of solutions.

The solution set is $\left\{ (x, y) \left| \dfrac{x}{2} = \dfrac{y+8}{3} \right. \right\}$ or

$\left\{ (x, y) \left| \dfrac{x+2}{2} = \dfrac{y+11}{3} \right. \right\}$.

69. $-0.45x + y = 0.8$

$-1.5x + y = 2.6$

Multiply the second equation by -1, then add the equations.

$-0.45x + y = 0.8$

$\underline{0.15x - y = -2.6}$

$-0.3x = -1.8$

$x = 6$

Back-substitute 6 for x and solve for y.

$-0.45x + y = 0.8$

$-0.45(6) + y = 0.8$

$-2.7 + y = 0.8$

$y = 3.5$

The solution is $(6, 3.5)$. This means that in week 6, procrastinating and non procrastinating students reported the same number of symptoms of physical illness, namely 3.5 symptoms.

71. – 75. Answers will vary.

77. does not make sense; Explanations will vary. Sample explanation: The addition method does not involve graphing.

79. does not make sense; Explanations will vary.
Sample explanation: When one of equations has a
variable on one side by itself, it is typically best to
use the substitution method.

81. false; Changes to make the statement true will vary.
A sample change is: If $(2, -2)$ satisfies

$Ax + 2y = 2$ and $2x + By = 10$, then A and B can be
found by substitution.
Find A.

$$Ax + 2y = 2$$
$$A(2) + 2(-2) = 2$$
$$2A - 4 = 2$$
$$2A = 6$$
$$A = 3$$

Find B.

$$2x + By = 10$$
$$2(2) + B(-2) = 10$$
$$4 - 2B = 10$$
$$-2B = 6$$
$$B = -3$$

83. false; Changes to make the statement true will vary.
A sample change is: After these multiplications, the
coefficients of x will be the same. Thus, adding
them will not eliminate them.

85.
$$Ax - 3y = 16$$
$$A(5) - 3(-2) = 16$$
$$5A + 6 = 16$$
$$5A = 10$$
$$A = 2$$

$$3x + By = 7$$
$$3(5) + B(-2) = 7$$
$$15 - 2B = 7$$
$$-2B = -8$$
$$B = 4$$

86. Let $x =$ the unknown number.

$$5x = x + 40$$
$$4x = 40$$
$$x = 10$$

The number is 10.

87. Because the x-coordinate is negative and the y-
coordinate is positive, $\left(-\dfrac{3}{2}, 15\right)$ is located in
quadrant II.

88.
$$29,700 + 150x = 5000 + 1100x$$
$$29,700 - 950x = 5000$$
$$-950x = -24,700$$
$$x = 26$$

The solution set is $\{26\}$.

89. a. $\begin{aligned} x + y &= 28 \\ x - y &= 6 \end{aligned}$

b. $\begin{aligned} x + y &= 28 \\ \underline{x - y} &= \underline{6} \\ 2x &= 34 \\ x &= 17 \end{aligned}$

Substitute 17 for x to find y.

$$x + y = 28$$
$$17 + y = 28$$
$$y = 11$$

The numbers are 17 and 11.

90. $3x + 2y$

91. a. $\$20 + \$0.05(200) = \$30$

b. $y = 20 + 0.05x$

Mid-Chapter Check Points

1. $\begin{aligned} 3x + 2y &= 6 \\ 2x - y &= 4 \end{aligned}$

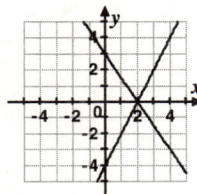

The lines intersect at (2, 0).
The solution set is $\left\{(2, 0)\right\}$.

2. $\begin{aligned} y &= 2x - 1 \\ y &= 3x - 2 \end{aligned}$

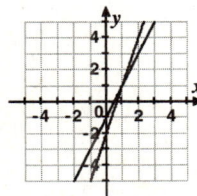

The lines intersect at (1, 1).
The solution set is $\left\{(1, 1)\right\}$.

3.
$$y = 2x - 1$$
$$6x - 3y = 12$$

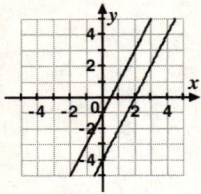

Since the lines are parallel, the system is inconsistent and there is no solution.
The solution set is $\{\ \}$.

4. $5x - 3y = 1$
$$y = 3x - 7$$

Substitute $3x - 7$ for y in the first equation and solve for x.
$$5x - 3y = 1$$
$$5x - 3(3x - 7) = 1$$
$$5x - 9x + 21 = 1$$
$$-4x + 21 = 1$$
$$-4x = -20$$
$$x = 5$$

Back-substitute to find y.
$$y = 3x - 7 = 3(5) - 7 = 15 - 7 = 8$$

The solution set is $\{(5, 8)\}$.

5. $6x + 5y = 7$
$$3x - 7y = 13$$

Multiply the second equation by -2 and add to eliminate x.
$$\begin{array}{r} 6x + 5y = 7 \\ -6x + 14y = -26 \\ \hline 19y = -19 \\ y = -1 \end{array}$$

Back-substitute -1 for y and solve for x.
$$6x + 5y = 7$$
$$6x + 5(-1) = 7$$
$$6x - 5 = 7$$
$$6x = 12$$
$$x = 2$$
The solution set is $\{(2, -1)\}$.

6. $x = \dfrac{y}{3} - 1$
$$6x + y = 21$$

Substitute $\dfrac{y}{3} - 1$ for x in the second equation and solve for y.
$$6x + y = 21$$
$$6\left(\dfrac{y}{3} - 1\right) + y = 21$$
$$2y - 6 + y = 21$$
$$3y - 6 = 21$$
$$3y = 27$$
$$y = 9$$

Back-substitute to find x.
$$x = \dfrac{y}{3} - 1 = \dfrac{9}{3} - 1 = 3 - 1 = 2$$
The solution set is $\{(2, 9)\}$.

7. $3x - 4y = 6$
$$5x - 6y = 8$$

Multiply the first equation by -5, the second equation by 3 and add to eliminate x.
$$\begin{array}{r} -15x + 20y = -30 \\ 15x - 18y = 24 \\ \hline 2y = -6 \\ y = -3 \end{array}$$

Back-substitute -3 for y and solve for x.
$$3x - 4y = 6$$
$$3x - 4(-3) = 6$$
$$3x + 12 = 6$$
$$3x = -6$$
$$x = -2$$
The solution set is $\{(-2, -3)\}$.

8. $3x - 2y = 32$
$$\dfrac{x}{5} + 3y = -1$$

Multiply the second equation by 5 to clear the fraction.
$$5\left(\dfrac{x}{5} + 3y\right) = 5(-1)$$
$$x + 15y = -5$$
The system is as follows.
$$3x - 2y = 32$$
$$x + 15y = -5$$

Multiply the second equation by -3 and solve by addition.

$$3x - 2y = 32$$
$$\underline{-3x - 45y = 15}$$
$$-47y = 47$$
$$y = -1$$

Back-substitute -1 for y and solve for x.

$$x + 15y = -5$$
$$x + 15(-1) = -5$$
$$x - 15 = -5$$
$$x = 10$$

The solution set is $\{(10, -1)\}$.

9. $x - y = 3$

$2x = 4 + 2y$

Solve the first equation for x.

$$x - y = 3$$
$$x = y + 3$$

Substitute $y + 3$ for x in the second equation and solve for y.

$$2x = 4 + 2y$$
$$2(y + 3) = 4 + 2y$$
$$2y + 6 = 4 + 2y$$
$$6 = 4$$

Notice that y has also been eliminated. The false statement $6 = 4$ indicates that the system is inconsistent and has no solution. The solution set is $\{\ \}$.

10. $x = 2(y - 5)$

$4x + 40 = y - 7$

Substitute $2(y - 5)$ for x in the second equation and solve for y.

$$4x + 40 = y - 7$$
$$4(2(y - 5)) + 40 = y - 7$$
$$4(2y - 10) + 40 = y - 7$$
$$8y - 40 + 40 = y - 7$$
$$8y = y - 7$$
$$7y = -7$$
$$y = -1$$

Back-substitute -1 for y and solve for x.

$$x = 2(y - 5)$$
$$= 2(-1 - 5) = 2(-6) = -12$$

The solution set is $\{(-12, -1)\}$.

11. $y = 3x - 2$

$y = 2x - 9$

Substitute $3x - 2$ for y in the second equation and solve for x.

$$y = 2x - 9$$
$$3x - 2 = 2x - 9$$
$$x - 2 = -9$$
$$x = -7$$

Back-substitute -7 for x and solve for y.

$$y = 3(-7) - 2 = -21 - 2 = -23$$

The solution set is $\{(-7, -23)\}$.

12. $2x - 3y = 4$

$3x + 4y = 0$

Multiply the first equation by 4, the second equation by 3 and add to eliminate y.

$$8x - 12y = 16$$
$$\underline{9x + 12y = 0}$$
$$17x \qquad = 16$$
$$x \qquad = \frac{16}{17}$$

Back-substitution of $\frac{16}{17}$ to find y would cause cumbersome arithmetic.

Instead, use the addition to eliminate x and find y.

$$2x - 3y = 4$$
$$3x + 4y = 0$$

Multiply the top equation by -3 and multiply the bottom equation by 2.

$$-6x + 9y = -12$$
$$\underline{6x + 8y = \ 0}$$
$$17y = -12$$
$$y = \frac{-12}{17}$$

The solution set is $\left\{ \left(\frac{16}{17}, -\frac{12}{17} \right) \right\}$.

13. $y - 2x = 7$

$4x = 2y - 14$

Solve the first equation for y.

$y - 2x = 7$

$y = 2x + 7$

Substitute $2x + 7$ for y and solve for x.

$4x = 2y - 14$

$4x = 2(2x + 7) - 14$

$4x = 4x + 14 - 14$

$0 = 14 - 14$

$0 = 0$

Notice that x has also been eliminated. The true statement $0 = 0$ indicates that the system is dependent and has infinitely many solutions. The solution set is $\{(x, y) | y - 2x = 7\}$ or

$\{(x, y) | 4x = 2y - 14\}$.

14. $4(x + 3) = 3y + 7$

$2(y - 5) = x + 5$

First, rewrite both equations in the form $Ax + By = C$.

$4(x + 3) = 3y + 7$

$4x + 12 = 3y + 7$

$4x - 3y + 12 = 7$

$4x - 3y = -5$

$2(y - 5) = x + 5$

$2y - 10 = x + 5$

$-x + 2y - 10 = 5$

$-x + 2y = 15$

The system is as follows.

$4x - 3y = -5$

$-x + 2y = 15$

Multiply the second equation by 4 and add to eliminate x.

$\begin{array}{r} 4x - 3y = -5 \\ \underline{-4x + 8y = 60} \\ 5y = 55 \\ y = 11 \end{array}$

Back-substitute 11 for y and solve for x.

$-x + 2y = 15$

$-x + 2(11) = 15$

$-x + 22 = 15$

$-x = -7$

$x = 7$

The solution set is $\{(7, 11)\}$.

15. $\dfrac{x}{2} - \dfrac{y}{5} = 1$

$y - \dfrac{x}{3} = 8$

Multiply the first equation by 10 and the second equation by 3 to clear fractions.

$10\left(\dfrac{x}{2}\right) - 10\left(\dfrac{y}{5}\right) = 10(1)$

$5x - 2y = 10$

$3y - 3\left(\dfrac{x}{3}\right) = 3(8)$

$3y - x = 24$

$-x + 3y = 24$

The system is as follows.

$5x - 2y = 10$

$-x + 3y = 24$

Multiply the second equation by 5 and solve by addition.

$\begin{array}{r} 5x - 2y = 10 \\ \underline{-5x + 15y = 120} \\ 13y = 130 \\ y = 10 \end{array}$

Back-substitute and solve for x.

$-x + 3y = 24$

$-x + 3(10) = 24$

$-x + 30 = 24$

$-x = -6$

$x = 6$

The solution set is $\{(6, 10)\}$.

4.4 Check Points

1. Let x = average time per day women spend socializing.
   Let y = average time per day men spend socializing.

$$x + y = 138$$
$$\underline{x - y = 8}$$
$$2x = 146$$
$$x = 73$$

Back-substitute 73 for x to find y.

$$x + y = 138$$
$$73 + y = 138$$
$$y = 65$$

Men average 65 minutes per day socializing and women average 73 minutes.

2. Let x = the number of calories in a Quarter Pounder.
   Let y = the number of calories in a Whopper with cheese.

$$2x + 3y = 2607$$
$$x + y = 1000 + 9$$

Solve the second equation for x.

$$x + y = 1000 + 9$$
$$x = -y + 1009$$

Substitute $-y + 1009$ for x to find y.

$$2x + 3y = 2607$$
$$2(\overbrace{-y + 1009}^{x}) + 3y = 2607$$
$$2(-y + 1009) + 3y = 2607$$
$$-2y + 2018 + 3y = 2607$$
$$y + 2018 = 2607$$
$$y = 589$$

Back substitute to find x.

$$x = -y + 1009 = -589 + 1009 = 420$$

There are 420 calories in a Quarter Pounder and 589 calories in a Whopper with cheese.

3. Let x = the length of the lot.
   Let y = the width of the lot.
   Use the formula for the perimeter of a rectangle to write the first equation.

$$P = 2l + 2w$$
$$360 = 2x + 2y$$

Use the other information in the problem to write the second equation.

$$20x + 8 \cdot 2y = 3280$$

The two equations form the system.

$$2x + 2y = 360$$
$$20x + 16y = 3280$$

Multiply the first equation by -8 and add the result to the second equation.

$$-16x - 16y = -2880$$
$$\underline{20x + 16y = 3280}$$
$$4x = 400$$
$$x = 100$$

Back-substitute to find y.

$$2x + 2y = 360$$
$$2(100) + 2y = 360$$
$$200 + 2y = 360$$
$$2y = 160$$
$$y = 80$$

The length is 100 feet and the width is 80 feet.

4. Let x = the number of years the heating system is used.
   Let y = the total cost of the heating system.

$$y = 5000 + 1100x$$
$$y = 12,000 + 700x$$

Solve by the substitution method.

$$y = 12,000 + 700x$$

$$\overbrace{5000 + 1100x}^{y} = 12,000 + 700x$$
$$5000 + 1100x = 12,000 + 700x$$
$$5000 + 400x = 12,000$$
$$400x = 7000$$
$$x = 17.5$$

Back-substitute to find y.

$$y = 5000 + 1100(17.5)$$
$$y = 24,250$$

After 17.5 years, the total costs for the two systems will be the same, namely $24,250.

5. Let x = the amount invested at 9%.
    Let y = the amount invested at 12%.
    $$x + y = 25,000$$
    $$0.09x + 0.12y = 2550$$
    This system can be solved by substitution.
    Solve for y in terms of x.
    $$x + y = 25,000$$
    $$y = -x + 25,000$$
    Substitute this value into the other equation.
    $$0.09x + 0.12y = 2550$$

    $$0.09x + 0.12(\overbrace{-x + 25,000}^{y}) = 2550$$
    $$0.09x - 0.12x + 3000 = 2550$$
    $$-0.03x + 3000 = 2550$$
    $$-0.03x = -450$$
    $$x = 15,000$$
    Back-substitute to find y.
    $$y = -x + 25,000$$
    $$y = -(15,000) + 25,000$$
    $$y = 10,000$$
    There was $15,000 invested at 9% and $10,000 invested at 12%.

6. Let x = the number of ounces of 10% acid solution.
    Let y = the number of ounces of 60% acid solution.
    $$x + y = 50$$
    $$0.10x + 0.60y = 0.30(50)$$
    This system can be solved by substitution.
    Solve for y in terms of x.
    $$x + y = 50$$
    $$y = -x + 50$$
    Substitute this value into the other equation.
    $$0.10x + 0.60y = 0.30(50)$$
    $$0.10x + 0.60y = 15$$

    $$0.10x + 0.60(\overbrace{-x + 50}^{y}) = 15$$
    $$0.10x - 0.60x + 30 = 15$$
    $$-0.50x + 30 = 15$$
    $$-0.50x = -15$$
    $$x = 30$$
    Back-substitute to find y.
    $$y = -x + 50$$
    $$y = -(30) + 50$$
    $$y = 20$$
    The chemist should mix 30 milliliters of the 10% acid solution and 20 milliliters of the 60% acid solution.

4.4 Concept and Vocabulary Check

1. $5x + 6y$

2. $10 \cdot 2x + 15 \cdot 2y$ or $20x + 30y$

3. $25,600 + 225x$

4. $0.4x + 0.05y$

5. $0.07x + 0.15y$

4.4 Exercise Set

1. Let x = one number.
    Let y = the other number.
    $$\begin{aligned} x + y &= 17 \\ \underline{x - y} &= \underline{-3} \\ 2x\phantom{{}+y} &= 14 \\ x\phantom{{}+2y} &= 7 \end{aligned}$$
    Back-substitute 7 for x to find y.
    $$x + y = 17$$
    $$7 + y = 17$$
    $$y = 10$$
    The numbers are 7 and 10.

3. Let x = one number.
    Let y = the other number.
    $$3x - y = -1$$
    $$x + 2y = 23$$
    Solve the second equation for x.
    $$x + 2y = 23$$
    $$x = -2y + 23$$
    Substitute $-2y + 23$ for x to find y.
    $$3x - y = -1$$
    $$3(-2y + 23) - y = -1$$
    $$-6y + 69 - y = -1$$
    $$-7y + 69 = -1$$
    $$-7y = -70$$
    $$y = 10$$
    Back substitute 10 for y to find x.
    $$x = -2y + 23 = -2(10) + 23 = 3$$
    The numbers are 3 and 10.

5. Let x = the average number of minutes 20- to 24-year-old women spend grooming.

Let y = the average number of minutes 20- to 24-year-old men spend grooming.

$x + y = 86$

$x - y = 12$

Add the equations to eliminate y.

$x + y = 86$

$\underline{x - y = 12}$

$2x \quad = 98$

$\quad x = 49$

Back-substitute to find y.

$x + y = 86$

$49 + y = 86$

$\quad y = 37$

20- to 24-year-old women averaged 49 minutes grooming and men averaged 37 minutes.

7. Let x = the number of calories in a Mr. Goodbar.

Let y = the number of calories in a Mounds bar.

$x + 2y = 780$

$2x + \ y = 786$

Multiply the bottom equation by –2 and then add the equations to eliminate y.

$\ x + 2y = \quad 780$

$\underline{-4x - 2y = -1572}$

$-3x \quad\quad = -792$

$\quad\quad x = \ 264$

Back-substitute to find y.

$\ x + 2y = 780$

$264 + 2y = 780$

$\quad\quad 2y = 516$

$\quad\quad y = 258$

There are 264 calories in a Mr. Goodbar and 258 calories in a Mounds bar.

9. Let x = the number of Mr. Goodbars.

Let y = the number of Mounds bars.

$x + y = 5$

$16.3x + 14.1y - 70 = 7.1$

Solve the first equation for y in terms of x.

$x + y = 5$

$\quad y = -x + 5$

Substitute $-x + 5$ for y in the second equation.

$16.3x + 14.1y - 70 = 7.1$

$\qquad\qquad\overbrace{\qquad}^{y}$

$16.3x + 14.1(-x + 5) - 70 = 7.1$

$16.3x + 14.1(-x + 5) - 70 = 7.1$

$16.3x - 14.1x + 70.5 - 70 = 7.1$

$\qquad\qquad 2.2x + 0.5 = 7.1$

$\qquad\qquad\quad 2.2x = 6.6$

$\qquad\qquad\qquad x = 3$

Back-substitute to find y.

$x + y = 5$

$3 + y = 5$

$\quad y = 2$

There are 3 Mr. Goodbars and 2 Mounds bars.

11. Let x = the price of one sweater.

Let y = the price of one shirt.

$x + 3y = 42$

$3x + 2y = 56$

Multiply the first equation by –3 and add the result to the second equation.

$-3x - 9y = -126$

$\underline{\ 3x + 2y = \ \ 56}$

$\qquad -7y = -70$

$\qquad\quad y = \ \ 10$

Back-substitute 10 for y and solve for x.

$\quad x + 3y = 42$

$x + 3(10) = 42$

$\quad x + 30 = 42$

$\qquad\quad x = 12$

The price of one sweater is $12 and the price of one shirt is $10.

13. Let x = the length of a badminton court.
Let y = the width of a badminton court.
Use the formula for the perimeter of a rectangle to write the first equation.
$$P = 2l + 2w$$
$$128 = 2x + 2y$$
Use the other information in the problem to write the second equation.
$$6x + 9y = 444$$
The two equations form the system.
$$2x + 2y = 128$$
$$6x + 9y = 444$$
Multiply the first equation by -3 and add the result to the second equation.
$$-6x - 6y = -384$$
$$\underline{6x + 9y = \ 444}$$
$$3y = 60$$
$$y = 20$$
Back-substitute 20 for y and solve for x.
$$2x + 2y = 128$$
$$2x + 2(20) = 128$$
$$2x + 40 = 128$$
$$2x = 88$$
$$x = 44$$
The length is 44 feet and the width is 20 feet, so the dimensions of a standard badminton court are 44 feet by 20 feet.

15. Let x = the length of the lot.
Let y = the width of the lot.
Use the formula for the perimeter of a rectangle to write the first equation.
$$2x + 2y = 320$$
Use the other information in the problem to write the second equation.
$$16x + 5(2y) = 2140$$
These two equations form the system.
$$2x + 2y = 320$$
$$16x + 10y = 2140$$
Multiply the first equation by -5 and add the result to the second equation.
$$-10x - 10y = -1600$$
$$\underline{16x + 10y = \ 2140}$$
$$6x \qquad = \quad 540$$
$$x \qquad = \quad 90$$

Back-substitute 90 for x to find y.
$$2(90) + 2y = 320$$
$$180 + 2y = 320$$
$$2y = 140$$
$$y = 70$$
The length is 90 feet and the width is 70 feet, so the dimensions of the lot are 90 feet by 70 feet.

17. **a.** Let x = the number of minutes of calls.
Let y = the monthly cost of a telephone plan.
Plan A: $y = 20 + 0.05x$
Plan B: $y = 5 + 0.10x$
Solve by substitution. Substitute $5 + 0.10x$ for y in the first equation.
$$5 + 0.10x = 20 + 0.05x$$
$$5 + 0.05x = 20$$
$$0.05x = 15$$
$$x = 300$$
Back-substitute 300 for x.
$$y = 20 + 0.05x = 20 + 0.05(300) = 35$$
The costs for the two plans will be equal for 300 minutes of calls per month. The cost for each plan will be $35.

b. $x = 10(20) = 200$
Plan A: $y = 20 + 0.05(200) = 30$
Plan B: $y = 5 + 0.10(200) = 25$
The monthly cost would be $30 for Plan A and $25 for Plan B, so Plan B should be selected to get the lower cost.

19. Let x = the number of dollars of merchandise purchased in a year.
Let y = the total cost for a year.
Plan A: $y = 100 + 0.80x$
Plan B: $y = 40 + 0.90x$
Substitute $40 = 0.90x$ for y in the first equation and solve for x.
$$40 + 0.90x = 100 + 0.80x$$
$$40 + 0.10x = 100$$
$$0.10x = 60$$
$$x = 600$$
Back-substitute 600 for x to find y.
$$y = 100 + 0.80(600) = 580.$$
If you purchase $600 worth of merchandise, you will pay the $580 under both plans.

21. Let x = the number of adult tickets sold.
Let y = the number of student tickets sold.
$$x + y = 301$$
$$3x + y = 487$$
Multiply the first equation by -1 and add it to the
second equation.
$$-x - y = -301$$
$$\underline{3x + y = 487}$$
$$2x = 186$$
$$x = 93$$
Back-substitute to find y.
$$x + y = 301$$
$$93 + y = 301$$
$$y = 208$$
There were 93 adult tickets sold and 208 student
tickets sold.

23. Let x = the cost of an item from column A.
Let y = the cost of an item from column B.
$$x + y = 5.49$$
$$x + 2y = 6.99$$
Multiply the first equation by -1 and add it to the
second equation.
$$-x - y = -5.49$$
$$\underline{x + 2y = 6.99}$$
$$y = 1.50$$
Back-substitute to find x.
$$x + y = 5.49$$
$$x + 1.50 = 5.49$$
$$x = 3.99$$
The cost of an item from column A is $3.99 and the
cost of an item from column B is $1.50.

25. Let x = the number of servings of macaroni.
Let y = the number of servings of broccoli.
$$3x + 2y = 14$$
$$16x + 4y = 48$$
Multiply the first equation by -2 and add to second
equation.
$$-6x - 4y = -28$$
$$\underline{16x + 4y = 48}$$
$$10x = 20$$
$$x = 2$$

Back-substitute 2 for x to find y.
$$3(2) + 2y = 14$$
$$2y = 8$$
$$y = 4$$
It would take 2 servings of macaroni and 4 servings
of broccoli to get 14 grams of protein and 48 grams
of carbohydrate.

27. The sum of the measures of the three angles of any
triangle is $180°$, so
$$(x + 8y - 1) + (3y + 4) + (7x + 5) = 180.$$
Simplify this equation.
$$8x + 11y + 8 = 180$$
$$8x + 11y = 172$$
The base angles of an isosceles triangle have equal
measures, so
$$3y + 4 = 7x + 5$$
Rewrite this equation in the form $Ax + By = C$.
$$7x + 5 = 3y + 4$$
$$7x - 3y = -1$$
Use the addition method to solve the system.
$$8x + 11y = 172$$
$$7x - 3y = -1$$
Multiply the first equation by 3 and the second
equation by 11; then add the results.
$$24x + 33y = 516$$
$$\underline{77x - 33y = -11}$$
$$101x = 505$$
$$x = 5$$
Back-substitute 5 for x to find y.
$$7(5) - 3y = -1$$
$$35 - 3y = -1$$
$$-3y = -36$$
$$y = 12$$
Use the values of x and y to find the angle measures.
Angle A: $(x + 8y - 1)° = (5 + 8 \cdot 12 - 1)° = 100°$
Angle B: $(3y + 4)° = (3 \cdot 12 + 4)° = 40°$
Angle C: $(7x + 5)° = (7 \cdot 5 + 5)° = 40°$

29.

	Principal ×	Rate =	Interest
7% Investment	x	0.07	$0.07x$
8% Investment	y	0.08	$0.08y$

Since the total investment is \$20,000 the first equation is $x+y=20,000$.

Since the total interest is \$1520 the second equation is $0.07x+0.08y=1520$.

System of equations: $\begin{cases} x+y=20,000 \\ 0.07x+0.08y=1520 \end{cases}$

Solve the first equation for y and substitute into the second equation.

$x+y=20,000$

$y=20,000-x$

Solve for x. $0.07x+0.08\overbrace{(20,000-x)}^{y}=1520$

$$0.07x+1600-0.08x=1520$$
$$-0.01x+1600=1520$$
$$-0.10x+1600-1600=1520-1600$$
$$-0.01x=-80$$
$$\frac{-0.01x}{-0.01}=\frac{80}{-0.01}$$
$$x=8000$$

Back-substitute to find y.

$x+y=20,000$

$8000+y=20,000$

$y=12,000$

\$8000 should be invested at 7% and \$12,000 should be invested at 8%.

31.

	Principal ×	Rate =	Interest
8% Loan	x	0.08	$0.08x$
18% Loan	y	0.18	$0.18y$

Since the total loaned is \$120,000 the first equation is $x+y=120,000$.

Since the total interest is \$10,000 the second equation is $0.08x+0.18y=10,000$.

System of equations: $\begin{cases} x+y=120,000 \\ 0.08x+0.18y=10,000 \end{cases}$

Solve the first equation for y and substitute into the second equation.

$x+y=120,000$

$y=120,000-x$

Solve for x. $0.08x+0.18\overbrace{(120,000-x)}^{y}=10,000$

$$0.08x+21,600-0.18x=10,000$$
$$-0.10x+21,600=10,000$$
$$-0.10x+21,600-21,600=10,000-21,600$$
$$-0.10x=-11,600$$
$$\frac{-0.10x}{-0.10}=\frac{-11,600}{-0.10}$$
$$x=116,000$$

Back-substitute to find y.

$$x+y=120,000$$

$$116,000+y=120,000$$

$$y=4000$$

$116,000 was loaned at 8% and $4000 was loaned at 18%.

33.

	Principal ×	Rate =	Interest
6% Investment	x	0.06	$0.06x$
9% Investment	y	0.09	$0.09y$

Since the total investment is $6000 the first equation is $x+y=6000$.

Since the interest is equal for both accounts, the second equation is $0.06x=0.09y$

System of equations: $\begin{cases} x+y=6000 \\ 0.06x=0.09y \end{cases}$

Solve the first equation for y and substitute into the second equation.

$$x+y=6000$$

$$y=6000-x$$

Solve for x.
$$0.06x=0.09\overbrace{(6000-x)}^{y}$$

$$0.06x=540-0.09x$$

$$0.06x+0.09x=540$$

$$0.15x=540$$

$$\frac{0.15x}{0.15}=\frac{540}{0.15}$$

$$x=3600$$

Back-substitute to find y.

$$x+y=6000$$

$$3600+y=6000$$

$$y=2400$$

$3600 was invested at 6% and $2400 was invested at 9%.

35.

	Principal ×	Rate =	Interest
15% Investment	x	0.15	$0.15x$
7% Investment	y	0.07	$0.07y$

Since the total investment is $50,000 the first equation is $x+y=50,000$.

Since the total interest is $6000 the second equation is $0.15x+0.07y=6000$.

System of equations: $\begin{cases} x+y=50,000 \\ 0.15x+0.07y=6000 \end{cases}$

Solve the first equation for y and substitute into the second equation.

$$x+y=20,000$$

$$y=20,000-x$$

Solve for x. $0.15x + 0.07\overbrace{(50,000 - x)}^{y} = 6000$

$$0.15x + 3500 - 0.07x = 6000$$

$$0.08x = 2500$$

$$\frac{0.08x}{0.08} = \frac{2500}{0.08}$$

$$x = 31,250$$

Back-substitute to find y.

$$x + y = 50,000$$

$$31,250 + y = 50,000$$

$$y = 18,750$$

$31,250 should be invested at 15% and $18,750 should be invested at 7%.

37.

	Number of Liters	× Percent Fungicide	= Amount of Fungicide
5% Fungicide Solution	x	0.05	$0.05x$
10% Fungicide Solution	y	0.10	$0.10y$
8% Fungicide Solution	50	0.08	$0.08(50)$

Since there are 50 total liters, the first equation is $x + y = 50$.

Since the total amount of fungicide is $0.08(50)$, the second equation is $0.05x + 0.10y = 0.08(50)$.

System of equations: $\begin{cases} x + y = 50 \\ 0.05x + 0.10y = 0.08(50) \end{cases}$

Solve the first equation for y and substitute into the second equation.

$$x + y = 50$$

$$y = 50 - x$$

Solve for x. $0.05x + 0.10\overbrace{(50 - x)}^{y} = 0.08(50)$

$$0.05x + 5 - 0.10x = 4$$

$$-0.05x + 5 = 4$$

$$-0.05x = -1$$

$$\frac{-0.05x}{-0.05} = \frac{-1}{-0.05}$$

$$x = 20$$

Back-substitute to find y.

$$x + y = 50$$

$$20 + y = 50$$

$$y = 30$$

20 liters of 5% fungicide solution and 30 liters of 10% fungicide solution should be used.

39.

	Number of Ounces	× Percent Alcohol	= Amount of Alcohol
15% Alcohol	x	0.15	$0.15x$
20% Alcohol	4	0.20	$0.20(4)$
17% Alcohol	y	0.17	$0.17y$

Since there are y total ounces, the first equation is $x + 4 = y$.

Since the total amount of alcohol is $0.17y$, the second equation is $0.15x - 0.20(4) = 0.17y$.

System of equations: $\begin{cases} x+4=y \\ 0.15x-0.20(4)=0.17y \end{cases}$

The first equation is solved for y, so we substitute $x+4$ for y in the second equation and solve for x.

$$0.15x+0.20(4)=0.17\overbrace{(x+4)}^{y}$$
$$0.15x+0.8=0.17x+0.68$$
$$0.15x+0.8-0.17x=0.17x+0.68-0.17x$$
$$-0.02x+0.8=0.68$$
$$-0.02x+0.8-0.8=0.68-0.8$$
$$-0.02x=-0.12$$
$$-0.02x=-0.12$$
$$\frac{-0.02x}{-0.02}=\frac{-0.12}{-0.02}$$
$$x=6$$

Back-substitute to find y.

$$x+4=y$$
$$6+4=y$$
$$10=y$$

To make a 17% alcohol solution (10 ounces), 6 ounces of 15% alcohol should be mixed with the 4 ounces of 20% alcohol solution.

41.

	Number of Students	× Percent Music Majors	= Music Majors
North Campus	x	0.10	$0.10x$
South Campus	y	0.90	$0.90y$
Merged Campus	1000	0.42	$0.42(1000)$

Since the east campus has 1000 students, the first equation is $x+y=1000$.

Since 42% of the east campus students are music majors, the second equation is $0.10x-0.90y=0.42(1000)$.

System of equations: $\begin{cases} x+y=1000 \\ 0.10x-0.90y=0.42(1000) \end{cases}$

Solve the first equation for y and substitute into the second equation.

$$x+y=1000$$
$$y=1000-x$$

Solve for x. $0.10x+0.90\overbrace{(1000-x)}^{y}=0.42(1000)$

$$0.10x+900-0.90x=420$$
$$-0.8x+900=420$$
$$-0.8x=-480$$
$$\frac{-0.8x}{-0.8}=\frac{-420}{-0.8}$$
$$x=600$$

Back-substitute to find y.

$$x+y=1000$$
$$600+y=1000$$
$$y=400$$

There were 600 students at the north campus and 400 students at the south campus.

43. – 47. Answers will vary.

49. does not make sense; Explanations will vary. Sample explanation: The model is $y = 4 + 0.07x$.

51. does not make sense; Explanations will vary. Sample explanation: The model is $(4x - 2y + 4) + (12x + 6y + 12) = 180$.

53. Let x = the number of birds.
Let y = the number of lions.
Since each bird has one head and each lion has one head, $x + y = 30$.
Since each bird has two feet and each lion has four feet, $2x + 4y = 100$.
Solve the first equation for y. $y = 30 - x$
Substitute $30 - x$ for y in the second equation.
$$2x + 4(30 - x) = 100$$
$$2x + 120 - 4x = 100$$
$$-2x + 120 = 100$$
$$-2x = -20$$
$$x = 10$$
Back-substitute 10 for x to find y.
$$10 + y = 30$$
$$y = 20$$
There were 10 birds and 20 lions in the zoo.

55. Let x = the number of people in the downstairs apartment.
Let y = the number of people in the upstairs apartment.
If one of the people in the upstairs apartment goes downstairs, there will be the same number of people in both apartments, so $y - 1 = x + 1$.
If one of the people in the downstairs apartment goes upstairs, there will be twice as many people upstairs as downstairs, so $y + 1 = 2(x - 1)$.
Solve the first equation for y.
$$y = x + 2$$
Also solve the second equation for y.
$$y + 1 = 2x - 2$$
$$y = 2x - 3$$
Substitute $x + 2$ for y in the last equation.
$$x + 2 = 2x - 3$$
$$-x + 2 = -3$$
$$-x = -5$$
$$x = 5$$
Back-substitute to find y.
$$y = 5 + 2 = 7$$
There are 5 people downstairs and 7 people upstairs.

57.

	Principal	× Rate	= Interest
8% Investment	x	0.08	$0.08x$
12% Investment	y	0.12	$0.12y$
9% Investment	70,000	0.09	0.09(70,000)

Since the total investment is $70,000 the first equation is $x + y = 70,000$.

Since the total interest is $0.09(70,000)$ the second equation is $0.08x + 0.12y = 0.09(70,000)$.

System of equations: $\begin{cases} x + y = 70,000 \\ 0.08x + 0.12y = 0.09(70,000) \end{cases}$

Solve the first equation for y and substitute into the second equation.

$x + y = 70,000$

$\quad y = 70,000 - x$

Solve for x. $0.08x + 0.12\overbrace{(70,000 - x)}^{y} = 0.09(70,000)$

$\qquad 0.08x + 8400 - 0.12x = 6300$

$\qquad\qquad -0.04x + 8400 = 6300$

$\qquad\qquad\qquad -0.04x = -2100$

$\qquad\qquad\qquad \dfrac{-0.04x}{-0.04} = \dfrac{-2100}{-0.04}$

$\qquad\qquad\qquad\qquad x = 52,500$

Back-substitute to find y.

$\qquad x + y = 70,000$

$52,500 + y = 70,000$

$\qquad\qquad y = 17,500$

$52,500 should be invested at 8% and $17,500 should be invested at 12% to obtain an overall return of 9%.

59. $2(x + 3) = 24 - 2(x + 4)$

$\quad 2x + 6 = 24 - 2x - 8$

$\quad 2x + 6 = 16 - 2x$

$\qquad 4x = 10$

$\qquad\ x = \dfrac{10}{4}$

$\qquad\ x = \dfrac{5}{2}$

The solution set is $\left\{\dfrac{5}{2}\right\}$.

60. $5 + 6(x + 1) = 5 + 6x + 6$

$\qquad\qquad\qquad = 6x + 11$

61. Find slope: $m = \dfrac{y_2 - y_1}{x_2 - x_1} = \dfrac{-10 - 6}{3 - (-5)} = \dfrac{-16}{8} = -2$

Use either point in the point-slope form and then solve for y.

$$y - y_1 = m(x - x_1)$$
$$y - (-10) = -2(x - 3)$$
$$y + 10 = -2x + 6$$
$$y = -2x - 4$$

62. $2x - y < 4$

Graph $2x - y = 4$ as a dashed line with x-intercept 2 and y-intercept -4. Use $(0,0)$ as a test point. Since $2(0) - 0 < 4$ is true, shade the half-plane containing $(0,0)$.

$2x - y < 4$

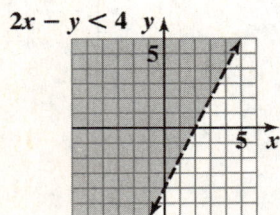

63. $y \geq x + 1$

Graph the line $y = x + 1$ using the slope of 1 and y-intercept of 1. Make the line solid because the inequality symbol is $\geq$. Use $(0,0)$ as a test point. Since $0 \geq 0 + 1$ is false, shade the half-plane *not* containing $(0,0)$.

$y \geq x + 1$

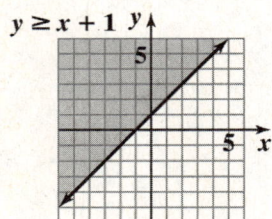

64. $x \geq 2$

Graph $x = 2$ as a solid vertical line. Use $(0,0)$ as a test point. Since $0 \geq 2$ is false, shade the half-plane *not* containing $(0,0)$. This is the region to the right of the line $x = 2$.

$x \geq 2$

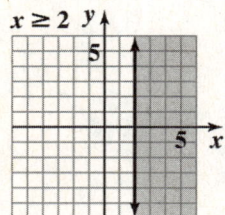

4.5 Check Points

1. Point $B = (66, 130)$

$$4.9x - y \geq 165$$
$$4.9(66) - 130 \geq 165$$
$$193.4 \geq 165, \text{ true}$$

$$3.7x - y \leq 125$$
$$3.7(66) - 130 \leq 125$$
$$114.2 \leq 125, \text{ true}$$

Point B is a solution of the system.

2. $x + 2y > 4$
$2x - 3y \leq -6$

Graph $x + 2y > 4$ by graphing $x + 2y = 4$ as a dashed line using the x-intercept, 4, and the y-intercept, 2.

Use $(0,0)$ as a test point. $x + 2y > 4$
$$0 + 2(0) > 4$$
$$0 > 4, \text{ false}$$

Because $0 + 2(0) > 4$ is false, shade the half-plane *not* containing $(0,0)$.

Graph $2x - 3y \leq -6$ by graphing $2x - 3y = -6$ as a solid line using the x-intercept, -3, and the y-intercept, 2.

Use $(0,0)$ as a test point. $2x - 3y \leq -6$
$$2(0) - 3(0) \leq -6$$
$$0 \leq -6, \text{ false}$$

Because $2(0) - 3(0) \leq -6$ is false, shade the half-plane *not* containing $(0,0)$.

The solution set of the system is the intersection of the two shaded regions.

$x + 2y > 4$
$2x - 3y \leq -6$

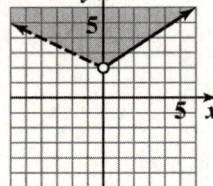

3. $y \geq x + 2$

$x \geq 1$

Graph $y \geq x + 2$ by graphing $y = x + 2$ as a solid line using the y-intercept, 2, and the slope, 1.
Use $(0,0)$ as a test point. $y \geq x + 2$

$$0 \geq 0 + 2$$

$$0 \geq 2, \text{ false}$$

Because $0 \geq 0 + 2$ is false, shade the half-plane *not* containing $(0,0)$.

Graph $x \geq 1$ by graphing $x = 1$ as a solid vertical line through $x = 1$.
Use $(0,0)$ as a test point. $x \geq 1$

$$0 \geq 1, \text{ false}$$

Because $0 \geq 1$ is false, shade the half-plane *not* containing $(0,0)$.

The solution set of the system is the intersection of the two shaded regions.

$y \geq x + 2$
$x \geq 1$

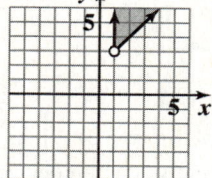

4.5 Concept and Vocabulary Check

1. $x + y \geq 4; \ x - y \leq 2$

2. $y \geq 2x + 1; \ y \leq 4$

3. false

4. true

4.5 Exercise Set

1. $x + y \leq 4$

$x - y \leq 2$

Graph $x + y \leq 4$ by graphing $x + y = 4$ as a solid line using the x-intercept, 4, and the y-intercept, 4. Because $(0,0)$ makes the inequality $x + y \leq 4$ true, shade the half-plane containing $(0,0)$.

Graph $x - y \leq 2$ by graphing $x - y = 2$ or $y = x - 2$ as a solid line with x-intercept 2 and y-intercept -2. Because $(0,0)$ makes the inequality $x - y \leq 2$ true, shade the half-plane containing $(0,0)$.

The solution set of the system is the intersection of the shaded regions.

$x + y \leq 4$
$x - y \leq 2$

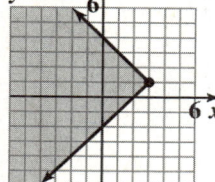

3. $2x - 4y \leq 8$

$x + y \geq -1$

Graph $2x - 4y \leq 8$ by graphing $2x - 4y = 8$ as a solid line using the x-intercept, 4, and the y-intercept, -2. Because $4(0) - 2(0) \leq 8$ is true, shade the half-plane containing $(0,0)$.

Graph $x + y \geq -1$ by graphing $x + y = -1$ as a solid line using the x-intercept, -1, and the y-intercept, -1. Because $0 + 0 \geq -1$ is true, shade the half-plane containing $(0,0)$.

The solution set of the system is the intersection of the two shaded regions.

$2x - 4y \leq 8$
$x + y \geq -1$

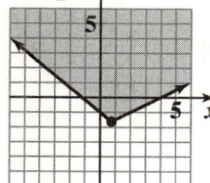

5. $x + 3y \leq 6$

$x - 2y \leq 4$

Graph $x + 3y \leq 6$ by graphing $x + 3y = 6$ as a solid line using the x-intercept, 6, and the y-intercept, 2. Because $0 + 3(0) \leq 6$ is true, shade the half-plane containing $(0,0)$.

Graph $x - 2y \leq 4$ by graphing $x - 2y = 4$ as a solid line using the x-intercept, 4, and the y-intercept, -2. Because $0 - 2(0) \leq 4$ is true, shade the half-plane containing $(0,0)$.

The solution set of the system is the intersection of the shaded regions.

$x + 3y \leq 6$
$x - 2y \leq 4$

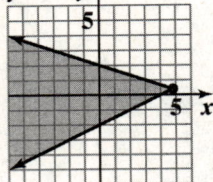

7. $x - 2y > 4$

$2x + y \geq 6$

Graph $x - 2y > 4$ by graphing $x - 2y = 4$ as a dashed line using the x-intercept, 4, and the y-intercept, -2. Because $0 - 2(0) > 4$ is false, shade the half-plane *not* containing $(0,0)$.

Graph $2x + y \geq 6$ by graphing $2x + y = 6$ as a solid line using the x-intercept, 3, and the y-intercept, 6. Because $2(0) + 0 \geq 6$ is false, shade the half-plane *not* containing $(0,0)$.

The solution set of the system is the intersection of the two shaded regions.

$x - 2y > 4$
$2x + y \geq 6$

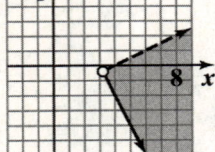

9. $x + y > 1$

$x + y < 4$

Graph $x + y > 1$ by graphing $x + y = 1$ as a dashed line using the x-intercept 1 and y-intercept 1. Because $0 + 0 > 1$ is false, *do not* shade the half-plane containing $(0,0)$.

Graph $x + y < 4$ by graphing $x + y = 4$ as a dashed line using the x-intercept 4 and y-intercept 4. Because $0 + 0 < 4$ is true, shade the half-plane containing $(0,0)$.

The solution set is the shaded region between the two dashed parallel lines.

$x + y > 1$
$x + y < 4$

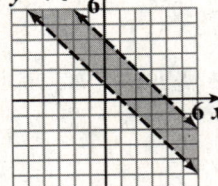

11. $y \geq 2x + 1$

$y \leq 4$

Graph $y \geq 2x + 1$ by graphing $y = 2x + 1$ as a solid line using the slope, 2, and the y-intercept, 1. Because $0 \geq 2(0) + 1$ is false, shade the half-plane *not* containing $(0,0)$.

Graph $y \leq 4$ by graphing $y = 4$ as a solid horizontal line with y-intercept 4. Because $0 \leq 4$ is true, shade the half-plane containing $(0,0)$.

The solution set of the system is the intersection of the shaded regions.

$y \geq 2x + 1$
$y \leq 4$

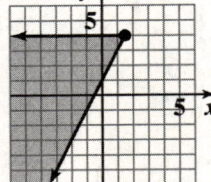

13. $y > x - 1$

$x > 5$

Graph $y > x - 1$ by graphing $y = x - 1$ as a dashed line using the slope, 1, and the y-intercept, -1. Because $0 > 0 - 1$ is true, shade the half-plane containing $(0, 0)$.

Graph $x > 5$ by graphing $x = 5$ as a dashed vertical line with x-intercept 5. Because $0 > 5$ is false, shade the half-plane not containing $(0, 0)$.

The solution set of the system is the intersection of the two shaded regions.

$y > x - 1$
$x > 5$

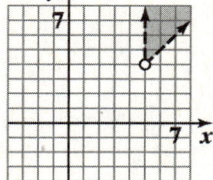

15. $y \geq 2x - 3$

$y \leq 2x + 1$

Graph $y \geq 2x - 3$ by graphing $y = 2x - 3$ as a solid line using slope, 2, and y-intercept, -3. Since $0 \geq 2(0) - 3$ is true, shade the half plane containing $(0, 0)$.

Graph $y \leq 2x + 1$ by graphing $y = 2x + 1$ as a solid line using its slope, 2, and y-intercept, 1. Since $0 \leq 2(0) + 1$ is true, shade the half-plane containing $(0, 0)$.

The solution set of the system is the intersection of the shaded regions.

$y \geq 2x - 3$
$y \leq 2x + 1$

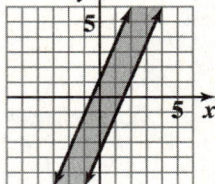

17. $y > 2x + 3$

$y \leq -x + 6$

Graph $y > 2x - 3$ by graphing $y = 2x - 3$ as a dashed line using its slope, 2, and y-intercept, -3. Since $0 > 2(0) - 3$ is true, shade the half-plane containing $(0, 0)$.

Graph $y \leq -x + 6$ by graphing $y = -x + 6$ as a solid line using its slope, -1, and y-intercept, 6. Since $0 \leq -0 + 6$ is true, shade the half-plane containing $(0, 0)$.

The solution set of the system is the intersection of the two shaded regions.

$y > 2x - 3$
$y \leq -x + 6$

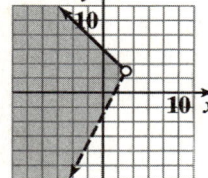

19. $x + 2y \leq 4$

$y \geq x - 3$

Graph $x + 2y \leq 4$ by graphing $x + 2y = 4$ as a solid line using its x-intercept, 4, and y-intercept, 2. Since $0 + 2(0) \leq 4$ is true, shade the half-plane containing $(0, 0)$.

Graph $y \geq x - 3$ by graphing $y = x - 3$ as a solid line, using its slope, 1, and y-intercept, -3. Since $0 \geq 0 - 3$ is true, shade the half-plane containing $(0, 0)$.

The solution set of the system is the intersection of the two shaded regions.

$x + 2y \leq 4$
$y \geq x - 3$

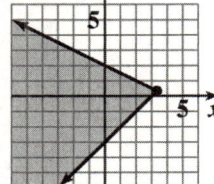

21. $x \leq 3$

$y \geq -2$

Graph $x \leq 3$ by graphing $x = 3$ as a solid vertical line. Since $0 \leq 3$ is true, shade the half-plane containing $(0,0)$.

Graph $y \geq -2$ by graphing $y = -2$ as a solid horizontal line. Since $0 \geq -2$ is true, shade the half-plane containing $(0,0)$.

The solution set of the system is the intersection of the shaded regions.

$x \leq 3$
$y \geq -2$

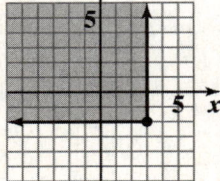

23. $x \geq 3$

$y < 2$

Graph $x \geq 3$ by graphing $x = 3$ as a solid vertical line. Since $0 \geq 3$ is false, shade the half-plane *not* containing $(0,0)$.

Graph $y < 2$ by graphing $y = 2$ as a dashed horizontal line. Since $0 < 2$ is true, shade the half-plane containing $(0,0)$.

The solution set of the system is the intersection of the two shaded regions.

$x \geq 3$
$y < 2$

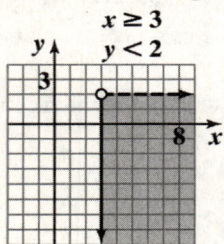

25. $x \geq 0$

$y \leq 0$

Graph $x \geq 0$ by graphing $x = 0$ (the y–axis) as a solid vertical line. Shade the half-plane to the right of the axis.

Graph $y \leq 0$ by graphing $y = 0$ (the x–axis) as a solid horizontal line. Shade the half-plane below the axis.

The solution set of the system is the intersection of the two shaded regions. This is all of quadrant IV, including the portions of the axes that are the boundaries of this region.

$x \geq 0$
$y \leq 0$

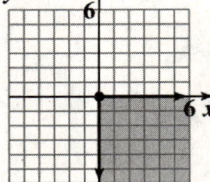

27. $x \geq 0$

$y > 0$

Graph $x \geq 0$ by graphing $x = 0$ as a solid vertical line. This is the y-axis. Shade the half-plane to the right of the y-axis.

Graph $y > 0$ by graphing $y = 0$ as a dashed solid line. This is the x-axis. Shade the half-plane above the x-axis.

The solution set of the system is the intersection of the two shaded regions. This is all of quadrant I, including the portion of the y-axis, but excluding the portion of the x-axis, that are boundaries of this region.

$x \geq 0$
$y > 0$

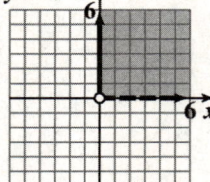

29. $x + y \leq 5$

$x \geq 0$

$y \geq 0$

Graph $x + y \leq 5$ by graphing $x + y = 5$ as a solid line using its x-intercept, 5, and y-intercept, 5. Since $0 + 0 \leq 5$ is true, shade the half-plane containing $(0,0)$.

Graph $x \geq 0$ by graphing $x = 0$ (the y-axis) as a solid line. Shade the half-plane to the right of the y-axis.

Graph $y \geq 0$ by graphing $y = 0$ (the x-axis) as a solid line. Shade the half-plane above the x-axis. The solution set of the system is the intersection of the shaded regions. This is the set of points satisfying $x + y \leq 5$ that lie in quadrant I, together with the portions of the axes that are boundaries of this region.

$x + y \leq 5$

$x \geq 0$

$y \geq 0$

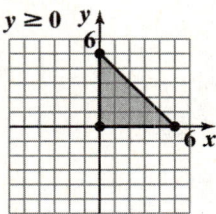

31. $4x - 3y > 12$

$x \geq 0$

$y \leq 0$

Graph $4x - 3y > 12$ by graphing $4x - 3y = 12$ as a dashed line using x-intercept, 3, and y-intercept, –4. Since $4(0) - 3(0) > 12$ is false, shade the half-plane *not* including $(0, 0)$.

Graph $x \geq 0$ by graphing $x = 0$ (the y-axis) as a solid line and shade the half-plane to the right of it.

Graph $y \leq 0$ by graphing $y = 0$ (the x-axis) as a solid line and shade the half-plane below it.

The solution set of the system is the intersection of the three shaded regions. Notice that this is the set of points satisfying $4x - 3y > 12$ that lie in quadrant IV, together with the portions of the axes that are

boundaries of these regions.

$4x - 3y > 12$

$x \geq 0$

$y \leq 0$

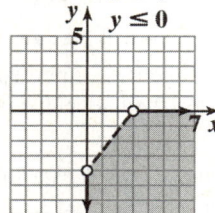

33. $0 \leq x \leq 3$

$0 \leq y \leq 3$

Graph $0 \leq x \leq 3$ by graphing the vertical lines $x = 0$ (the y-axis) and $x = 3$ as solid lines. Shade the region between the parallel lines.

Graph $0 \leq y \leq 3$ by graphing the horizontal lines $y = 0$ (the x-axis) and $y = 3$ as solid lines. Shade the region between these horizontal lines.

The solution set of the system is the intersection of the shaded regions.

$0 \leq x \leq 3$

$0 \leq y \leq 3$

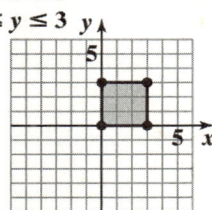

35. $x - y \leq 4$

$x + 2y \leq 4$

Graph $x - y \leq 4$ by graphing $x - y = 4$ as a solid line using the x-intercept, 4, and the y-intercept, –4. Since $0 - 0 \leq 4$ is true, shade the half-plane containing $(0, 0)$.

Graph $x + 2y \leq 4$ by graphing $x + 2y = 4$ as a solid line using the x-intercept, 4, and the y-intercept, 2. Since $0 + 2(0) \leq 4$ is true, shade the half-plane containing $(0, 0)$.

The solution set of the system is the intersection of the shaded regions.

$x - y \leq 4$

$x + 2y \leq 4$

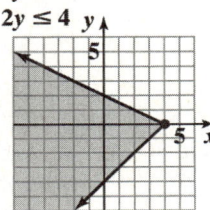

37. $x + y \geq 1$

$x - y \geq 1$

$x \geq 4$

Graph $x + y \geq 1$ by graphing $x + y = 1$ as a solid line using the x-intercept, 1, and the y-intercept, 1. Since $0 + 0 \geq 1$ is false, shade the half-plane *not* containing $(0, 0)$.

Graph $x - y \geq 1$ by graphing $x - y = 1$ as a solid line using the x-intercept, 1, and the y-intercept, -1. Since $0 - 0 \geq 1$ is false, shade the half-plane *not* containing $(0, 0)$.

Graph $x \geq 4$ by graphing $x = 4$ as a solid vertical line and shade the half-plane to the right of it.

The solution set of the system is the intersection of the shaded regions.

$\boldsymbol{x + y \geq 1}$

$\boldsymbol{x - y \geq 1}$

$\boldsymbol{x \geq 4}$

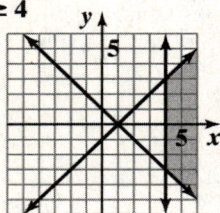

39. $x + 2y < 6$

$y > 2x - 2$

$y \geq 2$

Graph $x + 2y < 6$ by graphing $x + 2y < 6$ as a dashed line using the x-intercept, 6, and the y-intercept, 3. Since $0 + 2(0) < 6$ is true, shade the half-plane containing $(0, 0)$.

Graph $y > 2x - 2$ by graphing $y = 2x - 2$ as a dashed line using the slope, 2, and the y-intercept, -2. Since $0 > 2(0) - 2$ is true, shade the half-plane containing $(0, 0)$.

Graph $y \geq 2$ by graphing $y = 2$ as a solid horizontal line and shade the half-plane above it.

The solution set of the system is the intersection of the shaded regions.

$\boldsymbol{x + 2y < 6}$

$\boldsymbol{y > 2x - 2}$

$\boldsymbol{y \geq 2}$

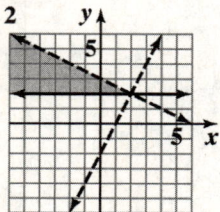

41. $y \leq -3x + 3$

$y > -x - 1$

$y < x + 7$

Graph $y \leq -3x + 3$ by graphing $y = -3x + 3$ as a solid line with slope, -3, and y-intercept, 3. Since $0 \leq -3(0) + 3$ is true, shade the half-plane containing $(0, 0)$.

Graph $y > -x - 1$ by graphing $y = -x - 1$ as a dashed line using the slope, -1, and the y-intercept, -1.

Since $0 > -0 - 1$ is true, shade the half-plane containing $(0, 0)$.

Graph $y < x + 7$ by graphing $y = x + 7$ as a dashed line with slope, 1, and y-intercept, 7. Since $0 < 0 + 7$ is true, shade the half-plane containing $(0, 0)$.

The solution set of the system is the intersection of the shaded regions.

$\boldsymbol{y \leq -3x + 3}$

$\boldsymbol{y \geq -x - 1}$

$\boldsymbol{y < x + 7}$

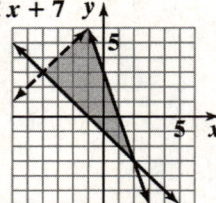

43. $y \geq 2x + 2$

$y < 2x - 3$

$x \geq 2$

Graph $y \geq 2x + 2$ by graphing $y = 2x + 2$ as a solid line with slope, 2, and y-intercept, 2. Since $0 \geq 2(0) + 2$ is false, shade the half-plane *not* containing $(0, 0)$.

Graph $y < 2x - 3$ by graphing $y = 2x - 3$ as a dashed line using the slope, 2, and the y-intercept, -3.

Since $0 < 2(0) - 3$ is false, shade the half-plane *not* containing $(0, 0)$.

Graph $x \geq 2$ by graphing $x = 2$ as a solid vertical line. Since $0 \geq 2$ is false, shade the half-plane containing $(0, 0)$.

The solution to the system is the intersection of the shaded regions. In this case, there is no region on the graph where the three shaded regions intersect, so there is no solution.

$y \geq 2x + 2$
$y < 2x - 3$
$x \geq 2$

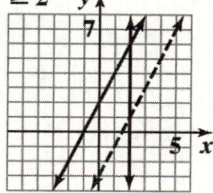

45. Point $A = (66, 160)$

$$5.3x - y \geq 180$$
$$5.3(66) - 160 \geq 180$$
$$189.8 \geq 180, \text{ true}$$

$$4.1x - y \leq 14$$
$$4.1(66) - 160 \leq 140$$
$$110.6 \leq 140, \text{ true}$$

Point A is a solution of the system.

47. Point $= (72, 205)$

$$5.3x - y \geq 180$$
$$5.3(72) - 205 \geq 180$$
$$176.6 \geq 180, \text{ false}$$

$$4.1x - y \leq 14$$
$$4.1(72) - 205 \leq 140$$
$$90.2 \leq 140, \text{ true}$$

The data does not satisfy both inequalities. The person is not within the healthy weight region.

49. $165x + 110y \leq 330$

$165x + 110y \geq 165$

Graph $165x + 110y = 330$ as a solid line with x-intercept 2 and y-intercept 3. Draw the graph in quadrant I only. Use $(1,1)$ as a test point. Since

$$165(1) + 110(1) = 275 < 330$$ is true, shade the

region in quadrant I containing $(1,1)$.

Graph $165x + 110y = 165$ as a solid line with x-intercept 1 and y-intercept 1.5. Draw the graph in quadrant I only. Use $(1,3)$ as a test point. Since

$$165(1) + 110(3) = 495 \geq 165$$

is true, shade the region in quadrant I containing $(1,3)$.

The graph of the system is the intersection of the two shaded regions.

$165x + 110y \leq 330$

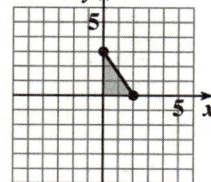

One point in the solution set is $(1,1)$. This means that the patient is allowed to have 1 egg and 1 ounce of meat per day.

Another point in the solution set is $(0,3)$ This means that the patient is also allowed to have no eggs and 3 ounces of meat per day.

51. Answers will vary.

53. makes sense

55. makes sense

57. $y \geq -x + 3$
$x > 5$

59. $x < 0$
$y < 0$

61. Let x = the number of \$35 tickets sold.
Let y = the number of \$50 tickets sold.
From the information given in the problem, we have the following inequalities.
$x + y \geq 25{,}000$
$35x + 50y \geq 1{,}025{,}000$
Also, the number of tickets sold and the amount of money in ticket sales cannot be negative, so we also have $x \geq 0$ and $y \geq 0$. so the system is as follows.
$x + y \geq 25{,}000$
$35x + 50y \geq 1{,}025{,}000$
$x \geq 0$
$y \geq 0$
Graph $x + y \geq 25{,}000$ by graphing $x + y = 25{,}000$ as a solid line using its x-intercept, 25,000, and y-intercept, 25,000. Since $0 + 0 \geq 25{,}000$ is false, shade the region *not* containing (0,0).
Graph $35x + 50y \geq 1{,}025{,}000$ by graphing $35x + 50y = 1{,}025{,}000$ using its x-intercept, approximately 29,286, and its y-intercept, 20,500. Since $35(0) + 50(0) \geq 1{,}025{,}000$ is false, shade the region *not* containing (0,0).
The inequalities $x \geq 0$ and $y \geq 0$ restrict the graph to quadrant I.

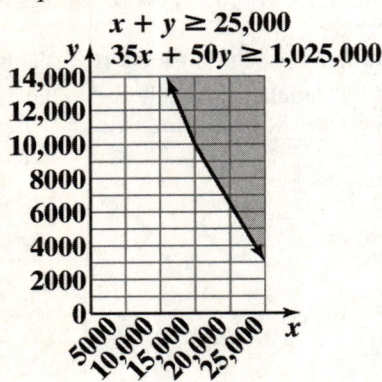

62. $(-6,1)$ and $(2,-1)$
$m = \dfrac{-1-1}{2-(-6)} = \dfrac{-2}{8} = -\dfrac{1}{4}$

63. $\dfrac{1}{5} + \left(-\dfrac{3}{4}\right) = \dfrac{4}{20} + \left(-\dfrac{15}{20}\right) = -\dfrac{11}{20}$

64. $7x = 10 + 6(11 - 2x)$
$7x = 10 + 66 - 12x$
$7x = 76 - 12x$
$19x = 76$
$x = \dfrac{76}{19}$
$x = 4$
The solution set is {4}.

65. $5x^3 + 12x^3 = (5 + 12)x^3 = 17x^3$

66. $-8x^2 + 6x^2 = (-8 + 6)x^2 = -2x^2$

67. $-9y^4 - (-2y^4) = -9y^4 + 2y^4$
$\qquad\qquad\qquad = (-9 + 2)y^4$
$\qquad\qquad\qquad = -7y^4$

Chapter 4 Review Exercises

1. $\quad 4x - y = 9$
$4(1) - (-5) = 9$
$\quad 4 + 5 = 9$
$\qquad 9 = 9$, true

$\quad 2x + 3y = -13$
$2(1) + 3(-5) = -13$
$\quad 2 - 15 = -13$
$\qquad -13 = -13$, true
Since the ordered pair (1, −5) satisfies both equations, it is a solution of the given system.

2. $\quad 2x + 3y = -4$
$2(-5) + 3(2) = -4$
$\quad -10 + 6 = -4$
$\qquad -4 = -4$, true

$\quad x - 4y = -10$
$-5 - 4(-2) = -10$
$\quad -5 + 8 = -10$
$\qquad 3 = -10$, false
Since (−5,2) fails to satisfy *both* equations, it is not a solution of the given system.

3. $x + y = 2$
  $-1 + 3 = 2$
  $ 2 = 2$, true

  $2x + y = -5$
  $2(-1) + 3 = -5$
  $ -2 + 3 = -5$
  $ 1 = -5$, false

Since $(-1, 3)$ fails to satisfy *both* equations, it is not a solution of the given system. Also, the second equation in the system, which can be rewritten as $y = -2x - 5$, is a line with slope -2 and y-intercept -5, while the graph shows a line with slope 2 and y-intercept 5.

4. $x + y = 2$
  $x - y = 6$

Graph both lines on the same axes.
$x + y = 2$: x-intercept = 2; y-intercept 2
$x - y = 6$: x-intercept = 6, y-intercept = -6

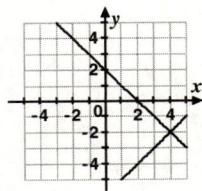

The solution set is $\{(4, -2)\}$.

5. $2x - 3y = 12$
  $-2x + y = -8$

Graph both equations.
$2x - 3y = 12$: x-intercept = 6; y-intercept = -4
$-2x + y = -8$: x-intercept = 4; y-intercept = -8

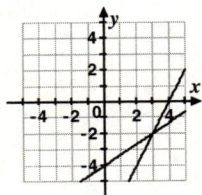

The solution set is $\{(3, -2)\}$.

6. $3x + 2y = 6$
  $3x - 2y = 6$

Graph both equations.
$3x + 2y = 6$: x-intercept = 2; y-intercept = 3
$3x - 2y = 6$: x-intercept = 2; y-intercept = -3

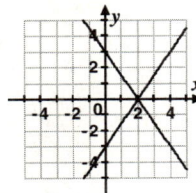

The solution set is $\{(2, 0)\}$.

7. $y = \dfrac{1}{2}x$
  $y = 2x - 3$

Graph both equations.

$y = \dfrac{1}{2}x$: slope = $\dfrac{1}{2}$; y-intercept = 0

$y = 2x - 3$: slope = 2; y-intercept = -3

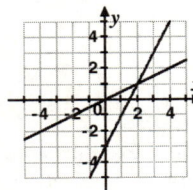

The solution set is $\{(2, 1)\}$.

8. $x + 2y = 2$
  $ y = x - 5$

Graph both equations.
$x + 2y = 2$: x-intercept = 2; y-intercept = 1
$y = x - 5$: slope = 1; y-intercept = -5

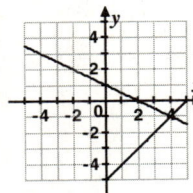

The solution set is $\{(4, -1)\}$.

9. $x + 2y = 8$

$3x + 6y = 12$

Graph both equations.

$x + 2y = 8$: x-intercept $= 8$; y-intercept $= 4$

$3x + 6y = 12$: x-intercept $= 4$; y-intercept $= 2$

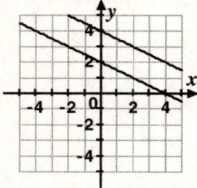

The lines are parallel. The system is inconsistent and has no solution. The solution set is $\{\ \}$.

10. $2x - 4y = 8$

$x - 2y = 4$

Graph both equations.

$2x - 4y = 8$: x-intercept $= 4$; y-intercept $= -2$

$x - 2y = 4$: x-intercept $= 4$; y-intercept $= -2$

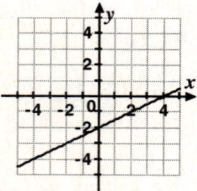

The graphs of the two equations are the same line. The system is dependent and has infinitely many solutions.

The solution set is $\{(x, y) | 2x - 4y = 8\}$ or $\{(x, y) | x - 2y = 4\}$.

11. $y = 3x - 1$

$y = 3x + 2$

Graph both equations.

$y = 3x - 1$: slope $= 3$; y-intercept $= -1$

$y = 3x + 2$: slope $= 3$; y-intercept $= 2$

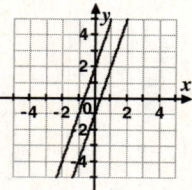

The lines are parallel, so the system is inconsistent and has no solution. The solution set is $\{\ \}$.

12. $x - y = 4$

$x = -2$

Graph both equations:

$x - y = 4$: x-intercept $= 4$; y-intercept $= -4$

$x = 2$: vertical line with x-intercept $= -2$

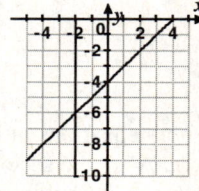

The solution set is $\{(-2, -6)\}$.

13. $x = 2$

$y = 5$

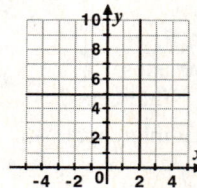

The solution set is $\{(2, 5)\}$.

14. $x = 2$

$x = 5$

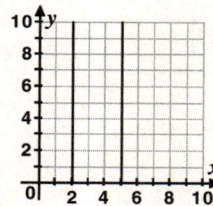

The lines are parallel, so the system inconsistent and has no solution.

The solution set is $\{\ \}$.

15.　$2x - 3y = 7$

　　　$y = 3x - 7$

Substitute $3x - 7$ for y in the first equation.

　　　$2x - 3y = 7$

$2x - 3(3x - 7) = 7$

$2x - 9x + 21 = 7$

$-7x + 21 = 7$

　　　$-7x = -14$

　　　　$x = 2$

Back-substitute 7 for x into the second equation and solve for y.

$y = 3x - 7$

$y = 3(2) - 7 = -1$

The solution set is $\{(2, -1)\}$.

16.　$2x - y = 6$

　　　$x = 13 - 2y$

Substitute $13 - 2y$ for x into the first equation.

　　　$2x - y = 6$

$2(13 - 2y) - y = 6$

　$26 - 4y - y = 6$

　　$26 - 5y = 6$

　　　$-5y = -20$

　　　　$y = 4$

Back-substitute 4 for y in the second equation.

$x = 13 - 2y$

$x = 13 - 2(4) = 5$

The solution set is $\{(5, 4)\}$.

17.　$2x - 5y = 1$

　　　$3x + y = -7$

Solve the second equation for y.

$3x + y = -7$

　　　$y = -3x - 7$

Substitute $-3x - 7$ in the first equation.

　　　$2x - 5y = 1$

$2x - 5(-3x - 7) = 1$

　$2x + 15x + 35 = 1$

　　$17x + 35 = 1$

　　　　$17x = -34$

　　　　　$x = -2$

Back-substitute in the equation $y = -3x - 7$.

$y = -3x - 7$

$y = -3(-2) - 7 = -1$

The solution set is $\{(-2, -1)\}$.

18.　$3x + 4y = -13$

　　　$5y - x = -21$

Solve the second equation for x.

$5y - x = -21$

　$-x = -5y - 21$

　　$x = 5y + 21$

Substitute $5y + 21$ for x in the first equation.

　　　$3x + 4y = -13$

$3(5y + 21) + 4y = -13$

　$15y + 63 + 4y = -13$

　　$19y + 63 = -13$

　　　$19y = -76$

　　　　$y = -4$

Back-substitute.

　　$3x + 4y = -13$

$3x + 4(-4) = -13$

　$3x - 16 = -13$

　　$3x = 3$

　　　$x = 1$

The solution set is $\{(1, -4)\}$.

19.　$y = 39 - 3x$

　　　$y = 2x - 61$

Substitute $2x - 61$ for y in the first equation.

$2x - 61 = 39 - 3x$

$5x - 61 = 39$

　$5x = 100$

　　$x = 20$

Back-substitute.

$y = 2x - 61 = 2(20) - 61 = -21$

The solution set is $\{(20, -21)\}$.

20. $4x + y = 5$

$12x + 3y = 15$

Solve the first equation for y.

$4x + y = 5$

$y = -4x + 5$

Substitute $-4x + 5$ for y in the second equation.

$12x + 3y = 15$

$12x + 3(-4x + 5) = 15$

$12x - 12x + 15 = 15$

$15 = 15$, true

The true statement indicates that the given system has infinitely many solutions.

The solution set is $\{(x, y) | 4x + y = 5\}$ or $\{(x, y) | 12x + 3y = 15\}$.

21. $4x - 2y = 10$

$y = 2x + 3$

Substitute $2x + 3$ for y in the first equation.

$4x - 2y = 10$

$4x - 2(2x + 3) = 10$

$4x - 4x - 6 = 10$

$-6 = 10$, false

The false statement $-6 = 10$ indicates that the system is inconsistent and has no solution.

The solution set is $\{\ \}$.

22. $x - 4 = 0$

$9x - 2y = 0$

Solve the first equation for x.

$x - 4 = 0$

$x = 4$

Substitute 4 for x in the second equation.

$9x - 2y = 0$

$9(4) - 2y = 0$

$36 - 2y = 0$

$-2y = -36$

$y = 18$

The solution set is $\{(4, 18)\}$.

23. $8y = 4x$

$7x + 2y = -8$

Solve the first equation for y.

$8y = 4x$

$y = \dfrac{1}{2}x$

Substitute $\dfrac{1}{2}x$ for y in the second equation.

$7x + 2y = -8$

$7x + 2\left(\dfrac{1}{2}x\right) = -8$

$7x + x = -8$

$8x = -8$

$x = -1$

Back-substitute.

$y = \dfrac{1}{2}x = \dfrac{1}{2}(-1) = -\dfrac{1}{2}$

The solution set is $\left\{\left(-1, -\dfrac{1}{2}\right)\right\}$.

24. a. Demand model: $p = -50x + 2000$

Supply model: $p = 50x$

Use the substitution method.

$p = -50x + 2000$

$\overbrace{50x}^{p} = -50x + 2000$

$50x = -50x + 2000$

$100x = 2000$

$x = 20$

Back-substitute 20 for x and find p.

$p = 50x$

$p = 50(20) = 1000$

The solution set is $\{(20, 1000)\}$.

The equilibrium quantity is 20,000 and the equilibrium price is $1000.

b. When rents are $\underline{\$1000}$ per month, consumers will demand $\underline{20{,}000}$ apartments and suppliers will offer $\underline{20{,}000}$ apartments for rent.

25. $x + y = 6$

$2x + y = 8$

Multiply the first equation by -1 and add the result to the second equation to eliminate the y-terms.

$-x - y = -6$

$\underline{2x + y = 8}$

$x = 2$

Back-substitute into either of the original equations to solve for y.

$x + y = 6$

$2 + y = 6$

$ y = 4$

The solution set is $\{(2, 4)\}$.

26. $3x - 4y = 1$

$12x - y = -11$

Multiply the first equation by -4 and add the result to the second equation.

$-12x + 16y = -4$

$\underline{12x - y = -11}$

$ 15y = -15$

$ y = -1$

Back-substitute.

$3x - 4y = 1$

$3x - 4(-1) = 1$

$3x + 4 = 1$

$3x = -3$

$x = -1$

The solution set is $\{(-1, -1)\}$.

27. $3x - 7y = 13$

$6x + 5y = 7$

Multiply the first equation by -2.

$-6x + 14y = -26$

$\underline{6x + 5y = 7}$

$ 19y = -19$

$ y = -1$

Back-substitute.

$3x - 7y = 13$

$3x - 7(-1) = 13$

$3x + 7 = 13$

$3x = 6$

$x = 2$

The solution set is $\{(2, -1)\}$.

28. $8x - 4y = 16$

$4x + 5y = 22$

Multiply the second equation by -2.

$8x - 4y = 16$

$\underline{-8x - 10y = -44}$

$ -14y = -28$

$ y = 2$

Back-substitute.

$8x - 4y = 16$

$8x - 4(2) = 16$

$8x - 8 = 16$

$8x = 24$

$x = 3$

The solution set is $\{(3, 2)\}$.

29. $5x - 2y = 8$

$3x - 5y = 1$

Multiply the first equation by 3.

Multiply the second equation by -5.

$15x - 6y = 24$

$\underline{-15x + 25y = -5}$

$ 19y = 19$

$ y = 1$

Back-substitute.

$5x - 2y = 8$

$5x - 2(1) = 8$

$5x - 2 = 8$

$5x = 10$

$x = 2$

The solution set is $\{(2, 1)\}$.

30. $2x + 7y = 0$

$7x + 2y = 0$

Multiply the first equation by 7.
Multiply the second equation by -2.

$14x + 49y = 0$

$\underline{-14x - 4y = 0}$

$\qquad 45y = 0$

$\qquad\quad y = 0$

Back-substitute.

$2x + 7y = 0$

$2x + 7(0) = 0$

$\qquad 2x = 0$

$\qquad\ x = 0$

The solution set is $\{(0,0)\}$.

31. $x + 3y = -4$

$3x + 2y = 3$

Multiply the first equation by -3.

$-3x - 9y = 12$

$\underline{\ 3x + 2y = \ 3}$

$\qquad -7y = 15$

$\qquad\quad y = -\dfrac{15}{7}$

Instead of back-substituting $-\dfrac{15}{7}$ and working with

fractions, go back to the original system. Multiply
the first equation by 2 and the second equation by
-3.

$\quad 2x + 6y = -8$

$\underline{-9x - 6y = -9}$

$-7x \qquad\ = -17$

$\qquad x\ \ =\ \dfrac{17}{7}$

The solution set is $\left\{\left(\dfrac{17}{7}, -\dfrac{15}{7}\right)\right\}$.

32. $2x + y = 5$

$2x + y = 7$

Multiply the first equation by -1.

$-2x - y = -5$

$\underline{\ 2x + y = \ 7}$

$\qquad 0 = 2,\ \text{false}$

The false statement indicates that the system has no
solution. The solution set is $\{\ \}$.

33. $3x - 4y = -1$

$-6x + 8y = 2$

Multiply the first equation by 2.

$6x - 8y = -2$

$\underline{-6x + 8y = \ 2}$

$\qquad\qquad 0 = 0,\ \text{true}$

The true statement indicates that the system is
dependent and has infinitely many solutions.
The solution set is $\{(x, y)\,|\,3x - 4y = -1\}$ or

$\{(x, y)\,|\,-6x + 8y = 2\}$.

34. $2x = 8y + 24$

$3x + 5y = 2$

Rewrite the first equation in the form $Ax + By = C$.

$2x - 8y = 24$

Multiply this equation by 3.
Multiply the second equation by -2.

$6x - 24y = 72$

$\underline{-6y - 10y = -4}$

$\qquad -34y = 68$

$\qquad\quad y = -2$

Back-substitute.

$3x + 5y = 2$

$3x + 5(-2) = 2$

$3x - 10 = 2$

$\qquad 3x = 12$

$\qquad\ x = 4$

The solution set is $\{(4, -2)\}$.

35. $5x - 7y = 2$

$3x = 4y$

Rewrite the second equation in the form
$Ax + By = C.$

$3x - 4y = 0$

Multiply this equation by -5.
Multiply the first equation by 3.

$15x - 21y = 6$

$\underline{-15x + 20y = 0}$

$-y = 6$

$y = -6$

Back-substitute.

$3x - 4y = 0$

$3x - 4(-6) = 0$

$3x + 24 = 0$

$3x = -24$

$x = -8$

The solution set is $\{(-8, -6)\}$.

36. $3x + 4y = -8$

$2x + 3y = -5$

Multiply the first equation by 2.
Multiply the second equation by -3.

$6x + 8y = -16$

$\underline{-6x - 9y = \ \ 15}$

$-y = -1$

$y = 1$

Back-substitute.

$3x + 4y = -8$

$3x + 4(1) = -8$

$3x + 4 = -8$

$3x = -12$

$x = -4$

The solution set is $\{(-4, 1)\}$.

37. $6x + 8y = 39$

$y = 2x - 2$

Substitute $2x - 2$ for y in the first equation.

$6x + 8y = 39$

$6x + 8(2x - 2) = 39$

$6x + 16x - 16 = 39$

$22x - 16 = 39$

$22x = 55$

$x = \dfrac{55}{22} = \dfrac{5}{2}$

Back-substitute $\dfrac{5}{2}$ for x into the second equation of
the system.

$y = 2x - 2$

$y = 2\left(\dfrac{5}{2}\right) - 2 = 5 - 2 = 3$

The solution set is $\left\{\left(\dfrac{5}{2}, 3\right)\right\}$.

38. $x + 2y = 7$

$2x + y = 8$

Multiply the first equation by -2.

$-2x - 4y = -14$

$\underline{2x + \ y = \ \ 8}$

$-3y = -6$

$y = 2$

Back-substitute.

$x + 2y = 7$

$x + 2(2) = 7$

$x + 4 = 7$

$x = 3$

The solution set is $\{(3, 2)\}$.

39. $y = 2x - 3$

$y = -2x - 1$

Substitute $-2x - 1$ for y in the first equation.

$-2x - 1 = 2x - 3$

$-4x - 1 = -3$

$-4x = -2$

$x = \dfrac{1}{2}$

Back-substitute.

$y = 2x - 3$

$y = 2\left(\dfrac{1}{2}\right) - 3 = -2$

The solution set is $\left\{\left(\dfrac{1}{2}, -2\right)\right\}$.

40. $3x - 6y = 7$

$3x = 6y$

Solve the second equation for x.

$3x = 6y$

$x = 2y$

Substitute $2y$ for x in the first equation.

$3x - 6y = 7$

$3(2y) - 6y = 7$

$6y - 6y = 7$

$0 = 7$

The false statement indicates that the system has no solution. The solution set is $\{\ \}$.

41. $y - 7 = 0$

$7x - 3y = 0$

Solve the first equation for y.

$y - 7 = 0$

$y = 7$

Substitute 7 for y in the second equation.

$7x - 3y = 0$

$7x - 3(7) = 0$

$7x - 21 = 0$

$7x = 21$

$x = 3$

The solution set is $\{(3, 7)\}$.

42. Let $x =$ the selling price for Klint's work.

Let $y =$ the selling price for Picasso's work.

$x + y = 239$

$x - y = 31$

Add the equations to eliminate y and solve for x.

$x + y = 239$

$\underline{x - y = 31}$

$2x = 270$

$x = 135$

Back-substitute to find y.

$x + y = 239$

$135 + y = 239$

$y = 104$

Klint's work sold for \$135 million and Picasso's work sold for \$104 million.

43. Let $x =$ the cholesterol content of one ounce of shrimp (in milligrams).

Let $y =$ the cholesterol content in one ounce of scallops.

$3x + 2y = 156$

$5x + 3y = 300 - 45$

Simplify the second equation.

$5x + 3y = 255$

Multiply this equation by 2.
Multiply the first equation by -3.

$-9x - 6y = -468$

$\underline{10x + 6y = 510}$

$x = 42$

Back-substitute to find y.

$3x + 2y = 156$

$3(42) + 2y = 156$

$126 + 2y = 156$

$2y = 30$

$y = 15$

There are 42 mg of cholesterol in an ounce of shrimp and 15 mg in an ounce of scallops.

44. Let x = the length of a tennis table top.
Let y = the width.
Use the formula for perimeter of a rectangle to write the first equation and the other information in the problem to write the second equation.
$2x + 2y = 28$
$4x - 3y = 21$
Multiply the first equation by –2.
$-4x + 4y = -56$
$\underline{4x - 3y = 21}$
$-7y = -35$
$y = 5$
Back-substitute to find x.
$2x + 2(5) = 28$
$2x + 10 = 28$
$2x = 18$
$x = 9$
The length is 9 feet and the width is 5 feet, so the dimensions of the table are 9 feet by 5 feet.

45. Let x = the length of the garden.
Let y = the width of the garden.
The perimeter of the garden is 24 yards, so
$2x + 2y = 24$.
Since there are two lengths and two widths to be fenced, the information about the cost of fencing leads to the equation $3(2x) + 2(2y) = 62$.
Simplify the second equation.
$6x + 4y = 62$.
Multiply the first equation by –2.
$-4x - 4y = -48$
$\underline{6x + 4y = 62}$
$2x = 14$
$x = 7$
Back-substitute to find y.
$2(7) + 2y = 24$
$14 + 2y = 24$
$2y = 10$
$y = 5$
The length of the garden is 7 yards and the width is 5 yards.

46. Let x = daily cost for room.
Let y = daily cost for car.
First plan: $3x + 2y = 360$
Second plan: $4x + 3y = 500$
Multiply the first equation by 3.
Multiply the second equation by –2.
$9x + 6y = 1080$
$\underline{-8x - 6y = -1000}$
$x = 80$
Back-substitute to find y.
$3(80) + 2y = 360$
$240 + 2y = 360$
$2y = 120$
$y = 60$
The cost per day is \$80 for the room and \$60 for the car.

47. Let x = the number of minutes of long-distance calls.
Let y = the monthly cost of a telephone plan.
Plan A: $y = 15 + 0.05x$
Plan B: $y = 10 + 0.075x$
To determine the amount of calling time that will result in the same cost for both plans, solve this system by the substitution method. Substitute $15 + 0.05x$ for y in the first equation.
$15 + 0.05x = 10 + 0.075x$
$15 - 0.025x = 10$
$-0.025x = -5$
$\dfrac{-0.025x}{-0.025} = \dfrac{-5}{-0.025}$
$x = 200$
Back-substitute to find y.
$y = 15 + 0.05(200) = 25$
The costs for the two plans will be equal for 200 minutes of long-distance calls per month. The cost of each plan will be \$25.

48. Let $x =$ the number orchestra tickets.
Let $y =$ the number balcony tickets.

$$x + y = 9$$
$$90x + 60y = 720$$

Solve the first equation for y.

$$x + y = 9$$
$$y = -x + 9$$

Use the substitution method.

$$90x + 60y = 720$$

$$90x + 60\overbrace{(-x + 9)}^{y} = 720$$

$$90x + 60(-x + 9) = 720$$

$$90x - 60x + 540 = 720$$

$$30x + 540 = 720$$

$$30x = 180$$

$$x = 6$$

Back-substitute to find y.

$$y = -x + 9$$

$$y = -6 + 9 = 3$$

You purchased 6 orchestra tickets and 3 balcony tickets.

49.

	Principal	× Rate	= Interest
8% Investment	x	0.08	$0.08x$
10% Investment	y	0.10	$0.10y$

Since the total investment is \$10,000 the first equation is $x + y = 10,000$.

Since the total interest is \$940, the second equation is $0.08x + 0.10y = 940$.

System of equations: $\begin{cases} x + y = 10,000 \\ 0.08x + 0.10y = 940 \end{cases}$

Solve the first equation for y and substitute into the second equation.

$$x + y = 10,000$$

$$y = 10,000 - x$$

Solve for x. $\quad 0.08x + 0.10\overbrace{(10,000 - x)}^{y} = 940$

$$0.08x + 1000 - 0.10x = 940$$

$$-0.02x + 1000 = 940$$

$$-0.02x + 1000 - 1000 = 940 - 1000$$

$$-0.02x = -60$$

$$\frac{-0.02x}{-0.02} = \frac{-60}{-0.02}$$

$$x = 3000$$

Back-substitute to find y.

$$x + y = 10,000$$

$$3000 + y = 10,000$$

$$y = 7000$$

\$3000 was invested at 8% and \$7000 was invested at 10%.

50.

	Gallons $\times$	Percent Salt $=$	Amount of Salt
75% Saltwater Solution	x	0.75	$0.75x$
50% Saltwater Solution	y	0.50	$0.50y$
60% Saltwater Solution	10	0.60	$0.60(10)$

Since there are a total of 10 gallons, the first equation is $x + y = 10$.

Since the total amount of salt is $0.60(10)$, the second equation is $0.75x + 0.50y = 0.60(10)$.

System of equations: $\begin{cases} x + y = 10 \\ 0.75x + 0.50y = 0.60(10) \end{cases}$

Solve the first equation for y and substitute into the second equation.

$x + y = 10$

$\quad y = 10 - x$

Solve for x. $0.75x + 0.50\overbrace{(10 - x)}^{y} = 0.60(10)$

$\qquad 0.75x + 5 - 0.50x = 6$

$\qquad\qquad 0.25x + 5 = 6$

$\qquad\qquad\quad 0.25x = 1$

$\qquad\qquad \dfrac{0.25x}{0.25} = \dfrac{1}{0.25}$

$\qquad\qquad\qquad x = 4$

Back-substitute to find y.

$x + y = 10$

$4 + y = 10$

$\quad y = 6$

To obtain 10 gallons of a 60% saltwater solution, 4 gallons of a 75% saltwater solution and 6 gallons of a 50% saltwater solution must be used.

51. $3x - y \le 6$

$\quad x + y \ge 2$

Graph $3x - y \le 6$ by graphing $3x - y = 6$ as a solid line using the x-intercept 2, and y-intercept -6. Because $(0,0)$ makes the inequality true, shade the half-plane containing $(0,0)$.

Graph $x + y \ge 2$ by graphing $x + y = 2$ as a solid line using the x-intercept 2, and y-intercept 2. Because $(0,0)$ makes the inequality false, shade the half-plane *not* containing $(0,0)$.

The solution set of the system is the intersection (overlap) of the two shaded regions.

$3x - y \le 6$
$x + y \ge 2$

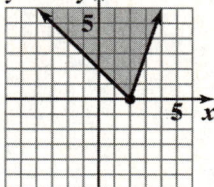

52. $x + y < 4$

$x - y < 4$

Graph $x + y < 4$ by graphing $x + y = 4$ as a dashed line using the *x*-intercept 4, and *y*-intercept 4. Because $0 + 0 < 4$ is true, shade the half-plane containing (0,0).

Graph $x - y < 4$ by graphing $x - y = 4$ as a dashed line using the *x*-intercept 4, and *y*-intercept −4. Because $0 - 0 < 4$ is true, shade the half-plane containing (0,0).

The solution set of the system is the intersection of the two shaded regions.

$x + y < 4$

$x - y < 4$

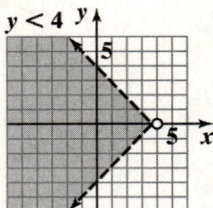

53. $y < 2x - 2$

$x \geq 3$

Graph $y < 2x - 2$ by graphing $y = 2x - 3$ as a dashed line using the slope 2 and *y*-intercept −2. Since $0 < 2(0) - 2$ is false, shade the half-plane *not* containing (0,0).

Graph $x \geq 3$ by graphing $x = 3$ as a solid vertical line with *x*-intercept 3. Since $0 \geq 3$ is false, shade the half-plane *not* containing (0,0).

The solution set of the system is the intersection of the two shaded regions.

$y < 2x - 2$

$x \geq 3$

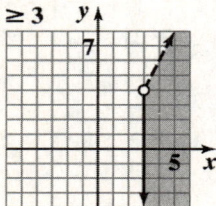

54. $4x + 6y \leq 24$

$y > 2$

Graph $4x + 6y \leq 24$ by graphing $4x + 6y = 24$ as a solid line with *x*-intercept 6 and *y*-intercept 4. Since $4(0) + 6(0) \leq 24$ is true, shade the half plane containing (0,0).

Graph $y > 2$ by graphing the dashed horizontal line at $y = 2$. Since $0 > 2$ is false, shade the half-plane *not* containing (0,0), the half-plane above the line. The solution set of the system is the intersection of the two shaded regions.

$4x + 6y \leq 24$

$y > 2$

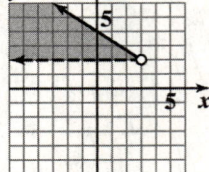

55. $x \leq 3$

$y \geq -2$

Graph $x \leq 3$ by graphing the solid vertical line at $x = 3$. Since $0 \leq 3$ is true, shade the half-plane containing (0,0).

Graph $y \geq -2$ by graphing $y = -2$ as a solid horizontal line. Since $0 \geq -2$ is true, shade the half-plane containing (0,0).

The solution set of the system is the intersection of the two shaded regions.

$x \leq 3$

$y \geq -2$

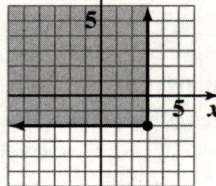

56. $y \ge \dfrac{1}{2}x - 2$

$y \le \dfrac{1}{2}x + 1$

Graph $y \ge \dfrac{1}{2}x - 2$ by graphing $y = \dfrac{1}{2}x - 2$ as a

solid line with slope $\dfrac{1}{2}$ and y-intercept -2. Since

$0 \ge \dfrac{1}{2}(0) - 2$ is true, shade the half-plane

containing $(0,0)$.

Graph $y \le \dfrac{1}{2}x + 1$ by graphing $y = \dfrac{1}{2}x + 1$ as a

solid line with slope $\dfrac{1}{2}$ and y-intercept 1. Since

$0 \le \dfrac{1}{2}(0) + 1$ is true, shade the half-plane containing

$(0,0)$.

The solution set of the system is the intersection of
the two shaded regions. This is the region between
and including the two parallel lines.

$y \ge \dfrac{1}{2}x - 2$

$y \le \dfrac{1}{2}x + 1$

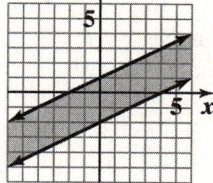

57. $x \le 0$

$y \ge 0$

Graph $x \le 0$ by graphing $x = 0$ (the y-axis) as a
solid line. Shade the half-plane to the left of the y-
axis.

Graph $y \ge 0$ by graphing $y = 0$ (the x-axis) as a
solid line. Shade the half-plane above the x-axis.
The solution set of the system is the intersection of
the two shaded regions. Notice that this is all of
quadrant II, including the portions of the axes that
are boundaries of the region.

$x \le 0$

$y \ge 0$

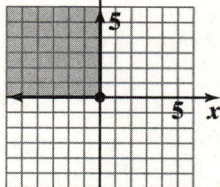

Chapter 4 Test

1. $2x + y = 5$

$2(5) + (-5) = 5$

$10 + (-5) = 5$

$5 = 5$, true

$x + 3y = -10$

$5 + 3(-5) = -10$

$5 + (-15) = -10$

$-10 = -10$, true

Since the ordered pair $(5, -5)$ satisfies both
equations, it is a solution of the given system.

2. $x + 5y = 7$

$-3 + 5(2) = 7$

$-3 + 10 = 7$

$7 = 7$, true

$3x - 4y = 1$

$3(-3) - 4(2) = 1$

$-9 - 8 = 1$

$-17 = 1$, false

Since the ordered pair $(-3,2)$ fails to satisfy *both*
equations, it is not a solution of the given system.

3. $x + y = 6$

$4x - y = 4$

Graph both lines on the same axes.
$x + y = 6$: x-intercept $= 6$; y-intercept $= 6$
$4x - y = 4$: x-intercept: 1; y-intercept $= -4$

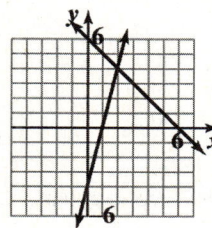

The solution set is $\{(2,4)\}$.

4. $2x + y = 8$

$\quad\quad y = 3x - 2$

Graph both lines on the same axes.

$2x + y = 8$: x-intercept = 4; y-intercept = 8

$y = 3x - 2$: slope = 3; y-intercept = -2

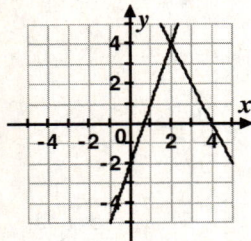

The solution set is $\{(2, 4)\}$.

5. $\quad\quad x = y + 4$

$\quad 3x + 7y = -18$

Substitute $y + 4$ for x in the second equation.

$3x + 7y = -18$

$3(y + 4) + 7y = -18$

$3y + 12 + 7y = -18$

$10y + 12 = -18$

$10y = -30$

$y = -3$

Back-substitute -3 for y in the first equation.

$x = y + 4$

$x = -3 + 4 = 1$

The solution set is $\{(1, -3)\}$.

6. $\quad 2x - y = 7$

$\quad 3x + 2y = 0$

Solve the first equation for y.

$2x - y = 7$

$-y = -2x + 7$

$y = 2x - 7$

Substitute $2x - 7$ for y in the second equation.

$3x + 2y = 0$

$3x + 2(2x - 7) = 0$

$3x + 4x - 14 = 0$

$7x - 14 = 0$

$7x = 14$

$x = 2$

Back-substitute 2 for x in the equation $3x + 2y = 0$.

$3x + 2y = 0$

$3(2) + 2y = 0$

$6 + 2y = 0$

$2y = -6$

$y = -3$

The solution set is $\{(2, -3)\}$.

7. $\quad 2x - 4y = 3$

$\quad\quad x = 2y + 4$

Substitute $2y + 4$ for x in the first equation.

$2x - 4y = 3$

$2(2y + 4) - 4y = 3$

$4y + 8 - 4y = 3$

$8 = 3$, false

The false statement indicates that the system has no solution. The solution set is $\{\ \}$.

8. $\quad 2x + y = 2$

$\quad \underline{4x - y = -8}$

$\quad 6x \quad\quad = -6$

$\quad\quad\quad x = -1$

Back-substitute to find y.

$2x + y = 2$

$2(-1) + y = 2$

$-2 + y = 2$

$y = 4$

The solution set is $\{(-1, 4)\}$.

9. $\quad 2x + 3y = 1$

$\quad 3x + 2y = -6$

Multiply the first equation by 3.

Multiply the second equation by -2.

$\quad 6x + 9y = 3$

$\quad \underline{-6x - 4y = 12}$

$\quad\quad\quad 5y = 15$

$\quad\quad\quad\quad y = 3$

Back-substitute to find x.

$2x + 3y = 1$

$2x + 3(3) = 1$

$2x + 9 = 1$

$2x = -8$

$x = -4$

The solution set is $\{(-4, 3)\}$.

10. $3x - 2y = 2$

$-9x + 6y = -6$

Multiply the first equation by 3.

$9x - 6y = 6$

$-9x + 6y = -6$

$0 = 0$, true

The true statement $0 = 0$ indicates that the system is dependent and the equation has infinitely many solutions.

The solution set is $\{(x, y) | 3x - 2y = 2\}$ or $\{(x, y) | -9x + 6y = -6\}$.

11. Let x = the percentage of females named Mary.
Let y = the percentage of females named Patricia.
The system is $x + y = 3.7$

$\qquad x - y = 1.5$

Add the two equations:

$x + y = 3.7$

$x - y = 1.5$

$2x \quad = 5.2$

$x \quad = 2.6$

Back substitute to find y.

$2.6 + y = 3.7$

$y = 1.1$

2.6% of females are named Mary and 1.1% are named Patricia.

12. Let x = the length of the garden.
Let y = the width of the garden.
The perimeter of the garden is 34 yards so

$2x + 2y = 34$.

Since there are two lengths and two widths to be fenced, the information about the cost of fencing leads to the equation $2(2x) + 1(2y) = 58$.

Simplify the second equation.

$4x + 2y = 58$

Multiply this equation by -1 and add the result to the first equation.

$2x + 2y = 34$

$-4x - 2y = -58$

$-2x \qquad = -24$

$x \qquad = 12$

Back substitute to find y.

$2(12) + 2y = 34$

$24 + 2y = 34$

$2y = 10$

$y = 5$

The length of the garden is 12 yards and the width is 5 yards.

13. Let x = the number of minutes of calls.
Let y = the monthly cost of a telephone plan.

Plan A: $y = 15 + 0.05x$

Plan B: $y = 5 + 0.07x$

To determine the amount of calling time that will result in the same cost for both plans, solve this system by the substitution method. Substitute $5 + 0.07x$ for y in the first equation.

$5 + 0.07x = 15 + 0.05x$

$5 + 0.02x = 15$

$0.02x = 10$

$\dfrac{0.02x}{0.02} = \dfrac{10}{0.02}$

$x = 500$

If $x = 500$, $y = 15 + 0.05(500) = 40$.

The cost of the two plans will be equal for 500 minutes per month. The cost of each plan will be $40.

14.

	Principal	$\times$ Rate	$=$ Interest
9% Investment	x	0.09	$0.09x$
6% Investment	y	0.06	$0.06y$

Since the total investment is $6000 the first equation is $x + y = 6000$.

Since the total interest is $940, the second equation is $0.09x + 0.06y = 480$.

System of equations: $\begin{cases} x + y = 6000 \\ 0.09x + 0.06y = 480 \end{cases}$

Solve the first equation for y and substitute into the second equation.

$x + y = 6000$

$y = 6000 - x$

Solve for x. $0.09x + 0.06\overbrace{(6000 - x)}^{y} = 480$

$0.09x + 360 - 0.06x = 480$

$0.03x + 360 = 480$

$0.03x = 120$

$\dfrac{0.03x}{0.03} = \dfrac{120}{0.03}$

$x = 4000$

Back-substitute to find y.

$x + y = 6000$

$4000 + y = 6000$

$y = 2000$

$4000 was invested at 9% at $2000 was invested at 6%.

15.

	Milliliters	× Percent of Acid	= Amount of Acid
50% Acid Solution	x	0.50	$0.50x$
80% Acid Solution	y	0.80	$0.80y$
68% Acid Mixture	100	0.68	$0.68(100)$

Since there are a total of 100 milliliters, the first equation is $x + y = 100$.

Since the total amount of salt is $0.68(100)$, the second equation is $0.50x + 0.80y = 0.68(100)$.

System of equations: $\begin{cases} x + y = 100 \\ 0.50x + 0.80y = 0.68(100) \end{cases}$

Solve the first equation for y and substitute into the second equation.

$x + y = 100$

$\quad y = 100 - x$

Solve for x. $\quad 0.50x + 0.80 \overbrace{(100 - x)}^{y} = 0.68(100)$

$\qquad\qquad 0.50x + 80 - 0.80x = 68$

$\qquad\qquad\qquad\quad -0.30x = -12$

$\qquad\qquad\qquad \dfrac{-0.30x}{-0.30} = \dfrac{-12}{-0.30}$

$\qquad\qquad\qquad\qquad\quad x = 40$

Back-substitute to find y.

$\quad x + y = 100$

$40 + y = 100$

$\qquad\quad y = 60$

To obtain 100 milliliters of 68% acid solution, 40 milliliters of 50% acid solution and 60 milliliters of 80% acid solution must be used.

16. $x - 3y > 6$

$2x + 4y \leq 8$

Graph $x - 3y > 6$ by graphing $x - 3y = 6$ as a dashed line using the x-intercept 6 and y-intercept −2. Because $0 - 3(0) > 6$ is false, shade the half-plane *not* containing (0,0).

Graph $2x + 4y \leq 8$ by graphing $2x + 4y = 8$ as a solid line using the x-intercept 4 and y-intercept 2. Because $2(0) + 4(0) \leq 8$ is true, shade the half-plane containing (0,0).

The solution set of the system is the intersection (overlap) of the two shaded regions.

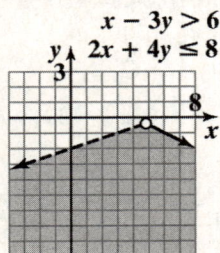

17. $y \geq 2x - 4$

$x < 2$

Graph $y \geq 2x - 4$ by graphing $y = 2x - 4$ as a solid line using the slope 2 and y-intercept –4. Since $0 \geq 2(0) - 4$ is true, shade the half-plane containing $(0,0)$.

Graph $x < 2$ by graphing $x = 2$ as a dashed vertical line with x-intercept 2. Since $0 < 2$ is true, shade the half-plane containing $(0,0)$.

The solution set of the system is the intersection (overlap) of the two shaded regions.

$y \geq 2x - 4$
$x < 2$

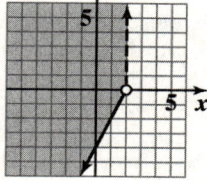

Cumulative Review Exercises (Chapters 1-4)

1. $-14 - \left[18 - (6 - 10) \right]$

$= -14 - \left[18 - (-4) \right]$

$= -14 - \left[18 + 4 \right]$

$= -14 - 22$

$= -14 + (-22)$

$= -36$

2. $6(3x - 2) - (x - 1) = 18x - 12 - x + 1$

$ = 17x - 11$

3. $17(x + 3) = 13 + 4(x - 10)$

$17x + 51 = 13 + 4x - 40$

$17x + 51 = 4x - 27$

$13x = -78$

$x = -6$

The solution set is {–6}.

4. $\dfrac{x}{4} - 1 = \dfrac{x}{5}$

To clear fractions, multiply both sides by 20.

$20\left(\dfrac{x}{4} - 1 \right) = 20\left(\dfrac{x}{5} \right)$

$5x - 20 = 4x$

$x - 20 = 0$

$x = 20$

The solution set is {20}.

5. $A = P + Prt$

$A - P = Prt$

$\dfrac{A - P}{Pr} = \dfrac{Prt}{Pr}$

$\dfrac{A - P}{Pr} = t$ or $t = \dfrac{A - P}{Pr}$

6. $2x - 5 < 5x - 11$

$-3x - 5 < -11$

$-3x < -6$

$\dfrac{-3x}{-3} > \dfrac{-6}{-3}$

$x > 2$

The solution set is $\{x \mid x > 2\}$.

7. $x - 3y = 6$

x-intercept:

$x - 3y = 6$

$x - 3(0) = 6$

$x = 6$

y-intercept:

$x - 3y = 6$

$0 - 3y = 6$

$-3y = 6$

$y = -2$

Check point:

$x - 3y = 6$

$3 - 3y = 6$

$-3y = 3$

$y = -1$

A check point is $(3, -1)$.

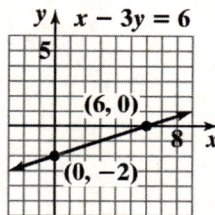

8.

x	$y = 0x + 4$	(x, y)
-6	$y = 0(-6) + 4 = 4$	$(-6, 4)$
-3	$y = 0(-3) + 4 = 4$	$(-3, 4)$
0	$y = 0(0) + 4 = 4$	$(0, 4)$
3	$y = 0(3) + 4 = 4$	$(3, 4)$
6	$y = 0(6) + 4 = 4$	$(6, 4)$

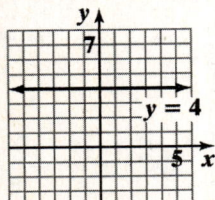

9. $y = -\dfrac{3}{5}x + 2$

slope $= -\dfrac{3}{5} = \dfrac{-3}{5}$; y-intercept $= 2$

Plot the point (0,2). From this point, move 3 units down (because -3 is negative) and 5 units to the right to reach the point (5, −1). Draw a line through (0,2) and (5, −1).

$y = -\dfrac{3}{5}x + 2$

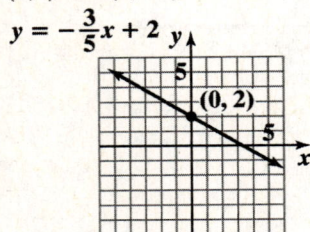

10. $3x - 4y = 8$

$4x + 5y = -10$

To solve this system by the addition method, multiply the first equation by 4 and the second equation by −3. Then add the results.

$\quad 12x - 16y = 32$

$\underline{-12x - 15y = 30}$

$\qquad -31y = 62$

$\qquad\quad y = -2$

Back-substitute to find x.

$3x - 4y = 8$

$3x - 4(-2) = 8$

$3x + 8 = 8$

$3x = 0$

$x = 0$

The solution set is $\{(0, -2)\}$.

11. $2x - 3y = 9$

$\quad\quad y = 4x - 8$

To solve this system by the substitution method, substitute $4x - 8$ for y in the first equation.

$2x - 3y = 9$

$2x - 3(4x - 8) = 9$

$2x - 12x + 24 = 9$

$-10x + 24 = 9$

$-10x = -15$

$x = \dfrac{3}{2}$

Back-substitute $\dfrac{3}{2}$ for x in the second equation.

$y = 4x - 8 = 4\left(\dfrac{3}{2}\right) - 8 = -2$

The solution set is $\left\{\left(\dfrac{3}{2}, -2\right)\right\}$.

12. $m = \dfrac{y_2 - y_1}{x_2 - x_1} = \dfrac{-5 - (-6)}{6 - 5} = \dfrac{1}{1} = 1$

13. $y - y_1 = m(x - x_1)$

$y - 6 = -4\left[x - (-1)\right]$

$y - 6 = -4(x + 1)$

slope-intercept form:

$y - 6 = -4x - 4$

$y = -4x + 2$

14. Use the formula for the area of a triangle.

$A = \dfrac{1}{2}bh$

$80 = \dfrac{1}{2} \cdot 16 \cdot h$

$80 = 8h$

$10 = h$

The height is 10 feet.

15. Let x = the cost of one pen.
Let y = the cost of one pad.
$$10x + 15y = 26$$
$$5x + 10y = 16$$
Multiply the second equation by –2, and add the result to the first equation.
$$10x + 15y = 26$$
$$\underline{-10x - 20y = -32}$$
$$-5y = -6$$
$$y = \frac{6}{5} = 1.20$$
Back-substitute 1.20 for y and solve for x.
$$10x + 15(1.20) = 26$$
$$10x + 18 = 26$$
$$10x = 8$$
$$x = \frac{8}{10} = 0.8$$
One pen costs $0.80 and one pad costs $1.20.

16. The integers in the given set are –93, 0, $\frac{7}{1}$ (=7) and $\sqrt{100}$ (=10).

17. The *living alone* line has a positive slope. This means that the percentage of U.S. adults living alone increased from 1960 through 2008.

18. The *married, living with kids* line has a negative slope. This means that the percentage of U.S. adults married living with kids decreased from 1960 through 2008.

19. If y is the percentage of U.S. adults married living with kids and x is the number of years after 1960, then this can be modeled by the equation
$$y = -0.4x + 47.$$
To find when the percentage of U.S. adults married living with kids will be 19%, let $y = 19$ and solve for x.
$$y = -0.4x + 47$$
$$19 = -0.4x + 47$$
$$-28 = -0.4x$$
$$\frac{-28}{-0.4} = \frac{-0.4x}{-0.4}$$
$$70 = x$$
According to the percentage of U.S. adults married living with kids will be 19%, 70 years after 1960, or 2030.

20. To find when the percentage of U.S. adults living alone will be 19%, let $y = 19$ and solve for x.
$$y = 0.2x + 5$$
$$19 = 0.2x + 5$$
$$14 = 0.2x$$
$$\frac{14}{0.2} = x$$
$$70 = x$$
According to the percentage of U.S. adults living alone will be 19%, 70 years after 1960, or 2030. Based on the previous exercise, in 2030 the same percentage of U.S. adults will be married, living with kids and living alone, namely 19%.

Chapter 5
Exponents and Polynomials

5.1 Check Points

1. $(-11x^3 + 7x^2 - 11x - 5) + (16x^3 - 3x^2 + 3x - 15)$
    $= -11x^3 + 7x^2 - 11x - 5 + 16x^3 - 3x^2 + 3x - 15$
    $= -11x^3 + 16x^3 + 7x^2 - 3x^2 - 11x + 3x - 5 - 15$
    $= 5x^3 + 4x^2 - 8x - 20$

2. $-11x^3 + 7x^2 - 11x - 5$
    $\underline{+16x^3 - 3x^2 + 3x - 15}$
    $5x^3 + 4x^2 - 8x - 20$

3. $(9x^2 + 7x - 2) - (2x^2 - 4x - 6)$
    $= 9x^2 + 7x - 2 - 2x^2 + 4x + 6$
    $= 9x^2 - 2x^2 + 7x + 4x - 2 + 6$
    $= 7x^2 + 11x + 4$

4. $(10x^3 - 5x^2 + 7x - 2) - (3x^3 - 8x^2 - 5x + 6)$
    $= 10x^3 - 5x^2 + 7x - 2 - 3x^3 + 8x^2 + 5x - 6$
    $= 10x^3 - 3x^3 - 5x^2 + 8x^2 + 7x + 5x - 2 - 6$
    $= 7x^3 + 3x^2 + 12x - 8$

5. $8y^3 - 10y^2 - 14y - 2$
    $-\left(5y^3 - 3y + 6\right)$

    To subtract, add the opposite of the polynomial being subtracted.

    $8y^3 - 10y^2 - 14y - 2$
    $\underline{-5y^3 + 3y - 6}$
    $3y^3 - 10y^2 - 11y - 8$

6. Make a table of values.

x	$y = x^2 - 1$	(x, y)
-3	$y = (-3)^2 - 1 = 8$	$(-3, 8)$
-2	$y = (-2)^2 - 1 = 3$	$(-2, 3)$
-1	$y = (-1)^2 - 1 = 0$	$(-1, 0)$
0	$y = (0)^2 - 1 = -1$	$(0, -1)$
1	$y = (1)^2 - 1 = 0$	$(1, 0)$
2	$y = (2)^2 - 1 = 3$	$(2, 3)$
3	$y = (3)^2 - 1 = 8$	$(3, 8)$

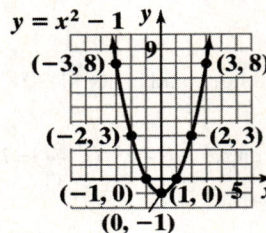

5.1 Concept and Vocabulary Check

1. whole

2. standard

3. monomial

4. binomial

5. trinomial

6. 0

7. greatest

8. like

9. opposite

5.1 Exercise Set

1. $3x + 7$ is a binomial of degree 1.

3. $x^2 - 2x$ is a binomial of degree 3.

5. $8x^2$ is a monomial of degree 2.

7. 5 is a monomial. Because it is a nonzero constant, its degree is 0.

9. $x^2 - 3x + 4$ is a trinomial of degree 2.

11. $7y^2 - 9y^4 + 5$ is a trinomial of degree 4.

13. $15x - 7x^3$ is a binomial of degree 3.

15. $-9y^{23}$ is a monomial of degree 23.

17. $(9x + 8) + (-17x + 5)$
$= 9x + 8 + (-17)x + 5$
$= 9x + 8 - 17x + 5$
$= (9x - 17x) + (8 + 5)$
$= -8x + 13$

19. $\left(4x^2 + 6x - 7\right) + \left(8x^2 + 9x - 2\right)$
$= 4x^2 + 6x - 7 + 8x^2 + 9x - 2$
$= \left(4x^2 + 8x^2\right) + (6x + 9x) + (-7 - 2)$
$= 12x^2 + 15x - 9$

21. $\left(7x^2 - 11x\right) + \left(3x^2 - x\right)$
$= 7x^2 - 11x + 3x^2 - x$
$= \left(7x^2 + 3x^2\right) + (-11x - x)$
$= 10x^2 - 12x$

23. $\left(4x^2 - 6x + 12\right) + \left(x^2 + 3x + 1\right)$
$= 4x^2 - 6x + 12 + x^2 + 3x + 1$
$= \left(4x^2 + x^2\right) + (-6x + 3x) + (12 + 1)$
$= 5x^2 - 3x + 13$

25. $\left(4y^3 + 7y - 5\right) + \left(10y^2 - 6y + 3\right)$
$= \left(4y^3 + 7y - 5\right) + \left(10y^2 - 6y + 3\right)$
$= 4y^3 + 7y - 5 + 10y^2 - 6y + 3$
$= 4y^3 + 10y^2 + (7y - 6y) + (-5 + 3)$
$= 4y^3 + 10y^2 + y - 2$

27. $= \left(2x^2 - 6x + 7\right) + \left(3x^3 - 3x\right)$

$= 2x^2 - 6x + 7 + 3x^3 - 3x$

$= 3x^3 + 2x^2 + \left(-6x - 3x\right) + 7$

$= 3x^3 + 2x^2 - 9x + 7$

29. $\left(4y^2 + 8y + 11\right) + \left(-2y^3 + 5y + 2\right)$

$= 4y^2 + 8y + 11 + (-2)y^3 + 5y + 2$

$= -2y^3 + 4y^2 + \left(8y + 5y\right) + \left(11 + 2\right)$

$= -2y^3 + 4y^2 + 13y + 13$

31. $\left(-2y^6 + 3y^4 - y^2\right) + \left(-y^6 + 5y^4 + 2y^2\right) = -2y^6 + 3y^4 - y^2 - y^6 + 5y^4 + 2y^2$

$= \left(-2y^6 - y^6\right) + \left(3y^4 + 5y^4\right) + \left(-y^2 + 2y^2\right)$

$= -3y^6 + 8y^4 + y^2$

33. $\left(9x^3 - x^2 - x - \dfrac{1}{3}\right) + \left(x^3 + x^2 + x + \dfrac{4}{3}\right) = \left(9x^3 + x^3\right) + \left(-x^2 + x^2\right) + \left(-x + x\right) + \left(-\dfrac{1}{3} + \dfrac{4}{3}\right)$

$= 10x^3 + \dfrac{3}{3}$

$= 10x^3 + 1$

35. $\left(\dfrac{1}{5}x^4 + \dfrac{1}{3}x^3 + \dfrac{3}{8}x^2 + 6\right) + \left(-\dfrac{3}{5}x^4 + \dfrac{2}{3}x^3 - \dfrac{1}{2}x^2 - 6\right)$

$= \left[\dfrac{1}{5}x^4 + \left(-\dfrac{3}{5}x^4\right)\right] + \left(\dfrac{1}{3}x^2 + \dfrac{2}{3}x^3\right) + \left[\dfrac{3}{8}x^2 + \left(-\dfrac{1}{2}x^2\right)\right] + \left[6 + (-6)\right]$

$= -\dfrac{2}{5}x^4 + x^3 - \dfrac{1}{8}x^2$

37. $\left(0.03x^5 - 0.1x^3 + x + 0.03\right) + \left(-0.02x^5 + x^4 - 0.7x + 0.3\right) = \left(0.03x^5 - 0.02x^5\right) + x^4 - 0.1x^3 + \left(x - 0.07x\right) + \left(0.03 + 0.3\right)$

$= 0.01x^5 + x^4 - 0.1x^3 + 0.3x + 0.33$

39. $\quad 5y^3 - 7y^2$

$\quad \underline{6y^3 + 4y^2}$

$\quad 11y^3 - 3y^2$

41. $\quad 3x^2 - 7x + 4$

$\quad \underline{-5x^2 + 6x - 3}$

$\quad -2x^2 - x + 1$

43. $\dfrac{1}{4}x^4 - \dfrac{2}{3}x^3 - 5$

$-\dfrac{1}{2}x^4 + \dfrac{1}{5}x^3 + 4.7$

To add, rewrite using common denominators for common terms.

$\dfrac{1}{4}x^4 - \dfrac{10}{15}x^3 - 5.0$

$-\dfrac{1}{2}x^4 + \dfrac{3}{15}x^3 + 4.7$

$-\dfrac{1}{4}x^4 - \dfrac{7}{15}x^3 - 0.3$

45. $y^3 + 5y^2 - 7y - 3$

$-2y^3 + 3y^2 + 4y - 11$

$-y^3 + 8y^2 - 3y - 14$

47. $4x^3 - 6x^2 + 5x - 7$

$-9x^3 \quad\quad - 4x + 3$

$-5x^3 - 6x^2 + x - 4$

49. $7x^4 - 3x^3 + x^2$

$\quad\quad x^3 - x^2 + 4x - 2$

$7x^4 - 2x^3 \quad\quad + 4x - 2$

51. $7x^2 - 9x + 3$

$4x^2 + 11x - 2$

$-3x^2 + 5x - 6$

$8x^2 + 7x - 5$

53. $1.2x^3 - 3x^2 + 9.1$

$7.8x^3 - 3.1x^2 + 8$

$\quad\quad\quad 1.2x^2 - 6$

$9x^3 - 4.9x^2 + 11.1$

55. $(x-8)-(3x+2)=(x-8)+(-3x-2)$
$\quad\quad\quad\quad = (x-3x)+(-8-2)$
$\quad\quad\quad\quad = -2x-10$

57. $\left(x^2 - 5x - 3\right) - \left(6x^2 - 4x + 9\right)$
$= \left(x^2 - 5x - 3\right) + \left(-6x^2 + 4x - 9\right)$
$= \left(x^2 - 6x^2\right) + \left(-5x - 4x\right) + \left(-3 - 9\right)$
$= -5x^2 - 9x - 12$

59. $\left(x^2 - 5x\right) - \left(6x^2 - 4x\right)$
$= \left(x^2 - 5x\right) + \left(-6x^2 + 4x\right)$
$= \left(x^2 - 6x^2\right) + \left(-5x + 4x\right)$
$= -5x^2 - x$

61. $\left(x^2 - 8x - 9\right) - \left(5x^2 - 4x - 3\right)$
$= \left(x^2 - 8x - 9\right) + \left(-5x^2 + 4x + 3\right)$
$= -4x^2 - 4x - 6$

63. $(y-8)-(3y-2)=(y-8)+(-3y+2)$
$\quad\quad\quad\quad\quad = -2y - 6$

65. $\left(6y^3 + 2y^2 - y - 11\right) - \left(y^2 - 8y + 9\right)$
$= \left(6y^3 + 2y^2 - y - 11\right) + \left(-y^2 + 8y - 9\right)$
$= 6y^3 + y^2 + 7y - 20$

67. $\left(7n^3 - n^7 - 8\right) - \left(6n^3 - n^7 - 10\right)$
$= \left(7n^3 - n^7 - 8\right) + \left(-6n^3 + n^7 + 10\right)$
$= \left(7n^3 - 6n^3\right) + \left(-n^7 + n^7\right) + \left(-8 + 10\right)$
$= n^3 + 2$

69. $\left(y^6 - y^3\right) - \left(y^2 - y\right)$
$= \left(y^6 - y^3\right) + \left(-y^2 + y\right)$
$= y^6 - y^3 - y^2 + y$

71. $\left(7x^4 + 4x^2 + 5x\right) - \left(-19x^4 - 5x^2 - x\right)$
$= \left(7x^4 + 4x^2 + 5x\right) + \left(19x^4 + 5x^2 + x\right)$
$= 26x^4 + 9x^2 + 6x$

73. $\left(\dfrac{3}{7}x^3 - \dfrac{1}{5}x - \dfrac{1}{3}\right) - \left(-\dfrac{2}{7}x^3 + \dfrac{1}{4}x - \dfrac{1}{3}\right)$

$= \left(\dfrac{3}{7}x^3 - \dfrac{1}{5}x - \dfrac{1}{3}\right) + \left(\dfrac{2}{7}x^3 - \dfrac{1}{4}x + \dfrac{1}{3}\right)$

$= \left(\dfrac{3}{7}x^3 + \dfrac{2}{7}x^3\right) + \left(-\dfrac{1}{5}x - \dfrac{1}{4}x\right) + \left(-\dfrac{1}{3} + \dfrac{1}{3}\right)$

$= \left(\dfrac{3}{7}x^3 + \dfrac{2}{7}x^3\right) + \left(-\dfrac{4}{20}x - \dfrac{5}{20}x\right)$

$= \dfrac{5}{7}x^3 - \dfrac{9}{20}x$

75. $7x + 1$

$\underline{-(3x - 5)}$

To subtract, add the opposite of the polynomial being subtracted.

$7x + 1$

$\underline{-3x + 5}$

$4x + 6$

77. $7x^2 - 3$

$\underline{-\left(-3x^2 + 4\right)}$

To subtract, add the opposite of the polynomial being subtracted.

$7x^2 - 3$

$\underline{3x^2 - 4}$

$10x^2 - 7$

79. $7y^2 - 5y + 2$

$\underline{-\left(11y^2 + 2y - 3\right)}$

To subtract, add the opposite of the polynomial being subtracted.

$7y^2 - 5y + 2$

$\underline{+ -11y^2 - 2y + 3}$

$-4y^2 - 7y + 5$

81. $7x^3 + 5x^2 - 3$

$\underline{-\left(-2x^3 - 6x^2 + 5\right)}$

To subtract, add the opposite of the polynomial being subtracted.

$7x^3 + 5x^2 - 3$

$\underline{2x^3 + 6x^2 - 5}$

$9x^3 + 11x^2 - 8$

83. $5y^3 + 6y^2 - 3y + 10$

$\underline{-\left(6y^3 - 2y^2 - 4y - 4\right)}$

To subtract, add the opposite of the polynomial being subtracted.

$5y^3 + 6y^2 - 3y + 10$

$\underline{+ -6y^3 + 2y^2 + 4y + 4}$

$-y^3 + 8y^2 + \ y + 14$

85. $7x^4 - 3x^3 + 2x^2$

$\underline{-\left(\quad - x^3 - x^2 + x - 2\right)}$

To subtract, add the opposite of the polynomial being subtracted.

$7x^4 - 3x^3 + 2x^2$

$\underline{+ \qquad x^3 + \ x^2 - x + 2}$

$7x^4 - 2x^3 + 3x^2 - x + 2$

87. $0.07x^3 - 0.01x^2 + 0.02x$

$\underline{-\left(0.02x^3 - 0.03x^2 - \quad x\right)}$

To subtract, add the opposite of the polynomial being subtracted.

$0.07x^3 - 0.01x^2 + 0.02x$

$\underline{-0.02x^3 + 0.03x^2 + \qquad x}$

$0.05x^3 + 0.02x^2 + 1.02x$

89. Table of values:

x	$y = x^2$	(x, y)
-3	$y = (-3)^2 = 9$	$(-3, 9)$
-2	$y = (-2)^2 = 4$	$(-2, 4)$
-1	$y = (-1)^2 = 1$	$(-1, 1)$
0	$y = (0)^2 = 0$	$(0, 0)$
1	$y = (1)^2 = 1$	$(1, 1)$
2	$y = (2)^2 = 4$	$(2, 4)$
3	$y = (3)^2 = 9$	$(3, 9)$

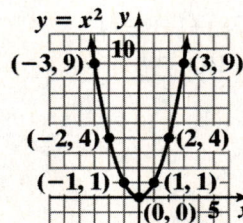

91. Table of values:

x	$y = x^2 + 1$	(x, y)
-3	$y = (-3)^2 + 1 = 10$	$(-3, 10)$
-2	$y = (-2)^2 + 1 = 5$	$(-2, 5)$
-1	$y = (-1)^2 + 1 = 2$	$(-1, 2)$
0	$y = (0)^2 + 1 = 1$	$(0, 1)$
1	$y = (1)^2 + 1 = 2$	$(1, 2)$
2	$y = (2)^2 + 1 = 5$	$(2, 5)$
3	$y = (3)^2 + 1 = 10$	$(3, 10)$

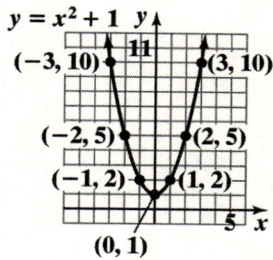

93. Table of values:

x	$y = 4 - x^2$	(x, y)
-3	$y = 4 - (-3)^2 = -5$	$(-3, -5)$
-2	$y = 4 - (-2)^2 = 0$	$(-2, 0)$
-1	$y = 4 - (-1)^2 = 3$	$(-1, 3)$
0	$y = 4 - (0)^2 = 4$	$(0, 4)$
1	$y = 4 - (1)^2 = 3$	$(1, 3)$
2	$y = 4 - (2)^2 = 0$	$(2, 0)$
3	$y = 4 - (3)^2 = -5$	$(3, -5)$

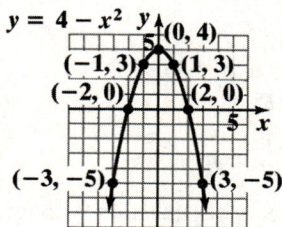

95. $\left[\left(4x^2 + 7x - 5 \right) - \left(2x^2 - 10x + 3 \right) \right] - \left(x^2 + 5x - 8 \right)$

$= \left[2x^2 + 17x - 8 \right] - x^2 - 5x + 8$

$= x^2 + 12x$

97. $\left[\left(4y^2 - 3y + 8\right) - \left(5y^2 + 7y - 4\right)\right] - \left[\left(8y^2 + 5y - 7\right) + \left(-10y^2 + 4y + 3\right)\right]$

$= \left[-y^2 - 10y + 12\right] - \left[-2y^2 + 9y - 4\right]$

$= y^2 - 19y + 16$

99. $\left[\left(4x^3 + x^2\right) + \left(-x^3 + 7x - 3\right)\right] - \left(x^3 - 2x^2 + 2\right)$

$= \left[3x^3 + x^2 + 7x - 3\right] + \left(-x^3 + 2x^2 - 2\right)$

$= 2x^3 + 3x^2 + 7x - 5$

101. $\left[\left(-5 + y^2 + 4y^3\right) - \left(-8 - y + 7y^3\right)\right] - \left(-y^2 + 7y^3\right)$

$= \left[-3y^3 + y^2 + y + 3\right] + \left(y^2 - 7y^3\right)$

$= -10y^3 + 2y^2 + y + 3$

103. a. $M - W = (-18x^3 + 923x^2 - 9603x + 48,446) - (17x^3 - 450x^2 + 6392x - 14,764)$

$M - W = -18x^3 + 923x^2 - 9603x + 48,446 - 17x^3 + 450x^2 - 6392x + 14,764$

$M - W = -18x^3 - 17x^3 + 923x^2 + 450x^2 - 9603x - 6392x + 48,446 + 14,764$

$M - W = -35x^3 + 1373x^2 - 15,995x + 63,210$

b. $M - W = -35x^3 + 1373x^2 - 15,995x + 63,210$

$M - W = -35(14)^3 + 1373(14)^2 - 15,995(14) + 63,210 = 12,348$

The difference in the median income between men and women with 14 years experience is $12,348.

c. $44,404 - 33,481 = 10,923$

The actual difference displayed in the graph in the median income between men and women with 14 years experience is $10,923.

The model overestimates this difference by $12,348 - $10,923 = $1425.

105. a. $M = 177x^2 + 288x + 7075$

$M = 177(16)^2 + 288(16) + 7075 = 56,995$

The model estimates the median annual income for a man with 16 years of education to be $56,995.

The model underestimates the actual value of $57,220 shown in the bar graph by $225.

b. The solution found in part (a) is represented by the point (16, 56,995) on the graph for men.

c. According to the graph in part (b) the median annual income for a woman with 16 years of education is about $42,000 (although answers will vary).

107. – 113. Answers will vary.

115. does not make sense; Explanations will vary. Sample explanation: The term of highest degree of a polynomial is not necessarily the first term. For example, the degree of $8x^2 - 7x + 9x^3 + 5$ is 3.

117. does not make sense; Explanations will vary. Sample explanation: $y = x^2 - 4$ is not a linear equation. Using two points and a checkpoint is an appropriate method for graphing a linear equation.

119. false; Changes to make the statement true will vary. A sample change is: The expression $\dfrac{1}{5x^2} + \dfrac{1}{3x}$ is not a polynomial expression (variables must not be in the denominator).

121. false; Changes to make the statement true will vary. A sample change is: The coefficient of x is -5.

123. $\left(5t - 3t^2 + t^3\right) - \left(t - t^2 + \dfrac{1}{3}t^3\right)$

$= \left(5t - 3t^2 + t^3\right) + \left(-t + t^2 - \dfrac{1}{3}t^3\right)$

$= \left(5t - t\right) + \left(-3t^2 + t^2\right) + \left(t^3 - \dfrac{1}{3}t^3\right)$

$= 4t - 2t^2 + \dfrac{2}{3}t^3$

$= \dfrac{2}{3}t^3 - 2t^2 + 4t$

125. $(-10)(-7) \div (1 - 8) = (-10)(-7) \div (-7)$

$= 70 \div (-7) = -10$

126. $-4.6 - (-10.2) = -4.6 + 10.2 = 5.6$

127. $3(x - 2) = 9(x + 2)$

$3x - 6 = 9x + 18$

$3x - 6 - 9x = 9x + 18 - 9x$

$-6x - 6 = 18$

$-6x - 6 + 6 = 18 + 6$

$-6x = 24$

$\dfrac{-6x}{-6} = \dfrac{24}{-6}$

$x = -4$

The solution set is $\{-4\}$.

128. $x^3 \cdot x^4 = (x \cdot x \cdot x) \cdot (x \cdot x \cdot x \cdot x) = x^7$

129. $3x(x + 5) = 3x \cdot x + 3x \cdot 5 = 3x^2 + 15x$

130. $x(x + 2) + 3(x + 2) = x^2 + 2x + 3x + 6 = x^2 + 5x + 6$

5.2 Check Points

1. a. $2^2 \cdot 2^4 = 2^{2+4} = 2^6$ or 64

   b. $x^6 \cdot x^4 = x^{6+4} = x^{10}$

   c. $y \cdot y^7 = y^{1+7} = y^8$

   d. $y^4 \cdot y^3 \cdot y^2 = y^{4+3+2} = y^9$

2. a. $\left(3^4\right)^5 = 3^{4 \cdot 5} = 3^{20}$

   b. $\left(x^9\right)^{10} = x^{9 \cdot 10} = x^{90}$

   c. $\left[(-5)^7\right]^3 = (-5)^{7 \cdot 3} = (-5)^{21}$

3. a. $(2x)^4 = 2^4 x^4 = 16x^4$

   b. $\left(-4y^2\right)^3 = (-4)^3 \left(y^2\right)^3 = (-4)^3 y^{2 \cdot 3} = -64y^6$

4. a. $(7x^2)(10x) = (7 \cdot 10)(x^2 \cdot x) = 70x^3$

   b. $(-5x^4)(4x^5) = (-5 \cdot 4)(x^4 \cdot x^5) = -20x^9$

5. a. $3x(x + 5) = 3x \cdot x + 3x \cdot 5 = 3x^2 + 15x$

   b. $6x^2(5x^3 - 2x + 3)$

$= 6x^2 \cdot 5x^3 - 6x^2 \cdot 2x + 6x^2 \cdot 3$

$= 30x^5 - 12x^3 + 18x^2$

6. a. $(x + 4)(x + 5) = x^2 + 5x + 4x + 20$

$= x^2 + 9x + 20$

   b. $(5x + 3)(2x - 7) = 10x^2 - 35x + 6x - 21$

$= 10x^2 - 29x - 21$

7. $(5x + 2)(x^2 - 4x + 3)$

$= 5x \cdot x^2 - 5x \cdot 4x + 5x \cdot 3 + 2 \cdot x^2 - 2 \cdot 4x + 2 \cdot 3$

$= 5x^3 - 20x^2 + 15x + 2x^2 - 8x + 6$

$= 5x^3 - 20x^2 + 2x^2 + 15x - 8x + 6$

$= 5x^3 - 18x^2 + 7x + 6$

8.

$$2x^3 - 5x^2 + 4x$$
$$3x^2 - 2x$$
$$\overline{-4x^4 + 10x^3 - 8x^2}$$
$$\underline{6x^5 - 15x^4 + 12x^3}$$
$$6x^5 - 19x^4 + 22x^3 - 8x^2$$

5.2 Concept and Vocabulary Check

1. b^{m+n}; add

2. b^{mn}; multiply

3. $a^n b^n$; factor

4. distributive; $x^2 + 5x + 7$; $2x^2$

5. $4x$; 7; like

5.2 Exercise Set

1. $x^{15} \cdot x^3 = x^{15+3} = x^{18}$

3. $y \cdot y^{11} = y^1 \cdot y^{11} = y^{1+11} = y^{12}$

5. $x^2 \cdot x^6 \cdot x^3 = x^{2+6+3} = x^{11}$

7. $7^9 \cdot 7^{10} = 7^{9+10} = 7^{19}$

9. $\left(6^9\right)^{10} = 6^{9 \cdot 10} = 6^{90}$

11. $\left(x^{15}\right)^3 = x^{15 \cdot 3} = x^{45}$

13. $\left[(-20)^3\right]^3 = (-20)^{3 \cdot 3} = (-20)^9$

15. $(2x)^3 = 2^3 \cdot x^3 = 8x^3$

17. $(-5x)^2 = (-5)^2 x^2 = 25x^2$

19. $\left(4x^3\right)^2 = 4^2 \left(x^3\right)^2 = 16x^6$

21. $\left(-2y^6\right)^4 = (-2)^4 \left(y^6\right)^4 = 16y^{24}$

23. $\left(-2x^7\right)^5 = (-2)^5 \left(x^7\right)^5 = -32x^{35}$

25. $(7x)(2x) = (7 \cdot 2)(x \cdot x) = 14x^2$

27. $(6x)\left(4x^2\right) = (6 \cdot 4)\left(x \cdot x^2\right) = 24x^3$

29. $\left(-5y^4\right)\left(3y^3\right) = (-5 \cdot 3)\left(y^4 \cdot y^3\right) = -15y^7$

31. $\left(-\dfrac{1}{2}a^3\right)\left(-\dfrac{1}{4}a^2\right) = \left(-\dfrac{1}{2} \cdot -\dfrac{1}{4}\right)\left(a^3 \cdot a^2\right)$
$$= \dfrac{1}{8}a^5$$

33. $\left(2x^2\right)(-3x)\left(8x^4\right)$
$$= (2 \cdot -3 \cdot 8)\left(x^2 \cdot x \cdot x^4\right) = -48x^7$$

35. $4x(x+3) = 4x \cdot x + 4x \cdot 3$
$$= 4x^2 + 12x$$

37. $x(x-3) = x \cdot x - x \cdot 3$
$$= x^2 - 3x$$

39. $2x(x-6) = 2x \cdot x - 2x \cdot 6$
$$= 2x^2 - 12x$$

41. $-4y(3y+5) = -4y \cdot 3y - 4y \cdot 5$
$$= -12y^2 - 20y$$

43. $4x^2(x+2) = 4x^2 \cdot x + 4x^2 \cdot 2$
$$= 4x^3 + 8x^2$$

45. $2y^2\left(y^2 + 3y\right) = 2y^2 \cdot y^2 + 2y^2 \cdot 3y$
$$= 2y^4 + 6y^3$$

47. $2y^2\left(3y^2 - 4y + 7\right)$
$$= 2y^2\left(3y^2\right) + 2y^2(-4y) + 2y^2(7)$$
$$= 6y^4 - 8y^3 + 14y^2$$

49. $\left(3x^3+4x^2\right)(2x)=3x^3\cdot 2x+4x^2\cdot 2x$

$\qquad\qquad\qquad =6x^4+8x^3$

51. $\left(x^2+5x-3\right)(-2x)$

$\qquad =x^2(-2x)+5x(-2x)-3(-2x)$

$\qquad =-2x^3-10x^2+6x$

53. $-3x^2\left(-4x^2+x-5\right)$

$\qquad =-3x^2\left(-4x^2\right)-3x^2(x)-3x^2(-5)$

$\qquad =12x^4-3x^3+15x^2$

55. $(x+3)(x+5)$

$\qquad =x(x+5)+3(x+5)$

$\qquad =x\cdot x+x\cdot 5+3\cdot x+3\cdot 5$

$\qquad =x^2+5x+3x+15$

$\qquad =x^2+8x+15$

57. $(2x+1)(x+4)$

$\qquad =2x(x+4)+1(x+4)$

$\qquad =2x^2+8x+x+4$

$\qquad =2x^2+9x+4$

59. $(x+3)(x-5)=x(x-5)+3(x-5)$

$\qquad\qquad\qquad =x^2-5x+3x-15$

$\qquad\qquad\qquad =x^2-2x-15$

61. $(x-11)(x+9)=x(x+9)-11(x+9)$

$\qquad\qquad\qquad\quad =x^2+9x-11x-99$

$\qquad\qquad\qquad\quad =x^2-2x-99$

63. $(2x-5)(x+4)$

$\qquad =2x(x+4)-5(x+4)$

$\qquad =2x^2+8x-5x-20$

$\qquad =2x^2+3x-20$

65. $\left(\dfrac{1}{4}x+4\right)\left(\dfrac{3}{4}x-1\right)$

$\qquad =\dfrac{1}{4}x\left(\dfrac{3}{4}x-1\right)+4\left(\dfrac{3}{4}x-1\right)$

$\qquad =\dfrac{1}{4}x\cdot\dfrac{3}{4}x+\dfrac{1}{4}x(-1)$

$\qquad\quad +4\left(\dfrac{3}{4}x\right)+4(-1)$

$\qquad =\dfrac{3}{16}x^2-\dfrac{1}{4}x+\dfrac{12}{4}x-4$

$\qquad =\dfrac{3}{16}x^2+\dfrac{11}{4}x-4$

67. $(x+1)\left(x^2+2x+3\right)$

$\qquad =x\left(x^2+2x+3\right)+1\left(x^2+2x+3\right)$

$\qquad =x^3+2x^2+3x+x^2+2x+3$

$\qquad =x^3+3x^2+5x+3$

69. $(y-3)\left(y^2-3y+4\right)$

$\qquad =y\left(y^2-3y+4\right)-3\left(y^2-3y+4\right)$

$\qquad =y^3-3y^2+4y-3y^2+9y-12$

$\qquad =y^3-6y^2+13y-12$

71. $(2a-3)\left(a^2-3a+5\right)$

$\qquad =2a\left(a^2-3a+5\right)-3\left(a^2-3a+5\right)$

$\qquad =2a^3-6a^2+10a-3a^2+9a-15$

$\qquad =2a^3-9a^2+19a-15$

73. $(x+1)\left(x^3+2x^2+3x+4\right)$

$\qquad =x\left(x^3+2x^2+3x+4\right)+1\left(x^3+2x^2+3x+4\right)$

$\qquad =x^4+2x^3+3x^2+4x+x^3+2x^2+3x+4$

$\qquad =x^4+\left(2x^3+x^3\right)+\left(3x^2+2x^2\right)+\left(4x+3x\right)+4$

$\qquad =x^4+3x^3+5x^2+7x+4$

75. $\left(x-\dfrac{1}{2}\right)\left(4x^3-2x^2+5x-6\right)$

$= x\left(4x^3-2x^2+5x-6\right)-\dfrac{1}{2}\left(4x^3-2x^2+5x-6\right)$

$= 4x^4-2x^3+5x^2-6x-2x^3+x^2-\dfrac{5}{2}x+3$

$= 4x^4-4x^3+6x^2-\dfrac{17}{2}x+3$

77. $\left(x^2+2x+1\right)\left(x^2-x+2\right)$

$= x^2\left(x^2-x+2\right)+2x\left(x^2-x+2\right)+1\left(x^2-x+2\right)$

$= x^4-x^3+2x^2+2x^3-2x^2+4x+x^2-x+2$

$= x^4+x^3+x^2+3x+2$

79.
$$
\begin{array}{r}
x^2-5x+3 \\
x+8 \\
\hline
8x^2-40x+24 \\
x^3-5x^2+3x \\
\hline
x^3+3x^2-37x+24
\end{array}
$$

81.
$$
\begin{array}{r}
x^2-3x+9 \\
2x-3 \\
\hline
-3x^2+9x-27 \\
2x^3-6x^2+18x \\
\hline
2x^3-9x^2+27x-27
\end{array}
$$

83.
$$
\begin{array}{r}
2x^3+x^2+2x+3 \\
x+4 \\
\hline
8x^3+4x^2+8x+12 \\
2x^4+x^3+2x^2+3x \\
\hline
2x^4+9x^3+6x^2+11x+12
\end{array}
$$

85.
$$
\begin{array}{r}
4z^3-2z^2+5z-4 \\
3z-2 \\
\hline
-8z^3+4z^2-10z+8 \\
12z^4-5z^3+15z^2-12z \\
\hline
12z^4-14z^3+19z^2-22z+8
\end{array}
$$

87.
$$
\begin{array}{r}
7x^3-5x^2+6x \\
3x^2-4x \\
\hline
-28x^4+20x^3-24x^2 \\
21x^5-15x^4+18x^3 \\
\hline
21x^5-43x^4+38x^3-24x^2
\end{array}
$$

89.
$$
\begin{array}{r}
2y^5-3y^3+y^2-2y+3 \\
2y-1 \\
\hline
-2y^5+3y^3-y^2+2y-3 \\
4y^6-6y^4+2y^3-4y^2+6y \\
\hline
4y^6-2y^5-6y^4+5y^3-5y^2+8y-3
\end{array}
$$

91.
$$
\begin{array}{r}
x^2+7x-3 \\
x^2-x-1 \\
\hline
-x^2-7x+3 \\
-x^3-7x^2+3x \\
x^4+7x^3-3x^2 \\
\hline
x^4+6x^3-11x^2-4x+3
\end{array}
$$

93. $(x+4)(x-5)-(x+3)(x-6)$

$= \left(x^2-x-20\right)-\left(x^2-3x-18\right)$

$= \left(x^2-x-20\right)+\left(-x^2+3x+18\right)$

$= 2x-2$

95. $4x^2\left(5x^3+3x-2\right)-5x^3\left(x^2-6\right)$

$= \left(20x^5+12x^3-8x^2\right)+\left(-5x^5+30x^3\right)$

$= 15x^5+42x^3-8x^2$

97. $(y+1)\left(y^2-y+1\right)+(y-1)\left(y^2+y+1\right)$

$= y\left(y^2-y+1\right)+1\left(y^2-y+1\right)$
$\quad +y\left(y^2+y+1\right)-1\left(y^2+y+1\right)$

$= y^3-y^2+y+y^2-y+1$
$\quad +y^3+y^2+y-y^2-y-1$

$= 2y^3$

99. $(y+6)^2 - (y-2)^2$

$= (y+6)(y+6) - (y-2)(y-2)$

$= (y^2 + 12y + 36) - (y^2 - 4y + 4)$

$= 16y + 32$

101. Use the formula for the area of a rectangle.

$A = l \cdot w$

$A = (x+5)(2x-3)$

$= x(2x-3) + 5(2x-3)$

$= 2x^2 - 3x + 10x - 15$

$= 2x^2 + 7x - 15$

The area of the rug is $2x^2 + 7x - 15$ square feet.

103. a. $(x+2)(2x+1)$

b. $x \cdot 2x + 2 \cdot 2x + x \cdot 1 + 2 \cdot 1$

$= 2x^2 + 4x + x + 2$

$= 2x^2 + 5x + 2$

c. $(x+2)(2x+1) = x(2x+1) + 2(2x+1)$

$= 2x^2 + x + 4x + 2$

$= 2x^2 + 5x + 2$

105. When multiplying numbers with the same base, keep the base and add the exponents.

Example: $2^3 \cdot 2^5 = 2^{3+5} = 2^8$

107. – 111. Answers will vary.

113. makes sense

115. makes sense

117. false; Changes to make the statement true will vary.

A sample change is: $4x^3 \cdot 3x^4 = 4 \cdot 3x^{3+4} = 12x^7$

119. true

121. The area of the outer square is

$(x+4)(x+4) = x(x+4) + 4(x+4)$

$= x^2 + 4x + 4x + 16$

$= x^2 + 8x + 16$

The area of the inner square is x^2. The area of the shaded region is the difference between the areas of the two squares, which is

$(x^2 + 8x + 16) - x^2 = 8x + 16$

123. $\left(-8x^4\right)\left(-\dfrac{1}{4}xy^3\right) = 2x^5y^3$, so the missing factor is $-8x^4$.

124.
$$4x - 7 > 9x - 2$$
$$4x - 7 - 9x > 9x - 2 - 9x$$
$$-5x - 7 > -2$$
$$-5x - 7 + 7 > -2 + 7$$
$$-5x > 5$$
$$\frac{-5x}{-5} < \frac{5}{-5}$$
$$x < -1$$

Solution: $(-\infty, -1)$

125. $3x - 2y = 6$

x-intercept:

$3x - 2y = 6$

$3x - 2(0) = 6$

$3x = 6$

$x = 2$

y-intercept:

$3x - 2y = 6$

$3(0) - 2y = 6$

$-2y = 6$

$y = -3$

checkpoint:

$3x - 2y = 6$

$3(4) - 2y = 6$

$12 - 2y = 6$

$-2y = -6$

$y = 3$

A checkpoint is (4,3).

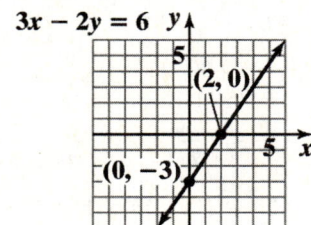

126. $m = \dfrac{y_2 - y_1}{x_2 - x_2} = \dfrac{6-8}{1-(-2)} = -\dfrac{2}{3}$

127. a. $(x+3)(x+4) = x^2 + 4x + 3x + 12$

$$= x^2 + 7x + 12$$

    b. $(x+5)(x+20) = x^2 + 5x + 20x + 100$

$$= x^2 + 25x + 100$$

A fast method is $(x+a)(x+b) = x^2 + (a+b)x + ab$.

128. a. $(x+3)(x-3) = x^2 + 3x - 3x - 9$

$$= x^2 - 9$$

    b. $(x+5)(x-5) = x^2 + 5x - 5x - 25$

$$= x^2 - 25$$

A fast method is $(x+a)(x-a) = x^2 - a^2$.

129. a. $(x+3)^2 = (x+3)(x+3)$

$$= x^2 + 3x + 3x + 9$$

$$= x^2 + 6x + 9$$

    b. $(x+5)^2 = (x+5)(x+5)$

$$= x^2 + 5x + 5x + 9$$

$$= x^2 + 10x + 25$$

A fast method is $(x+a)^2 = x^2 + 2ax + a^2$.

5.3 Check Points

1. $(x+5)(x+6) = \overset{\text{F}}{\overbrace{x \cdot x}} + \overset{\text{O}}{\overbrace{x \cdot 6}} + \overset{\text{I}}{\overbrace{5 \cdot x}} + \overset{\text{L}}{\overbrace{5 \cdot 6}}$

$$= x^2 + 6x + 5x + 30$$

$$= x^2 + 11x + 30$$

2. $(7x+5)(4x-3) = \overset{\text{F}}{\overbrace{7x \cdot 4x}} + \overset{\text{O}}{\overbrace{7x(-3)}} + \overset{\text{I}}{\overbrace{5 \cdot 4x}} + \overset{\text{L}}{\overbrace{5(-3)}}$

$$= 28x^2 - 21x + 20x - 15$$

$$= 28x^2 - x - 15$$

3. $(4-2x)(5-3x) = \overset{\text{F}}{\overbrace{4 \cdot 5}} + \overset{\text{O}}{\overbrace{4(-3x)}} + \overset{\text{I}}{\overbrace{(-2x)(5)}} + \overset{\text{L}}{\overbrace{(-2x)(-3x)}}$

$$= 20 - 12x - 10x + 6x^2$$

$$= 20 - 22x + 6x^2$$

$$= 6x^2 - 22x + 20$$

4. a. Since this product is of the form $(A+B)(A-B)$, use the special-product formula

$(A+B)(A-B) = A^2 - B^2$.

$$(7y+8)(7y-8) = \overbrace{(7y)^2}^{\substack{\text{first term}\\\text{squared}}} - \overbrace{8^2}^{\substack{\text{second term}\\\text{squared}}}$$

$$= 49y^2 - 64$$

b. Since this product is of the form $(A+B)(A-B)$, use the special-product formula

$(A+B)(A-B) = A^2 - B^2$.

$$(4x-5)(4x+5) = \overbrace{(4x)^2}^{\substack{\text{first term}\\\text{squared}}} - \overbrace{5^2}^{\substack{\text{second term}\\\text{squared}}}$$

$$= 16x^2 - 25$$

c. Since this product is of the form $(A+B)(A-B)$, use the special-product formula

$(A+B)(A-B) = A^2 - B^2$.

$$(2a^3+3)(2a^3-3) = \overbrace{(2a^3)^2}^{\substack{\text{first term}\\\text{squared}}} - \overbrace{3^2}^{\substack{\text{second term}\\\text{squared}}}$$

$$= 4a^6 - 9$$

5. a. Use the special-product formula

$(A+B)^2 = A^2 + 2AB + B^2$.

$$(x+10)^2 = \overbrace{x^2}^{\substack{\text{first term}\\\text{squared}}} + \overbrace{2 \cdot 10x}^{\substack{2 \cdot \text{product}\\\text{of the terms}}} + \overbrace{10^2}^{\substack{\text{last term}\\\text{squared}}}$$

$$= x^2 + 20x + 100$$

b. Use the special-product formula

$(A+B)^2 = A^2 + 2AB + B^2$.

$$(5x+4)^2 = \overbrace{(5x)^2}^{\substack{\text{first term}\\\text{squared}}} + \overbrace{2 \cdot 20x}^{\substack{2 \cdot \text{product}\\\text{of the terms}}} + \overbrace{4^2}^{\substack{\text{last term}\\\text{squared}}}$$

$$= 25x^2 + 40x + 16$$

6. a. Use the special-product formula

$(A-B)^2 = A^2 - 2AB + B^2$.

$$(x-9)^2 = \overbrace{x^2}^{\substack{\text{first term}\\\text{squared}}} - \overbrace{2 \cdot 9x}^{\substack{2 \cdot \text{product}\\\text{of the terms}}} + \overbrace{9^2}^{\substack{\text{last term}\\\text{squared}}}$$

$$= x^2 - 18x + 81$$

b. Use the special-product formula

$(A-B)^2 = A^2 - 2AB + B^2$.

$$(7x-3)^2 = \overbrace{(7x)^2}^{\substack{\text{first term}\\\text{squared}}} - \overbrace{2 \cdot 21x}^{\substack{2 \cdot \text{product}\\\text{of the terms}}} + \overbrace{3^2}^{\substack{\text{last term}\\\text{squared}}}$$

$$= 49x^2 - 42x + 9$$

5.3 Concept and Vocabulary Check

1. $2x^2$; $3x$; $10x$; 15

2. $A^2 - B^2$; minus

3. $A^2 + 2AB + B^2$; squared; product of the terms; squared

4. $A^2 - 2AB + B^2$; minus; product of the terms; squared

5. true

6. false

5.3 Exercise Set

1. $(x+4)(x+6) = x^2 + 6x + 4x + 24$
$$= x^2 + 10x + 24$$

3. $(y-7)(y+3) = y^2 + 3y - 7y - 21$
$$= y^2 - 4y - 21$$

5. $(2x-3)(x+5) = 2x^2 + 10x - 3x - 15$
$$= 2x^2 + 7x - 15$$

7. $(4y+3)(y-1) = 4y^2 - 4y + 3y - 3$
$$= 4y^2 - y - 3$$

9. $(2x-3)(5x+3) = 10x^2 + 6x - 15x - 9$
$$= 10x^2 - 9x - 9$$

11. $(3y-7)(4y-5) = 12y^2 - 15y - 28y + 35$
$$= 12y^2 - 43y + 35$$

13. $(7+3x)(1-5x) = 7 - 35x + 3x - 15x^2$
$$= -15x^2 - 32x + 7$$

15. $(5-3y)(6-2y) = 30 - 10y - 18y + 6y^2$
$$= 30 - 28y + 6y^2$$
$$= 6y^2 - 28y + 30$$

17. $\left(5x^2 - 4\right)\left(3x^2 - 7\right)$
$$= \left(5x^2\right)\left(3x^2\right) + \left(5x^2\right)(-7) + (-4)\left(3x^2\right) + (-4)(-7)$$
$$= 15x^4 - 35x^2 - 12x^2 + 28$$
$$= 15x^4 - 47x^2 + 28$$

19. $(6x-5)(2-x) = 12x - 6x^2 - 10 + 5x$
$$= -6x^2 + 17x - 10$$

21. $(x+5)\left(x^2 + 3\right) = x^3 + 3x + 5x^2 + 15$
$$= x^3 + 5x^2 + 3x + 15$$

23. $\left(8x^3 + 3\right)\left(x^2 + 5\right) = 8x^5 + 40x^3 + 3x^2 + 15$

25. $(x+3)(x-3) = x^2 - 3^2 = x^2 - 9$

27. $(3x+2)(3x-2) = (3x)^2 - 2^2 = 9x^2 - 4$

29. $(3r-4)(3r+4) = (3r)^2 - 4^2$
$$= 9r^2 - 16$$

31. $(3+r)(3-r) = 3^2 - r^2 = 9 - r^2$

33. $(5-7x)(5+7x) = 5^2 - \left(7x^2\right) = 25 - 49x^2$

35. $\left(2x + \dfrac{1}{2}\right)\left(2x - \dfrac{1}{2}\right) = (2x)^2 - \left(\dfrac{1}{2}\right)^2$
$$= 4x^2 - \dfrac{1}{4}$$

37. $\left(y^2 + 1\right)\left(y^2 - 1\right) = \left(y^2\right)^2 - 1^2 = y^4 - 1$

39. $\left(r^3 + 2\right)\left(r^3 - 2\right) = \left(r^3\right)^2 - 2^2 = r^6 - 4$

41. $\left(1 - y^4\right)\left(1 + y^4\right) = 1^2 - \left(y^4\right)^2 = 1 - y^8$

43. $\left(x^{10} + 5\right)\left(x^{10} - 5\right) = \left(x^{10}\right)^2 - 5^2$
$$= x^{20} - 25$$

45. $(x+2)^2 = x^2 + 2(2x) + 2^2$
$$= x^2 + 4x + 4$$

47. $(2x+5)^2 = (2x)^2 + 2(2x)(5) + 5^2$
$$= 4x^2 + 20x + 25$$

49. $(x-3)^2 = x^2 - 2(3x) + 3^2$
$$= x^2 - 6x + 9$$

51. $(3y-4)^2 = (3y)^2 - 2(3y)(4) + 4^2$
$$= 9y^2 - 24y + 16$$

53. $\left(4x^2 - 1\right)^2 = \left(4x^2\right)^2 - 2\left(4x^2\right)(1) + 1^2$
$$= 16x^4 - 8x^2 + 1$$

55. $(7-2x)^2 = 7^2 - 2(7)(2x) + (2x)^2$
$$= 49 - 28x + 4x^2$$

57. $\left(2x + \dfrac{1}{2}\right)^2 = 4x^2 + 2(2x)\left(\dfrac{1}{2}\right) + \left(\dfrac{1}{2}\right)^2$
$$= 4x^2 + 2x + \dfrac{1}{4}$$

59. $\left(4y - \dfrac{1}{4}\right)^2 = (4y)^2 - 2(4y)\left(\dfrac{1}{4}\right) + \left(\dfrac{1}{4}\right)^2$
$$= 16y^2 - 2y + \dfrac{1}{16}$$

61. $\left(x^8 + 3\right)^2 = \left(x^8\right)^2 + 2\left(x^8\right)(3) + 3^2$
$$= x^{16} + 6x^8 + 9$$

63. $(x-1)(x^2+x+1)$

$= x(x^2+x+1)-1(x^2+x+1)$

$= x^3+x^2+x-x^2-x-1$

$= x^3-1$

65. $(x-1)^2 = x^2 - 2(x)(1)+1^2$

$= x^2 - 2x+1$

67. $(3y+7)(3y-7)=(3y^2)-7^2$

$= 9y^2-49$

69. $3x^2(4x^2+x+9)$

$= 3x^2(4x^2)+3x^2(x)+3x^2(9)$

$= 12x^4+3x^3+27x^2$

71. $(7y+3)(10y-4)$

$= 70y^2-28y+30y-12$

$= 70y^2+2y-12$

73. $(x^2+1)^2 = (x^2)^2 + 2(x^2)(1)+1^2$

$= x^4+2x^2+1$

75. $(x^2+1)(x^2+2)$

$= x^2 \cdot x^2 + x^2 \cdot 2 + 1 \cdot x^2 + 1 \cdot 2$

$= x^4+3x^2+2$

77. $(x^2+4)(x^2-4)=(x^2)^2-4^2$

$= x^4-16$

79. $(2-3x^5)^2 = 2^2 - 2(2)(3x^5)+(3x^5)^2$

$= 4-12x^5+9x^{10}$

81. $\left(\dfrac{1}{4}x^2+12\right)\left(\dfrac{3}{4}x^2-8\right)$

$= \dfrac{1}{4}x^2\left(\dfrac{3}{4}x^2\right)+\dfrac{1}{4}x^2(-8)+12\left(\dfrac{3}{4}x^2\right)+12(-8)$

$= \dfrac{3}{16}x^4-2x^2+9x^2-96$

$= \dfrac{3}{16}x^2+7x^2-96$

83. $A=(x+1)^2 = x^2+2x+1$

85. $A=(2x-3)(2x+3)=(2x)^2-3^2$

$= 4x^2-9$

87. Area of outer rectangle:

$(x+9)(x+3)=x^2+12x+27$

Area of inner rectangle:

$(x+5)(x+1)=x^2+6x+5$

Area of shaded region:

$(x^2+12x+27)-(x^2+6x+5)=6x+22$

89. $\left[(2x+3)(2x-3)\right]^2$

$= \left[4x^2-9\right]^2$

$= 16x^4-72x^2+81$

91. $(4x^2+1)\left[(2x+1)(2x-1)\right]$

$= (4x^2+1)\left[4x^2-1\right]$

$= 16x^4-1$

93. $(x+2)^3$

$= (x+2)(x+2)^2$

$= (x+2)(x^2+4x+4)$

$= x(x^2+4x+4)+2(x^2+4x+4)$

$= x^3+4x^2+4x+2x^2+8x+8$

$= x^3+6x^2+12x+8$

95. $\left[(x+3)-y\right]\left[(x+3)+y\right]$

$= (x+3)^2-y^2$

$= x^2+6x+9-y^2$

97. $(x+2)(x+1)$

The area of the larger garden is given by $(x+2)(x+1)$ square yards.

99. If the original garden measures $x = 6$ yards on a side, the area of the larger garden would be

$(6)^2 + 3(6) + 2 = 56$ square yards.

This corresponds to the point $(6, 56)$ on the graph.

101. $A_{\text{total}} = (x+2)^2 = x^2 + 2 \cdot (2x) + 2^2 = x^2 + 4x + 4$

The total area is $(x^2 + 4x + 4)$ square inches.

103. – 107. Answers will vary.

109. makes sense

111. makes sense, although answers may vary

113. true

115. false; Changes to make the statement true will vary.

A sample change is: $(x-5)^2 = x^2 - 10x + 25$.

117. $V = l \cdot w \cdot h$

$ = (10 - 2x)(8 - 2x)(x)$

$ = \left(80 - 20x - 16x + 4x^2\right)(x)$

$ = \left(80 - 36x + 4x^2\right)(x)$

$ = 80x - 36x^2 + 4x^3$

$ = 4x^3 - 36x^2 + 80x$

The volume of the box is $(4x^3 - 36x^2 + 80x)$ cubic units.

119. Let $y_1 = (x+1)^2$ and $y_2 = x^2 + 1$.

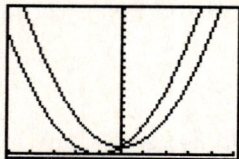

The graphs do not coincide so the multiplication is not correct.

To correct, let $y_2 = x^2 + 2x + 1$

121. Let $y_1 = (x+1)(x-1)$ and $y_2 = x^2 - 1$.

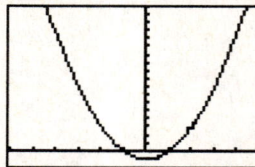

The graphs coincide so the multiplication is correct.

$(x+1)(x-1) = x^2 - x + x - 1 = x^2 - 1$

123. $2x + 3y = 1$

$y = 3x - 7$

The substitution method is a good choice because the second equation is already solved for y. Substitute $3x - 7$ for y into the first equation.

$2x + 3y = 1$

$2x + 3(3x - 7) = 1$

$2x + 9x - 21 = 1$

$11x - 21 = 1$

$11x = 22$

$x = 2$

Back-substitute to find y.

$y = 3x - 7$

$y = 3(2) - 7 = 6 + 7 = -1$

The solution set is $\{(2, -1)\}$.

124. $3x + 4y = 7$

$2x + 7y = 9$

The addition method is a good choice because both equations are written in the form $Ax + By = C$. To eliminate x, multiply the first equation by 2 and the second equation by -3. Then add the results.

$6x + 8y = 14$

$\underline{-6x - 21y = -27}$

$-13y = -13$

$y = 1$

Back-substitute 1 for y in either equation of the original system.

$3x + 4y = 7$

$3x + 4(1) = 7$

$3x + 4 = 7$

$3x = 3$

$x = 1$

The solution set is $\{(1, 1)\}$.

125. $y \le \dfrac{1}{3}x$

Graph $y = \dfrac{1}{3}x$ as a solid line using its slope, $\dfrac{1}{3}$, and y-intercept, 0. Since $(0,0)$ is on the line, chose a different point as a

test point, for example, $(3,2)$. Since $2 \le \dfrac{1}{3}(3)$ is false, shade the half-plane *not* containing $(3, 2)$.

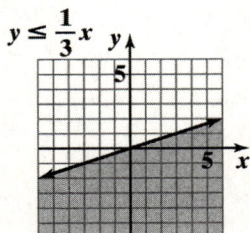

126.
$$\begin{aligned}
x^3y + 2xy^2 + 5x - 2 &= (-2)^3(3) + 2(-2)(3)^2 + 5(-2) - 2 \\
&= (-8)(3) + 2(-2)(9) + 5(-2) - 2 \\
&= -24 - 36 - 10 - 2 \\
&= -72
\end{aligned}$$

127. $5xy + 6xy = (5 + 6)xy = 11xy$

128.
$$\begin{aligned}
(x + 2y)(3x + 5y) &= x \cdot 3x + x \cdot 5y + 2y \cdot 3x + 2y \cdot 5y \\
&= 3x^2 + 5xy + 6xy + 10y^2 \\
&= 3x^2 + 11xy + 10y^2
\end{aligned}$$

5.4 Check Points

1. Begin by substituting -1 in for x and 5 in for y.
$$\begin{aligned}
3x^3y + xy^2 + 5y + 6 &= 3(-1)^3(5) + (-1)(5)^2 + 5(5) + 6 \\
&= 3(-1)(5) + (-1)(25) + 5(5) + 6 \\
&= -15 - 25 + 25 + 6 \\
&= -9
\end{aligned}$$

2. $8x^4y^5 - 7x^3y^2 - x^2y - 5x + 11$

Term	Coefficient	Degree
$8x^4y^5$	8	$4 + 5 = 9$
$-7x^3y^2$	-7	$3 + 2 = 5$
$-x^2y$	-1	$2 + 1 = 3$
$-5x$	-5	1
11	11	0

The degree of the polynomial is the highest degree of all its terms, which is 9.

3. $(-8x^2y - 3xy + 6) + (10x^2y + 5xy - 10)$

$= -8x^2y - 3xy + 6 + 10x^2y + 5xy - 10$

$= -8x^2y + 10x^2y - 3xy + 5xy + 6 - 10$

$= 2x^2y + 2xy - 4$

4. $(7x^3 - 10x^2y + 2xy^2 - 5) - (4x^3 - 12x^2y - 3xy^2 + 5)$

$= 7x^3 - 10x^2y + 2xy^2 - 5 - 4x^3 + 12x^2y + 3xy^2 - 5$

$= 7x^3 - 4x^3 - 10x^2y + 12x^2y + 2xy^2 + 3xy^2 - 5 - 5$

$= 3x^3 + 2x^2y + 5xy^2 - 10$

5. $(6xy^3)(10x^4y^2) = (6 \cdot 10)(x \cdot x^4)(y^3 \cdot y^2)$

$\qquad\qquad\qquad = 60x^{1+4}y^{3+2}$

$\qquad\qquad\qquad = 60x^5y^5$

6. $6xy^2(10x^4y^5 - 2x^2y + 3)$

$= 6xy^2 \cdot 10x^4y^5 - 6xy^2 \cdot 2x^2y + 6xy^2 \cdot 3$

$= 60x^{1+4}y^{2+5} - 12x^{1+2}y^{2+1} + 18xy^2$

$= 60x^5y^7 - 12x^3y^3 + 18xy^2$

7. a. $(7x - 6y)(3x - y)$

$$= \overbrace{(7x)(3x)}^{F} + \overbrace{(7x)(-y)}^{O} + \overbrace{(-6y)(3x)}^{I} + \overbrace{(-6y)(-y)}^{L}$$

$= 21x^2 - 7xy - 18xy + 6y^2$

$= 21x^2 - 25xy + 6y^2$

b. $\overbrace{(2x + 4y)^2}^{(A+B)^2} = \overbrace{(2x)^2}^{A^2} + \overbrace{2(2x)(4y)}^{2 \cdot A \cdot B} + \overbrace{(4y)^2}^{B^2}$

$\qquad\qquad = 4x^2 + 16xy + 16y^2$

$\qquad\qquad = 4x^2 + 16xy + 16y^2$

8. a. $\overbrace{(6xy^2 + 5x)(6xy^2 - 5x)}^{(A+B) \cdot (A-B)} = \overbrace{(6xy^2)^2}^{A^2} - \overbrace{(5x)^2}^{B^2}$

$\qquad\qquad\qquad\qquad = 36x^2y^4 - 25x^2$

b. $(x - y)(x^2 + xy + y^2) = x(x^2 + xy + y^2) - y(x^2 + xy + y^2)$

$\qquad\qquad\qquad\qquad = x \cdot x^2 + x \cdot xy + x \cdot y^2 - y \cdot x^2 - y \cdot xy - y \cdot y^2$

$\qquad\qquad\qquad\qquad = x^3 + x^2y + xy^2 - x^2y - xy^2 - y^3$

$\qquad\qquad\qquad\qquad = x^3 + x^2y - x^2y + xy^2 - xy^2 - y^3$

$\qquad\qquad\qquad\qquad = x^3 - y^3$

5.4 Concept and Vocabulary Check

1. -18

2. 6

3. $a; \ n+m$

4. $9; \ 9$

5. false

6. true

5.4 Exercise Set

1. $x^2 + 2xy + y^2 = 2^2 + 2(2)(-3) + (-3)^2$
$$= 4 - 12 + 9 = 1$$

3. $xy^3 - xy + 1 = 2(-3)^3 - 2(-3) + 1$
$$= 2(-27) + 6 + 1$$
$$= -54 + 6 + 1 = -47$$

5. $2x^2 y - 5y + 3 = 2(2^2)(-3) - 5(-3) + 3$
$$= 2(4)(-3) - 5(-3) + 3$$
$$= -24 + 15 + 3$$
$$= -6$$

7. $x^3 y^2 - 5x^2 y^7 + 6y^2 - 3$

Term	Coefficient	Degree
$x^3 y^2$	1	$3 + 2 = 5$
$-5x^2 y^7$	-5	$2 + 7 = 9$
$6y^2$	6	2
-3	-3	0

The degree of the polynomial is the highest degree of all its terms, which is 9.

9. $\left(5x^2 y - 3xy\right) + \left(2x^2 y - xy\right)$
$$= \left(5x^2 y + 2x^2 y\right) + \left(-3xy - xy\right)$$
$$= 7x^2 y - 4xy$$

11. $\left(4x^2 y + 8xy + 11\right) + \left(-2x^2 y + 5xy + 2\right)$
$$= \left(4x^2 y - 2x^2 y\right) + \left(8xy + 5xy\right) + \left(11 + 2\right)$$
$$= 2x^2 y + 13xy + 13$$

13. $\left(7x^4y^2 - 5x^2y^2 + 3xy\right) + \left(-18x^4y^2 - 6x^2y^2 - xy\right)$

$= \left(7x^4y^2 - 18x^4y^2\right) + \left(-5x^2y^2 - 6x^2y^2\right) + \left(3xy - xy\right)$

$= -11x^4y^2 - 11x^2y^2 + 2xy$

15. $\left(x^3 + 7xy - 5y^2\right) - \left(6x^3 - xy + 4y^2\right)$

$= \left(x^3 + 7xy - 5y^2\right) + \left(-6x^3 + xy - 4y^2\right)$

$= \left(x^3 - 6x^3\right) + \left(7xy + xy\right) + \left(-5y^2 - 4y^2\right)$

$= -5x^3 + 8xy - 9y^2$

17. $\left(3x^4y^2 + 5x^3y - 3y\right) - \left(2x^4y^2 - 3x^3y - 4y + 6x\right)$

$= \left(3x^4y^2 + 5x^3y - 3y\right) + \left(-2x^4y^2 + 3x^3y + 4y - 6x\right)$

$= \left(3x^4y^2 - 2x^4y^2\right) + \left(5x^3y + 3x^3y\right) + \left(-3y + 4y\right) + \left(-6x\right)$

$= x^4y^2 + 8x^3y + y - 6x$

19. $\left(x^3 - y^3\right) - \left(-4x^3 - x^2y + xy^2 + 3y^3\right)$

$= \left(x^3 - y^3\right) + \left(4x^3 + x^2y - xy^2 - 3y^3\right)$

$= \left(x^3 + 4x^3\right) + \left(-y^3 - 3y^3\right) + x^2y - xy^2$

$= 5x^3 - 4y^3 + x^2y - xy^2$

$= 5x^3 + x^2y - xy^2 - 4y^3$

21. $\quad 5x^2y^2 - 4xy^2 + 6y^2$

$\underline{-8x^2y^2 + 5xy^2 \ - y^2}$

$-3x^2y^2 + \ xy^2 \ + 5y^2$

23. $\quad 3a^2b^4 - 5ab^2 + 7ab$

$-\left(\underline{-5a^2b^4 - 8ab^2 \ - ab}\right)$

To subtract, add the opposite of the polynomial being subtracted.

$\quad 3a^2b^4 - 5ab^2 + \ 7ab$

$\underline{+5a^2b^4 + 8ab^2 \ + \ ab}$

$\quad 8a^2b^4 + 3ab^2 \ + 8ab$

25. $\left[\left(7x + 13y\right) + \left(-26x + 19y\right)\right] - \left(11x - 5y\right) = 7x + 13y - 26x + 19y - 11x + 5y$

$= 7x - 26x - 11x + 13y + 19y + 5y$

$= -30x + 37y$

27. $\left(5x^2 y\right)(8xy) = (5\cdot 8)\left(x^2\cdot x\right)(y\cdot y)$
$$= 40x^3 y^2$$

29. $\left(-8x^3 y^4\right)\left(3x^2 y^5\right) = (-8\cdot 3)\left(x^3\cdot x^2\right)\left(y^4\cdot y^5\right)$
$$= -24x^5 y^9$$

31. $9xy(5x+2y) = 9xy(5x) + 9xy(2y)$
$$= 45x^2 y + 18xy^2$$

33. $5xy^2\left(10x^2 - 3y\right) = 5xy^2\left(10x^2\right) - 5xy^2(3y)$
$$= 50x^3 y^2 - 15xy^3$$

35. $4ab^2\left(7a^2 b^3 + 2ab\right) = 4ab^2\left(7a^2 b^3\right) + 4ab^2(2ab)$
$$= 28a^3 b^5 + 8a^2 b^3$$

37. $-b\left(a^2 - ab + b^2\right) = -b\left(a^2\right) - b(-ab) - b\left(b^2\right)$
$$= -a^2 b + ab^2 - b^3$$

39. $(x+5y)(7x+3y)$
$= x(7x) + x(3y) + 5y(7x) + 5y(3y)$
$= 7x^2 + 3xy + 35xy + 15y^2$
$= 7x^2 + 38xy + 15y^2$

41. $(x-3y)(2x+7y)$
$= x(2x) + x(7y) - 3y(2x) - 3y(7y)$
$= 2x^2 + 7xy - 6xy - 21y^2$
$= 2x^2 + xy - 21y^2$

43. $(3xy-1)(5xy+2)$
$= 3xy(5xy) + 3xy(2) - 1(5xy) - 1(2)$
$= 15x^2 y^2 + 6xy - 5xy - 2$
$= 15x^2 y^2 + xy - 2$

45. $(2x+3y)^2 = (2x)^2 + 2(2x)(3y) + (3y)^2$
$$= 4x^2 + 12xy + 9y^2$$

47. $(xy-3)^2 = (xy)^2 - 2(xy)(3) + (-3)^2$
$$= x^2 y^2 - 6xy + 9$$

49. $\left(x^2+y^2\right)^2 = \left(x^2\right)^2 + 2\left(x^2\right)\left(y^2\right)+\left(y^2\right)^2$

$\qquad = x^4 + 2x^2y^2 + y^4$

51. $\left(x^2-2y^2\right)^2$

$\qquad = \left(x^2\right) - 2\left(x^2\right)\left(2y^2\right)+\left(-2y^2\right)^2$

$\qquad = x^4 - 4x^2y^2 + 4y^4$

53. $(3x+y)(3x-y) = (3x)^2 - y^2 = 9x^2 - y^2$

55. $(ab+1)(ab-1) = (ab)^2 - 1^2 = a^2b^2 - 1$

57. $\left(x+y^2\right)\left(x-y^2\right) = x^2 - \left(y^2\right)^2 = x^2 - y^4$

59. $\left(3a^2b+a\right)\left(3a^2b-a\right) = \left(3a^2b\right)^2 - a^2$

$\qquad\qquad = 9a^4b^2 - a^2$

61. $\left(3xy^2-4y\right)\left(3xy^2+4y\right) = \left(3xy^2\right)^2 - (4y)^2$

$\qquad\qquad = 9x^2y^4 - 16y^2$

63. $(a+b)\left(a^2-b^2\right)$

$\qquad = a\left(a^2\right)+a\left(-b^2\right)+b\left(a^2\right)+b\left(-b^2\right)$

$\qquad = a^3 - ab^2 + a^2b - b^3$

65. $(x+y)\left(x^2+3xy+y^2\right)$

$\qquad = x\left(x^2+3xy+y^2\right)+y\left(x^2+3xy+y^2\right)$

$\qquad = x^3 + 3x^2y + xy^2 + x^2y + 3xy^2 + y^3$

$\qquad = x^3 + 4x^2y + 4xy^2 + y^3$

67. $(x-y)\left(x^2-3xy+y^2\right)$

$\qquad = x\left(x^2-3xy+y^2\right)-y\left(x^2-3xy+y^2\right)$

$\qquad = x^3 - 3x^2y + xy^2 - x^2y + 3xy^2 - y^3$

$\qquad = x^3 - 4x^2y + 4xy^2 - y^3$

69. $(xy+ab)(xy-ab) = (xy)^2 - (ab)^2$

$\qquad\qquad = x^2y^2 - a^2b^2$

71. $\left(x^2+1\right)\left(x^4y+x^2+1\right)$

$\qquad = x^2\left(x^4y+x^2+1\right)+1\left(x^4y+x^2+1\right)$

$\qquad = x^6y + x^4 + x^2 + x^4y + x^2 + 1$

$\qquad = x^6y + x^4y + x^4 + 2x^2 + 1$

73. $\left(x^2y^2-3\right)^2$

$\qquad = \left(x^2y^2\right)^2 - 2\left(x^2y^2\right)(3)+(-3)^2$

$\qquad = x^4y^4 - 6x^2y^2 + 9$

75. $(x+y+1)(x+y-1)$

$\qquad = x(x+y-1)+y(x+y-1)+1(x+y-1)$

$\qquad = x^2 + xy - x + yx + y^2 - y + x + y - 1$

$\qquad = x^2 + 2xy + y^2 - 1$

77. $A = (3x+5y)(x+y)$

$\qquad = 3x(x)+3x(y)+5y(x)+5y(y)$

$\qquad = 3x^2 + 3xy + 5xy + 5y^2$

$\qquad = 3x^2 + 8xy + 5y^2$

The area of the shaded region is $3x^2 + 8xy + 5y^2$ square units.

79. Area of shaded region $=\ \underbrace{(x+y)^2}_{\substack{\text{Area of} \\ \text{larger square}}}\ -\ \underbrace{x^2}^{\substack{\text{Area of} \\ \text{smaller square}}}$

$\qquad\qquad = (x^2 + 2xy + y^2) - x^2$

$\qquad\qquad = 2xy + y^2$

81. $\left[\left(x^3y^3+1\right)\left(x^3y^3-1\right)\right]^2$

$\qquad = \left[\left(x^3y^3\right)^2 - (1)^2\right]^2$

$\qquad = \left[x^6y^6 - 1\right]^2$

$\qquad = \left(x^6y^6\right)^2 - 2\left(x^6y^6\right)(1)+(1)^2$

$\qquad = x^{12}y^{12} - 2x^6y^6 + 1$

83. $(xy-3)^2(xy+3)^2$

$=\left[(xy-3)(xy+3)\right]^2$

$=\left[(xy)^2-3^2\right]^2$

$=\left[x^2y^2-9\right]^2$

$=\left(x^2y^2\right)^2-2\left(x^2y^2\right)(9)+(9)^2$

$x^4y^4-18x^2y^2+81$

85. $[x+y+z][x-(y+z)]$

$=\left[x+(y+z)\right]\left[x-(y+z)\right]$

$=x^2-(y+z)^2$

$=x^2-\left(y^2+2yz+z^2\right)$

$=x^2-y^2-2yz-z^2$

87. $N=\frac{1}{4}x^2y-2xy+4y;\ x=10,\ y=16$

$N=\frac{1}{4}x^2y-2xy+4y$

$=\frac{1}{4}(10)^2(16)-2(10)(16)+4(16)$

$=\frac{1}{4}(100)(16)-2(10)(16)+4(16)$

$=400-320+64$

$=144$

Each tree provides 144 board feet of lumber, so 20 trees will provide 20(144) = 2880 board feet. This is not enough lumber to complete the job. Since 3000 − 2880 = 120, the contractor will need 120 more board feet.

89. $v_0=80,\ s_0=96$ and $t=2$

$s=-16t^2+v_0t+s_0$

$s=-16t^2+80t+96$

$s=-16(2)^2+80(2)+96$

$=-16(4)+80(6)+96$

$=-64+160+96$

$=192$

The ball will be 192 feet above the ground 2 seconds after being thrown.

91. $v_0=80,\ s_0=96$ and $t=6$

$s=-16t^2+v_0t+s_0$

$s=-16t^2+80t+96$

$s=-16(6)^2+80(6)+96$

$=-16(36)+80(6)+96$

$=-576+480+96=0$

The ball will be 0 feet above the ground after 6 seconds. This means that the ball hits the ground 6 seconds after being thrown.

93. The ball is falling from 2.5 seconds to 6 seconds. The graph is decreasing over this interval.

95. (2,192)

97. The ball reaches its maximum height 2.5 seconds after it is thrown.

$v_0=80,\ s_0=96$ and $t=2.5$

$s=-16t^2+v_0t+s_0$

$s=-16t^2+80t+96$

$s=-16(2.5)^2+80(2.5)+96$

$=-16(6.25)+80(2.5)+96$

$=-100+200+96$

$=196$

The maximum height is 196 feet.

99. Answers will vary.

101. makes sense

103. does not make sense; Explanations will vary. Sample explanation: FOIL is used to multiply a binomial by a binomial.

105. false; Changes to make the statement true will vary. A sample change is: The term $-3x^{16}y^9$ has degree $16+9=25$.

107. true

109. Area of rectangle:

$(x+10y)(x+8y)=x^2+18xy+80y^2$

Area of two corner squares: $x^2+x^2=2x^2$

Area of shaded region:

$\left(x^2+18xy+80y^2\right)-2x^2=-x^2+18xy+80y^2$

111. Note that the shed consists of a rectangular solid with half a cylinder on top.

Radius of cylinder: x

Length of cylinder: y

Width of base: $2x$

Length of base: y

Height of base: x

Volume of rectangular solid: $l \cdot w \cdot h$

Volume of cylinder: $\pi r^2 \cdot l$

Volume of half-cylinder: $\dfrac{1}{2}\pi r^2 \cdot l$

$$V_{\text{shed}} = V_{\text{base}} + V_{\text{top}}$$

$$= l \cdot w \cdot h + \frac{1}{2}\pi r^2 \cdot l$$

$$= y \cdot 2x \cdot x + \frac{1}{2}\pi \cdot (x)^2 \cdot y$$

$$= 2x^2 y + \frac{1}{2}\pi x^2 y$$

112. $R = \dfrac{L+3W}{2}$; for W

$$R = \frac{L+3W}{2}$$

$$2R = 2\left(\frac{L+3W}{2}\right)$$

$$2R = L+3W$$

$$2R - L = L + 3W - L$$

$$2R - L = 3W$$

$$\frac{2R-L}{3} = \frac{3W}{3}$$

$$\frac{2R-L}{3} = W \quad \text{or} \quad W = \frac{2R-L}{3}$$

113. $-6.4 - (-10.2) = -6.4 + 10.2 = 3.8$

114.

$$0.02(x-5) = 0.03 - 0.03(x+7)$$

$$0.02x - 0.1 = 0.03 - 0.03x - 0.21$$

$$0.02x - 0.1 = -0.03x - 0.18$$

$$100(0.02x - 0.1) = 100(-0.03x - 0.18)$$

$$2x - 10 = -3x - 18$$

$$2x = -3x - 8$$

$$5x = -8$$

$$\frac{5x}{5} = \frac{-8}{5}$$

$$x = -1.6$$

The solution set is $\{-1.6\}$.

115. $\dfrac{x^7}{x^3} = \dfrac{\cancel{x} \cdot \cancel{x} \cdot \cancel{x} \cdot x \cdot x \cdot x \cdot x}{\cancel{x} \cdot \cancel{x} \cdot \cancel{x}} = x^4$

116. $\dfrac{(x^2)^3}{5^3} = \dfrac{x^6}{125}$

117. $\dfrac{(2a^3)^5}{(b^4)^5} = \dfrac{2^5(a^3)^5}{b^{20}} = \dfrac{32a^{15}}{b^{20}}$

Chapter 5 Mid-Chapter Check Point

1. $\left(11x^2 y^3\right)\left(-5x^2 y^3\right)$

$$= -55x^{2+2}y^{3+3} = -55x^4 y^6$$

2. $11x^2 y^3 - 5x^2 y^3 = 6x^2 y^3$

3. $(3x+5)(4x-7)$

$$= 12x^2 - 21x + 20x - 35$$

$$= 12x^2 - x - 35$$

4. $(3x+5) - (4x-7)$

$$= (3x+5) + (-4x+7)$$

$$= -x + 12$$

5. $(2x-5)\left(x^2 - 3x + 1\right)$

$$= 2x\left(x^2 - 3x + 1\right) - 5\left(x^2 - 3x + 1\right)$$

$$= 2x^3 - 6x^2 + 2x - 5x^2 + 15x - 5$$

$$= 2x^3 - 11x^2 + 17x - 5$$

6. $(2x-5) + \left(x^2 - 3x + 1\right)$

$$= x^2 - x - 4$$

7. $(8x-3)^2$

$$= (8x)^2 - 2(8x)(3) + 3^2$$

$$= 64x^2 - 48x + 9$$

8. $\left(-10x^4\right)\left(-7x^5\right) = 70x^9$

9. $\left(x^2+2\right)\left(x^2-2\right)$

$=\left(x^2\right)^2-2^2$

$=x^4-4$

10. $\left(x^2+2\right)^2$

$=\left(x^2\right)^2+2\left(x^2\right)(2)+2^2$

$=x^4+4x^2+4$

11. $(9a-10b)(2a+b)$

$=18a^2+9ab-20ba-10b^2$

$=18a^2+9ab-20ab-10b^2$

$=18a^2-11ab-10b^2$

12. $7x^2\left(10x^3-2x+3\right)$

$=70x^5-14x^3+21x^2$

13. $\left(3a^2b^3-ab+4b^2\right)-\left(-2a^2b^3-3ab+5b^2\right)$

$=\left(3a^2b^3-ab+4b^2\right)+\left(2a^2b^3+3ab-5b^2\right)$

$=5a^2b^3+2ab-b^2$

14. $2(3y-5)(3y+5)=2\left(9y^2-25\right)$

$=18y^2-50$

15. $\left(-9x^3+5x^2-2x+7\right)$

$\quad +\left(11x^3-6x^2+3x-7\right)$

$=\left(-9x^3+11x^3\right)+\left(5x^2-6x^2\right)$

$\quad +(-2x+3x)+(7-7)$

$=2x^3-x^2+x$

16. $10x^2-8xy-3\left(y^2-xy\right)$

$=10x^2-8xy-3y^2+3xy$

$=10x^2-5xy-3y^2$

17. $\left(-2x^5+x^4-3x+10\right)$

$\quad -\left(2x^5-6x^4+7x-13\right)$

$=\left(-2x^5+x^4-3x+10\right)$

$\quad +\left(-2x^5+6x^4-7x+13\right)$

$=-4x^5+7x^4-10x+23$

18. $(x+3y)\left(x^2-3xy+9y^2\right)$

$=x\left(x^2-3xy+9y^2\right)+3y\left(x^2-3xy+9y^2\right)$

$=x^3-3x^2y+9xy^2+3x^2y-9xy^2+27y^3$

$=x^3+27y^3$

19. $\left(5x^4+4\right)\left(2x^3-1\right)$

$=10x^7-5x^4+8x^3-4$

20. $(y-6z)^2=y^2-2(y)(6z)+(6z)^2$

$=y^2-12yz+36z^2$

21. $(2x+3)(2x-3)-(5x+4)(5x-4)$

$=\left(4x^2-9\right)-\left(25x^2-16\right)$

$=\left(4x^2-9\right)+\left(-25x^2+16\right)$

$=-21x^2+7$

22. Make a table of values:

x	$y=1-x^2$	(x,y)
-3	$y=1-(-3)^2=-8$	$(-3,-8)$
-2	$y=1-(-2)^2=-3$	$(-2,-3)$
-1	$y=1-(-1)^2=0$	$(-1,0)$
0	$y=1-(0)^2=1$	$(0,1)$
1	$y=1-(1)^2=0$	$(1,0)$
2	$y=1-(2)^2=-3$	$(2,-3)$
3	$y=1-(3)^2=-8$	$(3,-8)$

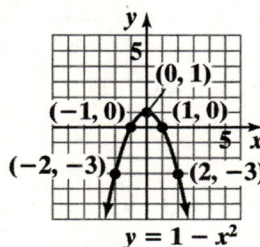

$y=1-x^2$

5.5 Check Points

1. a. $\dfrac{5^{12}}{5^4} = 5^{12-4} = 5^8$

   b. $\dfrac{x^9}{x^2} = x^{9-2} = x^7$

   c. $\dfrac{y^{20}}{y} = y^{20-1} = y^{19}$

2. a. $14^0 = 1$

   b. $(-10)^0 = 1$

   c. $-10^0 = -1 \cdot 10^0 = -1 \cdot 1 = -1$

   d. $20x^0 = 20 \cdot 1 = 20$

   e. $(20x)^0 = 1$

3. a. $\left(\dfrac{x}{5}\right)^2 = \dfrac{x^2}{5^2} = \dfrac{x^2}{25}$

   b. $\left(\dfrac{x^4}{2}\right)^3 = \dfrac{x^{4\cdot 3}}{2^3} = \dfrac{x^{12}}{8}$

   c. $\left(\dfrac{2a^{10}}{b^3}\right)^4 = \dfrac{2^4 (a^{10})^4}{(b^3)^4} = \dfrac{16a^{40}}{b^{12}}$

4. a. $\dfrac{-20x^{12}}{10x^4} = \dfrac{-20}{10}x^{12-4} = -2x^8$

   b. $\dfrac{3x^4}{15x^4} = \dfrac{3}{15}x^{4-4} = \dfrac{1}{5}x^0 = \dfrac{1}{5}$

   c. $\dfrac{9x^6 y^5}{3xy^2} = \dfrac{9}{3} \cdot x^{6-1} y^{5-2} = 3x^5 y^3$

5. $\dfrac{-15x^9 + 6x^5 - 9x^3}{3x^2} = \dfrac{-15x^9}{3x^2} + \dfrac{6x^5}{3x^2} - \dfrac{9x^3}{3x^2}$
   $$= -5x^7 + 2x^3 - 3x$$

6. $\dfrac{25x^9 - 7x^4 + 10x^3}{5x^3} = \dfrac{25x^9}{5x^3} - \dfrac{7x^4}{5x^3} + \dfrac{10x^3}{5x^3}$
   $$= 5x^6 - \dfrac{7}{5}x + 2$$

7. $\dfrac{18x^7 y^6 - 6x^2 y^3 + 60xy^2}{6xy^2}$
   $$= \dfrac{18x^7 y^6}{6xy^2} - \dfrac{6x^2 y^3}{6xy^2} + \dfrac{60xy^2}{6xy^2}$$
   $$= 3x^6 y^4 - xy + 10$$

5.5 Concept and Vocabulary Check

1. b^{m+n}; subtract

2. 1

3. $\dfrac{a^n}{b^n}$; numerator; denominator

4. divide; subtract

5. dividend; divisor; quotient

6. divisor; quotient; dividend

7. $20x^8 - 10x^4 + 6x^3$; $2x^3$

5.5 Exercise Set

1. $\dfrac{3^{20}}{3^5} = 3^{20-5} = 3^{15}$

3. $\dfrac{x^6}{x^2} = x^{6-2} = x^4$

5. $\dfrac{y^{13}}{y^5} = y^{13-5} = y^8$

7. $\dfrac{5^6 \cdot 2^8}{5^3 \cdot 2^4} = 5^{6-3} \cdot 2^{8-4} = 5^3 \cdot 2^4$

9. $\dfrac{x^{100} y^{50}}{x^{25} y^{10}} = x^{100-25} y^{50-10} = x^{75} y^{40}$

11. $2^0 = 1$

13. $(-2)^0 = 1$

15. $-2^0 = -(2^0) = -(1) = -1$

17. $100y^0 = 100 \cdot 1 = 100$

19. $(100y)^0 = 1$

21. $-5^0 + (-5)^0 = -1 + 1 = 0$

23. $-\pi^0 - (-\pi)^0 = -1 - 1 = -2$

25. $\left(\dfrac{x}{3}\right)^2 = \dfrac{x^2}{3^2} = \dfrac{x^2}{9}$

27. $\left(\dfrac{x^2}{4}\right)^3 = \dfrac{\left(x^2\right)^3}{4^3} = \dfrac{x^{2 \cdot 3}}{4^3} = \dfrac{x^6}{64}$

29. $\left(\dfrac{2x^3}{5}\right)^2 = \dfrac{2^2\left(x^3\right)^2}{5^2} = \dfrac{4x^6}{25}$

31. $\left(\dfrac{-4}{3a^3}\right)^3 = \dfrac{(-4)^3}{3^3\left(a^3\right)^3} = \dfrac{-64}{27a^9} = -\dfrac{64}{27a^9}$

33. $\left(\dfrac{-2a^7}{b^4}\right)^5 = \dfrac{\left(-2a^7\right)^5}{\left(b^4\right)^5} = \dfrac{(-2)^5\left(a^7\right)^5}{\left(b^4\right)^5}$

$$= \dfrac{-32a^{35}}{b^{20}} = -\dfrac{32a^{35}}{b^{20}}$$

35. $\left(\dfrac{x^2y^3}{2z}\right)^4 = \dfrac{\left(x^2\right)^4\left(y^3\right)^4}{2^4 z^4} = \dfrac{x^8 y^{12}}{16z^4}$

37. $\dfrac{30x^{10}}{10x^5} = \dfrac{30}{10}x^{10-5} = 3x^5$

39. $\dfrac{-8x^{22}}{4x^2} = \dfrac{-8}{4}x^{22-2} = -2x^{20}$

41. $\dfrac{-9y^8}{18y^5} = \dfrac{-9}{18}y^{8-5} = -\dfrac{1}{2}y^3$

43. $\dfrac{7y^{17}}{5y^5} = \dfrac{7}{5}y^{17-5} = \dfrac{7}{5}y^{12}$

45. $\dfrac{30x^7 y^5}{5x^2 y} = \dfrac{30}{5}x^{7-2}y^{5-1} = 6x^5 y^4$

47. $\dfrac{-18x^{14}y^2}{36x^2 y^2} = \dfrac{-18}{36}x^{14-2}y^{2-2}$

$$= -\dfrac{1}{2}x^{12}y^0 = -\dfrac{1}{2}x^{12} \cdot 1$$

$$= -\dfrac{1}{2}x^{12}$$

49. $\dfrac{9x^{20}y^{20}}{7x^{20}y^{20}} = \dfrac{9}{7}x^{20-20}y^{20-20}$

$$= \dfrac{9}{7}x^0 y^0 = \dfrac{9}{7} \cdot 1 \cdot 1 = \dfrac{9}{7}$$

51. $\dfrac{-5x^{10}y^{12}z^6}{50x^2 y^3 z^2} = -\dfrac{1}{10}x^{10-2}y^{12-3}z^{6-2}$

$$= -\dfrac{1}{10}x^8 y^9 z^4$$

53. $\dfrac{10x^4 + 2x^3}{2} = \dfrac{10x^4}{2} + \dfrac{2x^3}{2} = 5x^4 + x^3$

55. $\dfrac{14x^4 - 7x^3}{7x} = \dfrac{14x^4}{7x} - \dfrac{7x^3}{7x}$

$$= 2x^{4-1} - x^{3-1} = 2x^3 - x^2$$

57. $\dfrac{y^7 - 9y^2 + y}{y} = \dfrac{y^7}{y} - \dfrac{9y^2}{y} + \dfrac{y}{y}$

$$= y^{7-1} - 9y^{2-1} + y^{1-1}$$

$$= y^6 - 9y^1 + y^0$$

$$= y^6 - 9y + 1$$

59. $\dfrac{24x^3 - 15x^2}{-3x} = \dfrac{24x^3}{-3x} + \dfrac{-15x^2}{-3x}$

$$= -8x^{3-1} + 5x^{2-1} = -8x^2 + 5x$$

61. $\dfrac{18x^5 + 6x^4 + 9x^3}{3x^2} = \dfrac{18x^5}{3x^2} + \dfrac{6x^4}{3x^2} + \dfrac{9x^3}{3x^2}$

$\qquad\qquad\qquad = 6x^3 + 2x^2 + 3x$

63. $\dfrac{12x^4 - 8x^3 + 40x^2}{4x} = \dfrac{12x^4}{4x} - \dfrac{8x^3}{4x} + \dfrac{40x^2}{4x}$

$\qquad\qquad\qquad = 3x^3 - 2x^2 + 10x$

65. $\left(4x^2 - 6x\right) \div x = \dfrac{4x^2 - 6x}{x} = \dfrac{4x^2}{x} - \dfrac{6x}{x}$

$\qquad\qquad\qquad = 4x - 6$

67. $\dfrac{30z^3 + 10z^2}{-5z} = \dfrac{30z^3}{-5z} + \dfrac{10z^2}{-5z} = -6z^2 - 2z$

69. $\dfrac{8x^3 + 6x^2 - 2x}{2x} = \dfrac{8x^3}{2x} + \dfrac{6x^2}{2x} - \dfrac{2x}{2x}$

$\qquad\qquad\qquad = 4x^2 + 3x - 1$

71. $\dfrac{25x^7 - 15x^5 - 5x^4}{5x^3} = \dfrac{25x^7}{5x^3} - \dfrac{15x^5}{5x^3} - \dfrac{5x^4}{5x^3}$

$\qquad\qquad\qquad = 5x^4 - 3x^2 - x$

73. $\dfrac{18x^7 - 9x^6 + 20x^5 - 10x^4}{-2x^4}$

$\quad = \dfrac{18x^7}{-2x^4} - \dfrac{9x^6}{-2x^4} + \dfrac{20x^5}{-2x^4} - \dfrac{10x^4}{-2x^4}$

$\quad = -9x^3 + \dfrac{9}{2}x^2 - 10x + 5$

75. $\dfrac{12x^2 y^2 + 6x^2 y - 15xy^2}{3xy}$

$\quad = \dfrac{12x^2 y^2}{3xy} + \dfrac{6x^2 y}{3xy} - \dfrac{15xy^2}{3xy}$

$\quad = 4xy + 2x - 5y$

77. $\dfrac{20x^7 y^4 - 15x^3 y^2 - 10x^2 y}{-5x^2 y}$

$\quad = \dfrac{20x^7 y^4}{-5x^2 y} + \dfrac{-15x^3 y^2}{-5x^2 y} + \dfrac{-10x^2 y}{-5x^2 y}$

$\quad = -4x^5 y^3 + 3xy + 2$

79. $\dfrac{2x^3\left(4x + 2\right) - 3x^2\left(2x - 4\right)}{2x^2}$

$\quad = \dfrac{8x^4 + 4x^3 - 6x^3 + 12x^2}{2x^2}$

$\quad = \dfrac{8x^4 - 2x^3 + 12x^2}{2x^2}$

$\quad = \dfrac{8x^4}{2x^2} - \dfrac{2x^3}{2x^2} + \dfrac{12x^2}{2x^2}$

$\quad = 4x^2 - x + 6$

81. $\left(\dfrac{18x^2 y^4}{9xy^2}\right) - \left(\dfrac{15x^5 y^6}{5x^4 y^4}\right) = 2xy^2 - 3xy^2$

$\qquad\qquad\qquad\qquad\qquad = -xy^2$

83. $\dfrac{(y+5)^2 + (y+5)(y-5)}{2y}$

$\quad = \dfrac{\left(y^2 + 10y + 25\right) + \left(y^2 - 25\right)}{2y}$

$\quad = \dfrac{2y^2 + 10y}{2y} = \dfrac{2y^2}{2y} + \dfrac{10y}{2y} = y + 5$

85. $\dfrac{12x^{15n} - 24x^{12n} + 8x^{3n}}{4x^{3n}}$

$\quad = \dfrac{12x^{15n}}{4x^{3n}} - \dfrac{24x^{12n}}{4x^{3n}} + \dfrac{8x^{3n}}{4x^{3n}}$

$\quad = 3x^{12n} - 6x^{9n} + 2$

87. a. Average $= \dfrac{\text{Receipts}}{\text{Admissions}} = \dfrac{7661}{1421} = 5.39$

The average admission charge for a film in 2000 was $5.39.

b. Average $= \dfrac{\text{Receipts}}{\text{Admissions}} = \dfrac{3.6x^2 + 158x + 2790}{-0.2x^2 + 21x + 1015}$

c. $\overset{\text{2000 is 20 years after 1980.}}{\dfrac{3.6x^2 + 158x + 2790}{-0.2x^2 + 21x + 1015} = \dfrac{3.6(20)^2 + 158(20) + 2790}{-0.2(20)^2 + 21(20) + 1015}}$

$\qquad\qquad\qquad\qquad = \dfrac{7390}{1355}$

$\qquad\qquad\qquad\qquad = 5.45$

According to the model, the average admission charge for a film in 2000 was $5.45.
This overestimates the actual value by $0.06.

d. Polynomial division cannot be performed using the methods in this section because the divisor is not a monomial.

89. – 95. Answers will vary.

97. does not make sense; Explanations will vary. Sample explanation: The quotient rule involves subtracting exponents.

99. does not make sense; Explanations will vary. Sample explanation: Divide each term of the polynomial by the monomial.

101. false; Changes to make the statement true will vary. A sample change is:

$$\frac{12x^3 - 6x}{2x} = \frac{12x^3}{2x} - \frac{6x}{2x} = 6x^2 - 3$$

103. true

105. $\dfrac{? x^8 - ? x^6}{3x^?} = 3x^5 - 4x^3$

To get 3 as the coefficient of the first term of the quotient, the coefficient of the first term of the dividend must be 9. For the exponent in the first term of the quotient to be 5, the exponent in the divisor must be 3. Since now know that the divisor is $3x^3$, the coefficient of the second term in the dividend must be 12. Therefore,

$$\frac{? x^8 - ? x^6}{3x^?} = \frac{9x^8 - 12x^6}{3x^3}$$

107. $\left| -20.3 \right| = 20.3$

108.

```
    0.875
8 ) 7.000
    64
    ‾‾
     60
     56
     ‾‾
      40
      40
      ‾‾
       0
```

$$\frac{7}{8} = 0.875$$

109. $y = \dfrac{1}{3}x + 2$

slope $= \dfrac{1}{3}$; y-intercept $= 2$

Plot (0,2). From this point move 1 unit *up* and 3 units to the *right* to reach the point (3,3). Draw a line through (0,2) and (3,3).

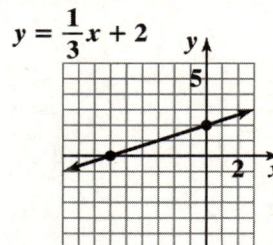

110.

```
      26
19 ) 494
     38
     ‾‾
     114
     114
     ‾‾‾
       0
```

The quotient is 26 and the remainder is 0.

111.

```
      123
24 ) 2958
     24
     ‾‾
      55
      48
      ‾‾
       78
       72
       ‾‾
        6
```

The quotient is 123 and the remainder is 6.

112.

```
        257
98 ) 25187
     196
     ‾‾‾
      558
      490
      ‾‾‾
       687
       686
       ‾‾‾
         1
```

The quotient is 257 and the remainder is 1.

5.6 Check Points

1.
$$x + 9 \overline{\smash{\big)}\,x^2 + 14x + 45} \quad\begin{array}{r} x+5 \end{array}$$

$$\underline{x^2 + 9x}$$
$$5x + 45$$
$$\underline{5x + 45}$$
$$0$$

$$\frac{x^2 + 14x + 45}{x + 9} = x + 5$$

2. $\dfrac{6x + 8x^2 - 12}{2x + 3} = \dfrac{8x^2 + 6x - 12}{2x + 3}$

$$2x + 3 \overline{\smash{\big)}\,8x^2 + 6x - 12} \quad\begin{array}{r} 4x-3 \end{array}$$

$$\underline{8x^2 + 12x}$$
$$-6x - 12$$
$$\underline{-6x - 9}$$
$$-3$$

$$\frac{6x + 8x^2 - 12}{2x + 3} = 4x - 3 - \frac{3}{2x + 3}$$

3. Rewrite $x^3 - 1$ using coefficients of 0 on the missing terms gives $x^3 + 0x^2 + 0x - 1$.

$$x - 1 \overline{\smash{\big)}\,x^3 + 0x^2 + 0x - 1} \quad\begin{array}{r} x^2 + x + 1 \end{array}$$

$$\underline{x^3 - x^2}$$
$$x^2 + 0x$$
$$\underline{x^2 - x}$$
$$x - 1$$
$$\underline{x - 1}$$
$$0$$

$$\frac{x^3 - 1}{x - 1} = x^2 + x + 1$$

5.6 Concept and Vocabulary Check

1. $10x^3 + 4x^2 + 0x + 9$

2. $8x^2$; $2x$; $4x$; $10x$

3. $5x$; $3x - 2$; $15x^2 - 10x$; $15x^2 - 22x$

4. $5x^2 - 10x$; $6x^2 + 8x$; $18x$; -4; $18x - 4$

5. 14; $x - 12$; 14; $x - 12 + \dfrac{14}{x - 5}$

5.6 Exercise Set

1.
$$x + 2 \overline{\smash{\big)}\,x^2 + 6x + 8} \quad\begin{array}{r} x+4 \end{array}$$

$$\underline{x^2 + 2x}$$
$$4x + 8$$
$$\underline{4x + 8}$$
$$0$$

$$\frac{x^2 + 6x + 8}{x + 2} = x + 4$$

3.
$$x - 2 \overline{\smash{\big)}\,2x^2 + x - 10} \quad\begin{array}{r} 2x+5 \end{array}$$

$$\underline{2x^2 - 4x}$$
$$5x - 10$$
$$\underline{5x - 10}$$
$$0$$

$$\frac{2x^2 + x - 10}{x - 2} = 2x + 5$$

5.
$$x - 3 \overline{\smash{\big)}\,x^2 - 5x + 6} \quad\begin{array}{r} x-2 \end{array}$$

$$\underline{x^2 - 3x}$$
$$-2x + 6$$
$$\underline{-2x + 6}$$
$$0$$

$$\frac{x^2 - 5x + 6}{x - 3} = x - 2$$

7.
$$y + 2 \overline{\smash{\big)}\,2y^2 + 5y + 2} \quad\begin{array}{r} 2y+1 \end{array}$$

$$\underline{2y^2 + 4y}$$
$$y + 2$$
$$\underline{y + 2}$$
$$0$$

$$\frac{2y^2 + 5y + 2}{y + 2} = 2y + 1$$

9.
$$x+2 \overline{)x^2 - 3x + 4} \quad \text{quotient } x-5$$

$$\underline{x^2 + 2x}$$
$$-5x + 4$$
$$\underline{-5x - 10}$$
$$14$$

$$\frac{x^2 - 3x + 4}{x+2} = x - 5 + \frac{14}{x+2}$$

11.
$$y+2 \overline{)y^2 + 5y + 10} \quad y+3$$

$$\underline{y^2 + 2y}$$
$$3y + 10$$
$$\underline{3y + 6}$$
$$4$$

$$\frac{5y + 10 + y^2}{y+2} = \frac{y^2 + 5y + 10}{y+2} = y + 3 + \frac{4}{y+2}$$

13.
$$x-1 \overline{)x^3 - 6x^2 + 7x - 2} \quad x^2 - 5x + 2$$

$$\underline{x^3 - x^2}$$
$$-5x^2 + 7x$$
$$\underline{-5x^2 + 5x}$$
$$2x - 2$$
$$\underline{2x - 2}$$
$$0$$

$$\frac{x^3 - 6x^2 + 7x - 2}{x-1} = x^2 - 5x + 2$$

15.
$$2y-3 \overline{)12y^2 - 20y + 3} \quad 6y - 1$$

$$\underline{12y^2 - 18y}$$
$$-2y + 3$$
$$\underline{-2y + 3}$$
$$0$$

$$\frac{12y^2 - 20y + 3}{2y-3} = 6y - 1$$

17.
$$2a-1 \overline{)4a^2 + 4a - 3} \quad 2a + 3$$

$$\underline{4a^2 - 2a}$$
$$6a - 3$$
$$\underline{6a - 3}$$
$$0$$

$$\frac{4a^2 + 4a - 3}{2a-1} = 2a + 3$$

19.
$$2y+1 \overline{)2y^3 - y^2 + 3y + 2} \quad y^2 - y + 2$$

$$\underline{2y^3 + y^2}$$
$$-2y^2 + 3y$$
$$\underline{-2y^2 - y}$$
$$4y + 2$$
$$\underline{4y + 2}$$
$$0$$

$$\frac{3y - y^2 + 2y^3 + 2}{2y + 1}$$

$$= \frac{2y^3 - y^2 + 3y + 2}{2y + 1} = y^2 - y + 2$$

21.
$$2x-5 \overline{)6x^2 - 5x - 30} \quad 3x + 5$$

$$\underline{6x^2 - 15x}$$
$$10x - 30$$
$$\underline{10x - 25}$$
$$-5$$

$$\frac{6x^2 - 5x - 30}{2x - 5} = 3x + 5 - \frac{5}{2x - 5}$$

23.

$$x-2\overline{\smash{\big)}\,x^3+0x^2+4x-3} \quad\text{with quotient } x^2+2x+8$$

$$\underline{x^3-2x^2}$$
$$2x^2+4x$$
$$\underline{2x^2-4x}$$
$$8x-3$$
$$\underline{8x-16}$$
$$13$$

$$\frac{x^2+4x-3}{x-2}=x^2+2x+8+\frac{13}{x-2}$$

25.

$$2y+3\overline{\smash{\big)}\,4y^3+8y^2+5y+9}\quad\text{with quotient } 2y^2+y+1$$

$$\underline{4y^3+6y^2}$$
$$2y^2+5y$$
$$\underline{2y^2+3y}$$
$$2y+9$$
$$\underline{2y+3}$$
$$6$$

$$\frac{4y^3+8y^2+5y+9}{2y+3}=2y^2+y+1+\frac{6}{2y+3}$$

27.

$$3y+2\overline{\smash{\big)}\,6y^3-5y^2+0y+5}\quad\text{with quotient } 2y^2-3y+2$$

$$\underline{6y^3+4y^2}$$
$$-9y^2+0y$$
$$\underline{-9y^2-6y}$$
$$6y+5$$
$$\underline{6y+4}$$
$$1$$

$$\frac{6y^3-5y^2+5}{3y+2}=2y^2-3y+2+\frac{1}{3y+2}$$

29.

$$3x-1\overline{\smash{\big)}\,27x^3+0x^2+0x-1}\quad\text{with quotient } 9x^2+3x+1$$

$$\underline{27x^3-9x^2}$$
$$9x^2+0x$$
$$\underline{9x^2-3x}$$
$$3x-1$$
$$\underline{3x-1}$$
$$0$$

$$\frac{27x^3-1}{3x-1}=9x^2+3x+1$$

31.

$$y-3\overline{\smash{\big)}\,y^4-12y^3+54y^2-108y+81}\quad\text{with quotient } y^3-9y^2+27y-27$$

$$\underline{y^4-3y^3}$$
$$-9y^3+54y^2$$
$$\underline{-9y^3+27y^2}$$
$$27y^2-108y$$
$$\underline{27y^2-81y}$$
$$-27y+81$$
$$\underline{-27y+81}$$
$$0$$

$$\frac{81-12y^3+54y^2+y^4-108y}{y-3}$$
$$=\frac{y^4-12y^3+54y^2-108y+81}{y-3}$$
$$=y^3-9y^2+27y-27$$

33.

$$2y-1\overline{\smash{\big)}\,4y^2+6y+0}\quad\text{with quotient } 2y+4$$

$$\underline{4y^2-2y}$$
$$8y+0$$
$$\underline{8y-4}$$
$$4$$

$$\frac{4y^2+6y}{2y-1}=2y+4+\frac{4}{2y-1}$$

35.
$$y-1 \overline{\smash{\big)}\,y^4+0y^3-2y^2+0y+5} \quad\quad \begin{array}{c} y^3+y^2-y-1 \end{array}$$

$$
\begin{array}{r}
y^4-y^3 \\ \hline
y^3-2y^2 \\
y^3-\ y^2 \\ \hline
-y^2+0y \\
-y^2+\ y \\ \hline
-y+5 \\
-y+1 \\ \hline
4
\end{array}
$$

$$\frac{y^4-2y^2+5}{y-1}=y^3+y^2-y-1+\frac{4}{y-1}$$

37.
$$x^2+0x+2 \overline{\smash{\big)}\,4x^3-3x^2+x+1} \quad\quad \begin{array}{c}4x-3\end{array}$$

$$
\begin{array}{r}
4x^3+0x^2+8x \\ \hline
-3x^2-7x+1 \\
-3x^2+0x-6 \\ \hline
-7x+7
\end{array}
$$

$$\frac{4x^3-3x^2+x+1}{x^2+2}=4x-3+\frac{-7x+7}{x^2+2}$$

39.
$$x-a \overline{\smash{\big)}\,x^3+0x^2+0x-a^3} \quad\quad \begin{array}{c}x^2+ax+a^2\end{array}$$

$$
\begin{array}{r}
x^3-ax^2 \\ \hline
ax^2+0x \\
ax^2-a^2x \\ \hline
a^2x-a^3 \\
a^2x-a^3 \\ \hline
0
\end{array}
$$

$$\frac{x^3+0x^2+0x-a^3}{x-a}=x^2+ax+a^2$$

41.
$$3x^2+2x-4 \overline{\smash{\big)}\,6x^4-5x^3-8x^2+16x-8} \quad\quad \begin{array}{c}2x^2-3x+2\end{array}$$

$$
\begin{array}{r}
6x^4+4x^3-8x^2 \\ \hline
-9x^3+0x^2+16x \\
-9x^3-6x^2+12x \\ \hline
6x^2+4x-8 \\
6x^2+4x-8 \\ \hline
0
\end{array}
$$

$$\frac{6x^4-5x^3-8x^2+16x-8}{3x^2+2x-4}=2x^2-3x+2$$

43. First, compute the difference:
$$\left(4x^3+x^2-2x+7\right)-\left(3x^3-2x^2-7x+4\right)$$
$$=x^3+3x^2+5x+3$$
Now, complete the division:
$$\frac{x^3+3x^2+5x+3}{x+1}$$

$$x+1 \overline{\smash{\big)}\,x^3+3x^2+5x+3} \quad\quad \begin{array}{c}x^2+2x+3\end{array}$$

$$
\begin{array}{r}
x^3+x^2 \\ \hline
2x^2+5x \\
2x^2+2x \\ \hline
3x+3 \\
3x+3 \\ \hline
0
\end{array}
$$

$$\frac{x^3+3x^2+5x+3}{x+1}=x^2+2x+3$$

45. $A=l\cdot w$ so $l=\dfrac{A}{w}=\dfrac{x^3+3x^2+5x+3}{x+1}$

$$x+1 \overline{\smash{\big)}\,x^3+3x^2+5x+3} \quad\quad \begin{array}{c}x^2+2x+3\end{array}$$

$$
\begin{array}{r}
x^3+x^2 \\ \hline
2x^2+5x \\
2x^2+2x \\ \hline
3x+3 \\
3x+3 \\ \hline
0
\end{array}
$$

The base of the parallelogram is (x^2+2x+3) units.

47. a. Substitute $n = 3$ into the formula:

$$\frac{30,000x^3 - 30,000}{x-1}$$

b. Factor out 30,000 from the numerator:

$$\frac{30,000x^3 - 30,000}{x-1} = 30,000\frac{x^3 - 1}{x-1}$$

Now use the formula from #39 with $a = 1$:

$$\frac{x^3 - a^3}{x-a} = x^2 + ax + a^2$$

$$\frac{x^3 - 1}{x-1} = \frac{x^3 - 1^3}{x-1}$$

$$= x^2 + 1 \cdot x + 1^2 = x^2 + x + 1$$

So $\dfrac{30,000x^3 - 30,000}{x-1} = 30,000 \cdot \dfrac{x^3 - 1}{x-1}$

$$= 30,000\left(x^2 + x + 1\right)$$

$$= 30,000x^2 + 30,000x + 30,000$$

c. Substitute in $x = 1.05$ into your formulas from parts (a) and (b) above:

$$\frac{30,000x^3 - 30,000}{x-1}$$

$$= \frac{30,000(1.05)^3 - 30,000}{(1.05) - 1} = 94,575$$

$$30,000x^2 + 30,000x + 30,000$$

$$= 30,000(1.05)^2 + 30,000(1.05)$$
$$+ 30,000$$

$$= 94,575$$

Total salary over three years is \$94,575.

49. – 51. Answers will vary.

53. makes sense

55. does not make sense; Explanations will vary.
Sample explanation: The correct answer is
$x^2 - x + 1$.

57. false; Changes to make the statement true will vary.
A sample change is: The remainder is –9.

59. true

61. We can find the polynomial by multiplying the divisor by the quotient and adding the remainder.

$$(x-3)(2x+4) + 17$$

$$= 2x^2 + 4x - 6x - 12 + 17$$

$$= 2x^2 - 2x + 5$$

63. Answers will vary. The quotient starts with x to a power that is one less than the power in the dividend. It is made up of terms that are all powers of x down to 1, but with alternating signs. The remainder is always -2. Following this pattern,

$$\frac{x^7 - 1}{x+1} = x^6 - x^5 + x^4 - x^3 + x^2 - x + 1 - \frac{2}{x+1}$$

$$\require{enclose}
\begin{array}{r}
x^6 - x^5 + x^4 - x^3 + x^2 - x + 1 \\
x+1 \enclose{longdiv}{x^7 + 0x^6 + 0x^5 + 0x^4 + 0x^3 + 0x^2 + 0x - 1}
\end{array}$$

$$\begin{array}{r}
\underline{x^7 + x^6} \\
-x^6 + 0x^5 \\
\underline{-x^6 - x^5} \\
x^5 + 0x^4 \\
\underline{x^5 + x^4} \\
-x^4 + 0x^3 \\
\underline{-x^4 - x^3} \\
x^3 + 0x^2 \\
\underline{x^3 + x^2} \\
-x^2 + 0x \\
\underline{-x^2 - x} \\
x - 1 \\
\underline{x + 1} \\
-2
\end{array}$$

Long division yields the same result.

65. Let $y_1 = \dfrac{x^2 - 25}{x - 5}$ and $y_2 = x - 5$.

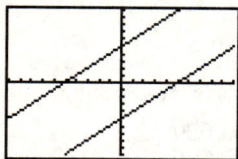

The graphs do not coincide so the division is incorrect.

$$x - 5 \overline{\smash{\big)}\, x^2 + 0x - 25} \quad \overset{\textstyle x+5}{}$$

$$\underline{x^2 - 5x}$$
$$5x - 25$$
$$\underline{5x - 25}$$
$$0$$

The right side should be $x + 5$.

67. Let $y_1 = \dfrac{6x^2 + 16x + 8}{3x + 2}$ and $y_2 = 2x - 4$.

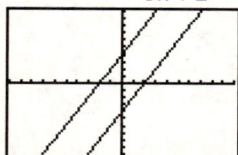

The graphs do not coincide so the division is incorrect.

$$3x + 2 \overline{\smash{\big)}\, 6x^2 + 16x + 8} \quad \overset{\textstyle 2x+4}{}$$

$$\underline{6x^2 + 4x}$$
$$12x + 8$$
$$\underline{12x + 8}$$
$$0$$

The right side should be $2x + 4$.

69. $2x - y \ge 4$

$x + y \le -1$

Graph $2x - y \ge 4$:

Graph $2x - y = 4$ as a solid line with x-intercept 2 and y-intercept -4. Since $2(0) - 0 \ge 4$ is false, shade the half-plane *not* containing $(0,0)$.
Graph $x + y \le -1$;

Graph $x + y = -1$ as a solid line with x-intercept 1, and y-intercept 1. Since $0 + 0 \le -1$ is false, shade the half-plane not containing $(0,0)$.

The solution set of the system is the intersection (overlap) of the two shaded regions.

$2x - y \ge 4$
$x + y \le -1$

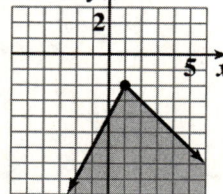

70. $P = 6\% = 0.06, \; B = 20$

$A = PB$

$A = (0.06)(20)$

$A = 1.2$

1.2 is 6% of 20.

71. $\dfrac{x}{3} + \dfrac{2}{5} = \dfrac{x}{5} - \dfrac{2}{5}$

To clear fractions, multiply by the LCD, 15.

$$15\left(\dfrac{x}{3} + \dfrac{2}{5}\right) = 15\left(\dfrac{x}{5} - \dfrac{2}{5}\right)$$

$$15\left(\dfrac{x}{3}\right) + 15\left(\dfrac{2}{5}\right) = 15\left(\dfrac{x}{5}\right) - 15\left(\dfrac{2}{5}\right)$$

$$5x + 6 = 3x - 6$$
$$2x + 6 = -6$$
$$2x = -12$$
$$x = -6$$

The solution set is $\{-6\}$.

72. a. $\dfrac{7^3}{7^5} = \dfrac{\cancel{7} \cdot \cancel{7} \cdot \cancel{7}}{\cancel{7} \cdot \cancel{7} \cdot \cancel{7} \cdot 7 \cdot 7} = \dfrac{1}{7^2}$

$\dfrac{7^3}{7^5} = 7^{3-5} = 7^{-2}$

b. $\dfrac{1}{7^2} = 7^{-2}$

73. $\dfrac{(2x^3)^4}{x^{10}} = \dfrac{2^4(x^3)^4}{x^{10}} = \dfrac{16x^{12}}{x^{10}} = 16x^{12-10} = 16x^2$

74. $\left(\dfrac{x^5}{x^2}\right)^3 = \left(x^{5-2}\right)^3 = \left(x^3\right)^3 = x^9$

5.7 Check Points

1. **a.** $6^{-2} = \dfrac{1}{6^2} = \dfrac{1}{36}$

   b. $5^{-3} = \dfrac{1}{5^3} = \dfrac{1}{125}$

   c. $(-3)^{-4} = \dfrac{1}{(-3)^4} = \dfrac{1}{81}$

   d. $-3^{-4} = -\dfrac{1}{3^4} = -\dfrac{1}{81}$

   e. $8^{-1} = \dfrac{1}{8^1} = \dfrac{1}{8}$

2. **a.** $\dfrac{2^{-3}}{7^{-2}} = \dfrac{7^2}{2^3} = \dfrac{49}{8}$

   b. $\left(\dfrac{4}{5}\right)^{-2} = \dfrac{5^2}{4^2} = \dfrac{25}{16}$

   c. $\dfrac{1}{7y^{-2}} = \dfrac{y^2}{7}$

   d. $\dfrac{x^{-1}}{y^{-8}} = \dfrac{y^8}{x^1} = \dfrac{y^8}{x}$

3. $x^{-12} \cdot x^2 = x^{-12+2} = x^{-10} = \dfrac{1}{x^{10}}$

4. **a.** $\dfrac{x^2}{x^{10}} = x^{2-10} = x^{-8} = \dfrac{1}{x^8}$

   b. $\dfrac{75x^3}{5x^9} = \dfrac{75}{5} \cdot \dfrac{x^3}{x^9} = 15x^{3-9} = 15x^{-6} = \dfrac{15}{x^6}$

   c. $\dfrac{50y^8}{-25y^{14}} = \dfrac{50}{-25} \cdot \dfrac{y^8}{y^{14}} = -2y^{8-14} = -2y^{-6} = -\dfrac{2}{y^6}$

5. $\dfrac{(6x^4)^2}{x^{11}} = \dfrac{6^2(x^4)^2}{x^{11}} = \dfrac{36x^{4\cdot2}}{x^{11}} = \dfrac{36x^8}{x^{11}}$
   $= 36x^{8-11} = 36x^{-3} = \dfrac{36}{x^3}$

6. $\left(\dfrac{x^8}{x^4}\right)^{-5} = \left(x^4\right)^{-5} = x^{-20} = \dfrac{1}{x^{20}}$

7. **a.** The exponent is positive so we move the decimal point eight places to the right.
   $7.4 \times 10^9 = 7,400,000,000$

   b. The exponent is negative so we move the decimal point six places to the left.
   $3.017 \times 10^{-6} = 0.000003017$

8. **a.** $7,410,000,000 = 7.41 \times 10^9$

   b. $0.000000092 = 9.2 \times 10^{-8}$

9. **a.** $(3 \times 10^8)(2 \times 10^2) = (3 \times 2) \times (10^8 \times 10^2)$
   $= 6 \times 10^{8+2}$
   $= 6 \times 10^{10}$

   b. $\dfrac{8.4 \times 10^7}{4 \times 10^{-4}} = \dfrac{8.4}{4} \cdot \dfrac{10^7}{10^{-4}}$
   $= 2.1 \times 10^{7-(-4)}$
   $= 2.1 \times 10^{11}$

   c. $(4 \times 10^{-2})^3 = 4^3 \times (10^{-2})^3$
   $= 64 \times 10^{-6}$
   $= 6.4 \times 10^{-5}$

10. $\dfrac{7.87 \times 10^{11}}{3.07 \times 10^8} = \dfrac{7.87}{3.07} \times \dfrac{10^{11}}{10^8} \approx 2.56 \times 10^3 = 2560$
    Each citizen would have to pay about $2560.

5.7 Concept and Vocabulary Check

1. $\dfrac{1}{b^n}$

2. false

3. true

4. b^n

5. true

6. false

7. a number greater than or equal to 1 and less than 10; integer

8. true

9. false

5.7 Exercise Set

1. $8^{-2} = \dfrac{1}{8^2} = \dfrac{1}{64}$

3. $5^{-3} = \dfrac{1}{5^3} = \dfrac{1}{125}$

5. $(-6)^{-2} = \dfrac{1}{(-6)^2} = \dfrac{1}{36}$

7. $-6^{-2} = -\dfrac{1}{6^2} = -\dfrac{1}{36}$

9. $4^{-1} = \dfrac{1}{4^1} = \dfrac{1}{4}$

11. $2^{-1} + 3^{-1} = \dfrac{1}{2^1} + \dfrac{1}{3^1} = \dfrac{1}{2} + \dfrac{1}{3}$

$\qquad = \dfrac{3}{6} + \dfrac{2}{6} = \dfrac{5}{6}$

13. $\dfrac{1}{3^{-2}} = 3^2 = 9$

15. $\dfrac{1}{(-3)^{-2}} = (-3)^2 = 9$

17. $\dfrac{2^{-3}}{8^{-2}} = \dfrac{8^2}{2^3} = \dfrac{64}{8} = 8$

19. $\left(\dfrac{1}{4}\right)^{-2} = \dfrac{1^{-2}}{4^{-2}} = \dfrac{4^2}{1^2} = \dfrac{16}{1} = 16$

21. $\left(\dfrac{3}{5}\right)^{-3} = \dfrac{3^{-3}}{5^{-3}} = \dfrac{5^3}{3^3} = \dfrac{125}{27}$

23. $\dfrac{1}{6x^{-5}} = \dfrac{1 \cdot x^5}{6} = \dfrac{x^5}{6}$

25. $\dfrac{x^{-8}}{y^{-1}} = \dfrac{y^1}{x^8} = \dfrac{y}{x^8}$

27. $\dfrac{3}{(-5)^{-3}} = 3 \cdot (-5)^3 = 5(-125) = -375$

29. $x^{-8} \cdot x^3 = x^{-8+3} = x^{-5} = \dfrac{1}{x^5}$

31. $(4x^{-5})(2x^2) = 8x^{-5+2} = 8x^{-3} = \dfrac{8}{x^3}$

33. $\dfrac{x^3}{x^9} = x^{3-9} = x^{-6} = \dfrac{1}{x^6}$

35. $\dfrac{y}{y^{100}} = \dfrac{y^1}{y^{100}} = y^{1-100} = y^{-99} = \dfrac{1}{y^{99}}$

37. $\dfrac{30z^5}{10z^{10}} = \dfrac{30}{10} \cdot \dfrac{z^5}{z^{10}} = 3z^{5-10}$

$\qquad = -3z^{-5} = \dfrac{3}{z^5}$

39. $\dfrac{-8x^3}{2x^7} = \dfrac{-8}{2} \cdot \dfrac{x^3}{x^7} = -4x^{-4} = -\dfrac{4}{x^4}$

41. $\dfrac{-9a^5}{27a^8} = \dfrac{-9}{27} \cdot \dfrac{a^5}{a^8} = -\dfrac{1}{3}a^{-3} = -\dfrac{1}{3a^3}$

43. $\dfrac{7w^5}{5w^{13}} = \dfrac{7}{5} \cdot \dfrac{w^5}{w^{13}} = \dfrac{7}{5}w^{-8} = \dfrac{7}{5w^8}$

45. $\dfrac{x^3}{(x^4)^2} = \dfrac{x^3}{x^{4\cdot2}} = \dfrac{x^3}{x^8} = x^{-5} = \dfrac{1}{x^5}$

47. $\dfrac{y^{-3}}{(y^4)^2} = \dfrac{y^{-3}}{y^8} = y^{-3-8} = y^{-11} = \dfrac{1}{y^{11}}$

49. $\dfrac{(4x^3)^2}{x^8} = \dfrac{4^2x^6}{x^8} = 16x^{-2} = \dfrac{16}{x^2}$

51. $\dfrac{(6y^4)^3}{y^{-5}} = \dfrac{6^3y^{12}}{y^{-5}} = 216y^{12-(-5)} = 216y^{17}$

53. $\left(\dfrac{x^4}{x^2}\right)^{-3} = \left(x^2\right)^{-3} = x^{-6} = \dfrac{1}{x^6}$

55. $\left(\dfrac{4x^5}{2x^2}\right)^{-4} = \left(2x^3\right)^{-4} = 2^{-4}x^{-12} = \dfrac{1}{2^4 x^{12}}$

$\qquad = \dfrac{1}{16x^{12}}$

57. $\left(3x^{-1}\right)^{-2} = 3^{-2}\left(x^{-1}\right)^{-2} = 3^{-2}x^2$

$\qquad = \dfrac{x^2}{3^2} = \dfrac{x^2}{9}$

59. $\left(-2y^{-1}\right)^{-3} = (-2)^{-3}\left(y^{-1}\right)^{-3} = \dfrac{y^3}{(-2)^3}$

$\qquad = \dfrac{y^3}{-8} = -\dfrac{y^3}{8}$

61. $\dfrac{2x^5 \cdot 3x^7}{15x^6} = \dfrac{6x^{12}}{15x^6} = \dfrac{6}{15} \cdot \dfrac{x^{12}}{x^6}$

$\qquad = \dfrac{2}{5} \cdot x^6 = \dfrac{2x^6}{5}$

63. $\left(x^3\right)^5 \cdot x^{-7} = x^{15} \cdot x^{-7} = x^{15+(-7)} = x^8$

65. $\left(2y^3\right)^4 y^{-6} = 2^4\left(y^3\right)^4 y^{-6} = 16y^{12}y^{-6}$

$\qquad = 16y^6$

67. $\dfrac{\left(y^3\right)^4}{\left(y^2\right)^7} = \dfrac{y^{12}}{y^{14}} = y^{-2} = \dfrac{1}{y^2}$

69. $\left(y^{10}\right)^{-5} = y^{(10)(-5)} = y^{-50} = \dfrac{1}{y^{50}}$

71. $\left(a^4 b^5\right)^{-3} = \left(a^4\right)^{-3}\left(b^5\right)^{-3} = a^{-12}b^{-15}$

$\qquad = \dfrac{1}{a^{12}b^{15}}$

73. $\left(a^{-2}b^6\right)^{-4} = a^8 b^{-24} = \dfrac{a^8}{b^{24}}$

75. $\left(\dfrac{x^2}{2}\right)^{-2} = \dfrac{x^{-4}}{2^{-2}} = \dfrac{2^2}{x^4} = \dfrac{4}{x^4}$

77. $\left(\dfrac{x^2}{y^3}\right)^{-3} = \dfrac{\left(x^2\right)^{-3}}{\left(y^3\right)^{-3}} = \dfrac{x^{-6}}{y^{-9}} = \dfrac{y^9}{x^6}$

79. The exponent is positive so we move the decimal point two places to the right.

$8.7 \times 10^2 = 870$

81. The exponent is positive so we move the decimal point five places to the right.

$9.23 \times 10^5 = 923,000$

83. $3.4 \times 10^0 = 3.4$ (Don't move decimal point.)

85. The exponent is negative so we move the decimal point one place to the left.

$7.9 \times 10^{-1} = 0.79$

87. The exponent is negative so we move the decimal point two places to the left.

$2.15 \times 10^{-2} = 0.0215$

89. The exponent is negative so we move the decimal point four places to the left.

$7.86 \times 10^{-4} = 0.000786$

91. $32,400 = 3.24 \times 10^4$

93. $220,000,000 = 2.2 \times 10^8$

95. $713 = 7.13 \times 10^2$

97. $6751 = 6.751 \times 10^3$

99. $0.0027 = 2.7 \times 10^{-3}$

101. $0.000020 = 2.02 \times 10^{-5}$

103. $0.005 = 5 \times 10^{-3}$

105. $3.14159 = 3.14159 \times 10^0$

107. $\left(2 \times 10^3\right)\left(3 \times 10^2\right) = 6 \times 10^{3+2} = 6 \times 10^5$

109. $\left(2\times10^5\right)\left(8\times10^3\right)=16\times10^{5+3}=16\times10^8$
$$=1.6\times10^9$$

111. $\dfrac{12\times10^6}{4\times10^2}=3\times10^{6-2}=3\times10^4$

113. $\dfrac{15\times10^4}{5\times10^{-2}}=3\times10^{4+2}=3\times10^6$

115. $\dfrac{15\times10^{-4}}{5\times10^2}=3\times10^{-4-2}=3\times10^{-6}$

117. $\dfrac{180\times10^6}{2\times10^3}=90\times10^{6-3}=90\times10^3$
$$=9\times10^4$$

119. $\dfrac{3\times10^4}{12\times10^{-3}}=0.25\times10^{4+3}=0.25\times10^7$
$$=2.5\times10^6$$

121. $\left(5\times10^2\right)^3=5^3\times10^{2(3)}=125\times10^6$
$$=1.25\times10^8$$

123. $\left(3\times10^{-2}\right)^4=3^4\times10^{-2(4)}=81\times10^{-8}$
$$=8.1\times10^{-7}$$

125. $\left(4\times10^6\right)^{-1}=4^{-1}\times10^{6(-1)}=0.25\times10^{-6}$
$$=2.5\times10^{-7}$$

127. $\dfrac{\left(x^{-2}y\right)^{-3}}{\left(x^2y^{-1}\right)^3}=\dfrac{x^6y^{-3}}{x^6y^{-3}}$
$$=x^{6-6}y^{-3-(-3)}=x^0y^0=1$$

129. $\left(2x^{-3}yz^{-6}\right)\left(2x\right)^{-5}=2x^{-3}yz^{-6}\cdot2^{-5}x^{-5}$
$$=2^{-4}x^{-8}yz^{-6}=\dfrac{y}{2^4x^8z^6}=\dfrac{y}{16x^8z^6}$$

131. $\left(\dfrac{x^3y^4z^5}{x^{-3}y^{-4}z^{-5}}\right)^{-2}=\left(x^6y^8z^{10}\right)^{-2}$
$$=x^{-12}y^{-16}z^{-20}=\dfrac{1}{x^{12}y^{16}z^{20}}$$

133. $\dfrac{\left(2^{-1}x^{-2}y^{-1}\right)^{-2}\left(2x^{-4}y^3\right)^{-2}\left(16x^{-3}y^3\right)^0}{\left(2x^{-3}y^{-5}\right)^2}$

$=\dfrac{\left(2^2x^2y^2\right)\left(2^{-2}x^8y^{-6}\right)(1)}{\left(2^2x^{-6}y^{-10}\right)}$

$=\dfrac{x^{18}y^6}{4}$

135. $\dfrac{\left(5\times10^3\right)\left(1.2\times10^{-4}\right)}{\left(2.4\times10^2\right)}=2.5\times10^{-3}$

137. $\dfrac{\left(1.6\times10^4\right)\left(7.2\times10^{-3}\right)}{\left(3.6\times10^8\right)\left(4\times10^{-3}\right)}=0.8\times10^{-4}$
$$=8\times10^{-5}$$

139. a. 1.35×10^{12}

    b. 3.07×10^8

    c. $\dfrac{1.35\times10^{12}}{3.07\times10^8}=\dfrac{1.35}{3.07}\times\dfrac{10^{12}}{10^8}$
$$=0.44\times10^4$$
$$=\$4.40\times10^3$$
$$=\$4400$$

141. $\dfrac{1.35\times10^{12}}{3.2\times10^7}=\dfrac{1.35}{3.2}\times\dfrac{10^{12}}{10^7}$
$$=0.42\times10^5$$
$$=4.2\times10^4$$
$$=42,000 \text{ years}$$

143. $\dfrac{2.325\times10^5}{1.86\times10^5}=1.25 \text{ seconds}$

145. – 151. Answers will vary.

153. makes sense

155. makes sense

157. true

159. false; Changes to make the statement true will vary.
A sample change is: $5^2 \cdot 5^{-2} = 5^{2-2} = 5^0 = 1$ and
$2^5 \cdot 2^{-5} = 2^{5-5} = 2^0 = 1$

161. false; Changes to make the statement true will vary.
A sample change is:
$$\frac{8 \times 10^{30}}{4 \times 10^{-5}} = 2 \times 10^{30-(-5)} = 2 \times 10^{35}$$

163. true

165. The calculator verifies your results.

167. The calculator verifies your results.

169.
$$8 - 6x > 4x - 12$$
$$8 - 8 - 6x > 4x - 12 - 8$$
$$-6x > 4x - 20$$
$$-6x - 4x > 4x - 4x - 20$$
$$-10x > -20$$
$$\frac{-10x}{-10} < \frac{-20}{-10}$$
$$x < 2$$
Interval notation: $(-\infty, 2)$

170. $24 \div 8 \cdot 3 + 28 \div (-7) = 3 \cdot 3 + 28 \div (-7)$
$$= 9 + (-4) = 5$$

171. The whole numbers in the given set are 0 and
$\sqrt{16} \, (= 4)$.

172. $4x^3(4x^2 - 3x + 1) = 4x^3 \cdot 4x^2 - 4x^3 \cdot 3x + 4x^3 \cdot 1$
$$= 16x^5 - 12x^4 + 4x^3$$

173. $9xy(3xy^2 - y + 9) = 9xy \cdot 3xy^2 - 9xy \cdot y + 9xy \cdot 9$
$$= 27x^2y^3 - 9xy^2 + 81xy$$

174. $(x+3)(x^2+5) = (x)(x^2) + (x)(5) + (3)(x^2) + (3)(5)$
$$= x^3 + 5x + 3x^2 + 15$$
$$= x^3 + 3x^2 + 5x + 15$$

Chapter 5 Review Exercises

1. $7x^4 + 9x$ is a binomial of degree 4.

2. $3x + 5x^2 - 2$ is a trinomial of degree 2.

3. $16x$ is a monomial of degree 1.

4. $\left(-6x^3 + 7x^2 - 9x + 3\right) + \left(14x^3 + 3x^2 - 11x - 7\right)$
$$= \left(-6x^3 + 14x^3\right) + \left(7x^2 + 3x^2\right) + \left(-9x - 11x\right) + \left(3 - 7\right)$$
$$= 8x^3 + 10x^2 - 20x - 4$$

5. $\left(9y^3 - 7y^2 + 5\right) + \left(4y^3 - y^2 + 7y - 10\right)$
$$= \left(9y^3 + 4y^3\right) + \left(-7y^2 - y^2\right) + 7y + \left(5 - 10\right)$$
$$= 13y^3 - 8y^2 + 7y - 5$$

6. $\left(5y^2 - y - 8\right) - \left(-6y^2 + 3y - 4\right)$
$$= \left(5y^2 - y - 8\right) + \left(6y^2 - 3y + 4\right)$$
$$= \left(5y^2 + 6y^2\right) + \left(-y - 3y\right) + \left(-8 + 4\right)$$
$$= 11y^2 - 4y - 4$$

7. $\left(13x^4 - 8x^3 + 2x^2\right) - \left(5x^4 - 3x^3 + 2x^2 - 6\right)$
$$= \left(13x^4 - 8x^3 + 2x^2\right)$$
$$\quad + \left(-5x^4 + 3x^3 - 2x^2 + 6\right)$$
$$= \left(13x^4 - 5x^4\right) + \left(-8x^3 + 3x^3\right)$$
$$\quad + \left(2x^2 - 2x^2\right) + 6$$
$$= 8x^4 - 5x^3 + 6$$

8. $\left(-13x^4 - 6x^2 + 5x\right) - \left(x^4 + 7x^2 - 11x\right)$
$$= \left(-13x^4 - 6x^2 + 5x\right) + \left(-x^4 - 7x^2 + 11x\right)$$
$$= \left(-13x^4 - x^4\right) + \left(-6x^2 - 7x^2\right) + \left(5x + 11x\right)$$
$$= -14x^4 - 13x^2 + 16x$$

9. $7y^4 - 6y^3 + 4y^2 - 4y$

$\underline{\quad\quad y^3 - \quad y^2 + 3y - 4}$

$7y^4 - 5y^3 + 3y^2 - \ y - 4$

10. $\quad 7x^2 - 9x + 2$

$\underline{-\left(4x^2 - 2x - 7\right)}$

To subtract, add the opposite of the polynomial being subtracted.

$7x^2 - 9x + 2$

$\underline{-4x^2 + 2x + 7}$

$3x^2 - 7x + 9$

11. $\quad 5x^3 - 6x^2 - \ 9x + 14$

$\underline{-\left(-5x^3 + 3x^2 - 11x + \quad 3\right)}$

To subtract, add the opposite of the polynomial being subtracted.

$5x^3 - 6x^2 - \ 9x + 14$

$\underline{5x^3 - 3x^2 + 11x - \ 3}$

$10x^3 - 9x^2 + \ 2x + 11$

12.

x	$y = x^2 + 3$	(x, y)
-3	$y = (-3)^2 + 3 = 12$	$(-3, 12)$
-2	$y = (-2)^2 + 3 = 7$	$(-2, 7)$
-1	$y = (-1)^2 + 3 = 4$	$(-1, 4)$
0	$y = (0)^2 + 3 = 3$	$(0, 3)$
1	$y = (1)^2 + 3 = 4$	$(1, 4)$
2	$y = (2)^2 + 3 = 7$	$(2, 7)$
3	$y = (3)^2 + 3 = 12$	$(3, 12)$

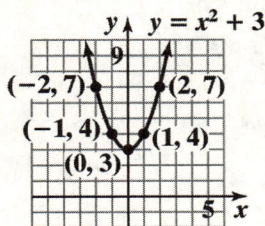

$y = x^2 + 3$

13.

x	$y = 1 - x^2$	(x, y)
-3	$y = 1 - (-3)^2 = -8$	$(-3, -8)$
-2	$y = 1 - (-2)^2 = -3$	$(-2, -3)$
-1	$y = 1 - (-1)^2 = 0$	$(-1, 0)$
0	$y = 1 - (0)^2 = 1$	$(0, 1)$
1	$y = 1 - (1)^2 = 0$	$(1, 0)$
2	$y = 1 - (2)^2 = -3$	$(2, -3)$
3	$y = 1 - (3)^2 = -8$	$(3, -8)$

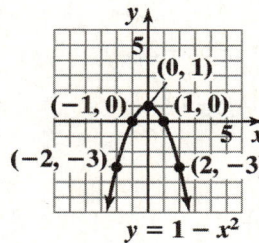

$y = 1 - x^2$

14. $x^{20} \cdot x^3 = x^{20+3} = x^{23}$

15. $y \cdot y^5 \cdot y^8 = y^1 \cdot y^5 \cdot y^8 = y^{1+5+8} = y^{14}$

16. $\left(x^{20}\right)^5 = x^{20 \cdot 5} = x^{100}$

17. $(10y)^2 = 10^2 y^2 = 100y^2$

18. $\left(-4x^{10}\right)^3 = (-4)^3 \left(x^{10}\right)^3 = -64x^{30}$

19. $(5x)\left(10x^3\right) = (5 \cdot 10)\left(x^1 \cdot x^3\right) = 50x^4$

20. $\left(-12y^7\right)\left(3y^4\right) = -36y^{11}$

21. $\left(-2x^5\right)\left(-3x^4\right)\left(5x^3\right) = 30x^{12}$

22. $7x\left(3x^2 + 9\right) = 7x\left(3x^2\right) + (7x)(9)$

$\qquad\qquad = 21x^3 + 63x$

23. $5x^3\left(4x^2 - 11x\right) = 5x^3\left(4x^2\right) - 5x^3(11x)$

$\qquad\qquad = 20x^5 - 55x^4$

24. $3y^2\left(-7y^2+3y-6\right)$

$\quad=3y^2\left(-7y^2\right)+3y^2\left(3y\right)+3y^2\left(-6\right)$

$\quad=-21y^4+9y^3-18y^2$

25. $2y^5\left(8y^3-10y^2+1\right)$

$\quad=2y^5\left(8y^3\right)+2y^5\left(-10y^2\right)+2y^5\left(1\right)$

$\quad=16y^8-20y^7+2y^5$

26. $(x+3)\left(x^2-5x+2\right)$

$\quad=x\left(x^2-5x+2\right)+3\left(x^2-5x+2\right)$

$\quad=x^3-5x^2+2x+3x^2-15x+6$

$\quad=x^3-2x^2-13x+6$

27. $(3y-2)\left(4y^2+3y-5\right)$

$\quad=3y\left(4y^2+3y-5\right)-2\left(4y^2+3y-5\right)$

$\quad=12y^3+9y^2-15y-8y^2-6y+10$

$\quad=12y^3+y^2-21y+10$

28. y^2-4y+7

$\quad\quad\underline{3y-5}$

$\quad-5y^2+20y-35$

$\quad\underline{3y^3-12y^2+21y}$

$\quad3y^3-17y^2+41y-35$

29. $4x^3-2x^2-6x-1$

$\quad\quad\quad\underline{2x+3}$

$\quad12x^3-6x^2-18x-3$

$\quad\underline{8x^4-4x^3-12x^2-2x}$

$\quad8x^4+8x^3-18x^2-20x-3$

30. $(x+6)(x+2)$

$\quad=x\cdot x+x\cdot 2+6\cdot x+6\cdot 2$

$\quad=x^2+2x+6x+12$

$\quad=x^2+8x+12$

31. $(3y-5)(2y+1)=6y^2+3y-10y-5$

$\quad\quad\quad\quad\quad=6y^2-7y-5$

32. $\left(4x^2-2\right)\left(x^2-3\right)$

$\quad=4x^2\cdot x^2+4x^2\left(-3\right)-2\cdot x^2-2\left(-3\right)$

$\quad=4x^4-12x^2-2x^2+6$

$\quad=4x^4-14x^2+6$

33. $(5x+4)(5x-4)=(5x)^2-4^2$

$\quad\quad\quad\quad\quad\quad=25x^2-16$

34. $(7-2y)(7+2y)=7^2-(2y)^2$

$\quad\quad\quad\quad\quad\quad=49-4y^2$

35. $\left(y^2+1\right)\left(y^2-1\right)=\left(y^2\right)^2-1^2=y^4-1$

36. $(x+3)^2=x^2+2(x)(3)+3^2$

$\quad\quad\quad\quad=x^2+6x+9$

37. $(3y+4)^2=(3y)^2+2(3y)(4)+16$

$\quad\quad\quad\quad=9y^2+24y+16$

38. $(y-1)^2=y^2-2y+1$

39. $(5y-2)^2=(5y)^2-2(5y)(2)+2^2$

$\quad\quad\quad\quad=25y^2-20y+4$

40. $\left(x^2+4\right)^2=\left(x^2\right)^2+2\left(x^2\right)(4)+4^2$

$\quad\quad\quad\quad=x^4+8x^2+16$

41. $\left(x^2+4\right)\left(x^2-4\right)=\left(x^2\right)^2-4^2=x^4-16$

42. $\left(x^2+4\right)\left(x^2-5\right)=\left(x^2\right)^2-5x^2+4x^2-20$

$\quad\quad\quad\quad\quad\quad=x^4-x^2-20$

43. $A=(x+3)(x+4)$

$\quad=x^2+4x+3x+12$

$\quad=x^2+7x+12$

44. $A = (x+30)(x+20)$

$\qquad = x^2 + 20x + 30x + 600$

$\qquad = x^2 + 50x + 600$

The area of the expanded garage is

$\left(x^2 + 50x + 600\right)$ yards^2.

45. $2x^3 y - 4xy^2 + 5y + 6$

$= 2(-1)^3 (2) - 4(-1)(2)^2 + 5(2) + 6$

$= 2(-1)(2) + 4(1)(4) + 5(2) + 6$

$= -4 + 16 + 10 + 6 = 28$

46. $4x^2 y + 9x^3 y^2 - 17x^4 - 12$

Term	Coefficient	Degree
$4x^2 y$	4	$2+1 = 3$
$9x^3 y^2$	9	$3+2 = 5$
$-17x^4$	-17	4
-12	-12	0

The degree of the polynomial is the highest degree of all its terms, which is 5.

47. $\left(7x^2 - 8xy + y^2\right) + \left(-8x^2 - 9xy + 4y^2\right)$

$= \left(7x^2 - 8x^2\right) + (-8xy - 9xy) + \left(y^2 + 4y^2\right)$

$= -x^2 - 17xy + 5y^2$

48. $\left(13x^3 y^2 - 5x^2 y - 9x^2\right) - \left(11x^3 y^2 - 6x^2 y - 3x^2 + 4\right)$

$= \left(13x^3 y^2 - 5x^2 y - 9x^2\right) + \left(-11x^3 y^2 + 6x^2 y + 3x^2 - 4\right)$

$= \left(13x^3 y^2 - 11x^3 y^2\right) + \left(-5x^2 y + 6x^2 y\right) + \left(-9x^2 + 3x^2\right) - 4$

$= 2x^3 y^2 + x^2 y - 6x^2 - 4$

49. $\left(-7x^2 y^3\right)\left(5x^4 y^6\right) = (-7)(-5)x^{2+4} y^{3+6}$

$\qquad\qquad\qquad = -35x^6 y^9$

50. $5ab^2 \left(3a^2 b^3 - 4ab\right)$

$= 5ab^2 \left(3a^2 b^3\right) + 5ab^2 (-4ab)$

$= 15a^3 b^5 - 20a^2 b^3$

51. $(x + 7y)(3x - 5y)$

$= x(3x) + x(-5y) + 7y(3x) + 7y(-5y)$

$= 3x^2 - 5xy + 21xy - 35y^2$

$= 3x^2 + 16xy - 35y^2$

52. $(4xy - 3)(9xy - 1)$

$= 4xy(9xy) + 4xy(-1) - 3(9xy) - 3(-1)$

$= 36x^2 y^2 - 4xy - 27xy + 3$

$= 36x^2 y^2 - 31xy + 3$

53. $(3x + 5y)^2 = (3x)^2 + 2(3x)(5y) + (5y)^2$

$\qquad\qquad\quad = 9x^2 + 30xy + 25y^2$

54. $(xy - 7)^2 = (xy)^2 - 2(xy)(7) + 7^2$

$\qquad\qquad\quad = x^2 y^2 - 14xy + 49$

55. $(7x + 4y)(7x - 4y) = (7x)^2 - (4y)^2$

$\qquad\qquad\qquad\qquad\quad = 49x^2 - 16y^2$

56. $(a - b)\left(a^2 + ab + b^2\right)$

$= a\left(a^2 + ab + b^2\right) - b\left(a^2 + ab + b^2\right)$

$= a^3 + a^2 b + ab^2 - a^2 b - ab^2 - b^3$

$= a^3 + \left(a^2 b - a^2 b\right) + \left(ab^2 - ab^2\right) - b^3$

$= a^3 - b^3$

57. $\dfrac{6^{40}}{6^{10}} = 6^{40-10} = 6^{30}$

58. $\dfrac{x^{18}}{x^3} = x^{18-3} = x^{15}$

59. $(-10)^0 = 1$

60. $-10^0 = -(1) = -1$

61. $400x^0 = 400 \cdot 1 = 400$

62. $\left(\dfrac{x^4}{2}\right)^3 = \dfrac{\left(x^4\right)^3}{2^3} = \dfrac{x^{4 \cdot 3}}{8} = \dfrac{x^{12}}{8}$

63. $\left(\dfrac{-3}{2y^6}\right)^4 = \dfrac{(-3)^4}{\left(2y^6\right)^4} = \dfrac{81}{\left(2^4 y^6\right)^4} = \dfrac{81}{16y^{24}}$

64. $\dfrac{-15y^8}{3y^2} = \dfrac{-15}{3} \cdot \dfrac{y^8}{y^2} = -5y^6$

65. $\dfrac{40x^8 y^6}{5xy^3} = \dfrac{40}{5} \cdot \dfrac{x^8}{x^1} \cdot \dfrac{y^6}{y^3} = 8x^7 y^3$

66. $\dfrac{18x^4 - 12x^2 + 36x}{6x} = \dfrac{18x^4}{6x} - \dfrac{12x^2}{6x} + \dfrac{36x}{6x}$

$\qquad\qquad\qquad\qquad = 3x^2 - 2x + 6$

67. $\dfrac{30x^8 - 25x^7 - 40x^5}{-5x^3}$

$\quad = \dfrac{30x^8}{-5x^3} - \dfrac{25x^7}{-5x^3} - \dfrac{40x^5}{-5x^3}$

$\quad = -6x^5 + 5x^4 + 8x^2$

68. $\dfrac{27x^3 y^2 - 9x^2 y - 18xy^2}{3xy}$

$\quad = \dfrac{27x^3 y^2}{3xy} - \dfrac{9x^2 y}{3xy} - \dfrac{18xy^2}{3xy}$

$\quad = 9x^2 y - 3x - 6y$

69.
$$\require{enclose}\begin{array}{r}2x+7\\ x-2\enclose{longdiv}{2x^2+3x-14}\\ \underline{2x^2-4x}\\ 7x-14\\ \underline{7x-14}\\ 0\end{array}$$

$\dfrac{2x^2 + 3x - 14}{x - 2} = 2x + 7$

70.
$$\begin{array}{r}x^2-3x+5\\ 2x+1\enclose{longdiv}{2x^3-5x^2+7x+5}\\ \underline{2x^3+x^2}\\ -6x^2+7x\\ \underline{-6x^2-3x}\\ 10x+5\\ \underline{10x+5}\\ 0\end{array}$$

$\dfrac{2x^3 - 5x^2 + 7x + 5}{2x + 1} = x^2 - 3x + 5$

71.
$$\begin{array}{r}x^2+5x+2\\ x-7\enclose{longdiv}{x^3-2x^2-33x-7}\\ \underline{x^3-7x^2}\\ 5x^2-33x\\ \underline{5x^2-35x}\\ 2x-7\\ \underline{2x-14}\\ 7\end{array}$$

$\dfrac{x^3 - 2x^2 - 33x - 7}{x - 7} = x^2 + 5x + 2 + \dfrac{7}{x - 7}$

72.
$$\begin{array}{r}y^2+3y+9\\ y-3\enclose{longdiv}{y^3+0y^2+0y-27}\\ \underline{y^3-3y^2}\\ 3y^2+0y\\ \underline{3y^2-9y}\\ 9y-27\\ \underline{9y-27}\\ 0\end{array}$$

$\dfrac{y^2 - 27}{y - 3} = y^2 + 3y + 9$

73. $7^{-2} = \dfrac{1}{7^2} = \dfrac{1}{49}$

74. $(-4)^{-3} = \dfrac{1}{(-4)^3} = \dfrac{1}{-64} = -\dfrac{1}{64}$

75. $2^{-1} + 4^{-1} = \dfrac{1}{2} + \dfrac{1}{4} = \dfrac{3}{4}$

76. $\dfrac{1}{5^{-2}} = 5^2 = 25$

77. $\left(\dfrac{2}{5}\right)^{-3} = \dfrac{2^{-3}}{5^{-3}} = \dfrac{5^3}{2^3} = \dfrac{125}{8}$

78. $\dfrac{x^3}{x^9} = x^{3-9} = x^{-6} = \dfrac{1}{x^6}$

79. $\dfrac{30y^6}{5y^8} = \dfrac{30}{5} \cdot \dfrac{y^6}{y^8} = 6y^{-2} = \dfrac{6}{y^2}$

80. $\left(5x^{-7}\right)\left(6x^2\right) = (5 \cdot 6)\left(x^{-7+2}\right)$
$$= 30x^{-5} = \dfrac{30}{x^5}$$

81. $\dfrac{x^4 \cdot x^{-2}}{x^{-6}} = \dfrac{x^{4+(-2)}}{x^{-6}} = \dfrac{x^2}{x^{-6}}$
$$= x^{2-(-6)} = x^8$$

82. $\dfrac{\left(3y^3\right)^4}{y^{10}} = \dfrac{3^4 y^{3(4)}}{y^{10}} = \dfrac{81y^{12}}{y^{10}}$
$$= 81y^{12-10} = 81y^2$$

83. $\dfrac{y^{-7}}{\left(y^4\right)^3} = \dfrac{y^{-7}}{y^{12}} = y^{-7-12} = y^{-19} = \dfrac{1}{y^{19}}$

84. $\left(2x^{-1}\right)^{-3} = 2^{-3}\left(x^{-1}\right)^{-3} = 2^{-3}x^3$
$$= \dfrac{x^3}{2^3} = \dfrac{x^3}{8}$$

85. $\left(\dfrac{x^7}{x^4}\right)^{-2} = \left(x^3\right)^{-2} = x^{-6} = \dfrac{1}{x^6}$

86. $\dfrac{\left(y^3\right)^4}{\left(y^{-2}\right)^4} = \dfrac{y^{12}}{y^{-8}} = y^{12-(-8)} = y^{20}$

87. $2.3 \times 10^4 = 23,000$

88. $1.76 \times 10^{-3} = 0.00176$

89. $9 \times 10^{-1} = 0.9$

90. $73,900,000 = 7.39 \times 10^7$

91. $0.00062 = 6.2 \times 10^{-4}$

92. $0.38 = 3.8 \times 10^{-1}$

93. $3.8 = 3.8 \times 10^0$

94. $\left(6 \times 10^{-3}\right)\left(1.5 \times 10^6\right) = 6(1.5) \times 10^{-3+6}$
$$= 9 \times 10^3$$

95. $\dfrac{2 \times 10^2}{4 \times 10^{-3}} = 0.5 \cdot 10^{2+3} = 0.5 \times 10^5$
$$= 5 \times 10^{-1} \times 10^5$$
$$= 5.0 \times 10^4$$

96. $\left(4 \times 10^{-2}\right)^2 = 4^2 \times 10^{-2(2)} = 16 \times 10^{-4}$
$$= 1.6 \times 10^1 \times 10^{-4} = 1.6 \times 10^{1-4}$$
$$= 1.6 \times 10^{-3}$$

97. $53.6 \times 10^9 = 5.36 \times 10^1 \times 10^9 = 5.36 \times 10^{10}$

98. $307 \times 10^6 = 3.07 \times 10^2 \times 10^6 = 3.07 \times 10^8$

99. $\dfrac{5.36 \times 10^{10}}{3.06 \times 10^8} \approx 1.75 \times 10^2 = \$175$

Chapter 5 Test

1. $9x + 6x^2 - 4$ is a trinomial of degree 2.

2. $\left(7x^3 + 3x^2 - 5x - 11\right) + \left(6x^3 - 2x^2 + 4x - 13\right)$
$$= \left(7x^3 + 6x^3\right) + \left(3x^2 - 2x^2\right) + (-5x + 4x) + (-11 - 13)$$
$$= 13x^3 + x^2 - x - 24$$

3. $\left(9x^3 - 6x^2 - 11x - 4\right) - \left(4x^3 - 8x^2 - 13x + 5\right)$

$= \left(9x^3 - 6x^2 - 11x - 4\right) + \left(-4x^3 + 8x^2 + 13x - 5\right)$

$= \left(9x^3 - 4x^3\right) + \left(-6x^2 + 8x^2\right) + \left(-11x + 13x\right) + \left(-4 - 5\right)$

$= 5x^3 + 2x^2 + 2x - 9$

4.

x	$y = x^2 - 3$	(x, y)
-3	$y = (-3)^2 - 3 = 6$	$(-3, 6)$
-2	$y = (-2)^2 - 3 = 1$	$(-2, 1)$
-1	$y = (-1)^2 - 3 = -2$	$(-1, -2)$
0	$y = (0)^2 - 3 = -3$	$(0, -3)$
1	$y = (1)^2 - 3 = -2$	$(1, -2)$
2	$y = (2)^2 - 3 = 1$	$(2, 1)$
3	$y = (3)^2 - 3 = 6$	$(3, 6)$

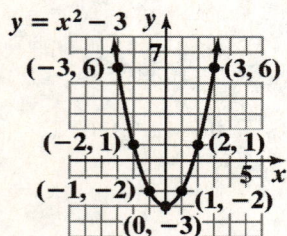

5. $\left(-7x^3\right)\left(5x^8\right) = (-7 \cdot 5)\left(x^{3+8}\right) = -35x^{11}$

6. $6x^2\left(8x^3 - 5x - 2\right)$

$= 6x^2\left(8x^3\right) + 6x^2\left(-5x\right) + 6x^2\left(-2\right)$

$= 48x^5 - 30x^3 - 12x^2$

7. $(3x + 2)\left(x^2 - 4x - 3\right)$

$= 3x\left(x^2 - 4x - 3\right) + 2\left(x^2 - 4x - 3\right)$

$= 3x^3 - 12x^2 - 9x + 2x^2 - 8x - 6$

$= 3x^3 - 10x^2 - 17x - 6$

8. $(3y + 7)(2y - 9)$

$= 6y^2 + 14y - 27y - 63$

$= 6y^2 - 13y - 63$

9. $(7x + 5)(7x - 5) = (7x)^2 - 5^2$

$= 49x^2 - 25$

10. $\left(x^2 + 3\right)^2 = \left(x^2\right)^2 + 2\left(x^2\right)(3) + 3^2$

$= x^4 + 6x^2 + 9$

11. $(5x - 3)^2 = (5x)^2 - 2(5x)(3) + 3^2$

$= 25x^2 - 30x + 9$

12. $4x^2 y + 5xy - 6x$

$= 4(-2)^2(3) + 5(-2)(3) - 6(-2)$

$= 4(4)(3) + 5(-2)(3) - 6(-2)$

$= 48 - 30 + 12 = 30$

13. $\left(8x^2 y^3 - xy + 2y^2\right) - \left(6x^2 y^3 - 4xy - 10y^2\right)$

$= \left(8x^2 y^3 - xy + 2y^2\right) + \left(-6x^2 y^3 + 4xy + 10y^2\right)$

$= \left(8x^2 y^3 - 6x^2 y^3\right) + \left(-xy + 4xy\right) + \left(2y^2 + 10y^2\right)$

$= 2x^2 y^3 + 3xy + 12y^2$

14. $(3a - 7b)(4a + 5b)$

$= (3a)(4a) + (3a)(5b) - (7b)(4a) - (7b)(5b)$

$= 12a^2 + 15ab - 28ab - 35b^2$

$= 12a^2 - 13ab - 35b^2$

15. $(2x + 3y)^2 = (2x)^2 + 2(2x)(3y) + (3y)^2$

$= 4x^2 + 12xy + 9y^2$

16. $\dfrac{-25x^{16}}{5x^4} = \dfrac{-25}{5} \cdot \dfrac{x^{16}}{x^4} = -5x^{16-4}$

$= -5x^{12}$

Check by multiplication:

$5x^4\left(-5x^{12}\right) = -25x^{4+12} = -25x^{16}$

17. $\dfrac{15x^4 - 10x^3 + 25x^2}{5x}$

$= \dfrac{15x^4}{5x} - \dfrac{10x^3}{5x} + \dfrac{25x^2}{5x}$

$= 3x^3 - 2x^2 + 5x$

Check by multiplication:

$5x\left(3x^3 - 2x^2 + 5x\right)$

$= 5x\left(3x^3\right) + 5x\left(-2x^2\right) + 5x(5x)$

$= 15x^4 - 10x^3 + 25x^2$

18.

$$
\begin{array}{r}
x^2 - 2x + 3 \\
2x+1\overline{\smash{)}\,2x^3 - 3x^2 + 4x + 4}
\end{array}
$$

$\underline{2x^3 + \ x^2}$

$\quad -4x^2 + 4x$

$\quad \underline{-4x^2 - 2x}$

$\qquad\quad 6x + 4$

$\qquad\quad \underline{6x + 3}$

$\qquad\qquad\ 1$

$\dfrac{2x^3 - 3x^2 + 4x + 4}{2x+1} = x^2 - 2x + 3 + \dfrac{1}{2x+1}$

Check by multiplication:

$(2x+1)\left(x^2 - 2x + 3\right) + 1$

$= \left[2x\left(x^2 - 2x + 3\right) + 1\left(x^2 - 2x + 3\right)\right] + 1$

$= \left(2x^3 - 4x^2 + 6x + x^2 - 2x + 3\right) + 1$

$= \left(2x^3 - 3x^2 + 4x + 3\right) + 1$

$= 2x^3 - 3x^2 + 4x + 4$

19. $10^{-2} = \dfrac{1}{10^2} = \dfrac{1}{100}$

20. $\dfrac{1}{4^{-3}} = 1 \cdot 4^3 = 4^3 = 64$

21. $\left(-3x^2\right)^3 = (-3)^3\left(x^2\right)^3 = -27x^6$

22. $\dfrac{20x^3}{5x^8} = \dfrac{20}{5} \cdot \dfrac{x^3}{x^8} = 4x^{3-8} = 4x^{-5} = \dfrac{4}{x^5}$

23. $\left(-7x^{-8}\right)\left(3x^2\right) = -21x^{-8+2} = -\dfrac{21}{x^6}$

24. $\dfrac{\left(2y^3\right)^4}{y^8} = \dfrac{2^4\left(y^3\right)^4}{y^8} = \dfrac{16y^{12}}{y^8} = 16y^4$

25. $\left(5x^{-4}\right)^{-2} = 5^{-2}\left(x^{-4}\right)^{-2} = 5^{-2}x^8$

$\qquad = \dfrac{x^8}{5^2} = \dfrac{x^8}{25}$

26. $\left(\dfrac{x^{10}}{x^5}\right)^{-3} = \left(x^{10-5}\right)^{-3} = \left(x^5\right)^{-3}$

$\qquad = x^{-15} = \dfrac{1}{x^{15}}$

27. $3.7 \times 10^{-4} = 0.00037$

28. $7{,}600{,}000 = 7.6 \times 10^6$

29. $\left(4.1 \times 10^2\right)\left(3 \times 10^{-5}\right)$

$= (4.1 \cdot 3)\left(10^2 \cdot 10^{-5}\right)$

$= 12.3 \times 10^{-3}$

$= 1.23 \times 10^{-2}$

30. $\dfrac{8.4 \times 10^6}{4 \times 10^{-2}} = \dfrac{8.4}{4} \times \dfrac{10^6}{10^{-2}}$

$\qquad = 2.1 \times 10^{6-(-2)}$

$\qquad = 2.1 \times 10^8$

31. $A = (x+8)(x+2)$

$\qquad = x^2 + 2x + 8x + 16$

$\qquad = x^2 + 10x + 16$

Cumulative Review Exercises (Chapters 1-5)

1. $(-7)(-5) \div (12-3) = (-7)(-5) \div 9$

$\qquad\qquad = 35 \div 9 = \dfrac{35}{9}$

2. $(3-7)^2(9-11)^3 = (-4)^2(-2)^3$

$\qquad\qquad = 16(-8) = -128$

3. $14,300 - (-750) = 14,300 + 750$
$$= 15,050$$

The difference in elevation between the plane and the submarine is 15,050 feet.

4. $2(x+3) + 2x = x + 4$
$$2x + 6 + 2x = x + 4$$
$$4x + 6 = x + 4$$
$$3x + 6 = 4$$
$$3x = -2$$
$$x = -\frac{2}{3}$$

The solution set is $\left\{-\dfrac{2}{3}\right\}$.

5. $\dfrac{x}{5} - \dfrac{1}{3} = \dfrac{x}{10} - \dfrac{1}{2}$

To clear fractions, multiply by the LCD = 30.
$$30\left(\frac{x}{5} - \frac{1}{3}\right) = 30\left(\frac{x}{10} - \frac{1}{2}\right)$$
$$30\left(\frac{x}{5}\right) - 30\left(\frac{1}{3}\right) = 30\left(\frac{x}{10}\right) - 30\left(\frac{1}{2}\right)$$
$$6x - 10 = 3x - 15$$
$$3x - 10 = -15$$
$$3x = -5$$
$$x = -\frac{5}{3}$$

The solution set is $\left\{-\dfrac{5}{3}\right\}$.

6. Let x = width of sign.
Then $3x - 2$ = length of sign.
$$2x + 2(3x - 2) = 28$$
$$2x + 6x - 4 = 28$$
$$8x - 4 = 28$$
$$8x = 32$$
$$x = 4$$
$$3x - 2 = 3(4) - 2 = 10$$

The length of the sign is 10 feet and the width is 4 feet, so the dimensions are 10 feet by 4 feet.

7. $\qquad 7 - 8x \le -6x - 5$
$$7 - 8x + 6x \le -6x - 5 + 6x$$
$$-2x + 7 \le -5$$
$$-2x + 7 - 7 \le -5 - 7$$
$$-2x \le -12$$
$$\frac{-2x}{-2} \ge \frac{-12}{-2}$$
$$x \ge 6$$

$[6, \infty)$

8.

	Principal	× Rate	= Interest
12% Investment	x	0.12	$0.12x$
14% Investment	y	0.14	$0.14y$

Since the total investment is \$6000 the first equation is $x + y = 6000$.

Since the total interest is \$6000 the second equation is $0.12x + 0.14y = 772$.

System of equations: $\begin{cases} x + y = 6000 \\ 0.12x + 0.14y = 772 \end{cases}$

Solve the first equation for y and substitute into the second equation.
$$x + y = 6000$$
$$y = 6000 - x$$

Solve for x. $\quad 0.12x + 0.14\overbrace{(6000 - x)}^{y} = 772$
$$0.12x + 840 - 0.14x = 772$$
$$-0.02x + 840 = 772$$
$$-0.02x = -68$$
$$\frac{-0.02x}{-0.02} = \frac{-68}{-0.02}$$
$$x = 3400$$

Back-substitute to find y.
$$x + y = 6000$$
$$3400 + y = 6000$$
$$y = 2600$$

\$3400 should be invested at 12% and \$2600 should be invested at 14%.

9.

	Number of Liters	×	Percent Antifreeze	=	Amount of Antifreeze
70% Antifreeze Solution	x		0.70		$0.70x$
30% Antifreeze Solution	y		0.30		$0.30y$
60% Antifreeze Solution	20		0.60		$0.60(20)$

Since there are 20 total liters, the first equation is $x + y = 20$.

Since the total amount of antifreeze is $0.60(20)$, the second equation is $0.70x + 0.30y = 0.60(20)$.

System of equations: $\begin{cases} x + y = 50 \\ 0.70x + 0.30y = 0.60(20) \end{cases}$

Solve the first equation for y and substitute into the second equation.

$x + y = 20$

$\quad y = 20 - x$

Solve for x. $\quad 0.70x + 0.30\overbrace{(20 - x)}^{y} = 0.60(20)$

$0.70x + 6 - 0.30x = 12$

$0.40x + 6 = 12$

$0.40x = 6$

$\dfrac{0.40x}{0.40} = \dfrac{6}{0.40}$

$x = 15$

Back-substitute to find y.

$x + y = 20$

$15 + y = 20$

$\quad y = 5$

15 liters of 70% antifreeze solution and 5 liters of 30% antifreeze solution should be used.

10. $y = -\dfrac{2}{5}x + 2$

slope $= -\dfrac{2}{5} = \dfrac{-2}{5}$; y-intercept $= 2$

Plot $(0,2)$. Move 2 units *down* (since -2 is negative) and 5 units to the *right* to reach the point $(5,0)$. Draw a line through $(0,2)$ and $(5,0)$.

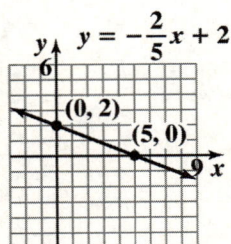

11. $x - 2y = 4$

x-intercept: 4
y-intercept: -2
checkpoint: $(-2, -3)$

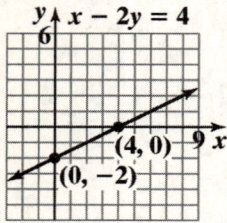

12. $m = \dfrac{y_2 - y_1}{x_2 - x_1} = \dfrac{-4 - 2}{2 - (-3)} = \dfrac{-6}{5} = -\dfrac{6}{5}$

Because the slope is negative, the line is falling.

13. $y - (-1) = -2(x - 3)$

$y + 1 = -2(x - 3)$ point-slope form

$y + 1 = -2x + 6$

$y = -2x + 5$ slope-intercept form

14. $3x + 2y = 10$

$4x - 3y = -15$

Multiply the first equation by 3 and the second equation by 2:

$9x + 6y = 30$

$\underline{8x - 6y = -30}$

$17x \qquad = 0$

$x = 0$

Back-substitute $x = 0$ to find y:

$3(0) + 2y = 10$

$2y = 10$

$y = 5$

The solution set is $\{(0, 5)\}$.

15. $2x + 3y = -6$

$y = 3x - 13$

Substitute the second equation in for y in the first equation:

$2x + 3y = -6$

$2x + 3(3x - 13) = -6$

$2x + 9x - 39 = -6$

$11x - 39 = -6$

$11x = 33$

$x = 3$

Back-substitute $x = 3$ to find y:

$y = 3x - 13 = 3(3) - 13 = -4$

The solution set is $\{(3, -4)\}$.

16. Let y = total charge.
Let x = # of minutes.
Plan A: $y = 0.05x + 15$

Plan B: $y = 0.07x + 5$

To find when the plans are the same, substitute the second equation into the first equation:

$0.07x + 5 = 0.05x + 15$

$0.02x + 5 = 15$

$0.02x = 10$

$x = 500$

Back-substitute to find y:

$y = 0.07x + 5 = 0.07(500) + 5 = 40$

The plans will be the same for 500 minutes at $40 a plan.

17. $2x + 5y \le 10$

$x - y \ge 4$

Graph $2x + 5y \le 10$:

Graph $2x + 5y = 10$ as a solid line using its x-intercept, 5 and y-intercept, 2.

Since $2(0) + 5(0) \le 10$ is true, shade the half-plane containing (0,0).

Graph $x - y \ge 4$:

Graph $x - y = 4$ as a solid line using its x-intercept, 4 and y-intercept, -4. Since $0 - 0 \ge 4$ is false, shade the half-plane *not* containing (0,0).

The solution set of the system is the intersection of the two shaded regions.

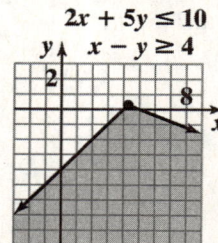

18. $\left(9x^5 - 3x^3 + 2x - 7\right) - \left(6x^5 + 3x^3 - 7x - 9\right)$

$= \left(9x^5 - 3x^3 + 2x - 7\right) + \left(-6x^5 - 3x^3 + 7x + 9\right)$

$= \left(9x^5 - 6x^5\right) + \left(-3x^3 - 3x^3\right) + \left(2x + 7x\right) + \left(-7 + 9\right)$

$= 3x^5 - 6x^3 + 9x + 2$

19.

$$
\begin{array}{r}
x^2 + 2x + 3 \\
x+1{\overline{\smash{\big)}\,x^3 + 3x^2 + 5x + 3}} \\
\underline{x^3 +x^2} \\
2x^2 + 5x \\
\underline{2x^2 + 2x} \\
3x + 3 \\
\underline{3x + 3} \\
0
\end{array}
$$

$\dfrac{x^3 + 3x^2 + 5x + 3}{x+1} = x^2 + 2x + 3$

20. $\dfrac{\left(3x^2\right)^4}{x^{10}} = \dfrac{3^4\left(x^2\right)^4}{x^{10}} = \dfrac{81x^8}{x^{10}}$

$= 81x^{8-10} = 81x^{-2} = \dfrac{81}{x^2}$

Chapter 6
Factoring Polynomials

6.1 Check Points

1. **a.** $18x^3 = 3x^2 \cdot 6x$

   $15x^2 = 3x^2 \cdot 5$

   The GCF is $3x^2$.

   b. $-20x^2 = 4x^2 \cdot (-5)$

   $12x^4 = 4x^2 \cdot 3x^2$

   $40x^3 = 4x^2 \cdot 10x$

   The GCF is $4x^2$.

   c. $x^4 y = x^2 y \cdot x^2$

   $x^3 y^2 = x^2 y \cdot xy$

   $x^2 y = x^2 y$

   The GCF is $x^2 y$.

2. $6x^2 + 18 = 6 \cdot x^2 + 6 \cdot 3 = 6(x^2 + 3)$

3. $25x^2 + 35x^3 = 5x^2 \cdot 5 + 5x^2 \cdot 7x = 5x^2(5 + 7x)$

4. $15x^5 + 12x^4 - 27x^3 = 3x^3 \cdot 5x^2 + 3x^3 \cdot 4x - 3x^3 \cdot 9$

   $\qquad\qquad\qquad\quad = 3x^3(5x^2 + 4x - 9)$

5. $8x^3 y^2 - 14x^2 y + 2xy = 2xy \cdot 4x^2 y - 2xy \cdot 7x + 2xy \cdot 1$

   $\qquad\qquad\qquad\qquad = 2xy(4x^2 y - 7x + 1)$

6. $-16a^4 b^5 + 24a^3 b^4 - 20ab^5 = -4ab^2 \cdot 4a^3 b^3 - 4ab^2 \cdot (-6a^2 b^2) - 4ab^2 \cdot 5$

   $\qquad\qquad\qquad\qquad\qquad\quad = -4ab^2(4a^3 b^3 - 6a^2 b^2 + 5)$

7. **a.** $x^2 \overbrace{(x+1)}^{\text{GCF}} + 7\overbrace{(x+1)}^{\text{GCF}} = (x+1)(x^2 + 7)$

   b. $x\overbrace{(y+4)}^{\text{GCF}} - 7\overbrace{(y+4)}^{\text{GCF}} = (y+4)(x-7)$

8. $x^3 + 5x^2 + 2x + 10 = (x^3 + 5x^2) + (2x + 10)$

   $\qquad\qquad\qquad\qquad = x^2(x+5) + 2(x+5)$

   $\qquad\qquad\qquad\qquad = (x+5)(x^2 + 2)$

9. $xy + 3x - 5y - 15 = x(y+3) - 5(y+3)$

   $\qquad\qquad\qquad\qquad = (y+3)(x-5)$

6.1 Concept and Vocabulary Check

1. factoring

2. greatest common factor; smallest/least

3. false

4. false

6.1 Exercise Set

1. The GCF of 4 and $8x$ is 4.

3. The GCF of $12x^2$ and $8x$ is $4x$.

5. The GCF of $-2x^4$ and $6x^3$ is $2x^3$.

7. The GCF of $9y^5, 18y^2$, and $-3y$ is $3y$.

9. The GCF of xy, xy^2, and xy^3 is xy.

11. The GCF of $16x^5y^4, 8x^6y^3$, and $20x^4y^5$ is $4x^4y^3$.

13. $8x+8 = 8\cdot x + 8\cdot 1$
    $= 8(x+1)$

15. $4y-4 = 4\cdot y - 4\cdot 1$
    $= 4(y-1)$

17. $5x+30 = 5\cdot x + 5\cdot 6$
    $= 5(x+6)$

19. $30x-12 = 6\cdot 5x - 6\cdot 2$
    $= 6(5x-2)$

21. $x^2 + 5x = x\cdot x + x\cdot 5$
    $= x(x+5)$

23. $18y^2 + 12 = 6\cdot 3y^2 + 6\cdot 2$
    $= 6(3y^2+2)$

25. $14x^3 + 21x^2 = 7x^2\cdot 2x + 7x^2\cdot 3$
    $= 7x^2(2x+3)$

27. $13y^2 - 25y = y\cdot 13y - y\cdot 25$
    $= y(13y-25)$

29. $9y^4 + 27y^6 = 9y^4\cdot 1 + 9y^4\cdot 3y^2$
    $= 9y^4(1+3y^2)$

31. $8x^2 - 4x^4 = 4x^2(2) - 4x^2(x^2)$
    $= 4x^2(2-x^2)$

33. $12y^2 + 16y - 8 = 4(3y^2) + 4(4y) - 4(2)$
    $= 4(3y^2 + 4y - 2)$

35. $9x^4 + 18x^3 + 6x^2$
    $= 3x^2(3x^2) + 3x^2(6x) + 3x^2(2)$
    $= 3x^2(3x^2 + 6x + 2)$

37. $100y^5 - 50y^3 + 100y^2$
    $= 50y^2(2y^3) - 50y^2(y) + 50y^2(2)$
    $= 50y^2(2y^3 - y + 2)$

39. $10x - 20x^2 + 5x^3$
    $= 5x(2) - 5x(4x) + 5x(x^2)$
    $= 5x(2 - 4x + x^2)$

41. $11x^2 - 23$ cannot be factored because the two terms have no common factor other than 1.

43. $6x^3y^2 + 9xy = 3xy(2x^2 + y) + 3xy(3)$
    $= 3xy(2x^2y + 3)$

45. $30x^2y^2 - 10xy^2 + 20xy$
    $= 10xy(3xy^2) - 10xy(y) + 10xy(2)$
    $= 10xy(3xy^2 - y + 2)$

47. $32x^3y^2 - 24x^3y - 16x^2y$
$= 8x^2y(4xy) - 8x^2y(3x) - 8x^2y(2)$
$= 8x^2y(4xy - 3x - 2)$

49. $-12x^2 + 18 = -6(2x^2 - 3)$

51. $-8x^4 + 32x^3 + 16x^2 = -8x^2(x^2 - 4x - 2)$

53. $-4a^3b^2 + 6ab = -2ab(2a^2b - 3)$

55. $-12x^3y^2 - 18x^3y + 24x^2y = -6x^2y(2xy + 3x - 4)$

57. $x(x+5) + 3(x+5) = (x+5)(x+3)$

59. $x(x+2) - 4(x+2) = (x+2)(x-4)$

61. $x(y+6) - 7(y+6) = (y+6)(x-7)$

63. $3x(x+y) - (x+y)$
$= 3x(x+y) - 1(x+y)$
$= (x+y)(3x-1)$

65. $4x(3x+1) + 3x + 1$
$= 4x(3x+1) + 1(3x+1)$
$= (3x+1)(4x+1)$

67. $7x^2(5x+4) + 5x + 4$
$= 7x^2(5x+4) + 1(5x+4)$
$= (5x+4)(7x^2+1)$

69. $x^2 + 2x + 4x + 8 = (x^2 + 2x) + (4x + 8)$
$= x(x+2) + 4(x+2)$
$= (x+2)(x+4)$

71. $x^2 + 3x - 5x - 15 = (x^2 + 3x) + (-5x - 15)$
$= x(x+3) - 5(x+3)$
$= (x+3)(x-5)$

73. $x^3 - 2x^2 + 5x - 10$
$= (x^3 - 2x^2) + (5x - 10)$
$= x^2(x-2) + 5(x-2)$
$= (x-2)(x^2+5)$

75. $x^3 - x^2 + 2x - 2 = x^2(x-1) + 2(x-1)$
$= (x-1)(x^2+2)$

77. $xy + 5x + 9y + 45 = x(y+5) + 9(y+5)$
$= (y+5)(x+9)$

79. $xy - x + 5y - 5 = x(y-1) + 5(y-1)$
$= (y-1)(x+5)$

81. $3x^2 - 6xy + 5xy - 10y^2$
$= 3x(x-2y) + 5y(x-2y)$
$= (x-2y)(3x+5y)$

83. $3x^3 - 2x^2 - 6x + 4$
$= x^2(3x-2) - 2(3x-2)$
$= (3x-2)(x^2-2)$

85. $x^2 - ax - bx + ab = x(x-a) - b(x-a)$
$= (x-a)(x-b)$

87. $24x^3y^3z^3 + 30x^2y^2z + 18x^2yz^2$
$= 6x^2yz(4xy^2z^2) + 6x^2yz(5y) + 6x^2yz(3z)$
$= 6x^2yz(4xy^2z^2 + 5y + 3z)$

89. $x^3 - 4 + 3x^3y - 12y = 1(x^3 - 4) + 3y(x^3 - 4)$
$= (x^3 - 4)(1 + 3y)$

91. $4x^5(x+1) - 6x^3(x+1) - 8x^2(x+1)$
$= 2x^2(x+1) \cdot 2x^3 - 2x^2(x+1) \cdot 3x - 2x^2(x+1) \cdot 4$
$= 2x^2(x+1)(2x^3 - 3x - 4)$

93. $3x^5 - 3x^4 + x^3 - x^2 + 5x - 5$

$= \left(3x^5 - 3x^4\right) + \left(x^3 - x^2\right) + \left(5x - 5\right)$

$= 3x^4\left(x-1\right) + x^2\left(x-1\right) + 5\left(x-1\right)$

$= \left(x-1\right)\left(3x^4 + x^2 + 5\right)$

95. The area of the square is $6x \cdot 6x = 36x^2$. The area of the circle is $\pi\left(2x\right)^2 = \pi \cdot 4x^2 = 4\pi x^2$. So the shaded area is the area of the square minus the area of the circle, which is $36x^2 - 4\pi x^2 = 4x^2\left(9 - \pi\right)$.

97. a. Use the formula, $64x - 16x^2$, for the height of the debris above the ground. Substitute 3 for x.

$64x - 16x^2 = 64\left(3\right) - 16\left(3\right)^2$

$= 192 - 16\left(9\right) = 192 - 144 = 48$

Therefore, the height of the debris after 3 seconds is 48 feet.

b. $64x - 16x^2 = 16x\left(4 - x\right)$

c. Substitute 3 for x in the factored polynomial.
$16 \cdot 3\left(4 - 3\right) = 48\left(1\right) = 48$

You do get the same answer as in part (a) but this does not prove your factorization is correct.

99. Use the formula for the area of a rectangle, $A = l \cdot w$. Substitute $5x^4 - 10x$ for A and $5x$ for w.

$A = l \cdot w$

$5x^4 - 10x = l\left(5x\right)$

$\dfrac{5x^4 - 10x}{5x} = \dfrac{l\left(5x\right)}{5x}$

$\dfrac{5x\left(x^3 - 2\right)}{5x} = l$

$x^3 - 2 = l$

The length, l, is $\left(x^3 - 2\right)$ units.

101. – 105. Answers will vary.

107. makes sense

109. does not make sense; Explanations will vary.
Sample explanation: The power must be the greatest that is *common* for this variable.

111. false; Changes to make the statement true will vary.
A sample change is: Since $\dfrac{3x}{3x} = 1$, it is necessary to write the 1.

113. false; Changes to make the statement true will vary.
A sample change is: $a^2 + b^2$ is not factorable.

115. The polynomial will be
$x + \left(x + 100\right) + \left(x + 200\right) + \left(x + 300\right)$
$= 4x + 600$
$= 4\left(x + 150\right)$

117. Answers will vary. One example is
$5x^2 + 10x - 4x - 8$.

119. The graphs do not coincide.

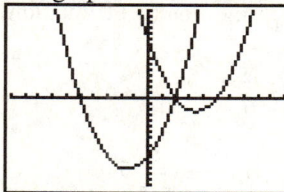

Factor by grouping.
$x^2 - 2x + 5x - 10 = x\left(x - 2\right) + 5\left(x - 2\right)$
$\qquad\qquad\qquad\qquad = \left(x - 2\right)\left(x + 5\right)$

Change the expression on the right side to $\left(x - 2\right)\left(x + 5\right)$.

121. $\left(x + 7\right)\left(x + 10\right) = x^2 + 10x + 7x + 70$
$\qquad\qquad\qquad\qquad\quad = x^2 + 17x + 70$

122. $2x - y = -4$
$x - 3y = 3$
Graph both equations on the same axes.
$2x - y = -4$: x-intercept: -2; y-intercept: 4
$x - 3y = 3$: x-intercept: 3; y-intercept: -1

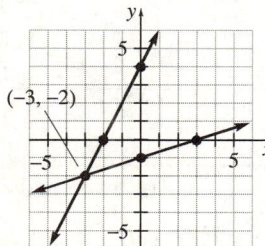

The lines intersect as $\left(-3, -2\right)$.
The solution set is $\left\{\left(-3, -2\right)\right\}$.

123. First, find the slope $m = \dfrac{5-2}{-4-(-7)} = \dfrac{3}{3} = 1$

Write the point-slope equation using
$m = 1$ and $(x_1, y_1) = (-7, 2)$.

$$y - y_1 = m(x - x_1)$$
$$y - 2 = 1\left[x - (-7)\right]$$
$$y - 2 = 1(x + 7)$$

Now rewrite this equation in slope-intercept form.
$$y - 2 = x + 7$$
$$y = x + 9$$

Note: If $(-4, 5)$ is used as $(x_1 y_1)$, the point-slope equation will be

$$y - 5 = 1\left[x - (-4)\right]$$
$$y - 5 = x + 4$$

This also leads to the slope-intercept equation $y = x + 9$.

124. $2 \times 4 = 8$ and $2 + 4 = 6$

125. $(-3)(-2) = 6$ and $(-3) + (-2) = -5$

126. $(-5)(7) = -35$ and $(-5) + 7 = 2$

6.2 Check Points

1. $x^2 + 5x + 6$

Factors of 6	6,1	-6,-1	2,3	-2,-3
Sum of Factors	7	-7	5	-5

The factors of 6 whose sum is 5, are 2 and 3.
Thus, $x^2 + 5x + 6 = (x + 2)(x + 3)$.
Check:
$$(x + 2)(x + 3) = x^2 + 3x + 2x + 6$$
$$= x^2 + 5x + 6$$

2. $x^2 - 6x + 8$

Factors of 8	8,1	-8,-1	2,4	-2,-4
Sum of Factors	9	-9	6	-6

The factors of 8 whose sum is -6, are -2 and -4.
Thus, $x^2 - 6x + 8 = (x - 2)(x - 4)$.
Check:
$$(x - 2)(x - 4) = x^2 - 4x - 2x + 8$$
$$= x^2 - 6x + 8$$

3. $x^2 + 3x - 10$

Factors of -10	-10,1	10,-1	-5,2	5,-2
Sum of Factors	-9	9	-3	3

The factors of –10 whose sum is 3, are 5 and –2.
Thus, $x^2 + 3x - 10 = (x + 5)(x - 2)$.
Check:
$$(x + 5)(x - 2) = x^2 - 2x + 5x - 10$$
$$= x^2 + 3x - 10$$

4. The factors of –27 whose sum is –6, are –9 and 3.
Thus, $y^2 - 6y - 27 = (y - 9)(y + 3)$.

5. No factor pair of –7 has a sum of 1.
Thus, $x^2 + x - 7$ is prime.

6. The factors of 3 whose sum is –4, are –3 and –1.
Thus, $x^2 - 4xy + 3y^2 = (x - 3y)(x - y)$.

7. First factor out the common factor of $2x$.
$$2x^3 + 6x^2 - 56x = 2x(x^2 + 6x - 28)$$
Continue by factoring the trinomial.
$$2x^3 + 6x^2 - 56x = 2x(x^2 + 6x - 28)$$
$$= 2x(x - 4)(x + 7)$$

8. First factor out the common factor of –2.
$$-2y^2 - 10y + 28 = -2(y^2 + 5y - 14)$$
Continue by factoring the trinomial.
$$-2y^2 - 10y + 28 = -2(y^2 + 5y - 14)$$
$$= -2(y - 2)(y + 7)$$

6.2 Concept and Vocabulary Check

1. 20; –12

2. completely

3. +10

4. –6

5. +5

6. –7

7. –2y

6.2 Exercise Set

1. $x^2 + 7x + 6$

Factors of 6	6,1	−6,−1	3,2	−3,−2
Sum of Factors	7	−7	5	−5

The factors of 6 whose sum is 7 are 6 and 1.
Thus, $x^2 + 7x + 6 = (x + 6)(x + 1)$.
Check:
$$(x + 6)(x + 1) = x^2 + 1x + 6x + 6$$
$$= x^2 + 7x + 6$$

3. $x^2 + 7x + 10 = (x + 5)(x + 2)$
   $5(2) = 10;\ 5 + 2 = 7$

5. $x^2 + 11x + 10 = (x + 10)(x + 1)$
   $10(1) = 10;\ 10 + 1 = 11$

7. $x^2 - 7x + 12 = (x - 4)(x - 3)$
   $-4(-3) = 12;\ -4 + -3 = -7$

9. $x^2 - 12x + 36 = (x - 6)(x - 6)$
   $-6(-6) = 36;\ -6 + -6 = -12$

11. $y^2 - 8y + 15 = (y - 5)(y - 3)$
    $-5(-3) = 15;\ -5 + -3 = -8$

13. $x^2 + 3x - 10 = (x + 5)(x - 2)$
    $5(-2) = -10;\ 5 + -2 = 3$

15. $y^2 + 10y - 39 = (y + 13)(y - 3)$
    $(13)(-3) = -39;\ 13 + -3 = 10$

17. $x^2 - 2x - 15 = (x - 5)(x + 3)$
    $(-5)(3) = -15;\ -5 + 3 = -2$

19. $x^2 - 2x - 8 = (x - 4)(x + 2)$
    $(-4)(2) = -8;\ -4 + 2 = -2$

21. $x^2 + 4x + 12$ is prime because there is no pair of integers whose product is 12 and whose sum is 4.

23. $y^2 - 16y + 48 = (y - 4)(y - 12)$
    $(-4)(-12) = 48;\ -4 + -12 = -16$

25. $x^2 - 3x + 6$ is prime because there is no pair of integers whose product is 6 and whose sum is −3.

27. $w^2 - 30w - 64 = (w - 32)(w + 2)$
    $(-32)(2) = -64;\ -32 + 2 = -30$

29. $y^2 - 18y + 65 = (y - 5)(y - 13)$
    $(-5)(-13) = 65;\ -5 + -13 = -18$

31. $r^2 + 12r + 27 = (r + 3)(r + 9)$
    $(3)(9) = 27;\ 3 + 9 = 12$

33. $y^2 - 7y + 5$ is prime because there is no pair of integers whose product is 5 and whose sum is −7.

35. $x^2 + 7xy + 6y^2 = (x + 6y)(x + y)$
    $(6)(1) = 6;\ 6 + 1 = 7$

37. $x^2 - 8xy + 15y^2 = (x - 3y)(x - 5y)$
    $(-3)(-5) = 15;\ -3 + -5 = -8$

39. $x^2 - 3xy - 18y^2 = (x - 6y)(x + 3y)$
    $(-6)(3) = -18;\ -6 + 3 = -3$

41. $a^2 - 18ab + 45b^2 = (a - 15b)(a - 3b)$
    $(-15)(-3) = 45;\ -15 + -3 = -18$

43. $3x^2 + 15x + 18$
    First factor out the GCF, 3. Then factor the resulting binomial.
    $$3x^2 + 15x + 18 = 3\left(x^2 + 5x + 6\right)$$
    $$= 3(x + 2)(x + 3)$$

45. $4y^2 - 4y - 8 = 4\left(y^2 - y - 2\right)$
    $$= 4(y - 2)(y + 1)$$

47. $10x^2 - 40x - 600 = 10\left(x^2 - 4x - 60\right)$
    $$= 10(x - 10)(x + 6)$$

49. $3x^2 - 33x + 54 = 3\left(x^2 - 11x + 18\right)$
    $$3(x - 2)(x - 9)$$

51. $2r^3 + 6r^2 + 4r = 2r\left(r^2 + 3r + 2\right)$

$= 2r(r+2)(r+1)$

53. $4x^3 + 12x^2 - 72x = 4x\left(x^2 + 3x - 18\right)$

$= 4x(x+6)(x-3)$

55. $2r^3 + 8r^2 - 64r = 2r\left(r^2 + 4r - 32\right)$

$= 2r(r+8)(r-4)$

57. $y^4 + 2y^3 - 80y^2 = y^2\left(y^2 + 2y - 80\right)$

$= y^2(y+10)(y-8)$

59. $x^4 - 3x^3 - 10x^2 = x^2\left(x^2 - 3x - 10\right)$

$= x^2(x-5)(x+2)$

61. $2w^4 - 26w^3 - 96w^2$

$= 2w^2\left(w^2 - 13w - 48\right)$

$= 2w^2(w-16)(w+3)$

63. $15xy^2 + 45xy - 60x = 15x\left(y^2 + 3y - 4\right)$

$= 15x(y+4)(y-1)$

65. $x^5 + 3x^4 y - 4x^3 y^2 = x^3\left(x^2 + 3xy - 4y^2\right)$

$= x^3(x+4y)(x-y)$

67. $-16t^2 + 64t + 80 = -16(t^2 - 4t - 5)$

$= -16(t-5)(t+1)$

69. $-5x^2 + 50x - 45 = -5(x^2 - 10x + 9)$

$= -5(x-9)(x-1)$

71. $-x^2 - 3x + 40 = -(x^2 + 3x - 40)$

$= -(x+8)(x-5)$

73. $-2x^3 - 6x^2 + 8x = -2x(x^2 + 3x - 4)$

$= -2x(x+4)(x-1)$

75. $2x^2 y^2 - 32x^2 yz + 30x^2 z^2$

$= 2x^2\left(y^2 - 16yz + 15z^2\right)$

$= 2x^2(y - 15z)(y - z)$

77. $(a+b)x^2 + (a+b)x - 20(a+b)$

$= (a+b)\left(x^2 + x - 20\right)$

$= (a+b)(x+5)(x-4)$

79. $x^2 + 0.5x + 0.06 = (x+0.2)(x+0.3)$

$0.2(0.3) = 0.06; \ 0.2 + 0.3 = 0.5$

81. $x^2 - \dfrac{2}{5}x + \dfrac{1}{25} = \left(x - \dfrac{1}{5}\right)\left(x - \dfrac{1}{5}\right)$

$\dfrac{1}{5}\left(\dfrac{1}{5}\right) = \dfrac{1}{25}; \ -\dfrac{1}{5} + -\dfrac{1}{5} = -\dfrac{2}{5}$

83. a. $-16t^2 + 16t + 32 = -16\left(t^2 - t - 2\right)$

$= -16(t-2)(t+1)$

b. Substitute 2 for t in the original polynomial:

$-16t^2 + 16t + 32 = -16(2)^2 + 16(2) + 32$

$= -16(4) + 32 + 32$

$= -64 + 64$

$= 0$

Substitute 2 for t in the factored polynomial:

$-16(t-2)(t+1) = -16(2-2)(2+1)$

$= -16(0)(3)$

$= 0$

The answers are the same.
This answer means that after 2 seconds you hit the water.

85. – 87. Answers will vary.

89. does not make sense; Explanations will vary.
Sample explanation: If the order of the terms is switched, the factors of an expression will not change.

91. does not make sense; Explanations will vary.
Sample explanation: $x^2 + x + 1$ is prime.

93. false; Changes to make the statement true will vary.
A sample change is: $x^2 + x + 20$ is prime.

95. true

97. In order for $x^2 + bx + 15$ to be factorable, b must be the sum of two integers that are positive factors of 15. The only positive factor pairs for 15 are $3 \cdot 5$ and $1 \cdot 15$. Since $3 + 5 = 8, 1 + 15 = 16$, the possible values of b are 8 and 16.

99. Answers will vary. An example is $x^2 + 14x + 2$.

101. $x^3 + 3x^2 + 2x = x\left(x^2 + 3x + 2\right)$
$$= x(x+1)(x+2)$$
The trinomial represents the product of three consecutive integers.

103. The graphs coincide.

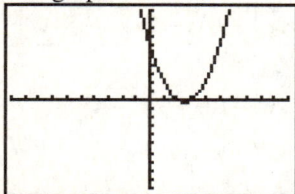

This verifies the factorization
$x^2 - 5x + 6 = (x-2)(x-3)$.

105. The graphs do not coincide.

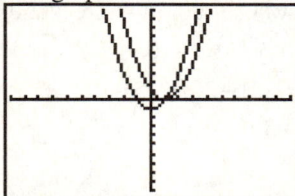

$x^2 - 2x + 1 = (x-1)(x-1)$
Change the polynomial on the right to $(x-1)(x-1)$.

107. $4(x-2) = 3x + 5$
$4x - 8 = 3x + 5$
$x - 8 = 5$
$x = 13$
The solution set is $\{13\}$.

108. Graph $6x - 5y = 30$ with a solid line. Since the test point $(0,0)$ makes the inequality true, shade the half-plane that contains the test point.

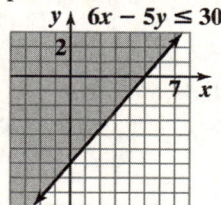

109 $y = -\dfrac{1}{2}x + 2$

The y-intercept is 2. Find an additional point by using the slope. From the y-intercept, move down one unit and to the right 2 units. Draw the line through these points.

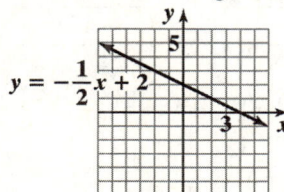

110. $(2x+3)(x-2) = 2x^2 - 4x + 3x - 6$
$$= 2x^2 - x - 6$$

111. $(3x+4)(3x+1) = 9x^2 + 3x + 12x + 4$
$$= 9x^2 + 15x + 4$$

112. $8x^2 - 2x - 20x + 5 = 2x(4x-1) - 5(4x-1)$
$$= (4x-1)(2x-5)$$

6.3 Check Points

1. Factor $5x^2 - 14x + 8$ by trial and error.

   Step 1 $5x^2 - 14x + 8 = (5x\quad)(x\quad)$

   Step 2 The number 8 has pairs of factors that are either both positive or both negative. Because the middle term, $-14x$, is negative, both factors must be negative.

   Step 3

Possible Factors of $5x^2 - 14x + 8$	Sum of Outside and Inside Products
	required middle term
$(5x-4)(x-2)$	$-10x - 4x = -14x$
$(5x-2)(x-4)$	$-20x - 2x = -22x$
$(5x-1)(x-8)$	$-40x - x = -41x$
$(5x-8)(x-1)$	$-5x - 8x = -13x$

   Check:

   $$(5x-4)(x-2) = 5x^2 - 10x - 4x + 8$$
   $$= 5x^2 - 14x + 8$$

   Thus, $5x^2 - 14x + 8 = (5x-4)(x-2)$.

2. Factor $6x^2 + 19x - 7$ by trial and error.

   Step 1 Find two First terms whose product is $6x^2$.

   $$6x^2 + 19x - 7 = (6x\quad)(x\quad)$$
   $$6x^2 + 19x - 7 = (3x\quad)(2x\quad)$$

   Step 2 The last term, -7, has possible factorizations of $1(-7)$ and $-1(7)$.

   Step 3

Possible Factors of $6x^2 + 19x - 7$	Sum of Outside and Inside Products
$(6x+1)(x-7)$	$-42x + x = -41x$
$(6x-7)(x+1)$	$6x - 7x = -x$
$(6x-1)(x+7)$	$42x - x = 41x$
$(6x+7)(x-1)$	$-6x + 7x = x$
$(3x+1)(2x-7)$	$-21x + 2x = -19x$
$(3x-7)(2x+1)$	$3x - 14x = -11x$
	required middle term
$(3x-1)(2x+7)$	$21x - 2x = 19x$
$(3x+7)(2x-1)$	$-3x + 14x = 11x$

   Check:

   $$(3x-1)(2x+7) = 6x^2 + 21x - 2x - 7$$
   $$= 6x^2 + 19x - 7$$

   Thus, $6x^2 + 19x - 7 = (3x-1)(2x+7)$

3. Factor $3x^2 - 13xy + 4y^2$ by trial and error.

   Step 1 Find two First terms whose product is $2x^2$.

   $$3x^2 - 13xy + 4y^2 = (3x\quad)(x\quad)$$

   Step 2 The last term, $4y^2$, has pairs of factors that are either both positive or both negative. Because the middle term, $-13xy$, is negative, both factors must be negative. Thus the last term has possible factorizations of $-2y(-2y)$ or $-y(-4y)$.

   Step 3

Possible Factors of $3x^2 - 13xy + 4y^2$	Sum of Outside and Inside Products
$(3x-4y)(x-y)$	$-3xy - 4xy = -7xy$
	required middle term
$(3x-y)(x-4y)$	$-12xy - xy = -13xy$
$(3x-2y)(x-2y)$	$-6xy - 2xy = -8xy$

   Check:

   $$(3x-y)(x-4y) = 3x^2 - 12xy - xy + 4y^2$$
   $$= 3x^2 - 13xy + 4y^2$$

   Thus, $3x^2 - 13xy + 4y^2 = (3x-y)(x-4y)$

4. Factor $3x^2 - x - 10$ by grouping.

   $a = 3$ and $c = -10$, so $ac = 3(-10) = -30$.

   The factors of -30 whose sum is -1 are 5 and -6.

   $$3x^2 - x - 10 = 3x^2 + 5x - 6x - 10$$
   $$= x(3x+5) - 2(3x+5)$$
   $$= (3x+5)(x-2)$$

5. Factor $8x^2 - 10x + 3$ by grouping.

   $a = 8$ and $c = 3$, so $ac = 8(3) = 24$.

   The factors of 24 whose sum is -10 are -6 and -4.

   $$8x^2 - 10x + 3 = 8x^2 - 4x - 6x + 3$$
   $$= 4x(2x-1) - 3(2x-1)$$
   $$= (2x-1)(4x-3)$$

6. First factor out the GCF.

   $$5y^4 + 13y^3 + 6y^2 = y^2(5y^2 + 13y + 6)$$

   Then factor the resulting trinomial.

   $$5y^4 + 13y^3 + 6y^2 = y^2(5y^2 + 13y + 6)$$
   $$= y^2(5y + 3)(y + 2)$$

6.3 Concept and Vocabulary Check

1. greatest common factor

2. -3

3. -4

4. $2x - 3$

5. $3x + 4$

6. $x - 2y$

6.3 Exercise Set

1. Factor $2x^2 + 5x + 3$ by trial and error.

   Step 1 $\quad 2x^2 + 5x + 3 = (2x \quad)(x \quad)$

   Step 2 The number 3 has pairs of factors that are either both positive or both negative. Because the middle term, $5x$, is positive, both factors must be positive. The only positive factorization is $(1)(3)$.

   Step 3

Possible Factors of $2x^2 + 5x + 3$	Sum of Outside and Inside Products
$(2x+1)(x+3)$	$6x + x = 7x$
$(2x+3)(x+1)$	$2x + 3x = 5x$

   Check:

   $$(2x+3)(x+1) = 2x^2 + 2x + 3x + 3$$
   $$= 2x^2 + 5x + 3$$

   Thus, $2x^2 + 5x + 3 = (2x+3)(x+1)$.

3. Factor $3x^2 + 13x + 4$ by trial and error. The only possibility for the first terms is $(3x)(x) = 3x^2$.

   Because the middle term is positive and the last term is also positive, the possible factorizations of 4 are $(1)(4)$ and $(2)(2)$.

Possible Factors of $3x^2 + 13x + 4$	Sum of Outside and Inside Products
$(3x+1)(x+4)$	$12x + x = 13x$
$(3x+4)(x+1)$	$3x + 4x = 7x$
$(3x+2)(x+2)$	$6x + 2x = 8x$

   Thus, $3x^2 + 13x + 4 = (3x+1)(x+4)$.

5. Factor $2x^2 + 11x + 12$ by grouping.

   $a = 2$ and $c = 12$, so $ac = 2(12) = 24$.

   The factors of 24 whose sum is 11 are 8 and 3.

   $$2x^2 + 11x + 12 = 2x^2 + 8x + 3x + 12$$
   $$= 2x(x+4) + 3(x+4)$$
   $$= (x+4)(2x+3)$$

7. Factor $5y^2 - 16y + 3$ by trial and error. The first terms must be $5y$ and y. Because the middle term is negative, the factors of 3 must be -3 and -1.

Possible Factors of $5y^2 - 16y + 3$	Sum of Outside and Inside Products
$(5y-1)(y-3)$	$-15y - y = -16y$
$(5y-3)(y-1)$	$-5y - 3y = -8y$

   Thus, $y^2 - 16y + 3 = (5y-1)(y-3)$.

9. Factor $3y^2 + y - 4$ by trial and error.

   $$(3y+1)(y-4) = 3y^2 - 11y - 4$$
   $$(3y-1)(y+4) = 3y^2 + 11y - 4$$
   $$(3y+4)(y-1) = 3y^2 + y - 4$$
   $$(3y-4)(y+1) = 3y^2 - y - 4$$
   $$(3y+2)(y-2) = 3y^2 - 4y - 4$$
   $$(3y-2)(y+2) = 3y^2 + 4y - 4$$

   Thus, $3y^2 + y - 4 = (3y+4)(y-1)$.

11. Factor $3x^2 + 13x - 10$ by grouping.

    $a = 3$ and $c = -10$, so $ac = -30$.

    The factors of -30 whose sum is 13 are 15 and -2.

    $$3x^2 + 13x - 10 = 3x^2 + 15x - 2x - 10$$
    $$= 3x(x+5) - 2(x+5)$$
    $$= (x+5)(3x-2)$$

13. Factor $3x^2 - 22x + 7$ by trial and error.

    $$(3x-7)(x-1) = 3x^2 - 10x + 7$$
    $$(3x-1)(x-7) = 3x^2 - 22x + 7$$

    Thus, $3x^2 - 22x + 7 = (3x-1)(x-7)$.

15. Factor $5y^2 - 16y + 3$ by trial and error.

$$(5y - 3)(y - 1) = 5y^2 - 8y + 3$$

$$(5y - 1)(y - 3) = 5y^2 - 16y + 3$$

Thus, $5y^2 - 16y + 3 = (5y - 1)(y - 3)$.

17. Factor $3x^2 - 17x + 10$ by grouping.

$a = 3$ and $c = 10$, so $ac = 30$.

The factors of 30 whose sum is -17 are -15 and -2.

$$3x^2 - 17x + 10 = 3x^2 - 15x - 2x + 10$$

$$= 3x(x - 5) - 2(x - 5)$$

$$= (x - 5)(3x - 2)$$

19. Factor $6w^2 - 11w + 4$ by grouping.

$a = 6$ and $c = 4$, so $ac = 24$.

The factors of 24 whose sum is -11 are -3 and -8.

$$6w^2 - 11w + 4 = 6w^2 - 3w - 8w + 4$$

$$= 3w(2w - 1) - 4(2w - 1)$$

$$= (2w - 1)(3w - 4)$$

21. Factor $8x^2 + 33x + 4$ by grouping.

$a = 8$ and $c = 4$, so $ac = 32$.

The factors of 32 whose sum is 33 are 32 and 1.

$$8x^2 + 33x + 4 = 8x^2 + 32x + x + 4$$

$$= 8x(x + 4) + 1(x + 4)$$

$$= (x + 4)(8x + 1)$$

23. Factor $5x^2 + 33x - 14$ by trial and error.

$$(5x - 7)(x + 2) = 5x^2 + 3x - 14$$

$$(5x + 7)(x - 2) = 5x^2 - 3x - 14$$

$$(5x - 2)(x + 7) = 5x^2 + 33x - 14$$

Because the correct factorization has been found, there is no need to try additional possibilities.

Thus, $5x^2 + 33x - 14 = (5x - 2)(x + 7)$.

25. Factor $14y^2 + 15y - 9$ by trial and error. The sign in one factor must be positive and the other negative.

$$(7y + 9)(2y - 1) = 14y^2 + 11y - 9$$

$$(7y + 1)(2y - 9) = 14y^2 - 61y - 9$$

$$(7y + 3)(2y - 3) = 14y^2 - 15y - 9$$

$$(7y - 3)(2y + 3) = 14y^2 + 15y - 9$$

Thus, $14y^2 + 15y - 9 = (7y - 3)(2y + 3)$.

27. Factor $6x^2 - 7x + 3$ by trial and error. List all the possibilities in which both signs are negative.

$$(6x - 1)(x - 3) = 6x^2 - 19x + 3$$

$$(6x - 3)(x - 1) = 6x^2 - 9x + 3$$

$$(3x - 1)(2x - 3) = 6x^2 - 11x + 3$$

$$(3x - 3)(2x - 1) = 6x^2 - 9x + 3$$

None of these possibilities gives the required middle term, $-7x$, and there are no more possibilities to try, so $6x^2 - 7x + 3$ is prime.

29. Factor $25z^2 - 30z + 9$ by trial and error until the correct factorization is obtained. The signs in both factors must be negative.

$$(5z - 1)(5z - 9) = 25z^2 - 50z + 9$$

$$(5z - 3)(5z - 3) = 25z^2 - 30z + 9$$

Thus, $25z^2 - 30z + 9 = (5z - 3)(5z - 3)$.

31. Factor $15y^2 - y - 2$ by grouping.

$a = 15$ and $c = -2$, so $ac = -30$.

The factors of -30 whose sum is -1 are -6 and 5.

$$15y^2 - y - 2 = 15y^2 - 6y + 5y - 2$$

$$= 3y(5y - 2) + 1(5y - 2)$$

$$= (5y - 2)(3y + 1)$$

33. Factor $5x^2 + 2x + 9$ by trial and error. The signs in both factors must be positive.

$$(5x + 3)(x + 3) = 5x^2 + 18x + 9$$

$$(5x + 9)(x + 1) = 5x^2 + 14x + 9$$

$$(5x + 1)(x + 9) = 5x^2 + 46x + 9$$

None of these possibilities gives the required middle term, $2x$, and there are no more possibilities to try, so $5x^2 + 2x + 9$ is prime.

35. Factor $10y^2 + 43y - 9$ by grouping.

$a = 10$ and $c = -9$ so $ac = -90$.

The factors of -90 whose sum is 43 are 45 and -2.

$$10y^2 + 43y - 9 = 10y^2 + 45y - 2y - 9$$

$$= 5y(2y + 9) - 1(2y + 9)$$

$$= (2y + 9)(5y - 1)$$

37. Factor $8x^2 - 2x - 1$ by trial and error until the correct factorization is obtained. The sign must be negative in one factor and positive in the other.

$$(4x-1)(2x+1) = 8x^2 + 2x - 1$$
$$(4x+1)(2x-1) = 8x^2 - 2x - 1$$

Thus, $8x^2 - 2x - 1 = (4x+1)(2x-1)$.

39. Factor $9y^2 - 9y + 2$ by grouping.

$a = 9$ and $c = 2$, so $ac = 18$.

The factors of 18 whose sum is -9 are -3 and -6.

$$9y^2 - 9y + 2 = 9y^2 - 3y - 6y + 2$$
$$= 3y(3y-1) - 2(3y-1)$$
$$= (3y-1)(3y-2)$$

41. Factor $20x^2 + 27x - 8$ by grouping.

$a = 20$ and $c = -8$, so $ac = -160$.

The factors of -160 whose sum is 27 are -5 and 32.

$$20x^2 + 27x - 8 = 20x^2 - 5x + 32x - 8$$
$$= 5x(4x-1) + 8(4x-1)$$
$$= (4x-1)(5x+8)$$

43. $2x^2 + 3xy + y^2 = (2x+y)(x+y)$

45. Factor $3x^2 + 5xy + 2y^2$ by trial and error.

$$(3x+y)(x+2y) = 3x^2 + 7xy + 2y^2$$
$$(3x+2y)(x+y) = 3x^2 + 5xy + 2y^2$$

Thus, $3x^2 + 5xy + 2y^2 = (3x+2y)(x+y)$.

47. Factor $2x^2 - 9xy + 9y^2$ by trial and error until the correct factorization is obtained. The signs in both factors must be negative.

$$(2x-9y)(x-y) = 2x^2 - 11xy + 9y^2$$
$$(2x+9y)(x+y) = 2x^2 + 11xy + 9y^2$$
$$(2x-3y)(x-3y) = 2x^2 - 9xy + 9y^2$$

Thus, $2x^2 - 9xy + 9y^2 = (2x-3y)(x-3y)$.

49. Factor $6x^2 - 5xy - 6y^2$ by grouping.

$a = 6$ and $c = -6$, so $ac = -36$.

The factors of -36 whose sum is -5 are -9 and 4.

$$6x^2 - 5xy - 6y^2$$
$$= 6x^2 - 9xy + 4xy - 6y^2$$
$$= 3x(2x-3y) + 2y(2x-3y)$$
$$= (2x-3y)(3x+2y)$$

51. Factor $15x^2 + 11xy - 14y^2$ by grouping.

$a = 15$ and $c = -14$, so $ac = -210$.

The factors of -210 whose sum is 11 are 21 and -10.

$$15x^2 + 11xy - 14y^2$$
$$= 15x^2 + 21xy - 10xy - 14y^2$$
$$= 3x(5x+7y) - 2y(5x+7y)$$
$$= (5x+7y)(3x-2y)$$

53. Factor $2a^2 + 7ab + 5b^2$ by trial and error.

$$(2a+5b)(a+b) = 2a^2 + 7ab + 5b^2$$
$$(2a+b)(a+5b) = 2a^2 + 11ab + 5b^2$$

Thus, $2a^2 + 7ab + 5b^2 = (2a+5b)(a+b)$.

55. Factor $15a^2 - ab - 6b^2$ by grouping.

$a = 15$ and $c = -6$, so $ac = -90$.

The factors of -90 whose sum is -1 are 9 and -10.

$$15a^2 - ab - 6b^2 = 15a^2 + 9ab - 10ab - 6b^2$$
$$= 3a(5a+3b) - 2b(5a+3b)$$
$$= (5a+3b)(3a-2b)$$

57. Factor $12x^2 - 25xy + 12y^2$ by grouping.

$a = 12$ and $c = 12$, so $ac = 144$.

The factors of 144 whose sum is -25 are -9 and -16.

$$12x^2 - 25xy + 12y^2 = 12x^2 - 9xy - 16xy + 12y^2$$
$$= 3x(4x-3y) - 4y(4x-3y)$$
$$= (4x-3y)(3x-4y)$$

59. $4x^2 + 26x + 30$

First factor out the GCF, 2. Then factor the resulting trinomial by trial and error or grouping.

$$4x^2 + 26x + 30 = 2(2x^2 + 13x + 15)$$
$$= 2(2x + 3)(x + 5)$$

61. $9x^2 - 6x - 24$

First factor out the GCF, 3. Then factor the resulting trinomial by trial and error or grouping.

$$9x^2 - 6x - 24 = 3(3x^2 - 2x - 8)$$
$$= 3(3x + 4)(x - 2)$$

63. $4y^2 + 2y - 30 = 2(2y^2 + y - 15)$
$$= 2(2y - 5)(y + 3)$$

65. $9y^2 + 33y - 60 = 3(3y^2 + 11y - 20)$
$$= 3(3y - 4)(y + 5)$$

67. $3x^3 + 4x^2 + x$

First factor out the GCF, x. Then factor the resulting trinomial by trial and error or grouping.

$$3x^3 + 4x^2 + x = x(3x^2 + 4x + 1)$$
$$= x(3x + 1)(x + 1)$$

69. $2x^3 - 3x^2 - 5x = x(2x^2 - 3x - 5)$
$$= x(2x - 5)(x + 1)$$

71. $9y^3 - 39y^2 + 12y$

First factor out the GCF, $3y$. Then factor the resulting trinomial by trial and error or grouping.

$$9y^3 - 39y^2 + 12y = 3y(3y^2 - 13y + 4)$$
$$= 3y(3y - 1)(y - 4)$$

73. $60z^3 + 40z^2 + 5z = 5z(12z^2 + 8z + 1)$
$$= 5z(6z + 1)(2z + 1)$$

75. $15x^4 - 39x^3 + 18x^2 = 3x^2(5x^2 - 13x + 6)$
$$= 3x^2(5x - 3)(x - 2)$$

77. $10x^5 - 17x^4 + 3x^3 = x^3(10x^2 - 17x + 3)$
$$= x^3(2x - 3)(5x - 1)$$

79. $6x^2 - 3xy - 18y^2 = 3(2x^2 - xy - 6y^2)$
$$= 3(2x + 3y)(x - 2y)$$

81. $12x^2 + 10xy - 8y^2 = 2(6x^2 + 5xy - 4y^2)$
$$= 2(2x - y)(3x + 4y)$$

83. $8x^2y + 34xy - 84y = 2y(4x^2 + 17x - 42)$
$$= 2y(4x - 7)(x + 6)$$

85. $12a^2b - 46ab^2 + 14b^3 = 2b(6a^2 - 23ab + 7b^2)$
$$= 2b(2a - 7b)(3a - b)$$

87. $-32x^2y^4 + 20xy^4 + 12y^4$
$$= -4y^4(8x^2 - 5x - 3)$$
$$= -4y^4(8x + 3)(x - 1)$$

89. $30(y + 1)x^2 + 10(y + 1)x - 20(y + 1)$
$$= 10(y + 1)(3x^2 + x - 2)$$
$$= 10(y + 1)(3x - 2)(x + 1)$$

91. a. $2x^2 - 5x - 3 = (2x + 1)(x - 3)$

b. $2(y + 1)^2 - 5(y + 1) - 3$
$$= [2(y + 1) + 1][(y + 1) - 3]$$
$$= [2y + 2 + 1][y + 1 - 3]$$
$$= (2y + 3)(y - 2)$$

93.
$$
\begin{array}{r}
3x^2 - 5x + 2 \\
x - 2 \overline{\smash{)}\ 3x^3 - 11x^2 + 12x - 4} \\
\underline{3x^3 - 6x^2} \\
-5x^2 + 12x \\
\underline{-5x^2 + 10x} \\
2x - 4 \\
\underline{2x - 4} \\
0
\end{array}
$$

The quotient $3x^2 - 5x + 2$ factors into $(3x - 2)(x - 1)$.

Thus, $3x^3 - 11x^2 + 12x - 4 = (x - 2)(3x - 2)(x - 1)$.

95. a. $x^2 + 3x + 2$

b. $(x+2)(x+1)$

c. Yes, the pieces are the same in both figures: one large square, three long rectangles, and two small squares. This geometric model illustrates the factorization:
$x^2 + 3x + 2 = (x+2)(x+1)$.

97. – 99. Answers will vary.

101. makes sense

103. does not make sense; Explanations will vary. Sample explanation: If a polynomial has a greatest common factor then it is not prime.

105. true

107. false; Changes to make the statement true will vary. A sample change is: The factorization of $4y^2 - 11y - 3$ is $(y-3)(4y+1)$.

109. $3x^2 + bx + 2$

The possible factorizations that will give $3x^2$ as the first term and 2 as the last term are:
$(3x+1)(x+2) = 3x^2 + 7x + 2$
$(3x+2)(x+1) = 3x^2 + 5x + 2$
$(3x-1)(x-2) = 3x^2 - 7x + 2$
$(3x-2)(x-1) = 3x^2 - 5x + 2$
The possible middle terms are
$5x, 7x, -5x$ and $-7x$, so $3x^2 + bx + 2$ can be factored if b is 5, 7, –5, or –7.

111. $3x^{10} - 4x^5 - 15$

Since $\left(x^5\right)^2 = x^{10}$, the first terms of the factors must be $3x^5$ and x^5 so the middle term will contain x^5. Use trial and error or grouping to obtain the correct factorization.
$3x^{10} - 4x^5 - 15 = \left(3x^5 + 5\right)\left(x^5 - 3\right)$

113. $4x - y = 105$
$x + 7y = -10$
Multiply the second equation by –4 and then add the equations.
$$4x - y = 105$$
$$\underline{-4x - 28y = 40}$$
$$-29y = 145$$
$$\frac{-29y}{-29} = \frac{145}{-29}$$
$$y = -5$$
Back-substitute to find x.
$$x + 7y = -10$$
$$x + 7(-5) = -10$$
$$x - 35 = -10$$
$$x = 25$$
The solution set is $\{(25, -5)\}$.

114. $0.00086 = 8.6 \times 10^{-4}$

115. $8x - \dfrac{x}{6} = \dfrac{1}{6} - 8$
Multiply both sides by the LCD of 6.
$$6\left(8x - \frac{x}{6}\right) = 6\left(\frac{1}{6} - 8\right)$$
$$48x - x = 1 - 48$$
$$47x = -47$$
$$x = -1$$
The solution set is $\{-1\}$.

116. $(9x+10)(9x-10) = (9x)^2 - 10^2$
$$= 81x^2 - 100$$

117. $(4x+5y)^2 = (4x)^2 + 2(4x)(5y) + (5y)^2$
$$= 16x^2 + 40xy + 25y^2$$

118. $(x+2)(x^2 - 2x + 4)$
$$= x(x^2 - 2x + 4) + 2(x^2 - 2x + 4)$$
$$= x^3 - 2x^2 + 4x + 2x^2 - 4x + 8$$
$$= x^3 + 8$$

Mid-Chapter Check Point

1. The GCF is x^4.
$$x^5 + x^4 = x^4(x+1)$$

2. $x^2 + 7x - 18 = (x-2)(x+9)$

3. The GCF is $x^2 y$. The polynomial in the parentheses is prime because there are no factors of 1 whose sum is -1.
$$x^2 y^3 - x^2 y^2 + x^2 y = x^2 y\left(y^2 - y + 1\right)$$

4. $x^2 - 2x + 4$ is prime because there are no factors of 4 whose sum is -2.

5. Factor $7x^2 - 22x + 3$ by grouping.
$a = 7$ and $c = 3$, so $ac = 21$.
The only factors of 21 whose sum is -22 are -21 and -1.
$$7x^2 - 22x + 3 = 7x^2 - 21x - x + 3$$
$$= 7x(x-3) - 1(x-3)$$
$$= (x-3)(7x-1)$$

6. Factor $x^3 + 5x^2 + 3x + 15$ by grouping.
$$x^3 + 5x^2 + 3x + 15 = \left(x^3 + 5x^2\right) + (3x+15)$$
$$= x^2(x+5) + 3(x+5)$$
$$= (x+5)\left(x^2 + 3\right)$$

7. The GCF is x.
$$2x^3 - 11x^2 + 5x = x\left(2x^2 - 11x + 5\right)$$
$$= x(2x-1)(x-5)$$

8. Factor $xy - 7x - 4y + 28$ by grouping.
$$xy - 7x - 4y + 28 = (xy - 7x) + (-4y + 28)$$
$$= x(y-7) - 4(y-7)$$
$$= (y-7)(x-4)$$

9. Factor $x^2 - 17xy + 30y^2$ by trial and error. The only factors of 30 whose sum is -17 are -15 and -2.
$$x^2 - 17xy + 30y^2 = (x-15y)(x-2y)$$

10. Factor $25x^2 - 25x - 14$ by trial and error.
$$(5x-2)(5x+7) = 25x^2 + 25x - 14$$
$$(5x+2)(5x-7) = 25x^2 - 25x - 14$$
Because the correct factorization has been found, there is no need to try additional possibilities.
$$25x^2 - 25x - 14 = (5x+2)(5x-7).$$

11. The GCF is 2.
$$16x^2 - 70x + 24 = 2\left(8x^2 - 35x + 12\right)$$
Factor the polynomial in parentheses by grouping.
$a = 8$ and $c = 12$, so $ac = 96$.
The only factors of 96 whose sum is -35 are -32 and -3.
$$16x^2 - 70x + 24 = 2(8x^2 - 35x + 12)$$
$$= 2(8x^2 - 32x - 3x + 12)$$
$$= 2\left[8x(x-4) - 3(x-4)\right]$$
$$= 2(x-4)(8x-3)$$

12. Factor $3x^2 + 10xy + 7y^2$ by grouping.
$a = 3$ and $c = 7$, so $ac = 21$.
The only factors of 21 whose sum is 10 are 3 and 7.
$$3x^2 + 10xy + 7y^2 = 3x^2 + 3xy + 7xy + 7y^2$$
$$= 3x(x+y) + 7y(x+y)$$
$$= (x+y)(3x+7y)$$

13. First, factor out the greatest common factor.
$$-6x^3 + 8x^2 + 30x = -2x(3x^2 - 4x - 15)$$
Then, factor the trinomial.
$$-6x^3 + 8x^2 + 30x = -2x(3x^2 - 4x - 15)$$
$$= -2x(3x+5)(x-3)$$

6.4 Check Points

1. **a.** Notice that the trinomial fits the form $A^2 - B^2$. Thus, factor using $A^2 - B^2 = (A+B)(A-B)$.
$$x^2 - 81 = x^2 - 9^2$$
$$= (x+9)(x-9)$$

   b. Notice that the trinomial fits the form $A^2 - B^2$. Thus, factor using $A^2 - B^2 = (A+B)(A-B)$.
$$36x^2 - 25 = (6x)^2 - 5^2$$
$$= (6x+5)(6x-5)$$

2. a. Notice that the trinomial fits the form $A^2 - B^2$.
Thus, factor using $A^2 - B^2 = (A + B)(A - B)$.
$$25 - 4x^{10} = 5^2 - (2x^5)^2$$
$$= (5 + 2x^5)(5 - 2x^5)$$

b. Notice that the trinomial fits the form $A^2 - B^2$.
Thus, factor using $A^2 - B^2 = (A + B)(A - B)$.
$$100x^2 - 9y^2 = (10x)^2 - (3y)^2$$
$$= (10x + 3y)(10x - 3y)$$

3. a. First factor out the GCF.
$$18x^3 - 2x = 2x(9x^2 - 1)$$
Next, factor the difference of two squares.
$$18x^3 - 2x = 2x(9x^2 - 1)$$
$$= 2x(3x + 1)(3x - 1)$$

b. First factor out the GCF.
$$72 - 18x^2 = 18(4 - x^2)$$
Next, factor the difference of two squares.
$$72 - 18x^2 = 18(4 - x^2)$$
$$= 18(2 + x)(2 - x)$$

4. First, factor the difference of two squares.
$$81x^4 - 16 = (9x^2 + 4)(9x^2 - 4)$$

The factor of $9x^2 - 4$ is the difference of two squares and can be factored.
$$81x^4 - 16 = (9x^2 + 4)(9x^2 - 4)$$
$$= (9x^2 + 4)(3x + 2)(3x - 2)$$

5. a. Notice that the trinomial fits the form $A^2 + 2AB + B^2$.
Thus, factor using $A^2 + 2AB + B^2 = (A + B)^2$.
$$x^2 + 14x + 49 = (x + 7)^2$$

b. Notice that the trinomial fits the form $A^2 - 2AB + B^2$.
Thus, factor using $A^2 - 2AB + B^2 = (A - B)^2$.
$$x^2 - 6x + 9 = (x - 3)^2$$

c. Notice that the trinomial fits the form $A^2 - 2AB + B^2$.
Thus, factor using $A^2 - 2AB + B^2 = (A - B)^2$.
$$16x^2 - 56x + 49 = (4x - 7)^2$$

6. Notice that the trinomial fits the form $A^2 + 2AB + B^2$.
Thus it factors as $(A + B)^2$.
$$4x^2 + 12xy + 9y^2 = (2x + 3y)^2$$

7. Notice that the polynomial fits the form $A^3 + B^3$.
Thus it factors as $(A + B)(A^2 - AB + B^2)$.
$$x^3 + 27 = x^3 + 3^3$$
$$= (x + 3)(x^2 - 3x + 3^2)$$
$$= (x + 3)(x^2 - 3x + 9)$$

8. Notice that the polynomial fits the form $A^3 - B^3$.
Thus it factors as $(A - B)(A^2 + AB + B^2)$.
$$1 - y^3 = 1^3 - y^3$$
$$= (1 - y)(1^2 + 1 \cdot y + y^2)$$
$$= (1 - y)(1 + y + y^2)$$

9. Notice that the polynomial fits the form $A^3 + B^3$.
Thus it factors as $(A + B)(A^2 - AB + B^2)$.
$$125x^3 + 8 = (5x)^3 + 2^3$$
$$= (5x + 2)\left[(5x)^2 - (5x)(2) + 2^2\right]$$
$$= (5x + 2)(25x^2 - 10x + 4)$$

6.4 Concept and Vocabulary Check

1. $(A + B)(A - B)$

2. $(A + B)^2$

3. $(A - B)^2$

4. $(A + B)(A^2 - AB + B^2)$

5. $(A - B)(A^2 + AB + B^2)$

6. $6x$; $6x$

7. -6

8. $4x$

9. $+2$

10. -3; $+9$

11. false

12. true

13. false

14. true

15. false

6.4 Exercise Set

1. $x^2 - 25 = x^2 - 5^2 = (x+5)(x-5)$

3. $y^2 - 1 = y^2 - 1^2 = (y+1)(y-1)$

5. $4x^2 - 9 = (2x)^2 - 3^2 = (2x+3)(2x-3)$

7. $25 - x^2 = 5^2 - x^2 = (5+x)(5-x)$

9. $1 - 49x^2 = 1^2 - (7x)^2 = (1+7x)(1-7x)$

11. $9 - 25y^2 = 3^2 - (5y)^2 = (3+5y)(3-5y)$

13. $x^4 - 9 = (x^2)^2 - 3^2 = (x^2+3)(x^2-3)$

15. $49y^4 - 16 = (7y^2)^2 - 4^2$
$$= (7y^2+4)(7y^2-4)$$

17. $x^{10} - 9 = (x^5)^2 - 3^2 = (x^5+3)(x^5-3)$

19. $25x^2 - 16y^2 = (5x)^2 - (4y)^2$
$$= (5x+4y)(5x-4y)$$

21. $x^4 - y^{10} = (x^2)^2 - (y^5)^2$
$$= (x^2+y^5)(x^2-y^5)$$

23. $x^4 - 16 = (x^2)^2 - 4^2 = (x^2+4)(x^2-4)$

Because $x^2 - 4$ is also the difference of two squares, the factorization must be continued.
$$x^4 - 16 = (x^2+4)(x^2-4)$$
$$= (x^2+4)(x^2-2^2)$$
$$= (x^2+4)(x+2)(x-2).$$

25. $16x^4 - 81 = (4x^2)^2 - 9^2$
$$= (4x^2+9)(4x^2-9)$$
$$= (4x^2+9)[(2x)^2 - 3^2]$$
$$= (4x^2+9)(2x+3)(2x-3)$$

27. $2x^2 - 18 = 2(x^2-9) = 2(x+3)(x-3)$

29. $2x^3 - 72x = 2x(x^2-36)$
$$= 2x(x+6)(x-6)$$

31. $x^2 + 36$ is prime because it is the sum of two squares with no common factor other than 1.

33. $3x^3 + 27x = 3x(x^2+9)$

35. $18 - 2y^2 = 2(9-y^2) = 2(3+y)(3-y)$

37. $3y^3 - 48y = 3y(y^2-16)$
$$= 3y(y+4)(y-4)$$

39. $18x^3 - 2x = 2x(9x^2-1)$
$$= 2x(3x+1)(3x-1)$$

41. $-3x^2 + 75 = -3(x^2-25)$
$$= -3(x+5)(x-5)$$

43. $-5y^3 + 20y = -5y(y^2-4)$
$$= -5y(y+2)(y-2)$$

45. $x^2 + 2x + 1 = x^2 + 2(1x) + 1^2$
$$= (x+1)^2$$

47. $x^2 - 14x + 49 = x^2 - 2(7x) + 7^2$
$$= (x-7)^2$$

49. $x^2 - 2x + 1 = x^2 - 2(1x) + 1^2$
$$= (x-1)^2$$

51. $x^2 + 22x + 121 = x^2 + 2(11x) + 11^2$
$$= (x+11)^2$$

53. $4x^2 + 4x + 1 = (2x)^2 + 2(2x) + 1^2$
$$= (2x+1)^2$$

55. $25y^2 - 10y + 1 = (5y)^2 - 2(5y) + 1^2$
$$= (5y-1)^2$$

57. $x^2 - 10x + 100$ is prime.
To be a perfect square trinomial, the middle term would have to be $2(-10x) = -20x$ rather than $-10x$.

59. $x^2 + 14xy + 49y^2 = x^2 + 2(7xy) + (7y)^2$
$$= (x+7y)^2$$

61. $x^2 - 12xy + 36y^2 = x^2 - 2(6xy) + (6y)^2$
$$= (x-6y)^2$$

63. $x^2 - 8xy + 64y^2$ is prime.
To be a perfect square trinomial, the middle term would have to be $2(-8xy) = -16xy$ rather than $-8xy$.

65. $16x^2 - 40xy + 25y^2$
$$= (4x)^2 - 2(4x \cdot 5y) + (5y)^2$$
$$= (4x-5y)^2$$

67. $12x^2 - 12x + 3 = 3(4x^2 - 4x + 1)$
$$= 3\left[(2x)^2 - 2(2x) + 1^2\right]$$
$$= 3(2x-1)^2$$

69. $9x^3 + 6x^2 + x$
$$= x(9x^2 + 6x + 1)$$
$$= x\left[(3x)^2 + 2(3x) + 1^2\right]$$
$$= x(3x+1)^2$$

71. $2y^2 - 4y + 2 = 2(y^2 - 2y + 1)$
$$= 2(y-1)^2$$

73. $2y^3 + 28y^2 + 98y = 2y(y^2 + 14y + 49)$
$$= 2y(y+7)^2$$

75. $-6x^2 + 24x - 24 = -6(x^2 - 4x + 4)$
$$= -6(x-2)^2$$

77. $-16y^3 - 16y^2 - 4y = -4y(4y^2 + 4y + 1)$
$$= -4y(2y+1)^2$$

79. $x^3 + 1 = x^3 + 1^3$
$$= (x+1)(x^2 - x \cdot 1 + 1^2)$$
$$= (x+1)(x^2 - x + 1)$$

81. $x^3 - 27 = x^3 - 3^3$
$$= (x-3)(x^2 + x \cdot 3 + 3^2)$$
$$= (x-3)(x^2 + 3x + 9)$$

83. $8y^3 - 1 = (2y)^3 - 1^3$
$$= (2y-1)\left[(2y)^2 + 2y \cdot 1 + 1\right]$$
$$= (2y-1)(4y^2 + 2y + 1)$$

85. $27x^3 + 8 = (3x)^3 + 2^3$

$\qquad = (3x+2)\left[(3x)^2 - 3x \cdot 2 + 2^2\right]$

$\qquad = (3x+2)\left(9x^2 - 6x + 4\right)$

87. $x^3 y^3 - 64 = (xy)^3 - 4^3$

$\qquad = (xy - 4)\left[(xy)^2 + xy \cdot 4 + 4^2\right]$

$\qquad = (xy - 4)\left(x^2 y^2 + 4xy + 16\right)$

89. $27y^4 + 8y = y\left(27y^3 + 8\right)$

$\qquad = y\left[(3y)^3 + 2^3\right]$

$\qquad = y(3y+2)\left[(3y)^2 - 3y \cdot 2 + 2^2\right]$

$\qquad = y(3y+2)\left(9y^2 - 6y + 4\right)$

91. $54 - 16y^3 = 2\left(27 - 8y^3\right)$

$\qquad = 2\left[3^3 - (2y)^3\right]$

$\qquad = 2(3 - 2y)\left[3^2 + 3 \cdot 2y + (2y)^2\right]$

$\qquad = 2(3 - 2y)\left(9 + 6y + 4y^2\right)$

93. $64x^3 + 27y^3 = (4x)^3 + (3y)^3$

$\qquad = (4x + 3y)\left[(4x)^2 - 4x \cdot 3y + (3y)^2\right]$

$\qquad = (4x + 3y)\left(16x^2 - 12xy + 9y^2\right)$

95. $125x^3 - 64y^3 = (5x)^3 - (4y)^3$

$\qquad = (5x - 4y)\left[(5x)^2 + 5x \cdot 4y + (4y)^2\right]$

$\qquad = (5x - 4y)\left(25x^2 + 20xy + 16y^2\right)$

97. $25x^2 - \dfrac{4}{49} = (5x)^2 - \left(\dfrac{2}{7}\right)^2$

$\qquad = \left(5x + \dfrac{2}{7}\right)\left(5x - \dfrac{2}{7}\right)$

99. $y^4 - \dfrac{y}{1000} = y\left(y^3 - \dfrac{1}{1000}\right)$

$\qquad = y\left[y^3 - \left(\dfrac{1}{10}\right)^3\right]$

$\qquad = y\left(y - \dfrac{1}{10}\right)\left[y^2 + y \cdot \dfrac{1}{10} + \left(\dfrac{1}{10}\right)^2\right]$

$\qquad = y\left(y - \dfrac{1}{10}\right)\left(y^2 + \dfrac{y}{10} + \dfrac{1}{100}\right)$

101. $0.25x - x^3 = x\left(0.25 - x^2\right)$

$\qquad = x\left[(0.5)^2 - x^2\right]$

$\qquad = x(0.5 + x)(0.5 - x)$

103. $(x+1)^2 - 25 = (x+1)^2 - 5^2$

$\qquad = \left[(x+1) + 5\right]\left[(x+1) - 5\right]$

$\qquad = (x + 6)(x - 4)$

105.
$$
\begin{array}{r}
x^2 + 2x + 1 \\
x - 3 \overline{\smash{\big)}\, x^3 - x^2 - 5x - 3} \\
\underline{x^3 - 3x^2} \\
2x^2 - 5x \\
\underline{2x^2 - 6x} \\
x - 3 \\
\underline{x - 3} \\
\end{array}
$$

The quotient $x^2 + 2x + 1$ factors further.

$x^2 + 2x + 1 = (x+1)^2$.

Thus, $x^3 - x^2 - 5x - 3 = (x - 3)(x + 1)^2$.

107. Area of outer square $= x^2$

Area of inner square $= 5^2 = 25$

Area of shaded region $= x^2 - 25 = (x + 5)(x - 5)$

109. Area of large square $= x^2$

Area of each small corner squares $= 2^2 = 4$

Area of four corner squares $= 4 \cdot 4 = 16$

Area of shaded region $= x^2 - 16$

$\qquad\qquad\qquad\qquad = (x + 4)(x - 4)$

111. – 113. Answers will vary.

115. does not make sense; Explanations will vary. Sample explanation: The original expression has a common factor of 9.

$$9x^2 - 36 = 9(x^2 - 4) = 9(x+2)(x-2)$$

117. does not make sense; Explanations will vary. Sample explanation: The second factor involves a subtraction. The commutative property does not apply to subtraction.

119. false; Changes to make the statement true will vary. A sample change is: $x^2 + 25$ is prime.

121. false; Changes to make the statement true will vary. A sample change is: $4x^2 + 36$ is a polynomial that is the sum of two squares. $4x^2 + 36$ factors into $4(x^2 + 9)$.

123. The error in the proof is in step 7 where you are asked to divide by $a - b = 0$.
Division by 0 is not permitted because division by zero is undefined.

125. $x^{2n} - 25y^{2n} = \left(x^n\right)^2 - \left(5y^n\right)^2$

$$= \left(x^n + 5y^n\right)\left(x^n - 5y^n\right)$$

127. $(x+3)^2 - 2(x+3) + 1$

$$= \left[(x+3) - 1\right]^2$$

$$= (x+2)^2$$

129. $64x^2 - 16x + k$

Let r be the number such that $r^2 = k$.

Then, $64x^2 - 16x + k = (8x)^2 - 2 \cdot 8x \cdot r + r^2$.

Comparing the middle terms, we see that
$-2 \cdot 8x \cdot r = -16x$
$\quad -16xr = -16x$
$\qquad\quad r = 1.$

Therefore, $k = r^2 = 1^2 = 1$.

131. The graphs coincide.

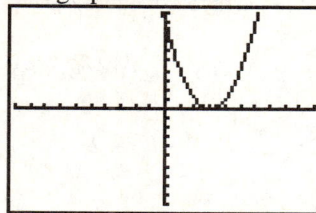

This verifies that $x^2 - 6x + 9 = (x-3)^2$.

133. The graphs do not coincide.

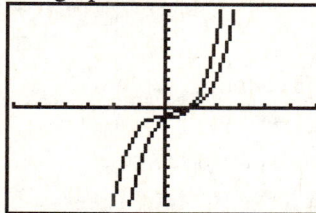

$$x^3 - 1 = x^3 - 1^3 = (x-1)\left(x^2 + x + 1\right)$$

The polynomial on the right side should be changed to $(x-1)\left(x^2 + x + 1\right)$.

134. $\left(2x^2 y^3\right)^4 \left(5xy^2\right)$

$$= \left[2^4 \left(x^2\right)^4 \left(y^3\right)^4\right] \cdot \left(5xy^2\right)$$

$$= \left(16x^8 y^{12}\right)\left(5xy^2\right)$$

$$= (16 \cdot 5)\left(x^8 \cdot x^1\right)\left(y^{12} \cdot y^2\right)$$

$$= 80x^9 y^{14}$$

135. $\left(10x^2 - 5x + 2\right) - \left(14x^2 - 5x - 1\right)$

$$= \left(10x^2 - 5x + 2\right) + \left(-14x^2 + 5x + 1\right)$$

$$= \left(10x^2 - 14x^2\right) + \left(-5x + 5x\right) + (2 + 1)$$

$$= -4x^2 + 3$$

136.
$$\require{enclose}\begin{array}{r} 2x+5 \\ 3x-2 \enclose{longdiv}{6x^2 + 11x - 10} \\ \underline{6x^2 - 4x} \\ 15x - 10 \\ \underline{15x - 10} \\ 0 \end{array}$$

$$\frac{6x^2 + 11x - 10}{3x - 2} = 2x + 5$$

137. $3x^3 - 75x = 3x(x^2 - 25) = 3x(x+5)(x-5)$

138. $2x^2 - 20x + 50 = 2(x^2 - 10x + 25) = 2(x-5)^2$

139. $x^3 - 2x^2 - x + 2 = x^2(x-2) - 1(x-2)$
$$= (x-2)(x^2 - 1)$$
$$= (x-2)(x+1)(x-1)$$

6.5 Check Points

1. First use common factoring.
$$5x^4 - 45x^2 = 5x^2(x^2 - 9)$$
Then use difference of two squares.
$$5x^4 - 45x^2 = 5x^2(x^2 - 9)$$
$$= 5x^2(x+3)(x-3)$$

2. First use common factoring.
$$4x^2 - 16x - 48 = 4(x^2 - 4x - 12)$$
Then use trial and error or grouping.
$$4x^2 - 16x - 48 = 4(x^2 - 4x - 12)$$
$$= 4(x-6)(x+2)$$

3. First use common factoring.
$$4x^5 - 64x = 4x(x^4 - 16)$$
Then use difference of two squares.
$$4x^5 - 64x = 4x(x^4 - 16)$$
$$= 4x(x^2 + 4)(x^2 - 4)$$
Finally, use difference of two squares again.
$$4x^5 - 64x = 4x(x^4 - 16)$$
$$= 4x(x^2 + 4)(x^2 - 4)$$
$$= 4x(x^2 + 4)(x+2)(x-2)$$

4. Use factor by grouping.
$$x^3 - 4x^2 - 9x + 36 = x^2(x-4) - 9(x-4)$$
$$= (x-4)(x^2 - 9)$$
Then use difference of two squares.
$$x^3 - 4x^2 - 9x + 36 = x^2(x-4) - 9(x-4)$$
$$= (x-4)(x^2 - 9)$$
$$= (x-4)(x+3)(x-3)$$

5. First use common factoring.
$$3x^3 - 30x^2 + 75x = 3x(x^2 - 10x + 25)$$
The trinomial is a perfect square trinomial.
$$3x^3 - 30x^2 + 75x = 3x(x^2 - 10x + 25)$$
$$= 3x(x-5)^2$$

6. First use common factoring.
$$2x^5 + 54x^2 = 2x^2(x^3 + 27)$$
The binomial is the sum of two cubes.
$$2x^5 + 54x^2 = 2x^2(x^3 + 27)$$
$$= 2x^2(x+3)(x^2 - 3x + 9)$$

7. First use common factoring.
$$3x^4 y - 48y^5 = 3y(x^4 - 16y^4)$$
Then use difference of two squares.
$$3x^4 y - 48y^5 = 3y(x^4 - 16y^4)$$
$$= 3y(x^2 + 4y^2)(x^2 - 4y^2)$$
Finally, use difference of two squares again.
$$3x^4 y - 48y^5 = 3y(x^4 - 16y^4)$$
$$= 3y(x^2 + 4y^2)(x^2 - 4y^2)$$
$$= 3y(x^2 + 4y^2)(x+2y)(x-2y)$$

8. First use common factoring.
$$12x^3 + 36x^2 y + 27xy^2 = 3x(4x^2 + 12xy + 9y^2)$$
The trinomial is a perfect square trinomial.
$$12x^3 + 36x^2 y + 27xy^2 = 3x(4x^2 + 12xy + 9y^2)$$
$$= 3x(2x + 3y)^2$$

6.5 Concept and Vocabulary Check

1. b

2. e

3. h

4. c

5. d

6. f

7. a

8. g

6.5 Exercise Set

1. $-7x^2 + 35x = -7x(x-5)$

3. $25x^2 - 49 = (5x+7)(5x-7)$

5. $27x^3 - 1 = (3x-1)(9x^2+3x+1)$

7. $5x + 5y + x^2 + xy = 5(x+y) + x(x+y)$
$$= (5+x)(x+y)$$

9. $14x^2 - 9x + 1 = (7x-1)(2x-1)$

11. $x^2 - 2x + 1 = (x-1)(x-1)$
$$= (x-1)^2$$

13. $27x^3y^3 + 8 = (3xy+2)(9x^2y^2 - 6xy + 4)$

15. $6x^2 + x - 15 = (3x+5)(2x-3)$

17. $5x^3 - 20x = 5x\left(x^2 - 4\right)$
$$= 5x(x+2)(x-2)$$

19. $7x^3 + 7x = 7x\left(x^2 + 1\right)$

21. $5x^2 - 5x - 30 = 5\left(x^2 - x - 6\right)$
$$= 5(x+2)(x-3)$$

23. $2x^4 - 162 = 2\left(x^4 - 81\right)$
$$= 2\left(x^2 + 9\right)\left(x^2 - 9\right)$$
$$= 2\left(x^2 + 9\right)(x+3)(x-3)$$

25. $x^3 + 2x^2 - 9x - 18 = \left(x^3 + 2x^2\right) + \left(-9x - 18\right)$
$$= x^2(x+2) - 9(x+2)$$
$$= (x+2)\left(x^2 - 9\right)$$
$$= (x+2)(x+3)(x-3)$$

27. $3x^3 - 24x^2 + 48x = 3x\left(x^2 - 8x + 16\right)$
$$= 3x(x-4)^2$$

29. $2x^5 + 2x^2 = 2x^2\left(x^3 + 1\right)$
$$= 2x^2(x+1)\left(x^2 - x + 1\right)$$

31. $6x^2 + 8x = 2x(3x+4)$

33. $2y^2 - 2y - 112 = 2\left(y^2 - y - 56\right)$
$$= 2(y-8)(y+7)$$

35. $7y^4 + 14y^3 + 7y^2 = 7y^2\left(y^2 + 2y + 1\right)$
$$= 7y^2(y+1)^2$$

37. $y^2 + 8y - 16$ is prime because there are no two integers whose product is -16 and whose sum is 8.

39. $16y^2 - 4y - 2 = 2\left(8y^2 - 2y - 1\right)$
$$= 2(4y+1)(2y-1)$$

41. $r^2 - 25r = r(r-25)$

43. $4w^2 + 8w - 5 = (2w+5)(2w-1)$

45. $x^3 - 4x = x\left(x^2 - 4\right) = x(x+2)(x-2)$

47. $x^2 + 64$ is prime because it is the sum of two squares with no common factor other than 1.

49. $9y^2 + 13y + 4 = (9y+4)(y+1)$

51. $y^3 + 2y^2 - 4y - 8 = \left(y^3 + 2y^2\right) + \left(-4y - 8\right)$
$$= y^2(y+2) - 4(y+2)$$
$$= (y+2)\left(y^2 - 4\right)$$
$$= (y+2)(y+2)(y-2)$$
$$= (y+2)^2(y-2)$$

53. $16y^2 + 24y + 9 = (4y)^2 + 2(4y \cdot 3) + 3^2$
$$= (4y+3)^2$$

55. $4y^3 - 28y^2 + 40y = 4y\left(y^2 - 7y + 10\right)$
$$= 4y(y-5)(y-2)$$

57. $y^5 - 81y = y\left(y^4 - 81\right)$

$\qquad = y\left(y^2 + 9\right)\left(y^2 - 9\right)$

$\qquad = y\left(y^2 + 9\right)(y + 3)(y - 3)$

59. $20a^4 - 45a^2 = 5a^2\left(4a^2 - 9\right)$

$\qquad = 5a^2(2a + 3)(2a - 3)$

61. $9x^4 + 18x^3 + 6x^2 = 3x^2(3x^2 + 6x + 2)$

63. $12y^2 - 11y + 2 = (4y - 1)(3y - 2)$

65. $9y^2 - 64 = (3y)^2 - 8^2$

$\qquad = (3y + 8)(3y - 8)$

67. $9y^2 + 64$ is prime because it is the sum of two squares with no common factor other than 1.

69. $2y^3 + 3y^2 - 50y - 75 = \left(2y^3 + 3y^2\right) + (-50y - 75)$

$\qquad = y^2(2y + 3) - 25(2y + 3)$

$\qquad = (2y + 3)\left(y^2 - 25\right)$

$\qquad = (2y + 3)(y + 5)(y - 5)$

71. $2r^3 + 30r^2 - 68r = 2r\left(r^2 + 15r - 34\right)$

$\qquad = 2r(r + 17)(r - 2)$

73. $8x^5 - 2x^3 = 2x^3\left(4x^2 - 1\right)$

$\qquad = 2x^3\left[(2x)^2 - 1^2\right]$

$\qquad = 2x^3(2x + 1)(2x - 1)$

75. $3x^2 + 243 = 3\left(x^2 + 81\right)$

77. $x^4 + 8x = x\left(x^3 + 8\right)$

$\qquad = x\left(x^3 + 2^3\right)$

$\qquad = x(x + 2)\left(x^2 - 2x + 4\right)$

79. $2y^5 - 2y^2 = 2y^2\left(y^3 - 1\right)$

$\qquad = 2y^2(y - 1)\left(y^2 + y + 1\right)$

81. $6x^2 + 8xy = 2x(3x + 4y)$

83. $xy - 7x + 3y - 21 = (xy - 7x) + (3y - 21)$

$\qquad = x(y - 7) + 3(y - 7)$

$\qquad = (y - 7)(x + 3)$

85. $x^2 - 3xy - 4y^2 = (x - 4y)(x + y)$

87. $72a^3b^2 + 12a^2 - 24a^4b^2$

$\qquad = 12a^2\left(6ab^2 + 1 - 2a^2b^2\right)$

89. $3a^2 + 27ab + 54b^2 = 3\left(a^2 + 9ab + 18b^2\right)$

$\qquad = 3(a + 6b)(a + 3b)$

91. $48x^4y - 3x^2y = 3x^2y\left(16x^2 - 1\right)$

$\qquad = 3x^2y(4x + 1)(4x - 1)$

93. $6a^2b + ab - 2b = b\left(6a^2 + a - 2\right)$

$\qquad = b(3a + 2)(2a - 1)$

95. $7x^5y - 7xy^5 = 7xy\left(x^4 - y^4\right)$

$\qquad = 7xy\left(x^2 + y^2\right)\left(x^2 - y^2\right)$

$\qquad = 7xy\left(x^2 + y^2\right)(x + y)(x - y)$

97. $10x^3y - 14x^2y^2 + 4xy^3 = 2xy\left(5x^2 - 7xy + 2y^2\right)$

$\qquad = 2xy(5x - 2y)(x - y)$

99. $2bx^2 + 44bx + 242b = 2b\left(x^2 + 22x + 121\right)$

$\qquad = 2b\left(x^2 + 2(11x) + 11^2\right)$

$\qquad = 2b(x + 11)^2$

101. $15a^2 + 11ab - 14b^2 = (5a + 7b)(3a - 2b)$

103. $36x^3y - 62x^2y^2 + 12xy^3$
$= 2xy\left(18x^2 - 31xy + 6y^2\right)$
$= 2xy(9x - 2y)(2x - 3y)$

105. $a^2y - b^2y - a^2x + b^2x$
$= \left(a^2y - b^2y\right) + \left(-a^2x + b^2x\right)$
$= y\left(a^2 - b^2\right) - x\left(a^2 - b^2\right)$
$= \left(a^2 - b^2\right)(y - x)$
$= (a + b)(a - b)(y - x)$

107. $9ax^3 + 15ax^2 - 14ax = ax\left(9x^2 + 15x - 14\right)$
$= ax(3x + 7)(3x - 2)$

109. $2x^4 + 6x^3y + 2x^2y^2 = 2x^2(x^2 + 3xy + y^2)$

111. $81x^4y - y^5 = y\left(81x^4 - y^4\right)$
$= y\left(9x^2 + y^2\right)\left(9x^2 - y^2\right)$
$= y\left(9x^2 + y^2\right)(3x + y)(3x - y)$

113. $10x^2(x + 1) - 7x(x + 1) - 6(x + 1)$
$= (x + 1)\left(10x^2 - 7x - 6\right)$
$= (x + 1)(5x - 6)(2x + 1)$

115. $6x^4 + 35x^2 - 6 = \left(x^2 + 6\right)\left(6x^2 - 1\right)$

117. $(x - 7)^2 - 4a^2 = (x - 7)^2 - (2a)^2$
$= [(x - 7) + 2a][(x - 7) - 2a]$
$= (x - 7 + 2a)(x - 7 - 2a)$

119. $x^2 + 8x + 16 - 25a^2$
$= \left(x^2 + 8x + 16\right) - (5a)^2$
$= (x + 4)^2 - (5a)^2$
$= [(x + 4) + 5a][(x + 4) - 5a]$
$= (x + 4 + 5a)(x + 4 - 5a)$

121. $y^7 + y = y\left(y^6 + 1\right) = y\left[\left(y^2\right)^3 + 1^3\right]$
$= y\left(y^2 + 1\right)\left(y^4 - y^2 + 1\right)$

123. $256 - 16t^2 = 16\left(16 - t^2\right)$
$= 16(4 + t)(4 - t)$

125. Area of outer circle $= \pi b^2$
Area of inner circle $= \pi a^2$
Area of shaded ring $= \pi b^2 - \pi a^2$
$\pi b^2 - \pi a^2 = \pi\left(b^2 - a^2\right)$
$= \pi(b + a)(b - a)$

127. Answers will vary.

129. makes sense

131. makes sense

133. false; Changes to make the statement true will vary. A sample change is: $x^2 - 9 = (x + 3)(x - 3)$ for any real number x.

135. false; Changes to make the statement true will vary. A sample change is: Some polynomials are completely factored after one step.

137. $3x^5 - 21x^3 - 54x = 3x\left(x^4 - 7x^2 - 18\right)$
$= 3x\left(x^2 + 2\right)\left(x^2 - 9\right)$
$= 3x\left(x^2 + 2\right)(x + 3)(x - 3)$

139. $4x^4 - 9x^2 + 5 = \left(4x^2 - 5\right)\left(x^2 - 1\right)$
$= \left(4x^2 - 5\right)(x + 1)(x - 1)$

141. $3x^{2n} - 27y^{2n} = 3\left(x^{2n} - 9y^{2n}\right)$
$= 3\left[\left(x^n\right)^2 - \left(3y^n\right)^2\right]$
$= 3\left(x^n + 3y^n\right)\left(x^n - 3y^n\right)$

143. The graphs do not coincide.

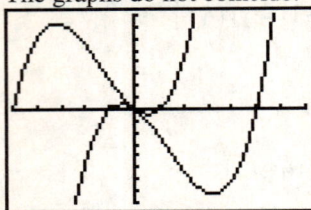

$$3x^3 - 12x^2 - 15x = 3x\left(x^2 - 4x - 5\right)$$
$$= 3x(x-5)(x+1)$$

Change the polynomial on the right side to $3x(x-5)(x+1)$.

145. The graphs coincide.

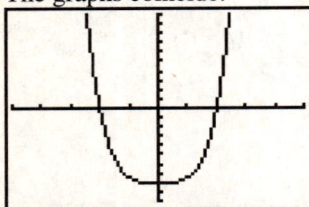

This verifies that the factorization
$x^4 - 16 = \left(x^2 + 4\right)(x+2)(x-2)$ is correct.

147. $9x^2 - 16 = (3x)^2 - 4^2$
$$= (3x+4)(3x-4)$$

148. $5x - 2y = 10$

To find the x-intercept, let $y = 0$.
$$5x - 2(0) = 10$$
$$5x = 10$$
$$x = 2$$

To find the y-intercept, let $x = 0$.
$$5(0) - 2y = 10$$
$$-2y = 10$$
$$y = -5$$

Checkpoint: $(4,5)$

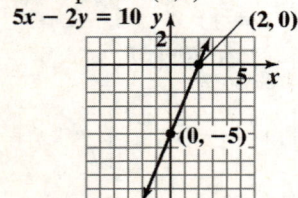

149. Let $x =$ the measure of the first angle.
Then $3x =$ the measure of the second angle,
$x + 80 =$ the measure of the third angle.
$$x + 3x + (x + 80) = 180$$
$$5x + 80 = 180$$
$$5x = 100$$
$$x = 20$$

Measure of first angle $= x = 20°$
Measure of second angle $= 3x = 60°$
Measure of third angle $= x + 80 = 100°$

150. $(3x-1)(x+2) = \left(3 \cdot \frac{1}{3} - 1\right)\left(\frac{1}{3} + 2\right)$
$$= (1-1)\left(2\frac{1}{3}\right)$$
$$= (0)\left(2\frac{1}{3}\right)$$
$$= 0$$

151. $2x^2 + 7x - 4 = 2\left(\frac{1}{2}\right)^2 + 7\left(\frac{1}{2}\right) - 4$
$$= \frac{1}{2} + \frac{7}{2} - 4$$
$$= \frac{8}{2} - 4$$
$$= 4 - 4$$
$$= 0$$

152. $(x-2)(x+3) - 6 = x^2 + 3x - 2x - 6 - 6$
$$= x^2 + x - 12$$
$$= (x+4)(x-3)$$

6.6 Check Points

1. $(2x+1)(x-4) = 0$

The equation is in factored form on the left side with zero on the right side.
Thus, set each factor equal to zero and solve the resulting equations.

$2x+1 = 0$ or $x-4 = 0$
$2x = -1$ $x = 4$
$x = -\frac{1}{2}$

The solution set is $\left\{-\frac{1}{2}, 4\right\}$.

2. All the terms are on one side and zero is on the other side. Thus, factor.

$$x^2 - 6x + 5 = 0$$

$$(x-1)(x-5) = 0$$

Next, set each factor equal to zero and solve the resulting equations.

$$x - 1 = 0 \quad \text{or} \quad x - 5 = 0$$
$$x = 1 \qquad\qquad x = 5$$

The solution set is $\{1, 5\}$.

3. Move all terms to one side and obtain zero on the other side. Then factor.

$$4x^2 = 2x$$

$$4x^2 - 2x = 0$$

$$2x(2x - 1) = 0$$

Next, set each factor equal to zero and solve the resulting equations.

$$2x = 0 \quad \text{or} \quad 2x - 1 = 0$$
$$x = 0 \qquad\qquad 2x = 1$$
$$x = \frac{1}{2}$$

The solution set is $\left\{0, \frac{1}{2}\right\}$.

4. Move all terms to one side and obtain zero on the other side. Then factor.

$$x^2 = 10x - 25$$

$$x^2 - 10x + 25 = 0$$

$$(x - 5)^2 = 0$$

Because both factors are the same, it is only necessary to set one of them equal to zero.

$$x - 5 = 0$$
$$x = 5$$

The solution set is $\{5\}$.

5. Move all terms to one side and obtain zero on the other side. Then factor.

$$16x^2 = 25$$

$$16x^2 - 25 = 0$$

$$(4x + 5)(4x - 5) = 0$$

Next, set each factor equal to zero and solve the resulting equations.

$$4x + 5 = 0 \quad \text{or} \quad 4x - 5 = 0$$
$$4x = -5 \qquad\qquad 4x = 5$$
$$x = -\frac{5}{4} \qquad\qquad x = \frac{5}{4}$$

The solution set is $\left\{-\frac{5}{4}, \frac{5}{4}\right\}$.

6. Write the equation in standard form by finding the product on the left side and then subtracting 28 from both sides.

$$(x - 5)(x - 2) = 28$$

$$x^2 - 7x + 10 = 28$$

$$x^2 - 7x - 18 = 0$$

Factor, and then set each factor equal to zero and solve the resulting equations.

$$x^2 - 7x - 18 = 0$$

$$(x - 9)(x + 2) = 0$$

$$x - 9 = 0 \quad \text{or} \quad x + 2 = 0$$
$$x = 9 \qquad\qquad x = -2$$

The solution set is $\{-2, 9\}$.

7.
$$h = -16t^2 + 48t + 160$$
$$192 = -16t^2 + 48t + 160$$
$$0 = -16t^2 + 48t - 32$$
$$0 = -16(t^2 - 3t + 2)$$
$$0 = -16(t - 1)(t - 2)$$
$$x - 1 = 0 \quad \text{or} \quad x - 2 = 0$$
$$x = 1 \qquad\qquad x = 2$$

The ball's height will be 192 feet at 1 second and 2 seconds after it is thrown.
This is represented on the graph by the points (1,192) and (2,192).

8. Let $x =$ the width of the sign.

Let $x + 3 =$ the length of the sign.

The area of 54 square units can be represented as follows.

$$A = l \cdot w$$

$$54 = (x + 3) \cdot x$$

$$54 = x^2 + 3x$$

$$0 = x^2 + 3x - 54$$

$$0 = (x - 6)(x + 9)$$

$$x - 6 = 0 \quad \text{or} \quad x + 9 = 0$$

$$x = 6 \qquad\qquad x = -9$$

Reject –9 because the width cannot be negative. The width of the sign is 6 units and the length is 9 units.

6.6 Concept and Vocabulary Check

1. quadratic equation

2. $A = 0$ or $B = 0$

3. x-intercepts

4. subtracting $5x$

5. subtracting $30x$; adding 25

6.6 Exercise Set

1. $x(x + 7) = 0$

$x = 0$ or $x + 7 = 0$

$\qquad\qquad x = -7$

The solution set is $\{-7, 0\}$.

3. $(x - 6)(x + 4) = 0$

$x - 6 = 0$ or $x + 4 = 0$

$x = 6 \qquad\quad x = -4$

The solution set is $\{-4, 6\}$.

5. $(x - 9)(5x + 4) = 0$

$x - 9 = 0$ or $5x + 4 = 0$

$\quad x = 9 \qquad\quad 5x = -4$

$$x = -\frac{4}{5}$$

The solution set is $\left\{-\dfrac{4}{5}, 9\right\}$.

7. $10(x - 4)(2x + 9) = 0$

$x - 4 = 0$ or $2x + 9 = 0$

$\quad x = 4 \qquad\qquad 2x = -9$

$$x = -\frac{9}{2}$$

The solution set is $\left\{-\dfrac{9}{2}, 4\right\}$.

9. $x^2 + 8x + 15 = 0$

$(x + 5)(x + 3) = 0$

$x + 5 = 0$ or $x + 3 = 0$

$\quad x = -5 \qquad\quad x = -3$

Check -5:

$$x^2 + 8x + 15 = 0$$

$$(-5)^2 + 8(-5) + 15 = 0$$

$$25 - 40 + 15 = 0$$

$$0 = 0 \text{ true}$$

Check -3:

$$x^2 + 8x + 15 = 0$$

$$(-3)^2 + 8(-3) + 15 = 0$$

$$9 - 24 + 15 = 0$$

$$0 = 0 \text{ true}$$

The solution set is $\{-5, -3\}$.

11. $x^2 - 2x - 15 = 0$

$(x + 3)(x - 5) = 0$

$x + 3 = 0$ or $x - 5 = 0$

$\quad x = -3 \qquad\quad x = 5$

The solution set is $\{-3, 5\}$.

13. $x^2 - 4x = 21$

$x^2 - 4x - 21 = 0$

$(x + 3)(x - 7) = 0$

$x + 3 = 0 \quad \text{or} \quad x - 7 = 0$

$\quad x = -3 \qquad\qquad x = 7$

The solution set is $\{-3, 7\}$.

15. $x^2 + 9x = -8$

$x^2 + 9x + 8 = 0$

$(x + 8)(x + 1) = 0$

$x + 8 = 0$ or $x + 1 = 0$

$x = -8$ $x = -1$

The solution set is $\{-8, -1\}$.

17. $x^2 + 4x = 0$

$x(x + 4) = 0$

$x = 0$ or $x + 4 = 0$

$x = -4$

The solution set is $\{-4, 0\}$.

19. $x^2 - 5x = 0$

$x(x - 5) = 0$

$x = 0$ or $x - 5 = 0$

$x = 5$

The solution set is $\{0, 5\}$.

21. $x^2 = 4x$

$x^2 - 4x = 0$

$x(x - 4) = 0$

$x = 0$ or $x - 4 = 0$

$x = 4$

The solution set is $\{0, 4\}$.

23. $2x^2 = 5x$

$2x^2 - 5x = 0$

$x(2x - 5) = 0$

$x = 0$ or $2x - 5 = 0$

$2x = 5$

$x = \dfrac{5}{2}$

The solution set is $\left\{0, \dfrac{5}{2}\right\}$.

25. $3x^2 = -5x$

$3x^2 + 5x = 0$

$x(3x + 5) = 0$

$x = 0$ or $3x + 5 = 0$

$3x = -5$

$x = -\dfrac{5}{3}$

The solution set is $\left\{-\dfrac{5}{3}, 0\right\}$.

27. $x^2 + 4x + 4 = 0$

$(x - 2)^2 = 0$

$x + 2 = 0$

$x = -2$

The solution set is $\{-2\}$.

29. $x^2 = 12x - 36$

$x^2 - 12x + 36 = 0$

$(x - 6)^2 = 0$

$x - 6 = 0$

$x = 6$

The solution set is $\{6\}$.

31. $4x^2 = 12x - 9$

$4x^2 - 12x + 9 = 0$

$(2x - 3)^2 = 0$

$2x - 3 = 0$

$2x = 3$

$x = \dfrac{3}{2}$

The solution set is $\left\{\dfrac{3}{2}\right\}$.

33.
$$2x^2 = 7x + 4$$
$$2x^2 - 7x - 4 = 0$$
$$(2x+1)(x-4) = 0$$
$$2x+1 = 0 \quad \text{or} \quad x-4 = 0$$
$$2x = -1 \qquad\qquad x = 4$$
$$x = -\frac{1}{2}$$

The solution set is $\left\{-\frac{1}{2}, 4\right\}$.

35.
$$5x^2 = 18 - x$$
$$5x^2 + x - 18 = 0$$
$$(5x-9)(x+2) = 0$$
$$5x-9 = 0 \quad \text{or} \quad x+2 = 0$$
$$5x = 9 \qquad\qquad x = -2$$
$$x = \frac{9}{5}$$

The solution set is $\left\{-2, \frac{9}{5}\right\}$.

37.
$$x^2 - 49 = 0$$
$$(x+7)(x-7) = 0$$
$$x+7 = 0 \quad \text{or} \quad x-7 = 0$$
$$x = -7 \qquad\qquad x = 7$$

The solution set is $\{-7, 7\}$.

39.
$$4x^2 - 25 = 0$$
$$(2x+5)(2x-5) = 0$$
$$2x+5 = 0 \quad \text{or} \quad 2x-5 = 0$$
$$2x = -5 \qquad\qquad 2x = 5$$
$$x = -\frac{5}{2} \qquad\qquad x = \frac{5}{2}$$

The solution set is $\left\{-\frac{5}{2}, \frac{5}{2}\right\}$.

41.
$$81x^2 = 25$$
$$81x^2 - 25 = 0$$
$$(9x+5)(9x-5) = 0$$
$$9x+5 = 0 \quad \text{or} \quad 9x-5 = 0$$
$$9x = -5 \qquad\qquad 9x = 5$$
$$x = -\frac{5}{9} \qquad\qquad x = \frac{5}{9}$$

The solution set is $\left\{-\frac{5}{9}, \frac{5}{9}\right\}$.

43.
$$x(x-4) = 21$$
$$x^2 - 4x = 21$$
$$x^2 - 4x - 21 = 0$$
$$(x+3)(x-7) = 0$$
$$x+3 = 0 \quad \text{or} \quad x-7 = 0$$
$$x = -3 \qquad\qquad x = 7$$

The solution set is $\{-3, 7\}$.

45.
$$4x(x+1) = 15$$
$$4x^2 + 4x = 15$$
$$4x^2 + 4x - 15 = 0$$
$$(2x+5)(2x-3) = 0$$
$$2x+5 = 0 \quad \text{or} \quad 2x-3 = 0$$
$$2x = -5 \qquad\qquad 2x = 3$$
$$x = -\frac{5}{2} \qquad\qquad x = \frac{3}{2}$$

The solution set is $\left\{-\frac{5}{2}, \frac{3}{2}\right\}$.

47.
$$(x-1)(x+4) = 14$$
$$x^2 + 3x - 4 = 14$$
$$x^2 + 3x - 18 = 0$$
$$(x+6)(x-3) = 0$$
$$x+6 = 0 \quad \text{or} \quad x-3 = 0$$
$$x = -6 \qquad\qquad x = 3$$

The solution set is $\{-6, 3\}$.

49. $(x+1)(2x+5)=-1$

$2x^2+7x+5=-1$

$2x^2+7x+6=0$

$(2x+3)(x+2)=0$

$2x+3=0 \quad$ or $\quad x+2=0$

$2x=-3 \qquad\qquad x=-2$

$x=-\dfrac{3}{2}$

The solution set is $\left\{-2,-\dfrac{3}{2}\right\}$.

51. $y(y+8)=16(y-1)$

$y^2+8y=16y-16$

$y^2-8y+16=0$

$(y-4)^2=0$

$y-4=0$

$y=4$

The solution set is $\{4\}$.

53. $4y^2+20y+25=0$

$(2y+5)^2=0$

$2y+5=0$

$2y=-5$

$y=-\dfrac{5}{2}$

The solution set is $\left\{-\dfrac{5}{2}\right\}$.

55. $64w^2=48w-9$

$64w^2-48w+9=0$

$(8w-3)^2=0$

$8w-3=0$

$8w=3$

$w=\dfrac{3}{8}$

The solution set is $\left\{\dfrac{3}{8}\right\}$.

57. $(x-4)(x^2+5x+6)=0$

$(x-4)(x+3)(x+2)=0$

$x-4=0$ or $x+3=0 \quad$ or $\quad x+2=0$

$x=4 \qquad x=-3 \qquad\quad x=-2$

The solution set is $\{-3,-2,4\}$.

59. $x^3-36x=0$

$x(x^2-36)=0$

$x(x+6)(x-6)=0$

$x=0$ or $x+6=0$ or $x-6=0$

$x=-6 \qquad x=6$

The solution set is $\{-6,0,6\}$.

61. $y^3+3y^2+2y=0$

$y(y^2+3y+2)=0$

$y(y+2)(y+1)=0$

$y=0$ or $y+2=0$ or $y+1=0$

$y=-2 \qquad y=-1$

The solution set is $\{-2,-1,0\}$.

63. $2(x-4)^2+x^2=x(x+50)-46x$

$2(x^2-8x+16)+x^2=x^2+50x-46x$

$2x^2-16x+32+x^2=x^2+4x$

$3x^2-16x+32=x^2+4x$

$2x^2-20x+32=0$

$2(x^2-10x+16)=0$

$2(x-8)(x-2)=0$

$x-8=0 \quad$ or $\quad x-2=0$

$x=8 \qquad\qquad x=2$

The solution set is $\{2,8\}$.

65. $(x-2)^2-5(x-2)+6=0$

$[(x-2)-3][(x-2)-2]=0$

$(x-5)(x-4)=0$

$x-5=0 \quad$ or $\quad x-4=0$

$x=5 \qquad\qquad x=4$

The solution set is $\{4,5\}$.

67. $h = -16t^2 + 20t + 300$
Substitute 0 for h and solve for t.

$$0 = -16t^2 + 20t + 300$$

$$-16t^2 + 20t + 300 = 0$$

$$-4t\left(4t^2 - 5t - 75\right) = 0$$

$$-4t\left(4t + 15\right)\left(t - 5\right) = 0$$

$-4t = 0$ or $4t + 15 = 0$ or $t - 5 = 0$

$t = 0$ $\qquad$ $4t = -15$ $\qquad$ $t = 5$

$$t = -\frac{15}{4} = -3.75$$

Reject $t = 0$ since this represents the time before the ball was thrown. Also discard $t = -3.75$ since time cannot be negative. The only solution that makes sense is 5. So it will take 5 seconds for the ball to hit the ground. Each tick mark represents one second.

69. Substitute 276 for h and solve for t.

$$276 = -16t^2 + 20t + 300$$

$$16t^2 - 20t - 24 = 0$$

$$4\left(4t^2 - 5t - 6\right) = 0$$

$$4\left(4t + 3\right)\left(t - 2\right) = 0$$

$4t + 3 = 0$ $\qquad$ or $\qquad$ $t - 2 = 0$

$4t = -3$ $\qquad\qquad$ $t = 2$

$$t = -\frac{3}{4}$$

Reject $t = -\dfrac{3}{4}$ since time cannot be negative. The ball's height will be 276 feet 2 seconds after it is thrown. This corresponds to the point $(2, 276)$ on the graph.

71. $h = -16t^2 + 72t$
Substitute 32 for h and solve for t.

$$32 = -16t^2 + 72t$$

$$16t^2 - 72t + 32 = 0$$

$$8\left(2t^2 - 9t + 4\right) = 0$$

$$8\left(2t - 1\right)\left(t - 4\right) = 0$$

$2t - 1 = 0$ or $t - 4 = 0$

$$t = \frac{1}{2} \qquad\qquad t = 4$$

The debris will be 32 feet above the ground $\dfrac{1}{2}$ second after the explosion and 4 seconds after the explosion.

73. $S = 2x^2 - 12x + 82$
Substitute 72 for S and solve for x.

$$S = 2x^2 - 12x + 82$$

$$66 = 2x^2 - 12x + 82$$

$$0 = 2x^2 - 12x + 16$$

$$0 = 2(x^2 - 6x + 8)$$

$$0 = x^2 - 6x + 8$$

$$0 = (x - 2)(x - 4)$$

$x - 2 = 0$ $\qquad$ or $\qquad$ $x - 4 = 0$

$x = 2$ $\qquad\qquad\qquad$ $x = 4$

International travelers spent \$66 billion 2 years and 4 years after 2000, or 2002 and 2004.

75. International travelers spent \$66 billion 2 years and 4 years after 2000. This corresponds to the point $(2, 66)$ and the point $(4, 66)$ on the graph.

77. $P = -10x^2 + 475x + 3500$
Substitute 7250 for P and solve for x.

$$7250 = -10x^2 + 475x + 3500$$

$$10x^2 - 475x + 3750 = 0$$

$$5\left(2x^2 - 95x + 750\right) = 0$$

$$5\left(x - 10\right)\left(2x - 75\right) = 0$$

$x - 10 = 0$ or $2x - 75 = 0$

$x = 10$ $\qquad\qquad$ $2x = 75$

$$x = \frac{75}{2} \text{ or } 37.5$$

The alligator population will have increased to 7250 after 10 years. (Discard 37.5 because this value is outside of $0 \le x \le 12$.)

79. The solution in Exercise 77 corresponds to the point $(10, 7250)$ on the graph.

81. $N = \dfrac{t^2 - t}{2}$

Substitute 45 for N and solve for t.

$$45 = \dfrac{t^2 - t}{2}$$

$$2 \cdot 45 = 2\left(\dfrac{t^2 - t}{2}\right)$$

$$90 = t^2 - t$$

$$0 = t^2 - t - 90$$

$$0 = (t - 10)(t + 9)$$

$$t - 10 = 0 \quad \text{or} \quad t + 9 = 0$$

$$t = 10 \qquad\qquad t = -9$$

Reject $t = -9$ since the number of teams cannot be negative. If 45 games are scheduled, there are 10 teams in the league.

83. Let x = the width of the parking lot.
Then $x + 3$ = the length.

$$l \cdot w = A$$

$$(x + 3)(x) = 180$$

$$x^2 + 3x = 180$$

$$x^2 + 3x - 180 = 0$$

$$(x + 15)(x - 12) = 0$$

$$x + 15 = 0 \quad \text{or} \quad x - 12 = 0$$

$$x = -15 \qquad\qquad x = 12$$

Reject $x = -15$ since the width cannot be negative. Then $x = 12$ and $x + 3 = 15$, so the length is 15 yards and the width is 12 yards.

85. Use the formula for the area of a triangle where x is the base and $x + 1$ is the height.

$$\dfrac{1}{2}bh = A$$

$$\dfrac{1}{2}x(x + 1) = 15$$

$$2\left[\dfrac{1}{2}x(x + 1)\right] = 2 \cdot 15$$

$$x(x + 1) = 30$$

$$x^2 + x = 30$$

$$x^2 + x - 30 = 0$$

$$(x + 6)(x - 5) = 0$$

$$x + 6 = 0 \quad \text{or} \quad x - 5 = 0$$

$$x = -6 \qquad\qquad x = 5$$

Reject $x = -6$ since the length of the base cannot be negative. Then $x = 5$ and $x + 1 = 6$, so the base is 5 centimeters and the height is 6 centimeters.

87. a. Area of border

$$= \overbrace{(2x + 12)(2x + 10)}^{\text{Area of a large rectangle}} - \overbrace{10 \cdot 12}^{\substack{\text{Area of}\\ \text{flower bed}}}$$

$$= 4x^2 + 20x + 24x + 120 - 120$$

$$= 4x^2 + 44x$$

b. Find the width of the border for which the area of the border would be 168 square feet.

$$4x^2 + 44x = 168$$

$$4x^2 + 44x - 168 = 0$$

$$4\left(x^2 + 11x - 42\right) = 0$$

$$4(x + 14)(x - 3) = 0$$

$$x + 14 = 0 \quad \text{or} \quad x - 3 = 0$$

$$x = -14 \qquad\qquad x = 3$$

Reject $x = -14$ since the width of the border cannot be negative. You should prepare a strip that is 3 feet wide for the border.

89. Answers will vary.

91. does not make sense; Explanations will vary. Sample explanation: Though 4 cannot equal 0 and can therefore be ignored, $4x$ will equal 0 when $x = 0$ and so it cannot be ignored.

93. makes sense

95. false; Changes to make the statement true will vary. A sample change is: If $(x + 3)(x - 4) = 0$, then $x + 3 = 0$ or $x - 4 = 0$.

97. false; Changes to make the statement true will vary. A sample change is: Some equations solved by factoring have more than 2 solutions and some have fewer than 2 solutions

99. If -3 and 5 are solutions of the quadratic equation, then $x - (-3) = x + 3$ and $x - 5$ must be factors of the polynomial on the left side when the quadratic equation is written in standard form.

$$(x + 3)(x - 5) = 0$$

$$x^2 - 5x + 3x - 15 = 0$$

$$x^2 - 2x - 15 = 0$$

Thus, $x^2 - 2x - 15 = 0$ is a quadratic equation in standard form whose solutions are -3 and 5.

101. $3^{x^2-9x+20} = 1$

Because $3^0 = 1$ (and there is no other power of 3 that is equal to 1), $x^2 - 9x + 20 = 0$. Solve this equation.

$$x^2 - 9x + 20 = 0$$
$$(x-4)(x-5) = 0$$
$$x - 4 = 0 \quad \text{or} \quad x - 5 = 0$$
$$x = 4 \qquad x = 5$$

The solution set is $\{4, 5\}$.

103. $y = x^2 - x - 2$

To match this equation with its graph, find the intercepts.
To find the y-intercept, let $x = 0$ and solve for y.

$$y = 0^2 - 0 - 2 = -2$$

The y-intercept is -2.
To find the x intercepts (if any), let $y = 0$ and solve for x.

$$0 = x^2 - x - 2$$
$$0 = (x+1)(x-2)$$

The x intercepts are -1 and 2.
The only graph with y-intercept -2 and x-intercepts -1 and 2 is graph, c, so this is the graph of $y = x^2 - x - 2$

105. $y = x^2 - 4$

To match this equation with its graph, find the intercepts.
To find the y-intercepts, let $x = 0$ and solve for y.

$$y = 0^2 - 4 = -4$$

The y-intercept is -4.
To find the x-intercepts (if any), let $y = 0$ and solve for x.

$$0 = x^2 - 4$$
$$0 = (x+2)(x-2)$$
$$x + 2 = 0 \quad \text{or} \quad x - 2 = 0$$
$$x = -2 \qquad x = 2$$

The x-intercepts are -2 and 2.
The only graph with y-intercept -4 and x-intercepts -2 and 2 is graph d, so this is the graph of $y = x^2 - 4$.

107. $y = x^2 + 3x - 4$

$$x^2 + 3x - 4 = 0$$

Graph $y = x^2 + 3x - 4$ and use the graph to find the x-intercepts.

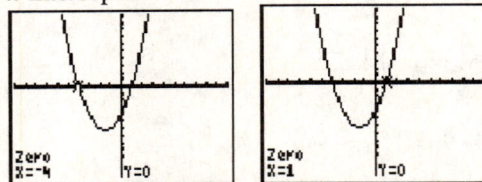

The x-intercepts of $y = x^2 + 3x - 4$ are -4 and 1, so the solution set of $x^2 + 3x - 4 = 0$ is $\{-4, 1\}$.
Check -4:

$$x^2 + 3x - 4 = 0$$
$$(-4)^2 + 3(-4) - 4 = 0$$
$$16 - 12 - 4 = 0$$
$$4 - 4 = 0$$
$$0 = 0, \text{ true}$$

Check 1:

$$x^2 + 3x - 4 = 0$$
$$1^2 + 3(1) - 4 = 0$$
$$1 + 3 - 4 = 0$$
$$4 - 4 = 0$$
$$0 = 0, \text{ true}$$

109. $y = (x-2)(x+3) - 6$

$$(x-2)(x+3) - 6 = 0$$

Graph $y = (x-2)(x-3) - 6$ and use the graph to find the x-intercepts.

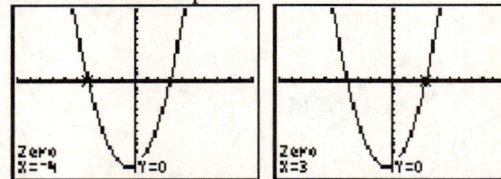

The x-intercepts of $y = (x-2)(x+3) - 6$ are -4 and 3, so the solution set of $(x-2)(x+3) - 6 = 0$ is $\{-4, 3\}$.

Check -4:

$$(x-2)(x+3)-6=0$$
$$(-4-2)(-4+3)-6=0$$
$$(-6)(-1)-6=0$$
$$6-6=0$$
$$=0, \text{ true}$$

Check 3:

$$(x-2)(x+3)-6=0$$
$$(3-1)(3+3)-6=0$$
$$(1)(6)-6=0$$
$$6-6=0$$
$$0=0, \text{ true}$$

111. Answers will vary depending on the exercises chosen.

113. $y > -\dfrac{2}{3}x+1$

Graph $y = -\dfrac{2}{3}x+1$ as a dashed line using the slope

$-\dfrac{2}{3} = \dfrac{-2}{3}$ and y-intercept 1. (Plot $(0,1)$ and move 2

units *down* and 3 units to the *right* to reach the point $(3, -1)$. Draw a line through $(0,1)$ and $(3, -1)$.)

Use $(0,0)$ as a test point. Since $0 > -\dfrac{2}{3}(0)+1$ is

false, shade the half-plane *not* containing $(0,0)$.

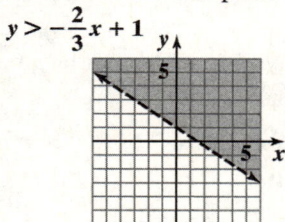

$y > -\dfrac{2}{3}x+1$

114. $\left(\dfrac{8x^4}{4x^7}\right)^2 = \left(\dfrac{8}{4} \cdot x^{4-7}\right) = \left(2x^{-3}\right)^2$

$= 2^2 \cdot \left(x^{-3}\right)^2 = 4x^{-6} = \dfrac{4}{x^6}$

115.
$$5x+28 = 6-6x$$
$$5x+6x+28 = 6-6x+6x$$
$$11x+28 = 6$$
$$11x+28-28 = 6-28$$
$$11x = -22$$
$$\dfrac{11x}{11} = \dfrac{-22}{11}$$
$$x = -2$$

The solution set is $\{-2\}$.

116. $\dfrac{250x}{100-x} = \dfrac{250(60)}{100-60} = \dfrac{15000}{40} = 375$

117. When x is replaced with 4, the denominator is 0. Division by zero is undefined.
$$7x-28 = 7(4)-28 = 28-28 = 0$$

118. $\dfrac{x^2+6x+5}{x^2-25} = \dfrac{(x+5)(x+1)}{(x+5)(x-5)}$

$= \dfrac{\cancel{(x+5)}(x+1)}{\cancel{(x+5)}(x-5)}$

$= \dfrac{x+1}{x-5}$

Chapter 6 Review Exercises

1. $30x-45 = 15(2x-3)$

2. $12x^3 + 16x^2 - 400x = 4x\left(3x^2+4x-100\right)$

3. $30x^4y + 15x^3y + 5x^2y = 5x^2y\left(6x^2+3x+1\right)$

4. $7(x+3)-2(x+3) = (x+3)(7-2)$
$= (x+3) \cdot 5 \text{ or } 5(x+3)$

5. $7x^2(x+y)-(x+y) = 7x^2(x+y)-1(x+y)$
$= (x+y)(7x^2-1)$

6. $x^3+3x^2+2x+6 = \left(x^3+3x^2\right)+(2x+6)$
$= x^2(x+3)+2(x+3)$
$= (x+3)\left(x^2+2\right)$

7. $xy + y + 4x + 4 = (xy + y) + (4x + 4)$
$= y(x+1) + 4(x+1)$
$= (x+1)(y+4)$

8. $x^3 + 5x + x^2 + 5 = (x^3 + 5x) + (x^2 + 5)$
$= x(x^2 + 5) + 1(x^2 + 5)$
$= (x^2 + 5)(x+1)$

9. $xy + 4x - 2y - 8 = (xy + 4x) + (-2y - 8)$
$= x(y+4) - 2(y+4)$
$= (y+4)(x-2)$

10. $x^2 - 3x + 2 = (x-2)(x-1)$

11. $x^2 - x - 20 = (x-5)(x+4)$

12. $x^2 + 19x + 48 = (x+3)(x+16)$

13. $x^2 - 6xy + 8y^2 = (x-4y)(x-2y)$

14. $x^2 + 5x - 9$ is prime because there is no pair of integers whose product is -9 and whose sum is 5.

15. $x^2 + 16xy - 17y^2 = (x+17y)(x-y)$

16. $3x^2 + 6x - 24 = 3(x^2 + 2x - 8)$
$= 3(x+4)(x-2)$

17. $3x^3 - 36x^2 + 33x = 3x(x^2 - 12x + 11)$
$= 3x(x-11)(x-1)$

18. Factor $3x^2 + 17x + 10$ by trial and error or by grouping. To factor by grouping, find two integers whose product is $ac = 3 \cdot 10 = 30$ and whose sum is $b = 17$. These integers are 15 and 2.
$3x^2 + 17x + 10 = 3x^2 + 15x + 2x + 10$
$= 3x(x+5) + 2(x+5)$
$= (x+5)(3x+2)$

19. Factor $5y^2 - 17y + 6$ by trial and error or by grouping. To factor by trial and error, start with the First terms, which must be $5y$ and y. Because the middle term is negative, the factors of 6 must both be negative. Try various combinations until the correct middle term is obtained.
$(5y-1)(y-6) = 5y^2 - 31y + 6$
$(5y-6)(y-1) = 5y^2 - 11y + 6$
$(5y-3)(y-2) = 5y^2 - 13y + 6$
$(5y-2)(y-3) = 5y^2 - 17y + 6$
Thus, $5y^2 - 17y + 6 = (5y-2)(y-3)$.

20. $4x^2 + 4x - 15 = (2x+5)(2x-3)$

21. Factor $5y^2 + 11y + 4$ by trial and error. The first terms must be $5y$ and y. Because the middle term is positive, the factors of 4 must both be positive. Try all the combinations.
$(5y+2)(y+2) = 5y^2 + 12y + 4$
$(5y+4)(y+1) = 5y^2 + 9y + 4$
$(5y+1)(y+4) = 5y^2 + 21y + 4$
None of these possibilities gives the required middle term, $11x$, and there are no more possibilities to try, so $5y^2 + 11y + 4$ is prime.

22. First factor out the GCF, -2. Then factor the resulting trinomial by trial and error or by grouping.
$-8x^2 - 8x + 6 = -2(4x^2 + 4x - 3)$
$= -2(2x+3)(2x-1)$

23. $2x^3 + 7x^2 - 72x = x(2x^2 + 7x - 72)$
$= x(2x-9)(x+8)$

24. $12y^3 + 28y^2 + 8y = 4y(3y^2 + 7y + 2)$
$= 4y(3y+1)(y+2)$

25. $2x^2 - 7xy + 3y^2 = (2x-y)(x-3y)$

26. $5x^2 - 6xy - 8y^2 = (5x+4y)(x-2y)$

27. $4x^2 - 1 = (2x)^2 - 1^2 = (2x+1)(2x-1)$

28. $81 - 100y^2 = 9^2 - (10y)^2$
$$= (9 + 10y)(9 - 10y)$$

29. $25a^2 - 49b^2 = (5a)^2 - (7b)^2$
$$= (5a + 7b)(5a - 7b)$$

30. $z^4 - 16 = (z^2)^2 - 4^2$
$$= (z^2 + 4)(z^2 - 4)$$
$$= (z^2 + 4)(z + 2)(z - 2)$$

31. $2x^2 - 18 = 2(x^2 - 9) = 2(x + 3)(x - 3)$

32. $x^2 + 1$ is prime because it is the sum of two squares with no common factor other than 1.

33. $9x^3 - x = x(9x^2 - 1) = x(3x + 1)(3x - 1)$

34. $18xy^2 - 8x = 2x(9y^2 - 4)$
$$= 2x(3y + 2)(3y - 2)$$

35. $x^2 + 22x + 121 = x^2 + 2(11x) + 11^2$
$$= (x + 11)^2$$

36. $x^2 - 16x + 64 = x^2 - 2(8 \cdot x) + 8^2$
$$= (x - 8)^2$$

37. $9y^2 + 48y + 64 = (3y)^2 + 2(3y \cdot 8) + 8^2$
$$= (3y + 8)^2$$

38. $16x^2 - 40x + 25 = (4x)^2 - 2(4x \cdot 5) + 5^2$
$$= (4x - 5)^2$$

39. $25x^2 + 15x + 9$ is prime.
Note that to be a perfect square trinomial, the middle term would have to be $2(5x \cdot 3) = 30x$.

40. $36x^2 + 60xy + 25y^2$
$$= (6x)^2 + 2(6x \cdot 5y) + (5y)^2$$
$$= (6x + 5y)^2$$

41. $25x^2 - 40xy + 16y^2$
$$= (5x)^2 - 2(5x \cdot 4y) + (4y)^2$$
$$= (5x - 4y)^2$$

42. $x^3 - 27 = x^3 - 3^2 = (x - 3)(x^2 + 3x + 9)$

43. $64x^3 + 1 = (4x)^3 + 1^3$
$$= (4x + 1)\left[(4x)^2 - 4x \cdot 1 + 1^2\right]$$
$$= (4x + 1)(16x^2 - 4x + 1)$$

44. $54x^3 - 16y^3$
$$= 2(27x^3 - 8y^3)$$
$$= 2\left[(3x)^3 - (2y)^3\right]$$
$$= 2(3x - 2y)\left[(3x)^2 + 3x \cdot 2y + (2y)^2\right]$$
$$= 2(3x - 2y)(9x^2 + 6xy + 4y^2)$$

45. $27x^3y + 8y = y(27x^3 + 8)$
$$= y\left[(3x)^3 + 2^3\right]$$
$$= y(3x + 2)\left[(3x)^2 - 3x \cdot 2 + 2^2\right]$$
$$= y(3x + 2)(9x^2 - 6x + 4)$$

46. Area of outer square = a^2
Area of inner square = $3^2 = 9$
Area of shaded region = $a^2 - 9$
$$= (a + 3)(a - 3)$$

47. Area of large square = a^2
Area of each small corner square = b^2
Area of four corner squares = $4b^2$
Area of shaded region = $a^2 - 4b^2$
$$= (a + 2b)(a - 2b)$$

48. Area on the left:

Area of large square = A^2
Area of each rectangle: $A \cdot 1 = A$
Area of two rectangles = $2A$
Area of small square = $1^2 = 1$
Area on the right:
Area of square = $(A+1)^2$

This geometric model illustrates the factorization $A^2 + 2A - 1 = (A+1)^2$

49. $x^3 - 8x^2 + 7x = x\left(x^2 - 8x + 7\right) = x(x-7)(x-1)$

50. $10y^2 + 9y + 2 = (5y+2)(2y+1)$

51. $128 - 2y^2 = 2\left(64 - y^2\right) = 2(8+y)(8-y)$

52. $9x^2 + 6x + 1 = (3x)^2 + 2(3x) + 1^2 = (3x+1)^2$

53. $-20x^7 + 36x^3 = -4x^3(5x^4 - 9)$

54. $x^3 - 3x^2 - 9x + 27$
$= \left(x^3 - 3x^2\right) + \left(-9x + 27\right)$
$= x^2(x-3) - 9(x-3)$
$= (x-3)\left(x^2 - 9\right)$
$= (x-3)(x+3)(x-3)$
or $(x-3)^2(x+3)$

55. $y^2 + 16$ is prime because it is the sum of two squares with no common factor other than 1.

56. $2x^3 + 19x^2 + 35x = x\left(2x^2 + 19x + 35\right)$
$= x(2x+5)(x+7)$

57. $3x^3 - 30x^2 + 75x = 3x\left(x^2 - 10x + 25\right)$
$= 3x(x-5)^2$

58. $3x^5 - 24x^2 = 3x^2\left(x^3 - 8\right)$
$= 3x^2\left(x^3 - 2^3\right)$
$= 3x^2(x-2)\left(x^2 + 2x + 4\right)$

59. $4y^4 - 36y^2 = 4y^2\left(y^2 - 9\right)$
$= 4y^2(y+3)(y-3)$

60. $5x^2 + 20x - 105 = 5\left(x^2 + 4x - 21\right)$
$= 5(x+7)(x-3)$

61. $9x^2 + 8x - 3$ is prime because there are no two integers whose product is $ac = -27$ and whose sum is 8.

62. $-10x^5 + 44x^4 - 16x^3 = -2x^3(5x^2 - 22x + 8)$
$= -2x^3(5x-2)(x-4)$

63. $100y^2 - 49 = (10y)^2 - 7^2$
$= (10y+7)(10y-7)$

64. $9x^5 - 18x^4 = 9x^4(x-2)$

65. $x^4 - 1 = \left(x^2\right)^2 - 1^2$
$= \left(x^2 + 1\right)\left(x^2 - 1\right)$
$= \left(x^2 + 1\right)(x+1)(x-1)$

66. $2y^3 - 16 = 2\left(y^3 - 8\right)$
$= 2\left(y^3 - 2^3\right)$
$= 2(y-2)\left(y^2 + 2y + 2^2\right)$
$= 2(y-2)\left(y^2 + 2y + 4\right)$

67. $x^3 + 64 = x^3 + 4^3$
$= (x+4)\left(x^2 - 4x + 4^2\right)$
$= (x+4)\left(x^2 - 4x + 16\right)$

68. $6x^2 + 11x - 10 = (3x-2)(2x+5)$

69. $3x^4 - 12x^2 = 3x^2\left(x^2 - 4\right)$
$= 3x^2(x+2)(x-2)$

70. $x^2 - x - 90 = (x-10)(x+9)$

71. $25x^2 + 25xy + 6y^2 = (5x + 2y)(5x + 3y)$

72. $x^4 + 125x = x(x^3 + 125)$
$$= x(x^3 + 5^3)$$
$$= x(x + 5)(x^2 - 5x + 5^2)$$
$$= x(x + 5)(x^2 - 5x + 25)$$

73. $32y^3 + 32y^2 + 6y = 2y(16y^2 + 16y + 3)$
$$= 2y(4y + 3)(4y + 1)$$

74. $-2y^2 + 16y - 32 = -2(y^2 - 8y + 16)$
$$= -2(y - 4)^2$$

75. $x^2 - 2xy - 35y^2 = (x + 5y)(x - 7y)$

76. $x^2 + 7x + xy + 7y = x(x + 7) + y(x + 7)$
$$= (x + 7)(x + y)$$

77. $9x^2 + 24xy + 16y^2$
$$= (3x)^2 + 2(3x \cdot 4y) + (4y)^2$$
$$= (3x + 4y)^2$$

78. $2x^4y - 2x^2y = 2x^2y(x^2 - 1)$
$$= 2x^2y(x + 1)(x - 1)$$

79. $100y^2 - 49z^2 = (10y)^2 - (7z)^2$
$$= (10y + 7z)(10y - 7z)$$

80. $x^2 + xy + y^2$ is prime.

Note that to be a perfect square trinomial, the middle term would have to be $2xy$.

81. $3x^4y^2 - 12x^2y^4 = 3x^2y^2(x^2 - 4y^2)$
$$= 3x^2y^2(x + 2y)(x - 2y)$$

82. $x(x - 12) = 0$
$$x = 0 \quad \text{or} \quad x - 12 = 0$$
$$x = 12$$
The solution set is $\{0, 12\}$.

83. $3(x - 7)(4x + 9) = 0$
$$x - 7 = 0 \quad \text{or} \quad 4x + 9 = 0$$
$$x = 7 \qquad\qquad 4x = -9$$
$$x = -\frac{9}{4}$$
The solution set is $\left\{-\frac{9}{4}, 7\right\}$.

84. $x^2 + 5x - 14 = 0$
$$(x + 7)(x - 2) = 0$$
$$x + 7 = 0 \quad \text{or} \quad x - 2 = 0$$
$$x = -7 \qquad\qquad x = 2$$
The solution set is $\{-7, 2\}$.

85. $5x^2 + 20x = 0$
$$5x(x + 4) = 0$$
$$5x = 0 \quad \text{or} \quad x + 4 = 0$$
$$x = 0 \qquad\qquad x = -4$$
The solution set is $\{-4, 0\}$.

86. $2x^2 + 15x = 8$
$$2x^2 + 15x - 8 = 0$$
$$(2x - 1)(x + 8) = 0$$
$$2x - 1 = 0 \quad \text{or} \quad x + 8 = 0$$
$$2x = 1 \qquad\qquad x = -8$$
$$x = \frac{1}{2}$$
The solution set is $\left\{-8, \frac{1}{2}\right\}$.

87. $x(x - 4) = 32$
$$x^2 - 4x = 32$$
$$x^2 - 4x - 32 = 0$$
$$(x + 4)(x - 8) = 0$$
$$x + 4 = 0 \quad \text{or} \quad x - 8 = 0$$
$$x = -4 \qquad\qquad x = 8$$
The solution set is $\{-4, 8\}$.

88. $(x+3)(x-2)=50$

$$x^2+x-6=50$$

$$x^2+x-56=0$$

$$(x+8)(x-7)=0$$

$x+8=0$ or $x-7=0$

$\quad x=-8 \qquad\quad x=7$

The solution set is $\{-8,7\}$.

89. $\qquad\qquad x^2=14x-49$

$$x^2-14x+49=0$$

$$(x-7)^2=0$$

$$x-7=0$$

$$x=7$$

The solution set is $\{7\}$.

90. $\qquad\qquad 9x^2=100$

$$9x^2-100=0$$

$$(3x+10)(3x-10)=0$$

$3x+10=0$ or $3x-10=0$

$\quad 3x=-10 \qquad\quad 3x=10$

$\quad x=-\dfrac{10}{3} \qquad\quad x=\dfrac{10}{3}$

The solution set is $\left\{-\dfrac{10}{3},\dfrac{10}{3}\right\}$.

91. $3x^2+21x+30=0$

$$3\left(x^2+7x+10\right)=0$$

$$3(x+5)(x+2)=0$$

$x+5=0$ or $x+2=0$

$\quad x=-5 \qquad\quad x=-2$

The solution set is $\{-5,-2\}$.

92. $\qquad\qquad 3x^2=22x-7$

$$3x^2-22x+7=0$$

$$(3x-1)(x-7)=0$$

$3x-1=0$ or $x-7=0$

$\quad 3x=1 \qquad\qquad x=7$

$\quad x=\dfrac{1}{3}$

The solution set is $\left\{\dfrac{1}{3},7\right\}$.

93. $h=-16t^2+16t+32$

Substitute 0 for h and solve for t.

$$0=-16t^2+16t+32$$

$$0=-16(t^2-t-2)$$

$$0=-16(t+1)(t-2)$$

$t+1=0$ or $t-2=0$

$\quad t=-1 \qquad\quad t=2$

Because time cannot be negative, reject the solution $t=-1$.

It will take you 2 seconds to hit the water.

94. Let x = the width of the sign.

Then $x+3$ = the length of the sign.

Use the formula for the area of a rectangle.

$$l\cdot w=A$$

$$(x+3)(x)=40$$

$$x^2+3x=40$$

$$x^2+3x-40=0$$

$$(x+8)(x-5)=0$$

$x+8=0$ or $x-5=0$

$\quad x=-8 \qquad\quad x=5$

A rectangle cannot have a negative length. Thus $x=5$, and $x+3=8$.

The length of the sign is 8 feet and the width is 5 feet.

95. Area of garden $=x(x-3)=88$

$$x(x-3)=88$$

$$x^2-3x=88$$

$$x^2-3x-88=0$$

$$(x-11)(x+8)=0$$

$x-11=0$ or $x+8=0$

$\quad x=11 \qquad\quad x=-8$

Because a length cannot be negative, reject $x=-8$. Each side of the square lot is 11 meters, that is, the dimensions of the square lot are 11 meters by 11 meters.

Chapter 6 Test

1. $x^2 - 9x + 18 = (x-3)(x-6)$

2. $x^2 - 14x + 49 = x^2 - 2(x \cdot 7) + 7^2$
$$= (x-7)^2$$

3. $15y^4 - 35y^3 + 10y^2 = 5y^2(3y^2 - 7y + 2)$
$$= 5y^2(3y-1)(y-2)$$

4. $x^3 + 2x^2 + 3x + 6 = (x^3 + 2x^2) + (3x+6)$
$$= x^2(x+2) + 3(x+2)$$
$$= (x+2)(x^2+3)$$

5. $x^2 - 9x = x(x-9)$

6. $x^3 + 6x^2 - 7x = x(x^2 + 6x - 7)$
$$= x(x+7)(x-1)$$

7. $14x^2 + 64x - 30 = 2(7x^2 + 32x - 15)$
$$= 2(7x-3)(x+5)$$

8. $25x^2 - 9 = (5x)^2 - 3^2$
$$= (5x+3)(5x-3)$$

9. $x^3 + 8 = x^3 + 2^3 = (x+2)(x^2 - 2x + 2^2)$
$$= (x+2)(x^2 - 2x + 4)$$

10. $x^2 - 4x - 21 = (x+3)(x-7)$

11. $x^2 + 4$ is prime.

12. $6y^3 + 9y^2 + 3y = 3y(2y^2 + 3y + 1)$
$$= 3y(2y+1)(y+1)$$

13. $4y^2 - 36 = 4(y^2 - 9) = 4(y+3)(y-3)$

14. $16x^2 + 48x + 36$
$$= 4(4x^2 + 12x + 9)$$
$$= 4\left[(2x)^2 + 2(2x \cdot 3) + 3^2\right]$$
$$= 4(2x+3)^2$$

15. $2x^4 - 32 = 2(x^4 - 16)$
$$= 2(x^2 + 4)(x^2 - 4)$$
$$= 2(x^2 + 4)(x+2)(x-2)$$

16. $36x^2 - 84x + 49 = (6x)^2 - 2(6x \cdot 7) + 7^2$
$$= (6x-7)^2$$

17. $7x^2 - 50x + 7 = (7x-1)(x-7)$

18. $x^3 + 2x^2 - 5x - 10$
$$= (x^3 + 2x^2) + (-5x - 10)$$
$$= x^2(x+2) - 5(x+2)$$
$$= (x+2)(x^2 - 5)$$

19. $-12y^3 + 12y^2 + 45y = -3y(4y^2 - 4y - 15)$
$$= -3y(2y+3)(2y-5)$$

20. $y^3 - 125 = y^3 - 5^3$
$$= (y-5)(y^2 + 5y + 5^2)$$
$$= (y-5)(y^2 + 5y + 25)$$

21. $5x^2 - 5xy - 30y^2 = 5(x^2 - xy - 6y^2)$
$$= 5(x-3y)(x+2y)$$

22. $x^2 + 2x - 24 = 0$
$(x+6)(x-4) = 0$
$x+6 = 0$ or $x - 4 = 0$
$\;x = -6 \quad x = 4$
The solution set is $\{-6, 4\}$.

23.
$$3x^2 - 5x = 2$$
$$3x^2 - 5x - 2 = 0$$
$$(3x+1)(x-2) = 0$$
$$3x+1 = 0 \quad \text{or} \quad x-2 = 0$$
$$3x = -1 \qquad\qquad x = 2$$
$$x = -\frac{1}{3}$$
The solution set is $\left\{-\frac{1}{3}, 2\right\}$.

24.
$$x(x-6) = 16$$
$$x^2 - 6x = 16$$
$$x^2 - 6x - 16 = 0$$
$$(x+2)(x-8) = 0$$
$$x+2 = 0 \quad \text{or} \quad x-8 = 0$$
$$x = -2 \qquad\qquad x = 8$$
The solution set is $\{-2, 8\}$.

25.
$$6x^2 = 21x$$
$$6x^2 - 21x = 0$$
$$3x(2x-7) = 0$$
$$3x = 0 \quad \text{or} \quad 2x-7 = 0$$
$$x = 0 \qquad\qquad 2x = 7$$
$$x = \frac{7}{2}$$
The solution set is $\left\{0, \frac{7}{2}\right\}$.

26.
$$16x^2 = 81$$
$$16x^2 - 81 = 0$$
$$(4x+9)(4x-9) = 0$$
$$4x+9 = 0 \quad \text{or} \quad 4x-9 = 0$$
$$4x = -9 \qquad\qquad 4x = 9$$
$$x = -\frac{9}{4} \qquad\qquad x = \frac{9}{4}$$
The solution set is $\left\{-\frac{9}{4}, \frac{9}{4}\right\}$.

27.
$$(5x+4)(x-1) = 2$$
$$5x^2 - x - 4 = 2$$
$$5x^2 - x - 6 = 0$$
$$(5x-6)(x+1) = 0$$
$$5x-6 = 0 \quad \text{or} \quad x+1 = 0$$
$$5x = 6 \qquad\qquad x = -1$$
$$x = \frac{6}{5}$$
The solution set is $\left\{-1, \frac{6}{5}\right\}$.

28. Area of large square = x^2

Area of each small (corner) square = $1^2 = 1$
Area of four corner squares = $4 \cdot 1 = 4$
Area of shaded region = $x^2 - 4$
$$= (x+2)(x-2)$$

29. $h = -16t^2 + 80t + 96$
Substitute 0 for h and solve for t.
$$0 = -16t^2 + 80t + 96$$
$$0 = -16(t^2 - 5t - 6)$$
$$0 = -16(t-6)(t+1)$$
$$t - 6 = 0 \quad \text{or} \quad t + 1 = 0$$
$$t = 6 \qquad\qquad t = -1$$
Since time cannot be negative, reject $t = -1$. The rocket will reach the ground after 6 seconds.

30. Let x = the width of the garden.
Then $x + 6$ = the length of the garden.
$$(x+6)(x) = 55$$
$$x^2 + 6x = 55$$
$$x^2 + 6x - 55 = 0$$
$$(x+11)(x-5) = 0$$
$$x+11 = 0 \quad \text{or} \quad x-5 = 0$$
$$x = -11 \qquad\qquad x = 5$$
Since the width cannot be negative, reject $x = -11$. Then $x = 5$ and $x + 6 = 11$, so the width is 5 feet and the length is 11 feet.

Cumulative Review Exercises (Chapters 1-6)

1. $6\left[5+2(3-8)-3\right]=6\left[5+2(-5)-3\right]$
$$=6\left[5-10-3\right]$$
$$=6(-8)=-48$$

2. $4(x-2)=2(x-4)+3x$
$$4x-8=2x-8+3x$$
$$4x-8=5x-8$$
$$-x=0$$
$$x=0$$
The solution set is $\{0\}$.

3. $\dfrac{x}{2}-1=\dfrac{x}{3}+1$
$$6\left(\dfrac{x}{2}-1\right)=6\left(\dfrac{x}{3}+1\right)$$
$$3x-6=2x+6$$
$$x=12$$
The solution set is $\{12\}$.

The solution is 12.

4. $5-5x>2(5-x)+1$
$$5-5x>10-2x+1$$
$$5-5x>11-2x$$
$$5-5x+2x>11-2x+2x$$
$$5-3x>11$$
$$5-3x-5>11-5$$
$$-3x>6$$
$$\dfrac{-3x}{-3}<\dfrac{6}{-3}$$
$$x<-2$$
$(-\infty,-2)$

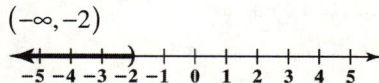

5. Let $x=$ the measure of each of the two base angles. Then $3x-10=$ the measure of the third angle.
$$x+x+(3x-10)=180$$
$$5x-10=180$$
$$5x=190$$
$$x=38$$
$$3x-10=104$$
The measures of the three angles of the triangle are 38°, 38°, and 104°.

6. Let $x=$ the cost of the dinner before tax.
$$x+0.06x=159$$
$$1.06x=159$$
$$\dfrac{1.06x}{1.06}=\dfrac{159}{1.06}$$
$$x\approx150$$
The cost of the dinner before tax was $150.

7. $y=-\dfrac{3}{5}x+3$

slope $=-\dfrac{3}{5}=\dfrac{-3}{5}$; *y*-intercept $=3$

Plot (0,3). From this point, move 3 units *down* (because −3 is negative) and 5 units to the *right* to reach the point (5,0). Draw a line through (0,3) and (5,0).

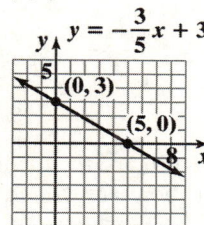

8. First, find slope $m=\dfrac{1-(-4)}{3-2}=\dfrac{5}{1}=5$.

Use the point (2, −4) in the point-slope equation.
$$y-y_1=m(x-x_1)$$
$$y-(-4)=5(x-2)$$
$$y+4=5(x-2)$$
Rewrite this equation in slope-intercept form.
$$y+4=5x-10$$
$$y=5x-14$$

9. $5x-6y>30$

Graph $5x-6y=30$ as a dashed line through (6,0) and (0, −5). Use (0,0) as a test point. Since $0-0>30$ is false, shade the half-plane *not* containing (0,0).

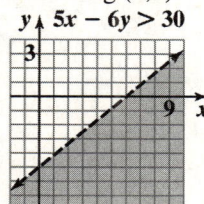

10. $5x + 2y = 13$

$y = 2x - 7$

The substitution method is a good choice for solving this system because the second equation is already solved for y.

Substitute $2x - 7$ for y in the first equation.

$$5x + 2y = 13$$
$$5x + 2(2x - 7) = 13$$
$$5x + 4x - 14 = 13$$
$$9x - 14 = 13$$
$$9x = 27$$
$$x = 3$$

Back-substitute into the second given equation.

$y = 2x - 7$

$y = 2(3) - 7 = -1$

The solution set is $\{(3,-1)\}$.

11. $2x + 3y = 5$

$3x - 2y = -4$

The addition method is a good choice for solving this system because both equations are written in the form $Ax + By = C$.

Multiply the first equation by 2 and the second equation by 3; then add the results.

$$\begin{array}{r} 4x + 6y = 10 \\ 9x - 6y = -12 \\ \hline 13x \quad\quad = -2 \end{array}$$

$$x = -\frac{2}{13}$$

Instead of back-substituting $-\dfrac{2}{13}$ and working with fractions, go back to the original system and eliminate x. Multiply the first equation by 3 and the second equation by -2; then add the results.

$$\begin{array}{r} 6x + 9y = 15 \\ -6x + 4y = 8 \\ \hline 13y = 23 \end{array}$$

$$y = \frac{23}{13}$$

The solution set is $\left\{\left(-\dfrac{2}{13}, \dfrac{23}{13}\right)\right\}$.

12. $\dfrac{4}{5} - \dfrac{9}{8} = \dfrac{4}{5} \cdot \dfrac{8}{8} - \dfrac{9}{8} \cdot \dfrac{5}{5}$

$= \dfrac{32}{40} - \dfrac{45}{40} = -\dfrac{13}{40}$

13. $\dfrac{6x^5 - 3x^4 + 9x^2 + 27x}{3x}$

$= \dfrac{6x^5}{3x} - \dfrac{3x^4}{3x} + \dfrac{9x^2}{3x} + \dfrac{27x}{3x}$

$= 2x^4 - x^3 + 3x + 9$

14. $(3x - 5y)(2x + 9y)$

$= 6x^2 + 27xy - 10xy - 45y^2$

$= 6x^2 + 17xy - 45y^2$

15.
$$\require{enclose}\begin{array}{r} 2x^2 + 5x - 3 \\ 3x-5 \enclose{longdiv}{6x^3 + 5x^2 - 34x + 13} \\ \underline{6x^3 - 10x^2} \\ 15x^2 - 34x \\ \underline{15x^2 - 25x} \\ -9x + 13 \\ \underline{-9x + 15} \\ -2 \end{array}$$

$\dfrac{6x^3 + 5x^2 - 34x + 13}{3x - 5} = 2x^2 + 5x - 3 + \dfrac{-2}{3x - 5}$

or $2x^2 + 5x - 3 - \dfrac{2}{3x - 5}$

16. $0.0071 = 7.1 \times 10^{-3}$

To write 0.0071 in scientific notation, move the decimal point 3 places to the right. Because the given number is between 0 and 1, the exponent will be negative.

17. Factor $3x^2 + 11x + 6$ by trial and error or by grouping. To Factor by grouping, find two integers whose product is $ac = 3 \cdot 6 = 18$ and whose sum is $b = 11$. These integers are 9 and 2.

$3x^2 + 11x + 6 = 3x^2 + 9x + 2x + 6$

$= 3x(x + 3) + 2(x + 3)$

$= (x + 3)(3x + 2)$

18. $y^5 - 16y = y(y^4 - 16)$

$= y(y^2 + 4)(y^2 - 4)$

$= y(y^2 + 4)(y + 2)(y - 2)$

19. $4x^2 + 12x + 9 = (2x)^2 + 2(2x \cdot 3) + 3x^2$

$\qquad\qquad\qquad = (2x+3)^2$

20. Let x = the width of the rectangle.

Then $x + 2$ = the length of the rectangle.

Use the formula for the area of a rectangle.

$$l \cdot w = A$$

$$(x+2)(x) = 24$$

$$x^2 + 2x = 24$$

$$x^2 + 2x - 24 = 0$$

$$(x+6)(x-4) = 0$$

$x+6=0$ $\qquad$ or $\qquad$ $x-4=0$

$\qquad x=-6$ $\qquad\qquad\qquad x=4$

Reject −6 because the width cannot be negative.

$x = 4$ and $x + 2 = 6$.

The dimensions of the rectangle are 6 feet by 4 feet.

Chapter 7
Rational Expressions

7.1 Check Points

1. a. $\dfrac{7x-28}{8x-40}$

Set the denominator equal to 0 and solve for x.

$8x - 40 = 0$

$\qquad 8x = 40$

$\qquad\ x = 5$

The rational expression is undefined for $x = 5$.

b. $\dfrac{8x-40}{x^2+3x-28}$

Set the denominator equal to 0 and solve for x.

$x^2 + 3x - 28 = 0$

$(x+7)(x-4) = 0$

$x + 7 = 0 \quad$ or $\quad x - 4 = 0$

$\quad x = -7 \qquad\qquad x = 4$

The rational expression is undefined for $x = -7$ and $x = 4$.

2. $\dfrac{7x+28}{21x} = \dfrac{7(x+4)}{7\cdot 3x} = \dfrac{x+4}{3x}$

3. $\dfrac{x^3-x^2}{7x-7} = \dfrac{x^2(x-1)}{7(x-1)} = \dfrac{x^2}{7}$

4. $\dfrac{x^2-1}{x^2+2x+1} = \dfrac{(x+1)(x-1)}{(x+1)^2} = \dfrac{\cancel{(x+1)}(x-1)}{\cancel{(x+1)}(x+1)} = \dfrac{x-1}{x+1}$

5. $\dfrac{9x^2-49}{28-12x} = \dfrac{(3x+7)(3x-7)}{4(7-3x)}$

$\qquad = \dfrac{(3x+7)\overset{-1}{\cancel{(3x-7)}}}{4\cancel{(7-3x)}}$

$\qquad = \dfrac{-(3x+7)}{4} \quad$ or $\quad -\dfrac{3x+7}{4} \quad$ or $\quad \dfrac{-3x-7}{4}$

7.1 Concept and Vocabulary Check

1. polynomials

2. 0

3. factoring; common factors

4. 1

5. -1

6. false

7. false

8. true

9. false

7.1 Exercise Set

1. $\dfrac{5}{2x}$

Set the denominator equal to 0 and solve for x.

$2x = 0$

$\ x = 0$

The rational expression is undefined for $x = 0$.

3. $\dfrac{x}{x-8}$

Set the denominator equal to 0 and solve for x.

$x - 8 = 0$

$\quad x = 8$

The rational expression is undefined for $x = 8$.

5. $\dfrac{13}{5x-20}$

Set the denominator equal to 0 and solve for x.

$5x - 20 = 0$

$\qquad 5x = 20$

$\qquad\ x = 4$

The rational expression is undefined for $x = 4$.

7. $\dfrac{x+3}{(x+9)(x-2)}$

Set the denominator equal to 0 and solve for x.

$(x+9)(x-2) = 0$

$x + 9 = 0 \quad$ or $\quad x - 2 = 0$

$\quad x = -9 \qquad\qquad x = 2$

The rational expression is undefined for $x = -9$ and $x = 2$.

9. $\dfrac{4x}{(3x-17)(x+3)}$

Set the denominator equal to 0 and solve for x.

$(3x-17)(x+3)=0$

$3x-17=0$　or　$x+3=0$

$\quad 3x=17 \qquad\qquad x=-3$

$\quad x=\dfrac{17}{3}$

The rational expression is undefined for $x=\dfrac{17}{3}$

and $x=-3$.

11. $\dfrac{x+5}{x^2+x-12}$

Set the denominator equal to 0 and solve for x.

$x^2+x-12=0$

$(x+4)(x-3)=0$

$x+4=0$　or　$x-3=0$

$\quad x=-4 \qquad\qquad x=3$

The rational expression is undefined for $x=-4$ and $x=3$.

13. $\dfrac{x+5}{5}$

Because the denominator, 5, is not zero for any value of x, the rational expression is defined for all real numbers.

15. $\dfrac{y+3}{4y^2+y-3}$

Set the denominator equal to 0 and solve for x.

$4y^2+y-3=0$

$(y+1)(4y-3)=0$

$y+1=0$　or　$4y-3=0$

$\quad y=-1 \qquad\qquad 4y=3$

$\qquad\qquad\qquad\qquad y=\dfrac{3}{4}$

The rational expression is undefined for $y=-1$ and

$y=\dfrac{3}{4}$.

17. $\dfrac{y+5}{y^2-25}$

Set the denominator equal to 0 and solve for x.

$y^2-25=0$

$(y+5)(y-5)=0$

$y+5=0$　or　$y-5=0$

$\quad y=-5 \qquad\qquad y=5$

The rational expression is undefined for $y=-5$ and $y=5$.

19. $\dfrac{5}{x^2+1}$

The smallest possible value of x^2 is 0, so $x^2+1\geq 1$ for all real numbers of x. This means that there is no real number x for which $x^2+1=0$. Thus, the rational expression is defined for all real numbers.

21. $\dfrac{14x^2}{7x}=\dfrac{2\cdot 7\cdot x\cdot x}{7\cdot x}=\dfrac{2x}{1}=2x$

23. $\dfrac{5x-15}{25}=\dfrac{5(x-3)}{5\cdot 5}=\dfrac{x-3}{5}$

25. $\dfrac{2x-8}{4x}=\dfrac{2(x-4)}{2\cdot 2x}=\dfrac{x-4}{2x}$

27. $\dfrac{3}{3x-9}=\dfrac{3}{3(x-3)}=\dfrac{1}{x-3}$

29. $\dfrac{-15}{3x-9}=\dfrac{-15}{3(x-3)}=\dfrac{-5}{x-3}$ or $-\dfrac{5}{x-3}$

31. $\dfrac{3x+9}{x+3}=\dfrac{3(x+3)}{x+3}=\dfrac{3}{1}=3$

33. $\dfrac{x+5}{x^2-25}=\dfrac{x+5}{(x+5)(x-5)}=\dfrac{1}{x-5}$

35. $\dfrac{2y-10}{3y-15}=\dfrac{2(y-5)}{3(y-5)}=\dfrac{2}{3}$

37. $\dfrac{x+1}{x^2-2x-3}=\dfrac{x+1}{(x+1)(x-3)}=\dfrac{1}{x-3}$

39. $\dfrac{4x-8}{x^2-4x+4}=\dfrac{4(x-2)}{(x-2)(x-2)}=\dfrac{4}{x-2}$

41. $\dfrac{y^2-3y+2}{y^2+7y-18}=\dfrac{(y-1)(y-2)}{(y+9)(y-2)}=\dfrac{y-1}{y+9}$

43. $\dfrac{2y^2-7y+3}{2y^2-5y+2}=\dfrac{(2y-1)(y-3)}{(2y-1)(y-2)}=\dfrac{y-3}{y-2}$

45. $\dfrac{2x+3}{2x+5}$

The numerator and denominator have no common factor (other than 1), so this rational expression cannot be simplified.

47. $\dfrac{x^2+12x+36}{x^2-36}=\dfrac{(x+6)(x+6)}{(x+6)(x-6)}=\dfrac{x+6}{x-6}$

49. $\dfrac{x^3-2x^2+x-2}{x-2}=\dfrac{x^2(x-2)+1(x-2)}{x-2}$

$=\dfrac{(x-2)(x^2+1)}{x-2}$

$=x^2+1$

51. $\dfrac{x^3-8}{x-2}=\dfrac{(x-2)(x^2+2x+4)}{x-2}$

$=x^2+2x+4$

53. $\dfrac{(x-4)^2}{x^2-16}=\dfrac{(x-4)(x-4)}{(x+4)(x-4)}=\dfrac{x-4}{x+4}$

55. $\dfrac{x}{x+1}$; The numerator and denominator have no common factor (other than 1), so this rational expression cannot be simplified.

57. $\dfrac{x+4}{x^2+16}$; The numerator and denominator are both prime polynomials. They have no common factor (other than 1), so this rational expression cannot be simplified.

59. $\dfrac{x-5}{5-x}=\dfrac{-1(5-x)}{5-x}=-1$

Notice that the numerator and denominator of the given rational expression are additive inverses.

61. The numerator and denominator of this rational expression are additive inverses, so $\dfrac{2x-3}{3-2x}=-1$.

63. $\dfrac{x-5}{x+5}$; The numerator and denominator have no common factor and they are not additive inverses, so this rational expression cannot be simplified.

65. $\dfrac{4x-6}{3-2x}=\dfrac{2(2x-3)}{3-2x}=\dfrac{-2(3-2x)}{3-2x}=-2$

67. $\dfrac{4-6x}{3x^2-2x}=\dfrac{2(2-3x)}{x(3x-2)}$

$=\dfrac{-2(3x-2)}{x(3x-2)}$

$=-\dfrac{2}{x}$

69. $\dfrac{x^2-1}{1-x}=\dfrac{(x+1)(x-1)}{1-x}$

$=\dfrac{(x+1)\cdot-1(1-x)}{1-x}$

$=-1(x+1)=-x-1$

71. $\dfrac{y^2-y-12}{4-y}=\dfrac{(y-4)(y+3)}{4-y}$

$=\dfrac{-1(4-y)(y+3)}{4-y}$

$=-1(y+3)=-y-3$

73. $\dfrac{x^2y-x^2}{x^3-x^3y}=\dfrac{x^2(y-1)}{x^3(1-y)}$

$=\dfrac{x^2\cdot-1(1-y)}{x^3(1-y)}$

$=-\dfrac{1}{x}$

75. $\dfrac{x^2+2xy-3y^2}{2x^2+5xy-3y^2}=\dfrac{(x-y)(x+3y)}{(2x-y)(x+3y)}$

$=\dfrac{x-y}{2x-y}$

77. $\dfrac{x^2-9x+18}{x^3-27}=\dfrac{(x-3)(x-6)}{(x-3)(x^2+3x+9)}$

$=\dfrac{x-6}{x^2+3x+9}$

79. $\dfrac{9-y^2}{y^2-3(2y-3)} = \dfrac{(3+y)(3-y)}{y^2-6y+9}$

$= \dfrac{(3+y)(3-y)}{(y-3)(y-3)} = \dfrac{(3+y)\cdot-1(y-3)}{(y-3)(y-3)}$

$= \dfrac{-1(3+y)}{y-3}$ or $\dfrac{3+y}{-1(y-3)} = \dfrac{3+y}{3-y}$

81. $\dfrac{xy+2y+3x+6}{x^2+5x+6} = \dfrac{y(x+2)+3(x+2)}{(x+2)(x+3)}$

$= \dfrac{(x+2)(y+3)}{(x+2)(x+3)} = \dfrac{y+3}{x+3}$

83. $\dfrac{8x^2+4x+2}{1-8x^3} = \dfrac{2(4x^2+2x+1)}{(1-2x)(1+2x+4x^2)}$

$= \dfrac{2}{1-2x}$

85. $\dfrac{130x}{100-x}$

a. $x = 40$:

$\dfrac{130x}{100-x} = \dfrac{130(40)}{100-40}$

$= \dfrac{5200}{60}$

≈ 86.67

This means it costs about $86.67 million to inoculate 40% of the population.

$x = 80$:

$\dfrac{130x}{100-x} = \dfrac{130(80)}{100-80}$

$= \dfrac{10,400}{20}$

$= 520$

This means it costs $520 million to inoculate 80% of the population.

$x = 90$:

$\dfrac{130x}{100-x} = \dfrac{130(90)}{100-90}$

$= \dfrac{11,700}{10}$

$= 1170$

This means it costs $1170 million ($1,170,000,000) to inoculate 90% of the population.

b. Set the denominator equal to 0 and solve for x.

$100 - x = 0$

$100 = x$

The rational expression is undefined for $x = 100$.

c. The cost keeps rising as x approaches 100. No amount of money will be enough to inoculate 100% of the population.

87. $D = 1000, A = 8$

$\dfrac{DA}{A+12} = \dfrac{1000 \cdot 8}{8+12}$

$= \dfrac{8000}{20} = 400$

The correct dosage for an 8-year old is 400 milligrams.

89. $C = \dfrac{100x+100,000}{x}$

a. $x = 500$

$C = \dfrac{100(500)+100,000}{500}$

$= \dfrac{150,000}{500} = 300$

The cost per bicycle when manufacturing 500 bicycles is $300.

b. $x = 4000$

$C = \dfrac{100(4000)+100,000}{4000}$

$= \dfrac{400,000+100,000}{4000}$

$= \dfrac{500,000}{4000} = 125$

The cost per bicycle when manufacturing 4000 bicycles is $125.

c. The cost per bicycle decreases as more bicycles are manufactured. One possible reason for this is that there could be fixed costs for equipment, so the more the equipment is used, the lower the cost per bicycle.

91. $y = \dfrac{5x}{x^2+1}; x = 3$

$y = \dfrac{5 \cdot 3}{3^2+1} = \dfrac{15}{10} = 1.5$

The equation indicates that the drug's concentration after 3 hours is 1.5 milligram per liter. This is represented on the graph by the point $(3, 1.5)$.

93. – 97. Answers will vary.

99. makes sense

101. does not make sense; Explanations will vary. Sample explanation: 1 makes the denominator equal to 0 and thus the expression is undefined at 1.

103. false; Changes to make the statement true will vary.

A sample change is: 3 is not a factor of $x^2 + 3$.

105. false; Changes to make the statement true will vary. A sample change is: x is not a factor of the numerator or the denominator.

107. Answers will vary. The denominator should be $x + 4$ or contain a factor of $x + 4$.

109. The graphs coincide.

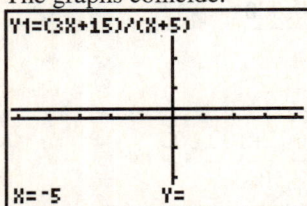

This verifies that the

simplification $\dfrac{3x+15}{x+5} = 3, x \neq -5$, is correct. Notice

the screen shows no y – value for $x = -5$.

111. The graphs do not coincide.

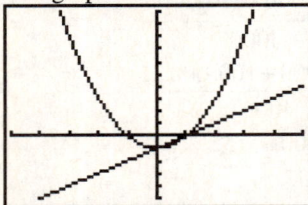

$$\frac{x^2 - x}{x} = \frac{x(x-1)}{x}$$
$$= x - 1, x \neq 0$$

Change the expression on the right from $x^2 - 1$ to $x - 1$.

113. $\dfrac{5}{6} \cdot \dfrac{9}{25} = \dfrac{\overset{1}{\cancel{5}}}{\underset{2}{\cancel{6}}} \cdot \dfrac{\overset{3}{\cancel{9}}}{\underset{5}{\cancel{25}}} = \dfrac{3}{10}$

114. $\dfrac{2}{3} \div 4 = \dfrac{2}{3} \cdot \dfrac{1}{4} = \dfrac{2}{12} = \dfrac{2 \cdot 1}{2 \cdot 6} = \dfrac{1}{6}$

115. $2x - 5y = -2$

$3x + 4y = 20$

Multiply the first equation by 3 and the second equation by -2; then add the results.

$\quad 6x - 15y = -6$

$\underline{-6x - \ 8y = -40}$

$\qquad - 23y = -46$

$\qquad \quad y = 2$

Back-substitute to find x.

$\quad 2x - 5y = -2$

$2x - 5(2) = -2$

$\ 2x - 10 = -2$

$\qquad 2x = 8$

$\qquad \ x = 4$

The solution set is $\{(4,2)\}$.

116. $\dfrac{2}{5} \cdot \dfrac{3}{7} = \dfrac{6}{35}$

117. $\dfrac{3}{4} \div \dfrac{1}{2} = \dfrac{3}{4} \cdot \dfrac{2}{1} = \dfrac{6}{4} = \dfrac{2 \cdot 3}{2 \cdot 2} = \dfrac{3}{2}$

118. $\dfrac{5}{4} \div \dfrac{15}{8} = \dfrac{5}{4} \cdot \dfrac{8}{15} = \dfrac{\overset{1}{\cancel{5}}}{\underset{1}{\cancel{4}}} \cdot \dfrac{\overset{2}{\cancel{8}}}{\underset{3}{\cancel{15}}} = \dfrac{2}{3}$

7.2 Check Points

1. $\dfrac{9}{x+4} \cdot \dfrac{x-5}{2} = \dfrac{9(x-5)}{(x+4)2} = \dfrac{9x-45}{2x+8}$

2. $\dfrac{x+4}{x-7} \cdot \dfrac{3x-21}{8x+32} = \dfrac{x+4}{x-7} \cdot \dfrac{3(x-7)}{8(x+4)}$

$\qquad = \dfrac{\overset{1}{\cancel{x+4}}}{\underset{1}{\cancel{x-7}}} \cdot \dfrac{3\,\overset{1}{\cancel{(x-7)}}}{8\,\underset{1}{\cancel{(x+4)}}}$

$\qquad = \dfrac{3}{8}$

3. $\dfrac{x-5}{x-2} \cdot \dfrac{x^2-4}{9x-45} = \dfrac{x-5}{x-2} \cdot \dfrac{(x+2)(x-2)}{9(x-5)}$

$$= \dfrac{\cancel{x-5}}{\cancel{x-2}} \cdot \dfrac{(x+2)\,\cancel{(x-2)}}{9\,\cancel{(x-5)}}$$

$$= \dfrac{x+2}{9}$$

4. $\dfrac{5x+5}{7x-7x^2} \cdot \dfrac{2x^2+x-3}{4x^2-9} = \dfrac{5(x+1)}{7x(1-x)} \cdot \dfrac{(2x+3)(x-1)}{(2x+3)(2x-3)}$

$$= \dfrac{5(x+1)}{7x(1-x)} \cdot \dfrac{\cancel{(2x+3)}\,\overset{-1}{\cancel{(x-1)}}}{\cancel{(2x+3)}(2x-3)}$$

$$= \dfrac{-5(x+1)}{7x(2x-3)} \ \text{ or } \ -\dfrac{5(x+1)}{7x(2x-3)}$$

5. $(x+3) \div \dfrac{x-4}{x+7} = \dfrac{x+3}{1} \cdot \dfrac{x+7}{x-4}$

$$= \dfrac{(x+3)(x+7)}{x-4}$$

6. $\dfrac{x^2+5x+6}{x^2-25} \div \dfrac{x+2}{x+5} = \dfrac{x^2+5x+6}{x^2-25} \cdot \dfrac{x+5}{x+2}$

$$= \dfrac{(x+3)(x+2)}{(x+5)(x-5)} \cdot \dfrac{x+5}{x+2}$$

$$= \dfrac{(x+3)\,\cancel{(x+2)}}{\cancel{(x+5)}(x-5)} \cdot \dfrac{\cancel{x+5}}{\cancel{x+2}}$$

$$= \dfrac{x+3}{x-5}$$

7. $\dfrac{y^2+3y+2}{y^2+1} \div \left(5y^2+10y\right) = \dfrac{y^2+3y+2}{y^2+1} \div \dfrac{5y^2+10y}{1}$

$$= \dfrac{y^2+3y+2}{y^2+1} \cdot \dfrac{1}{5y^2+10y}$$

$$= \dfrac{(y+2)(y+1)}{y^2+1} \cdot \dfrac{1}{5y(y+2)}$$

$$= \dfrac{\cancel{(y+2)}(y+1)}{y^2+1} \cdot \dfrac{1}{5y\,\cancel{(y+2)}}$$

$$= \dfrac{y+1}{5y(y^2+1)}$$

7.2 Concept and Vocabulary Check

1. numerators; denominators; $\dfrac{PR}{QS}$

2. multiplicative inverse/reciprocal

3. $\dfrac{x^2}{15}$

4. $\dfrac{3}{5}$

7.2 Exercise Set

1. $\dfrac{4}{x+3} \cdot \dfrac{x-5}{9} = \dfrac{4(x-5)}{(x+3)9} = \dfrac{4x-20}{9x+27}$

3. $\dfrac{x}{3} \cdot \dfrac{12}{x+5} = \dfrac{3 \cdot 4x}{3(x+5)} = \dfrac{4x}{x+5}$

5. $\dfrac{3}{x} \cdot \dfrac{4x}{15} = \dfrac{3 \cdot 4x}{3 \cdot 5x} = \dfrac{4}{5}$

7. $\dfrac{x-3}{x+5} \cdot \dfrac{4x+20}{9x-27} = \dfrac{x-3}{x+5} \cdot \dfrac{4(x+5)}{9(x-3)} = \dfrac{4}{9}$

9. $\dfrac{x^2+9x+14}{x+7} \cdot \dfrac{1}{x+2} = \dfrac{(x+7)(x+2)\cdot1}{(x+7)(x+2)} = 1$

11. $\dfrac{x^2-25}{x^2-3x-10} \cdot \dfrac{x+2}{x} = \dfrac{(x+5)(x-5)}{(x+2)(x-5)} \cdot \dfrac{(x+2)}{x}$

$= \dfrac{x+5}{x}$

13. $\dfrac{4y+30}{y^2-3y} \cdot \dfrac{y-3}{2y+15} = \dfrac{2(2y+15)}{y(y-3)} \cdot \dfrac{(y-3)}{(2y+15)}$

$= \dfrac{2}{y}$

15. $\dfrac{y^2-7y-30}{y^2-6y-40} \cdot \dfrac{2y^2+5y+2}{2y^2+7y+3}$

$= \dfrac{(y+3)(y-10)}{(y+4)(y-10)} \cdot \dfrac{(2y+1)(y+2)}{(2y+1)(y+3)}$

$= \dfrac{y+2}{y+4}$

17. $(y^2-9) \cdot \dfrac{4}{y-3} = \dfrac{y^2-9}{1} \cdot \dfrac{4}{y-3}$

$= \dfrac{(y+3)(y-3)}{1} \cdot \dfrac{4}{y-3}$

$= 4(y+3) \text{ or } 4y+12$

19. $\dfrac{x^2-5x+6}{x^2-2x-3} \cdot \dfrac{x^2-1}{x^2-4}$

$= \dfrac{(x-2)(x-3)}{(x+1)(x-3)} \cdot \dfrac{(x+1)(x-1)}{(x+2)(x-2)}$

$= \dfrac{x-1}{x+2}$

21. $\dfrac{x^3-8}{x^2-4} \cdot \dfrac{x+2}{3x}$

$= \dfrac{(x-2)(x^2+2x+4)}{(x+2)(x-2)} \cdot \dfrac{(x+2)}{3x}$

$= \dfrac{x^2+2x+4}{3x}$

23. $\dfrac{(x-2)^3}{(x-1)^3} \cdot \dfrac{x^2-2x+1}{x^2-4x+4}$

$= \dfrac{(x-2)^3}{(x-1)^3} \cdot \dfrac{(x-1)^2}{(x-2)^2}$

$= \dfrac{x-2}{x-1}$

25. $\dfrac{6x+2}{x^2-1} \cdot \dfrac{1-x}{3x^2+x}$

$= \dfrac{2(3x+1)}{(x+1)(x-1)} \cdot \dfrac{(1-x)}{x(3x+1)}$

$= \dfrac{2(3x+1)}{(x+1)(x-1)} \cdot \dfrac{-1(x-1)}{x(3x+1)}$

$= \dfrac{-2}{x(x+1)} \text{ or } -\dfrac{2}{x(x+1)}$

27. $\dfrac{25-y^2}{y^2-2y-35} \cdot \dfrac{y^2-8y-20}{y^2-3y-10}$

$= \dfrac{(5+y)(5-y)}{(y+5)(y-7)} \cdot \dfrac{(y-10)(y+2)}{(y-5)(y+2)}$

$= \dfrac{-(y-10)}{y-7} \text{ or } -\dfrac{y-10}{y-7}$

29. $\dfrac{x^2-y^2}{x}\cdot\dfrac{x^2+xy}{x+y}$

$=\dfrac{(x+y)(x-y)}{x}\cdot\dfrac{x(x+y)}{(x+y)}$

$=(x-y)(x+y)\quad\text{or}\quad x^2-y^2$

31. $\dfrac{x^2+2xy+y^2}{x^2-2xy+y^2}\cdot\dfrac{4x-4y}{3x+3y}$

$=\dfrac{(x+y)(x+y)}{(x-y)(x-y)}\cdot\dfrac{4(x-y)}{3(x+y)}$

$=\dfrac{4(x+y)}{3(x-y)}$

33. $\dfrac{x}{7}\div\dfrac{5}{3}=\dfrac{x}{7}\cdot\dfrac{3}{5}=\dfrac{3x}{35}$

35. $\dfrac{3}{x}\div\dfrac{12}{x}=\dfrac{3}{x}\cdot\dfrac{x}{12}=\dfrac{1}{4}$

37. $\dfrac{15}{x}\div\dfrac{3}{2x}=\dfrac{15}{x}\cdot\dfrac{2x}{3}=10$

39. $\dfrac{x+1}{3}\div\dfrac{3x+3}{7}=\dfrac{x+1}{3}\cdot\dfrac{7}{3x+3}$

$=\dfrac{x+1}{3}\cdot\dfrac{7}{3(x+1)}$

$=\dfrac{7}{9}$

41. $\dfrac{7}{x-5}\div\dfrac{28}{3x-15}=\dfrac{7}{x-5}\cdot\dfrac{3x-15}{28}$

$=\dfrac{7}{(x-5)}\cdot\dfrac{3(x-5)}{7\cdot4}$

$=\dfrac{3}{4}$

43. $\dfrac{x^2-4}{x}\div\dfrac{x+2}{x-2}=\dfrac{x^2-4}{x}\cdot\dfrac{x-2}{x+2}$

$=\dfrac{(x+2)(x-2)}{x}\cdot\dfrac{x-2}{x+2}$

$=\dfrac{(x-2)^2}{x}$

45. $\left(y^2-16\right)\div\dfrac{y^2+3y-4}{y^2+4}$

$=\dfrac{y^2-16}{1}\cdot\dfrac{y^2+4}{y^2+3y-4}$

$=\dfrac{(y+4)(y-4)}{1}\cdot\dfrac{y^2+4}{(y+4)(y-1)}$

$=\dfrac{(y-4)\left(y^2+4\right)}{y-1}$

47. $\dfrac{y^2-y}{15}\div\dfrac{y-1}{5}=\dfrac{y^2-y}{15}\cdot\dfrac{5}{y-1}$

$=\dfrac{y(y-1)}{15}\cdot\dfrac{5}{(y-1)}$

$=\dfrac{y}{3}$

49. $\dfrac{4x^2+10}{x-3}\div\dfrac{6x^2+15}{x^2-9}$

$=\dfrac{4x^2+10}{x-3}\cdot\dfrac{x^2-9}{6x^2+15}$

$=\dfrac{2\left(2x^2+5\right)}{(x-3)}\cdot\dfrac{(x+3)(x-3)}{3\left(2x^2+5\right)}$

$=\dfrac{2(x+3)}{3}\quad\text{or}\quad\dfrac{2x+6}{3}$

51. $\dfrac{x^2-25}{2x-2}\div\dfrac{x^2+10x+25}{x^2+4x-5}$

$=\dfrac{x^2-25}{2x-2}\cdot\dfrac{x^2+4x-5}{x^2+10x+25}$

$=\dfrac{(x+5)(x-5)}{2(x-1)}\cdot\dfrac{(x+5)(x-1)}{(x+5)(x+5)}$

$=\dfrac{x-5}{2}$

53. $\dfrac{y^3+y}{y^2-y} \div \dfrac{y^3-y^2}{y^2-2y+1}$

$= \dfrac{y^3+y}{y^2-y} \cdot \dfrac{y^2-2y+1}{y^3-y^2}$

$= \dfrac{y\left(y^2+1\right)}{y(y-1)} \cdot \dfrac{(y-1)(y-1)}{y^2(y-1)}$

$= \dfrac{y^2+1}{y^2}$

55. $\dfrac{y^2+5y+4}{y^2+12y+32} \div \dfrac{y^2-12y+35}{y^2+3y-40}$

$= \dfrac{y^2+5y+4}{y^2+12y+32} \cdot \dfrac{y^2+3y-40}{y^2-12y+35}$

$= \dfrac{(y+4)(y+1)}{(y+4)(y+8)} \cdot \dfrac{(y+8)(y-5)}{(y-7)(y-5)}$

$= \dfrac{y+1}{y-7}$

57. $\dfrac{2y^2-128}{y^2+16y+64} \div \dfrac{y^2-6y-16}{3y^2+30y+48}$

$= \dfrac{2y^2-128}{y^2+16y+64} \cdot \dfrac{3y^2+30y+48}{y^2-6y-16}$

$= \dfrac{2\left(y^2-64\right)}{(y+8)(y+8)} \cdot \dfrac{3\left(y^2+10y+16\right)}{(y+2)(y-8)}$

$= \dfrac{2(y+8)(y-8)}{(y+8)(y+8)} \cdot \dfrac{3(y+2)(y+8)}{(y+2)(y-8)}$

$= 6$

59. $\dfrac{2x+2y}{3} \div \dfrac{x^2-y^2}{x-y}$

$= \dfrac{2x+2y}{3} \cdot \dfrac{x-y}{x^2-y^2}$

$= \dfrac{2(x+y)}{3} \cdot \dfrac{x-y}{(x+y)(x-y)}$

$= \dfrac{2}{3}$

61. $\dfrac{x^2-y^2}{8x^2-16xy+8y^2} \div \dfrac{4x-4y}{x+y}$

$= \dfrac{x^2-y^2}{8x^2-16xy+8y^2} \cdot \dfrac{x+y}{4x-4y}$

$= \dfrac{(x+y)(x-y)}{8\left(x^2-2xy+y^2\right)} \cdot \dfrac{x+y}{4(x-y)}$

$= \dfrac{(x+y)(x-y)}{8(x-y)(x-y)} \cdot \dfrac{x+y}{4(x-y)}$

$= \dfrac{(x+y)^2}{32(x-y)^2}$

63. $\dfrac{xy-y^2}{x^2+2x+1} \div \dfrac{2x^2+xy-3y^2}{2x^2+5xy+3y^2}$

$= \dfrac{xy-y^2}{x^2+2x+1} \cdot \dfrac{2x^2+5xy+3y^2}{2x^2+xy-3y^2}$

$= \dfrac{y(x-y)}{(x+1)(x+1)} \cdot \dfrac{(2x+3y)(x+y)}{(2x+3y)(x-y)}$

$= \dfrac{y(x+y)}{(x+1)^2}$

65. $\left(\dfrac{y-2}{y^2-9y+18}\cdot\dfrac{y^2-4y-12}{y+2}\right)\div\dfrac{y^2-4}{y^2+5y+6}=\left(\dfrac{y-2}{y^2-9y+18}\cdot\dfrac{y^2-4y-12}{y+2}\right)\cdot\dfrac{y^2+5y+6}{y^2-4}$

$=\left(\dfrac{y-2}{(y-6)(y-3)}\cdot\dfrac{(y-6)(y+2)}{y+2}\right)\cdot\dfrac{(y+2)(y+3)}{(y+2)(y-2)}=\left(\dfrac{y-2}{y-3}\right)\cdot\dfrac{y+3}{y-2}=\dfrac{y+3}{y-3}$

67. $\dfrac{3x^2+3x-60}{2x-8}\div\left(\dfrac{30x^2}{x^2-7x+10}\cdot\dfrac{x^3+3x^2-10x}{25x^3}\right)$

$=\dfrac{3x^2+3x-60}{2x-8}\div\left(\dfrac{30x^2}{(x-2)(x-5)}\cdot\dfrac{x\left(x^2+3x-10\right)}{25x^3}\right)$

$=\dfrac{3\left(x^2+x-20\right)}{2x-8}\div\left(\dfrac{30x^2}{(x-2)(x-5)}\cdot\dfrac{x(x+5)(x-2)}{25x^3}\right)$

$=\dfrac{3(x+5)(x-4)}{2(x-4)}\div\dfrac{6(x+5)}{5(x-5)}=\dfrac{3(x+5)(x-4)}{2(x-4)}\cdot\dfrac{5(x-5)}{6(x+5)}=\dfrac{5(x-5)}{4}$

69. $\dfrac{x^2+xz+xy+yz}{x-y}\div\dfrac{x+z}{x+y}=\dfrac{x(x+z)+y(x+z)}{x-y}\cdot\dfrac{x+y}{x+z}=\dfrac{(x+z)(x+y)}{x-y}\cdot\dfrac{x+y}{x+z}=\dfrac{(x+y)^2}{x-y}$

71. $\dfrac{3xy+ay+3xb+ab}{9x^2-a^2}\div\dfrac{y^3+b^3}{6x-2a}=\dfrac{3xy+ay+3xb+ab}{9x^2-a^2}\cdot\dfrac{6x-2a}{y^3+b^3}$

$=\dfrac{y(3x+a)+b(3x+a)}{(3x+a)(3x-a)}\cdot\dfrac{2(3x-a)}{(y+b)\left(y^2-by+b^2\right)}$

$=\dfrac{(3x+a)(y+b)}{(3x+a)(3x-a)}\cdot\dfrac{2(3x-a)}{(y+b)\left(y^2-by+b^2\right)}$

$=\dfrac{2}{y^2-by+b^2}$

73. $A=l\cdot w$

$A=\dfrac{x+1}{x^2+2x}\cdot\dfrac{x^2-4}{x^2-1}$

$=\dfrac{x+1}{x(x+2)}\cdot\dfrac{(x+2)(x-2)}{(x+1)(x-1)}$

$=\dfrac{\cancel{x+1}}{x\,\cancel{(x+2)}}\cdot\dfrac{\cancel{(x+2)}\,(x-2)}{\cancel{(x+1)}\,(x-1)}$

$=\dfrac{x-2}{x(x-1)}\text{in.}^2$

75. $A = \frac{1}{2} \cdot b \cdot h$

$A = \frac{1}{2} \cdot \frac{18x}{x^2 + 3x + 2} \cdot \frac{x+1}{3}$

$= \frac{1}{2} \cdot \frac{2 \cdot 3 \cdot 3x}{(x+2)(x+1)} \cdot \frac{x+1}{3}$

$= \frac{1}{\cancel{2}} \cdot \frac{\cancel{2} \cdot \cancel{3} \cdot 3x}{(x+2)\cancel{(x+1)}} \cdot \frac{\cancel{x+1}}{\cancel{3}}$

$= \frac{3x}{x+2}$ in.^{2}

77. – 79. Answers will vary.

81. makes sense

83. makes sense

85. true

87. false; Changes to make the statement true will vary. A sample change is: The quotient of two rational expressions can be found by inverting the divisor and then multiplying .

89. $-\frac{1}{2x-3} \div \frac{?}{?} = \frac{1}{3}$

The numerator of the unknown rational expression must contain a factor of -3 . The denominator of the unknown rational expression must contain a factor of $(2x-3)$. Therefore, the simplest pair of polynomials that will work are -3 in the numerator and $2x-3$ in the denominator, to give the rational expression $\frac{-3}{2x-3}$.

Check:

$-\frac{1}{2x-3} \div \frac{-3}{2x-3} = -\frac{1}{2x-3} \cdot \frac{2x-3}{-3} = \frac{1}{3}$

91. The graph coincides.

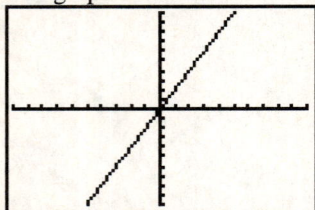

This verifies that $\frac{x^3+x}{3x} \cdot \frac{6x}{x+1} = 2x$.

93. The graphs do not coincide.

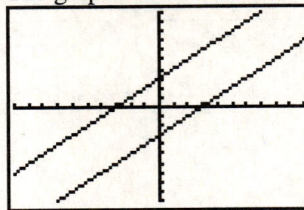

$\frac{x^2-9}{x+4} \div \frac{x-3}{x+4} = \frac{x^2-9}{x+4} \cdot \frac{x+4}{x-3}$

$= \frac{(x+3)(x-3)}{(x+4)} \cdot \frac{(x+4)}{(x-3)}$

$= x+3$

Change the expression on the right from $(x-3)$ to $(x+3)$.

95. $2x+3 < 3(x-5)$

$2x+3 < 3x-15$

$-x+3 < -15$

$-x < -18$

$x > 18$

$(18, \infty)$

96. $3x^2 - 15x - 42 = 3(x^2 - 5x - 14)$

$= 3(x-7)(x+2)$

97. $x(2x+9) = 5$

$2x^2 + 9x = 5$

$2x^2 + 9x - 5 = 0$

$(2x-1)(x+5) = 0$

$2x-1 = 0$ or $x+5 = 0$

$2x = 1 \qquad\qquad x = -5$

$x = \frac{1}{2}$

The solution set is $\left\{-5, \frac{1}{2}\right\}$.

98. $\frac{7}{9} - \frac{1}{9} = \frac{6}{9} = \frac{2}{3}$

99. $\frac{2x}{3} + \frac{x}{3} = \frac{3x}{3} = x$

100. $\frac{x^2-6x+9}{x^2-9} = \frac{(x-3)^2}{(x+3)(x-3)} = \frac{(x-3)(x-3)}{(x+3)(x-3)} = \frac{x-3}{x+3}$

7.3 Check Points

1. $\dfrac{3x-2}{5}+\dfrac{2x+12}{5}=\dfrac{3x-2+2x+12}{5}$

$\qquad\qquad\qquad\quad =\dfrac{5x+10}{5}$

$\qquad\qquad\qquad\quad =\dfrac{\overset{1}{\cancel{5}}(x+2)}{\underset{1}{\cancel{5}}}$

$\qquad\qquad\qquad\quad =x+2$

2. $\dfrac{x^2}{x^2-25}+\dfrac{25-10x}{x^2-25}=\dfrac{x^2-10x+25}{x^2-25}$

$\qquad\qquad\qquad\qquad =\dfrac{(x-5)^2}{(x+5)(x-5)}$

$\qquad\qquad\qquad\qquad =\dfrac{(x-5)\,\cancel{(x-5)}}{(x+5)\,\cancel{(x-5)}}$

$\qquad\qquad\qquad\qquad =\dfrac{x-5}{x+5}$

3. a. $\dfrac{4x+5}{x+7}-\dfrac{x}{x+7}=\dfrac{4x+5-x}{x+7}=\dfrac{3x+5}{x+7}$

b. $\dfrac{3x^2+4x}{x-1}-\dfrac{11x-4}{x-1}=\dfrac{3x^2+4x-(11x-4)}{x-1}$

$\qquad\qquad\qquad\qquad\quad =\dfrac{3x^2+4x-11x+4}{x-1}$

$\qquad\qquad\qquad\qquad\quad =\dfrac{3x^2-7x+4}{x-1}$

$\qquad\qquad\qquad\qquad\quad =\dfrac{(3x-4)(x-1)}{x-1}$

$\qquad\qquad\qquad\qquad\quad =3x-4$

4. $\dfrac{y^2+3y-6}{y^2-5y+4}-\dfrac{4y-4-2y^2}{y^2-5y+4}$

$\quad =\dfrac{y^2+3y-6-(4y-4-2y^2)}{y^2-5y+4}$

$\quad =\dfrac{y^2+3y-6-4y+4+2y^2}{y^2-5y+4}$

$\quad =\dfrac{3y^2-y-2}{y^2-5y+4}$

$\quad =\dfrac{(3y+2)(y-1)}{(y-4)(y-1)}$

$\quad =\dfrac{3y+2}{y-4}$

5. $\dfrac{x^2}{x-7}+\dfrac{4x+21}{7-x}=\dfrac{x^2}{x-7}+\dfrac{(-1)}{(-1)}\cdot\dfrac{4x+21}{7-x}$

$\qquad\qquad\qquad\quad =\dfrac{x^2}{x-7}+\dfrac{-4x-21}{x-7}$

$\qquad\qquad\qquad\quad =\dfrac{x^2-4x-21}{x-7}$

$\qquad\qquad\qquad\quad =\dfrac{(x+3)(x-7)}{x-7}$

$\qquad\qquad\qquad\quad =x+3$

6. $\dfrac{7x-x^2}{x^2-2x-9}-\dfrac{5x-3x^2}{9+2x-x^2}$

$\quad =\dfrac{7x-x^2}{x^2-2x-9}-\dfrac{3x^2-5x}{x^2-2x-9}$

$\quad =\dfrac{7x-x^2-(3x^2-5x)}{x^2-2x-9}$

$\quad =\dfrac{7x-x^2-3x^2+5x}{x^2-2x-9}$

$\quad =\dfrac{-4x^2+12x}{x^2-2x-9}$

7.3 Concept and Vocabulary Check

1. $\dfrac{P+Q}{R}$; numerators; common denominator

2. $\dfrac{P-Q}{R}$; numerations; common denominator

3. $\dfrac{-1}{-1}$

4. $\dfrac{x+5}{3}$

5. $\dfrac{x-5}{3}$

6. $\dfrac{x-5+y}{3}$

7.3 Exercise Set

1. $\dfrac{7x}{13} + \dfrac{2x}{13} = \dfrac{9x}{13}$

3. $\dfrac{8x}{15} + \dfrac{x}{15} = \dfrac{9x}{15} = \dfrac{3x}{5}$

5. $\dfrac{x-3}{12} + \dfrac{5x+21}{12} = \dfrac{6x+18}{12}$

$\qquad = \dfrac{6(x+3)}{12}$

$\qquad = \dfrac{x+3}{2}$

7. $\dfrac{4}{x} + \dfrac{2}{x} = \dfrac{6}{x}$

9. $\dfrac{8}{9x} + \dfrac{13}{9x} = \dfrac{21}{9x} = \dfrac{7}{3x}$

11. $\dfrac{5}{x+3} + \dfrac{4}{x+3} = \dfrac{9}{x+3}$

13. $\dfrac{x}{x-3} + \dfrac{4x+5}{x-3} = \dfrac{5x+5}{x-3}$

15. $\dfrac{4x+1}{6x+5} + \dfrac{8x+9}{6x+5} = \dfrac{12x+10}{6x+5}$

$\qquad = \dfrac{2(6x+5)}{6x+5} = 2$

17. $\dfrac{y^2+7y}{y^2-5y} + \dfrac{y^2-4y}{y^2-5y} = \dfrac{y^2+7y+y^2-4y}{y^2-5y}$

$\qquad = \dfrac{2y^2+3y}{y^2-5y}$

$\qquad = \dfrac{y(2y+3)}{y(y-5)}$

$\qquad = \dfrac{2y+3}{y-5}$

19. $\dfrac{4y-1}{5y^2} + \dfrac{3y+1}{5y^2} = \dfrac{4y-1+3y+1}{5y^2}$

$\qquad = \dfrac{7y}{5y^2} = \dfrac{7}{5y}$

21. $\dfrac{x^2-2}{x^2+x-2} + \dfrac{2x-x^2}{x^2+x-2}$

$\qquad = \dfrac{x^2-2+2x-x^2}{x^2+x-2}$

$\qquad = \dfrac{2x-2}{x^2+x-2} = \dfrac{2(x-1)}{(x+2)(x-1)} = \dfrac{2}{x+2}$

23. $\dfrac{x^2-4x}{x^2-x-6} + \dfrac{4x-4}{x^2-x-6}$

$\qquad = \dfrac{x^2-4x+4x-4}{x^2-x-6}$

$\qquad = \dfrac{x^2-4}{x^2-x-6} = \dfrac{(x+2)(x-2)}{(x-3)(x+2)} = \dfrac{x-2}{x-3}$

25. $\dfrac{3x}{5x-4} - \dfrac{4}{5x-4} = \dfrac{3x-4}{5x-4}$

27. $\dfrac{4x}{4x-3} - \dfrac{3}{4x-3} = \dfrac{4x-3}{4x-3} = 1$

29. $\dfrac{14y}{7y+2} - \dfrac{7y-2}{7y+2} = \dfrac{14y-(7y-2)}{7y+2}$

$\qquad = \dfrac{14y-7y+2}{7y+2}$

$\qquad = \dfrac{7y+2}{7y+2} = 1$

31. $\dfrac{3x+1}{4x-2} - \dfrac{x+1}{4x-2} = \dfrac{(3x+1)-(x+1)}{4x-2}$

$\qquad = \dfrac{3x+1-x-1}{4x-2}$

$\qquad = \dfrac{2x}{4x-2}$

$\qquad = \dfrac{2x}{2(2x-1)}$

$\qquad = \dfrac{x}{2x-1}$

33. $\dfrac{3y^2-1}{3y^3} - \dfrac{6y^2-1}{3y^3}$

$\qquad = \dfrac{(3y^2-1)-(6y^2-1)}{3y^3}$

$\qquad = \dfrac{3y^2-1-6y^2+1}{3y^3} = \dfrac{-3y^2}{3y^3} = -\dfrac{1}{y}$

35. $\dfrac{4y^2+5}{9y^2-64}-\dfrac{y^2-y+29}{9y^2-64}$

$=\dfrac{\left(4y^2+5\right)-\left(y^2-y+29\right)}{9y^2-64}$

$=\dfrac{4y^2+5-y^2+y-29}{9y^2-64}$

$=\dfrac{3y^2+y-24}{9y^2-64}$

$=\dfrac{(3y-8)(y+3)}{(3y+8)(3y-8)}=\dfrac{y+3}{3y+8}$

37. $\dfrac{6y^2+y}{2y^2-9y+9}-\dfrac{2y+9}{2y^2-9y+9}-\dfrac{4y-3}{2y^2-9y+9}$

$=\dfrac{\left(6y^2+y\right)-(2y+9)-(4y-3)}{2y^2-9y+9}$

$=\dfrac{6y^2+y-2y-9-4y+3}{2y^2-9y+9}$

$=\dfrac{6y^2-5y-6}{2y^2-9y+9}$

$=\dfrac{(2y-3)(3y+2)}{(2y-3)(y-3)}$

$=\dfrac{3y+2}{y-3}$

39. $\dfrac{4}{x-3}+\dfrac{2}{3-x}=\dfrac{4}{x-3}+\dfrac{(-1)}{(-1)}\cdot\dfrac{2}{3-x}$

$=\dfrac{4}{x-3}+\dfrac{-2}{x-3}$

$=\dfrac{2}{x-3}$

41. $\dfrac{6x+7}{x-6}+\dfrac{3x}{6-x}=\dfrac{6x+7}{x-6}+\dfrac{(-1)}{(-1)}\cdot\dfrac{3x}{6-x}$

$=\dfrac{6x+7}{x-6}+\dfrac{-3x}{x-6}$

$=\dfrac{3x+7}{x-6}$

43. $\dfrac{5x-2}{3x-4}+\dfrac{2x-3}{4-3x}=\dfrac{5x-2}{3x-4}+\dfrac{(-1)}{(-1)}\cdot\dfrac{2x-3}{4-3x}$

$=\dfrac{5x-2}{3x-4}+\dfrac{-2x+3}{3x-4}$

$=\dfrac{5x-2-2x+3}{3x-4}$

$=\dfrac{3x+1}{3x-4}$

45. $\dfrac{x^2}{x-2}+\dfrac{4}{2-x}=\dfrac{x^2}{x-2}+\dfrac{(-1)}{(-1)}\cdot\dfrac{4}{2-x}$

$=\dfrac{x^2}{x-2}+\dfrac{-4}{x-2}$

$=\dfrac{x^2-4}{x-2}$

$=\dfrac{(x+2)(x-2)}{x-2}$

$=x+2$

47. $\dfrac{y-3}{y^2-25}+\dfrac{y-3}{25-y^2}$

$=\dfrac{y-3}{y^2-25}+\dfrac{(-1)}{(-1)}\cdot\dfrac{y-3}{25-y^2}$

$=\dfrac{y-3}{y^2-25}+\dfrac{-y+3}{y^2-25}$

$=\dfrac{y-3-y+3}{y^2-25}=\dfrac{0}{y^2-25}=0$

49. $\dfrac{6}{x-1}-\dfrac{5}{1-x}=\dfrac{6}{x-1}-\dfrac{(-1)}{(-1)}\cdot\dfrac{5}{1-x}$

$=\dfrac{6}{x-1}-\dfrac{-5}{x-1}$

$=\dfrac{6+5}{x-1}=\dfrac{11}{x-1}$

51. $\dfrac{10}{x+3}-\dfrac{2}{-x-3}=\dfrac{10}{x+3}-\dfrac{(-1)}{(-1)}\cdot\dfrac{2}{-x-3}$

$=\dfrac{10}{x+3}-\dfrac{-2}{x+3}$

$=\dfrac{10+2}{x+3}=\dfrac{12}{x+3}$

53. $\dfrac{y}{y-1} - \dfrac{1}{1-y} = \dfrac{y}{y-1} - \dfrac{(-1)}{(-1)} \cdot \dfrac{1}{1-y}$

$\qquad\qquad\qquad = \dfrac{y}{y-1} - \dfrac{-1}{y-1}$

$\qquad\qquad\qquad = \dfrac{y+1}{y-1}$

55. $\dfrac{3-x}{x-7} - \dfrac{2x-5}{7-x} = \dfrac{3-x}{x-7} - \dfrac{(-1)}{(-1)} \cdot \dfrac{2x-5}{7-x}$

$\qquad\qquad\qquad = \dfrac{3-x}{x-7} - \dfrac{-2x+5}{x-7}$

$\qquad\qquad\qquad = \dfrac{(3-x)-(-2x+5)}{x-7}$

$\qquad\qquad\qquad = \dfrac{3-x+2x-5}{x-7}$

$\qquad\qquad\qquad = \dfrac{x-2}{x-7}$

57. $\dfrac{x-2}{x^2-25} - \dfrac{x-2}{25-x^2}$

$\quad = \dfrac{x-2}{x^2-25} - \dfrac{(-1)}{(-1)} \cdot \dfrac{x-2}{25-x^2}$

$\quad = \dfrac{x-2}{x^2-25} - \dfrac{-x+2}{x^2-25}$

$\quad = \dfrac{(x-2)-(-x+2)}{x^2-25}$

$\quad = \dfrac{x-2+x-2}{x^2-25} = \dfrac{2x-4}{x^2-25}$

59. $\dfrac{x}{x-y} + \dfrac{y}{y-x} = \dfrac{x}{x-y} + \dfrac{(-1)}{(-1)} \cdot \dfrac{y}{y-x}$

$\qquad\qquad\qquad = \dfrac{x}{x-y} + \dfrac{-y}{x-y}$

$\qquad\qquad\qquad = \dfrac{x-y}{x-y} = 1$

61. $\dfrac{2x}{x^2-y^2} + \dfrac{2y}{y^2-x^2}$

$\quad = \dfrac{2x}{x^2-y^2} + \dfrac{(-1)}{(-1)} \cdot \dfrac{2y}{y^2-x^2}$

$\quad = \dfrac{2x}{x^2-y^2} + \dfrac{-2y}{x^2-y^2}$

$\quad = \dfrac{2x-2y}{x^2-y^2} = \dfrac{2(x-y)}{(x+y)(x-y)} = \dfrac{2}{x+y}$

63. $\dfrac{x^2-2}{x^2+6x-7} + \dfrac{19-4x}{7-6x-x^2}$

$\quad = \dfrac{x^2-2}{x^2+6x-7} + \dfrac{(-1)}{(-1)} \cdot \dfrac{19-4x}{7-6x-x^2}$

$\quad = \dfrac{x^2-2}{x^2+6x-7} + \dfrac{-19+4x}{-7+6x+x^2}$

$\quad = \dfrac{x^2-2}{x^2+6x-7} + \dfrac{-19+4x}{x^2+6x-7}$

$\quad = \dfrac{x^2-2-19+4x}{x^2+6x-7}$

$\quad = \dfrac{x^2+4x-21}{x^2+6x-7}$

$\quad = \dfrac{(x+7)(x-3)}{(x+7)(x-1)} = \dfrac{x-3}{x-1}$

65. $\dfrac{6b^2-10b}{16b^2-48b+27}+\dfrac{7b^2-20b}{16b^2-48b+27}-\dfrac{6b-3b^2}{16b^2-48b+27}$

$=\dfrac{6b^2-10b+7b^2-20b-6b+3b^2}{16b^2-48b+27}$

$=\dfrac{16b^2-36b}{16b^2-48b+27}=\dfrac{4b(4b-9)}{(4b-9)(4b-3)}$

$=\dfrac{4b}{4b-3}$

67. $\dfrac{2y}{y-5}-\left(\dfrac{2}{y-5}+\dfrac{y-2}{y-5}\right)$

$=\dfrac{2y}{y-5}-\left(\dfrac{2+y-2}{y-5}\right)=\dfrac{2y}{y-5}-\dfrac{y}{y-5}$

$=\dfrac{y}{y-5}$

69. $\dfrac{b}{ac+ad-bc-bd}-\dfrac{a}{ac+ad-bc-bd}$

$=\dfrac{b-a}{ac+ad-bc-bd}$

$=\dfrac{b-a}{a(c+d)-b(c+d)}$

$=\dfrac{b-a}{(c+d)(a-b)}=\dfrac{(-1)}{(-1)}\cdot\dfrac{b-a}{(c+d)(a-b)}$

$=\dfrac{a-b}{-(c+d)(a-b)}=\dfrac{-1}{c+d}$ or $-\dfrac{1}{c+d}$

71. $\dfrac{(y-3)(y+2)}{(y+1)(y-4)}-\dfrac{(y+2)(y+3)}{(y+1)(4-y)}-\dfrac{(y+5)(y-1)}{(y+1)(4-y)}=\dfrac{y^2-y-6}{(y+1)(y-4)}-\dfrac{y^2+5y+6}{(y+1)(4-y)}-\dfrac{y^2+4y-5}{(y+1)(4-y)}$

$=\dfrac{y^2-y-6}{(y+1)(y-4)}-\dfrac{(-1)}{(-1)}\cdot\dfrac{y^2+5y+6}{(y+1)(4-y)}-\dfrac{(-1)}{(-1)}\cdot\dfrac{y^2+4y-5}{(y+1)(4-y)}$

$=\dfrac{y^2-y-6}{(y+1)(y-4)}+\dfrac{y^2+5y+6}{(y+1)(y-4)}+\dfrac{y^2+4y-5}{(y+1)(y-4)}=\dfrac{y^2-y-6+y^2+5y+6+y^2+4y-5}{(y+1)(y-4)}$

$=\dfrac{3y^2+8y-5}{(y+1)(y-4)}$

73. a. $\dfrac{L+60W}{L}-\dfrac{L-40W}{L}$

$=\dfrac{(L+60W)-(L-40W)}{L}$

$=\dfrac{L+60W-L+40W}{L}$

$=\dfrac{100W}{L}$

b. $\dfrac{100W}{L}$; $W = 5, L = 6$

$$\dfrac{100W}{L} = \dfrac{100 \cdot 5}{6} \approx 83.3$$

Since this value is over 80, the skull is round.

75. $P = 2L + 2W$

$$= 2\left(\dfrac{5x+10}{x+3}\right) + 2\left(\dfrac{5}{x+3}\right)$$

$$= \dfrac{10x+20}{x+3} + \dfrac{10}{x+3}$$

$$= \dfrac{10x+30}{x+3} = \dfrac{10(x+3)}{x+3} = 10$$

The perimeter is 10 meters.

77. – 79. Answers will vary.

81. does not make sense; Explanations will vary.

Sample explanation: $\dfrac{3x+1}{4} + \dfrac{x+2}{4} = \dfrac{4x+3}{4}$ and

$4x+3$ is not divisible by 4.

83. makes sense

85. false; Changes to make the statement true will vary. A sample change is: You do not add the common denominators. You just keep the common denominator in your answer.

87. false; Changes to make the statement true will vary. A sample change is: Some such rational expressions cannot be simplified.

89. $\left(\dfrac{3x-1}{x^2+5x-6} - \dfrac{2x-7}{x^2+5x-6}\right) \div \dfrac{x+2}{x^2-1}$

$$= \left(\dfrac{(3x-1)-(2x-7)}{x^2+5x-6}\right) \div \dfrac{x+2}{x^2-1}$$

$$= \dfrac{3x-1-2x+7}{x^2+5x-6} \div \dfrac{x+2}{x^2-1}$$

$$= \dfrac{x+6}{x^2+5x-6} \div \dfrac{x+2}{x^2-1}$$

$$= \dfrac{x+6}{x^2+5x-6} \cdot \dfrac{x^2-1}{x+2}$$

$$= \dfrac{(x+6)}{(x+6)(x-1)} \cdot \dfrac{(x+1)(x-1)}{(x+2)}$$

$$= \dfrac{x+1}{x+2}$$

91. $\dfrac{2x}{x+3} + \dfrac{?}{x+3} = \dfrac{4x+1}{x+3}$

The sum of numerators on the left side must be $4x+1$, so the missing expression is $2x+1$.

Check:

$$\dfrac{2x}{x+3} + \dfrac{2x+1}{x+3} = \dfrac{2x+2x+1}{x+3}$$

$$= \dfrac{4x+1}{x+3}$$

93. $\dfrac{6}{x-2} + \dfrac{?}{2-x} = \dfrac{13}{x-2}$

$$\dfrac{6}{x-2} + \dfrac{(-1)}{(-1)} \cdot \dfrac{?}{2-x} = \dfrac{13}{x-2}$$

$$\dfrac{6}{x-2} + \dfrac{(-1)?}{x-2} = \dfrac{13}{x-2}$$

Since $6 + 7 = 13$, the opposite of the missing expression must be 7, so the missing expression is -7.

Check:

$$\dfrac{6}{x-2} + \dfrac{-7}{2-x} = \dfrac{6}{x-2} + \dfrac{(-1) \cdot -7}{(-1)(2-x)}$$

$$= \dfrac{6}{x-2} + \dfrac{7}{x-2} = \dfrac{13}{x-2}$$

95. $\dfrac{3x}{x-5} + \dfrac{?}{5-x} = \dfrac{7x+1}{x-5}$

$$\dfrac{3x}{x-5} + \dfrac{(-1)}{(-1)} \cdot \dfrac{?}{5-x} = \dfrac{7x+1}{x-5}$$

$$\dfrac{3x}{x-5} + \dfrac{(-1)?}{x-5} = \dfrac{7x+1}{x-5}$$

Since $3x + (4x+1) = 7x+1$, the opposite of the missing expression must be $4x+1$, so the missing expression is $-4x-1$.

Check:

$$\dfrac{3x}{x-5} + \dfrac{-4x-1}{5-x} = \dfrac{3x}{x-5} + \dfrac{4x+1}{x-5}$$

$$= \dfrac{7x+1}{x-5}$$

97. The graphs do not coincide.

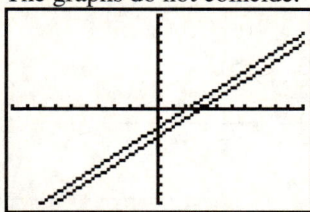

$$\frac{x^2+4x+3}{x+2}-\frac{5x+9}{x+2}$$

$$=\frac{\left(x^2+4x+3\right)-\left(5x+9\right)}{x+2}$$

$$=\frac{x^2+4x+3-5x-9}{x+2}$$

$$=\frac{x^2-x-6}{x+2}$$

$$=\frac{\left(x+2\right)\left(x-3\right)}{x+2}$$

$$=x-3$$

Change $x-2$ to $x-3$.

99. $\dfrac{13}{15}-\dfrac{8}{45}=\dfrac{13}{15}\cdot\dfrac{3}{3}-\dfrac{8}{45}$

$$=\frac{39}{45}-\frac{8}{45}=\frac{31}{45}$$

100. $81x^4-1=\left(9x^2+1\right)\left(9x^2-1\right)$

$$=\left(9x^2+1\right)\left(3x+1\right)\left(3x-1\right)$$

101.

$$
\begin{array}{r}
3x^2-7x-5 \\
x+3\overline{\smash{\big)}\,3x^3+2x^2-26x-15} \\
\underline{3x^3+9x^2} \\
-7x^2-26x \\
\underline{-7x^2-21x} \\
-5x-15 \\
\underline{-5x-15} \\
0
\end{array}
$$

$$\frac{3x^3+2x^2-26x-15}{x+3}=3x^2-7x-5$$

102. $\dfrac{1}{2}+\dfrac{2}{3}=\dfrac{3}{6}+\dfrac{4}{6}=\dfrac{7}{6}$

103. $\dfrac{1}{8}-\dfrac{5}{6}=\dfrac{3}{24}-\dfrac{20}{24}=-\dfrac{17}{24}$

104. $\dfrac{(y+2)y-2\cdot4}{4y(y+4)}=\dfrac{y^2+2y-8}{4y(y+4)}$

$$=\frac{(y+4)(y-2)}{4y(y+4)}$$

$$=\frac{y-2}{4y}$$

7.4 Check Points

1. $\dfrac{3}{10x^2}$ and $\dfrac{7}{15x}$

List the factors for each denominator.

$10x^2=2\cdot5x^2$

$15x=3\cdot5x$

LCD $=2\cdot3\cdot5\cdot x^2=30x^2$

2. $\dfrac{2}{x+3}$ and $\dfrac{4}{x-3}$

List the factors for each denominator.

$x+3=1(x+3)$

$x-3=1(x-3)$

LCD $=\left(x+3\right)\left(x-3\right)$

3. $\dfrac{9}{7x^2+28x}$ and $\dfrac{11}{x^2+8x+16}$

List the factors for each denominator.

$7x^2+28x=7x(x+4)$

$x^2+8x+16=(x+4)^2$

LCD $=7x(x+4)^2$

4. $\dfrac{3}{10x^2}+\dfrac{7}{15x}$

LCD $=30x^2$

$$\frac{3}{10x^2}+\frac{7}{15x}=\frac{3}{3}\cdot\frac{3}{10x^2}+\frac{2x}{2x}\cdot\frac{7}{15x}$$

$$=\frac{9}{30x^2}+\frac{14x}{30x^2}$$

$$=\frac{9+14x}{30x^2}$$

5. $\dfrac{2}{x+3} + \dfrac{4}{x-3}$

$\text{LCD} = (x+3)(x-3)$

$\dfrac{2}{x+3} + \dfrac{4}{x-3} = \dfrac{x-3}{x-3} \cdot \dfrac{2}{x+3} + \dfrac{x+3}{x+3} \cdot \dfrac{4}{x-3}$

$= \dfrac{2x-6}{(x+3)(x-3)} + \dfrac{4x+12}{(x+3)(x-3)}$

$= \dfrac{2x-6+4x+12}{(x+3)(x-3)}$

$= \dfrac{6x+6}{(x+3)(x-3)}$

6. $\dfrac{x}{x+5} - 1$

$\text{LCD} = x+5$

$\dfrac{x}{x+5} - 1 = \dfrac{x}{x+5} - \dfrac{x+5}{x+5}$

$= \dfrac{x-(x+5)}{x+5}$

$= \dfrac{x-x-5}{x+5}$

$= \dfrac{-5}{x+5} \text{ or } -\dfrac{5}{x+5}$

7. $\dfrac{5}{y^2-5y} - \dfrac{y}{5y-25} = \dfrac{5}{y(y-5)} - \dfrac{y}{5(y-5)}$

$\text{LCD} = 5y(y-5)$

$\dfrac{5}{y^2-5y} - \dfrac{y}{5y-25} = \dfrac{5}{y(y-5)} - \dfrac{y}{5(y-5)}$

$= \dfrac{5}{5} \cdot \dfrac{5}{y(y-5)} - \dfrac{y}{y} \cdot \dfrac{y}{5(y-5)}$

$= \dfrac{25}{5y(y-5)} - \dfrac{y^2}{5y(y-5)}$

$= \dfrac{25-y^2}{5y(y-5)}$

$= \dfrac{(5+y)(5-y)}{5y(y-5)}$

$= \dfrac{(5+y)\overset{-1}{\cancel{(5-y)}}}{5y\underset{1}{\cancel{(y-5)}}}$

$= -\dfrac{5+y}{5y}$

8. $\dfrac{4x}{x^2-25} + \dfrac{3}{5-x} = \dfrac{4x}{(x+5)(x-5)} + \dfrac{3}{5-x}$

$\text{LCD} = (x+5)(x-5)$

$\dfrac{4x}{x^2-25} + \dfrac{3}{5-x} = \dfrac{4x}{(x+5)(x-5)} + \dfrac{3}{5-x}$

$= \dfrac{4x}{(x+5)(x-5)} + \dfrac{-1(x+5)}{-1(x+5)} \cdot \dfrac{3}{5-x}$

$= \dfrac{4x}{(x+5)(x-5)} + \dfrac{-1(x+5)\cdot 3}{(x+5)(x-5)}$

$= \dfrac{4x}{(x+5)(x-5)} + \dfrac{-3x-15}{(x+5)(x-5)}$

$= \dfrac{4x-3x-15}{(x+5)(x-5)}$

$= \dfrac{x-15}{(x+5)(x-5)}$

7.4 Concept and Vocabulary Check

1. factor denominators

2. x and $x+3$; $x+3$ and $x+5$; $x(x+3)(x+5)$

3. 3

4. $x-5$

5. -1

7.4 Exercise Set

1. $\dfrac{7}{15x^2}$ and $\dfrac{13}{24x}$

$15x^2 = 3 \cdot 5x^2$

$24x = 2^3 \cdot 3x$

$\text{LCD} = 2^3 \cdot 3 \cdot 5x^2 = 120x^2$

3. $\dfrac{8}{15x^2}$ and $\dfrac{5}{6x^5}$

$15x^2 = 3 \cdot 5x^2$

$6x^5 = 2 \cdot 3x^5$

$\text{LCD} = 2 \cdot 3 \cdot 5 \cdot x^5 = 30x^5$

5. $\dfrac{4}{x-3}$ and $\dfrac{7}{x+1}$

LCD $= (x-3)(x+1)$

7. $\dfrac{5}{7(y+2)}$ and $\dfrac{10}{y}$

LCD $= 7y(y+2)$

9. $\dfrac{17}{x+4}$ and $\dfrac{18}{x^2-16}$

$x+4 = 1(x+4)$

$x^2 - 16 = (x+4)(x-4)$

LCD $= (x+4)(x-4)$

11. $\dfrac{8}{y^2-9}$ and $\dfrac{14}{y(y+3)}$

$y^2 - 9 = (y+3)(y-3)$

$y(y+3) = y(y+3)$

LCD $= y(y+3)(y-3)$

13. $\dfrac{7}{y^2-1}$ and $\dfrac{y}{y^2-2y+1}$

$y^2 - 1 = (y+1)(y-1)$

$y^2 - 2y + 1 = (y-1)(y-1)$

LCD $= (y+1)(y-1)(y-1)$

15. $\dfrac{3}{x^2-x-20}$ and $\dfrac{x}{2x^2+7x-4}$

$x^2 - x - 20 = (x-5)(x+4)$

$2x^2 + 7x - 4 = (2x-1)(x+4)$

LCD $= (x-5)(x+4)(2x-1)$

17. $\dfrac{3}{x}+\dfrac{5}{x^2}$

LCD $= x^2$

$\dfrac{3}{x}+\dfrac{5}{x^2} = \dfrac{3}{x}\cdot\dfrac{x}{x}+\dfrac{5}{x^2} = \dfrac{3x+5}{x^2}$

19. $\dfrac{2}{9x}+\dfrac{11}{6x}$

LCD $= 18x$

$\dfrac{2}{9x}+\dfrac{11}{6x} = \dfrac{2}{9x}\cdot\dfrac{2}{2}+\dfrac{11}{6x}\cdot\dfrac{3}{3}$

$= \dfrac{4}{18x}+\dfrac{33}{18x} = \dfrac{37}{18x}$

21. $\dfrac{4}{x}+\dfrac{7}{2x^2}$

LCD $= 2x^2$

$\dfrac{4}{x}+\dfrac{7}{2x^2} = \dfrac{4}{x}\cdot\dfrac{2x}{2x}+\dfrac{7}{2x^2} = \dfrac{8x}{2x^2}+\dfrac{7}{2x^2}$

$= \dfrac{8x+7}{2x^2}$

23. $6+\dfrac{1}{x}$

LCD $= x$

$6+\dfrac{1}{x} = \dfrac{6}{1}\cdot\dfrac{x}{x}+\dfrac{1}{x} = \dfrac{6x}{x}+\dfrac{1}{x} = \dfrac{6x+1}{x}$

25. $\dfrac{2}{x}+9$

LCD $= x$

$\dfrac{2}{x}+9 = \dfrac{2}{x}+\dfrac{9}{1}\cdot\dfrac{x}{x} = \dfrac{2}{x}+\dfrac{9x}{x} = \dfrac{2+9x}{x}$

27. $\dfrac{x-1}{6}+\dfrac{x+2}{3}$

LCD $= 6$

$\dfrac{x-1}{6}+\dfrac{x+2}{3} = \dfrac{x-1}{6}+\dfrac{(x+2)}{3}\cdot\dfrac{2}{2}$

$= \dfrac{x-1}{6}+\dfrac{2x+4}{6} = \dfrac{3x+3}{6} = \dfrac{3(x+1)}{6}$

$= \dfrac{x+1}{2}$

29. $\dfrac{4}{x} + \dfrac{3}{x-5}$

LCD $= x(x-5)$

$\dfrac{4}{x} + \dfrac{3}{x-5} = \dfrac{4(x-5)}{x(x-5)} + \dfrac{3}{x-5} \cdot \dfrac{x}{x}$

$\qquad = \dfrac{4(x-5)}{x(x-5)} + \dfrac{3x}{x(x-5)}$

$\qquad = \dfrac{4x-20+3x}{x(x-5)}$

$\qquad = \dfrac{7x-20}{x(x-5)}$

31. $\dfrac{2}{x-1} + \dfrac{3}{x+2}$

LCD $= (x-1)(x+2)$

$\dfrac{2}{x-1} + \dfrac{3}{x+2}$

$\quad = \dfrac{2(x+2)}{(x-1)(x+2)} + \dfrac{3(x-1)}{(x-1)(x+2)}$

$\quad = \dfrac{2x+4+3x-3}{(x-1)(x+2)}$

$\quad = \dfrac{5x+1}{(x-1)(x+2)}$

33. $\dfrac{2}{y+5} + \dfrac{3}{4y}$

LCD $= 4y(y+5)$

$\dfrac{2}{y+5} + \dfrac{3}{4y} = \dfrac{2(4y)}{(4y)(y+5)} + \dfrac{3(y+5)}{4y(y+5)}$

$\qquad = \dfrac{2(4y)+3(y+5)}{4y(y+5)}$

$\qquad = \dfrac{8y+3y+15}{4y(y+5)}$

$\qquad = \dfrac{11y+15}{4y(y+5)}$

35. $\dfrac{x}{x+7} - 1$

LCD $= x+7$

$\dfrac{x}{x+7} - 1 = \dfrac{x}{x+7} - \dfrac{x+7}{x+7}$

$\qquad = \dfrac{x-(x+7)}{x+7}$

$\qquad = \dfrac{x-x-7}{x+7}$

$\qquad = \dfrac{-7}{x+7} \text{ or } -\dfrac{7}{x+7}$

37. $\dfrac{7}{x+5} - \dfrac{4}{x-5}$

LCD $= (x+5)(x-5)$

$\dfrac{7}{x+5} - \dfrac{4}{x-5}$

$= \dfrac{7(x-5)}{(x+5)(x-5)} - \dfrac{4(x+5)}{(x+5)(x-5)}$

$= \dfrac{7(x-5)-4(x+5)}{(x+5)(x-5)}$

$= \dfrac{7x-35-4x-20}{(x+5)(x-5)}$

$= \dfrac{3x-55}{(x+5)(x-5)}$

39. $\dfrac{2x}{x^2-16} + \dfrac{x}{x-4}$

$x^2-16 = (x+4)(x-4)$

$x-4 = 1(x-4)$

LCD $= (x+4)(x-4)$

$\dfrac{2x}{x^2-16} + \dfrac{x}{x-4}$

$= \dfrac{2x}{(x+4)(x-4)} + \dfrac{x}{x-4}$

$= \dfrac{2x}{(x+4)(x-4)} + \dfrac{x(x+4)}{(x+4)(x-4)}$

$= \dfrac{2x+x(x+4)}{(x+4)(x-4)}$

$= \dfrac{2x+x^2+4x}{(x+4)(x-4)}$

$= \dfrac{x^2+6x}{(x+4)(x-4)}$

41. $\dfrac{5y}{y^2-9}-\dfrac{4}{y+3}$

LCD $=(y+3)(y-3)$

$\dfrac{5y}{y^2-9}-\dfrac{4}{y+3}$

$=\dfrac{5y}{(y+3)(y-3)}-\dfrac{4}{y+3}$

$=\dfrac{5y}{(y+3)(y-3)}-\dfrac{4(y-3)}{(y+3)(y-3)}$

$=\dfrac{5y-4(y-3)}{(y+3)(y-3)}$

$=\dfrac{5y-4y+12}{(y+3)(y-3)}$

$=\dfrac{y+12}{(y+3)(y-3)}$

43. $\dfrac{7}{x-1}-\dfrac{3}{(x-1)(x-1)}$

LCD $=(x-1)(x-1)$

$\dfrac{7}{x-1}-\dfrac{3}{(x-1)(x-1)}$

$=\dfrac{7(x-1)}{(x-1)(x-1)}-\dfrac{3}{(x-1)(x-1)}$

$=\dfrac{7x-7-3}{(x-1)(x-1)}$

$=\dfrac{7x-10}{(x-1)(x-1)}$ or $\dfrac{7x-10}{(x-1)^2}$

45. $\dfrac{3y}{4y-20}+\dfrac{9y}{6y-30}$

$4y-20=4(y-5)$

$6y-30=6(y-5)$

LCD $=12(y-5)$

$\dfrac{3y}{4y-20}+\dfrac{9y}{6y-30}$

$=\dfrac{4y}{4(y-5)}+\dfrac{9y}{6(y-5)}$

$=\dfrac{4y}{4(y-5)}\cdot\dfrac{3}{3}+\dfrac{9y}{6(y-5)}\cdot\dfrac{2}{2}$

$=\dfrac{12y}{12(y-5)}+\dfrac{18y}{12(y-5)}$

$=\dfrac{9y+18y}{12(y-5)}=\dfrac{27y}{12(y-5)}$

$=\dfrac{9y}{4(y-5)}$

47. $\dfrac{y+4}{y}-\dfrac{y}{y+4}$

LCD $=y(y+4)$

$\dfrac{y+4}{y}-\dfrac{y}{y+4}$

$=\dfrac{(y+4)(y+4)}{y(y+4)}-\dfrac{y\cdot y}{y(y+4)}$

$=\dfrac{y^2+8y+16-y^2}{y(y+4)}$

$=\dfrac{8y+16}{y(y+4)}$

49. $\dfrac{2x+9}{x^2-7x+12}-\dfrac{2}{x-3}$

$x^2-7x+12=(x-3)(x-4)$

$x-3=1(x-3)$

$\text{LCD}=(x-3)(x-4)$

$\dfrac{2x+9}{x^2-7x+12}-\dfrac{2}{x-3}$

$=\dfrac{2x+9}{(x-3)(x-4)}-\dfrac{2}{x-3}$

$=\dfrac{2x+9}{(x-3)(x-4)}-\dfrac{2(x-4)}{(x-3)(x-4)}$

$=\dfrac{2x+9-2(x-4)}{(x-3)(x-4)}$

$=\dfrac{2x+9-2x+8}{(x-3)(x-4)}$

$=\dfrac{17}{(x-3)(x-4)}$

51. $\dfrac{3}{x^2-1}+\dfrac{4}{(x+1)^2}$

$x^2-1=(x+1)(x-1)$

$(x+1)^2=(x+1)(x+1)$

$\text{LCD}=(x+1)(x+1)(x-1)$

$\dfrac{3}{x^2-1}+\dfrac{4}{(x+1)^2}$

$=\dfrac{3}{(x+1)(x-1)}+\dfrac{4}{(x+1)(x+1)}$

$=\dfrac{3(x+1)}{(x+1)(x+1)(x-1)}+\dfrac{4(x-1)}{(x+1)(x+1)(x-1)}$

$=\dfrac{3(x+1)+4(x-1)}{(x+1)(x+1)(x-1)}$

$=\dfrac{3x+3+4x-4}{(x+1)(x+1)(x-1)}$

$=\dfrac{7x-1}{(x+1)(x+1)(x-1)}$

53. $\dfrac{3x}{x^2+3x-10}-\dfrac{2x}{x^2+x-6}$

$x^2+3x-10=(x-2)(x+5)$

$x^2+x-6=(x+3)(x-2)$

$\text{LCD}=(x+3)(x-2)(x+5)$

$\dfrac{3x}{x^2+3x-10}-\dfrac{2x}{x^2+x-6}$

$=\dfrac{3x}{(x-2)(x+5)}-\dfrac{2x}{(x+3)(x-2)}$

$=\dfrac{3x(x+3)}{(x+3)(x-2)(x+5)}-\dfrac{2x(x+5)}{(x+3)(x-2)(x+5)}$

$=\dfrac{3x(x+3)-2x(x+5)}{(x+3)(x-2)(x+5)}$

$=\dfrac{3x^2+9x-2x^2-10x}{(x+3)(x-2)(x+5)}$

$=\dfrac{x^2-x}{(x+3)(x-2)(x+5)}$

55. $\dfrac{y}{y^2+2y+1}+\dfrac{4}{y^2+5y+4}$

$y^2+2y+1=(y+1)(y+1)$

$y^2+5y+4=(y+4)(y+1)$

$\text{LCD}=(y+4)(y+1)(y+1)$

$\dfrac{y}{y^2+2y+1}+\dfrac{4}{y^2+5y+4}$

$=\dfrac{y}{(y+1)(y+1)}+\dfrac{4}{(y+4)(y+1)}$

$=\dfrac{y(y+4)}{(y+4)(y+1)(y+1)}+\dfrac{4(y+1)}{(y+4)(y+1)(y+1)}$

$=\dfrac{y(y+4)+4(y+1)}{(y+4)(y+1)(y+1)}$

$=\dfrac{y^2+4y+4y+4}{(y+4)(y+1)(y+1)}$

$=\dfrac{y^2+8y+4}{(y+4)(y+1)(y+1)}$

57. $\dfrac{x-5}{x+3}+\dfrac{x+3}{x-5}$

$\text{LCD} = (x+3)(x-5)$

$\dfrac{x-5}{x+3}+\dfrac{x+3}{x-5}$

$=\dfrac{(x-5)(x-5)}{(x+3)(x-5)}+\dfrac{(x+3)(x+3)}{(x-5)(x+3)}$

$=\dfrac{(x-5)(x-5)+(x+3)(x+3)}{(x+3)(x-5)}$

$=\dfrac{\left(x^2-10x+25\right)+\left(x^2+6x+9\right)}{(x+3)(x-5)}$

$=\dfrac{2x^2-4x+34}{(x+3)(x-5)}$

59. $\dfrac{5}{2y^2-2y}-\dfrac{3}{2y-2}$

$2y^2-2y=2y(y-1)$

$2y-2=2(y-1)$

$\text{LCD} = 2y(y-1)$

$\dfrac{5}{2y^2-2y}-\dfrac{3}{2y-2}$

$=\dfrac{5}{2y(y-1)}-\dfrac{3}{2(y-1)}$

$=\dfrac{5}{2y(y-1)}-\dfrac{3\cdot y}{2y(y-1)}$

$=\dfrac{5-3y}{2y(y-1)}$

61. $\dfrac{4x+3}{x^2-9}-\dfrac{x+1}{x-3}$

$\text{LCD} = (x+3)(x-3)$

$\dfrac{4x+3}{x^2-9}-\dfrac{x+1}{x-3}$

$=\dfrac{4x+3}{(x+3)(x-3)}-\dfrac{(x+1)(x+3)}{(x+3)(x-3)}$

$=\dfrac{(4x+3)-(x+1)(x+3)}{(x+3)(x-3)}$

$=\dfrac{(4x+3)-\left(x^2+4x+3\right)}{(x+3)(x-3)}$

$=\dfrac{4x+3-x^2-4x-3}{(x+3)(x-3)}$

$=\dfrac{-x^2}{(x+3)(x-3)}=-\dfrac{x^2}{(x+3)(x-3)}$

63. $\dfrac{y^2-39}{y^2+3y-10}-\dfrac{y-7}{y-2}$

$y^2+3y-10=(y-2)(y+5)$

$y-2=1(y-2)$

$\text{LCD} = (y-2)(y+5)$

$\dfrac{y^2-39}{y^2+3y-10}-\dfrac{y-7}{y-2}$

$=\dfrac{y^2-39}{(y-2)(y+5)}-\dfrac{y-7}{y-2}$

$=\dfrac{y^2-39}{(y-2)(y+5)}-\dfrac{(y-7)(y+5)}{(y-2)(y+5)}$

$=\dfrac{\left(y^2-39\right)-(y-7)(y+5)}{(y-2)(y+5)}$

$=\dfrac{\left(y^2-39\right)-\left(y^2-2y-35\right)}{(y-2)(y+5)}$

$=\dfrac{y^2-39-y^2+2y+35}{(y-2)(y+5)}$

$=\dfrac{2y-4}{(y-2)(y+5)}=\dfrac{2(y-2)}{(y-2)(y+5)}$

$=\dfrac{2}{y+5}$

65. $4 + \dfrac{1}{x-3}$

LCD = $x-3$

$4 + \dfrac{1}{x-3} = \dfrac{4(x-3)}{x-3} + \dfrac{1}{x-3}$

$= \dfrac{4(x-3)+1}{x-3}$

$= \dfrac{4x-12+1}{x-3}$

$= \dfrac{4x-11}{x-3}$

67. $3 - \dfrac{3y}{y+1}$

LCD = $y+1$

$3 - \dfrac{3y}{y+1} = \dfrac{3(y+1)}{y+1} - \dfrac{3y}{y+1}$

$= \dfrac{3(y+1)-3y}{y+1}$

$= \dfrac{3y+3-3y}{y+1}$

$= \dfrac{3}{y+1}$

69. $\dfrac{9x+3}{x^2-x-6} + \dfrac{x}{3-x}$

$x^2-x-6 = (x-3)(x+2)$

$3-x = -1(x-3)$

LCD = $(x-3)(x+2)$

$\dfrac{9x+3}{x^2-x-6} + \dfrac{x}{3-x}$

$= \dfrac{9x+3}{(x-3)(x+2)} + \dfrac{(-1)}{(-1)} \cdot \dfrac{x}{3-x}$

$= \dfrac{9x+3}{(x-3)(x+2)} + \dfrac{-x}{x-3}$

$= \dfrac{9x+3}{(x-3)(x+2)} + \dfrac{-x(x+2)}{(x-3)(x+2)}$

$= \dfrac{9x+3-x(x+2)}{(x-3)(x+2)}$

$= \dfrac{9x+3-x^2-2x}{(x-3)(x+2)}$

$= \dfrac{-x^2+7x+3}{(x-3)(x+2)}$

71. $\dfrac{x+3}{x^2+x-2} - \dfrac{2}{x^2-1}$

$x^2+x-2 = (x-1)(x+2)$

$x^2-1 = (x+1)(x-1)$

LCD = $(x+1)(x-1)(x+2)$

$\dfrac{x+3}{x^2+x-2} - \dfrac{2}{x^2-1}$

$= \dfrac{x+3}{(x-1)(x+2)} - \dfrac{2}{(x+1)(x-1)}$

$= \dfrac{(x+3)(x+1)}{(x+1)(x-1)(x+2)}$

$\quad - \dfrac{2(x+2)}{(x+1)(x-1)(x+2)}$

$= \dfrac{(x+3)(x+1)-2(x+2)}{(x+1)(x-1)(x+2)}$

$= \dfrac{x^2+4x+3-2x-4}{(x+1)(x-1)(x+2)}$

$= \dfrac{x^2+2x-1}{(x+1)(x-1)(x+2)}$

73. $\dfrac{y+3}{5y^2} - \dfrac{y-5}{15y}$

LCD = $15y^2$

$\dfrac{y+3}{5y^2} - \dfrac{y-5}{15y}$

$= \dfrac{(y+3)(3)}{5y^2(3)} - \dfrac{(y-5)(y)}{15y(y)}$

$= \dfrac{(3y+9)-\left(y^2-5y\right)}{15y^2}$

$= \dfrac{3y+9-y^2+5y}{15y^2}$

$= \dfrac{-y^2+8y+9}{15y^2}$

75. $\dfrac{x+3}{3x+6}+\dfrac{x}{4-x^2}$

$3x+6=3(x+2)$

$4-x^2=(2+x)(2-x)$

Note that $-1(2-x)=x-2$

LCD $=3(x+2)(x-2)$

$\dfrac{x+3}{3x+6}+\dfrac{x}{4-x^2}$

$=\dfrac{x+3}{3(x+2)}+\dfrac{x}{(2+x)(2-x)}$

$=\dfrac{x+3}{3(x+2)}+\dfrac{(-1)}{(-1)}\cdot\dfrac{x}{(2+x)(2-x)}$

$=\dfrac{x+3}{3(x+2)}+\dfrac{-x}{(x+2)(x-2)}$

$=\dfrac{(x+3)(x-2)}{3(x+2)(x-2)}+\dfrac{-x(3)}{3(x+2)(x-2)}$

$=\dfrac{x^2+x-6-3x}{3(x+2)(x-2)}$

$=\dfrac{x^2-2x-6}{3(x+2)(x-2)}$

77. $\dfrac{y}{y^2-1}+\dfrac{2y}{y-y^2}$

$y^2-1=(y+1)(y-1)$

$y-y^2=y(1-y)$

Note that $-1(1-y)=y-1$

LCD $=y(y+1)(y-1)$

$\dfrac{y}{y^2-1}+\dfrac{2y}{y-y^2}$

$=\dfrac{y}{(y+1)(y-1)}+\dfrac{2y}{y(1-y)}$

$=\dfrac{y}{(y+1)(y-1)}+\dfrac{(-1)}{(-1)}\cdot\dfrac{2y}{y(1-y)}$

$=\dfrac{y}{(y+1)(y-1)}+\dfrac{-2y}{y(y-1)}$

$=\dfrac{y\cdot y}{y(y+1)(y-1)}+\dfrac{-2y(y+1)}{y(y+1)(y-1)}$

$=\dfrac{y^2-2y(y+1)}{y(y+1)(y-1)}=\dfrac{y^2-2y^2-2y}{y(y+1)(y-1)}$

$=\dfrac{-y^2-2y}{y(y+1)(y-1)}=\dfrac{-y(y+2)}{y(y+1)(y-1)}$

$=\dfrac{-1(y+2)}{(y+1)(y-1)}=\dfrac{-y-2}{(y+1)(y-1)}$

79. $\dfrac{x-1}{x}+\dfrac{y+1}{y}$

LCD $=xy$

$\dfrac{x-1}{x}+\dfrac{y+1}{y}$

$=\dfrac{(x-1)(y)}{xy}+\dfrac{(y+1)(x)}{xy}$

$=\dfrac{xy-y+xy+x}{xy}$

$=\dfrac{x+2xy-y}{xy}$

81. $\dfrac{3x}{x^2-y^2}-\dfrac{2}{y-x}$

$x^2-y^2=(x+y)(x-y)$

Note that $y-x=-1(x-y)$

LCD $=(x+y)(x-y)$

$\dfrac{3x}{x^2-y^2}-\dfrac{2}{y-x}$

$=\dfrac{3x}{(x+y)(x-y)}-\dfrac{(-1)}{(-1)}\cdot\dfrac{2}{y-x}$

$=\dfrac{3x}{(x+y)(x-y)}-\dfrac{-2}{x-y}$

$=\dfrac{3x}{(x+y)(x-y)}-\dfrac{-2(x+y)}{(x+y)(x-y)}$

$=\dfrac{3x+2(x+y)}{(x+y)(x-y)}=\dfrac{3x+2x+2y}{(x+y)(x-y)}$

$=\dfrac{5x+2y}{(x+y)(x-y)}$

83. $\dfrac{x+6}{x^2-4}-\dfrac{x+3}{x+2}+\dfrac{x-3}{x-2}$

$\text{LCD}=(x+2)(x-2)$

$\dfrac{x+6}{x^2-4}-\dfrac{x+3}{x+2}+\dfrac{x-3}{x-2}=\dfrac{x+6}{(x+2)(x-2)}-\dfrac{x+3}{x+2}+\dfrac{x-3}{x-2}$

$=\dfrac{x+6}{(x+2)(x-2)}-\dfrac{(x+3)(x-2)}{(x+2)(x-2)}+\dfrac{(x-3)(x+2)}{(x-2)(x+2)}$

$=\dfrac{x+6-(x+3)(x-2)+(x-3)(x+2)}{(x+2)(x-2)}=\dfrac{x+6-\left(x^2+x-6\right)+\left(x^2-x-6\right)}{(x+2)(x-2)}$

$=\dfrac{x+6-x^2-x+6+x^2-x-6}{(x+2)(x-2)}=\dfrac{-x+6}{(x+2)(x-2)}$

85. $\dfrac{5}{x^2-25}+\dfrac{4}{x^2-11x+30}-\dfrac{3}{x^2-x-30}$

$x^2-25=(x+5)(x-5)$

$x^2-11x+30=(x-6)(x-5)$

$x^2-x-30=(x-6)(x+5)$

$\text{LCD}=(x+5)(x-5)(x-6)$

$\dfrac{5}{x^2-25}+\dfrac{4}{x^2-11x+30}-\dfrac{3}{x^2-x-30}$

$=\dfrac{5}{(x+5)(x-5)}+\dfrac{4}{(x-6)(x-5)}-\dfrac{3}{(x-6)(x+5)}$

$=\dfrac{5(x-6)}{(x+5)(x-5)(x-6)}+\dfrac{4(x+5)}{(x-6)(x-5)(x+5)}-\dfrac{3(x-5)}{(x-6)(x+5)(x-5)}$

$=\dfrac{5(x-6)+4(x+5)-3(x-5)}{(x+5)(x-5)(x-6)}=\dfrac{5x-30+4x+20-3x+15}{(x+5)(x-5)(x-6)}$

$=\dfrac{6x+5}{(x+5)(x-5)(x-6)}$

87. $\dfrac{x+6}{x^3-27} - \dfrac{x}{x^3+3x^2+9x}$

$x^3-27 = (x-3)\left(x^2+3x+9\right)$

$x^3+3x^2+9x = x\left(x^2+3x+9\right)$

LCD $= x(x-3)\left(x^2+3x+9\right)$ $\dfrac{x+6}{x^3-27} - \dfrac{x}{x^3+3x^2+9x} = \dfrac{x+6}{(x-3)\left(x^2+3x+9\right)} - \dfrac{x}{x\left(x^2+3x+9\right)}$

$= \dfrac{(x+6)x}{(x-3)\left(x^2+3x+9\right)x} - \dfrac{x(x-3)}{x\left(x^2+3x+9\right)(x-3)}$

$= \dfrac{x(x+6)-x(x-3)}{x(x-3)\left(x^2+3x+9\right)} = \dfrac{x^2+6x-x^2+3x}{x(x-3)\left(x^2+3x+9\right)}$

$= \dfrac{9x}{x(x-3)\left(x^2+3x+9\right)} = \dfrac{9}{(x-3)\left(x^2+3x+9\right)}$

89. $\dfrac{9y+3}{y^2-y-6} + \dfrac{y}{3-y} + \dfrac{y-1}{y+2}$

$y^2-y-6 = (y-3)(y+2)$

$3-y = -1(y-3)$

$y+2 = 1(y+2)$

LCD $= (y-3)(y+2)$

$\dfrac{9y+3}{y^2-y-6} + \dfrac{y}{3-y} + \dfrac{y-1}{y+2} = \dfrac{9y+3}{(y-3)(y+2)} + \dfrac{y}{-1(y-3)} + \dfrac{y-1}{y+2}$

$= \dfrac{9y+3}{(y-3)(y+2)} + \dfrac{-y(y+2)}{(y-3)(y+2)} + \dfrac{(y-1)(y-3)}{(y+2)(y-3)}$

$= \dfrac{9y+3+-y(y+2)+(y-1)(y-3)}{(y-3)(y+2)} = \dfrac{9y+3-y^2-2y+y^2-4y+3}{(y-3)(y+2)}$

$= \dfrac{3y+6}{(y-3)(y+2)} = \dfrac{3(y+2)}{(y-3)(y+2)} = \dfrac{3}{y-3}$

91. $\dfrac{3}{x^2+4xy+3y^2} - \dfrac{5}{x^2-2xy-3y^2} + \dfrac{2}{x^2-9y^2}$

$x^2+4xy+3y^2 = (x+y)(x+3y)$

$x^2-2xy-3y^2 = (x-3y)(x+y)$

$x^2-9y^2 = (x+3y)(x-3y)$

LCD $= (x+y)(x+3y)(x-3y)$

$$\frac{3}{x^2+4xy+3y^2}-\frac{5}{x^2-2xy-3y^2}+\frac{2}{x^2-9y^2}$$

$$=\frac{3}{(x+y)(x+3y)}-\frac{5}{(x-3y)(x+y)}+\frac{2}{(x+3y)(x-3y)}$$

$$=\frac{3(x-3y)}{(x+y)(x+3y)(x-3y)}-\frac{5(x+3y)}{(x-3y)(x+y)(x+3y)}+\frac{2(x+y)}{(x+3y)(x-3y)(x+y)}$$

$$=\frac{3(x-3y)-5(x+3y)+2(x+y)}{(x+y)(x+3y)(x-3y)}=\frac{3x-9y-5x-15y+2x+2y}{(x+y)(x+3y)(x-3y)}$$

$$=\frac{-22y}{(x+y)(x+3y)(x-3y)}$$

93. Young's Rule: $C=\dfrac{DA}{A+12}$

$A=8;$

$C=\dfrac{D\cdot 8}{8+12}=\dfrac{8D}{20}=\dfrac{2D}{5}$

$A=3;$

$C=\dfrac{D\cdot 3}{3+12}=\dfrac{3D}{15}=\dfrac{D}{5}$

Difference:

$\dfrac{2D}{5}-\dfrac{D}{5}=\dfrac{D}{5}$

The difference in dosages for an 8-year-old child and a 3-year-old child is $\dfrac{D}{5}$. This means that an 8-year-old should be given $\dfrac{1}{5}$ of the adult dosage more than a 3-year-old.

95. Young's Rule: $C=\dfrac{DA}{A+12}$

Cowling's Rule: $C=\dfrac{D(A+1)}{24}$

For $A=12$, Young's Rule gives $C=\dfrac{D\cdot 12}{12+12}=\dfrac{12D}{24}=\dfrac{D}{2}$ and Cowling's Rule gives $C=\dfrac{D(12+1)}{24}=\dfrac{13D}{24}$.

The difference between the dosages given by Cowling's Rule and Young's Rule is $\dfrac{13D}{24}-\dfrac{12D}{24}=\dfrac{D}{24}$.

This means that Cowling's Rule says to give a 12-year-old $\dfrac{1}{24}$ of the adult dose more than Young's Rule says the dosage should be.

97. No, because the graphs cross, neither formula gives a consistently smaller dosage.

99. The difference in dosage is greatest at 5 years. This is where the graphs are farthest apart.

101. $P = 2L + 2W$

$= 2\left(\dfrac{x}{x+3}\right) + 2\left(\dfrac{x}{x-4}\right)$

$= \dfrac{2x}{x+3} + \dfrac{2x}{x+4}$

$= \dfrac{2x(x+4)}{(x+3)(x+4)} + \dfrac{2x(x+3)}{(x+3)(x+4)}$

$= \dfrac{2x^2 + 8x + 2x^2 + 6x}{(x+3)(x+4)}$

$= \dfrac{4x^2 + 14x}{(x+3)(x+4)}$

103. Answers will vary.

105. Explanations will vary. The right side of the equation should be charged from $\dfrac{3}{x+5}$ to $\dfrac{5+2x}{5x}$.

107. Answers will vary.

109. makes sense

111. does not make sense; Explanations will vary. Sample explanation: It is acceptable to leave the numerator in factored form.

113. false; Changes to make the statement true will vary. A sample change is: The LCD is $x(x-1)$ or $x^2 - x$.

115. true

117. $\left(\dfrac{1}{x+h} - \dfrac{1}{x}\right) \div h$

$= \left(\dfrac{1(x)}{(x+h)(x)} - \dfrac{1(x+h)}{x(x+h)}\right) \div h$

$= \left(\dfrac{x - (x+h)}{x(x+h)}\right) \div h$

$= \left(\dfrac{x - x - h}{x(x+h)}\right) \div h = \dfrac{-h}{x(x+h)} \div h$

$= \dfrac{-h}{x(x+h)} \cdot \dfrac{1}{h} = \dfrac{-1}{x(x+h)}$

119. $\dfrac{4}{x-2} - \dfrac{?}{?} = \dfrac{2x+8}{(x-2)(x+1)}$

The missing rational expression must have $(x+1)$ as a factor in its denominator or as the complete denominator. Let y = the numerator of the missing rational expression.

$\dfrac{4}{x-2} - \dfrac{y}{x+1} = \dfrac{2x+8}{(x-2)(x+1)}$

Then, $\dfrac{4(x+1) - y(x-2)}{(x-2)(x+1)} - \dfrac{2x+8}{(x-2)(x+1)}$

So $4x + 4 - yx + 2y = 2x + 8$, which implies that $4x - yx = 2x$ and $4 + 2y = 8$. Both of these equations give $y = 2$. Thus, the missing rational expression is $\dfrac{2}{x+1}$.

120. $(3x+5)(2x-7)$

$= 6x^2 - 21x + 10x - 35$

$= 6x^2 - 11x - 35$

121. $3x - y < 3$

Graph $3x - y = 3$ as a dashed line using the x-intercept, 1, and the y-intercept -3. Use $(9,0)$ as a test point. Since $3 \cdot 0 - 9 < 3$ is a true statement, shade the half-plane including $(0,0)$.

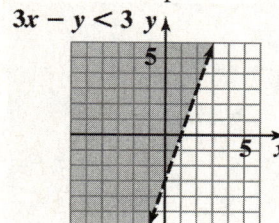

$3x - y < 3$

122. First find the slope $m = \dfrac{0 - (-4)}{1 - (-3)} = \dfrac{4}{4} = 1$

Use $m = 1$ and $(x_1, y_1) = (1, 0)$ in the point-slope form and simplify to find the slope-intercept form.

$y - y_1 = m(x - x_1)$

$y - 0 = 1(x - 1)$

$y = x - 1$

123. a. $\dfrac{1}{3}+\dfrac{2}{5}=\dfrac{1\cdot 5}{3\cdot 5}+\dfrac{2\cdot 3}{5\cdot 3}=\dfrac{5}{15}+\dfrac{6}{15}=\dfrac{11}{15}$

b. $\dfrac{2}{5}-\dfrac{1}{3}=\dfrac{2\cdot 3}{5\cdot 3}-\dfrac{1\cdot 5}{3\cdot 5}=\dfrac{6}{15}-\dfrac{5}{15}=\dfrac{1}{15}$

c. $\left(\dfrac{1}{3}+\dfrac{2}{5}\right)\div\left(\dfrac{2}{5}-\dfrac{1}{3}\right)=\dfrac{11}{15}\div\dfrac{1}{15}=\dfrac{11}{15}\cdot\dfrac{15}{1}=11$

124. a. $\dfrac{1}{x}+\dfrac{1}{y}=\dfrac{1\cdot y}{x\cdot y}+\dfrac{1\cdot x}{y\cdot x}=\dfrac{y}{xy}+\dfrac{x}{xy}=\dfrac{y+x}{xy}$

b. $\dfrac{1}{xy}\div\left(\dfrac{1}{x}+\dfrac{1}{y}\right)=\dfrac{1}{xy}\div\left(\dfrac{y+x}{xy}\right)$

$=\dfrac{1}{xy}\cdot\dfrac{xy}{y+x}$

$=\dfrac{1}{y+x}$

125. $xy\left(\dfrac{1}{x}+\dfrac{1}{y}\right)=\dfrac{xy}{x}+\dfrac{xy}{y}=y+x$

Mid-Chapter Check Point

1. $\dfrac{x^2-4}{x^2-2x-8}$

$x^2-2x-8=0$

$(x-4)(x+2)=0$

$x-4=0 \quad$ or $\quad x+2=0$

$x=4 \qquad\qquad x=-2$

The rational expression is undefined for $x=4$ and $x=-2$.

2. $\dfrac{3x^2-7x+2}{6x^2+x-1}=\dfrac{(3x-1)(x-2)}{(3x-1)(2x+1)}$

$=\dfrac{x-2}{2x+1}$

3. $\dfrac{9-3y}{y^2-5y+6}=\dfrac{3(3-y)}{(y-3)(y-2)}$

$=\dfrac{-3(y-3)}{(y-3)(y-2)}=\dfrac{-3}{y-2}$

4. $\dfrac{16w^3-24w^2}{8w^4-12w^3}=\dfrac{8w^2(2w-3)}{4w^3(2w-3)}=\dfrac{2}{w}$

5. $\dfrac{7x-3}{x^2+3x-4}-\dfrac{3x+1}{x^2+3x-4}=\dfrac{7x-3-3x-1}{(x-1)(x+4)}$

$=\dfrac{4x-4}{(x-1)(x+4)}=\dfrac{4(x-1)}{(x-1)(x+4)}$

$=\dfrac{4}{x+4}$

6. $\dfrac{x+2}{2x-4}\cdot\dfrac{8}{x^2-4}$

$=\dfrac{x+2}{2(x-2)}\cdot\dfrac{8}{(x+2)(x-2)}$

$=\dfrac{4}{(x-2)(x-2)}$ or $\dfrac{4}{(x-2)^2}$

7. $1+\dfrac{7}{x-2}=\dfrac{x-2}{x-2}+\dfrac{7}{x-2}=\dfrac{x+5}{x-2}$

8. $\dfrac{2x^2+x-1}{2x^2-7x+3}\div\dfrac{x^2-3x-4}{x^2-x-6}$

$=\dfrac{2x^2+x-1}{2x^2-7x+3}\cdot\dfrac{x^2-x-6}{x^2-3x-4}$

$=\dfrac{(2x-1)(x+1)}{(2x-1)(x-3)}\cdot\dfrac{(x-3)(x+2)}{(x-4)(x+1)}$

$=\dfrac{x+2}{x-4}$

9.

$$\frac{1}{x^2+2x-3}+\frac{1}{x^2+5x+6}$$

$$x^2+2x-3=(x+3)(x-1)$$

$$x^2+5x+6=(x+2)(x+3)$$

$$\text{LCD}=(x+3)(x-1)(x+2)$$

$$\frac{1}{x^2+2x-3}+\frac{1}{x^2+5x+6}=\frac{1}{(x+3)(x-1)}+\frac{1}{(x+2)(x+3)}$$

$$=\frac{1(x+2)}{(x+3)(x-1)(x+2)}+\frac{1(x-1)}{(x+2)(x+3)(x-1)}$$

$$=\frac{x+2+x-1}{(x+3)(x-1)(x+2)}=\frac{2x+1}{(x+3)(x-1)(x+2)}$$

10.

$$\frac{17}{x-5}+\frac{x+8}{5-x}$$

Note: $5-x=-1(x-5)$

$$\text{LCD}=x-5$$

$$\frac{17}{x-5}+\frac{-1(x+8)}{-1(5-x)}=\frac{17}{x-5}+\frac{-x-8}{x-5}=\frac{17-x-8}{x-5}=\frac{9-x}{x-5}$$

11.

$$\frac{4y^2-1}{9y-3y^2}\cdot\frac{y^2-7y+12}{2y^2-7y-4}=\frac{(2y+1)(2y-1)}{3y(3-y)}\cdot\frac{(y-4)(y-3)}{(2y+1)(y-4)}$$

$$=\frac{-1(2y+1)(2y-1)}{-1\cdot3y(3-y)}\cdot\frac{(y-4)(y-3)}{(2y+1)(y-4)}=\frac{-1(2y+1)(2y-1)}{3y(y-3)}\cdot\frac{(y-4)(y-3)}{(2y+1)(y-4)}$$

$$=\frac{-(2y-1)}{3y}=\frac{-2y+1}{3y}$$

12.

$$\frac{y}{y+1}-\frac{2y}{y+2}$$

$$\text{LCD}=(y+1)(y+2)$$

$$\frac{y(y+2)}{(y+1)(y+2)}-\frac{2y(y+1)}{(y+2)(y+1)}=\frac{y^2+2y-2y^2-2y}{(y+1)(y+2)}=\frac{-y^2}{(y+1)(y+2)}$$

13.

$$\frac{w^2+6w+5}{7w^2-63}\div\frac{w^2+10w+25}{7w+21}=\frac{w^2+6w+5}{7w^2-63}\cdot\frac{7w+21}{w^2+10w+25}$$

$$=\frac{(w+5)(w+1)}{7(w^2-9)}\cdot\frac{7(w+3)}{(w+5)(w+5)}=\frac{(w+5)(w+1)}{7(w+3)(w-3)}\cdot\frac{7(w+3)}{(w+5)(w+5)}=\frac{w+1}{(w-3)(w+5)}$$

14. $\dfrac{2z}{z^2-9} - \dfrac{5}{z^2+4z+3}$

$z^2-9 = (z+3)(z-3)$

$z^2+4z+3 = (z+3)(z+1)$

$\text{LCD} = (z+3)(z-3)(z+1)$

$\dfrac{2z}{z^2-9} - \dfrac{5}{z^2+4z+3} = \dfrac{2z}{(z+3)(z-3)} - \dfrac{5}{(z+3)(z+1)}$

$= \dfrac{2z(z+1)}{(z+3)(z-3)(z+1)} - \dfrac{5(z-3)}{(z+3)(z+1)(z-3)} = \dfrac{2z^2+2z-5z+15}{(z+3)(z-3)(z+1)} = \dfrac{2z^2-3z+15}{(z+3)(z-3)(z+1)}$

15. $\dfrac{z+2}{3z-1} + \dfrac{5}{(3z-1)^2}$

$\text{LCD} = (3z-1)(3z-1)$

$\dfrac{(z+2)(3z-1)}{(3z-1)(3z-1)} + \dfrac{5}{(3z-1)^2} = \dfrac{3z^2+5z-2+5}{(3z-1)^2} = \dfrac{3z^2+5z+3}{(3z-1)^2}$

16. $\dfrac{8}{x^2+4x-21} + \dfrac{3}{x+7}$

$x^2+4x-21 = (x+7)(x-3)$

$x+7 = 1(x+7)$

$\text{LCD} = (x+7)(x-3)$

$\dfrac{8}{x^2+4x-21} + \dfrac{3}{x+7}$

$= \dfrac{8}{(x+7)(x-3)} + \dfrac{3}{x+7}$

$= \dfrac{8}{(x+7)(x-3)} + \dfrac{3(x-3)}{(x+7)(x-3)}$

$= \dfrac{8+3x-9}{(x+7)(x-3)} = \dfrac{3x-1}{(x+7)(x-3)}$

17. $\dfrac{x^4-27x}{x^2-9} \cdot \dfrac{x+3}{x^2+3x+9}$

$= \dfrac{x(x^3-27)}{(x+3)(x-3)} \cdot \dfrac{x+3}{x^2+3x+9}$

$= \dfrac{x(x-3)(x^2+3x+9)}{(x+3)(x-3)} \cdot \dfrac{x+3}{x^2+3x+9}$

$= \dfrac{x}{1} = x$

18. $\dfrac{x-1}{x^2-x-2} - \dfrac{x+2}{x^2+4x+3}$

$x^2-x-2 = (x-2)(x+1)$

$x^2+4x+3 = (x+3)(x+1)$

LCD $= (x-2)(x+1)(x+3)$

$\dfrac{x-1}{x^2-x-2} - \dfrac{x+2}{x^2+4x+3}$

$= \dfrac{x-1}{(x-2)(x+1)} - \dfrac{x+2}{(x+1)(x+3)}$

$= \dfrac{(x-1)(x+3)}{(x-2)(x+1)(x+3)} - \dfrac{(x+2)(x-2)}{(x+1)(x+3)(x-2)}$

$= \dfrac{x^2+2x-3-(x^2-4)}{(x-2)(x+1)(x+3)}$

$= \dfrac{x^2+2x-3-x^2+4}{(x-2)(x+1)(x+3)}$

$= \dfrac{2x+1}{(x-2)(x+1)(x+3)}$

19. $\dfrac{x^2-2xy+y^2}{x+y} \div \dfrac{x^2-xy}{5x+5y}$

$= \dfrac{x^2-2xy+y^2}{x+y} \cdot \dfrac{5x+5y}{x^2-xy}$

$= \dfrac{(x-y)(x-y)}{x+y} \cdot \dfrac{5(x+y)}{x(x-y)}$

$= \dfrac{5(x-y)}{x} = \dfrac{5x-5y}{x}$

20. $\dfrac{5}{x+5} + \dfrac{x}{x-4} - \dfrac{11x-8}{x^2+x-20}$

$x^2+x-20 = (x+5)(x-4)$

LCD $= (x+5)(x-4)$

$\dfrac{5}{x+5} + \dfrac{x}{x-4} - \dfrac{11x-8}{x^2+x-20}$

$= \dfrac{5}{x+5} + \dfrac{x}{x-4} - \dfrac{11x-8}{(x+5)(x-4)}$

$= \dfrac{5(x-4)}{(x+5)(x-4)} + \dfrac{x(x+5)}{(x-4)(x+5)}$

$\qquad - \dfrac{11x-8}{(x+5)(x-4)}$

$= \dfrac{5x-20+x^2+5x-11x+8}{(x+5)(x-4)}$

$= \dfrac{x^2-x-12}{(x+5)(x-4)} = \dfrac{(x+3)(x-4)}{(x+5)(x-4)}$

$= \dfrac{x+3}{x+5}$

7.5 Check Points

1. $\dfrac{\dfrac{1}{4}+\dfrac{2}{3}}{\dfrac{2}{3}-\dfrac{1}{4}}$

Add to get a single rational expression in the numerator.

$$\frac{1}{4}+\frac{2}{3}=\frac{3}{12}+\frac{8}{12}=\frac{11}{12}$$

Subtract to get a single rational expression in the denominator.

$$\frac{2}{3}-\frac{1}{4}=\frac{8}{12}-\frac{3}{12}=\frac{5}{12}$$

Perform the division indicated by the fraction bar. Invert and multiply.

$$\frac{\dfrac{1}{4}+\dfrac{2}{3}}{\dfrac{2}{3}-\dfrac{1}{4}}=\frac{\dfrac{11}{12}}{\dfrac{5}{12}}=\frac{11}{12}\cdot\frac{12}{5}=\frac{11}{5}$$

2. $\dfrac{2-\dfrac{1}{x}}{2+\dfrac{1}{x}}$

Subtract to get a single rational expression in the numerator.

$$2-\frac{1}{x}=\frac{2x}{x}-\frac{1}{x}=\frac{2x-1}{x}$$

Add to get a single rational expression in the denominator.

$$2+\frac{1}{x}=\frac{2x}{x}+\frac{1}{x}=\frac{2x+1}{x}$$

Perform the division indicated by the fraction bar. Invert and multiply.

$$\frac{2-\dfrac{1}{x}}{2+\dfrac{1}{x}}=\frac{2x-1}{2x+1}$$

3. $\dfrac{\dfrac{1}{x}-\dfrac{1}{y}}{\dfrac{1}{xy}}$

Subtract to get a single rational expression in the numerator.

$$\frac{1}{x}-\frac{1}{y}=\frac{y}{xy}-\frac{x}{xy}=\frac{y-x}{xy}$$

Perform the division indicated by the fraction bar. Invert and multiply.

$$\frac{\dfrac{y-x}{xy}}{\dfrac{1}{xy}}=\frac{y-x}{xy}\cdot\frac{xy}{1}=y-x$$

4. $\dfrac{\dfrac{1}{4}+\dfrac{2}{3}}{\dfrac{2}{3}-\dfrac{1}{4}}$

Multiply the numerator and the denominator by the LCD of 12.

$$\dfrac{\dfrac{1}{4}+\dfrac{2}{3}}{\dfrac{2}{3}-\dfrac{1}{4}}=\dfrac{12\left(\dfrac{1}{4}+\dfrac{2}{3}\right)}{12\left(\dfrac{2}{3}-\dfrac{1}{4}\right)}=\dfrac{12\cdot\dfrac{1}{4}+12\cdot\dfrac{2}{3}}{12\cdot\dfrac{2}{3}-12\cdot\dfrac{1}{4}}=\dfrac{3+8}{8-3}=\dfrac{11}{5}$$

5. $\dfrac{2-\dfrac{1}{x}}{2+\dfrac{1}{x}}$

Multiply the numerator and the denominator by the LCD of x.

$$\dfrac{2-\dfrac{1}{x}}{2+\dfrac{1}{x}}=\dfrac{x\left(2-\dfrac{1}{x}\right)}{x\left(2+\dfrac{1}{x}\right)}=\dfrac{x\cdot 2-x\cdot\dfrac{1}{x}}{x\cdot 2+x\cdot\dfrac{1}{x}}=\dfrac{2x-1}{2x+1}$$

6. $\dfrac{\dfrac{1}{x}-\dfrac{1}{y}}{\dfrac{1}{xy}}$

Multiply the numerator and the denominator by the LCD of xy.

$$\dfrac{\dfrac{1}{x}-\dfrac{1}{y}}{\dfrac{1}{xy}}=\dfrac{xy\left(\dfrac{1}{x}-\dfrac{1}{y}\right)}{xy\left(\dfrac{1}{xy}\right)}=\dfrac{xy\cdot\dfrac{1}{x}-xy\cdot\dfrac{1}{y}}{xy\cdot\dfrac{1}{xy}}=\dfrac{y-x}{1}=y-x$$

7.5 Concept and Vocabulary Check

1. complex; complex

2. 18; 12; 30; 9; 4; 5; 6

3. 30; $2x^2$; 24; x^2

4. $2x$; 3; $5x$; x^2

7.5 Exercise Set

1. $\dfrac{\dfrac{1}{2}+\dfrac{1}{4}}{\dfrac{1}{2}+\dfrac{1}{3}}$

Add to get a single rational expression in the numerator.

$$\frac{1}{2}+\frac{1}{4}=\frac{2}{4}+\frac{1}{4}=\frac{3}{4}$$

Add to get a single rational expression in the denominator.

$$\frac{1}{2}+\frac{1}{3}=\frac{3}{6}+\frac{2}{6}=\frac{5}{6}$$

Perform the division indicated by the fraction bar. Invert and multiply.

$$\frac{\dfrac{1}{2}+\dfrac{1}{4}}{\dfrac{1}{2}+\dfrac{1}{3}}=\frac{\dfrac{3}{4}}{\dfrac{5}{6}}=\frac{3}{4}\cdot\frac{6}{5}=\frac{9}{10}$$

3. $\dfrac{5+\dfrac{2}{5}}{7-\dfrac{1}{10}}=\dfrac{\dfrac{25}{5}+\dfrac{2}{5}}{\dfrac{70}{10}-\dfrac{1}{10}}$

$$=\frac{\dfrac{27}{5}}{\dfrac{69}{10}}=\frac{27}{5}\cdot\frac{10}{69}=\frac{9\cdot3\cdot2\cdot5}{5\cdot3\cdot23}=\frac{18}{23}$$

5. $\dfrac{\dfrac{2}{5}-\dfrac{1}{3}}{\dfrac{2}{3}-\dfrac{3}{4}}$

LCD = 60

$$\frac{\dfrac{2}{5}-\dfrac{1}{3}}{\dfrac{2}{3}-\dfrac{3}{4}}=\frac{60\cdot\left(\dfrac{2}{5}-\dfrac{1}{3}\right)}{60\cdot\left(\dfrac{2}{3}-\dfrac{3}{4}\right)}$$

$$=\frac{60\cdot\dfrac{2}{5}-60\cdot\dfrac{1}{3}}{60\cdot\dfrac{2}{3}-60\cdot\dfrac{3}{4}}$$

$$=\frac{24-20}{40-45}=\frac{4}{-5}=-\frac{4}{5}$$

7. $\dfrac{\dfrac{3}{4}-x}{\dfrac{3}{4}+x}=\dfrac{\dfrac{3}{4}-\dfrac{4x}{4}}{\dfrac{3}{4}+\dfrac{4x}{4}}$

$$=\frac{\dfrac{3-4x}{4}}{\dfrac{3+4x}{4}}$$

$$=\frac{3-4x}{4}\cdot\frac{4}{3+4x}=\frac{3-4x}{3+4x}$$

9. $\dfrac{7-\dfrac{2}{x}}{5+\dfrac{1}{x}}=\dfrac{\dfrac{7x-2}{x}}{\dfrac{5x+1}{x}}=\dfrac{7x-2}{x}\cdot\dfrac{x}{5x+1}=\dfrac{7x-2}{5x+1}$

11. $\dfrac{2+\dfrac{3}{y}}{1-\dfrac{7}{y}}=\dfrac{\dfrac{2y+3}{y}}{\dfrac{y-7}{y}}$

$$=\frac{2y+3}{y}\cdot\frac{y}{y-7}=\frac{2y+3}{y-7}$$

13. $\dfrac{\dfrac{1}{y}-\dfrac{3}{2}}{\dfrac{1}{y}+\dfrac{3}{4}}=\dfrac{\dfrac{2-3y}{2y}}{\dfrac{4+3y}{4y}}$

$$=\frac{2-3y}{2y}\cdot\frac{4y}{4+3y}$$

$$=\frac{2(2-3y)}{4+3y}=\frac{4-6y}{4+3y}$$

15. $\dfrac{\dfrac{x}{5}-\dfrac{5}{x}}{\dfrac{1}{5}+\dfrac{1}{x}}$

LCD $= 5x$

$\dfrac{\dfrac{x}{5}-\dfrac{5}{x}}{\dfrac{1}{5}+\dfrac{1}{x}}=\dfrac{5x\cdot\left(\dfrac{x}{5}-\dfrac{5}{x}\right)}{5x\cdot\left(\dfrac{1}{5}+\dfrac{1}{x}\right)}$

$=\dfrac{5x\cdot\dfrac{x}{5}-5x\cdot\dfrac{5}{x}}{5x\cdot\dfrac{1}{5}+5x\cdot\dfrac{1}{x}}$

$=\dfrac{x^2-25}{x+5}$

$=\dfrac{(x+5)(x-5)}{x+5}=x-5$

17. $\dfrac{1+\dfrac{1}{x}}{1-\dfrac{1}{x^2}}=\dfrac{\dfrac{x+1}{x}}{\dfrac{x^2-1}{x^2}}$

$=\dfrac{x+1}{x}\cdot\dfrac{x^2}{x^2-1}$

$=\dfrac{x+1}{x}\cdot\dfrac{x^2}{(x+1)(x-1)}$

$=\dfrac{x}{x-1}$

19. $\dfrac{\dfrac{1}{7}-\dfrac{1}{y}}{\dfrac{7-y}{7}}$

LCD $= 7y$

$\dfrac{\dfrac{1}{7}-\dfrac{1}{y}}{\dfrac{7-y}{7}}=\dfrac{7y\left(\dfrac{1}{7}-\dfrac{1}{y}\right)}{7y\left(\dfrac{7-y}{7}\right)}$

$=\dfrac{7y\left(\dfrac{1}{7}\right)-7y\left(\dfrac{1}{y}\right)}{7y\left(\dfrac{7-y}{7}\right)}$

$=\dfrac{y-7}{y(7-y)}$

$=\dfrac{-1(7-y)}{y(7-y)}=-\dfrac{1}{y}$

21. $\dfrac{x+\dfrac{2}{y}}{\dfrac{x}{y}}=\dfrac{\dfrac{xy+2}{y}}{\dfrac{x}{y}}=\dfrac{xy+2}{y}\cdot\dfrac{y}{x}=\dfrac{xy+2}{x}$

23. $\dfrac{\dfrac{1}{x}+\dfrac{1}{y}}{xy}$

LCD $= xy$

$\dfrac{\dfrac{1}{x}+\dfrac{1}{y}}{xy}=\dfrac{xy\left(\dfrac{1}{x}+\dfrac{1}{y}\right)}{xy(xy)}=\dfrac{y+x}{x^2y^2}$

25. $\dfrac{\dfrac{x}{y}+\dfrac{1}{x}}{\dfrac{y}{x}+\dfrac{1}{x}}=\dfrac{\dfrac{x^2+y}{xy}}{\dfrac{y+1}{x}}=\dfrac{x^2+y}{xy}\cdot\dfrac{x}{y+1}$

$=\dfrac{x^2+y}{y(y+1)}$

27. $\dfrac{\dfrac{1}{y}+\dfrac{2}{y^2}}{\dfrac{2}{y}+1}$

LCD $= y^2$

$\dfrac{\dfrac{1}{y}+\dfrac{2}{y^2}}{\dfrac{2}{y}+1}=\dfrac{y^2\left(\dfrac{1}{y}+\dfrac{2}{y^2}\right)}{y^2\left(\dfrac{2}{y}+1\right)}$

$=\dfrac{y^2\left(\dfrac{1}{y}\right)+y^2\left(\dfrac{2}{y^2}\right)}{y^2\left(\dfrac{2}{y}\right)+y^2(1)}$

$=\dfrac{y+2}{2y+y^2}$

$=\dfrac{(y+2)}{y(2+y)}=\dfrac{1}{y}$

29.
$$\frac{\dfrac{12}{x^2}-\dfrac{3}{x}}{\dfrac{15}{x}-\dfrac{9}{x^2}}=\frac{\dfrac{12}{x^2}-\dfrac{3x}{x^2}}{\dfrac{15x}{x^2}-\dfrac{9}{x^2}}=\frac{\dfrac{12-3x}{x^2}}{\dfrac{15x-9}{x^2}}$$

$$=\frac{12-3x}{x^2}\cdot\frac{x^2}{15x-9}=\frac{12-3x}{15x-9}$$

$$=\frac{3(4-x)}{3(5x-3)}=\frac{4-x}{5x-3}$$

31.
$$\frac{2+\dfrac{6}{y}}{1-\dfrac{9}{y^2}}$$

$\text{LCD}=\ y^2$

$$\frac{2+\dfrac{6}{y}}{1-\dfrac{9}{y^2}}=\frac{y^2\left(2+\dfrac{6}{y}\right)}{y^2\left(1-\dfrac{9}{y^2}\right)}$$

$$=\frac{2y^2+6y}{y^2-9}$$

$$=\frac{2y(y+3)}{(y+3)(y-3)}=\frac{2y}{y-3}$$

33.
$$\frac{\dfrac{1}{x+2}}{1+\dfrac{1}{x+2}}$$

$\text{LCD}=\ x+2$

$$\frac{\dfrac{1}{x+2}}{1+\dfrac{1}{x+2}}=\frac{(x+2)\left(\dfrac{1}{x+2}\right)}{(x+2)\left(1+\dfrac{1}{x+2}\right)}$$

$$=\frac{1}{x+2+1}=\frac{1}{x+3}$$

35.
$$\frac{x-5+\dfrac{3}{x}}{x-7+\dfrac{2}{x}}$$

$\text{LCD}=x$

$$\frac{x-5+\dfrac{3}{x}}{x-7+\dfrac{2}{x}}=\frac{x\left(x-5+\dfrac{3}{x}\right)}{x\left(x-7+\dfrac{2}{x}\right)}$$

$$=\frac{x^2-5x+3}{x^2-7x+2}$$

37.
$$\frac{\dfrac{3}{xy^2}+\dfrac{2}{x^2y}}{\dfrac{1}{x^2y}+\dfrac{2}{xy^3}}=\frac{\dfrac{3x}{x^2y^2}+\dfrac{2y}{x^2y^2}}{\dfrac{y^2}{x^2y^3}+\dfrac{2x}{x^2y^3}}$$

$$=\frac{\dfrac{3x+2y}{x^2y^2}}{\dfrac{y^2+2x}{x^2y^3}}$$

$$=\frac{3x+2y}{x^2y^2}\cdot\frac{x^2y^3}{y^2+2x}$$

$$=\frac{3x+2y}{x^2y^2}\cdot\frac{x^2y^3}{y^2+2x}$$

$$=\frac{(3x+2y)(y)}{y^2+2x}$$

$$=\frac{3xy+2y^2}{y^2+2x}$$

39.

$$\dfrac{\dfrac{3}{x+1}-\dfrac{3}{x-1}}{\dfrac{5}{x^2-1}}$$

$$=\dfrac{\dfrac{3(x-1)-3(x+1)}{(x+1)(x-1)}}{\dfrac{5}{x^2-1}}$$

$$=\dfrac{\dfrac{3x-3-3x-3}{(x+1)(x-1)}}{\dfrac{5}{x^2-1}}$$

$$=\dfrac{\dfrac{-6}{(x+1)(x-1)}}{\dfrac{5}{x^2-1}}$$

$$=\dfrac{-6}{(x+1)(x-1)}\cdot\dfrac{x^2-1}{5}$$

$$=\dfrac{-6}{(x+1)(x-1)}\cdot\dfrac{(x+1)(x-1)}{5}$$

$$=-\dfrac{6}{5}$$

41. $\dfrac{\dfrac{6}{x^2+2x-15}-\dfrac{1}{x-3}}{\dfrac{1}{x+5}+1}=\dfrac{\dfrac{6}{(x+5)(x-3)}-\dfrac{1}{x-3}}{\dfrac{1}{x+5}+1}$

$\text{LCD}=(x+5)(x-3)$

$$\dfrac{\dfrac{6}{(x+5)(x-3)}-\dfrac{1}{x-3}}{\dfrac{1}{x+5}+1}=\dfrac{(x+5)(x-3)\left[\dfrac{6}{(x+5)(x-3)}-\dfrac{1}{x-3}\right]}{(x+5)(x-3)\left[\dfrac{1}{x+5}+1\right]}$$

$$=\dfrac{6-(x+5)}{x-3+(x+5)(x-3)}=\dfrac{6-x-5}{x-3+x^2+2x-15}=\dfrac{-x+1}{x^2+3x-18}=\dfrac{1-x}{(x-3)(x+6)}$$

43. $\dfrac{y^{-1}-(y+5)^{-1}}{5}=\dfrac{\dfrac{1}{y}-\dfrac{1}{y+5}}{5}$

$\text{LCD}=y(y+5)$

$$\dfrac{\dfrac{1}{y}-\dfrac{1}{y+5}}{5}=\dfrac{y(y+5)\left(\dfrac{1}{y}-\dfrac{1}{y+5}\right)}{y(y+5)(5)}=\dfrac{y+5-y}{5y(y+5)}=\dfrac{5}{5y(y+5)}=\dfrac{1}{y(y+5)}$$

45. $\dfrac{1}{1-\dfrac{1}{x}}-1=\dfrac{x(1)}{x\left(1-\dfrac{1}{x}\right)}-1=\dfrac{x}{x-1}-1=\dfrac{x}{x-1}-\dfrac{x-1}{x-1}=\dfrac{x-x+1}{x-1}=\dfrac{1}{x-1}$

47. $\dfrac{1}{1+\dfrac{1}{1+\dfrac{1}{x}}}=\dfrac{1}{1+\dfrac{x(1)}{x\left(1+\dfrac{1}{x}\right)}}=\dfrac{1}{1+\dfrac{x}{x+1}}=\dfrac{(x+1)(1)}{(x+1)\left(1+\dfrac{x}{x+1}\right)}=\dfrac{x+1}{x+1+x}=\dfrac{x+1}{2x+1}$

49. $\dfrac{2d}{\dfrac{d}{r_1}+\dfrac{d}{r_2}}$

LCD = $r_1 r_2$

$\dfrac{2d}{\dfrac{d}{r_1}+\dfrac{d}{r_2}}=\dfrac{r_1 r_2 (2d)}{r_1 r_2\left(\dfrac{d}{r_1}+\dfrac{d}{r_2}\right)}$

$=\dfrac{2r_1 r_2 d}{r_2 d+r_1 d}$

$=\dfrac{2r_1 r_2 d}{d(r_2+r_1)}=\dfrac{2r_1 r_2}{r_2+r_1}$

If $r_1=40$ and $r_2=30$, the value of this expression will be $\dfrac{2\cdot 40\cdot 30}{30+40}=\dfrac{2400}{70}=34\dfrac{2}{7}$.

Your average speed will be $34\dfrac{2}{7}$ miles per hour.

51. a. $\dfrac{175-\dfrac{1585}{x}+\dfrac{15{,}993}{x^2}}{206-\dfrac{349}{x}+\dfrac{25{,}984}{x^2}}=\dfrac{x^2\left(175-\dfrac{1585}{x}+\dfrac{15{,}993}{x^2}\right)}{x^2\left(206-\dfrac{349}{x}+\dfrac{25{,}984}{x^2}\right)}$

$=\dfrac{x^2\cdot 175-x^2\cdot\dfrac{1585}{x}+x^2\cdot\dfrac{15{,}993}{x^2}}{x^2\cdot 206-x^2\cdot\dfrac{349}{x}+x^2\cdot\dfrac{25{,}984}{x^2}}$

$=\dfrac{175x^2-1585x+15{,}993}{206x^2-349x+25{,}984}$

b. According to the data in the bar graph, $\dfrac{2504.0}{3720.7}$ or about 67% was spent on human resources in 2010.

c. $\dfrac{175x^2-1585x+15{,}993}{206x^2-349x+25{,}984}=\dfrac{175(40)^2-1585(40)+15{,}993}{206(40)^2-349(40)+25{,}984}\approx 0.68$

According to the model about 68% was spent on human resources in 2010.
This overestimates the actual percent by 1%.

53. – 55. Answers will vary.

57. makes sense

59. does not make sense; Explanations will vary.
Sample explanation: The expression simplifies to
$\dfrac{x+y}{x-y}$.

61. true

63. false; Changes to make the statement true will vary.
A sample change is: All complex rational
expressions can

65.
$$\dfrac{\dfrac{2y}{2+\dfrac{2}{y}}+\dfrac{y}{1+\dfrac{1}{y}}}{} = \dfrac{\dfrac{y}{y}\left(\dfrac{2y}{2+\dfrac{2}{y}}\right)+\dfrac{y}{y}\left(\dfrac{y}{1+\dfrac{1}{y}}\right)}{}$$

$$= \dfrac{2y^2}{2y+2}+\dfrac{y^2}{y+1} = \dfrac{2y^2}{2(y+1)}+\dfrac{y^2}{y+1}$$

$$= \dfrac{y^2}{y+1}+\dfrac{y^2}{y+1} = \dfrac{2y^2}{y+1}$$

67. The graphs coincide.

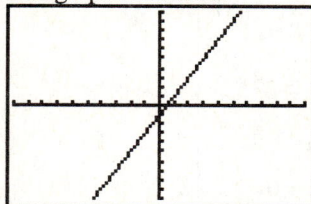

69. The graphs do not coincide.

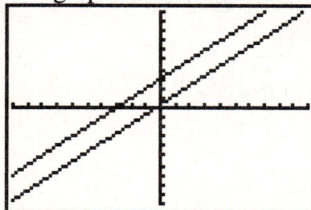

$$\dfrac{\dfrac{1}{x}+\dfrac{1}{3}}{\dfrac{1}{3x}} = \dfrac{3x\left(\dfrac{1}{x}+\dfrac{1}{3}\right)}{3x\left(\dfrac{1}{3x}\right)} = \dfrac{3+x}{1} = 3+x$$

Therefore, $x+\dfrac{1}{3}$ should be $3+x$.

70. $2x^3 - 20x^2 + 50x$

$= 2x\left(x^2 - 10x + 25\right)$

$= 2x(x-5)^2$

71. $2-3(x-2) = 5(x+5)-1$

$2-3x+6 = 5x+25-1$

$8-3x = 5x+24$

$8-3x-5x = 5x+24-5x$

$8-8x = 24$

$8-8x-8 = 24-8$

$-8x = 16$

$\dfrac{-8x}{-8} = \dfrac{16}{-8}$

$x = -2$

The solution set is $\{-2\}$.

72. $(x+y)\left(x^2 - xy + y^2\right)$

$= x\left(x^2 - xy + y^2\right) + y\left(x^2 - xy + y^2\right)$

$= x^3 - x^2y + xy^2 + x^2y - xy^2 + y^3$

$= x^3 + y^3$

73. $\dfrac{x}{3}+\dfrac{x}{2} = \dfrac{5}{6}$

$6\left(\dfrac{x}{3}+\dfrac{x}{2}\right) = 6\left(\dfrac{5}{6}\right)$

$\dfrac{6x}{3}+\dfrac{6x}{2} = \dfrac{6 \cdot 5}{6}$

$2x+3x = 5$

$5x = 5$

$x = 1$

The solution set is $\{1\}$.

74. $\dfrac{2x}{3} = \dfrac{14}{3}-\dfrac{x}{2}$

$6\left(\dfrac{2x}{3}\right) = 6\left(\dfrac{14}{3}-\dfrac{x}{2}\right)$

$\dfrac{6 \cdot 2x}{3} = \dfrac{6 \cdot 14}{3}-\dfrac{6 \cdot x}{2}$

$4x = 28-3x$

$7x = 28$

$x = 4$

The solution set is $\{4\}$.

75.
$$2x^2 + 2 = 5x$$
$$2x^2 - 5x + 2 = 0$$
$$(2x - 1)(x - 2) = 0$$
$$2x - 1 = 0 \quad \text{or} \quad x - 2 = 0$$
$$2x = 1 \qquad\qquad x = 2$$
$$x = \frac{1}{2}$$

The solution set is $\left\{\dfrac{1}{2}, 2\right\}$.

7.6 Check Points

1. $\dfrac{x}{6} = \dfrac{1}{6} + \dfrac{x}{8}$

There are no restrictions on the variable because the variable does not appear in any denominator.
The LCD is 24.
$$\frac{x}{6} = \frac{1}{6} + \frac{x}{8}$$
$$24\left(\frac{x}{6}\right) = 24\left(\frac{1}{6} + \frac{x}{8}\right)$$
$$24 \cdot \frac{x}{6} = 24 \cdot \frac{1}{6} + 24 \cdot \frac{x}{8}$$
$$4x = 4 + 3x$$
$$x = 4$$

The solution set is $\{4\}$.
Check:
$$\frac{x}{6} = \frac{1}{6} + \frac{x}{8}$$
$$\frac{4}{6} = \frac{1}{6} + \frac{4}{8}$$
$$\frac{16}{24} = \frac{4}{24} + \frac{12}{24}$$
$$\frac{16}{24} = \frac{16}{24}$$

2. $\dfrac{5}{2x} = \dfrac{17}{18} - \dfrac{1}{3x}$

The restriction is $x \neq 0$.
The LCD is $18x$.
$$\frac{5}{2x} = \frac{17}{18} - \frac{1}{3x}$$
$$18x\left(\frac{5}{2x}\right) = 18x\left(\frac{17}{18} - \frac{1}{3x}\right)$$
$$18x \cdot \frac{5}{2x} = 18x \cdot \frac{17}{18} - 18x \cdot \frac{1}{3x}$$
$$45 = 17x - 6$$
$$51 = 17x$$
$$3 = x$$

The solution set is $\{3\}$.

3. $x + \dfrac{6}{x} = -5$

The restriction is $x \neq 0$.
The LCD is x.
$$x + \frac{6}{x} = -5$$
$$x\left(x + \frac{6}{x}\right) = x(-5)$$
$$x \cdot x + x \cdot \frac{6}{x} = -5x$$
$$x^2 + 6 = -5x$$
$$x^2 + 5x + 6 = 0$$
$$(x + 3)(x + 2) = 0$$
$$x + 3 = 0 \quad \text{or} \quad x + 2 = 0$$
$$x = -3 \qquad\qquad x = -2$$

The solution set is $\{-3, -2\}$.

4.
$$\frac{11}{x^2-25}+\frac{4}{x+5}=\frac{3}{x-5}$$

$$\frac{11}{(x+5)(x-5)}+\frac{4}{x+5}=\frac{3}{x-5}$$

The restrictions are $x \neq -5$ and $x \neq 5$.

The LCD is $(x+5)(x-5)$.

$$\frac{11}{(x+5)(x-5)}+\frac{4}{x+5}=\frac{3}{x-5}$$

$$(x+5)(x-5)\left(\frac{11}{(x+5)(x-5)}+\frac{4}{x+5}\right)=(x+5)(x-5)\left(\frac{3}{x-5}\right)$$

$$\frac{11(x+5)(x-5)}{(x+5)(x-5)}+\frac{4(x+5)(x-5)}{x+5}=\frac{3(x+5)(x-5)}{x-5}$$

$$11+4(x-5)=3(x+5)$$

$$11+4x-20=3x+15$$

$$4x-9=3x+15$$

$$x-9=15$$

$$x=24$$

The solution set is $\{24\}$.

5.
$$\frac{x}{x-3}=\frac{3}{x-3}+9$$

The restriction is $x \neq 3$.

The LCD is $x-3$.

$$\frac{x}{x-3}=\frac{3}{x-3}+9$$

$$(x-3)\left(\frac{x}{x-3}\right)=(x-3)\left(\frac{3}{x-3}+9\right)$$

$$\frac{x(x-3)}{x-3}=\frac{3(x-3)}{x-3}+9(x-3)$$

$$x=3+9(x-3)$$

$$x=3+9x-27$$

$$x=9x-24$$

$$-8x=-24$$

$$x=3$$

The proposed solution, 3, is not a solution because of the restriction $x \neq 3$.

The solution set is $\{\ \}$.

6. $y = \dfrac{250x}{100 - x}$

The restriction is $x \neq 100$.

$$750 = \frac{250x}{100 - x}$$

$$(100 - x)(750) = (100 - x)\left(\frac{250x}{100 - x}\right)$$

$$75,000 - 750x = 250x$$

$$75,000 = 1000x$$

$$75 = x$$

If government funding is increased to \$750 million, then 75% of pollutants can be removed.

7.6 Concept and Vocabulary Check

1. LCD

2. 0

3. $8x$

4. $12(x + 3)$

5. $1 - x = 3x^2$

6. $6 - 5(x + 1) + 3(x - 1)$

7. $x \neq -2$

8. $x \neq 3; \quad x \neq 2$

9. true

10. false

7.6 Exercise Set

1. $\dfrac{x}{3} = \dfrac{x}{2} - 2$

There are no restrictions on the variable because the variable does not appear in any denominator.
The LCD is 6.

$$\frac{x}{3} = \frac{x}{2} - 2$$

$$6\left(\frac{x}{3}\right) = 6\left(\frac{x}{2} - 2\right)$$

$$6 \cdot \frac{x}{3} = 6 \cdot \frac{x}{2} - 6 \cdot 2$$

$$2x = 3x - 12$$

$$0 = x - 12$$

$$12 = x$$

The solution set is $\{12\}$.

3. $\dfrac{4x}{3} = \dfrac{x}{18} - \dfrac{x}{6}$

There are no restrictions.
The LCD is 18.

$$\frac{4x}{3} = \frac{x}{18} - \frac{x}{6}$$

$$18\left(\frac{4x}{3}\right) = 18\left(\frac{x}{18} - \frac{x}{6}\right)$$

$$18 \cdot \frac{4x}{3} = 18 \cdot \frac{x}{18} - 18 \cdot \frac{x}{6}$$

$$24x = x - 3x$$

$$24x = -2x$$

$$26x = 0$$

$$x = 0$$

The solution set is $\{0\}$.

5. $2 - \dfrac{8}{x} = 6$

The restriction is $x \neq 0$.
The LCD is x.

$$2 - \frac{8}{x} = 6$$

$$x\left(2 - \frac{8}{x}\right) = x \cdot 6$$

$$x \cdot 2 - x \cdot \frac{8}{x} = x \cdot 6$$

$$2x - 8 = 6x$$

$$-8 = 4x$$

$$-2 = x$$

The solution set is $\{-2\}$.

7. $\dfrac{2}{x} + \dfrac{1}{3} = \dfrac{4}{x}$

The restriction is $x \neq 0$.
The LCD is $3x$.

$$\frac{2}{x} + \frac{1}{3} = \frac{4}{x}$$

$$3x\left(\frac{2}{x} + \frac{1}{3}\right) = 3x\left(\frac{4}{x}\right)$$

$$3x \cdot \frac{2}{x} + 3x \cdot \frac{1}{3} = 3x \cdot \frac{4}{x}$$

$$6 + x = 12$$

$$x = 6$$

The solution set is $\{6\}$.

9. $\dfrac{2}{x} + 3 = \dfrac{5}{2x} + \dfrac{13}{4}$

The restriction is $x \neq 0$
The LCD is $4x$.

$$\frac{2}{x} + 3 = \frac{5}{2x} + \frac{13}{4}$$

$$4x\left(\frac{2}{x} + 3\right) = 4x\left(\frac{5}{2x} + \frac{13}{4}\right)$$

$$8 + 12x = 10 + 13x$$

$$8 = 10 + x$$

$$-2 = x$$

The solution set is $\{-2\}$.

11. $\dfrac{2}{3x} + \dfrac{1}{4} = \dfrac{11}{6x} - \dfrac{1}{3}$

The restriction is $x \neq 0$.
The LCD is $12x$.

$$\frac{2}{3x} + \frac{1}{4} = \frac{11}{6x} - \frac{1}{3}$$

$$12x\left(\frac{2}{3x} + \frac{1}{4}\right) = 12x\left(\frac{11}{6x} - \frac{1}{3}\right)$$

$$8 + 3x = 22 - 4x$$

$$8 + 7x = 22$$

$$7x = 14$$

$$x = 2$$

The solution set is $\{2\}$.

13. $\dfrac{6}{x+3} = \dfrac{4}{x-3}$

Restrictions: $x \neq -3, x \neq 3$

$\text{LCD} = (x+3)(x-3)$

$$\frac{6}{x+3} = \frac{4}{x-3}$$

$$(x+3)(x-3) \cdot \frac{6}{x+3} = (x+3)(x-3) \cdot \frac{4}{x-3}$$

$$(x-3) \cdot 6 = (x+3) \cdot 4$$

$$6x - 18 = 4x + 12$$

$$2x - 18 = 12$$

$$2x = 30$$

$$x = 15$$

The solution set is $\{15\}$.

15. $\dfrac{x-2}{2x} + 1 = \dfrac{x+1}{x}$

Restriction: $x \neq 0$

$\text{LCD} = 2x$.

$$\frac{x-2}{2x} + 1 = \frac{x+1}{x}$$

$$2x\left(\frac{x-2}{2x} + 1\right) = 2x\left(\frac{x+1}{x}\right)$$

$$x - 2 + 2x = 2(x+1)$$

$$3x - 2 = 2x + 2$$

$$x - 2 = 2$$

$$x = 4$$

The solution set is $\{4\}$.

17. $x + \dfrac{6}{x} = -7$

Restriction: $x \neq 0$

LCD = x

$$x + \dfrac{6}{x} = -7$$

$$x\left(x + \dfrac{6}{x}\right) = x(-7)$$

$$x^2 + 6 = -7x$$

$$x^2 + 7x + 6 = 0$$

$$(x+6)(x+1) = 0$$

$$x + 6 = 0 \quad \text{or} \quad x + 1 = 0$$

$$x = -6 \qquad x = -1$$

The solution set is $\{-6, -1\}$.

19. $\dfrac{x}{5} - \dfrac{5}{x} = 0$

Restriction: $x \neq 0$

LCD = $5x$

$$\dfrac{x}{5} - \dfrac{5}{x} = 0$$

$$5x\left(\dfrac{x}{5} - \dfrac{5}{x}\right) = 5x \cdot 0$$

$$5x \cdot \dfrac{x}{5} - 5x \cdot \dfrac{5}{x} = 0$$

$$x^2 - 25 = 0$$

$$(x+5)(x-5) = 0$$

$$x + 5 = 0 \quad \text{or} \quad x - 5 = 0$$

$$x = -5 \qquad x = 5$$

The solution set is $\{-5, 5\}$.

21. $x + \dfrac{3}{x} = \dfrac{12}{x}$

Restriction: $x \neq 0$

LCD = x

$$x + \dfrac{3}{x} = \dfrac{12}{x}$$

$$x\left(x + \dfrac{3}{x}\right) = x\left(\dfrac{12}{x}\right)$$

$$x^2 + 3 = 12$$

$$x^2 - 9 = 0$$

$$(x+3)(x-3) = 0$$

$$x + 3 = 0 \quad \text{or} \quad x - 3 = 0$$

$$x = -3 \qquad x = 3$$

The solution set is $\{-3, 3\}$.

23. $\dfrac{4}{y} - \dfrac{y}{2} = \dfrac{7}{2}$

Restrictions: $y \neq 0$

LCD = $2y$

$$\dfrac{4}{y} - \dfrac{y}{2} = \dfrac{7}{2}$$

$$2y\left(\dfrac{4}{y} - \dfrac{y}{2}\right) = 2y\left(\dfrac{7}{2}\right)$$

$$8 - y^2 = 7y$$

$$0 = y^2 + 7y - 8$$

$$0 = (y+8)(y-1)$$

$$y + 8 = 0 \quad \text{or} \quad y - 1 = 0$$

$$y = -8 \qquad y = 1$$

The solution set is $\{-8, 1\}$.

25. $\dfrac{x-4}{x} = \dfrac{15}{x+4}$

Restrictions: $x \neq 0, x \neq -4$

LCD = $x(x+4)$

$$\dfrac{x-4}{x} = \dfrac{15}{x+4}$$

$$x(x+4)\left(\dfrac{x-4}{x}\right) = x(x+4)\left(\dfrac{15}{x+4}\right)$$

$$(x+4)(x-4) = x \cdot 15$$

$$x^2 - 16 = 15x$$

$$x^2 - 15x - 16 = 0$$

$$(x+1)(x-16) = 0$$

$$x + 1 = 0 \quad \text{or} \quad x - 16 = 0$$

$$x = -1 \qquad x = 16$$

The solution set is $\{-1, 16\}$.

27.
$$\frac{2}{x^2-1} = \frac{4}{x+1}$$

$$\frac{2}{(x+1)(x-1)} = \frac{4}{x+1}$$

Restrictions: $x \neq -1, x \neq 1$

LCD $= (x+1)(x-1)$

$$\frac{2}{(x+1)(x-1)} = \frac{4}{x+1}$$

$$(x+1)(x-1)\frac{2}{(x+1)(x-1)} = (x+1)(x-1)\frac{4}{x+1}$$

$$2 = 4(x-1)$$

$$2 = 4x-4$$

$$6 = 4x$$

$$\frac{6}{4} = x$$

$$\frac{3}{2} = x$$

The solution set is $\left\{\dfrac{3}{2}\right\}$.

29.
$$\frac{1}{x-1} + 5 = \frac{11}{x-1}$$

Restriction: $x \neq 1$

LCD $= x-1$

$$\frac{1}{x-1} + 5 = \frac{11}{x-1}$$

$$(x-1)\left(\frac{1}{x-1} + 5\right) = (x-1)\left(\frac{11}{x-1}\right)$$

$$1 + (x-1)\cdot 5 = 11$$

$$1 + 5x - 5 = 11$$

$$5x - 4 = 11$$

$$5x = 15$$

$$x = 3$$

The solution set is $\{3\}$.

31.
$$\frac{8y}{y+1} = 4 - \frac{8}{y+1}$$

Restriction: $y \neq -1$

LCD $= y+1$

$$\frac{8y}{y+1} = 4 - \frac{8}{y+1}$$

$$(y+1)\left(\frac{8y}{y+1}\right) = (y+1)\left(4 - \frac{8}{y+1}\right)$$

$$8y = (y+1)\cdot 4 - 8$$

$$8y = 4y + 4 - 8$$

$$8y = 4y - 4$$

$$4y = -4$$

$$y = -1$$

The proposed solution, -1, is *not* a solution because of the restriction $x \neq -1$. Notice that -1 makes two of the denominators zero in the original equation. Therefore, the equation has no solution. The solution set is $\{\ \}$.

33.
$$\frac{3}{x-1} + \frac{8}{x} = 3$$

Restrictions: $x \neq 1, x \neq 0$

LCD $= x(x-1)$

$$\frac{3}{x-1} + \frac{8}{x} = 3$$

$$x(x-1)\left(\frac{3}{x-1} + \frac{8}{x}\right) = x(x-1)\cdot 3$$

$$x(x-1)\left(\frac{3}{x-1} + \frac{8}{x}\right) = 3x(x-1)$$

$$3x + 8(x-1) = 3x^2 - 3x$$

$$3x + 8x - 8 = 3x^2 - 3x$$

$$11x - 8 = 3x^2 - 3x$$

$$0 = 3x^2 - 14x + 8$$

$$0 = (3x-2)(x-4)$$

$$3x - 2 = 0 \quad \text{or} \quad x - 4 = 0$$

$$3x = 2 \qquad\qquad x = 4$$

$$x = \frac{2}{3}$$

The solution set is $\left\{\dfrac{2}{3}, 4\right\}$.

35. $\dfrac{3y}{y-4} - 5 = \dfrac{12}{y-4}$

Restriction: $y \neq 4$

LCD $= y-4$

$$\frac{3y}{y-4} - 5 = \frac{12}{y-4}$$

$$(y-4)\left(\frac{3y}{y-4} - 5\right) = (y-4)\left(\frac{12}{y-4}\right)$$

$$3y - 5(y-4) = 12$$

$$3y - 5y + 20 = 12$$

$$-2y + 20 = 12$$

$$-2y = -8$$

$$y = 4$$

The proposed solution, 4, is *not* a solution because of the restriction $y \neq 4$. Therefore, this equation has no solution. The solution set is $\{\ \}$.

37. $\dfrac{1}{x} + \dfrac{1}{x-3} = \dfrac{x-2}{x-3}$

Restrictions: $x \neq 0, x \neq 3$

LCD $= x(x-3)$

$$\frac{1}{x} + \frac{1}{x-3} = \frac{x-2}{x-3}$$

$$x(x-3)\left(\frac{1}{x} + \frac{1}{x-3}\right) = x(x-3) \cdot \frac{x-2}{x-3}$$

$$x - 3 + x = x(x-2)$$

$$2x - 3 = x^2 - 2x$$

$$0 = x^2 - 4x + 3$$

$$0 = (x-3)(x-1)$$

$x - 3 = 0$ or $x - 1 = 0$

$x = 3 \qquad\quad x = 1$

The proposed solution, 3, is *not* a solution because of the restriction $x \neq 3$.

The solution set is $\{1\}$.

39. $\dfrac{x+1}{3x+9} + \dfrac{x}{2x+6} = \dfrac{2}{4x+12}$

To find any restrictions and the LCD, factor the denominators.

$$\dfrac{x+1}{3(x+3)} + \dfrac{x}{2(x+3)} = \dfrac{2}{4(x+3)}$$

Restriction: $x \neq -3$

LCD $= 12(x+3)$

$$12(x+3)\left[\dfrac{x+1}{3(x+3)} + \dfrac{x}{2(x+3)}\right] = 12(x+3)\left[\dfrac{2}{4(x+3)}\right]$$

$$4(x+1) + 6x = 6$$
$$4x + 4 + 6x = 6$$
$$10x + 4 = 6$$
$$10x = 2$$
$$x = \dfrac{2}{10}$$
$$x = \dfrac{1}{5}$$

The solution set is $\dfrac{1}{5}$.

41. $\dfrac{4y}{y^2 - 25} + \dfrac{2}{y-5} = \dfrac{1}{y+5}$

To find any restrictions and the LCD, factor the first denominator.

$$\dfrac{4y}{(y+5)(y-5)} + \dfrac{2}{y-5} = \dfrac{1}{y+5}$$

Restrictions: $y \neq -5, y \neq 5$

LCD $= (y+5)(y-5)$

$$(y+5)(y-5)\left[\dfrac{4y}{(y+5)(y-5)} + \dfrac{2}{y-5}\right] = (y+5)(y-5) \cdot \dfrac{1}{y+5}$$

$$4y + 2(y+5) = y - 5$$
$$4y + 2y + 10 = y - 5$$
$$6y + 10 = y - 5$$
$$5y + 10 = -5$$
$$5y = -15$$
$$y = -3$$

The solution set is $\{-3\}$.

43. $\dfrac{1}{x-4} - \dfrac{5}{x+2} = \dfrac{6}{x^2 - 2x - 8}$

Factor the last denominator.

$\dfrac{1}{x-4} - \dfrac{5}{x+2} = \dfrac{6}{(x-4)(x+2)}$

Restrictions: $x \neq 4, x \neq -2$

$LCD = (x-4)(x+2)$

$(x-4)(x+2)\left[\dfrac{1}{x-4} - \dfrac{5}{x+2}\right] = (x-4)(x+2)\left[\dfrac{6}{(x-4)(x+2)}\right]$

$(x+2) \cdot 1 - (x-4) \cdot 5 = 6$

$x + 2 - 5x + 20 = 6$

$-4x + 22 = 6$

$-4x = -16$

$x = 4$

The proposed solution, 4, is *not* a solution because of the restriction $x \neq 4$. Therefore, the given equation has no solution. The solution set is $\{\ \}$.

45. $\dfrac{2}{x+3} - \dfrac{2x+3}{x-1} = \dfrac{6x-5}{x^2 + 2x - 3}$

Factor the denominators.

$\dfrac{2}{x+3} - \dfrac{2x+3}{x-1} = \dfrac{6x-5}{(x+3)(x-1)}$

Restrictions: $x \neq -3, x \neq 1$

$LCD = (x+3)(x-1)$

$(x+3)(x-1)\left[\dfrac{2}{x+3} - \dfrac{2x+3}{x-1}\right] = (x+3)(x-1)\left[\dfrac{6x-5}{(x+3)(x-1)}\right]$

$(x-1) \cdot 2 - (x+3)(2x+3) = 6x - 5$

$2x - 2 - \left(2x^2 + 9x + 9\right) = 6x - 5$

$2x - 2 - 2x^2 - 9x - 9 = 6x - 5$

$-2x^2 - 7x - 11 = 6x - 5$

$0 = 2x^2 + 13x + 6$

$0 = (x+6)(2x+1)$

$x + 6 = 0 \quad \text{or} \quad 2x + 1 = 0$

$x = -6 \qquad\qquad 2x = -1$

$x = -\dfrac{1}{2}$

The solution set is $\left\{-6, -\dfrac{1}{2}\right\}$.

47. Solve $\dfrac{x^2-10}{x^2-x-20}=1+\dfrac{7}{x-5}-\dfrac{1}{2}$.

Factor the first denominator.

$$\dfrac{x^2-10}{(x-5)(x+4)}=1+\dfrac{7}{x-5}-\dfrac{1}{2}.$$

Restrictions: $x\neq 5, x\neq -4$

$LCD = (x-5)(x+4)$

$$(x-5)(x+4)\left(\dfrac{x^2-10}{(x-5)(x+4)}\right)=(x-5)(x+4)\cdot 1+(x-5)(x+4)\left(\dfrac{7}{x-5}\right)$$

$$x^2-10=(x-5)(x+4)+(x+4)\cdot 7$$

$$x^2-10=x^2-x-20+7x+28$$

$$x^2-10=x^2+6x+8$$

$$-10=6x+8$$

$$-18=6x$$

$$-3=x$$

The solution set is $\{-3\}$.

49. Simplify $\dfrac{x^2-10}{x^2-x-20}-1-\dfrac{7}{x-5}$.

Factor the first denominator.

$$\dfrac{x^2-10}{(x-5)(x+4)}-1-\dfrac{7}{x-5}$$

$LCD = (x-5)(x+4)$

$$\dfrac{x^2-10}{(x-5)(x+4)}-\dfrac{(x-5)(x+4)}{(x-5)(x+4)}-\dfrac{7(x+4)}{(x-5)(x+4)}=\dfrac{x^2-10-(x-5)(x+4)-7(x+4)}{(x-5)(x+4)}$$

$$=\dfrac{x^2-10-\left(x^2-x-20\right)-7x-28}{(x-5)(x+4)}$$

$$=\dfrac{x^2-10-x^2+x+20-7x-28}{(x-5)(x+4)}$$

$$=\dfrac{-6x-18}{(x-5)(x+4)}$$

51. Solve $5y^{-2} + 1 = 6y^{-1}$

$$\frac{5}{y^2} + 1 = \frac{6}{y}$$

Restrictions: $y \neq 0$

LCD $= y^2$

$$y^2 \left(\frac{5}{y^2} + 1 \right) = y^2 \left(\frac{6}{y} \right)$$

$$y^2 \cdot \frac{5}{y^2} + y^2 \cdot 1 = 6y$$

$$5 + y^2 = 6y$$

$$y^2 - 6y + 5 = 0$$

$$(y-5)(y-1) = 0$$

$$y - 5 = 0 \quad \text{or} \quad y - 1 = 0$$

$$y = 5 \qquad\quad y = 1$$

The solution set is $\{1, 5\}$.

53. Solve $\dfrac{3}{y+1} - \dfrac{1}{1-y} = \dfrac{10}{y^2 - 1}$.

Factor the denominators.

$$\frac{3}{y+1} - \frac{(-1) \cdot 1}{(-1)(1-y)} = \frac{10}{(y+1)(y-1)}$$

$$\frac{3}{y+1} - \frac{-1}{y-1} = \frac{10}{(y+1)(y-1)}$$

$$\frac{3}{y+1} + \frac{1}{y-1} = \frac{10}{(y+1)(y-1)}$$

Restrictions: $y \neq -1, y \neq 1$

LCD $= (y+1)(y-1)$

$$(y+1)(y-1)\left(\frac{3}{y+1} + \frac{1}{y-1} \right) = (y+1)(y-1)\left(\frac{10}{(y+1)(y-1)} \right)$$

$$(y-1) \cdot 3 + (y+1) \cdot 1 = 10$$

$$3y - 3 + y + 1 = 10$$

$$4y - 2 = 10$$

$$4y = 12$$

$$y = 3$$

The solution set is $\{3\}$.

55. $C = \dfrac{400x + 500{,}000}{x}; C = 450$

$450 = \dfrac{400x + 500{,}000}{x}$

LCD $= x$

$x \cdot 450 = x\left(\dfrac{400x + 500{,}000}{x}\right)$

$450x = 400x + 500{,}000$

$50x = 500{,}000$

$x = 10{,}000$

At an average cost of \$450 per wheelchair, 10,000 wheelchairs can be produced.

57. $C = \dfrac{2x}{100 - x}; C = 2$

$2 = \dfrac{2x}{100 - x}$

LCD $= 100 - x$

$(100 - x) \cdot 2 = (100 - x) \cdot \dfrac{2x}{100 - x}$

$200 - 2x = 2x$

$200 = 4x$

$50 = x$

For \$2 million, 50% of the contaminants can be removed.

59. $C = \dfrac{DA}{A + 12}; C = 300, D = 1000$

$300 = \dfrac{1000A}{A + 12}$

LCD $= A + 12$

$(A + 12) \cdot 300 = (A + 12)\left(\dfrac{1000A}{A + 12}\right)$

$300A + 3600 = 1000A$

$3600 = 700A$

$\dfrac{3600}{700} = A$

$A = \dfrac{36}{7} \approx 5.14$

To the nearest year, the child is 5 years old.

61. $C = \dfrac{10{,}000}{x} + 3x; C = 350$

$350 = \dfrac{10{,}000}{x} + 3x$

LCD $= x$

$x \cdot 350 = x\left(\dfrac{10{,}000}{x} + 3x\right)$

$350x = 10{,}000 + 3x^2$

$0 = 3x^2 - 350x + 10{,}000$

$0 = (3x - 200)(x - 50)$

$3x - 200 = 0 \quad \text{or} \quad x - 50 = 0$

$3x = 200 \qquad\qquad x = 50$

$x = \dfrac{200}{3}$

$= 66\dfrac{2}{3} \approx 67$

For yearly inventory costs to be \$350, the owner should order either 50 or approximately 67 cases. These solutions correspond to the points $(50, 350)$ and $\left(66\dfrac{2}{3}, 350\right)$ on the graph.

63. – 67. Answers will vary.

69. makes sense

71. does not make sense; Explanations will vary. Sample explanation: If all potential solutions make any of the denominators of the rational equation equal to zero, then the equation will have no solution.

73. true

75. false; Changes to make the statement true will vary. A sample change is: You could begin by multiplying both sides by the LCD, $3x$.

77. $f = \dfrac{f_1 f_2}{f_1 + f_2}$ for f_2

$f = \dfrac{f_1 f_2}{f_1 + f_2}$

Multiply both sides by the LCD, $f_1 + f_2$.

$(f_1 + f_2) \cdot f = (f_1 + f_2)\left(\dfrac{f_1 f_2}{f_1 + f_2}\right)$

$ff_1 + ff_2 = f_1 f_2$

Get all terms containing f_2 on one side and all terms not containing f_2 on the other side.

$ff_2 - f_1 f_2 = -f_1 f$

Factor out the common factor f_2 on the left side.

$f_2(f - f_1) = -f_1 f$

Divide both sides by $f - f_1$.

$$\frac{f_2(f - f_1)}{f - f_1} = \frac{-f_1 f}{f - f_1}$$

$$f_2 = \frac{-f_1 f}{f - f_1} \text{ or } \frac{ff_1}{f_1 - f}$$

79. $\left(\dfrac{x+1}{x+7}\right)^2 \div \left(\dfrac{x+1}{x+7}\right)^4 = 0$

$\left(\dfrac{x+1}{x+7}\right)^2 \cdot \left(\dfrac{x+7}{x+1}\right)^4 = 0$

$\dfrac{(x+1)^2}{(x+7)^2} \cdot \dfrac{(x+7)^4}{(x+1)^4} = 0$

Restrictions: $x \neq -7, x \neq -1$

$\dfrac{(x+7)^2}{(x+1)^2} = 0$

Multiply both sides by $(x+1)^2$.

$(x+1)^2 \left[\dfrac{(x+7)^2}{(x+1)^2} \right] = (x+1)^2 \cdot 0$

$(x+7)^2 = 0$

$x + 7 = 0$

$x = -7$

The proposed solution, -7, is *not* a solution of the original equation because it is on the list of restrictions. Therefore, the given equation has no solution. The solution set is $\{\ \}$.

81. $\dfrac{x}{2} + \dfrac{x}{4} = 6$

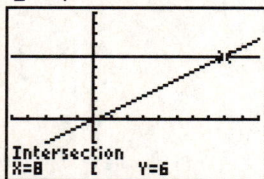

The solution set is $\{8\}$.

Check $x = 8$:

$\dfrac{8}{2} + \dfrac{8}{4} = 6$

$4 + 2 = 6$

$6 = 6$ true

83. $x + \dfrac{6}{x} = -5$

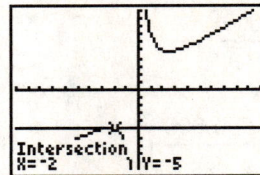

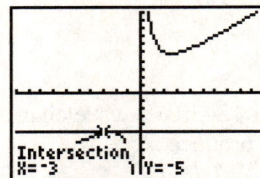

The solution set is $\{-3, -2\}$.

Check -3:

$x + \dfrac{6}{x} = -5$

$-3 + \dfrac{6}{-3} = -5$

$-3 + (-2) = -5$

$-5 = -5$, true

Check -2:

$x + \dfrac{6}{x} = -5$

$-2 + \dfrac{6}{-2} = -5$

$-2 + (-3) = -5$

$-5 = -5$, true

84. $x^4 + 2x^3 - 3x - 6$

Factor by grouping.

$x^4 + 2x^3 - 3x - 6$

$= \left(x^4 + 2x^3\right) + \left(-3x - 6\right)$

$= x^3(x + 2) - 3(x + 2)$

$= (x + 2)\left(x^3 - 3\right)$

85. $\left(3x^2\right)\left(-4x^{-10}\right)$

$= (3 \cdot -4)\left(x^2 \cdot x^{-10}\right) = -12x^{2 + (-10)}$

$= -12x^{-8} = -\dfrac{12}{x^8}$

86. $-5\left[4(x - 2) - 3\right] = -5\left[4x - 8 - 3\right]$

$= -5\left[4x - 11\right]$

$= -20x + 55$

87.
$$\frac{15}{8+x} = \frac{9}{8-x}$$

$$(8+x)(8-x)\frac{15}{8+x} = (8+x)(8-x)\frac{9}{8-x}$$

$$15(8-x) = 9(8+x)$$

$$120-15x = 72+9x$$

$$-24x = -48$$

$$x = 2$$

The solution set is $\{2\}$.

88. In 1 hour you can complete $\dfrac{1}{5}$ of the job.

In 3 hours you can complete $\dfrac{3}{5}$ of the job.

In x hours you can complete $\dfrac{x}{5}$ of the job.

89. $\dfrac{63}{x} = \dfrac{7}{5}$

7.7 Check Points

1. Let x = the rate of the current.
Then $3+x$ = the canoe's rate with the current.
and $3-x$ = the canoe's rate against the current.

	Distance	Rate	Time $= \dfrac{\text{Distance}}{\text{Rate}}$
With the current	10	$3+x$	$\dfrac{10}{3+x}$
Against the current	2	$3-x$	$\dfrac{2}{3-x}$

$$\frac{10}{3+x} = \frac{2}{3-x}$$

Use the cross-products principle to solve this equation.

$$\frac{10}{3+x} = \frac{2}{3-x}$$

$$10(3-x) = 2(3+x)$$

$$30-10x = 6+2x$$

$$30-12x = 6$$

$$-12x = -24$$

$$x = 2$$

The rate of the current is 2 miles per hour.

2. Let x = the number of hours for both people to paint a house together.

	Fractional part of job completed in 1 hour	Time working together	Fractional part of job completed in x hours
First person	$\dfrac{1}{8}$	x	$\dfrac{x}{8}$
Second person	$\dfrac{1}{4}$	x	$\dfrac{x}{4}$

Working together, the two people can complete the whole job, so $\dfrac{x}{8}+\dfrac{x}{4}=1$.

Multiply both sides by the LCD, 8.

$$8\left(\frac{x}{8}+\frac{x}{4}\right)=8\cdot 1$$
$$x+2x=8$$
$$3x=8$$
$$x=\frac{8}{3}$$
$$x=2\frac{2}{3}$$

It will take $2\dfrac{2}{3}$ hours (or 2 hours 40 minutes) if they work together.

3. Let x = the property tax on the \$420,000 house.

$$\frac{\text{Tax on \$250,000 house}}{\text{Assessed value (\$250,000)}}=\frac{\text{Tax on \$420,000 house}}{\text{Assessed value (\$420,000)}}$$
$$\frac{\$3500}{\$250,000}=\frac{\$x}{\$420,000}$$
$$\frac{3500}{250,000}=\frac{x}{420,000}$$
$$250,000x=(3500)(420,000)$$
$$250,000x=1,470,000,000$$
$$\frac{250,000x}{250,000}=\frac{1,470,000,000}{250,000}$$
$$x=5880$$

The property tax is \$5880.

4. Let x = the total number of deer in the refuge.

$$\frac{120}{x}=\frac{25}{150}$$
$$25x=(120)(150)$$
$$25x=18,000$$
$$\frac{25x}{25}=\frac{18,000}{25}$$
$$x=720$$

There are about 720 deer in the refuge.

5. $\dfrac{3}{8} = \dfrac{12}{x}$

$3x = 8 \cdot 12$

$3x = 96$

$x = 32$

The missing length is 32 inches.

6. $\dfrac{h}{2} = \dfrac{56}{3.5}$

$3.5h = 2 \cdot 56$

$3.5h = 112$

$h = \dfrac{112}{3.5}$

$h = 32$

The height of the tower is 32 yards.

7.7 Concept and Vocabulary Check

1. distance traveled; rate of travel

2. 1

3. $\dfrac{x}{5}$

4. $ad = bc$

5. similar

7.7 Exercise Set

1. The times are equal, so $\dfrac{10}{x} = \dfrac{15}{x+3}$

To solve this equation, multiply both sides by the LCD, $x(x+3)$.

$x(x+3) \cdot \dfrac{10}{x} = x(x+3) \cdot \dfrac{15}{x+3}$

$10(x+3) = 15x$

$10x + 30 = 15x$

$30 = 5x$

$6 = x$

If $x = 6, x+3 = 9$.

Note: The equation $\dfrac{10}{x} = \dfrac{15}{x+3}$ is a proportion, so it can also be solved by using the cross-products principle.

$10(x+3) = 15x$

This allows you to skip the first step of the solution process shown above.

The walking rate is 6 miles per hour and the car's rate is 9 miles per hour.

3. Let x = the jogger's rate running uphill.
Then $x + 4$ = the jogger's rate running downhill.

	Distance	Rate	Time = $\dfrac{\text{Distance}}{\text{Rate}}$
Downhill	5	$x + 4$	$\dfrac{5}{x+4}$
Uphill	3	x	$\dfrac{3}{x}$

The times are equal, so $\dfrac{5}{x+4} = \dfrac{3}{x}$.

Use the cross-products principle to solve this equation.

$5x = 3(x+4)$

$5x = 3x + 12$

$2x = 12$

$x = 6$

If $x = 6$, $x + 4 = 10$.

The jogger runs 10 miles per hour downhill and 6 miles per hour uphill.

5. Let x = the rate of the current.
Then $15 + x$ = the boat's rate with the current.
and $15 - x$ = the boat's rate against the current.

	Distance	Rate	Time = $\dfrac{\text{Distance}}{\text{Rate}}$
With the current	20	$15 + x$	$\dfrac{20}{15+x}$
Against the current	10	$15 - x$	$\dfrac{10}{15-x}$

$\dfrac{20}{15+x} = \dfrac{10}{15-x}$

Use the cross-products principle to solve this equation.

$20(15 - x) = 10(15 + x)$

$300 - 20x = 150 + 10x$

$300 = 150 + 30x$

$150 = 30x$

$5 = x$

The rate of the current is 5 miles per hour.

7. Let x = walking rate.
Then $2x$ = jogging rate.

	Distance	Rate	Time = $\dfrac{\text{Distance}}{\text{Rate}}$
Walking	2	x	$\dfrac{2}{x}$
Jogging	2	$2x$	$\dfrac{2}{2x}$

The total time is 1 hour, so

$$\frac{2}{x} + \frac{2}{2x} = 1$$

$$\frac{2}{x} + \frac{1}{x} = 1.$$

To solve this equation, multiply both sides by the LCD, x.

$$x\left(\frac{2}{x} + \frac{1}{x}\right) = x \cdot 1$$

$$2 + 1 = x$$

$$3 = x$$

If $x = 3$, $2x = 6$.

The walking rate is 3 miles per hour and the jogging rate is 6 miles per hour.

9. Let $x =$ the boat's average rate in still water.
Then $x + 2 =$ the boat's rate with the current (downstream).
and $x - 2 =$ the boat's rate against the current (upstream).

	Distance	Rate	Time $= \dfrac{\text{Distance}}{\text{Rate}}$
Downstream	6	$x + 2$	$\dfrac{6}{x+2}$
Upstream	4	$x - 2$	$\dfrac{4}{x-2}$

The times are equal so solve the following equation.

$$\frac{6}{x+2} = \frac{4}{x-2}$$

$$6(x-2) = 4(x+2)$$

$$6x - 12 = 4x + 8$$

$$2x - 12 = 8$$

$$2x = 20$$

$$x = 10$$

The boat's average rate in still water is 10 miles per hour.

11. Let $x =$ the time in minutes, for both people to shovel the driveway together.

	Fractional part of job completed in 1 minute	Time working together	Fractional part of job completed in x minutes
You	$\dfrac{1}{20}$	x	$\dfrac{x}{20}$
Your brother	$\dfrac{1}{15}$	x	$\dfrac{x}{15}$

Working together, you and your brother complete the whole job, so $\dfrac{x}{20} + \dfrac{x}{15} = 1$.

Multiply both sides by the LCD, 60.

$$60\left(\frac{x}{20}+\frac{x}{15}\right)=60\cdot 1$$

$$3x+4x=60$$

$$7x=60$$

$$x=\frac{60}{7}\approx 8.6$$

It will take about 8.6 minutes, which is enough time.

13. Let x = the time, in hours, for both teams to clean the streets working together.

	Fractional part of job completed in 1 hour	Time working together	Fractional part of job completed in x hours
First team	$\dfrac{1}{400}$	x	$\dfrac{x}{400}$
Second team	$\dfrac{1}{300}$	x	$\dfrac{x}{300}$

Working together, the two teams complete one whole job, so $\dfrac{x}{400}+\dfrac{x}{300}=1$.

Multiply both sides by the LCD, 1200.

$$1200\left(\frac{x}{400}+\frac{x}{300}\right)=1200\cdot 1$$

$$3x+4x=1200$$

$$7x=1200$$

$$x=\frac{1200}{7}\approx 171.4$$

It will take about 171.4 hours for both teams to clean the streets working together. One week is $7\cdot 24=168$ hours, so even if both crews work 24 hours a day, there is not enough time.

15. Let x = the time, in hours, for both pipes to fill in the pool.

	Fractional part of job completed in 1 hour	Time working together	Fractional part of job completed in x hours
First pipe	$\dfrac{1}{4}$	x	$\dfrac{x}{4}$
Second pipe	$\dfrac{1}{6}$	x	$\dfrac{x}{6}$

$$\frac{x}{4}+\frac{x}{6}=1$$

$$12\left(\frac{x}{4}+\frac{x}{6}\right)=12\cdot 1$$

$$3x+2x=12$$

$$5x=12$$

$$x=\frac{12}{5}=2.4$$

Using both pipes, it will take 2.4 hours or 2 hours 24 minutes to fill the pool.

17. Let $x =$ the tax on a property with an assessed value of \$162,500.

$$\frac{\text{Tax on \$62,000 house}}{\text{Assessed value (\$62,000)}} = \frac{\text{Tax on \$162,500 house}}{\text{Assessed value (\$162,500)}}$$

$$\frac{\$720}{\$65,000} = \frac{\$x}{\$162,500}$$

$$\frac{720}{65,000} = \frac{x}{162,500}$$

$$65,000x = (720)(162,500)$$

$$65,000x = 117,000,000$$

$$\frac{65,000x}{65,000} = \frac{117,000,000}{65,000}$$

$$x = 1800$$

The tax on a property assessed at \$162,500 is \$1800.

19. Let $x =$ the total number of fur seal pups in the rookery.

$$\frac{\text{Original \# tagged}}{\text{Total \# fur seal pups}} = \frac{\text{\# tagged in sample}}{\text{\# in sample}}$$

$$\frac{4963}{x} = \frac{218}{900}$$

$$218x = (4963)(900)$$

$$218x = 4,466,700$$

$$\frac{218x}{218} = \frac{4,466,700}{218}$$

$$x \approx 20,489$$

There were approximately 20,489 fur seal pups in the rookery.

21. Let $x =$ the number of people, in billions, suffering from malnutrition (in 2010).

$$\frac{28}{200} = \frac{x}{6.9}$$

$$200x = (28)(6.9)$$

$$200x = 193.2$$

$$\frac{200x}{200} = \frac{193.2}{200}$$

$$x \approx 0.97$$

0.97 billion people suffered from malnutrition in 2010.

23. Let $x =$ the height of the critter.

$$\frac{\text{foot length person}}{\text{height of person}} = \frac{\text{foot length critter}}{\text{height of critter}}$$

$$\frac{10 \text{ inches}}{67 \text{ inches}} = \frac{23 \text{ inches}}{x}$$

$$\frac{10}{67} = \frac{23}{x}$$

$$10x = (67)(23)$$

$$10x = 1541$$

$$x = 154.1$$

The height of the critter is 154.1 in.

25. $\dfrac{18}{9} = \dfrac{10}{x}$

$18x = 9 \cdot 10$

$18x = 90$

$x = 5$

The length of the side marked x is 5 inches.

27. $\dfrac{10}{30} = \dfrac{x}{18}$

$30x = 10 \cdot 18$

$30x = 180$

$x = 6$

The length of the side marked x is 6 meters.

29. $\dfrac{20}{15} = \dfrac{x}{12}$

$15x = 12 \cdot 20$

$15x = 240$

$x = 16$

The length of the side marked x is 16 inches.

31. $\dfrac{8}{6} = \dfrac{x}{12}$

$6x = 8 \cdot 12$

$6x = 96$

$x = 16$

The tree is 16 feet tall.

33. – 41. Answers will vary.

43. does not make sense; Explanations will vary. Sample explanation: In the same amount of time, you will go further with the current than against it.

45. makes sense

47. Let x = the usual average rate.
Then $x - 15$ = the average rate, in miles per hour, of the bus in the snowstorm.

	Distance	Rate	Time = $\dfrac{\text{Distance}}{\text{Rate}}$
Usual conditions	60	x	$\dfrac{60}{x}$
Snowstorm conditions	60	$x - 15$	$\dfrac{60}{x-15}$

Since the time during the snowstorm is 2 hours longer than the usual time, solve the following equation.

$\dfrac{60}{x} + 2 = \dfrac{60}{x-15}$

Multiply both sides of the equation by the LCD, $x(x-15)$

$$x(x-15)\ \frac{60}{x}+2\ =x(x-15)\ \frac{60}{x-15}$$

$$(x-15)\cdot 60+x(x-15)\cdot 2=60x$$

$$60x-900+2x^2-30x=60x$$

$$2x^2+30x-900=60x$$

$$2x^2-30x-900=0$$

$$x^2-15x-450=0$$

$$(x-30)(x+15)=0$$

$$x-30=0 \quad \text{or} \quad x+15=0$$

$$x=30 \qquad\qquad x=-15$$

Since the rate cannot be negative, the solution is 30. Therefore, the usual average rate of the bus is 30 miles per hour.

49. Let x = the time, in hours, it takes to prepare one report working together.

	Fractional part of job completed in 1 hour	Time working together	Fractional part of job completed in x hours
Ben	$\dfrac{1}{3}$	x	$\dfrac{x}{3}$
Shane	$\dfrac{1}{4.2}=\dfrac{10}{42}=\dfrac{5}{21}$	x	$\dfrac{5x}{21}$

Working together, Ben and Shane prepare one report, so $\dfrac{x}{3}+\dfrac{5x}{21}=1$.

To solve this equation, multiply both sides by the LCD, 21.

$$21\cdot\left(\frac{x}{3}+\frac{5x}{21}\right)=21\cdot 1$$

$$7x+5x=21$$

$$12x=21$$

$$x=\frac{21}{12}=1.75$$

Ben and Shane can prepare one report in 1.75 hours. Multiply this by four to determine how many hours it takes to prepare four reports. $4(1.75)=7$. Therefore, working together Ben and Shane can prepare four reports in 7 hours.

51. Let x = time, in hours, to fill empty swimming pool.

	Fractional part of job completed in 1 hour	Time working together	Fractional part of job completed in x hours
Normal filling of pool	$\dfrac{1}{2}$	x	$\dfrac{x}{2}$
Leak	$-\dfrac{1}{10}$	x	$-\dfrac{x}{10}$

The situation can be modeled by the equation $\dfrac{x}{2}-\dfrac{x}{10}=1$.

Multiply both sides of the equation by the LCD, 20.

$$20 \cdot \left(\frac{x}{2} - \frac{x}{10} \right) = 20 \cdot 1$$

$$10x - 2x = 20$$

$$8x = 20$$

$$x = 2.5$$

It will take 2.5 hours to fill the empty swimming pool.

53. $25x^2 - 81 = (5x)^2 - 9^2$

$\qquad = (5x + 9)(5x - 9)$

54. $x^2 - 12x + 36 = 0$

$\qquad (x - 6)^2 = 0$

$\qquad x - 6 = 0$

$\qquad x = 6$

The solution set is {6}.

55. $y = -\frac{2}{3}x + 4$

slope $= -\frac{2}{3} = \frac{-2}{3}$

y-intercept $= 4$

Plot (0,4). From this point, move 2 units *down* and 3 units to the *right* to reach the point (3,2). Draw a line through (0,4) and (3,2).

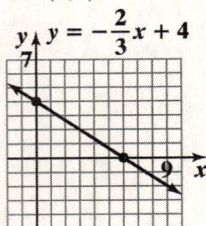

56. a. Substitute to find k.

$\qquad y = kx$

$\qquad 64 = k \cdot 2$

$\qquad 64 = 2k$

$\qquad k = 32$

b. $y = kx$

$\qquad y = 32x$

c. $y = 32x$

$\qquad y = 32 \cdot 5$

$\qquad y = 160$

57. Substitute to find k.

$$B = kW$$

$$5 = k \cdot 160$$

$$\frac{5}{160} = \frac{k \cdot 160}{160}$$

$$k = 0.03125$$

58. a. Substitute to find k.

$$y = \frac{k}{x}$$

$$12 = \frac{k}{8}$$

$$k = 12 \cdot 8$$

$$k = 96$$

b. $y = \frac{k}{x}$

$$y = \frac{96}{x}$$

c. $y = \frac{96}{x}$

$$y = \frac{96}{3}$$

$$y = 32$$

7.8 Check Points

1. *Step 1* Write an equation. $W = kt$

Step 2 Use the given values to find k.
Substitute 30 for W and 5 for t.

$$W = kt$$

$$30 = k \cdot 5$$

$$6 = k$$

Step 3 Substitute the value of k into the equation
$W = 6t$

Step 4 Answer the problem's question. Substitute 8
for t and solve for W.

$$W = 6t$$

$$W = 6 \cdot 8$$

$$W = 48$$

A shower lasting 8 minutes will use 48 gallons of
water.

2. *Step 1* Write an equation. $P = \dfrac{k}{V}$

Step 2 Use the given values to find k.
Substitute 12 for P and 8 for V.

$$P = \frac{k}{V}$$

$$12 = \frac{k}{8}$$

$$12 \cdot 8 = k$$

$$96 = k$$

Step 3 Substitute the value of k into the equation

$$P = \frac{96}{V}$$

Step 4 Answer the problem's question. Substitute
20 for V and solve for P.

$$P = \frac{96}{V}$$

$$P = \frac{96}{20}$$

$$P = 4.8$$

If the sample expands to a volume of 20 cubic
inches, the new pressure of the gas will be 4.8
pounds per square inch.

7.8 Concept and Vocabulary Check

1. $y = kx$;   constant of variation

2. $y = \dfrac{k}{x}$;   constant of variation

3. inversely

4. directly

7.8 Exercise Set

1. $g = kh$

3. $w = \dfrac{k}{v}$

5. $y = kx$

To find k, substitute 80 for y and 4 for x.

$$80 = k \cdot 4$$

$$20 = k$$

The constant of variation is 20.

7. $W = \dfrac{k}{r}$

To find k, substitute 600 for W and 10 for r.

$$600 = \dfrac{k}{10}$$

$$600 \cdot 10 = \dfrac{k}{10} \cdot 10$$

$$6000 = k$$

The constant of variation is 6000.

9. *Step 1* Write an equation. $y = kx$

Step 2 Use the given values to find k.
Substitute 35 for y and 5 for x.

$$35 = k \cdot 5$$

$$7 = k$$

Step 3 Substitute the value of k into the equation

$$y = 7k$$

Step 4 Answer the question. Substitute 12 for x and solve for y.

$$y = 7 \cdot 12 = 84$$

11. *Step 1* $y = \dfrac{k}{x}$

Step 2 Substitute 10 for y and 5 for x.

$$10 = \dfrac{k}{5}$$

$$50 = k$$

Step 3 $y = \dfrac{50}{x}$

Step 4 Substitute 2 for x and solve for y.

$$y = \dfrac{50}{2}$$

$$y = 25$$

13. **a.** $G = kW$

b. $G = 0.02W$

c. $G = 0.02(52) = 1.04$

Your fingernails would grow 1.04 inches in one year.

15. *Step 1* Write an equation. $C = kM$

Step 2 Use the given values to find k.

$$400 = k \cdot 3000$$

$$\dfrac{400}{3000} = k$$

$$\dfrac{2}{15} = k$$

Step 3 Substitute the value of k into the equation.

$$C = \dfrac{2}{15}M$$

Step 4 Answer the problem's question. Substitute 450 for M and solve for C.

$$C = \dfrac{2}{15}(450) = 60$$

The cost of a 450-mile trip is $60.

17. Let s = speed.
Then M = Mach number.

Step 1 $s = kM$

Step 2 To find k, substitute 1502.2 for s and 2.03 for M.

$$1502.2 = k \cdot 2.03$$

$$\dfrac{150.2}{2.03} = k$$

$$740 = k$$

Step 3 Substitute 740 for k.

$$s = 740M$$

Step 4 Substitute 3.3 for M and solve for s.

$$s = 740(3.3) = 2442$$

The speed of the Blackbird is 2442 miles per hour.

19. Let t = driving time.
Let r = driving rate.

Step 1 $t = \dfrac{k}{r}$

Step 2 $1.5 = \dfrac{k}{20}$

$$1.5 \cdot 20 = \dfrac{k}{20} \cdot 20$$

$$30 = k$$

Step 3 $t = \dfrac{30}{r}$

Step 4 $t = \dfrac{30}{60} = 0.5$

The trip would take 0.5 hour (or 30 minutes) at an average rate of 60 miles per hour.

21. Let V = volume and P = pressure.

Step 1 $V = \dfrac{k}{P}$

Step 2 $32 = \dfrac{k}{8}$

$256 = k$

Step 3 $V = \dfrac{256}{P}$

Step 4 $40 = \dfrac{256}{P}$

$40P = 256$

$P = 6.4$

When the volume is 40 cubic centimeters, the pressure is 6.4 pounds per square centimeter.

23. Let n = the number of pens.
Then p = the price per pen.

Step 1 $n = \dfrac{k}{p}$

Step 2 To find k, substitute 4000 for n and 1.50 for p.

$4000 = \dfrac{k}{1.50}$

$4000 \cdot 1.50 = \dfrac{k}{1.50} \cdot 150$

$6000 = k$

Step 3 $n = \dfrac{6000}{p}$

Step 4 Substitute 1.20 for p and solve for n.

$n = \dfrac{6000}{1.20} = 5000$

There will be 5000 pens sold at \$1.20 each.

25. – 29. Answers will vary.

31. makes sense

33. makes sense

35. $8(2-x) = -5x$

$16 - 8x = -5x$

$16 = 3x$

$\dfrac{16}{3} = x$

The solution set is $\left\{ \dfrac{16}{3} \right\}$.

36.

$$\begin{array}{r}
9x^2 - 6x + 4 \\
3x+2\overline{)27x^3 + \ 0x^2 + 0x - 8} \\
\underline{27x^3 + 18x^2} \\
-18x^2 + \ 0x \\
\underline{-18x^2 - 12x} \\
12x - 8 \\
\underline{12x + 8} \\
-16
\end{array}$$

$\dfrac{27x^3 + 8}{3x+2} = 9x^2 - 6x + 4 - \dfrac{16}{3x+2}$

37. $6x^3 - 6x^2 - 120x = 6x\left(x^2 - x - 20 \right)$

$= 6x(x-5)(x+4)$

38. $\sqrt{x-1} = \sqrt{17-1} = \sqrt{16} = 4$

39. $4\sqrt{x} + 30 = 4\sqrt{25} + 30 = 4 \cdot 5 + 30 = 20 + 30 = 50$

40. $(-2)^5 - (-1)^3 = (-1)^5 \cdot 2^5 - (-1)^3$

$= -1 \cdot 32 - (-1)$

$= -32 + 1$

$= -31$

Chapter 7 Review Exercises

1. $\dfrac{5x}{6x-24}$

Set the denominator equal to 0 and solve for x.

$6x - 24 = 0$

$6x = 24$

$x = 4$

The rational expression is undefined for $x = 4$.

2. $\dfrac{x+3}{(x-2)(x+5)}$

Set the denominator equal to 0 and solve for x.

$(x-2)(x+5) = 0$

$x - 2 = 0 \quad$ or $\quad x + 5 = 0$

$x = 2 \qquad\qquad x = -5$

The rational expression is undefined for $x = 2$ and $x = -5$.

3. $\dfrac{x^2+3}{x^2-3x+2}$

$x^2-3x+2=0$

$(x-1)(x-2)=0$

$x-1=0$ or $x-2=0$

$\qquad x=1 \qquad\qquad x=2$

The rational expression is undefined for $x=1$ and $x=2$.

4. $\dfrac{7}{x^2+81}$

The smallest possible value of x^2 is 0, so

$x^2+81 \geq 81$ for all real numbers x. This means that there is no real number for x for which

$x^2+81=0$. Thus, the rational expression is defined for all real numbers.

5. $\dfrac{16x^2}{12x}=\dfrac{4\cdot 4\cdot x\cdot x}{4\cdot 3\cdot x}=\dfrac{4x}{3}$

6. $\dfrac{x^2-4}{x-2}=\dfrac{(x+2)(x-2)}{(x-2)}=x+2$

7. $\dfrac{x^3+2x^2}{x+2}=\dfrac{x^2(x+2)}{(x+2)}=x^2$

8. $\dfrac{x^2+3x-18}{x^2-36}=\dfrac{(x+6)(x-3)}{(x+6)(x-6)}$

$\qquad =\dfrac{x-3}{x-6}$

9. $\dfrac{x^2-4x-5}{x^2+8x+7}=\dfrac{(x+1)(x-5)}{(x+1)(x+7)}$

$\qquad =\dfrac{x-5}{x+7}$

10. $\dfrac{y^2+2y}{y^2+4y+4}=\dfrac{y(y+2)}{(y+2)(y+2)}$

$\qquad =\dfrac{y}{y+2}$

11. $\dfrac{x^2}{x^2+4}$

The numerator and denominator have no common factor, so this rational expression cannot be simplified.

12. $\dfrac{2x^2-18y^2}{3y-x}=\dfrac{2\left(x^2-9y^2\right)}{3y-x}$

$\qquad =\dfrac{2(x+3y)(x-3y)}{(3y-x)}$

$\qquad =\dfrac{2(x+3y)(-1)(3y-x)}{(3y-x)}$

$\qquad =-2(x+3y)$ or $-2x-6y$

13. $\dfrac{x^2-4}{12x}\cdot\dfrac{3x}{x+2}=\dfrac{(x+2)(x-2)}{12x}\cdot\dfrac{3x}{(x+2)}$

$\qquad =\dfrac{x-2}{4}$

14. $\dfrac{5x+5}{6}\cdot\dfrac{3x}{x^2+x}=\dfrac{5(x+1)}{6}\cdot\dfrac{3x}{x(x+1)}$

$\qquad =\dfrac{5}{2}$

15. $\dfrac{x^2+6x+9}{x^2-4}\cdot\dfrac{x-2}{x+3}$

$\qquad =\dfrac{(x+3)(x+3)}{(x+2)(x-2)}\cdot\dfrac{x-2}{x+3}$

$\qquad =\dfrac{x+3}{x+2}$

16. $\dfrac{y^2-2y+1}{y^2-1}\cdot\dfrac{2y^2+y-1}{5y-5}$

$\qquad =\dfrac{(y-1)(y-1)}{(y+1)(y-1)}\cdot\dfrac{(2y-1)(y+1)}{5(y-1)}$

$\qquad =\dfrac{2y-1}{5}$

17. $\dfrac{2y^2+y-3}{4y^2-9}\cdot\dfrac{3y+3}{5y-5y^2}$

$\qquad =\dfrac{(2y+3)(y-1)}{(2y+3)(2y-3)}\cdot\dfrac{3(y+1)}{5y(1-y)}$

$\qquad =\dfrac{-3(y+1)}{5y(2y-3)}$ or $-\dfrac{3(y+1)}{5y(2y-3)}$

18. $\dfrac{x^2+x-2}{10} \div \dfrac{2x+4}{5}$

$= \dfrac{x^2+x-2}{10} \cdot \dfrac{5}{2x+4}$

$= \dfrac{(x-1)(x+2)}{10} \cdot \dfrac{5}{2(x+2)}$

$= \dfrac{x-1}{4}$

19. $\dfrac{6x+2}{x^2-1} \div \dfrac{3x^2+x}{x-1}$

$= \dfrac{6x+2}{x^2-1} \cdot \dfrac{x-1}{3x^2+x}$

$= \dfrac{2(3x+1)}{(x+1)(x-1)} \cdot \dfrac{(x-1)}{x(3x+1)}$

$= \dfrac{2}{x(x+1)}$

20. $\dfrac{1}{y^2+8y+15} \div \dfrac{7}{y+5}$

$= \dfrac{1}{y^2+8y+15} \cdot \dfrac{y+5}{7}$

$= \dfrac{1}{(y+3)(y+5)} \cdot \dfrac{(y+5)}{7}$

$= \dfrac{1}{7(y+3)}$

21. $\dfrac{y^2+y-42}{y-3} \div \dfrac{y+7}{(y-3)^2}$

$= \dfrac{y^2+y-42}{y-3} \cdot \dfrac{(y-3)^2}{y+7}$

$= \dfrac{(y+7)(y-6)}{(y-3)} \cdot \dfrac{(y-3)(y-3)}{y+7}$

$= (y-6)(y-3)$ or $y^2-9y+18$

22. $\dfrac{8x+8y}{x^2} \div \dfrac{x^2-y^2}{x^2}$

$= \dfrac{8x+8y}{x^2} \cdot \dfrac{x^2}{x^2-y^2}$

$= \dfrac{8(x+y)}{x^2} \cdot \dfrac{x^2}{(x+y)(x-y)}$

$= \dfrac{8}{x-y}$

23. $\dfrac{4x}{x+5} + \dfrac{20}{x+5} = \dfrac{4x+20}{x+5} = \dfrac{4(x+5)}{x+5} = 4$

24. $\dfrac{8x-5}{3x-1} + \dfrac{4x+1}{3x-1} = \dfrac{8x-5+4x+1}{3x-1}$

$= \dfrac{12x-4}{3x-1}$

$= \dfrac{4(3x-1)}{3x-1} = 4$

25. $\dfrac{3x^2+2x}{x-1} - \dfrac{10x-5}{x-1}$

$= \dfrac{\left(3x^2+2x\right)-\left(10x-5\right)}{x-1}$

$= \dfrac{3x^2+2x-10x+5}{x-1}$

$= \dfrac{3x^2-8x+5}{x-1}$

$= \dfrac{(3x-5)(x-1)}{x-1}$

$= 3x-5$

26. $\dfrac{6y^2-4y}{2y-3} - \dfrac{12-3y}{2y-3}$

$= \dfrac{\left(6y^2-4y\right)-\left(12-3y\right)}{2y-3}$

$= \dfrac{6y^2-4y-12+3y}{2y-3}$

$= \dfrac{6y^2-y-12}{2y-3}$

$= \dfrac{(2y-3)(3y+4)}{2y-3}$

$= 3y+4$

27. $\dfrac{x}{x-2}+\dfrac{x-4}{2-x}=\dfrac{x}{x-2}+\dfrac{(-1)}{(-1)}\cdot\dfrac{x-4}{x-2}$

$\qquad\qquad =\dfrac{x}{x-2}+\dfrac{-x+4}{x-2}$

$\qquad\qquad =\dfrac{x-x+4}{x-2}=\dfrac{4}{x-2}$

28. $\dfrac{x+5}{x-3}-\dfrac{x}{3-x}=\dfrac{x+5}{x-3}-\dfrac{(-1)}{(-1)}\cdot\dfrac{x}{3-x}$

$\qquad\qquad =\dfrac{x+5}{x-3}+\dfrac{x}{x-3}$

$\qquad\qquad =\dfrac{x+5+x}{x-3}=\dfrac{2x+5}{x-5}$

29. $\dfrac{7}{9x^3}$ and $\dfrac{5}{12x}$

$9x^3=3^2x^3$

$12x=2^2\cdot 3x$

LCD $=2^2\cdot 3^2\cdot x^3=36x^3$

30. $\dfrac{3}{x^2(x-1)}$ and $\dfrac{11}{x(x-1)^2}$

LCD $=x^2(x-1)^2$

31. $\dfrac{x}{x^2+4x+3}$ and $\dfrac{17}{x^2+10x+21}$

$x^2+4x+3=(x+3)(x+1)$

$x^2+10x+21=(x+3)(x+7)$

LCD $=(x+3)(x+1)(x+7)$

32. $\dfrac{7}{3x}+\dfrac{5}{2x^2}$

LCD $=6x^2$

$\dfrac{7}{3x}+\dfrac{6}{2x^2}=\dfrac{7}{3x}\cdot\dfrac{2x}{2x}+\dfrac{5}{2x^2}\cdot\dfrac{3}{3}$

$\qquad\qquad =\dfrac{14x+15}{6x^2}$

33. $\dfrac{5}{x+1}+\dfrac{2}{x}$

LCD $=x(x+1)$

$\dfrac{5}{x+1}+\dfrac{2}{x}=\dfrac{5x}{x(x+1)}+\dfrac{2(x+1)}{x(x+1)}$

$\qquad\qquad =\dfrac{5x+2(x+1)}{x(x+1)}=\dfrac{5x+2x+2}{x(x+1)}$

$\qquad\qquad =\dfrac{7x+2}{x(x+1)}$

34. $\dfrac{7}{x+3}+\dfrac{4}{(x+3)^2}$

LCD $=(x+3)^2$ or $(x+3)(x+3)$

$\dfrac{7}{x+3}+\dfrac{4}{(x+3)^2}$

$=\dfrac{7}{x+3}+\dfrac{4}{(x+3)(x+3)}$

$=\dfrac{7(x+3)}{(x+3)(x+3)}+\dfrac{4}{(x+3)(x+3)}$

$=\dfrac{7(x+3)+4}{(x+3)(x+3)}=\dfrac{7x+21+4}{(x+3)(x+3)}$

$=\dfrac{7x+25}{(x+3)(x+3)}$ or $\dfrac{7x+25}{(x+3)^2}$

35. $\dfrac{6y}{y^2-4}-\dfrac{3}{y+2}$

$y^2-4=(y+2)(y-2)$

$y+2=1(y+2)$

LCD $=(y+2)(y-2)$

$\dfrac{6y}{y^2-4}-\dfrac{3}{y+2}=\dfrac{6y}{(y+2)(y-2)}-\dfrac{3}{y+2}$

$=\dfrac{6y}{(y+2)(y-2)}-\dfrac{3(y-2)}{(y+2)(y-2)}$

$=\dfrac{6y-3(y-2)}{(y+2)(y-2)}=\dfrac{6y-3y+6}{(y+2)(y-2)}$

$=\dfrac{3y+6}{(y+2)(y-2)}=\dfrac{3(y+2)}{(y+2)(y-2)}$

$=\dfrac{3}{y-2}$

36. $\dfrac{y-1}{y^2-2y+1}-\dfrac{y+1}{y-1}$

$=\dfrac{y-1}{(y-1)(y-1)}-\dfrac{y+1}{y-1}$

$=\dfrac{1}{y-1}-\dfrac{y+1}{y-1}$

$=\dfrac{1-(y+1)}{y-1}=\dfrac{1-y-1}{y-1}$

$=\dfrac{-y}{y-1}$ or $-\dfrac{y}{y-1}$

37. $\dfrac{x+y}{y}-\dfrac{y-x}{x}$

LCD = xy

$\dfrac{x+y}{y}-\dfrac{y-x}{x}=\dfrac{(x+y)}{y}\cdot\dfrac{x}{x}-\dfrac{(x-y)}{x}\cdot\dfrac{y}{y}$

$=\dfrac{x^2+xy}{xy}-\dfrac{xy-y^2}{xy}$

$=\dfrac{(x^2+xy)-(xy-y^2)}{xy}$

$=\dfrac{x^2+xy-xy+y^2}{xy}$

$=\dfrac{x^2+y^2}{xy}$

38. $\dfrac{2x}{x^2+2x+1}+\dfrac{x}{x^2-1}$

$x^2+2x+1=(x+1)(x+1)$

$x^2-1=(x+1)(x-1)$

LCD $=(x+1)(x+1)(x-1)$

$\dfrac{2x}{x^2+2x+1}+\dfrac{x}{x^2-1}$

$=\dfrac{2x}{(x+1)(x+1)}+\dfrac{x}{(x+1)(x-1)}$

$=\dfrac{2x(x-1)}{(x+1)(x+1)(x-1)}+\dfrac{x(x+1)}{(x+1)(x-1)(x+1)}$

$=\dfrac{2x(x-1)+x(x+1)}{(x+1)(x+1)(x-1)}$

$=\dfrac{2x^2-2x+x^2+x}{(x+1)(x+1)(x-1)}$

$=\dfrac{3x^2-x}{(x+1)(x+1)(x-1)}$

39. $\dfrac{5x}{x+1}-\dfrac{2x}{1-x^2}$

$x+1=1(x+1)$

$1-x^2=-1(x^2-1)=-(x+1)(x-1)$

LCD $=(x+1)(x-1)$

$\dfrac{5x}{x+1}-\dfrac{2x}{1-x^2}$

$=\dfrac{5x}{x+1}-\dfrac{(-1)}{(-1)}\cdot\dfrac{2x}{1-x^2}=\dfrac{5x}{x+1}-\dfrac{-2x}{x^2-1}$

$=\dfrac{5x(x-1)}{(x+1)(x-1)}-\dfrac{-2x}{(x+1)(x-1)}$

$=\dfrac{5x(x-1)+2x}{(x+1)(x-1)}=\dfrac{5x^2-5x+2x}{(x+1)(x-1)}$

$=\dfrac{5x^2-3x}{(x+1)(x-1)}$

40. $\dfrac{4}{x^2-x-6}-\dfrac{4}{x^2-4}$

$x^2-x-6=(x+2)(x-3)$

$x^2-4=(x+2)(x-2)$

LCD $=(x+2)(x-3)(x-2)$

$\dfrac{4}{x^2-x-6}-\dfrac{4}{x^2-4}$

$=\dfrac{4}{(x+2)(x-3)}-\dfrac{4}{(x+2)(x-2)}$

$=\dfrac{4(x-2)}{(x+2)(x-3)(x-2)}-\dfrac{4(x-3)}{(x+2)(x-3)(x-2)}$

$=\dfrac{4(x-2)-4(x-3)}{(x+2)(x-3)(x-2)}$

$=\dfrac{4x-8-4x+12}{(x+2)(x-3)(x-2)}$

$=\dfrac{4}{(x+2)(x-3)(x-2)}$

41. $\dfrac{7}{x+3}+2$

LCD $=x+3$

$\dfrac{7}{x+3}+2=\dfrac{7}{x+3}+\dfrac{2(x+3)}{x+3}$

$=\dfrac{7+2(x+3)}{x+3}$

$=\dfrac{7+2x+6}{x+3}$

$=\dfrac{2x+13}{x+3}$

42. $\dfrac{2y-5}{6y+9}-\dfrac{4}{2y^2+3y}$

$6y+9=3(2y+3)$

$2y^2+3y=y(2y+3)$

LCD $=3y(2y+3)$

$\dfrac{2y-5}{6y+9}-\dfrac{4}{2y^2+3y}$

$=\dfrac{2y-5}{3(2y+3)}-\dfrac{4}{y(2y+3)}$

$=\dfrac{(2y-5)(y)}{3(2y+3)(y)}-\dfrac{4(3)}{y(2y+3)(3)}$

$=\dfrac{2y^2-5y-12}{3y(2y+3)}=\dfrac{(2y+3)(y-4)}{3y(2y+3)}$

$=\dfrac{y-4}{3y}$

43. $\dfrac{\frac{1}{2}+\frac{3}{8}}{\frac{3}{4}-\frac{1}{2}}=\dfrac{\frac{4}{8}+\frac{3}{8}}{\frac{3}{4}-\frac{2}{4}}=\dfrac{\frac{7}{8}}{\frac{1}{4}}=\dfrac{7}{8}\cdot\dfrac{4}{1}=\dfrac{7}{2}$

44. $\dfrac{\frac{1}{x}}{1-\frac{1}{x}}$

LCD $=x$

$\dfrac{\frac{1}{x}}{1-\frac{1}{x}}=\dfrac{x}{x}\cdot\dfrac{\left(\frac{1}{x}\right)}{\left(1-\frac{1}{x}\right)}$

$=\dfrac{x\cdot\frac{1}{x}}{x\cdot1-x\cdot\frac{1}{x}}=\dfrac{1}{x-1}$

45. $\dfrac{\frac{1}{x}+\frac{1}{y}}{\frac{1}{xy}}$

LCD $=xy$

$\dfrac{\frac{1}{x}+\frac{1}{y}}{\frac{1}{xy}}=\dfrac{xy}{xy}\cdot\dfrac{\left(\frac{1}{x}+\frac{1}{y}\right)}{\left(\frac{1}{xy}\right)}$

$=\dfrac{xy\cdot\frac{1}{x}+xy\cdot\frac{1}{y}}{xy\cdot\frac{1}{xy}}$

$=\dfrac{y+x}{1}=y+x\text{ or }x+y$

46.
$$\frac{\frac{1}{x}-\frac{1}{2}}{\frac{1}{3}-\frac{x}{6}}=\frac{\frac{2}{2x}-\frac{x}{2x}}{\frac{2}{6}-\frac{x}{6}}=\frac{\frac{2-x}{2x}}{\frac{2-x}{6}}$$
$$=\frac{2-x}{2x}\cdot\frac{6}{2-x}=\frac{3}{x}$$

47.
$$\frac{3+\frac{12}{x}}{1-\frac{16}{x^2}}$$
LCD $=x^2$
$$\frac{3+\frac{12}{x}}{1-\frac{16}{x^2}}=\frac{x^2}{x^2}\cdot\frac{\left(3+\frac{12}{x}\right)}{\left(1-\frac{16}{x^2}\right)}$$
$$=\frac{x^2\cdot 3+x^2\cdot\frac{12}{x}}{x^2\cdot 1-x^2\cdot\frac{16}{x^2}}=\frac{3x^2+12x}{x^2-16}$$
$$=\frac{3x(x+4)}{(x+4)(x-4)}=\frac{3x}{x-4}$$

48. $\frac{3}{x}-\frac{1}{6}=\frac{1}{x}$

The restriction is $x\neq 0$.
The LCD is $6x$.
$$\frac{3}{x}-\frac{1}{6}=\frac{1}{x}$$
$$6x\left(\frac{3}{x}-\frac{1}{6}\right)=6x\left(\frac{1}{x}\right)$$
$$6x\cdot\frac{3}{x}-6x\cdot\frac{1}{6}=6x\cdot\frac{1}{x}$$
$$18-x=6$$
$$-x=-12$$
$$x=12$$
The solution set is {12}.

49. $\frac{3}{4x}=\frac{1}{x}+\frac{1}{4}$

The restriction is $x\neq 0$.
The LCD is $4x$.
$$\frac{3}{4x}=\frac{1}{x}+\frac{1}{4}$$
$$4x\left(\frac{3}{4x}\right)=4x\left(\frac{1}{x}+\frac{1}{4}\right)$$
$$3=4+x$$
$$-1=x$$
The solution set is $\{-1\}$.

50. $x+5=\frac{6}{x}$

The restriction is $x\neq 0$.
The LCD is x.
$$x+5=\frac{6}{x}$$
$$x(x+5)=x\left(\frac{6}{x}\right)$$
$$x^2+5x=6$$
$$x^2+5x-6=0$$
$$(x+6)(x-1)=0$$
$$x+6=0\quad\text{or}\quad x-1=0$$
$$x=-6\qquad x=1$$
The solution set is $\{-6,1\}$.

51. $4-\frac{x}{x+5}=\frac{5}{x+5}$

The restriction is $x\neq -5$.
The LCD is $x+5$.
$$(x+5)\left(4-\frac{x}{x+5}\right)=(x+5)\left(\frac{5}{x+5}\right)$$
$$(x+5)\cdot 4-(x+5)\left(\frac{x}{x+5}\right)=(x+5)\left(\frac{5}{x+5}\right)$$
$$4x+20-x=5$$
$$3x+20=5$$
$$3x=-15$$
$$x=-5$$
The only proposed solution, −5, is *not* a solution because of the restriction $x\neq -5$. Notice that −5 makes two of the denominators zero in the original equation. The solution set is { }.

52. $\dfrac{2}{x-3} = \dfrac{4}{x+3} + \dfrac{8}{x^2-9}$

To find any restrictions and the LCD, all denominators should be written in factored form.

$\dfrac{2}{x-3} = \dfrac{4}{x+3} + \dfrac{8}{(x+3)(x-3)}$

Restrictions: $x \neq 3, x \neq -3$

LCD $= (x+3)(x-3)$

$$(x+3)(x-3) \cdot \dfrac{2}{x-3} = (x+3)(x-3)\left(\dfrac{4}{x+3} + \dfrac{8}{(x+3)(x-3)}\right)$$

$$2(x+3) = 4(x-3) + 8$$

$$2x + 6 = 4x - 12 + 8$$

$$2x + 6 = 4x - 4$$

$$6 = 2x - 4$$

$$10 = 2x$$

$$5 = x$$

The solution set is $\{5\}$.

53. $\dfrac{2}{x} = \dfrac{2}{3} + \dfrac{x}{6}$

Restriction: $x \neq 0$

LCD $= 6x$

$$6x\left(\dfrac{2}{x}\right) = 6x\left(\dfrac{2}{3} + \dfrac{x}{6}\right)$$

$$12 = 4x + x^2$$

$$0 = x^2 + 4x - 12$$

$$0 = (x+6)(x-2)$$

$$x + 6 = 0 \quad \text{or} \quad x - 2 = 0$$

$$x = -6 \qquad\qquad x = 2$$

The solution set is $\{-6, 2\}$.

54. $\dfrac{13}{y-1} - 3 = \dfrac{1}{y-1}$

Restriction: $y \neq 1$

LCD $= y - 1$

$$(y-1)\left(\dfrac{13}{y-1} - 3\right) = (y-1)\left(\dfrac{1}{y-1}\right)$$

$$13 - 3(y-1) = 1$$

$$13 - 3y + 3 = 1$$

$$16 - 3y = 1$$

$$-3y = -15$$

$$y = 5$$

The solution set is $\{5\}$.

55.

$$\frac{1}{x+3} - \frac{1}{x-1} = \frac{x+1}{x^2 + 2x - 3}$$

$$\frac{1}{x+3} - \frac{1}{x-1} = \frac{x+1}{(x+3)(x-1)}$$

Restrictions: $x \neq -3, x \neq 1$

LCD $= (x+3)(x-1)$

$$(x+3)(x-1)\left[\frac{1}{x+3} - \frac{1}{x-1}\right] = (x+3)(x-1)\cdot\left[\frac{x+1}{(x+3)(x-1)}\right]$$

$$(x-1) - (x+3) = x+1$$

$$x - 1 - x - 3 = x + 1$$

$$-4 = x + 1$$

$$-5 = x$$

The solution set is $\{-5\}$.

56.

$$P = \frac{250(3t+5)}{t+25}$$

$$125 = \frac{250(3t+5)}{t+25}$$

$$125(t+25) = \frac{250(3t+5)}{t+25}\cdot(t+25)$$

$$125t + 3125 = 250(3t+5)$$

$$125t + 3125 = 750t + 1250$$

$$3125 = 625t + 1250$$

$$1875 = 625t$$

$$3 = t$$

It will take 3 years for the population to reach 125 elk.

57.

$$S = \frac{C}{1-r}$$

$$200 = \frac{140}{1-r}$$

$$200(1-r) = \frac{140}{1-r}\cdot 1-r$$

$$200 - 200r = 140$$

$$-200r = -60$$

$$r = \frac{-60}{-200} = \frac{3}{10} = 30\%$$

The markup is 30%.

58. Let $x =$ the rate of the current. Then $20 + x =$ the rate of the boat with the current and $20 - x =$ the rate of the boat against the current.

	Distance	Rate	Time $= \dfrac{\text{Distance}}{\text{Rate}}$
Downstream	72	$20 + x$	$\dfrac{72}{20 + x}$
Upstream	48	$20 - x$	$\dfrac{48}{20 - x}$

The times are equal, so

$$\frac{72}{20 + x} = \frac{48}{20 - x}$$
$$72(20 - x) = 48(20 + x)$$
$$1440 - 72x = 960 + 48x$$
$$1440 = 960 + 120x$$
$$480 = 120x$$
$$4 = x$$

The rate of the current is 4 miles per hour.

59. Let $x =$ the rate of the slower car. Then $x + 10 =$ the rate of the faster car.

	Distance	Rate	Time $= \dfrac{\text{Distance}}{\text{Rate}}$
Slow car	60	x	$\dfrac{60}{x}$
Faster Car	90	$x + 10$	$\dfrac{90}{x + 10}$

$$\frac{60}{x} = \frac{90}{x + 10}$$
$$60(x + 10) = 90x$$
$$60x + 600 = 90x$$
$$600 = 30x$$
$$20 = x$$

If $x = 20$, $x + 10 = 30$.

The rate of the slower car is 20 miles per hour and the rate of the faster car is 30 miles per hour.

60. Let $x =$ the time, in hours, for both people to paint the fence together.

	Fractional part of job completed in 1 hour	Time working together	Fractional part of job completed in x hours
Painter	$\dfrac{1}{6}$	x	$\dfrac{x}{6}$
Apprentice	$\dfrac{1}{12}$	x	$\dfrac{x}{12}$

Working together, the two people complete one whole job, so $\dfrac{x}{6} + \dfrac{x}{12} = 1$.

Multiply both sides by the LCD, 12.

$$12\left(\frac{x}{6}+\frac{x}{12}\right)=12\cdot 1$$

$$2x+x=12$$

$$3x=12$$

$$x=4$$

It would take them 4 hours to paint the fence working together.

61. Let x = number of teachers needed for 5400 students.

$$\frac{3}{50}=\frac{x}{5400}$$

$$50x=3\cdot 5400$$

$$50x=16,200$$

$$\frac{50x}{50}=\frac{16,200}{50}$$

$$x=324$$

For an enrollment of 5400 students, 324 teachers are needed.

62. Let x = number of trout in the lake.

$$\frac{\text{Original Number Tagged Deer}}{\text{Total Number of Deer}}=\frac{\text{Number Tagged Deer in Sample}}{\text{Total Number Deer in Sample}}$$

$$\frac{112}{x}=\frac{32}{82}$$

$$32x=112\cdot 82$$

$$32x=9184$$

$$\frac{32x}{32}=\frac{9184}{32}$$

$$x=287$$

There are 287 trout in the lake.

63. $\dfrac{8}{4}=\dfrac{10}{x}$

$$8x=40$$

$$x=5$$

The length of the side marked with an x is 5 feet.

64. Write a proportion relating the corresponding sides of the large and small triangle. Notice that the length of the base of the larger triangle is $9 \text{ ft}+6 \text{ ft}=15$ ft.

$$\frac{x}{5}=\frac{15}{6}$$

$$6x=5\cdot 15$$

$$6x=75$$

$$x=\frac{75}{6}=12.5$$

The height of the lamppost is 12.5 feet.

Chapter 7: Rational Expressions

65. Let b = electric bill (in dollars).
Then e = number of kilowatts of electricity used.
Step 1 $b = ke$
Step 2 To find k, substitute 98 for b and 1400 for e.
$$98 = k \cdot 1400$$
$$\frac{98}{1400} = k$$
$$0.07 = k$$
Step 3 $b = 0.07e$
Step 4 Substitute 2200 for e and solve for b.
$$b = 0.07(2200) = 154$$
The bill is $154 for 2200 kilowatts of electricity.

66. *Step 1* $t = \dfrac{k}{r}$
Step 2 To find k, substitute k, substitute 4 for t and 50 for r.
$$4 = \frac{k}{50}$$
$$200 = k$$
Step 3 $t = \dfrac{200}{r}$
Step 4 Substitute 40 for r and solve for t.
$$t = \frac{200}{40} = 5$$
At 40 miles per hour, the trip will take 5 hours.

Chapter 7 Test

1. $\dfrac{x+7}{x^2+5x-36}$
Set the denominator equal to 0 and solve for x.
$$x^2 + 5x - 36 = 0$$
$$(x+9)(x-4) = 0$$
$$x + 9 = 0 \quad \text{or} \quad x - 4 = 0$$
$$x = -9 \qquad\qquad x = 4$$
The rational expression is undefined for $x = -9$ and $x = 4$.

2. $\dfrac{x^2+2x-3}{x^2-3x+2} = \dfrac{(x-1)(x+3)}{(x-1)(x-2)} = \dfrac{x+3}{x-2}$

3. $\dfrac{4x^2-20x}{x^2-4x-5} = \dfrac{4x(x-5)}{(x+1)(x-5)} = \dfrac{4x}{x+1}$

4. $\dfrac{x^2-16}{10} \cdot \dfrac{5}{x+4} = \dfrac{(x+4)(x-4)}{10} \cdot \dfrac{5}{(x+4)}$
$$= \dfrac{x-4}{2}$$

5. $\dfrac{x^2-7x+12}{x^2-4x} \cdot \dfrac{x^2}{x^2-9}$
$$= \dfrac{(x-3)(x-4)}{x(x-4)} \cdot \dfrac{x^2}{(x+3)(x-3)}$$
$$= \dfrac{x}{x+3}$$

6. $\dfrac{2x+8}{x-3} \div \dfrac{x^2+5x+4}{x^2-9}$
$$= \dfrac{2x+8}{x-3} \cdot \dfrac{x^2-9}{x^2+5x+4}$$
$$= \dfrac{2(x+4)}{(x-3)} \cdot \dfrac{(x+3)(x-3)}{(x+4)(x+1)}$$
$$= \dfrac{2(x+3)}{x+1} = \dfrac{2x+6}{x+1}$$

7. $\dfrac{5y+5}{(y-3)^2} \div \dfrac{y^2-1}{y-3}$
$$= \dfrac{5y+5}{(y-3)^2} \cdot \dfrac{y-3}{y^2-1}$$
$$= \dfrac{5(y+1)}{(y-3)(y-3)} \cdot \dfrac{(y-3)}{(y+1)(y-1)}$$
$$= \dfrac{5}{(y-3)(y-1)}$$

8. $\dfrac{2y^2+5}{y+3} + \dfrac{6y-5}{y+3}$
$$= \dfrac{(2y^2+5)+(6y-5)}{y+3}$$
$$= \dfrac{2y^2+5+6y-5}{y+3}$$
$$= \dfrac{2y^2+6y}{y+3}$$
$$= \dfrac{2y(y+3)}{y+3} = 2y$$

390　Copyright © 2013 Pearson Education, Inc.

9. $\dfrac{y^2-2y+3}{y^2+7y+12}-\dfrac{y^2-4y-5}{y^2+7y+12}$

$=\dfrac{\left(y^2-2y+3\right)-\left(y^2-4y-5\right)}{y^2+7y+12}$

$=\dfrac{y^2-2y+3-y^2+4y+5}{y^2+7y+12}$

$=\dfrac{2y+8}{y^2+7y+12}$

$=\dfrac{2(y+4)}{(y+3)(y+4)}$

$=\dfrac{2}{y+3}$

10. $\dfrac{x}{x+3}+\dfrac{5}{x-3}$

LCD $=(x+3)(x-3)$

$\dfrac{x}{x+3}+\dfrac{5}{x-3}$

$=\dfrac{x(x-3)}{(x+3)(x-3)}+\dfrac{5(x+3)}{(x+3)(x-3)}$

$=\dfrac{x(x-3)+5(x+3)}{(x+3)(x-3)}$

$=\dfrac{x^2-3x+5x+15}{(x+3)(x-3)}$

$=\dfrac{x^2+2x+15}{(x+3)(x-3)}$

11. $\dfrac{2}{x^2-4x+3}+\dfrac{6}{x^2+x-2}$

$x^2-4x+3=(x-1)(x-3)$

$x^2+x-2=(x-1)(x+2)$

LCD $=(x-1)(x-3)(x+2)$

$\dfrac{2}{x^2-4x+3}+\dfrac{6}{x^2+x-2}$

$=\dfrac{2}{(x-1)(x-3)}+\dfrac{6}{(x-1)(x+2)}$

$=\dfrac{2(x+2)}{(x-1)(x-3)(x+2)}$

$\quad+\dfrac{6(x-3)}{(x-1)(x-3)(x+2)}$

$=\dfrac{2(x+2)+6(x-3)}{(x-1)(x-3)(x+2)}$

$=\dfrac{2x+4+6x-18}{(x-1)(x-3)(x+2)}$

$=\dfrac{8x-14}{(x-1)(x-3)(x+2)}$

12. $\dfrac{4}{x-3}+\dfrac{x+5}{3-x}$

$3-x=-1(x-3)$

LCD $=x-3$

$\dfrac{4}{x-3}+\dfrac{x+5}{3-x}$

$=\dfrac{4}{x-3}+\dfrac{(-1)}{(-1)}\cdot\dfrac{(x+5)}{(3-x)}$

$=\dfrac{4}{x-3}+\dfrac{-x-5}{x-3}$

$=\dfrac{4-x-5}{x-3}=\dfrac{-x-1}{x-3}$

13. $1+\dfrac{3}{x-1}$

LCD $=x-1$

$1+\dfrac{3}{x-1}=\dfrac{1(x-1)}{x-1}+\dfrac{3}{x-1}$

$=\dfrac{x-1+3}{x-1}=\dfrac{x+2}{x-1}$

14. $\dfrac{2x+3}{x^2-7x+12}-\dfrac{2}{x-3}$

$x^2-7x+12=(x-3)(x-4)$

$x-3=1(x-3)$

LCD $=(x-3)(x-4)$

$\dfrac{2x+3}{x^2-7x+12}-\dfrac{2}{x-3}$

$=\dfrac{2x+3}{(x-3)(x-4)}-\dfrac{2(x-4)}{(x-3)(x-4)}$

$=\dfrac{2x+3-2(x-4)}{(x-3)(x-4)}$

$=\dfrac{2x+3-2x+8}{(x-3)(x-4)}$

$=\dfrac{11}{(x-3)(x-4)}$

15. $\dfrac{8y}{y^2-16}-\dfrac{4}{y-4}$

$y^2-16=(y+4)(y-4)$

$y-4=1(y-4)$

LCD $=(y+4)(y-4)$

$\dfrac{8y}{y^2-16}-\dfrac{4}{y-4}$

$=\dfrac{8y}{(y+4)(y-4)}-\dfrac{4}{y-4}$

$=\dfrac{8y}{(y+4)(y-4)}-\dfrac{4(y+4)}{(y+4)(y-4)}$

$=\dfrac{8y-4(y+4)}{(y+4)(y-4)}$

$=\dfrac{8y-4y-16}{(y+4)(y-4)}$

$=\dfrac{4y-16}{(y+4)(y-4)}$

$=\dfrac{4(y-4)}{(y+4)(y-4)}$

$=\dfrac{4}{y+4}$

16. $\dfrac{(x-y)^2}{x+y}\div\dfrac{x^2-xy}{3x+3y}=\dfrac{(x-y)^2}{x+y}\cdot\dfrac{3x+3y}{x^2-xy}$

$\phantom{\dfrac{(x-y)^2}{x+y}\div\dfrac{x^2-xy}{3x+3y}}=\dfrac{(x-y)(x-y)}{(x+y)}\cdot\dfrac{3(x+y)}{x(x-y)}$

$\phantom{\dfrac{(x-y)^2}{x+y}\div\dfrac{x^2-xy}{3x+3y}}=\dfrac{3(x-y)}{x}=\dfrac{3x-3y}{x}$

17. $\dfrac{5+\dfrac{5}{x}}{2+\dfrac{1}{x}}=\dfrac{\dfrac{5x}{x}+\dfrac{5}{x}}{\dfrac{2x}{x}+\dfrac{1}{x}}=\dfrac{\dfrac{5x+5}{x}}{\dfrac{2x+1}{x}}$

$\phantom{\dfrac{5+\dfrac{5}{x}}{2+\dfrac{1}{x}}}=\dfrac{5x+5}{x}\cdot\dfrac{x}{2x+1}$

$\phantom{\dfrac{5+\dfrac{5}{x}}{2+\dfrac{1}{x}}}=\dfrac{5x+5}{2x+1}$

18. $\dfrac{\dfrac{1}{x}-\dfrac{1}{y}}{\dfrac{1}{x}}$

LCD $=xy$

$\dfrac{\dfrac{1}{x}-\dfrac{1}{y}}{\dfrac{1}{x}}=\dfrac{xy}{xy}\cdot\dfrac{\left(\dfrac{1}{x}-\dfrac{1}{y}\right)}{\left(\dfrac{1}{x}\right)}$

$\phantom{\dfrac{\dfrac{1}{x}-\dfrac{1}{y}}{\dfrac{1}{x}}}=\dfrac{xy\cdot\dfrac{1}{x}-xy\cdot\dfrac{1}{y}}{xy\cdot\dfrac{1}{x}}$

$\phantom{\dfrac{\dfrac{1}{x}-\dfrac{1}{y}}{\dfrac{1}{x}}}=\dfrac{y-x}{y}$

19. $\dfrac{5}{x}+\dfrac{2}{3}=2-\dfrac{2}{x}-\dfrac{1}{6}$

Restriction: $x\neq 0$

LCD $=6x$

$6x\left(\dfrac{5}{x}+\dfrac{2}{3}\right)=6x\left(2-\dfrac{2}{x}-\dfrac{1}{6}\right)$

$6x\cdot\dfrac{5}{x}+6x\cdot\dfrac{2}{3}=6x\cdot 2-6x\cdot\dfrac{2}{x}-6x\cdot\dfrac{1}{6}$

$30+4x=12x-12-x$

$30+4x=11x-12$

$30=7x-12$

$42=7x$

$6=x$

The solution set is $\{6\}$.

20. $\dfrac{3}{y+5} - 1 = \dfrac{4-y}{2y+10}$

$\dfrac{3}{y+5} - 1 = \dfrac{4-y}{2(y+5)}$

Restriction: $y \neq -5$

LCD $= 2(y+5)$

$2(y+5)\left(\dfrac{3}{y+5} - 1\right) = 2(y+5)\left[\dfrac{4-y}{2(y+5)}\right]$

$\qquad 6 - 2(y+5) = 4 - y$

$\qquad 6 - 2y - 10 = 4 - y$

$\qquad\qquad -4 - 2y = 4 - y$

$\qquad\qquad\qquad -4 = 4 + y$

$\qquad\qquad\qquad -8 = y$

The solution set is $\{-8\}$.

21. $\dfrac{2}{x-1} = \dfrac{3}{x^2-1} + 1$

$\dfrac{2}{x-1} = \dfrac{3}{(x+1)(x-1)} + 1$

Restrictions: $x \neq 1, x \neq -1$

LCD $= (x+1)(x-1)$

$(x+1)(x-1)\left(\dfrac{2}{x-1}\right) = (x+1)(x-1)\left[\dfrac{3}{(x+1)(x-1)} + 1\right]$

$\qquad\qquad 2(x+1) = 3 + (x+1)(x-1)$

$\qquad\qquad 2x + 2 = 3 + x^2 - 1$

$\qquad\qquad 2x + 2 = 2 + x^2$

$\qquad\qquad\qquad 0 = x^2 - 2x$

$\qquad\qquad\qquad 0 = x(x-2)$

$\qquad x = 0 \ \text{ or } \ x - 2 = 0$

$\qquad\qquad\qquad\qquad\qquad x = 2$

The solution set is $\{0, 2\}$.

22. Let x = the rate of the current.
Then $30 + x$ = the rate of the boat with the current and $30 - x$ = the rate of the boat against the current.

	Distance	Rate	Time = $\dfrac{\text{Distance}}{\text{Rate}}$
Downstream	16	$30 + x$	$\dfrac{16}{30 + x}$
Upstream	14	$30 - x$	$\dfrac{14}{30 - x}$

$$\frac{16}{30+x} = \frac{14}{30-x}$$
$$16(30-x) = 14(30+x)$$
$$480 - 16x = 420 + 14x$$
$$480 = 420 + 30x$$
$$60 = 30x$$
$$2 = x$$

The rate of the current is 2 miles per hour.

23. Let x = the time (in minutes) for both pipes to fill the hot tub.
$$\frac{x}{20} + \frac{x}{30} = 1$$
$$\text{LCD} = 60$$
$$60\left(\frac{x}{20} + \frac{x}{30}\right) = 60 \cdot 1$$
$$3x + 2x = 60$$
$$5x = 60$$
$$x = 12$$

It will take 12 minutes for both pipes to fill the hot tub.

24. Let x = number of tule elk in the park.
$$\frac{200}{x} = \frac{5}{150}$$
$$5x = 30,000$$
$$x = 6000$$

There are 6000 tule elk in the park.

25. $\dfrac{10}{4} = \dfrac{8}{x}$
$$10x = 8 \cdot 4$$
$$10x = 32$$
$$x = 3.2$$

The length of the side marked with an x is 3.2 inches.

26. Let $C =$ the current (in amperes).
Then $R =$ the resistance (in ohms).

Step 1 $C = \dfrac{k}{R}$

Step 2 To find k, substitute 42 for C and 5 for R.

$42 = \dfrac{k}{5}$

$42 \cdot 5 = \dfrac{k}{5} \cdot 5$

$210 = k$

Step 3 $C = \dfrac{210}{R}$

Step 4 Substitute 4 for R and solve for C.

$C = \dfrac{210}{4} = 52.5$

When the resistance is 4 ohms, the current is 52.5 amperes.

Cumulative Review Exercises (Chapters 1-7)

1. $2(x-3)+5x = 8(x-1)$
$2x-6+5x = 8x-8$
$7x-6 = 8x-8$
$-6 = x-8$
$2 = x$

The solution set is $\{2\}$.

2. $-3(2x-4) > 2(6x-12)$
$-6x+12 > 12x-24$
$-18x+12 > -24$
$-18x > -36$
$\dfrac{-18x}{-18} < \dfrac{-36}{-18}$
$x < 2$

$(-\infty, 2)$

3. $x^2 + 3x = 18$
$x^2 + 3x - 18 = 0$
$(x+6)(x-3) = 0$
$x+6 = 0$ or $x-3 = 0$
$x = -6 \qquad x = 3$

The solution set is $\{-6, 3\}$.

4. $\dfrac{2x}{x^2-4}+\dfrac{1}{x-2}=\dfrac{2}{x+2}$

$$x^2-4=(x+2)(x-2)$$

Restrictions: $x\neq 2, x\neq -2$

$\text{LCD}=(x+2)(x-2)$

$$(x+2)(x-2)\left[\dfrac{2x}{(x+2)(x-2)}+\dfrac{1}{x-2}\right]=(x+2)(x-2)\cdot\dfrac{2}{x+2}$$

$$2x+(x+2)=2(x-2)$$
$$3x+2=2x-4$$
$$x=-6$$

The solution set is $\{-6\}$.

5. $y=2x-3$

$x+2y=9$

To solve this system by the substitution method, substitute $2x-3$ for y in the second equation.

$$x+2y=9$$
$$x+2(2x-3)=9$$
$$x+4x-6=9$$
$$5x-6=9$$
$$5x=15$$
$$x=3$$

Back-substitute 3 for x into the first equation.

$y=2x-3$

$y=2\cdot 3-3=3$

The solution set is $\{(3,3)\}$.

6. $3x+2y=-2$

$-4x+5y=18$

To solve this system by the addition method, multiply the first equation by 4 and the second equation by 3. Then add the equations.

$$12x+\;\;8y=-8$$
$$\underline{-12x+15y=54}$$
$$23y=46$$
$$y=2$$

Back-substitute 2 for y in the first equation of the original system.

$3x+2y=-2$

$3x+2(2)=-2$

$3x+4=-2$

$3x=-6$

$x=-2$

The solution set is $\{(-2,2)\}$.

7. $3x - 2y = 6$

x-intercept: 2

y-intercept: -3

checkpoint: $(4,3)$

Draw a line through $(2,0)$, $(0,-3)$ and $(4,3)$.

$3x - 2y = 6$

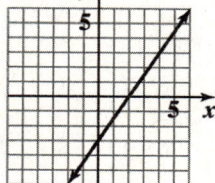

8. $y > -2x + 3$

Graph $y = -2x + 3$ as a dashed line using its slope,

$-2 = \dfrac{-2}{1}$, and its y-intercept, 3. Use $(0,0)$ as a test

point. Because $0 > 2 \cdot 0 + 3$ is false, shade the half-plane *not* containing $(0,0)$.

$y > -2x + 3$

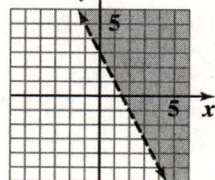

9. $y = -3$

The graph is a horizontal line with y-intercept -3.

$y = -3$

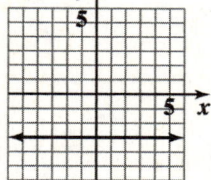

10.
$$\begin{aligned}
-21 - 16 - 3(2-8) &= -21 - 16 - 3(-6) \\
&= -21 - 16 + 18 \\
&= -37 + 18 = -19
\end{aligned}$$

11. $\left(\dfrac{4x^5}{2x^2}\right)^3 = \left(2x^3\right)^3 = 2^3 \cdot \left(x^3\right)^3 = 8x^9$

12. $\dfrac{\dfrac{1}{x} - 2}{4 - \dfrac{1}{x}}$

LCD $= x$

$$\dfrac{\dfrac{1}{x} - 2}{4 - \dfrac{1}{x}} = \dfrac{x\left(\dfrac{1}{x} - 2\right)}{x\left(4 - \dfrac{1}{x}\right)}$$

$$= \dfrac{x \cdot \dfrac{1}{x} - x \cdot 2}{x \cdot 4 - x \cdot \dfrac{1}{x}}$$

$$= \dfrac{1 - 2x}{4x - 1}$$

13. $4x^2 - 13x + 3 = (4x - 1)(x - 3)$

14.
$$\begin{aligned}
4x^2 - 20x + 25 &= (2x)^2 - 2(2x \cdot 5) + 5^2 \\
&= (2x - 5)^2
\end{aligned}$$

15.
$$\begin{aligned}
3x^2 - 75 &= 3\left(x^2 - 25\right) \\
&= 3(x + 5)(x - 5)
\end{aligned}$$

16.
$$\begin{aligned}
&\left(4x^2 - 3x + 2\right) - \left(5x^2 - 7x - 6\right) \\
&= \left(4x^2 - 3x + 2\right) + \left(-5x^2 + 7x + 6\right) \\
&= -x^2 + 4x + 8
\end{aligned}$$

17.
$$\begin{aligned}
\dfrac{-8x^6 + 12x^4 - 4x^2}{4x^2} &= \dfrac{-8x^6}{4x^2} + \dfrac{12x^4}{4x^2} - \dfrac{4x^2}{4x^2} \\
&= -2x^4 + 3x^2 - 1
\end{aligned}$$

18. $\dfrac{x+6}{x-2}+\dfrac{2x+1}{x+3}$

$LCD = (x-2)(x+3)$

$\dfrac{x+6}{x-2}+\dfrac{2x+1}{x+3}$

$= \dfrac{(x+6)(x+3)}{(x-2)(x+3)}+\dfrac{(2x+1)(x-2)}{(x-2)(x+3)}$

$= \dfrac{(x+6)(x+3)+(2x+1)(x-2)}{(x-2)(x+3)}$

$= \dfrac{x^2+9x+18+2x^2-3x-2}{(x-2)(x+3)}$

$= \dfrac{3x^2+6x+16}{(x-2)(x+3)}$

19. Let x = the amount invested at 5%.
Then $4000-x$ = the amount invested at 9%.

$0.05x+0.09(4000-x)=311$

$0.05x+360-0.09x=311$

$-0.04x+360=311$

$-0.04x=-49$

$x=\dfrac{-49}{-0.04}$

$x=1225$

If $x = 1225$, then $4000-x = 2775$.
$1225 was invested at 5% and $2775 at 9%.

20. Let x = the length of the shorter piece.
Then $3x$ = the length of the larger piece.

$x+3x=68$

$4x=68$

$x=17$

If $x = 17$, then $3x = 51$.
The lengths of the pieces are 17 inches and 51 inches.

Chapter 8
Roots and Radicals

8.1 Check Points

1. a. $\sqrt{81} = 9$

The principal square root of 81 is 9.

b. $-\sqrt{9} = -3$

The negative square root of 9 is -3.

c. $\sqrt{\dfrac{1}{25}} = \dfrac{1}{5}$ because $\left(\dfrac{1}{5}\right)^2 = \dfrac{1}{25}$.

d. $\sqrt{36 + 64} = \sqrt{100} = 10$

e. $\sqrt{36} + \sqrt{64} = 6 + 8 = 14$

2. $E = \dfrac{w}{20\sqrt{a}}$

$E = \dfrac{20}{20\sqrt{4}} = \dfrac{1}{2}$

The surface evaporation is $\dfrac{1}{2}$ of an inch on this day.

3. 1995 is 20 years after 1975. Therefore, substitute 20 for t into the formula.

$P = 2.2\sqrt{t} + 45$

$P = 2.2\sqrt{20} + 45 \approx 55$

According to the model, 55% of degrees were awarded to women in 1995.

This is the same as the actual percentage shown in the figure.

4. a. $\sqrt[3]{1} = 1$ because $1^3 = 1$.

b. $\sqrt[3]{-27} = -3$ because $(-3)^3 = -27$.

c. $\sqrt[3]{\dfrac{1}{125}} = \dfrac{1}{5}$ because $\left(\dfrac{1}{5}\right)^3 = \dfrac{1}{125}$.

5. a. $\sqrt[4]{81} = 3$ because $3^4 = 81$.

b. $\sqrt[4]{-81}$ is not a real number because the index, 4, is even and the radicand, -81, is negative.

c. $-\sqrt[4]{81} = -3$ because $\sqrt[4]{81} = 3$.

d. $\sqrt[5]{-\dfrac{1}{32}} = -\dfrac{1}{2}$ because $\left(-\dfrac{1}{2}\right)^5 = -\dfrac{1}{32}$.

8.1 Concept and Vocabulary Check

1. principal

2. 9^2

3. index, radicand

4. 4^3

5. 2^4

6. true

7. true

8. false

9. true

10. false

8.1 Exercise Set

1. $\sqrt{36} = 6$

The principal square root of 36 is 6.

3. $-\sqrt{36} = -6$

The negative square root of 36 is -6.

5. $\sqrt{-36}$ is not a real number.

There is no real number whose square is -36.

7. $\sqrt{\dfrac{1}{9}} = \dfrac{1}{3}$ because $\left(\dfrac{1}{3}\right)^2 = \dfrac{1}{9}$.

9. $\sqrt{\dfrac{1}{100}} = \dfrac{1}{10}$ because $\left(\dfrac{1}{10}\right)^2 = \dfrac{1}{100}$.

11. $-\sqrt{\dfrac{1}{36}} = -\dfrac{1}{6}$ because $\sqrt{\dfrac{1}{36}} = \dfrac{1}{6}$.

13. $\sqrt{-\dfrac{1}{36}}$ is not a real number.

15. $\sqrt{0.04} = 0.2$ because $(0.2)^2 = 0.04$ and 0.2 is positive.

17. $\sqrt{33-8} = \sqrt{25} = 5$

19. $\sqrt{2 \cdot 32} = \sqrt{64} = 8$

21. $\sqrt{144+25} = \sqrt{169} = 13$

23. $\sqrt{144} + \sqrt{25} = 12 + 5 = 17$

25. $\sqrt{25-144} = \sqrt{-119}$ which is not a real number.

27.

x	$y = \sqrt{x-1}$	(x, y)
1	$y = \sqrt{1-1} = 0$	$(1,0)$
2	$y = \sqrt{2-1} = 1$	$(2,1)$
5	$y = \sqrt{5-1} = 2$	$(5,2)$
10	$y = \sqrt{10-1} = 3$	$(10,3)$
17	$y = \sqrt{17-1} = 4$	$(17,4)$

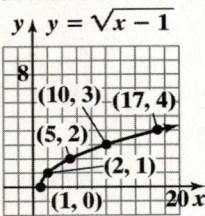

29. Answers will vary. The graph in exercise 27 is shaped like $y = \sqrt{x}$, but shifted 1 unit to the right.

31. $\sqrt{7} \approx 2.646$

33. $\sqrt{23} \approx 4.796$

35. $-\sqrt{65} \approx -8.062$

37. $12 + \sqrt{11} \approx 15.317$

39. $\dfrac{12 + \sqrt{11}}{2} \approx 7.658$

41. $\dfrac{-5 + \sqrt{321}}{6} \approx 2.153$

43. $\sqrt{13-5} = \sqrt{8} \approx 2.828$

45. $\sqrt{5-13} = \sqrt{-8}$, which is not a real number.

47. $\sqrt[3]{64} = 4$ because $4^3 = 64$.

49. $\sqrt[3]{-27} = -3$ because $(-3)^3 = -27$.

51. $-\sqrt[3]{8} = -2$ because $(2)^3 = 8$.

53. $\sqrt[3]{\dfrac{1}{125}} = \dfrac{1}{5}$ because $\left(\dfrac{1}{5}\right)^3 = \dfrac{1}{125}$.

55. $\sqrt[3]{-1000} = -10$ because $(-10)^3 = -1000$.

57. $\sqrt[4]{1} = 1$ because $1^4 = 1$.

59. $\sqrt[4]{16} = 2$ because $2^4 = 16$.

61. $-\sqrt[4]{16} = -2$ because $(2)^4 = 16$.

63. $\sqrt[4]{-16}$ is not a real number because the index, 4, is even and the radicand, -16, is negative.

65. $\sqrt[5]{-1} = -1$ because $(-1)^5 = -1$.

67. $\sqrt[6]{-1}$ is not a real number because the index, 6, is even and the radicand, -1, is negative.

69. $-\sqrt[4]{256} = -4$ because $\sqrt[4]{256} = 4$.

71. $\sqrt[6]{64} = 2$ because $2^6 = 64$.

73. $-\sqrt[5]{32} = -2$ because $\sqrt[5]{32} = 2$.

75. $\sqrt{2x}$ yields a real number if $2x \ge 0$
$$x \ge 0.$$

77. $\sqrt{x-2}$ yields a real number if $x - 2 \ge 0$
$$x \ge 2.$$

79. $\sqrt{2-x}$ yields a real number if $2 - x \ge 0$
$$-x \ge -2$$
$$x \le 2.$$

81. $\sqrt{x^2 + 2}$ yields a real number for all real numbers x since $x^2 + 2$ is always positive.

83. $\sqrt{12-2x}$ yields a real number if $12 - 2x \ge 0$
$$-2x \ge -12$$
$$x \le 6.$$

85. $v = 4\sqrt{r};\ r = 9$

$$v = 4\sqrt{9} = 4 \cdot 3 = 12$$

The maximum velocity is 12 miles per hour.

87. $v = \sqrt{20L} = \sqrt{20 \cdot 245} = \sqrt{4900} = 70$

The motorist was traveling 70 miles per hour, so she was speeding.

89. a. At birth we have $x = 0$.

$$y = 2.9\sqrt{x} + 36$$
$$= 2.9\sqrt{0} + 36$$
$$= 2.9(0) + 36$$
$$= 36$$

According to the model, the head circumference at birth is 36 cm.

b. At 9 months we have $x = 9$.

$$y = 2.9\sqrt{x} + 36$$
$$= 2.9\sqrt{9} + 36$$
$$= 2.9(3) + 36$$
$$= 44.7$$

According to the model, the head circumference at 9 months is 44.7 cm.

c. At 14 months we have $x = 14$.

$$y = 2.9\sqrt{x} + 36$$
$$= 2.9\sqrt{14} + 36$$
$$\approx 46.9$$

According to the model, the head circumference at 14 months is roughly 46.9 cm.

d. The model describes healthy children.

91. a. $\quad y = a\sqrt{x} + b$

$$68.6 = a\sqrt{0} + b$$
$$68.6 = b$$

Therefore, $y = a\sqrt{x} + 68.6$.

b. $\quad y = a\sqrt{x} + 68.6$

$$86.6 = a\sqrt{28} + 68.6$$
$$18 = a\sqrt{28}$$
$$\frac{18}{\sqrt{28}} = a$$
$$3.4 \approx a$$

Therefore, $y = 3.4\sqrt{x} + 68.6$.

c. In 2003, we have $x = 23$.

$$y = 3.4\sqrt{23} + 68.6. \approx 84.9$$

According to the model, 84.9% of Americans ages 25 and over completed at least four years of high school in 2003.
This overestimates the actual value shown in the bar graph by 0.3%.

93. – 97. Answers will vary.

99. does not make sense; Explanations will vary. Sample explanation: $-\sqrt{9} = -3$ and -3 is a real number.

101. makes sense

103. false; Changes to make the statement true will vary. A sample change is: $\sqrt{9} + \sqrt{16} = 3 + 4 = 7$.

105. false; Changes to make the statement true will vary. A sample change is: $\sqrt[3]{-27} = \sqrt[3]{(-3)^3} = -3$. The expression is a real number.

107. $\sqrt{\sqrt[3]{64}} = \sqrt{4} = 2$

109. $\sqrt{\sqrt{16}} - \sqrt[3]{\sqrt{64}} = \sqrt{4} - \sqrt[3]{8}$
$$= 2 - 2$$
$$= 0$$

111. $y_1 = \sqrt{x}$
$y_2 = \sqrt{x+4}$
$y_3 = \sqrt{x-3}$

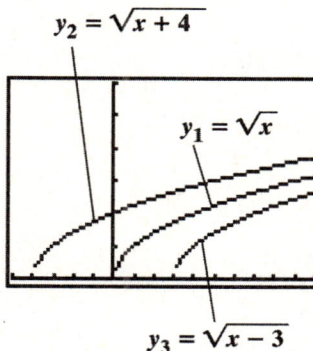

All three graphs are increasing and have a range of $y \geq 0$. The graphs of y_2 and y_3 look like the graph of y_1 but are shifted left 4 units and shifted right 3 units, respectively.

113. $4x - 5y = 20$

$$-5y = -4x + 20$$

$$y = \frac{4}{5}x - 4$$

x-intercept: 5
y-intercept: -4
checkpoint: $(-5, -8)$

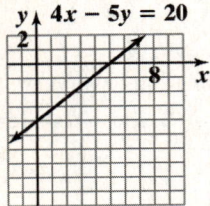

114. $2(x - 3) > 4x + 10$

$$2x - 6 > 4x + 10$$

$$-2x - 6 > 10$$

$$-2x > 16$$

$$x < -8$$

$$(-\infty, 8)$$

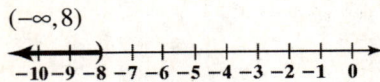

115. $\dfrac{1}{x^2 - 17x + 30} \div \dfrac{1}{x^2 + 7x - 18}$

$$= \frac{1}{x^2 - 17x + 30} \cdot \frac{x^2 + 7x - 18}{1}$$

$$= \frac{1}{(x - 15)(x - 2)} \cdot \frac{(x + 9)(x - 2)}{1}$$

$$= \frac{x + 9}{x - 15}$$

116. **a.** $\sqrt{25} \cdot \sqrt{4} = 5 \cdot 2 = 10$

b. $\sqrt{25 \cdot 4} = \sqrt{100} = 10$

c. $\sqrt{25} \cdot \sqrt{4} = \sqrt{25 \cdot 4}$

117. **a.** $\sqrt{500} \approx 22.361$

b. $10\sqrt{5} \approx 22.361$

c. $\sqrt{500} = 10\sqrt{5}$

118. **a.** $\dfrac{\sqrt{64}}{\sqrt{4}} = \dfrac{8}{2} = 4$

b. $\sqrt{\dfrac{64}{4}} = \sqrt{16} = 4$

c. $\dfrac{\sqrt{64}}{\sqrt{4}} = \sqrt{\dfrac{64}{4}}$

8.2 Check Points

1. **a.** $\sqrt{3} \cdot \sqrt{10} = \sqrt{3 \cdot 10} = \sqrt{30}$

b. $\sqrt{2x} \cdot \sqrt{13y} = \sqrt{2x \cdot 13y} = \sqrt{26xy}$

c. $\sqrt{5} \cdot \sqrt{5} = \sqrt{25} = 5$

d. $\sqrt{\dfrac{3}{2}} \cdot \sqrt{\dfrac{5}{11}} = \sqrt{\dfrac{3}{2} \cdot \dfrac{5}{11}} = \sqrt{\dfrac{15}{22}}$

2. **a.** $\sqrt{12} = \sqrt{4 \cdot 3} = \sqrt{4}\sqrt{3} = 2\sqrt{3}$

b. $\sqrt{60} = \sqrt{4 \cdot 15} = \sqrt{4}\sqrt{15} = 2\sqrt{15}$

c. $\sqrt{55}$ cannot be simplified because it has no perfect square factors other than 1.

3. $\sqrt{40x^{16}} = \sqrt{4x^{16} \cdot 10} = 2x^8\sqrt{10}$

4. $\sqrt{27x^9} = \sqrt{9x^8 \cdot 3x} = 3x^4\sqrt{3x}$

5. $\sqrt{15x^6} \cdot \sqrt{3x^7} = \sqrt{15x^6 \cdot 3x^7}$

$$= \sqrt{45x^{13}}$$

$$= \sqrt{9x^{12} \cdot 5x}$$

$$= 3x^6\sqrt{5x}$$

6. **a.** $\sqrt{\dfrac{49}{25}} = \dfrac{\sqrt{49}}{\sqrt{25}} = \dfrac{7}{5}$

b. $\dfrac{\sqrt{48x^5}}{\sqrt{3x}} = \sqrt{\dfrac{48x^5}{3x}} = \sqrt{16x^4} = 4x^2$

7. **a.** $\sqrt[3]{40} = \sqrt[3]{8 \cdot 5} = 2\sqrt[3]{5}$

b. $\sqrt[5]{8} \cdot \sqrt[5]{8} = \sqrt[5]{64} = \sqrt[5]{32 \cdot 2} = \sqrt[5]{32} \cdot \sqrt[5]{2} = 2\sqrt[5]{2}$

c. $\sqrt[3]{\dfrac{125}{27}} = \dfrac{5}{3}$

8.2 Concept and Vocabulary Check

1. $\sqrt{a} \cdot \sqrt{b}$

2. $49; 6; 7; 6; 7\sqrt{6}$

3. x^n; one-half

4. $12; 6$

5. $\dfrac{\sqrt{a}}{\sqrt{b}}$

6. $25; 4; \dfrac{5}{2}$

7. $4; 16; 3; 2; 3; 2\sqrt[4]{3}$

8.2 Exercise Set

1. $\sqrt{2} \cdot \sqrt{7} = \sqrt{2 \cdot 7} = \sqrt{14}$

3. $\sqrt{3x} \cdot \sqrt{5y} = \sqrt{3x \cdot 5y} = \sqrt{15xy}$

5. $\sqrt{5} \cdot \sqrt{5} = \sqrt{25} = 5$

7. $\sqrt{\dfrac{2}{3}} \cdot \sqrt{\dfrac{5}{7}} = \sqrt{\dfrac{2}{3} \cdot \dfrac{5}{7}} = \sqrt{\dfrac{10}{21}}$

9. $\sqrt{0.1x} \cdot \sqrt{5y} = \sqrt{0.5xy}$

11. $\sqrt{\dfrac{1}{5}a} \cdot \sqrt{\dfrac{1}{5}b} = \sqrt{\dfrac{1}{25}ab} = \sqrt{\dfrac{1}{25}} \cdot \sqrt{ab}$
$= \dfrac{1}{5}\sqrt{ab}$

13. $\sqrt{\dfrac{2x}{9}} \cdot \sqrt{\dfrac{9}{2}} = \sqrt{\dfrac{2x \cdot 9}{9 \cdot 2}} = \sqrt{\dfrac{18x}{18}} = \sqrt{x}$

15. $\sqrt{50} = \sqrt{25 \cdot 2} = \sqrt{25}\sqrt{2} = 5\sqrt{2}$

17. $\sqrt{45} = \sqrt{9 \cdot 5} = \sqrt{9}\sqrt{5} = 3\sqrt{5}$

19. $\sqrt{200} = \sqrt{100 \cdot 2} = \sqrt{100}\sqrt{2} = 10\sqrt{2}$

21. $\sqrt{75x} = \sqrt{25 \cdot 3x} = \sqrt{25}\sqrt{3x} = 5\sqrt{3x}$

23. $\sqrt{9x} = \sqrt{9}\sqrt{x} = 3\sqrt{x}$

25. $\sqrt{35}$ cannot be simplified because 35 has no perfect square factors other than 1.

27. $\sqrt{y^2} = y$

29. $\sqrt{64x^2} = 8x$

31. $\sqrt{11x^2} = \sqrt{x^2}\sqrt{11} = x\sqrt{11}$

33. $\sqrt{8x^2} = \sqrt{4x^2}\sqrt{2} = 2x\sqrt{2}$

35. $\sqrt{x^{20}} = x^{10}$ because $\left(x^{10}\right)^2 = x^{20}$.

37. $\sqrt{25y^{10}} = 5y^5$

39. $\sqrt{20x^6} = \sqrt{4x^6}\sqrt{5} = 2x^3\sqrt{5}$

41. $\sqrt{72y^{100}} = \sqrt{36y^{100}}\sqrt{2} = 6y^{50}\sqrt{2}$

43. $\sqrt{x^3} = \sqrt{x^2}\sqrt{x} = x\sqrt{x}$

45. $\sqrt{x^7} = \sqrt{x^6}\sqrt{x} = x^3\sqrt{x}$

47. $\sqrt{y^{17}} = \sqrt{y^{16}}\sqrt{y} = y^8\sqrt{y}$

49. $\sqrt{25x^5} = \sqrt{25x^4}\sqrt{x} = 5x^2\sqrt{x}$

51. $\sqrt{8x^{17}} = \sqrt{4x^{16}}\sqrt{2x} = 2x^8\sqrt{2x}$

53. $\sqrt{90y^{19}} = \sqrt{9y^{18}}\sqrt{10y} = 3y^9\sqrt{10y}$

55. $\sqrt{3} \cdot \sqrt{15} = \sqrt{45} = \sqrt{9}\sqrt{5} = 3\sqrt{5}$

57. $\sqrt{5x} \cdot \sqrt{10y} = \sqrt{50xy} = \sqrt{25}\sqrt{2xy}$
$= 5\sqrt{2xy}$

59. $\sqrt{12x} \cdot \sqrt{3x} = \sqrt{36x^2} = 6x$

61. $\sqrt{15x^2} \cdot \sqrt{3x} = \sqrt{45x^3} = \sqrt{9x^2}\sqrt{5x}$
$= 3x\sqrt{5x}$

63. $\sqrt{15x^4} \cdot \sqrt{5x^9} = \sqrt{75x^{13}}$
$= \sqrt{25x^{12}}\sqrt{3x} = 5x^6\sqrt{3x}$

65. $\sqrt{7x} \cdot \sqrt{3y} = \sqrt{21xy}$

67. $\sqrt{50xy} \cdot \sqrt{4xy^2} = \sqrt{200x^2y^3}$
$= \sqrt{100x^2y^2}\sqrt{2y} = 10xy\sqrt{2y}$

69. $\sqrt{\dfrac{49}{16}} = \dfrac{\sqrt{49}}{\sqrt{16}} = \dfrac{7}{4}$

71. $\sqrt{\dfrac{3}{4}} = \dfrac{\sqrt{3}}{\sqrt{4}} = \dfrac{\sqrt{3}}{2}$

73. $\sqrt{\dfrac{x^2}{36}} = \dfrac{\sqrt{x^2}}{6} = \dfrac{x}{6}$

75. $\sqrt{\dfrac{7}{x^4}} = \dfrac{\sqrt{7}}{\sqrt{x^4}} = \dfrac{\sqrt{7}}{x^2}$

77. $\sqrt{\dfrac{72}{y^{20}}} = \dfrac{\sqrt{72}}{\sqrt{y^{20}}} = \dfrac{\sqrt{36}\sqrt{2}}{y^{10}} = \dfrac{6\sqrt{2}}{y^{10}}$

79. $\dfrac{\sqrt{54}}{\sqrt{6}} = \sqrt{\dfrac{54}{6}} = \sqrt{9} = 3$

81. $\dfrac{\sqrt{24}}{\sqrt{3}} = \sqrt{\dfrac{24}{3}} = \sqrt{8} = \sqrt{4}\sqrt{2} = 2\sqrt{2}$

83. $\dfrac{\sqrt{75}}{\sqrt{15}} = \sqrt{\dfrac{75}{15}} = \sqrt{5}$

85. $\dfrac{\sqrt{48x}}{\sqrt{3x}} = \sqrt{\dfrac{48x}{3x}} = \sqrt{16} = 4$

87. $\dfrac{\sqrt{32x^3}}{\sqrt{8x}} = \sqrt{\dfrac{32x^3}{8x}} = \sqrt{4x^2} = 2x$

89. $\dfrac{\sqrt{150x^4}}{\sqrt{3x}} = \sqrt{\dfrac{150x^4}{3x}} = \sqrt{50x^3}$
$= \sqrt{25x^2}\sqrt{2x} = 5x\sqrt{2x}$

91. $\dfrac{\sqrt{400x^{10}}}{\sqrt{10x^3}} = \sqrt{\dfrac{400x^{10}}{10x^3}} = \sqrt{40x^7}$
$= \sqrt{4x^6}\sqrt{10x} = 2x^3\sqrt{10x}$

93. $\sqrt[3]{16} = \sqrt[3]{8 \cdot 2} = \sqrt[3]{8} \cdot \sqrt[3]{2} = 2\sqrt[3]{2}$

95. $\sqrt[3]{54} = \sqrt[3]{27 \cdot 2} = \sqrt[3]{27} \cdot \sqrt[3]{2} = 3\sqrt[3]{2}$

97. $\sqrt[4]{32} = \sqrt[4]{16 \cdot 2} = \sqrt[4]{16} \cdot \sqrt[4]{2} = 2\sqrt[4]{2}$

99. $\sqrt[3]{4} \cdot \sqrt[3]{2} = \sqrt[3]{8} = 2$

101. $\sqrt[3]{9} \cdot \sqrt[3]{6} = \sqrt[3]{54} = \sqrt[3]{27}\sqrt[3]{2} = 3\sqrt[3]{2}$

103. $\sqrt[4]{4} \cdot \sqrt[4]{8} = \sqrt[4]{32} = \sqrt[4]{16} \cdot \sqrt[4]{2} = 2\sqrt[4]{2}$

105. $\sqrt[3]{\dfrac{27}{8}} = \dfrac{\sqrt[3]{27}}{\sqrt[3]{8}} = \dfrac{3}{2}$

107. $\sqrt[3]{\dfrac{3}{8}} = \dfrac{\sqrt[3]{3}}{\sqrt[3]{8}} = \dfrac{\sqrt[3]{3}}{2}$

109. $\sqrt{90(x+4)^3} = \sqrt{9(x+4)^2} \cdot \sqrt{10(x+4)}$
$= 3(x+4)\sqrt{10(x+4)}$

111. $\sqrt{x^2 - 6x + 9} = \sqrt{(x-3)^2} = x - 3$

113. $\sqrt{2^{43}x^{104}y^{13}} = \sqrt{2^{42}x^{104}y^{12}} \cdot \sqrt{2y}$
$= 2^{21}x^{52}y^6\sqrt{2y}$

115. $\sqrt[3]{24x^5} = \sqrt[3]{8x^3} \cdot \sqrt[3]{3x^2}$
$= 2x\sqrt[3]{3x^2}$

117. $A = l \cdot w$
$= \sqrt{15} \cdot \sqrt{5} = \sqrt{75}$
$= \sqrt{25}\sqrt{3} = 5\sqrt{3}$
The area is $5\sqrt{3}$ square feet.

119. a. $A = \dfrac{\sqrt{h} \cdot \sqrt{w}}{56}$

$A = \dfrac{\sqrt{68} \cdot \sqrt{200}}{56}$

$= \dfrac{\sqrt{4 \cdot 17} \cdot \sqrt{100 \cdot 2}}{56}$

$= \dfrac{2\sqrt{17} \cdot 10\sqrt{2}}{56}$

$= \dfrac{20\sqrt{34}}{56}$

$= \dfrac{5\sqrt{34}}{14}$

The surface area is $\dfrac{5\sqrt{34}}{14}$ square meters.

b. $\dfrac{5\sqrt{34}}{14} \approx 2.08$

The surface area is approximately 2.08 square meters.

121. The message is "Paige Fox is bad at math."

123. – 127. Answers will vary.

129. does not make sense; Explanations will vary. Sample explanation: A simplified radical expression can contain no radicals.

131. makes sense

133. false; Changes to make the statement true will vary. A sample change is: $\sqrt{20} = \sqrt{4 \cdot 5} = 2\sqrt{5}$.

135. true

137. $\sqrt{\square x^{\square}} = 5x^7$

Since $\left(5x^7\right)^2 = 5^2 \left(x^7\right)^2 = 25x^{14}$, the radicand is $25x^{14}$. The missing coefficient is 25 and the missing exponent is 14.

139. Answers will vary.

141. The graphs do not coincide.

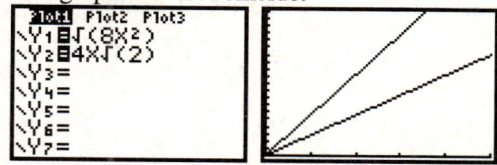

$\sqrt{8x^2} = \sqrt{4x^2 \cdot 2} = \sqrt{4x^2}\sqrt{2} = 2x\sqrt{2}$

Change y_2 to $2x\sqrt{2}$.

143. $4x + 3y = 18$

$5x - 9y = 48$

To solve this system by the addition method, multiply the first equation by 3 and add the result to the second equation.

$12x + 9y = 54$

$\underline{5x - 9y = 48}$

$17x = 102$

$x = 6$

Back-substitute into the first equation of the original system.

$4x + 3y = 18$

$4(6) + 3y = 18$

$24 + 3y = 18$

$3y = -6$

$y = -2$

The solution set is $\{(6, -2)\}$.

144. $\dfrac{6x}{x^2 - 4} - \dfrac{3}{x + 2}$

Factor the first denominator.

$x^2 - 4 = (x + 2)(x - 2)$

The LCD $= (x + 2)(x - 2)$.

$\dfrac{6x}{x^2 - 4} - \dfrac{3}{x + 2}$

$= \dfrac{6}{(x + 2)(x - 2)} - \dfrac{3(x - 2)}{(x + 2)(x - 2)}$

$= \dfrac{6x - 3(x - 2)}{(x + 2)(x - 2)} = \dfrac{6x - 3x + 6}{(x + 2)(x - 2)}$

$= \dfrac{3x + 6}{(x + 2)(x - 2)} = \dfrac{3(x + 2)}{(x + 2)(x - 2)}$

$= \dfrac{3}{x - 2}$

145. $2x^3 - 16x^2 + 30x = 2x\left(x^2 - 8x + 15\right)$
$$= 2x(x-3)(x-5)$$

146. a. $7x + 5x = (7+5)x = 12x$

b. $7\sqrt{2} + 5\sqrt{2} = (7+5)\sqrt{2} = 12\sqrt{2}$

147. $4\sqrt{50x} = 4\sqrt{25 \cdot 2x} = 4 \cdot 5\sqrt{2x} = 20\sqrt{2x}$

148. a. $3(x+5) = 3x + 3 \cdot 5 = 3x + 15$

b. $\sqrt{3}(\sqrt{7} + \sqrt{5}) = \sqrt{3}\sqrt{7} + \sqrt{3}\sqrt{5} = \sqrt{21} + \sqrt{15}$

8.3 Check Points

1. a. $8\sqrt{13} + 9\sqrt{13} = (8+9)\sqrt{13} = 17\sqrt{13}$

b. $\sqrt{17x} - 20\sqrt{17x} = (1-20)\sqrt{17x} = -19\sqrt{17x}$

2. a. $5\sqrt{27} + \sqrt{12} = 5\sqrt{9 \cdot 3} + \sqrt{4 \cdot 3}$
$$= 5 \cdot 3\sqrt{3} + 2\sqrt{3}$$
$$= 15\sqrt{3} + 2\sqrt{3}$$
$$= 17\sqrt{3}$$

b. $6\sqrt{18x} - 4\sqrt{8x} = 6\sqrt{9 \cdot 2x} - 4\sqrt{4 \cdot 2x}$
$$= 6 \cdot 3\sqrt{2x} - 4 \cdot 2\sqrt{2x}$$
$$= 18\sqrt{2x} - 8\sqrt{2x}$$
$$= 10\sqrt{2x}$$

3. a. $\sqrt{2}(\sqrt{5} + \sqrt{11}) = \sqrt{2}\sqrt{5} + \sqrt{2}\sqrt{11} = \sqrt{10} + \sqrt{22}$

b. $(4+\sqrt{3})(2+\sqrt{3}) = 4 \cdot 2 + 4\sqrt{3} + 2\sqrt{3} + \left(\sqrt{3}\right)^2$
$$= 8 + 4\sqrt{3} + 2\sqrt{3} + 3$$
$$= 11 + 6\sqrt{3}$$

c. $(3+\sqrt{5})(8-4\sqrt{5})$
$$= 3 \cdot 8 - 3 \cdot 4\sqrt{5} + 8\sqrt{5} - 4\left(\sqrt{5}\right)^2$$
$$= 24 - 12\sqrt{5} + 8\sqrt{5} - 4 \cdot 5$$
$$= 24 - 12\sqrt{5} + 8\sqrt{5} - 20$$
$$= 4 - 4\sqrt{5}$$

4. a. $(3+\sqrt{11})(3-\sqrt{11}) = 3^2 - \left(\sqrt{11}\right)^2$
$$= 9 - 11$$
$$= -2$$

b. $(\sqrt{7} - \sqrt{2})(\sqrt{7} + \sqrt{2}) = \left(\sqrt{7}\right)^2 - \left(\sqrt{2}\right)^2$
$$= 7 - 2$$
$$= 5$$

8.3 Concept and Vocabulary Check

1. like radicals

2. $8; 10; 18\sqrt{3}$

3. $7; 1; 8\sqrt{5}$

4. $5; 4; 9$

5. $21; 7\sqrt{2}; 3\sqrt{2}; 2$

6. $8 - \sqrt{6}$

7. $6^2; \sqrt{5}; 36; 5; 31$

8.3 Exercise Set

1. $8\sqrt{3} + 5\sqrt{3} = (8+5)\sqrt{3} = 13\sqrt{3}$

3. $17\sqrt{6} - 2\sqrt{6} = (17-2)\sqrt{6} = 15\sqrt{6}$

5. $3\sqrt{13} - 8\sqrt{13} = (3-8)\sqrt{13} = -5\sqrt{13}$

7. $12\sqrt{x} + 3\sqrt{x} = (12+3)\sqrt{x} = 15\sqrt{x}$

9. $70\sqrt{y} - 76\sqrt{y} = (70-76)\sqrt{y}$
$$= -6\sqrt{y}$$

11. $7\sqrt{10x} + 2\sqrt{10x} = (7+2)\sqrt{10x}$
$$= 9\sqrt{10x}$$

13. $7\sqrt{5y} - \sqrt{5y} = 7\sqrt{5y} - 1\sqrt{5y}$
$$= (7-1)\sqrt{5y} = 6\sqrt{5y}$$

15. $\sqrt{5}+\sqrt{5}=1\sqrt{5}+1\sqrt{5}=(1+1)\sqrt{5}$
$\qquad\qquad =2\sqrt{5}$

17. $4\sqrt{2}+3\sqrt{2}+5\sqrt{2}=(4+3+5)\sqrt{2}$
$\qquad\qquad\qquad\quad =12\sqrt{2}$

19. $4\sqrt{7}-5\sqrt{7}+8\sqrt{7}=(4-5+8)\sqrt{7}$
$\qquad\qquad\qquad\quad =7\sqrt{7}$

21. $4\sqrt{11}-6\sqrt{11}+2\sqrt{11}=(4-6+2)\sqrt{11}$
$\qquad\qquad\qquad\qquad =0\sqrt{11}=0$

23. $\sqrt{5}+\sqrt{20}=\sqrt{5}+\sqrt{4\cdot 5}=\sqrt{5}+2\sqrt{5}$
$\qquad\qquad =1\sqrt{5}+2\sqrt{5}=3\sqrt{5}$

25. $\sqrt{8}-\sqrt{2}=\sqrt{4\cdot 2}-\sqrt{2}=\sqrt{4}\sqrt{2}-\sqrt{2}$
$\qquad\qquad =2\sqrt{2}-1\sqrt{2}=(2-1)\sqrt{2}$
$\qquad\qquad =\sqrt{2}$

27. $\sqrt{50}+\sqrt{18}=\sqrt{25}\sqrt{2}+\sqrt{9}\sqrt{2}$
$\qquad\qquad =5\sqrt{2}+3\sqrt{2}=8\sqrt{2}$

29. $7\sqrt{12}+\sqrt{75}=7\sqrt{4}\sqrt{3}+\sqrt{25}\sqrt{3}$
$\qquad\qquad =7\cdot 2\sqrt{3}+5\sqrt{3}$
$\qquad\qquad =14\sqrt{3}+5\sqrt{3}=19\sqrt{3}$

31. $3\sqrt{27}-2\sqrt{18}=3\sqrt{9\cdot 3}-2\sqrt{9\cdot 2}$
$\qquad\qquad =3\cdot 3\sqrt{3}-2\cdot 3\sqrt{2}$
$\qquad\qquad =9\sqrt{3}-6\sqrt{2}$

Because $\sqrt{3}$ and $\sqrt{2}$ are unlike radicals, it is not possible to combine terms and simplify further.

33. $2\sqrt{45x}-2\sqrt{20x}$
$\quad =2\sqrt{9}\sqrt{5x}-2\sqrt{4}\sqrt{5x}$
$\quad =2\cdot 3\sqrt{5x}-2\cdot 2\sqrt{5x}$
$\quad =6\sqrt{5x}-4\sqrt{5x}$
$\quad =(6-4)\sqrt{5x}=2\sqrt{5x}$

35. $\sqrt{8}+\sqrt{16}+\sqrt{18}+\sqrt{25}$
$\quad =\sqrt{4}\sqrt{2}+4+\sqrt{9}\sqrt{2}+5$
$\quad =2\sqrt{2}+4+3\sqrt{2}+5$
$\quad =(4+5)+(2\sqrt{2}+3\sqrt{2})$
$\quad =9+5\sqrt{2}$

37. $\sqrt{2}+\sqrt{11}$
These are unlike radicals, so the terms cannot be combined.

39. $2\sqrt{80}+3\sqrt{75}=2\sqrt{16}\sqrt{5}+3\sqrt{25}\sqrt{3}$
$\qquad\qquad =2\cdot 4\sqrt{5}+3\cdot 5\sqrt{3}$
$\qquad\qquad =8\sqrt{5}+15\sqrt{3}$

Because $\sqrt{5}$ and $\sqrt{3}$ are unlike radicals, it is not possible to combine terms.

41. $3\sqrt{54}-2\sqrt{20}+4\sqrt{45}-\sqrt{24}$
$\quad =3\sqrt{9}\sqrt{6}-2\sqrt{4}\sqrt{5}+4\sqrt{9}\sqrt{5}-\sqrt{4}\sqrt{6}$
$\quad =3\cdot 3\sqrt{6}-2\cdot 2\sqrt{5}+4\cdot 3\sqrt{5}-2\sqrt{6}$
$\quad =9\sqrt{6}-4\sqrt{5}+12\sqrt{5}-2\sqrt{6}$
$\quad =(9-2)\sqrt{6}+(-4+12)\sqrt{5}$
$\quad =7\sqrt{6}+8\sqrt{5}$

43. $\sqrt{2}(\sqrt{3}+\sqrt{5})=\sqrt{2}\cdot\sqrt{3}+\sqrt{2}\cdot\sqrt{5}$
$\qquad\qquad\quad =\sqrt{6}+\sqrt{10}$

45. $\sqrt{7}(\sqrt{6}-\sqrt{10})=\sqrt{7}\cdot\sqrt{6}-\sqrt{7}\cdot\sqrt{10}$
$\qquad\qquad\qquad =\sqrt{42}-\sqrt{70}$

47. $\sqrt{3}(5+\sqrt{3})=\sqrt{3}\cdot 5+\sqrt{3}\cdot\sqrt{3}$
$\qquad\qquad\quad =5\sqrt{3}+3$

49. $\sqrt{3}(\sqrt{6}-\sqrt{3})=\sqrt{3}\cdot\sqrt{6}-\sqrt{3}\cdot\sqrt{3}$
$\qquad\qquad\qquad =\sqrt{18}-3=\sqrt{9}\sqrt{2}-3$
$\qquad\qquad\qquad =3\sqrt{2}-3$

51. $\left(5+\sqrt{2}\right)\left(6+\sqrt{2}\right)$

Use the FOIL method.

$\left(5+\sqrt{2}\right)\left(6+\sqrt{2}\right)$

$=5\cdot6+5\sqrt{2}+6\sqrt{2}+\sqrt{2}\cdot\sqrt{2}$

$=30+5\sqrt{2}+6\sqrt{2}+2$

$=\left(30+2\right)+\left(5+6\right)\sqrt{2}$

$=32+11\sqrt{2}$

53. $\left(4+\sqrt{5}\right)\left(10-3\sqrt{5}\right)$

$=4\cdot10+4\left(-3\sqrt{5}\right)+\sqrt{5}\cdot10+\sqrt{5}\left(-3\sqrt{5}\right)$

$=40+12\sqrt{5}+10\sqrt{5}-3\cdot5$

$=40-12\sqrt{5}+10\sqrt{5}-15$

$=\left(40-15\right)+\left(-12+10\right)\sqrt{5}$

$=25-2\sqrt{5}$

55. $\left(6-3\sqrt{7}\right)\left(2-5\sqrt{7}\right)$

$=6\cdot2+6\left(-5\sqrt{7}\right)-3\sqrt{7}\left(2\right)-3\sqrt{7}\left(-5\sqrt{7}\right)$

$=12-30\sqrt{7}-6\sqrt{7}+15\cdot7$

$=12-30\sqrt{7}-6\sqrt{7}+105$

$=\left(12+105\right)+\left(-30-6\right)\sqrt{7}$

$=117-36\sqrt{7}$

57. $\left(\sqrt{10}-3\right)\left(\sqrt{10}-5\right)$

$=\sqrt{10}\cdot\sqrt{10}+\sqrt{10}\left(-5\right)-3\sqrt{10}-3\left(-5\right)$

$=10-5\sqrt{10}-3\sqrt{10}+15$

$=25-8\sqrt{10}$

59. $\left(\sqrt{3}+\sqrt{6}\right)\left(\sqrt{3}+2\sqrt{6}\right)$

$=\sqrt{3}\cdot\sqrt{3}+\sqrt{3}\cdot2\sqrt{6}+\sqrt{6}\cdot3+\sqrt{6}\cdot2\sqrt{6}$

$=3+2\sqrt{18}+\sqrt{18}+2\cdot6$

$=3+2\sqrt{18}+\sqrt{18}+12$

$=15+3\sqrt{18}=15+3\sqrt{9}\sqrt{2}$

$=15+3\cdot3\sqrt{2}$

$=15+9\sqrt{2}$

61. $\left(\sqrt{2}+1\right)\left(\sqrt{3}-6\right)$

$=\sqrt{2}\cdot\sqrt{3}+\sqrt{2}\left(-6\right)+1\cdot\sqrt{3}+1\left(-6\right)$

$=\sqrt{6}-6\sqrt{2}+\sqrt{3}-6$

63. $\left(3+\sqrt{5}\right)\left(3-\sqrt{5}\right)$

These two radical expressions are conjugates. Use the special-product formula.

$\left(A+B\right)\left(A-B\right)=A^2-B^2$

$\left(3+\sqrt{5}\right)\left(3-\sqrt{5}\right)=3^2-\left(\sqrt{5}\right)^2$

$=9-5$

$=4$

65. $\left(1-\sqrt{6}\right)\left(1+\sqrt{6}\right)=1^2-\left(\sqrt{6}\right)^2$

$=1-6$

$=-5$

67. $\left(\sqrt{11}+5\right)\left(\sqrt{11}-5\right)=\left(\sqrt{11}\right)^2-5^2$

$=11-25=-14$

69. $\left(\sqrt{7}-\sqrt{5}\right)\left(\sqrt{7}+\sqrt{5}\right)=\left(\sqrt{7}\right)^2-\left(\sqrt{5}\right)^2$

$=7-5=2$

71. $\left(2\sqrt{3}+7\right)\left(2\sqrt{3}-7\right)=\left(2\sqrt{3}\right)^2-7^2$

$=12-49$

$=-37$

73. $\left(2\sqrt{3}+\sqrt{5}\right)\left(2\sqrt{3}-\sqrt{5}\right)$

$=\left(2\sqrt{3}\right)^2-\left(\sqrt{5}\right)^2$

$=12-5$

$=7$

75. Use the special-product formula.

$\left(A+B\right)^2=A^2+2AB+B^2$

$\left(\sqrt{2}+\sqrt{3}\right)^2$

$=\left(\sqrt{2}\right)^2+2\cdot\sqrt{2}\cdot\sqrt{3}+\left(\sqrt{3}\right)^2$

$=2+2\sqrt{6}+3=5+2\sqrt{6}$

77. Use the special-product formula
$$\left(A+B\right)^2 = A^2 - 2AB + B^2$$
$$\left(\sqrt{x}-\sqrt{10}\right)^2$$
$$= \left(\sqrt{x}\right)^2 - 2 \cdot \sqrt{x} \cdot \sqrt{10} + \left(\sqrt{10}\right)^2$$
$$= x - 2\sqrt{10x} + 10$$

79. $5\sqrt{27x^3} - 3x\sqrt{12x}$
$$= 5\sqrt{9x^2} \cdot \sqrt{3x} - 3x\sqrt{4} \cdot \sqrt{3x}$$
$$= 5(3x)\sqrt{3x} - 3x(2)\sqrt{3x}$$
$$= 15x\sqrt{3x} - 6x\sqrt{3x}$$
$$= 9x\sqrt{3x}$$

81. $6y^2\sqrt{x^5 y} + 2x^2\sqrt{xy^5}$
$$= 6y^2\sqrt{x^4} \cdot \sqrt{xy} + 2x^2\sqrt{y^4} \cdot \sqrt{xy}$$
$$= 6y^2\left(x^2\right)\sqrt{xy} + 2x^2\left(y^2\right)\sqrt{xy}$$
$$= 6x^2 y^2\sqrt{xy} + 2x^2 y^2\sqrt{xy}$$
$$= 8x^2 y^2\sqrt{xy}$$

83. $3\sqrt[3]{54} - 4\sqrt[3]{16}$
$$= 3\sqrt[3]{27} \cdot \sqrt[3]{2} - 4\sqrt[3]{8} \cdot \sqrt[3]{2}$$
$$= 3(3)\sqrt[3]{2} - 4(2)\sqrt[3]{2}$$
$$= 9\sqrt[3]{2} - 8\sqrt[3]{2}$$
$$= \sqrt[3]{2}$$

85. $x\sqrt[3]{32x} + 9\sqrt[3]{4x^4}$
$$= x\sqrt[3]{8}\sqrt[3]{4x} + 9\sqrt[3]{x^3} \cdot \sqrt[3]{4x}$$
$$= x(2)\sqrt[3]{4x} + 9(x)\sqrt[3]{4x}$$
$$= 2x\sqrt[3]{4x} + 9x\sqrt[3]{4x}$$
$$= 11x\sqrt[3]{4x}$$

87. Use the formulas for perimeter and area of a square with $s = \sqrt{3} + \sqrt{5}$.
$$P = 4x = 4\left(\sqrt{3} + \sqrt{5}\right)$$
$$= 4\sqrt{3} + 4\sqrt{5}$$
The perimeter is $\left(4\sqrt{3} + 4\sqrt{5}\right)$ inches.
$$A = s^2$$
$$A = \left(\sqrt{3} + \sqrt{5}\right)^2$$
$$= \left(\sqrt{3}\right)^2 + 2\sqrt{3}\sqrt{5} + \left(\sqrt{5}\right)^2$$
$$= 3 + 2\sqrt{15} + 5$$
$$= 8 + 2\sqrt{15}$$
The area is $\left(8 + 2\sqrt{15}\right)$ square inches.

89. Use the formulas for the perimeter and area of a rectangle with $l = \sqrt{6} + 1$ and $w = \sqrt{6} - 1$.
$$P = 2l + 2w = 2\left(\sqrt{6} + 1\right) + 2\left(\sqrt{6} - 1\right)$$
$$= 2\sqrt{6} + 2 + 2\sqrt{6} - 2$$
$$= 4\sqrt{6}$$
The perimeter is $4\sqrt{6}$ inches.
$$A = lw$$
$$A = \left(\sqrt{6} + 1\right)\left(\sqrt{6} - 1\right)$$
$$= \left(\sqrt{6}\right)^2 - 1^2$$
$$= 6 - 1$$
$$= 5$$
The area is 5 square inches.

91. To find the perimeter of a triangle, add the lengths of the three sides.
$$P = \sqrt{2} + \sqrt{2} + 2 = 2 + 2\sqrt{2}$$
The perimeter is $\left(2 + 2\sqrt{2}\right)$ inches.

Use the formula for the area of a triangle with $b = \sqrt{2}$ and $h = \sqrt{2}$.
$$A = \frac{1}{2}bh$$
$$A = \frac{1}{2}\left(\sqrt{2}\right)\left(\sqrt{2}\right) = \frac{1}{2} \cdot 2 = 1$$
The area is 1 square inch.

93. $\sqrt{2}+\sqrt{8}; a=2, b=8$

$$\sqrt{a}+\sqrt{b}=\sqrt{(a+b)+2\sqrt{ab}}$$

$$\sqrt{2}+\sqrt{8}=\sqrt{(2+8)+2\sqrt{2\cdot8}}$$

$$=\sqrt{10+2\sqrt{16}}$$

$$=\sqrt{10+2\cdot4}$$

$$=\sqrt{18}=\sqrt{9}\sqrt{2}=3\sqrt{2}$$

$$\sqrt{2}+\sqrt{8}=\sqrt{2}+\sqrt{4\cdot2}$$

$$=\sqrt{2}+\sqrt{4}\sqrt{2}$$

$$=\sqrt{2}+2\sqrt{2}$$

$$=(1+2)\sqrt{2}=3\sqrt{2}$$

Explanations of preferences will vary.

95. a. $J=1.4\sqrt{x}+55-(20-1.2\sqrt{x})$

$$=1.4\sqrt{x}+55-20+1.2\sqrt{x}$$

$$=1.4\sqrt{x}+1.2\sqrt{x}+55-20$$

$$=2.6\sqrt{x}+35$$

b. $J=2.6\sqrt{x}+35$

$J=2.6\sqrt{30}+35\approx49$

49% of full-time college students had jobs in 2005.
This is the same as the actual percentage displayed in the bar graph.

97. a. 49% of full-time college students had jobs in 30 years after 1975, or 2005. This is represented by the point $(30, 49)$.

b. According to the graph, about 51% or 52% of full-time college students will have jobs in 2015.

99. – 105. Answers will vary.

107. does not make sense; Explanations will vary.
Sample explanation: After they are simplified, the terms are not like radicals.

$$2\sqrt{20}+4\sqrt{75}=2\sqrt{4\cdot5}+4\sqrt{25\cdot3}$$

$$=2\cdot2\sqrt{5}+4\cdot5\sqrt{3}$$

$$=4\sqrt{5}+20\sqrt{3}$$

109. does not make sense; Explanations will vary.
Sample explanation: This product will eliminate the radicals.

111. false; Changes to make the statement true will vary.
A sample change is:

$$\left(\sqrt{5}+\sqrt{3}\right)^2=\left(\sqrt{5}\right)^2+2\sqrt{5}\sqrt{3}+\left(\sqrt{3}\right)^2$$

$$=5+2\sqrt{15}+3$$

$$=8+2\sqrt{15}.$$

113. false; Changes to make the statement true will vary.
A sample change is:

$$\left(\sqrt{7}+\sqrt{3}\right)\left(\sqrt{7}-\sqrt{3}\right)=\left(\sqrt{7}\right)^2-\left(\sqrt{3}\right)^2$$

$$=7-3$$

$$=4.$$

115. $\left(\sqrt[3]{4}+1\right)\left(\sqrt[3]{2}-3\right)$

$$=\sqrt[3]{4}\cdot\sqrt[3]{2}-3\cdot\sqrt[3]{4}+1\cdot\sqrt[3]{2}-1\cdot3$$

$$=\sqrt[3]{8}-3\sqrt[3]{4}+\sqrt[3]{2}-3$$

$$=2-3\sqrt[3]{4}+\sqrt[3]{2}-3$$

$$=-1-3\sqrt[3]{4}+\sqrt[3]{2}$$

117. $\left(4\sqrt{3x}+\sqrt{2y}\right)\left(4\sqrt{3x}-\sqrt{2y}\right)$

$$=\left(4\sqrt{3x}\right)^2-\left(\sqrt{2y}\right)^2$$

$$=4^2\left(\sqrt{3x}\right)^2-\left(\sqrt{2y}\right)^2$$

$$=16\cdot3x-2y$$

$$=48x-2y$$

119. $\sqrt{16x}-\sqrt{9x}=\sqrt{7x}$

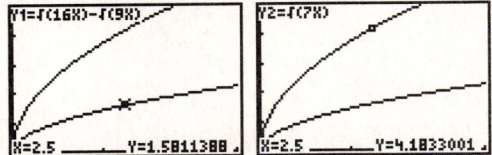

The graphs do not coincide so the simplification is incorrect. The right side of the equation should be $\sqrt{x}$.

$$\sqrt{16x}-\sqrt{9x}=\sqrt{16}\sqrt{x}-\sqrt{9}\sqrt{x}$$

$$=4\sqrt{x}-3\sqrt{x}$$

$$=(4-3)\sqrt{x}$$

$$=1\sqrt{x}=\sqrt{x}$$

121. $\left(\sqrt{x}+2\right)\left(\sqrt{x}-2\right)=x^2-4$

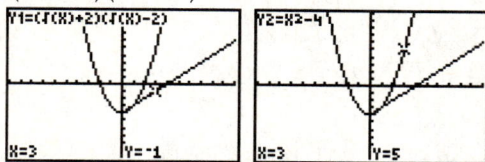

The graphs do not coincide so the simplification is incorrect. The right side of the equation should be $x-4$.

$$\left(\sqrt{x}+2\right)\left(\sqrt{x}-2\right)=\left(\sqrt{x}\right)^2-2^2=x-4$$

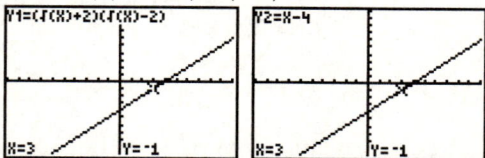

Note: the domain of the left side of the equation is $x\geq 0$ while the domain of the right side is any real number. Therefore, the two graphs are only identical for $x\geq 0$.

122. $(5x+3)(5x-3)=(5x)^2-3^2$
$$=25x^2-9$$

123. $64x^3-x=x\left(64x^2-1\right)$
$$=x\left[(8x)^2-1^2\right]$$
$$=x(8x+1)(8x-1)$$

124. $y=-\dfrac{1}{4}x+3$

slope $=-\dfrac{1}{4}=\dfrac{-1}{4}$; y-intercept $=3$

Plot $(0,3)$. From this point, move 1 unit *down* and 4 units to the *right* to reach the point $(4,2)$. Draw a line through $(0,3)$ and $(4,2)$.

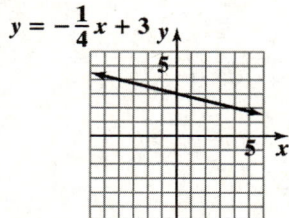

125. $\dfrac{\sqrt{3}}{\sqrt{5}}\cdot\dfrac{\sqrt{5}}{\sqrt{5}}=\dfrac{\sqrt{15}}{\sqrt{25}}=\dfrac{\sqrt{15}}{5}$

126. $\left(5+\sqrt{3}\right)\left(5-\sqrt{3}\right)=5^2-\left(\sqrt{3}\right)^2=25-3=22$

127. $\left(\sqrt{6}-\sqrt{2}\right)\left(\sqrt{6}+\sqrt{2}\right)=\left(\sqrt{6}\right)^2-\left(\sqrt{2}\right)^2=6-2=4$

Chapter 8 Mid-Chapter Check Points

1. $\sqrt{50}\cdot\sqrt{6}=\sqrt{50\cdot 6}=\sqrt{300}$
$$=\sqrt{100}\cdot\sqrt{3}=10\sqrt{3}$$

2. $\sqrt{6}+9\sqrt{6}=(1+9)\sqrt{6}=10\sqrt{6}$

3. $\sqrt{96x^3}=\sqrt{16x^2}\cdot\sqrt{6x}=4x\sqrt{6x}$

4. $\sqrt[5]{\dfrac{4}{32}}=\dfrac{\sqrt[5]{4}}{\sqrt[5]{32}}=\dfrac{\sqrt[5]{4}}{2}$

5. $\sqrt{27}+3\sqrt{12}=\sqrt{9}\cdot\sqrt{3}+3\sqrt{4}\cdot\sqrt{3}$
$$=3\sqrt{3}+3(2)\sqrt{3}$$
$$=3\sqrt{3}+6\sqrt{3}$$
$$=9\sqrt{3}$$

6. $\left(\sqrt{10}+\sqrt{3}\right)\left(\sqrt{10}-\sqrt{3}\right)$
$$=\left(\sqrt{10}\right)^2-\left(\sqrt{3}\right)^2=10-3=7$$

7. $\dfrac{\sqrt{5}}{2}\left(4\sqrt{3}-6\sqrt{20}\right)$
$$=\dfrac{4}{2}\sqrt{5\cdot 3}-\dfrac{6}{2}\sqrt{5\cdot 20}$$
$$=2\sqrt{15}-3\sqrt{100}$$
$$=2\sqrt{15}-3(10)$$
$$=2\sqrt{15}-30$$

8. $-\sqrt{32x^{21}}=-\sqrt{16x^{20}}\cdot\sqrt{2x}$
$$=-4x^{10}\sqrt{2x}$$

9. $\sqrt{6x^3}\cdot\sqrt{2x^4}=\sqrt{6x^3\cdot 2x^4}$
$$=\sqrt{12x^7}$$
$$=\sqrt{4x^6}\cdot\sqrt{3x}$$
$$=2x^3\sqrt{3x}$$

10. $\dfrac{\sqrt[3]{32}}{\sqrt[3]{2}}=\sqrt[3]{\dfrac{32}{2}}=\sqrt[3]{16}=\sqrt[3]{8}\cdot\sqrt[3]{2}=2\sqrt[3]{2}$

11. $-3\sqrt{90} - 5\sqrt{40}$

$= -3\sqrt{9} \cdot \sqrt{10} - 5\sqrt{4} \cdot \sqrt{10}$

$= -3(3)\sqrt{10} - 5(2)\sqrt{10}$

$= -9\sqrt{10} - 10\sqrt{10}$

$= -19\sqrt{10}$

12. $\left(2 - \sqrt{3}\right)\left(5 + 2\sqrt{3}\right)$

$= 10 + 4\sqrt{3} - 5\sqrt{3} - 2\left(\sqrt{3}\right)^2$

$= 10 - \sqrt{3} - 6$

$= 4 - \sqrt{3}$

13. $\dfrac{\sqrt{56x^5}}{\sqrt{7x^3}} = \sqrt{\dfrac{56x^5}{7x^3}} = \sqrt{8x^2}$

$= \sqrt{4x^2} \cdot \sqrt{2} = 2x\sqrt{2}$

14. $-\sqrt[4]{32} = -\sqrt[4]{16} \cdot \sqrt[4]{2} = -2\sqrt[4]{2}$

15. $\left(\sqrt{2} + \sqrt{7}\right)^2$

$= \left(\sqrt{2}\right)^2 + 2\left(\sqrt{2}\right)\left(\sqrt{7}\right) + \left(\sqrt{7}\right)^2$

$= 2 + 2\sqrt{2 \cdot 7} + 7$

$= 9 + 2\sqrt{14}$

16. $\sqrt[3]{\dfrac{1}{2}} \cdot \sqrt[3]{32} = \sqrt[3]{\dfrac{1}{2} \cdot 32} = \sqrt[3]{16}$

$= \sqrt[3]{8} \cdot \sqrt[3]{2} = 2\sqrt[3]{2}$

17. $\sqrt{5} + \sqrt{20} + \sqrt{45}$

$= \sqrt{5} + \sqrt{4} \cdot \sqrt{5} + \sqrt{9} \cdot \sqrt{5}$

$= 1\sqrt{5} + 2\sqrt{5} + 3\sqrt{5}$

$= 6\sqrt{5}$

18. $\dfrac{1}{3}\sqrt{\dfrac{90}{16}} = \dfrac{1}{3}\dfrac{\sqrt{90}}{\sqrt{16}} = \dfrac{\sqrt{9} \cdot \sqrt{10}}{3 \cdot 4}$

$= \dfrac{3\sqrt{10}}{12} = \dfrac{\sqrt{10}}{4}$

19. $3\sqrt{2}\left(\sqrt{2} + \sqrt{5}\right)$

$= 3\sqrt{2} \cdot \sqrt{2} + 3\sqrt{2} \cdot \sqrt{5}$

$= 3\sqrt{2 \cdot 2} + 3\sqrt{2 \cdot 5}$

$= 3\sqrt{4} + 3\sqrt{10}$

$= 3(2) + 3\sqrt{10}$

$= 6 + 3\sqrt{10}$

20. $\left(5 - \sqrt{2}\right)\left(5 + \sqrt{2}\right) = 5^2 - \left(\sqrt{2}\right)^2$

$= 25 - 2$

$= 23$

8.4 Check Points

1. a. $\dfrac{25}{\sqrt{10}} \cdot \dfrac{\sqrt{10}}{\sqrt{10}} = \dfrac{25\sqrt{10}}{10}$

$= \dfrac{5\sqrt{10}}{2}$

b. $\sqrt{\dfrac{2}{7}} = \dfrac{\sqrt{2}}{\sqrt{7}}$

$= \dfrac{\sqrt{2}}{\sqrt{7}} \cdot \dfrac{\sqrt{7}}{\sqrt{7}}$

$= \dfrac{\sqrt{14}}{7}$

2. a. $\dfrac{15}{\sqrt{18}} = \dfrac{15}{3\sqrt{2}} = \dfrac{15}{3\sqrt{2}} \cdot \dfrac{\sqrt{2}}{\sqrt{2}} = \dfrac{15\sqrt{2}}{3 \cdot 2} = \dfrac{5\sqrt{2}}{2}$

b. $\sqrt{\dfrac{7x}{20}} = \dfrac{\sqrt{7x}}{\sqrt{20}} = \dfrac{\sqrt{7x}}{2\sqrt{5}} \cdot \dfrac{\sqrt{5}}{\sqrt{5}} = \dfrac{\sqrt{35x}}{2 \cdot 5} = \dfrac{\sqrt{35x}}{10}$

3. $\dfrac{8}{4 + \sqrt{5}} = \dfrac{8}{4 + \sqrt{5}} \cdot \dfrac{4 - \sqrt{5}}{4 - \sqrt{5}}$

$= \dfrac{32 - 8\sqrt{5}}{16 - 5}$

$= \dfrac{32 - 8\sqrt{5}}{11}$

4. $\dfrac{8}{\sqrt{7}-\sqrt{3}} = \dfrac{8}{\sqrt{7}-\sqrt{3}} \cdot \dfrac{\sqrt{7}+\sqrt{3}}{\sqrt{7}+\sqrt{3}}$

$\qquad = \dfrac{8\sqrt{7}+8\sqrt{3}}{\left(\sqrt{7}\right)^2 - \left(\sqrt{3}\right)^2}$

$\qquad = \dfrac{8\sqrt{7}+8\sqrt{3}}{7-3}$

$\qquad = \dfrac{4\left(2\sqrt{7}+2\sqrt{3}\right)}{4}$

$\qquad = 2\sqrt{7}+2\sqrt{3}$

8.4 Concept and Vocabulary Check

1. rationalizing the denominator

2. $\sqrt{7}$

3. $2\sqrt{3}$; $\sqrt{3}$

4. $7-\sqrt{5}$

5. $\sqrt{11}+\sqrt{3}$

8.4 Exercise Set

1. $\dfrac{1}{\sqrt{10}} = \dfrac{1}{\sqrt{10}} \cdot \dfrac{\sqrt{10}}{\sqrt{10}} = \dfrac{\sqrt{10}}{10}$

3. $\dfrac{5}{\sqrt{5}} = \dfrac{5}{\sqrt{5}} \cdot \dfrac{\sqrt{5}}{\sqrt{5}} = \dfrac{5\sqrt{5}}{5} = \sqrt{5}$

5. $\dfrac{2}{\sqrt{6}} = \dfrac{2}{\sqrt{6}} \cdot \dfrac{\sqrt{6}}{\sqrt{6}} = \dfrac{2\sqrt{6}}{6} = \dfrac{\sqrt{6}}{3}$

7. $\dfrac{28}{\sqrt{7}} = \dfrac{28}{\sqrt{7}} \cdot \dfrac{\sqrt{7}}{\sqrt{7}} = \dfrac{28\sqrt{7}}{7} = 4\sqrt{7}$

9. $\sqrt{\dfrac{3}{5}} = \dfrac{\sqrt{3}}{\sqrt{5}} = \dfrac{\sqrt{3}}{\sqrt{5}} \cdot \dfrac{\sqrt{5}}{\sqrt{5}} = \dfrac{\sqrt{15}}{5}$

11. $\sqrt{\dfrac{7}{3}} = \dfrac{\sqrt{7}}{\sqrt{3}} = \dfrac{\sqrt{7}}{\sqrt{3}} \cdot \dfrac{\sqrt{3}}{\sqrt{3}} = \dfrac{\sqrt{21}}{3}$

13. $\sqrt{\dfrac{x^2}{3}} = \dfrac{\sqrt{x^2}}{\sqrt{3}} = \dfrac{x}{\sqrt{3}} \cdot \dfrac{\sqrt{3}}{\sqrt{3}} = \dfrac{x\sqrt{3}}{3}$

15. $\sqrt{\dfrac{11}{x}} = \dfrac{\sqrt{11}}{\sqrt{x}} = \dfrac{\sqrt{11}}{\sqrt{x}} \cdot \dfrac{\sqrt{x}}{\sqrt{x}} = \dfrac{\sqrt{11x}}{x}$

17. $\sqrt{\dfrac{x}{y}} = \dfrac{\sqrt{x}}{\sqrt{y}} = \dfrac{\sqrt{x}}{\sqrt{y}} \cdot \dfrac{\sqrt{y}}{\sqrt{y}} = \dfrac{\sqrt{xy}}{y}$

19. $\sqrt{\dfrac{x^4}{2}} = \dfrac{\sqrt{x^4}}{\sqrt{2}} = \dfrac{x^2}{\sqrt{2}}$

$\qquad = \dfrac{x^2}{\sqrt{2}} \cdot \dfrac{\sqrt{2}}{\sqrt{2}} = \dfrac{x^2\sqrt{2}}{2}$

21. $\dfrac{\sqrt{7}}{\sqrt{5}} = \dfrac{\sqrt{7}}{\sqrt{5}} \cdot \dfrac{\sqrt{5}}{\sqrt{5}} = \dfrac{\sqrt{35}}{5}$

23. $\dfrac{\sqrt{3x}}{\sqrt{14}} = \dfrac{\sqrt{3x}}{\sqrt{14}} \cdot \dfrac{\sqrt{14}}{\sqrt{14}} = \dfrac{\sqrt{42x}}{14}$

25. $\dfrac{1}{\sqrt{20}} = \dfrac{1}{\sqrt{4}\sqrt{5}} = \dfrac{1}{2\sqrt{5}} = \dfrac{1}{2\sqrt{5}} \cdot \dfrac{\sqrt{5}}{\sqrt{5}}$

$\qquad = \dfrac{\sqrt{5}}{2\cdot 5} = \dfrac{\sqrt{5}}{10}$

27. $\dfrac{12}{\sqrt{32}} = \dfrac{12}{\sqrt{6}\sqrt{2}} = \dfrac{12}{4\sqrt{2}} = \dfrac{3}{\sqrt{2}}$

$\qquad = \dfrac{3}{\sqrt{2}} \cdot \dfrac{\sqrt{2}}{\sqrt{2}} = \dfrac{3\sqrt{2}}{2}$

29. $\dfrac{15}{\sqrt{12}} = \dfrac{15}{\sqrt{4}\sqrt{3}} = \dfrac{15}{2\sqrt{3}} = \dfrac{15}{2\sqrt{3}} \cdot \dfrac{\sqrt{3}}{\sqrt{3}}$

$\qquad = \dfrac{15\sqrt{3}}{2\cdot 3} = \dfrac{15\sqrt{3}}{6} = \dfrac{5\sqrt{3}}{2}$

31. $\sqrt{\dfrac{5}{18}} = \dfrac{\sqrt{5}}{\sqrt{18}} = \dfrac{\sqrt{5}}{\sqrt{9}\sqrt{2}} = \dfrac{\sqrt{5}}{3\sqrt{2}}$

$\qquad = \dfrac{\sqrt{5}}{3\sqrt{2}} \cdot \dfrac{\sqrt{2}}{\sqrt{2}} = \dfrac{\sqrt{10}}{3\cdot 2} = \dfrac{\sqrt{10}}{6}$

33. $\sqrt{\dfrac{x}{32}} = \dfrac{\sqrt{x}}{\sqrt{32}} = \dfrac{\sqrt{x}}{\sqrt{16}\sqrt{2}} = \dfrac{\sqrt{x}}{4\sqrt{2}}$

$\qquad = \dfrac{\sqrt{x}}{4\sqrt{2}} \cdot \dfrac{\sqrt{2}}{\sqrt{2}} = \dfrac{\sqrt{2x}}{4\cdot 2} = \dfrac{\sqrt{2x}}{8}$

35. $\sqrt{\dfrac{1}{45}} = \dfrac{\sqrt{1}}{\sqrt{45}} = \dfrac{1}{\sqrt{45}} = \dfrac{1}{\sqrt{9}\sqrt{5}} = \dfrac{1}{3\sqrt{5}}$

$\qquad = \dfrac{1}{3\sqrt{5}} \cdot \dfrac{\sqrt{5}}{\sqrt{5}} = \dfrac{\sqrt{5}}{3 \cdot 5} = \dfrac{\sqrt{5}}{15}$

37. $\dfrac{\sqrt{7}}{\sqrt{12}} = \dfrac{\sqrt{7}}{\sqrt{4}\sqrt{3}} = \dfrac{\sqrt{7}}{2\sqrt{3}} \cdot \dfrac{\sqrt{3}}{\sqrt{3}}$

$\qquad = \dfrac{\sqrt{21}}{2 \cdot 3} = \dfrac{\sqrt{21}}{6}$

39. $\dfrac{8x}{\sqrt{8}} = \dfrac{8x}{\sqrt{4}\sqrt{2}} = \dfrac{8x}{2\sqrt{2}} = \dfrac{4x}{\sqrt{2}}$

$\qquad = \dfrac{4x}{\sqrt{2}} \cdot \dfrac{\sqrt{2}}{\sqrt{2}} = \dfrac{4x\sqrt{2}}{2}$

$\qquad = 2x\sqrt{2}$

41. $\dfrac{\sqrt{7y}}{\sqrt{8}} = \dfrac{\sqrt{7y}}{\sqrt{4}\sqrt{2}} = \dfrac{\sqrt{7y}}{2\sqrt{2}} \cdot \dfrac{\sqrt{2}}{\sqrt{2}}$

$\qquad = \dfrac{\sqrt{14y}}{2 \cdot 2} = \dfrac{\sqrt{14y}}{4}$

43. $\sqrt{\dfrac{7x}{12}} = \dfrac{\sqrt{7x}}{\sqrt{12}} = \dfrac{\sqrt{7x}}{\sqrt{4}\sqrt{3}} = \dfrac{\sqrt{7x}}{2\sqrt{3}}$

$\qquad = \dfrac{\sqrt{7x}}{2\sqrt{3}} \cdot \dfrac{\sqrt{3}}{\sqrt{3}} = \dfrac{\sqrt{21x}}{2 \cdot 3}$

$\qquad = \dfrac{\sqrt{21x}}{6}$

45. $\sqrt{\dfrac{45}{x}} = \dfrac{\sqrt{45}}{\sqrt{x}} = \dfrac{\sqrt{9}\sqrt{5}}{\sqrt{x}} = \dfrac{3\sqrt{5}}{\sqrt{x}}$

$\qquad = \dfrac{3\sqrt{5}}{\sqrt{x}} \cdot \dfrac{\sqrt{x}}{\sqrt{x}} = \dfrac{3\sqrt{5x}}{x}$

47. $\dfrac{5}{\sqrt{x^3}} = \dfrac{5}{\sqrt{x^2}\sqrt{x}} = \dfrac{5}{x\sqrt{x}} = \dfrac{5}{x\sqrt{x}} \cdot \dfrac{\sqrt{x}}{\sqrt{x}}$

$\qquad = \dfrac{5\sqrt{x}}{x \cdot x} = \dfrac{5\sqrt{x}}{x^2}$

49. $\sqrt{\dfrac{27}{y^3}} = \dfrac{\sqrt{27}}{\sqrt{y^3}} = \dfrac{\sqrt{9}\sqrt{3}}{\sqrt{y^2}\sqrt{y}}$

$\qquad = \dfrac{3\sqrt{3}}{y\sqrt{y}} = \dfrac{3\sqrt{3}}{y\sqrt{y}} \cdot \dfrac{\sqrt{y}}{\sqrt{y}}$

$\qquad = \dfrac{3\sqrt{3y}}{y \cdot y} = \dfrac{3\sqrt{3y}}{y^2}$

51. $\dfrac{\sqrt{50x^2}}{\sqrt{12y^3}} = \dfrac{\sqrt{25x^2}\sqrt{2}}{\sqrt{4y^2}\sqrt{3y}} = \dfrac{5x\sqrt{2}}{2y\sqrt{3y}}$

$\qquad = \dfrac{5x\sqrt{2}}{2y\sqrt{3y}} \cdot \dfrac{\sqrt{3y}}{\sqrt{3y}} = \dfrac{5x\sqrt{6y}}{2y \cdot 3y}$

$\qquad = \dfrac{5x\sqrt{6y}}{6y^2}$

53. Multiply the numerator and denominator by the conjugate of the denominator, $4 - \sqrt{3}$.

$\qquad \dfrac{1}{4+\sqrt{3}} \cdot \dfrac{4-\sqrt{3}}{4-\sqrt{3}} = \dfrac{1\left(4-\sqrt{3}\right)}{4^2 - \left(\sqrt{3}\right)^2}$

$\qquad\qquad = \dfrac{4-\sqrt{3}}{16-3}$

$\qquad\qquad = \dfrac{4-\sqrt{3}}{13}$

55. $\dfrac{9}{2-\sqrt{7}} = \dfrac{9}{2-\sqrt{7}} \cdot \dfrac{2+\sqrt{7}}{2+\sqrt{7}}$

$\qquad = \dfrac{9\left(2+\sqrt{7}\right)}{2^2 - \left(\sqrt{7}\right)^2} = \dfrac{9\left(2+\sqrt{7}\right)}{4-7}$

$\qquad = \dfrac{9\left(2+\sqrt{7}\right)}{-3}$

$\qquad = -3\left(2+\sqrt{7}\right)$

$\qquad -6 - 3\sqrt{7}$

57. $\dfrac{16}{\sqrt{11}+3} = \dfrac{16}{\sqrt{11}+3} \cdot \dfrac{\sqrt{11}-3}{\sqrt{11}-3}$

$\qquad = \dfrac{16\left(\sqrt{11}-3\right)}{\left(\sqrt{11}\right)^2 - 3^2} = \dfrac{16\left(\sqrt{11}-3\right)}{11-9}$

$\qquad = \dfrac{16\left(\sqrt{11}-3\right)}{2}$

$\qquad = 8\left(\sqrt{11}-3\right)$

$\qquad = 8\sqrt{11}-24$

59. $\dfrac{18}{3-\sqrt{3}} = \dfrac{18}{3-\sqrt{3}} \cdot \dfrac{3+\sqrt{3}}{3+\sqrt{3}}$

$\qquad = \dfrac{18\left(3+\sqrt{3}\right)}{3^2 - \left(\sqrt{3}\right)^2} = \dfrac{18\left(3+\sqrt{3}\right)}{9-3}$

$\qquad = \dfrac{18\left(3+\sqrt{3}\right)}{6}$

$\qquad = 3\left(3+\sqrt{3}\right)$

$\qquad = 9+3\sqrt{3}$

61. $\dfrac{\sqrt{2}}{\sqrt{2}+1} = \dfrac{\sqrt{2}}{\sqrt{2}+1} \cdot \dfrac{\sqrt{2}-1}{\sqrt{2}-1}$

$\qquad = \dfrac{\sqrt{2}\left(\sqrt{2}-1\right)}{\left(\sqrt{2}\right)^2 - 1^2} = \dfrac{\sqrt{2}\left(\sqrt{2}-1\right)}{2-1}$

$\qquad = \dfrac{\sqrt{2}\left(\sqrt{2}-1\right)}{1}$

$\qquad = 2-\sqrt{2}$

63. $\dfrac{\sqrt{10}}{\sqrt{10}-\sqrt{7}} = \dfrac{\sqrt{10}}{\sqrt{10}-\sqrt{7}} \cdot \dfrac{\sqrt{10}+\sqrt{7}}{\sqrt{10}+\sqrt{7}}$

$\qquad = \dfrac{\sqrt{10}\left(\sqrt{10}+\sqrt{7}\right)}{\left(\sqrt{10}\right)^2 - \left(\sqrt{7}\right)^2}$

$\qquad = \dfrac{\sqrt{10}\left(\sqrt{10}+\sqrt{7}\right)}{10-7}$

$\qquad = \dfrac{\sqrt{10}\left(\sqrt{10}+\sqrt{7}\right)}{3}$

$\qquad = \dfrac{10+\sqrt{70}}{3}$

65. $\dfrac{6}{\sqrt{6}+\sqrt{3}} = \dfrac{6}{\sqrt{6}+\sqrt{3}} \cdot \dfrac{\sqrt{6}-\sqrt{3}}{\sqrt{6}-\sqrt{3}}$

$\qquad = \dfrac{6\left(\sqrt{6}-\sqrt{3}\right)}{\left(\sqrt{6}\right)^2 - \left(\sqrt{3}\right)^2}$

$\qquad = \dfrac{6\left(\sqrt{6}-\sqrt{3}\right)}{6-3}$

$\qquad = \dfrac{6\left(\sqrt{6}-\sqrt{3}\right)}{3}$

$\qquad = 2\left(\sqrt{6}-\sqrt{3}\right)$

$\qquad = 2\sqrt{6}-2\sqrt{3}$

67. $\dfrac{2}{\sqrt{5}-\sqrt{3}} = \dfrac{2}{\sqrt{5}-\sqrt{3}} \cdot \dfrac{\sqrt{5}+\sqrt{3}}{\sqrt{5}+\sqrt{3}}$

$\qquad = \dfrac{2\left(\sqrt{5}+\sqrt{3}\right)}{\left(\sqrt{5}\right)^2 - \left(\sqrt{3}\right)^2}$

$\qquad = \dfrac{2\left(\sqrt{5}+\sqrt{3}\right)}{5-3}$

$\qquad = \dfrac{2\left(\sqrt{5}+\sqrt{3}\right)}{2}$

$\qquad = \sqrt{5}+\sqrt{3}$

69. $\dfrac{2}{4+\sqrt{x}} = \dfrac{2}{4+\sqrt{x}} \cdot \dfrac{4-\sqrt{x}}{4-\sqrt{x}}$

$\qquad = \dfrac{2\left(4-\sqrt{x}\right)}{4^2 - \left(\sqrt{x}\right)^2} = \dfrac{2\left(4-\sqrt{x}\right)}{16-x}$

$\qquad = \dfrac{8-2\sqrt{x}}{16-x}$

71.
$$\frac{2\sqrt{3}}{\sqrt{15}+2} = \frac{2\sqrt{3}}{\sqrt{15}+2} \cdot \frac{\sqrt{15}-2}{\sqrt{15}-2}$$

$$= \frac{2\sqrt{3}\left(\sqrt{15}-2\right)}{\left(\sqrt{15}\right)^2 - 2^2}$$

$$= \frac{2\sqrt{3}\left(\sqrt{15}-2\right)}{15-4}$$

$$= \frac{2\sqrt{3}\left(\sqrt{15}-2\right)}{11}$$

$$= \frac{2\sqrt{45}-4\sqrt{3}}{11}$$

$$= \frac{2\sqrt{9}\sqrt{5}-4\sqrt{3}}{11}$$

$$= \frac{2\cdot 3\sqrt{5}-4\sqrt{3}}{11}$$

$$= \frac{6\sqrt{5}-4\sqrt{3}}{11}$$

73.
$$\frac{\sqrt{5}+\sqrt{2}}{\sqrt{5}-\sqrt{2}} = \frac{\sqrt{5}+\sqrt{2}}{\sqrt{5}-\sqrt{2}} \cdot \frac{\sqrt{5}+\sqrt{2}}{\sqrt{5}+\sqrt{2}}$$

$$= \frac{\left(\sqrt{5}+\sqrt{2}\right)^2}{\left(\sqrt{5}\right)^2 - \left(\sqrt{2}\right)^2}$$

$$= \frac{\left(\sqrt{5}\right)^2 + 2\sqrt{5}\sqrt{2} + \left(\sqrt{2}\right)^2}{\left(\sqrt{5}\right)^2 - \left(\sqrt{2}\right)^2}$$

$$= \frac{5+2\sqrt{10}+2}{5-2}$$

$$= \frac{7+2\sqrt{10}}{3}$$

75.
$$\frac{\sqrt{36x^2y^5}}{\sqrt{2x^3y}} = \sqrt{\frac{36x^2y^5}{2x^3y}} = \sqrt{\frac{18y^4}{x}}$$

$$= \frac{\sqrt{9y^4}\sqrt{2}}{\sqrt{x}} = \frac{3y^2\sqrt{2}}{\sqrt{x}}$$

$$= \frac{3y^2\sqrt{2}}{\sqrt{x}} \cdot \frac{\sqrt{x}}{\sqrt{x}} = \frac{3y^2\sqrt{2x}}{x}$$

77.
$$\frac{2}{\sqrt{x+2}-\sqrt{x}}$$

$$= \frac{2}{\sqrt{x+2}-\sqrt{x}} \cdot \frac{\sqrt{x+2}+\sqrt{x}}{\sqrt{x+2}+\sqrt{x}}$$

$$= \frac{2\left(\sqrt{x+2}+\sqrt{x}\right)}{\left(\sqrt{x+2}\right)^2 - \left(\sqrt{x}\right)^2}$$

$$= \frac{2\left(\sqrt{x+2}+\sqrt{x}\right)}{x+2-x}$$

$$= \frac{2\left(\sqrt{x+2}+\sqrt{x}\right)}{2}$$

$$= \sqrt{x+2}+\sqrt{x}$$

79.
$$\frac{\sqrt{2}}{\sqrt{3}} + \frac{\sqrt{3}}{\sqrt{2}} = \frac{\sqrt{2}}{\sqrt{3}} \cdot \frac{\sqrt{3}}{\sqrt{3}} + \frac{\sqrt{3}}{\sqrt{2}} \cdot \frac{\sqrt{2}}{\sqrt{2}}$$

$$= \frac{\sqrt{6}}{3} + \frac{\sqrt{6}}{2} = \frac{2\sqrt{6}}{6} + \frac{3\sqrt{6}}{6}$$

$$= \frac{5\sqrt{6}}{6}$$

81.
$$\frac{2x+4-2h}{\sqrt{x+2-h}} = \frac{2(x+2-h)}{\sqrt{x+2-h}} \cdot \frac{\sqrt{x+2-h}}{\sqrt{x+2-h}}$$

$$= \frac{2(x+2-h)\sqrt{x+2-h}}{x+2-h}$$

$$= 2\sqrt{x+2-h}$$

83. a.
$$P = \frac{x\left(13+\sqrt{x}\right)}{5\sqrt{x}}; x = 25$$

$$P = \frac{25\left(13+\sqrt{25}\right)}{5\sqrt{25}}$$

$$= \frac{25 + \left(13+5\right)}{5\cdot 5}$$

$$= \frac{25\cdot 18}{25} = 18$$

According to the formula, 18% of 25-year-olds must pay more taxes.

b.
$$\frac{x(13+\sqrt{x})}{5\sqrt{x}} \cdot \frac{\sqrt{x}}{\sqrt{x}} = \frac{x\sqrt{x}(13+\sqrt{x})}{5\sqrt{x}\sqrt{x}}$$

$$= \frac{x\sqrt{x}\cdot 13 + x\sqrt{x}\sqrt{x}}{5\sqrt{x}\sqrt{x}}$$

$$= \frac{13x\sqrt{x} + x\cdot x}{5x}$$

$$= \frac{x(13\sqrt{x}+x)}{5x}$$

$$= \frac{13\sqrt{x}+x}{5}$$

c.
$$\frac{13\sqrt{25}+25}{5} = \frac{13(5)+25}{5} = 18$$

According to the formula, 18% of 25-year-olds must pay more taxes. This is the same result as in part (a).

85.
$$\frac{w}{h} = \frac{2}{\sqrt{5}-1}$$

$$= \frac{2}{\sqrt{5}-1} \cdot \frac{\sqrt{5}+1}{\sqrt{5}+1}$$

$$= \frac{2(\sqrt{5}+1)}{5-1}$$

$$= \frac{\sqrt{5}+1}{2} \approx 1.62$$

87. Answers will vary.

89.
$$P = \frac{x(13+\sqrt{x})}{5\sqrt{x}}$$

$$x = 30: \quad P = \frac{30(13+\sqrt{30})}{5\sqrt{30}} \approx 20$$

$$x = 40: \quad P = \frac{40(13+\sqrt{40})}{5\sqrt{40}} \approx 24$$

$$x = 50: \quad P = \frac{50(13+\sqrt{50})}{5\sqrt{50}} \approx 28$$

These results show that about 20% of 30-year-old taxpayers, 24% of 40-year-old taxpayers, and 28% of 50-year-old taxpayers must pay more taxes. The trend is for an increasing percentage of taxpayers to pay more taxes as the taxpayers get older. Explanations may vary.

91. makes sense

93. makes sense

95. true

97. false; Changes to make the statement true will vary. A sample change is: They may require more space.

99.
$$\sqrt{2} + \sqrt{\frac{1}{2}} = \sqrt{2} + \frac{\sqrt{1}}{\sqrt{2}} = \sqrt{2} + \frac{1}{\sqrt{2}}$$

$$= \sqrt{2} + \frac{1}{\sqrt{2}} \cdot \frac{\sqrt{2}}{\sqrt{2}}$$

$$= \sqrt{2} + \frac{\sqrt{2}}{2}$$

$$= \frac{2\sqrt{2}}{2} + \frac{\sqrt{2}}{2} = \frac{3\sqrt{2}}{2}$$

101.
$$\frac{4}{2+\sqrt{\Box}} = 8 - 4\sqrt{3}$$

$$\frac{4}{2+\sqrt{\Box}} \cdot \frac{2-\sqrt{\Box}}{2-\sqrt{\Box}} = 8 - 7\sqrt{3}$$

$$\frac{4(2-\sqrt{\Box})}{2^2 - (\sqrt{\Box})^2} = 8 - 4\sqrt{3}$$

$$\frac{8-4\sqrt{\Box}}{4-\Box} = 8 - 4\sqrt{3}$$

If the missing number is 3, the last equation becomes $\frac{8-4\sqrt{3}}{4-3} = 8-4\sqrt{3}$, which is true. Thus, 3 goes in the box.

102.
$$6x^2 - 11x + 5 = 0$$
$$(6x-5)(x-1) = 0$$
$$6x - 5 = 0 \quad \text{or} \quad x - 1 = 0$$
$$6x = 5 \qquad\qquad x = 1$$
$$x = \frac{5}{6}$$

The solution set is $\left\{\frac{5}{6}, 1\right\}$.

103.
$$(2x^2)^{-3} = \frac{1}{(2x^2)^3} = \frac{1}{2^3(x^2)^3} = \frac{1}{8x^6}$$

104. $\dfrac{x^2-6x+9}{12} \cdot \dfrac{3}{x^2-9}$

$= \dfrac{(x-3)(x-3)}{12} \cdot \dfrac{3}{(x+3)(x-3)}$

$= \dfrac{x-3}{4(x+3)}$

105. $\quad 7 = -2.5x+17$

$\quad -10 = -2.5x$

$\quad \dfrac{-10}{-2.5} = \dfrac{-2.5x}{-2.5}$

$\quad\quad 4 = x$

The solution set is $\{4\}$.

106. $\quad 2x-1 = x^2-4x+4$

$\quad\quad 0 = x^2-6x+5$

$\quad\quad 0 = (x-5)(x-1)$

$\quad x-5 = 0 \quad$ or $\quad x-1=0$

$\quad\quad x = 5 \quad\quad\quad\quad x = 1$

The solution set is $\{1,5\}$.

107. The solutions of the equation in exercise 106 are 1 and 5.

Check 1:

$\sqrt{2x-1}+2 = x$

$\sqrt{2(1)-1}+2 = 1$

$\quad\quad \sqrt{1}+2 = 1$

$\quad\quad\quad 1+2 = 1$

$\quad\quad\quad\quad\quad 3 = 1, \text{ false}$

Check 5:

$\sqrt{2x-1}+2 = x$

$\sqrt{2(5)-1}+2 = 5$

$\quad\quad \sqrt{9}+2 = 5$

$\quad\quad\quad 3+2 = 5$

$\quad\quad\quad\quad\quad 5 = 5, \text{ true}$

5 is a solution of $\sqrt{2x-1}+2 = x$.

8.5 Check Points

1. $\quad\quad \sqrt{2x+3} = 5$

$\quad \left(\sqrt{2x+3}\right)^2 = 5^2$

$\quad\quad\quad 2x+3 = 25$

$\quad\quad\quad\quad 2x = 22$

$\quad\quad\quad\quad\quad x = 11$

Check 11:

$\quad \sqrt{2x+3} = 5$

$\quad \sqrt{2\cdot 11+3} = 5$

$\quad\quad \sqrt{25} = 5$

$\quad\quad\quad\quad 5 = 5, \text{ true}$

The solution set is $\{11\}$.

2. $\quad \sqrt{x+32}-3\sqrt{x} = 0$

$\quad\quad \sqrt{x+32} = 3\sqrt{x}$

$\quad \left(\sqrt{x+32}\right)^2 = \left(3\sqrt{x}\right)^2$

$\quad\quad\quad x+32 = 9x$

$\quad\quad\quad -8x = -32$

$\quad\quad\quad\quad\quad x = 4$

Check 4:

$\sqrt{x+32}-3\sqrt{x} = 0$

$\sqrt{4+32}-3\sqrt{4} = 0$

$\quad \sqrt{36}-3\cdot 2 = 0$

$\quad\quad\quad 6-6 = 0$

$\quad\quad\quad\quad\quad 0 = 0, \text{ true}$

The solution set is $\{4\}$.

3. $\quad \sqrt{x}+1 = 0$

$\quad\quad \sqrt{x} = -1$

$\quad \left(\sqrt{x}\right)^2 = (-1)^2$

$\quad\quad\quad x = 1$

Check 1:

$\quad \sqrt{x}+1 = 0$

$\quad \sqrt{1}+1 = 0$

$\quad\quad 1+1 = 0$

$\quad\quad\quad 2 = 0, \text{ false}$

The false statement indicates that 1 is not a solution. The solution set is $\{\ \}$.

4. $\sqrt{x+3}+3=x$

$\sqrt{x+3}=x-3$

$\left(\sqrt{x+3}\right)^2=(x-3)^2$

$x+3=x^2-6x+9$

$0=x^2-7x+6$

$0=(x-1)(x-6)$

$x-1=0$ or $x-6=0$

$x=1$ $x=6$

Check 1:   Check 6:

$\sqrt{x+3}+3=x$ $\sqrt{x+3}+3=x$

$\sqrt{1+3}+3=1$ $\sqrt{6+3}+3=6$

$\sqrt{4}+3=1$ $\sqrt{9}+3=6$

$2+3=1$ $3+3=6$

$5=1$, false $6=6$, true

The false statement indicates that 1 is not a solution. The solution set is $\{6\}$.

5. $H=-2.3\sqrt{I}+67.6$

$33.1=-2.3\sqrt{I}+67.6$

$-34.5=-2.3\sqrt{I}$

$\dfrac{-34.5}{-2.3}=\dfrac{-2.3\sqrt{I}}{-2.3}$

$15=\sqrt{I}$

$15^2=\left(\sqrt{I}\right)^2$

$225=I$

The model indicates that an annual income of 225 thousand dollars, or $225,000, corresponds to 33.1 hours per week watching TV.

8.5 Concept and Vocabulary Check

1. radical

2. extraneous

3. $x+3;\ x^2-6x+9$

4. false

5. true

6. false

8.5 Exercise Set

1. $\sqrt{x}=5$

$\left(\sqrt{x}\right)^2=5^2$

$x=25$

Check:

$\sqrt{x}=5$

$\sqrt{25}=5$

$5=5$, true

The solution set is $\{25\}$.

3. $\sqrt{x}-4=0$

$\sqrt{x}=4$

$\left(\sqrt{x}\right)^2=4^2$

$x=16$

Check:

$\sqrt{x}-4=0$

$\sqrt{16}-4=0$

$4-4=0$

$0=0$, true

The solution set is $\{16\}$.

5. $\sqrt{x+2}=3$

$\left(\sqrt{x+2}\right)^2=3^2$

$x+2=9$

$x=7$

Substitution confirms the solution set is $\{7\}$.

7. $\sqrt{x-3}-11=0$

$\sqrt{x-3}=11$

$\left(\sqrt{x-3}\right)^2=11^2$

$x-3=121$

$x=124$

Substitution confirms the solution set is $\{124\}$.

9. $\sqrt{3x-5} = 4$

$\left(\sqrt{3x-5}\right)^2 = 4^2$

$3x - 5 = 16$

$3x = 21$

$x = 7$

Substitution confirms the solution set is $\{7\}$.

11. $\sqrt{x+5} + 2 = 5$

$\sqrt{x+5} = 3$

$\left(\sqrt{x+5}\right)^2 = 3^2$

$x + 5 = 9$

$x = 4$

Substitution confirms the solution set is $\{4\}$.

13. $\sqrt{x+3} = \sqrt{4x-3}$

$\left(\sqrt{x+3}\right)^2 = \left(\sqrt{4x-3}\right)^2$

$x + 3 = 4x - 3$

$-3x + 3 = -3$

$-3x = -6$

$x = 2$

Substitution confirms the solution set is $\{2\}$.

15. $\sqrt{6x-2} = \sqrt{4x+4}$

$\left(\sqrt{6x-2}\right)^2 = \left(\sqrt{4x+4}\right)$

$6x - 2 = 4x + 4$

$2x - 2 = 4$

$2x = 6$

$x = 3$

Substitution confirms the solution set is $\{3\}$.

17. $11 = 6 + \sqrt{x+1}$

$5 = \sqrt{x+1}$

$5^2 = \left(\sqrt{x+1}\right)^2$

$25 = x + 1$

$24 = x$

Substitution confirms the solution set is $\{24\}$.

19. $\sqrt{x} + 10 = 0$

$\sqrt{x} = -10$

$\left(\sqrt{x}\right)^2 = (-10)^2$

$x = 100$

Check 100:

$\sqrt{x} + 10 = 0$

$\sqrt{100} + 10 = 0$

$10 + 10 = 0$

$20 = 0$, false

This false statement indicates that 100 is not a solution. Since the only proposed solution is extraneous, the given equation has no solution. The solution set is $\{\ \}$.

21. $\sqrt{x-1} = -3$

$\left(\sqrt{x-1}\right)^2 = (-3)^2$

$x - 1 = 9$

$x = 10$

Check 10:

$\sqrt{x-1} = -3$

$\sqrt{10-1} = -3$

$\sqrt{9} = -3$

$3 = -3$, false

The false statement shows that 10 is an extraneous solution. Since it is the only proposed solution, the given equation has no solution. The solution set is $\{\ \}$.

23. $3\sqrt{x} = \sqrt{8x+16}$

$\left(3\sqrt{x}\right)^2 = \left(\sqrt{8x+16}\right)^2$

$9x = 8x + 16$

$x = 16$

Check 16:

$3\sqrt{x} = \sqrt{8x+16}$

$3\sqrt{16} = \sqrt{8 \cdot 16 + 16}$

$3 \cdot 4 = \sqrt{128 + 16}$

$12 = \sqrt{144}$

$12 = 12$, true

The solution set is $\{16\}$.

25. $\sqrt{2x-3}+5=0$

$\sqrt{2x-3}=-5$

$\left(\sqrt{2x-3}\right)^2=(-5)^2$

$2x-3=25$

$2x=28$

$x=14$

Check 14:

$\sqrt{2x-3}+5=0$

$\sqrt{2\cdot14-3}+5=0$

$\sqrt{28-3}+5=0$

$\sqrt{25}+5=0$

$5+5=0$

$10=0$, false

The false statement shows that 14 is an extraneous solution. The solution set is { }.

27. $\sqrt{3x+4}-2=3$

$\sqrt{3x+4}=5$

$\left(\sqrt{3x+4}\right)^2=5^2$

$3x+4=25$

$3x=21$

$x=7$

Substitution confirms the solution set is $\{7\}$.

29. $3\sqrt{x-1}=\sqrt{3x+3}$

$\left(3\sqrt{x-1}\right)^2=\left(\sqrt{3x+3}\right)^2$

$9(x-1)=3x+3$

$9x-9=3x+3$

$6x=12$

$x=2$

Substitution confirms the solution set is $\{2\}$.

31. $\sqrt{x+7}=x+5$

Square both sides,

$\left(\sqrt{x+7}\right)^2=(x+5)^2$

Square the binomial on the right.

$x+7=x^2+10x+25$

Simplify and solve this quadratic equation.

$0=x^2+9x+18$

$0=(x+3)(x+6)$

$x+3=0$ or $x+6=0$

$x=-3$ $x=-6$

There are two proposed solutions. Each must be checked in the original equation.

Check -3:   Check -6:

$\sqrt{x+7}=x+5$ $\sqrt{x+7}=x+5$

$\sqrt{-3+7}=-3+5$ $\sqrt{-6+7}=-6+5$

$\sqrt{4}=2$ $\sqrt{1}=-1$

$2=2$, true $1=-1$, false

Thus -6 is an extraneous solution, while -3 satisfies the equation. The solution set is $\{-3\}$.

33. $\sqrt{2x+13}=x+7$

$\left(\sqrt{2x+13}\right)^2=(x+7)^2$

$2x+13=x^2+14x+49$

$0=x^2+12x+36$

$0=(x+6)^2$

$0=x+6$

$-6=x$

Check -6:

$\sqrt{2x+13}=x+7$

$\sqrt{2(-6)+13}=-6+7$

$\sqrt{-12+13}=1$

$\sqrt{1}=1$

$1=1$, true

The solution set is $\{-6\}$.

35. $\sqrt{9x^2+2x-4}=3x$

$\left(\sqrt{9x^2+2x-4}\right)^2=(3x)^2$

$9x^2+2x-4=9x^2$

$2x-4=0$

$2x=4$

$x=2$

Substitution confirms the solution set is $\{2\}$.

37.
$$x = \sqrt{2x-2} + 1$$
$$x - 1 = \sqrt{2x-2}$$
$$(x-1)^2 = \left(\sqrt{2x-2}\right)^2$$
$$x^2 + 2x + 1 = 2x - 2$$
$$x^2 - 4x + 3 = 0$$
$$(x-1)(x-3) = 0$$
$$x - 1 = 0 \text{ or } x - 3 = 0$$
$$x = 1 \qquad\qquad x = 3$$

Check 1:
$$x = \sqrt{2x-2} + 1$$
$$1 = \sqrt{2 \cdot 1 - 2} + 1$$
$$1 = \sqrt{2 - 2} + 1$$
$$1 = 0 + 1$$
$$1 = 1, \text{ true}$$

Check 3:
$$x = \sqrt{2x-2} + 1$$
$$3 = \sqrt{2 \cdot 3 - 2} + 1$$
$$3 = \sqrt{6 - 2} + 1$$
$$3 = \sqrt{4} + 1$$
$$3 = 2 + 1$$
$$3 = 3, \text{ true}$$

Both proposed solutions, 1 and 3, satisfy the original equation. The solution set is $\{1,3\}$.

39.
$$x = \sqrt{8-7x} + 2$$
$$x - 2 = \sqrt{8-7x}$$
$$(x-2)^2 = \left(\sqrt{8-7x}\right)^2$$
$$x^2 - 4x + 4 = 8 - 7x$$
$$x^2 + 3x - 4 = 0$$
$$(x+4)(x-1) = 0$$
$$x + 4 = 0 \text{ or } x - 1 = 0$$
$$x = -4 \qquad\qquad x = 1$$

Check −4:
$$x = \sqrt{8-7x} + 2$$
$$-4 = \sqrt{8-7(-4)} + 2$$
$$-4 = \sqrt{8+28} + 2$$
$$-4 = \sqrt{36} + 2$$
$$-4 = 6 + 2$$
$$-4 = 8, \text{ false}$$

Check 1:
$$x = \sqrt{8-7x} + 2$$
$$1 = \sqrt{8-7 \cdot 1} + 2$$
$$1 = \sqrt{8-7} + 2$$
$$1 = \sqrt{1} + 2$$
$$1 = 1 + 2$$
$$1 = 3, \text{ false}$$

Both of the proposed solutions are extraneous so the given equation has no solution. The solution set is $\{ \ \}$.

41.
$$\sqrt{3x} + 10 = x + 4$$
$$\sqrt{3x} = x - 6$$
$$\left(\sqrt{3x}\right)^2 = (x-6)^2$$
$$3x = x^2 - 12x + 36$$
$$0 = x^2 - 15x + 36$$
$$0 = (x-3)(x-12)$$
$$x - 3 = 0 \text{ or } x - 12 = 0$$
$$x = 3 \qquad\qquad x = 12$$

Check 3:
$$\sqrt{3x} + 10 = x + 4$$
$$\sqrt{3 \cdot 3} + 10 = 3 + 4$$
$$\sqrt{9} + 10 = 7$$
$$3 - 10 = 7$$
$$13 = 7, \text{ false}$$

Check 12:
$$\sqrt{3x} + 10 = x + 4$$
$$\sqrt{3 \cdot 12} + 10 = 12 + 4$$
$$\sqrt{36} + 10 = 16$$
$$6 + 10 = 16$$
$$16 = 16, \text{ true}$$

The proposed solution, 3, is extraneous, while 12 satisfies the equation. The solution set is $\{12\}$.

43.
$$3\sqrt{x} + 5 = 2$$
$$3\sqrt{x} = -3$$
$$\left(3\sqrt{x}\right)^2 = (-3)^2$$
$$9x = 9$$
$$x = 1$$

Check 1:
$$3\sqrt{x} + 5 = 2$$
$$3\sqrt{1} + 5 = 2$$
$$3 \cdot 1 + 5 = 2$$
$$3 + 5 = 2$$
$$8 = 2, \text{ false}$$

The proposed solution, 1, is extraneous, so the equation has no solution. The solution set is $\{ \ \}$.

45. Let x = the number.

$$2+\sqrt{4x}=10$$

$$\sqrt{4x}=8$$

$$\left(\sqrt{4x}\right)^2=8^2$$

$$4x=64$$

$$x=16$$

Check 16:

$$2+\sqrt{4(16)}=10$$

$$2+\sqrt{64}=10$$

$$2+8=10$$

$$10=10,\ \text{true}$$

The number is 16.

47. Let x = the number.

$$x=4+\sqrt{2x}$$

$$x-4=\sqrt{2x}$$

$$(x-4)^2=\left(\sqrt{2x}\right)^2$$

$$x^2-8x+16=2x$$

$$x^2-10x+16=0$$

$$(x-8)(x-2)=0$$

$$x-8=0\ \text{ or }\ x-2=0$$

$$x=8\qquad\qquad x=2$$

Check 8:　　　　　　Check 2

$$8=4+\sqrt{2(8)}\qquad 2=4+\sqrt{2(2)}$$

$$8=4+\sqrt{16}\qquad 2=4+\sqrt{4}$$

$$8=4+4\qquad\quad 2=4+2$$

$$8=8\ \text{true}\qquad 2=6\ \text{false}$$

Discard 2. The number is 8.

49.

$$v=\sqrt{2gh}$$

$$v^2=\left(\sqrt{2gh}\right)^2$$

$$v^2=2gh$$

$$\frac{v^2}{2g}=\frac{2gh}{2g}$$

$$\frac{v^2}{2g}=h$$

51.

$$\sqrt{x}+2=\sqrt{x+8}$$

$$\left(\sqrt{x}+2\right)^2=\left(\sqrt{x+8}\right)^2$$

$$\left(\sqrt{x}\right)^2+2(2)\left(\sqrt{x}\right)+2^2=x+8$$

$$x+4\sqrt{x}+4=x+8$$

$$4\sqrt{x}=4$$

$$\sqrt{x}=1$$

$$\left(\sqrt{x}\right)^2=1^2$$

$$x=1$$

Check:　$\sqrt{1}+2=\sqrt{1+8}$

$$1+2=\sqrt{9}$$

$$3=3,\ \text{true}$$

The solution set is $\{1\}$.

53.

$$\sqrt{x-8}=\sqrt{x}-2$$

$$\left(\sqrt{x-8}\right)^2=\left(\sqrt{x}-2\right)^2$$

$$x-8=\left(\sqrt{x}\right)^2-2\left(\sqrt{x}\right)(2)+2^2$$

$$x-8=x-4\sqrt{x}+4$$

$$4\sqrt{x}=12$$

$$\sqrt{x}=3$$

$$\left(\sqrt{x}\right)^2=3^2$$

$$x=9$$

Check:　$\sqrt{9-8}=\sqrt{9}-2$

$$\sqrt{1}=3-2$$

$$1=1,\ \text{true}$$

The solution set is $\{9\}$.

55.

$$f=120\sqrt{p}$$

$$840=120\sqrt{p}$$

$$\frac{840}{120}=\frac{120\sqrt{p}}{120}$$

$$7=\sqrt{p}$$

$$7^2=\left(\sqrt{p}\right)^2$$

$$49=p$$

A nozzle pressure of 49 pounds per square inch is needed.

57. The answer to Exercise 55 is represented by the point (49, 840).

59. a. $p = -2.5\sqrt{t} + 17$

$p = -2.5\sqrt{9} + 17 = 9.5$

According to the formula, 9.5% of Americans were in favor of laws prohibiting interracial marriage in 2002.

b. The formula underestimates the actual value shown in the bar graph by 0.1%.

c.
$$p = -2.5\sqrt{t} + 17$$
$$7 = -2.5\sqrt{t} + 17$$
$$-10 = -2.5\sqrt{t}$$
$$\frac{-10}{-2.5} = \frac{-2.5\sqrt{t}}{-2.5}$$
$$4 = \sqrt{t}$$
$$4^2 = \left(\sqrt{t}\right)^2$$
$$16 = t$$

7% of Americans were in favor of laws prohibiting interracial marriage 16 years after 1993, or 2009.

61. – 63. Answers will vary.

65. does not make sense; Explanations will vary. Sample explanation: Extraneous solutions can be generated as early as the first step. You should check the proposed solution in the original equation.

67. does not make sense; Explanations will vary. Sample explanation: One, both, or neither of the solutions may satisfy the equation.

69. false; Changes to make the statement true will vary. A sample change is: –5 is a solution of the first equation only.

71. false; Changes to make the statement true will vary. A sample change is: The first step is to isolate the radical.

73. Let x = the smaller integer.
Then $x + 1$ = the larger integer.
$$\sqrt{x + (x+1)} = x - 1$$
$$\sqrt{2x+1} = x - 1$$
$$\left(\sqrt{2x+1}\right)^2 = (x-1)^2$$
$$2x + 1 = x^2 - 2x + 1$$
$$0 = x^2 - 4x$$
$$0 = x(x-4)$$

$x = 0$ or $x - 4 = 0$

$x = 4$

Check 0: | Check 4:
$\sqrt{x + (x+1)} = x - 1$ | $\sqrt{x + (x+1)} = x - 1$
$\sqrt{0+1} = 0 - 1$ | $\sqrt{x+5} = 4 - 1$
$\sqrt{1} = -1$ | $\sqrt{9} = 3$
$1 = -1$, false | $3 = 3$, true

The only solution to the equation is 4.
If $x = 4$, $x + 1 = 5$, so the consecutive integers are 4 and 5.

75. $\sqrt{2x+2} = \sqrt{3x-5}$

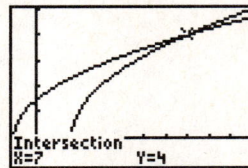

The graphs intersect at one point, $(7, 4)$, so the x-coordinate of the intersection point is 7.
Check 7:
$$\sqrt{2x+2} = \sqrt{3x-5}$$
$$\sqrt{2 \cdot 7 + 2} = \sqrt{3 \cdot 7 - 5}$$
$$\sqrt{14+2} = \sqrt{21-5}$$
$$\sqrt{16} = \sqrt{16}, \text{ true}$$
The solution set is $\{7\}$.

77. $\sqrt{x^2 + 3} = x + 1$

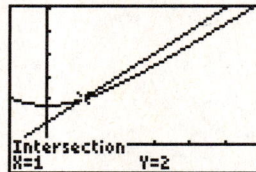

The graphs intersect at one point, $(1, 2)$. Therefore, the x-coordinate of the intersection point is 1.
Check 1:
$$\sqrt{x^2 + 3} = x + 1$$
$$\sqrt{1^2 + 3} = 1 + 1$$
$$\sqrt{4} = 2$$
$$2 = 2, \text{ true}$$
The solution set is $\{1\}$.

79. $\sqrt{x} + 4 = 2$

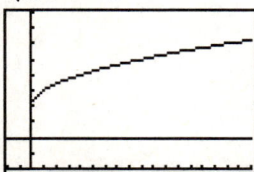

The graphs do not intersect. This indicates that the equation has no solution.

81. Let $x =$ the amount invested at 6%.
Then $9000 - x =$ the amount invested at 4%.
The investments earned a total of $500 in interest, so $0.06x + 0.04(9000 - x) = 500$.

$$0.06x + 0.04(9000 - x) = 500$$
$$0.06x + 360 - 0.04x = 500$$
$$0.02x + 360 = 500$$
$$0.02x = 140$$
$$\frac{0.02x}{0.02} = \frac{140}{0.02}$$
$$x = 7000$$
$$9000 - x = 2000$$

$7000 was invested at 6% and $2000 was invested at 4%.

82. Let $x =$ the price of an orchestra seat.
Then $y =$ the price of a mezzanine seat.
The given information leads to the system

$$4x + 2y = 22$$
$$2x + 3y = 16$$

Because both equations are written in the form $Ax + By = C$, the addition method is a good choice for solving this system. Because the problem only asks for the price of an orchestra seat, it is only necessary to solve the system for x. To do this, it is necessary to eliminate y.
Multiply the first equation by 3 and the second equation by -2. Then add the results.

$$12x + 6y = 66$$
$$\underline{-4x - 6y = -32}$$
$$8x \quad\quad = 34$$
$$x = \frac{34}{8} = 4.25$$

The price of an orchestra seat is $4.25.

83. $2x + y = -4$
$x + y = -3$
Graph $2x + y = -4$ using its x-intercept -2 and its y-intercept -4.
Graph $x + y = -3$ using its x-intercept -3 and y-intercept -3.

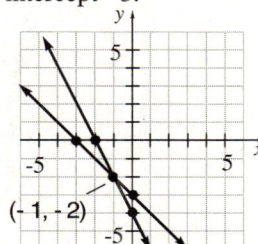

The two lines intersect at the point $(-1, -2)$ so the solution set is $\{(-1, -2)\}$.

84. $-\left(\sqrt[5]{32}\right)^4 = -2^4 = -16$

85. $\left(\sqrt[3]{27}\right)^2 = 3^2 = 9$

86. $\dfrac{1}{\left(\sqrt[4]{81}\right)^3} = \dfrac{1}{(3)^3} = \dfrac{1}{27}$

8.6 Check Points

1. a. $25^{\frac{1}{2}} = \sqrt{25} = 5$

b. $8^{\frac{1}{3}} = \sqrt[3]{8} = 2$

c. $-81^{\frac{1}{4}} = -\sqrt[4]{81} = -3$

d. $(-8)^{\frac{1}{3}} = \sqrt[3]{-8} = -2$

2. a. $27^{\frac{4}{3}} = \left(\sqrt[3]{27}\right)^4 = 3^4 = 81$

b. $4^{\frac{3}{2}} = \left(\sqrt{4}\right)^3 = 2^3 = 8$

c. $-16^{\frac{3}{4}} = -\left(\sqrt[4]{16}\right)^3 = -2^3 = -8$

3. a. $25^{-\frac{1}{2}} = \dfrac{1}{25^{\frac{1}{2}}} = \dfrac{1}{\sqrt{25}} = \dfrac{1}{5}$

b. $64^{-\frac{1}{3}} = \dfrac{1}{64^{\frac{1}{3}}} = \dfrac{1}{\sqrt[3]{64}} = \dfrac{1}{4}$

c. $32^{-\frac{4}{5}} = \dfrac{1}{32^{\frac{4}{5}}} = \dfrac{1}{\left(\sqrt[5]{32}\right)^4} = \dfrac{1}{2^4} = \dfrac{1}{16}$

4. a. Since this is group 3, substitute 3 for x.

$W = 62x^{-\frac{7}{5}}$

$W = 62(3)^{-\frac{7}{5}} \approx 13$

13% of women ages of 20-24 have never engaged in sexual activity with another person.

b. The answer in part (a) overestimates the actual percentage given by the bar graph by 1%.

c. $W = 62x^{-\frac{7}{5}}$

$W = \dfrac{62}{x^{7/5}}$

$W = \dfrac{62}{\left(\sqrt[5]{x}\right)^7}$ or $W = \dfrac{62}{\sqrt[5]{x^7}}$

8.6 Concept and Vocabulary Check

1. 6

2. $\sqrt{a}$

3. 2

4. $\sqrt[3]{a}$

5. 16; 2; 8

6. $\left(\sqrt[5]{a}\right)^4$

7. 3; 4; 3; $\dfrac{1}{64}$

8.6 Exercise Set

1. $49^{\frac{1}{2}} = \sqrt{49} = 7$

3. $121^{\frac{1}{2}} = \sqrt{121} = 11$

5. $27^{\frac{1}{3}} = \sqrt[3]{27} = 3$

7. $-125^{\frac{1}{3}} = -\left(\sqrt[3]{125}\right) = -5$

9. $16^{\frac{1}{4}} = \sqrt[4]{16} = 2$

11. $-32^{\frac{1}{5}} = -\left(\sqrt[5]{32}\right) = -2$

13. $\left(\dfrac{1}{9}\right)^{\frac{1}{2}} = \sqrt{\dfrac{1}{9}} = \dfrac{1}{3}$

15. $\left(\dfrac{27}{64}\right)^{\frac{1}{3}} = \sqrt[3]{\dfrac{27}{64}} = \dfrac{\sqrt[3]{27}}{\sqrt[3]{64}} = \dfrac{3}{4}$

17. $81^{\frac{3}{2}} = \left(\sqrt{81}\right)^3 = 9^3 = 729$

19. $125^{\frac{2}{3}} = \left(\sqrt[3]{125}\right)^2 = 5^2 = 25$

21. $9^{\frac{3}{2}} = \left(\sqrt{9}\right)^3 = 3^3 = 27$

23. $(-32)^{\frac{3}{5}} = \left(\sqrt[5]{-32}\right)^3 = (-2)^3 = -8$

25. $9^{-\frac{1}{2}} = \dfrac{1}{9^{\frac{1}{2}}} = \dfrac{1}{\sqrt{9}} = \dfrac{1}{3}$

27. $125^{-\frac{1}{3}} = \dfrac{1}{125^{\frac{1}{3}}} = \dfrac{1}{\sqrt[3]{125}} = \dfrac{1}{5}$

29. $32^{-\frac{1}{5}} = \dfrac{1}{32^{\frac{1}{5}}} = \dfrac{1}{\sqrt[5]{32}} = \dfrac{1}{2}$

31. $\left(\dfrac{1}{4}\right)^{-\frac{1}{2}} = \dfrac{1}{\left(\frac{1}{4}\right)^{\frac{1}{2}}} = \dfrac{1}{\sqrt{\frac{1}{4}}} = \dfrac{1}{\frac{1}{2}} = 2$

33. $16^{-\frac{3}{4}} = \dfrac{1}{16^{\frac{3}{4}}} = \dfrac{1}{\left(\sqrt[4]{16}\right)^3} = \dfrac{1}{2^3} = \dfrac{1}{8}$

35. $81^{-\frac{5}{4}} = \dfrac{1}{81^{\frac{5}{4}}} = \dfrac{1}{\left(\sqrt[4]{81}\right)^5} = \dfrac{1}{3^5} = \dfrac{1}{243}$

37. $8^{-\frac{2}{3}} = \dfrac{1}{8^{\frac{2}{3}}} = \dfrac{1}{\left(\sqrt[3]{8}\right)^2} = \dfrac{1}{2^2} = \dfrac{1}{4}$

39. $\left(\dfrac{4}{25}\right)^{-\frac{1}{2}} = \dfrac{1}{\left(\frac{4}{25}\right)^{\frac{1}{2}}} = \dfrac{1}{\sqrt{\frac{4}{25}}} = \dfrac{1}{\frac{2}{5}} = \dfrac{5}{2}$

41. $\left(\dfrac{8}{125}\right)^{-\frac{1}{3}} = \dfrac{1}{\left(\frac{8}{125}\right)^{\frac{1}{3}}} = \dfrac{1}{\sqrt[3]{\frac{8}{125}}} = \dfrac{1}{\frac{2}{5}} = \dfrac{5}{2}$

43. $(-8)^{-\frac{2}{3}} = \dfrac{1}{(-8)^{\frac{2}{3}}} = \dfrac{1}{\left(\sqrt[3]{-8}\right)^2}$

$\qquad = \dfrac{1}{(-2)^2} = \dfrac{1}{4}$

45. $27^{\frac{2}{3}} + 16^{\frac{3}{4}} = \left(\sqrt[3]{27}\right)^2 + \left(\sqrt[4]{16}\right)^3$

$\qquad = 3^2 + 2^3 = 9 + 8 = 17$

47. $25^{\frac{3}{2}} \cdot 81^{\frac{1}{4}} = \left(\sqrt{25}\right)^3 \cdot \left(\sqrt[4]{81}\right)$

$\qquad = 5^3 \cdot 3 = 125 \cdot 3$

$\qquad = 375$

49. $x^{\frac{1}{3}} \cdot x^{\frac{1}{4}} = x^{\frac{1}{3}+\frac{1}{4}} = x^{\frac{4}{12}+\frac{3}{12}} = x^{\frac{7}{12}}$

51. $\dfrac{x^{\frac{1}{6}}}{x^{\frac{5}{6}}} = x^{\frac{1}{6}-\frac{5}{6}} = x^{-\frac{4}{6}} = x^{-\frac{2}{3}} = \dfrac{1}{x^{\frac{2}{3}}}$

53. $\left(x^{1/4} y^3\right)^{2/3} = \left(x^{1/4}\right)^{2/3} \cdot \left(y^3\right)^{2/3}$

$\qquad = x^{\frac{1}{4} \cdot \frac{2}{3}} \cdot y^{3 \cdot \frac{2}{3}}$

$\qquad = x^{\frac{1}{6}} y^2$

55. $\left(\dfrac{x^{2/5}}{x^{6/5} \cdot x^{3/5}}\right)^5 = \dfrac{\left(x^{2/5}\right)^5}{\left(x^{6/5}\right)^5 \cdot \left(x^{3/5}\right)^5}$

$\qquad = \dfrac{x^2}{x^6 \cdot x^3} = \dfrac{x^2}{x^9}$

$\qquad = x^{2-9} = x^{-7} = \dfrac{1}{x^7}$

57. a. $h = 0.84 d^{\frac{2}{3}}$

$\qquad h = 0.84(985)^{\frac{2}{3}} \approx 83.2$

The tree is approximately 83.2 meters.

b. $h = 0.84 d^{\frac{2}{3}}$

$\qquad h = 0.84\left(\sqrt[3]{d}\right)^2$ or $h = 0.84\sqrt[3]{d^2}$

59. a. $N = \dfrac{a^{\frac{1}{2}}}{2\pi r^{\frac{1}{2}}}$

$\qquad N = \dfrac{(9.8)^{\frac{1}{2}}}{2\pi(1.7)^{\frac{1}{2}}} \approx 0.382$

It would need to rotate about 0.382 revolutions per second.

b. $0.382 \cdot 60 = 22.92$

0.382 revolutions per second is equivalent to 22.92 revolutions per minute.

c. $N = \dfrac{a^{\frac{1}{2}}}{2\pi r^{\frac{1}{2}}} \cdot \dfrac{r^{\frac{1}{2}}}{r^{\frac{1}{2}}}$

$= \dfrac{(ar)^{\frac{1}{2}}}{2\pi r}$

$= \dfrac{\sqrt{ar}}{2\pi r}$

61. – 65. Answers will vary.

67. does not make sense; Explanations will vary.

Sample explanation: $5^{\frac{1}{2}} \cdot 5^{\frac{1}{2}} = 5^{\frac{1}{2}+\frac{1}{2}} = 5^1 = 5.$

69. does not make sense; Explanations will vary.
Sample explanation: Because of the even index, the second equation is not correct.

71. false; Changes to make the statement true will vary.

A sample change is: $8^{-\frac{2}{3}} = \dfrac{1}{4}.$

73. false; Changes to make the statement true will vary.

A sample change is: $-3^{-2} = -\dfrac{1}{3^2} = -\dfrac{1}{9}.$

75. true

77. true

79. $\dfrac{3^{-1} \cdot 3^{\frac{1}{2}}}{3^{-\frac{3}{2}}} = \dfrac{3^{-1+\frac{1}{2}}}{3^{-\frac{3}{2}}}$

$= \dfrac{3^{-\frac{2}{2}+\frac{1}{2}}}{3^{-\frac{3}{2}}} = \dfrac{3^{-\frac{1}{2}}}{3^{-\frac{3}{2}}} = \dfrac{3^{\frac{3}{2}}}{3^{\frac{1}{2}}}$

$= 3^{\frac{3}{2}-\frac{1}{2}} = 3^{\frac{2}{2}} = 3^1$

$= 3$

81. a. $A = 6V^{\frac{2}{3}}$

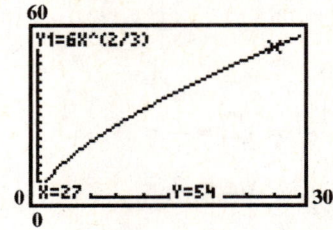

b. The graph shows that a cube whose volume is 27 cubic units has a surface area of 54 square units.

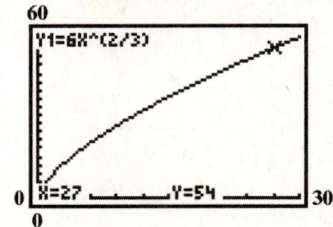

c. The graph shows that the volume of a cube whose volume is 8 cubic units is 24 square units.

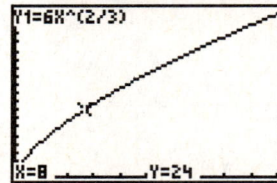

82. $7x - 3y = -14$

$\qquad y = 3x + 6$

To solve this system by the substitution method, substitute $3x + 6$ for y in the first equation and solve for x.

$7x - 3(3x + 6) = -14$

$7x - 9x - 18 = -14$

$-2x - 18 = -14$

$-2x = 4$

$x = -2$

Back-substitute -2 for x into the second equation.

$y = 3x + 6$

$y = 3(-2) + 6 = -6 + 6 = 0$

The solution set is $\{(-2, 0)\}.$

83. $-3x + 4y \leq 12$

$\qquad\quad x \geq 2$

To graph the solutions of this system of inequalities, first graph $-3x + 4y = 12$ as a solid line using its x-intercept, -4, and its y-intercept 3. Use $(0,0)$ as a test point. Since $-3(0) + 4(0) \leq 12$ is true, shade the half-plane containing $(0,0)$. Now graph $x = 2$ as a solid vertical line. Since $0 \geq 2$ is false, shade the half-plane *not* containing $(0,0)$, which is the half-plane to the right of the vertical line. The solution set of the system is the intersection (overlap) of the two shaded regions.

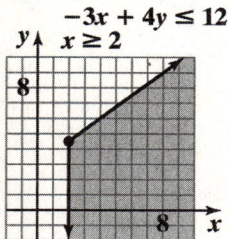

$-3x + 4y \leq 12$
$x \geq 2$

84. $\dfrac{(2x)^5}{x^3} = \dfrac{2^5 x^5}{x^3} = \dfrac{32x^5}{x^3} = 32x^2$

85. $\qquad\qquad 4x^2 = 20$

$\qquad\qquad 4\left(-\sqrt{5}\right)^2 = 20$

$\qquad\quad 4(-1)^2 \left(\sqrt{5}\right)^2 = 20$

$\qquad\qquad\quad 4 \cdot 1 \cdot 5 = 20$

$\qquad\qquad\qquad 20 = 20, \ \text{true}$

$-\sqrt{5}$ is a solution.

86. $\qquad (x-1)^2 = 5$

$\qquad \left(1 + \sqrt{3} - 1\right)^2 = 5$

$\qquad\qquad \left(\sqrt{3}\right)^2 = 5$

$\qquad\qquad\qquad 3 = 5, \ \text{false}$

$1 + \sqrt{3}$ is not a solution.

87. $\sqrt{[6-(-4)]^2 + [2-(-3)]^2} = \sqrt{[6+4]^2 + [2+3]^2}$

$\qquad\qquad\qquad\qquad\qquad = \sqrt{10^2 + 5^2}$

$\qquad\qquad\qquad\qquad\qquad = \sqrt{100 + 25}$

$\qquad\qquad\qquad\qquad\qquad = \sqrt{125}$

$\qquad\qquad\qquad\qquad\qquad = \sqrt{25 \cdot 5}$

$\qquad\qquad\qquad\qquad\qquad = 5\sqrt{5}$

Chapter 8 Review Exercises

1. $\sqrt{121} = 11$
The principal square root of 121 is 11.

2. $-\sqrt{121} = -11$
The negative square root of 121 is -11.

3. $\sqrt{-121}$ is not a real number because the square of a real number is never negative.

4. $\sqrt[3]{\dfrac{8}{125}} = \dfrac{2}{5}$ because $\left(\dfrac{2}{5}\right)^3 = \dfrac{8}{125}$.

5. $\sqrt[5]{-32} = -2$ because $(-2)^5 = -32$.

6. $-\sqrt[4]{81} = -3$ because $3^4 = 81$.

7. $\sqrt{75} \approx 8.660$

8. $\sqrt{398 - 5} = \sqrt{393} \approx 19.824$

9. $M = 0.7\sqrt{t} + 12.5$
$M = 0.7\sqrt{9} + 12.5 = 14.6$
According to the model, 14.6 non-program minutes disrupted an hour of cable TV action in 2005.

10. Answers will vary. Example: The number of disruptive minutes is increasing, but the rate of increase is slowing down.

11. $d = \sqrt{\dfrac{3h}{2}}$

$d = \sqrt{\dfrac{3 \cdot 1450}{2}} \approx 47$

Visitors will be able to see about 47 miles from the top of the Sear's Tower.

12. $\sqrt{54} = \sqrt{9 \cdot 6} = \sqrt{9}\sqrt{6} = 3\sqrt{6}$

13. $6\sqrt{20} = 6\sqrt{4 \cdot 5} = 6\sqrt{4}\sqrt{5}$
$\qquad\qquad = 6 \cdot 2\sqrt{5} = 12\sqrt{5}$

14. $\sqrt{63x^2} = \sqrt{9x^2}\sqrt{7} = 3x\sqrt{7}$

15. $\sqrt{48x^3} = \sqrt{16x^2}\sqrt{3x} = 4x\sqrt{3x}$

16. $\sqrt{x^8} = x^4$ because $\left(x^4\right)^2 = x^8$.

17. $\sqrt{75x^9} = \sqrt{25x^8}\sqrt{3x} = 5x^4\sqrt{3x}$

18. $\sqrt{45x^{23}} = \sqrt{9x^{22}}\sqrt{5x} = 3x^{11}\sqrt{5x}$

19. $\sqrt[3]{24} = \sqrt[3]{8 \cdot 3} = \sqrt[3]{8}\sqrt[3]{3} = 2\sqrt[3]{3}$

20. $\sqrt{7} \cdot \sqrt{11} = \sqrt{7 \cdot 11} = \sqrt{77}$

21. $\sqrt{3} \cdot \sqrt{12} = \sqrt{36} = 6$

22. $\sqrt{5x} \cdot \sqrt{10x} = \sqrt{50x^2} = \sqrt{25x^2}\sqrt{2}$
$\qquad = 5x\sqrt{2}$

23. $\sqrt{3x^2} \cdot \sqrt{4x^3} = \sqrt{12x^5} = \sqrt{4x^4}\sqrt{3x}$
$\qquad = 2x^2\sqrt{3x}$

24. $\sqrt[3]{6} \cdot \sqrt[3]{9} = \sqrt[3]{6 \cdot 9} = \sqrt[3]{54} = \sqrt[3]{27} \cdot \sqrt[3]{2}$
$\qquad = 3\sqrt[3]{2}$

25. $\sqrt{\dfrac{5}{2}} \cdot \sqrt{\dfrac{3}{8}} = \sqrt{\dfrac{5 \cdot 3}{2 \cdot 8}} = \sqrt{\dfrac{15}{16}}$
$\qquad = \dfrac{\sqrt{15}}{\sqrt{16}} = \dfrac{\sqrt{15}}{4}$

26. $\sqrt{\dfrac{121}{4}} = \dfrac{\sqrt{121}}{\sqrt{4}} = \dfrac{11}{2}$

27. $\sqrt{\dfrac{7x}{25}} = \dfrac{\sqrt{7x}}{\sqrt{25}} = \dfrac{\sqrt{7x}}{5}$

28. $\sqrt{\dfrac{18}{x^2}} - \dfrac{\sqrt{18}}{\sqrt{x^2}} = \dfrac{\sqrt{9}\sqrt{2}}{x} = \dfrac{3\sqrt{2}}{x}$

29. $\dfrac{\sqrt{200}}{\sqrt{2}} = \sqrt{\dfrac{200}{2}} = \sqrt{100} = 10$

30. $\dfrac{\sqrt{96}}{\sqrt{3}} = \sqrt{32} = \sqrt{16}\sqrt{2} = 4\sqrt{2}$

31. $\dfrac{\sqrt{72x^8}}{\sqrt{x^3}} = \sqrt{\dfrac{72x^8}{x^3}} = \sqrt{72x^5}$
$\qquad = \sqrt{36x^4}\sqrt{2x} = 6x^2\sqrt{2x}$

32. $\sqrt[3]{\dfrac{5}{64}} = \dfrac{\sqrt[3]{5}}{\sqrt[3]{64}} = \dfrac{\sqrt[3]{5}}{4}$

33. $\sqrt[3]{\dfrac{40}{27}} = \dfrac{\sqrt[3]{40}}{\sqrt[3]{27}} = \dfrac{\sqrt[3]{8}\sqrt[3]{5}}{3} = \dfrac{2\sqrt[3]{5}}{3}$

34. $7\sqrt{5} + 13\sqrt{5} = (7+13)\sqrt{5} = 20\sqrt{5}$

35. $\sqrt{8} + \sqrt{50} = \sqrt{4}\sqrt{2} + \sqrt{25}\sqrt{2}$
$\qquad = 2\sqrt{2} + 5\sqrt{2}$
$\qquad = (2+5)\sqrt{2} = 7\sqrt{2}$

36. $\sqrt{75} - \sqrt{48} = \sqrt{25}\sqrt{3} - \sqrt{16}\sqrt{3}$
$\qquad = 5\sqrt{3} - 4\sqrt{3}$
$\qquad = (5-4)\sqrt{3} = \sqrt{3}$

37. $2\sqrt{80} + 3\sqrt{45} = 2\sqrt{16}\sqrt{5} + 3\sqrt{9}\sqrt{5}$
$\qquad = 2 \cdot 4\sqrt{5} + 3 \cdot 3\sqrt{5}$
$\qquad = 8\sqrt{5} + 9\sqrt{5} = 17\sqrt{5}$

38. $4\sqrt{72} - 2\sqrt{48} = 4\sqrt{36}\sqrt{2} - 2\sqrt{16}\sqrt{3}$
$\qquad = 4 \cdot 6\sqrt{2} - 2 \cdot 4\sqrt{3}$
$\qquad = 24\sqrt{2} - 8\sqrt{3}$

39. $2\sqrt{18} + 3\sqrt{27} - \sqrt{12}$
$\qquad = 2\sqrt{9}\sqrt{2} + 3\sqrt{9}\sqrt{3} - \sqrt{4}\sqrt{3}$
$\qquad = 2 \cdot 3\sqrt{2} + 3 \cdot 3\sqrt{3} - 2\sqrt{3}$
$\qquad = 6\sqrt{2} + 9\sqrt{3} - 2\sqrt{3}$
$\qquad = 6\sqrt{2} + 7\sqrt{3}$

40. $\sqrt{10}\left(\sqrt{5} + \sqrt{6}\right) = \sqrt{10} \cdot \sqrt{5} + \sqrt{10} \cdot \sqrt{6}$
$\qquad = \sqrt{50} + \sqrt{60}$
$\qquad = \sqrt{25}\sqrt{2} + \sqrt{4}\sqrt{15}$
$\qquad = 5\sqrt{2} + 2\sqrt{15}$

41. $\sqrt{3}\left(\sqrt{6} - \sqrt{12}\right) = \sqrt{3} \cdot \sqrt{6} - \sqrt{3} \cdot \sqrt{12}$
$\qquad = \sqrt{18} - \sqrt{36}$
$\qquad = \sqrt{9}\sqrt{2} - 6 = 3\sqrt{2} - 6$

42. $\left(9+\sqrt{2}\right)\left(10+\sqrt{2}\right)$

$=9\cdot 10+9\sqrt{2}+10\sqrt{2}+\sqrt{2}\cdot\sqrt{2}$

$=90+19\sqrt{2}+2=92+19\sqrt{2}$

43. $\left(1+3\sqrt{7}\right)\left(4-\sqrt{7}\right)$

$=1\cdot 4-1\sqrt{7}+3\sqrt{7}\cdot 4-3\cdot\sqrt{7}\sqrt{7}$

$=4-\sqrt{7}+12\sqrt{7}-3\cdot 7$

$=4+11\sqrt{7}-21$

$=-17+11\sqrt{7}$

44. $\left(\sqrt{3}+2\right)\left(\sqrt{6}-4\right)$

$=\sqrt{3}\cdot\sqrt{6}+4\sqrt{3}+2\sqrt{6}-2\cdot 4$

$=\sqrt{18}-4\sqrt{3}+2\sqrt{6}-8$

$=\sqrt{9}\sqrt{2}-4\sqrt{3}+2\sqrt{6}-8$

$=3\sqrt{2}-4\sqrt{3}+2\sqrt{6}-8$

45. $\left(2+\sqrt{7}\right)\left(2-\sqrt{7}\right)=2^2-\left(\sqrt{7}\right)^2$

$=4-7=-3$

46. $\left(\sqrt{11}-\sqrt{5}\right)\left(\sqrt{11}+\sqrt{5}\right)$

$=\left(\sqrt{11}\right)^2-\left(\sqrt{5}\right)^2$

$=11-5=6$

47. $\left(1+\sqrt{2}\right)^2=1^2+2\cdot 1\cdot\sqrt{2}+\left(\sqrt{2}\right)^2$

$=1+2\sqrt{2}+2$

$=3+2\sqrt{2}$

48. $\dfrac{30}{\sqrt{5}}=\dfrac{30}{\sqrt{5}}\cdot\dfrac{\sqrt{5}}{\sqrt{5}}=\dfrac{30\sqrt{5}}{5}=6\sqrt{5}$

49. $\dfrac{13}{\sqrt{50}}=\dfrac{13}{\sqrt{25}\sqrt{2}}=\dfrac{13}{5\sqrt{2}}\cdot\dfrac{\sqrt{2}}{\sqrt{2}}$

$=\dfrac{13\sqrt{2}}{5\cdot 2}=\dfrac{13\sqrt{2}}{10}$

50. $\sqrt{\dfrac{2}{3}}=\dfrac{\sqrt{2}}{\sqrt{3}}=\dfrac{\sqrt{2}}{\sqrt{3}}\cdot\dfrac{\sqrt{3}}{\sqrt{3}}=\dfrac{\sqrt{6}}{3}$

51. $\sqrt{\dfrac{3}{8}}=\dfrac{\sqrt{3}}{\sqrt{8}}=\dfrac{\sqrt{3}}{\sqrt{4}\sqrt{2}}=\dfrac{\sqrt{3}}{2\sqrt{2}}$

$=\dfrac{\sqrt{3}}{2\sqrt{2}}\cdot\dfrac{\sqrt{2}}{\sqrt{2}}=\dfrac{\sqrt{6}}{2\cdot 2}=\dfrac{\sqrt{6}}{4}$

52. $\sqrt{\dfrac{17}{x}}=\dfrac{\sqrt{17}}{\sqrt{x}}=\dfrac{\sqrt{17}}{\sqrt{x}}\cdot\dfrac{\sqrt{x}}{\sqrt{x}}=\dfrac{\sqrt{17x}}{x}$

53. $\dfrac{11}{\sqrt{5}+2}=\dfrac{11}{\sqrt{5}+2}\cdot\dfrac{\sqrt{5}-2}{\sqrt{5}-2}$

$=\dfrac{11\left(\sqrt{5}-2\right)}{\left(\sqrt{5}\right)^2-2^2}$

$=\dfrac{11\left(\sqrt{5}-2\right)}{5-4}$

$=\dfrac{11\left(\sqrt{5}-2\right)}{1}$

$=11\sqrt{5}-22$

54. $\dfrac{21}{4-\sqrt{3}}=\dfrac{21}{4-\sqrt{3}}\cdot\dfrac{4+\sqrt{3}}{4+\sqrt{3}}$

$=\dfrac{21\left(4+\sqrt{3}\right)}{4^2-\left(\sqrt{3}\right)^2}$

$=\dfrac{21\left(4+\sqrt{3}\right)}{16-3}$

$=\dfrac{21\left(4+\sqrt{3}\right)}{13}$

$=\dfrac{84+21\sqrt{3}}{13}$

55. $\dfrac{12}{\sqrt{5}+\sqrt{3}}=\dfrac{12}{\sqrt{5}+\sqrt{3}}\cdot\dfrac{\sqrt{5}-\sqrt{3}}{\sqrt{5}-\sqrt{3}}$

$=\dfrac{12\left(\sqrt{5}-\sqrt{3}\right)}{\left(\sqrt{5}\right)^2-\left(\sqrt{3}\right)^2}$

$=\dfrac{12\left(\sqrt{5}-\sqrt{3}\right)}{5-3}$

$=\dfrac{12\left(\sqrt{5}-\sqrt{3}\right)}{2}$

$=6\left(\sqrt{5}-\sqrt{3}\right)$

$=6\sqrt{5}-6\sqrt{3}$

56. $\dfrac{7\sqrt{2}}{\sqrt{2}-4} = \dfrac{7\sqrt{2}}{\sqrt{2}-4} \cdot \dfrac{\sqrt{2}+4}{\sqrt{2}+4}$

$= \dfrac{7\sqrt{2}\left(\sqrt{2}+4\right)}{\left(\sqrt{2}\right)^2 - 4^2} = \dfrac{7\sqrt{2}\left(\sqrt{2}+4\right)}{2-16}$

$= \dfrac{7\sqrt{2}\left(\sqrt{2}+4\right)}{-14} = \dfrac{7 \cdot \sqrt{2}\sqrt{2} + 7 \cdot \sqrt{2} \cdot 4}{-14}$

$= \dfrac{7 \cdot 2 + 28\sqrt{2}}{-14} = \dfrac{14 + 28\sqrt{2}}{-14}$

$= \dfrac{14\left(1 + 2\sqrt{2}\right)}{14(-1)} = -1\left(1 + 2\sqrt{2}\right)$

$= -1 - 2\sqrt{2}$

57. $\sqrt{x+3} = 4$

$\left(\sqrt{x+3}\right) = 4^2$

$x + 3 = 16$

$x = 13$

Check 13:

$\sqrt{x+3} = 4$

$\sqrt{13+3} = 4$

$\sqrt{16} = 4$

$4 = 4, \text{ true}$

The solution set is $\{13\}$.

58. $\sqrt{2x+3} = 5$

$\left(\sqrt{2x+3}\right)^2 = 5^2$

$2x + 3 = 25$

$2x + 22$

$x = 11$

Check 11:

$\sqrt{2x+3} = 5$

$\sqrt{2(11)+3} = 5$

$\sqrt{22+3} = 5$

$\sqrt{25} = 5$

$5 = 5, \text{ true}$

The solution set is $\{11\}$.

59. $3\sqrt{x} = \sqrt{6x+15}$

$\left(3\sqrt{x}\right)^2 = \left(\sqrt{6x+15}\right)^2$

$9x = 6x + 15$

$3x = 15$

$x = 5$

Substitution confirms the solution set is $\{5\}$.

60. $\sqrt{5x+1} = x + 1$

$\left(\sqrt{5x+1}\right)^2 = (x+1)^2$

$5x + 1 = x^2 + 2x + 1$

$0 = x^2 - 3x$

$0 = x(x-3)$

$x = 0 \quad \text{or} \quad x - 3 = 0$

$\qquad\qquad\qquad x = 3$

Each of the proposed solutions must be checked separately in the original equation.

Check 0:   Check 3:

$\sqrt{5x+1} = x+1 \qquad \sqrt{5x+1} = x+1$

$\sqrt{5 \cdot 0 + 1} = 0 + 1 \qquad \sqrt{5 \cdot 3 + 1} = 3 + 1$

$\sqrt{1} = 0 + 1 \qquad\qquad \sqrt{15+1} = 4$

$1 = 1, \text{ true} \qquad\qquad \sqrt{16} = 4$

$\qquad\qquad\qquad\qquad\qquad 4 = 4, \text{ true}$

The solution set is $\{0, 3\}$.

61. $\sqrt{x+1} + 5 = x$

$\sqrt{x+1} = x - 5$

$\left(\sqrt{x+1}\right)^2 = (x-5)^2$

$x + 1 = x^2 - 10x + 25$

$0 = x^2 - 11x + 24$

$0 = (x-3)(x-8)$

$x - 3 = 0 \quad \text{or} \quad x - 8 = 0$

$x = 3 \qquad\qquad x = 8$

Check 3:   Check 8:

$\sqrt{x+1} + 5 = x \qquad \sqrt{x+1} + 5 = x$

$\sqrt{3+1} + 5 = 3 \qquad \sqrt{8+1} + 5 = 8$

$\sqrt{4} + 5 = 3 \qquad\qquad \sqrt{9} + 5 = 8$

$2 + 5 = 3 \qquad\qquad\qquad 3 + 5 = 8$

$\qquad 7 = 3, \text{ false} \qquad\qquad 8 = 8, \text{ true}$

The false statement indicates that 3 does not satisfy the original equation; it is an extraneous solution. The solution set is $\{8\}$.

62. $\sqrt{x-2}+5=1$

$\sqrt{x-2}=-4$

$\left(\sqrt{x-2}\right)^2=(-4)^2$

$x-2=16$

$x=18$

Check 18:

$\sqrt{x-2}+5=1$

$\sqrt{18-2}+5=1$

$\sqrt{16}+5=1$

$4+5=1$

$9=1$, false

The only proposed solution, 18, is extraneous, so the equation has no solution. The solution set is $\{\ \}$.

63. $x=\sqrt{x^2+4x+4}$

$x^2=\left(\sqrt{x^2+4x+4}\right)^2$

$x^2=x^2+4x+4$

$0=4x+4$

$-4=4x$

$-1=x$

Check:

$x=\sqrt{x^2+4x+4}$

$-1=\sqrt{(-1)^2+4(-1)+4}$

$-1=\sqrt{1-4+4}$

$-1=\sqrt{1}$

$-1=1$, false

The only proposed solution, -1, is extraneous, so the equation has no solution. The solution set is $\{\ \}$.

64. $W=4\sqrt{2x}$

$16=4\sqrt{2x}$

$\dfrac{16}{4}=\dfrac{4\sqrt{2x}}{4}$

$4=\sqrt{2x}$

$4^2=\left(\sqrt{2x}\right)^2$

$16=2x$

$8=x$

The leg length is 8 feet.

65. $p=3.5\sqrt{t}+38$

$66=3.5\sqrt{t}+38$

$28=3.5\sqrt{t}$

$\dfrac{28}{3.5}=\dfrac{3.5\sqrt{t}}{3.5}$

$8=\sqrt{t}$

$64=t$

According to the formula, 66% of U.S. women will participate in the labor force 64 years after 1960, or 2024.

65. a. According to the line graph, about 48% of U.S. women participated in the labor force in 2000.

b. $p=2.2\sqrt{t}+36.2$

$p=2.2\sqrt{28}+36.2\approx 47.8$

According to the formula, 47.8% of U.S. women participated in the labor force in 2000.

c. $p=2.2\sqrt{t}+36.2$

$52=2.2\sqrt{t}+36.2$

$15.8=2.2\sqrt{t}$

$\dfrac{15.8}{2.2}=\dfrac{2.2\sqrt{t}}{2.2}$

$\dfrac{15.8}{2.2}=\sqrt{t}$

$\left(\dfrac{15.8}{2.2}\right)^2=\left(\sqrt{t}\right)^2$

$52\approx t$

According to the formula, 52% of U.S. women will participate in the labor force 52 years after 1972, or 2024.

66. $16^{\frac{1}{2}}=\sqrt{16}=4$

67. $125^{\frac{1}{3}}=\sqrt[3]{125}=5$

68. $64^{\frac{2}{3}}=\left(\sqrt[3]{64}\right)^2=4^2=16$

69. $25^{-\frac{1}{2}}=\dfrac{1}{25^{\frac{1}{2}}}=\dfrac{1}{\sqrt{25}}=\dfrac{1}{5}$

70. $27^{-\frac{1}{3}} = \dfrac{1}{27^{\frac{1}{3}}} = \dfrac{1}{\sqrt[3]{27}} = \dfrac{1}{3}$

71. $(-8)^{-\frac{4}{3}} = \dfrac{1}{(-8)^{\frac{4}{3}}} = \dfrac{1}{\left(\sqrt[3]{-8}\right)^4}$

$= \dfrac{1}{(-2)^4} = \dfrac{1}{16}$

72. $S = 28.6A^{\frac{1}{3}}; \ A = 8$

$S = 28.6 \cdot 8^{\frac{1}{3}} = 28.6\sqrt[3]{8}$

$= 28.6 \cdot 2 = 57.2 \approx 57$

There are approximately 57 species on a Galapagos island whose area is 8 square miles.

Chapter 8 Test

1. $-\sqrt{64} = -8$
The negative square root of 64 is -8.

2. $\sqrt[3]{64} = 4$ because $4^3 = 64$.

3. $\sqrt{48} = \sqrt{16 \cdot 3} = \sqrt{16}\sqrt{3} = 4\sqrt{3}$

4. $\sqrt{72x^3} = \sqrt{36x^2}\sqrt{2x} = 6x\sqrt{2x}$

5. $\sqrt{x^{29}} = \sqrt{x^{28} \cdot x} = \sqrt{x^{28}}\sqrt{x} = x^{14}\sqrt{x}$

6. $\sqrt{\dfrac{25}{x^2}} = \dfrac{\sqrt{25}}{\sqrt{x^2}} = \dfrac{5}{x}$

7. $\sqrt{\dfrac{75}{27}} = \dfrac{\sqrt{75}}{\sqrt{27}} = \dfrac{\sqrt{25}\sqrt{3}}{\sqrt{9}\sqrt{3}}$

$= \dfrac{5\sqrt{3}}{3\sqrt{3}} = \dfrac{5}{3}$

8. $\sqrt[3]{\dfrac{5}{8}} = \dfrac{\sqrt[3]{5}}{\sqrt[3]{8}} = \dfrac{\sqrt[3]{5}}{2}$

9. $\dfrac{\sqrt{80x^4}}{\sqrt{2x^2}} = \sqrt{\dfrac{80x^4}{2x^2}} = \sqrt{40x^2}$

$= \sqrt{4x^2}\sqrt{10} = 2x\sqrt{10}$

10. $\sqrt{10} \cdot \sqrt{5} = \sqrt{50} = \sqrt{25 \cdot 2}$

$= \sqrt{25}\sqrt{2} = 5\sqrt{2}$

11. $\sqrt{6x} \cdot \sqrt{6y} = \sqrt{6x \cdot 6y}$

$= \sqrt{36xy}$

$= \sqrt{36} \cdot \sqrt{xy}$

$= 6\sqrt{xy}$

12. $\sqrt{10x^2} \cdot \sqrt{2x^3} = \sqrt{20x^5} = \sqrt{4x^4}\sqrt{5x}$

$= 2x^2\sqrt{5x}$

13. $\sqrt{24} + 3\sqrt{54} = \sqrt{4}\sqrt{6} + 3\sqrt{9}\sqrt{6}$

$= 2\sqrt{6} + 3 \cdot 3\sqrt{6}$

$= 2\sqrt{6} + 9\sqrt{6}$

$= 11\sqrt{6}$

14. $7\sqrt{8} - 2\sqrt{32} = 7\sqrt{4}\sqrt{2} - 2\sqrt{16}\sqrt{2}$

$= 7 \cdot 2\sqrt{2} - 2 \cdot 4\sqrt{2}$

$= 14\sqrt{2} - 8\sqrt{2} = 6\sqrt{2}$

15. $\sqrt{3}\left(\sqrt{10} + \sqrt{3}\right) = \sqrt{3} \cdot \sqrt{10} + \sqrt{3} \cdot \sqrt{3}$

$= \sqrt{30} + 3$

16. $\left(7 - \sqrt{5}\right)\left(10 + 3\sqrt{5}\right)$

$= 7 \cdot 10 + 7 \cdot 3\sqrt{5} - 10\sqrt{5} - 3\sqrt{5} \cdot \sqrt{5}$

$= 70 + 21\sqrt{5} - 10\sqrt{5} - 3 \cdot 5$

$= 70 + 11\sqrt{5} - 15 = 55 + 11\sqrt{5}$

17. $\left(\sqrt{6} + 2\right)\left(\sqrt{6} - 2\right) = \left(\sqrt{6}\right)^2 - 2^2$

$= 6 - 4 = 2$

18. $\left(3 + \sqrt{7}\right)^2 = 3^2 + 2 \cdot 3 \cdot \sqrt{7} + \left(\sqrt{7}\right)^2$

$= 9 + 6\sqrt{7} + 7 = 16 + 6\sqrt{7}$

19. $\dfrac{4}{\sqrt{5}} = \dfrac{4}{\sqrt{5}} \cdot \dfrac{\sqrt{5}}{\sqrt{5}} = \dfrac{4\sqrt{5}}{5}$

20.
$$\frac{5}{4+\sqrt{3}} = \frac{5}{4+\sqrt{3}} \cdot \frac{4-\sqrt{3}}{4-\sqrt{3}}$$
$$= \frac{5\left(4-\sqrt{3}\right)}{4^2-\left(\sqrt{3}\right)^2}$$
$$= \frac{5\left(4-\sqrt{3}\right)}{16-3}$$
$$= \frac{5\left(4-\sqrt{3}\right)}{13}$$
$$= \frac{20-5\sqrt{3}}{13}$$

21.
$$\sqrt{3x}+5=11$$
$$\sqrt{3x}=6$$
$$\left(\sqrt{3x}\right)^2=6^2$$
$$3x=36$$
$$x=12$$
Check the proposed solution in the original equation.
$$\sqrt{3x}+5=11$$
$$\sqrt{3\cdot12}+5=11$$
$$\sqrt{36}+5=11$$
$$6+5=11$$
$$11=11$$
The solution set is {12}.

22.
$$\sqrt{2x-1}=x-2$$
$$\left(\sqrt{2x-1}\right)^2=\left(x-2\right)^2$$
$$2x-1=x^2-4x+4$$
$$0=x^2-6x+5$$
$$0=\left(x-1\right)\left(x-5\right)$$
$$x-1=0 \text{ or } x-5=0$$
$$x=1 \qquad x=5$$
Check each proposed solution in the original equation.

Check 1:
$$\sqrt{2x-1}=x-2$$
$$\sqrt{2\cdot1-1}=1-2$$
$$\sqrt{2-1}=1-2$$
$$\sqrt{1}=-1$$
$$1=-1, \text{ false}$$

Check 5:
$$\sqrt{2x-1}=x-2$$
$$\sqrt{2\cdot5-1}=5-2$$
$$\sqrt{10-1}=3$$
$$\sqrt{9}=3$$
$$3=3, \text{ true}$$

The check shows that 1 is an extraneous solution. The solution set is {5}.

23.
$$p=-4.4\sqrt{x}+38$$
$$16=-4.4\sqrt{x}+38$$
$$-22=-4.4\sqrt{x}$$
$$\frac{-22}{-4.4}=\frac{-4.4\sqrt{x}}{-4.4}$$
$$5=\sqrt{x}$$
$$25=x$$
According to the formula, 16% of Americans with an income of $25 thousand report that their health is fair or poor.

24. $8^{\frac{2}{3}}=\left(\sqrt[3]{8}\right)^2=2^2=4$

25. $9^{-\frac{1}{2}}=\frac{1}{9^{\frac{1}{2}}}=\frac{1}{\sqrt{9}}=\frac{1}{3}$

Cumulative Review Exercises

(Chapters 1-8)

1. $2x+3x-5+7=10x+3-6x-4$
$$5x+2=4x-1$$
$$x+2=-1$$
$$x=-3$$
The solution set is { −3 }.

2.
$$2x^2 + 5x = 12$$
$$2x^2 + 5x - 12 = 0$$
$$(2x-3)(x+4) = 0$$
$$2x - 3 = 0 \quad \text{or} \quad x + 4 = 0$$
$$2x = 3 \qquad\qquad x = -4$$
$$x = \frac{3}{2}$$

The solution set is $\left\{-4, \dfrac{3}{2}\right\}$.

3.
$$8x - 5y = -4$$
$$2x + 15y = -66$$

Multiply the first equation by 3 and add the two equations.
$$24x - 15y = -12$$
$$\underline{2x + 15y = -66}$$
$$26x \qquad = -78$$
$$x \qquad = -3$$

Back-substitute $x = -3$ into the first equation and solve for *y*.
$$8(-3) - 5y = -4$$
$$-24 - 5y = -4$$
$$-5y = 20$$
$$y = -4$$

The solution set is $\{(-3, -4)\}$.

4. $\dfrac{15}{x} - 4 = \dfrac{6}{x} + 3$

Restriction: $x \neq 0$

Multiply both sides by the LCD, which is *x*.
$$x\left(\frac{15}{x} - 4\right) = x\left(\frac{6}{x} + 3\right)$$
$$x \cdot \frac{15}{x} - x \cdot 4 = x \cdot \frac{6}{x} + x \cdot 3$$
$$15 - 4x = 6 + 3x$$
$$15 - 7x = 6$$
$$-7x = -9$$
$$x = \frac{9}{7}$$

The solution set is $\left\{\dfrac{9}{7}\right\}$.

5.
$$-3x - 7 = 8$$
$$-3x = 15$$
$$x = -5$$

The solution set is $\{-5\}$.

6.
$$\sqrt{2x-1} - x = -2$$
$$\sqrt{2x-1} = x - 2$$
$$\left(\sqrt{2x-1}\right)^2 = (x-2)^2$$
$$2x - 1 = x^2 - 4x + 4$$
$$0 = x^2 - 6x + 5$$
$$0 = (x-1)(x-5)$$
$$x - 1 = 0 \quad \text{or} \quad x - 5 = 0$$
$$x = 1 \qquad\qquad x = 5$$

Check each proposed solution in the original equation.

Check 1: $\sqrt{2x-1} - x = -2$
$$\sqrt{2 \cdot 1 - 1} - 1 = -2$$
$$\sqrt{2-1} - 1 = -2$$
$$\sqrt{1} - 1 = -2$$
$$1 - 1 = -2$$
$$0 = -2, \text{ false}$$

Check 5: $\sqrt{2x-1} - x = -2$
$$\sqrt{2 \cdot 5 - 1} - 5 = -2$$
$$\sqrt{10-1} - 5 = -2$$
$$\sqrt{9} - 5 = -2$$
$$3 - 5 = -2$$
$$-2 = -2, \text{ true}$$

The check shows that 1 is an extraneous solution. The solution set is $\{5\}$.

7.
$$\frac{8x^3}{-4x^7} = \frac{8}{-4} \cdot x^{3-7} = -2x^{-4}$$
$$= \frac{-2}{x^4} = -\frac{2}{x^4}$$

8.
$$6\sqrt{75} - 4\sqrt{12} = 6\sqrt{25}\sqrt{3} - 4\sqrt{4}\sqrt{3}$$
$$= 6 \cdot 5\sqrt{3} - 4 \cdot 2\sqrt{3}$$
$$= 30\sqrt{3} - 8\sqrt{3}$$
$$= 22\sqrt{3}$$

9. $\dfrac{\dfrac{1}{x}-\dfrac{1}{2}}{\dfrac{1}{3}-\dfrac{x}{6}}$

LCD $= 6x$

$\dfrac{\dfrac{1}{x}-\dfrac{1}{2}}{\dfrac{1}{3}-\dfrac{x}{6}} = \dfrac{6x}{6x}\cdot\dfrac{\left(\dfrac{1}{x}-\dfrac{1}{2}\right)}{\left(\dfrac{1}{3}-\dfrac{x}{6}\right)}$

$= \dfrac{6x\cdot\dfrac{1}{x}-6x\cdot\dfrac{1}{2}}{6x\cdot\dfrac{1}{3}-6x\cdot\dfrac{x}{6}} = \dfrac{6-3x}{2x-x^2}$

$= \dfrac{3(2-x)}{x(2-x)} = \dfrac{3}{x}$

10. $\dfrac{4-x^2}{3x^2-5x-2} = \dfrac{(2+x)(2-x)}{(3x+1)(x-2)}$

$= \dfrac{(2+x)(-1)(x-2)}{(3x+1)(x-2)}$

$= \dfrac{-1(2+x)}{3x+1}$

$= -\dfrac{2+x}{3x+1}$

11. $-5-(-8)-(4-6)$

$= -5-(-8)-(-2)$

$= -5+8+2$

$= 3+2$

$= 5$

12. To factor this trinomial, find two integers whose product is 77 and whose sum is -18. These integers are -7 and -11.

$x^2-18x+77 = (x-7)(x-11)$

13. $x^3-25x = x\left(x^2-25\right)$

$= x(x+5)(x-5)$

14. $\begin{array}{r} 6x^2-7x+2 \\ x-2\overline{)6x^3-19x^2+16x-4} \end{array}$

$\underline{6x^3-12x^2}$

$\quad -7x^2+16x$

$\quad \underline{-7x^2+14x}$

$\qquad\quad 2x-4$

$\qquad\quad \underline{2x-4}$

$\qquad\qquad\quad 0$

$\dfrac{6x^3-19x^2+16x-4}{x-2} = 6x^2-7x+2$

15. $(2x-3)\left(4x^2+6x+9\right)$

$= 2x\left(4x^2+6x+9\right)-3\left(4x^2+6x+9\right)$

$= 8x^3+12x^2+18x-12x^2-18x-27$

$= 8x^3-27$

16. $\dfrac{3x}{x^2+x-2}-\dfrac{2}{x+2}$

$= \dfrac{3x}{(x+2)(x-1)}-\dfrac{2}{x+2}$

$= \dfrac{3x}{(x+2)(x-1)}-\dfrac{2(x-1)}{(x+2)(x-1)}$

$= \dfrac{3x-2(x-1)}{(x+2)(x-1)}$

$= \dfrac{3x-2x+2}{(x+2)(x-1)}$

$= \dfrac{x+2}{(x+2)(x-1)}$

$= \dfrac{1}{x-1}$

17. $\dfrac{5x^2-6x+1}{x^2-1}\div\dfrac{16x^2-9}{4x^2+7x+3}$

$= \dfrac{5x^2-6x+1}{x^2-1}\cdot\dfrac{4x^2+7x+3}{16x^2-9}$

$= \dfrac{(5x-1)(x-1)}{(x+1)(x-1)}\cdot\dfrac{(4x+3)(x+1)}{(4x+3)(4x-3)}$

$= \dfrac{5x-1}{4x-3}$

18. $\sqrt{12} - 4\sqrt{75} = \sqrt{4} \cdot \sqrt{3} - 4\sqrt{25} \cdot \sqrt{3}$

$\qquad\qquad\qquad = 2\sqrt{3} - 4(5)\sqrt{3}$

$\qquad\qquad\qquad = 2\sqrt{3} - 20\sqrt{3}$

$\qquad\qquad\qquad = -18\sqrt{3}$

19. $2x - y = 4$

The graph is a line with x-intercept 2 and y-intercept -4. The point $(1, -2)$ can be used as a checkpoint. Draw a line through $(2,0)$, $(0, -4)$, and $(1, -2)$.

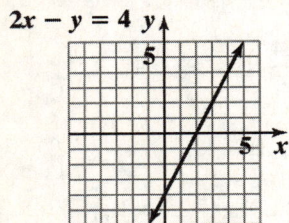

20. $y = -\dfrac{2}{3}x$

The graph is a line with slope $-\dfrac{2}{3} = \dfrac{-2}{3}$ and y-intercept 0. Plot $(0,0)$. From the origin, move 2 units *down* and 3 units to the *right* to reach the point $(3, -2)$. Draw a line through $(0,0)$ and $(3, -2)$.

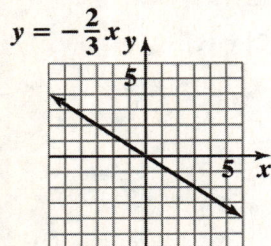

21. $x \geq -1$

Graph $x = 1$ as a solid vertical line. Use $(0,0)$ as a test point. Since $0 \geq -1$ is true, shade the half-plane that contains $(0,0)$. This is the region to the right of the vertical line.

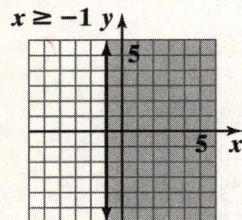

22. $m = \dfrac{y_2 - y_1}{x_2 - x_1} = \dfrac{-3 - 5}{2 - (-1)} = \dfrac{-8}{3} = -\dfrac{8}{3}$

23. Slope 5, passing through $(-2, -3)$.

First, substitute 5 for m, -2 for x_1, and -3 for y_1 in the point-slope form.

$y - y_1 = m(x - x_1)$

$y - (-3) = 5[x - (-2)]$

$\qquad y + 3 = 5(x + 2)$

Now rewrite the equation in slope-intercept form.

$y + 3 = 5x + 10$

$\qquad y = 5x + 7$

24. Let $x =$ the number.

$5x - 7 = 208$

$\qquad 5x = 215$

$\qquad\ x = 43$

The number is 43.

25. Let $x =$ the number of deer in the park.

$$\frac{\substack{\text{Original number} \\ \text{of tagged deer}}}{\substack{\text{Total number} \\ \text{of deer}}} = \frac{\substack{\text{Number of tagged} \\ \text{deer in sample}}}{\substack{\text{Number of deer} \\ \text{in sample}}}$$

$$\frac{318}{x} = \frac{56}{168}$$

$(318)(168) = 56x$

$\qquad 53,424 = 56x$

$\qquad\quad\ 954 = x$

There are approximately 954 deer in the park.

Chapter 9
Quadratic Equations and Introduction to Functions

9.1 Check Points

1. a. $x^2 = 36$

$x = \sqrt{36}$ or $x = -\sqrt{36}$

$x = 6 \qquad x = -6$

The solution set is $\{\pm 6\}$.

b. $5x^2 = 15$

$x^2 = 3$

$x = \sqrt{3}$ or $x = -\sqrt{3}$

The solution set is $\left\{\pm\sqrt{3}\right\}$.

c. $2x^2 - 7 = 0$

$2x^2 = 7$

$x^2 = \dfrac{7}{2}$

$x = \sqrt{\dfrac{7}{2}}$ or $x = -\sqrt{\dfrac{7}{2}}$

$\pm\sqrt{\dfrac{7}{2}} = \pm\dfrac{\sqrt{7}}{\sqrt{2}} \cdot \dfrac{\sqrt{2}}{\sqrt{2}} = \pm\dfrac{\sqrt{14}}{2}$

The solution set is $\left\{\pm\dfrac{\sqrt{14}}{2}\right\}$.

2. $(x-3)^2 = 25$

$x - 3 = \sqrt{25}$ or $x - 3 = -\sqrt{25}$

$x - 3 = 5 \quad$ or $\quad x - 3 = -5$

$x = 8 \qquad\qquad x = -2$

The solution set is $\{-2, 8\}$.

3. $(x-2)^2 = 7$

$x - 2 = \sqrt{7} \qquad$ or $\quad x - 2 = -\sqrt{7}$

$x = 2 + \sqrt{7} \qquad\qquad x = 2 - \sqrt{7}$

The solution set is $\left\{2 \pm \sqrt{7}\right\}$.

4. $w^2 + 20^2 = 50^2$

$w^2 + 400 = 2500$

$w^2 = 2100$

$w = \sqrt{2100} \quad$ or $\quad w = -\sqrt{2100}$

$w = 10\sqrt{21} \qquad\qquad w = -10\sqrt{21}$

$w \approx 45.8 \qquad\qquad w \approx -45.8$

The dimension must be positive. Reject –45.8.
The width of the rectangle is $10\sqrt{21}$ inches or about 45.8 inches.

5. $d = \sqrt{(x_2 - x_1)^2 + (y_2 - y_1)^2}$

$ = \sqrt{(1 - (-4))^2 + (-3 - 9)^2}$

$ = \sqrt{5^2 + (-12)^2}$

$ = \sqrt{25 + 144}$

$ = \sqrt{169}$

$ = 13$

The distance between the two points is 13 units.

9.1 Concept and Vocabulary Check

1. $\pm\sqrt{d}$

2. right; hypotenuse; legs

3. right; legs; the square of the length of the hypotenuse

4. $\sqrt{(x_2 - x_1)^2 + (y_2 - y_1)^2}$

9.1 Exercise Set

1. $x^2 = 16$

$x = \sqrt{16}$ or $x = -\sqrt{16}$

$x = 4 \qquad x = -4$

The solution set is $\{\pm 4\}$.

3. $y^2 = 81$

$y = \sqrt{81}$ or $y = -\sqrt{81}$

$y = 9 \qquad y = -9$

The solution set is $\{\pm 9\}$.

5. $x^2 = 7$

$x = \sqrt{7}$ or $x = -\sqrt{7}$

The solution set is $\left\{\pm\sqrt{7}\right\}$.

7. $x^2 = 50$

$x = \sqrt{50}$ or $x = -\sqrt{50}$

Simplify $\sqrt{50}$:

$\sqrt{50} = \sqrt{25}\sqrt{2} = 5\sqrt{2}$

$x = 5\sqrt{2}$ or $x = -5\sqrt{2}$

The solution set is $\left\{\pm 5\sqrt{2}\right\}$.

9. $5x^2 = 20$

$x^2 = 4$

$x = \sqrt{4}$ or $x = -\sqrt{4}$

$x = 2$ or $x = -2$

The solution set is $\{\pm 2\}$.

11. $4y^2 = 49$

$y^2 = \dfrac{49}{4}$

$y = \sqrt{\dfrac{49}{4}}$ or $y = -\sqrt{\dfrac{49}{4}}$

$y = \dfrac{7}{2}$ or $y = -\dfrac{7}{2}$

The solution set is $\left\{\pm\dfrac{7}{2}\right\}$.

13. $2x^2 + 1 = 51$

$2x^2 = 50$

$x^2 = 25$

$x = \sqrt{25}$ or $x = -\sqrt{25}$

$x = 5$ or $x = -5$

The solution set is $\{\pm 5\}$.

15. $3x^2 - 2 = 0$

$3x = 2$

$x^2 = \dfrac{2}{3}$

$x = \sqrt{\dfrac{2}{3}}$ or $x = -\sqrt{\dfrac{2}{3}}$

Rationalize the denominators.

$\sqrt{\dfrac{2}{3}} = \dfrac{\sqrt{2}}{\sqrt{3}} = \dfrac{\sqrt{2}}{\sqrt{3}}\cdot\dfrac{\sqrt{3}}{\sqrt{3}} = \dfrac{\sqrt{6}}{3}$

$x = \dfrac{\sqrt{6}}{3}$ or $x = -\dfrac{\sqrt{6}}{3}$

The solution set is $\left\{\pm\dfrac{\sqrt{6}}{3}\right\}$.

17. $5z^2 - 7 = 0$

$5z^2 = 7$

$z^2 = \dfrac{7}{5}$

$z = \sqrt{\dfrac{7}{5}}$ or $z = -\sqrt{\dfrac{7}{5}}$

Rationalize the denominators.

$\sqrt{\dfrac{7}{5}} = \dfrac{\sqrt{7}}{\sqrt{5}} = \dfrac{\sqrt{7}}{\sqrt{5}}\cdot\dfrac{\sqrt{5}}{\sqrt{5}} = \dfrac{\sqrt{35}}{5}$

$z = \dfrac{\sqrt{35}}{5}$ or $z = -\dfrac{\sqrt{35}}{5}$

The solution set is $\left\{\pm\dfrac{\sqrt{35}}{5}\right\}$.

19. $(x-3)^2 = 16$

$x - 3 = \sqrt{16}$ or $x - 3 = -\sqrt{16}$

$x - 3 = 4$ or $x - 3 = -4$

$x = 7$ $x = -1$

The solution set is $\{-1, 7\}$.

21. $(x+5)^2 = 121$

$x + 5 = \sqrt{121}$ or $x + 5 = -\sqrt{121}$

$x + 5 = 11$ or $x + 5 = -11$

$x = 6$ $x = -16$

The solution set is $\{-16, 6\}$.

23. $(3x+2)^2 = 9$

$3x+2 = \sqrt{9}$ or $3x+2 = -\sqrt{9}$

$3x+2 = 3$ or $3x+2 = -3$

$3x = 1$ or $3x = -5$

$x = \dfrac{1}{3}$ $x = -\dfrac{5}{3}$

The solution set is $\left\{ -\dfrac{5}{3}, \dfrac{1}{3} \right\}$.

25. $(x-5)^2 = 3$

$x-5 = \sqrt{3}$ or $x-5 = -\sqrt{3}$

$x = 5+\sqrt{3}$ $x = 5-\sqrt{3}$

The solution set is $\left\{ 5 \pm \sqrt{3} \right\}$.

27. $(y+8)^2 = 11$

$y+8 = \sqrt{11}$ or $y+8 = -\sqrt{11}$

$y = -8+\sqrt{11}$ $y = -8-\sqrt{11}$

The solution set is $\left\{ -8 \pm \sqrt{11} \right\}$.

29. $(z-4)^2 = 18$

$z-4 = \sqrt{18}$ or $z-4 = -\sqrt{18}$

$z-4 = 3\sqrt{2}$ $z-4 = -3\sqrt{2}$

$z = 4+3\sqrt{2}$ $z = 4-3\sqrt{2}$

The solution set is $\left\{ 4 \pm 3\sqrt{2} \right\}$.

31. $x^2 + 4x + 4 = 16$

$(x+2)^2 = 16$

$x+2 = \sqrt{16}$ or $x+2 = -\sqrt{16}$

$x+2 = 4$ or $x+2 = -4$

$x = 2$ $x = -6$

The solution set is $\left\{ -6, 2 \right\}$.

33. $x^2 - 6x + 9 = 36$

$(x-3)^2 = 36$

$x-3 = \sqrt{36}$ or $x-3 = -\sqrt{36}$

$x-3 = 6$ or $x-3 = -6$

$x = 9$ $x = -3$

The solution set is $\left\{ -3, 9 \right\}$.

35. $x^2 - 10x + 25 = 2$

$(x-5)^2 = 2$

$x-5 = \sqrt{2}$ or $x-5 = -\sqrt{2}$

$x = 5+\sqrt{2}$ $x = 5-\sqrt{2}$

The solution set is $\left\{ 5 \pm \sqrt{2} \right\}$.

37. $x^2 + 2x + 1 = 5$

$(x+1)^2 = 5$

$x+1 = \sqrt{5}$ or $x+1 = -\sqrt{5}$

$x = -1+\sqrt{5}$ $x = -1-\sqrt{5}$

The solution set is $\left\{ -1 \pm \sqrt{5} \right\}$.

39. $y^2 - 14y + 49 = 12$

$(y-7)^2 = 12$

$y-7 = \sqrt{12}$ or $y-7 = -\sqrt{12}$

$y-7 = 2\sqrt{3}$ or $y-7 = -2\sqrt{3}$

$y = 7+2\sqrt{3}$ $y = 7-2\sqrt{3}$

The solution set is $\left\{ 7 \pm 2\sqrt{3} \right\}$.

41. $8^2 + 15^2 = c^2$

$64 + 225 = c^2$

$289 = c^2$

$c = \sqrt{289} = 17$

The missing length is 17 meters.

43. $15^2 + 36^2 = c^2$

$225 + 1296 = c^2$

$1521 = c^2$

$c = \sqrt{1521}$

$= 39$

The missing length is 39 meters.

45. $a^2 + 16^2 = 20^2$

$a^2 + 256 = 440$

$a^2 = 144$

$a = \sqrt{144}$

$= 12$

The missing length is 12 centimeters.

47. $9^2 + b^2 = 16^2$

$81 + b^2 = 256$

$b^2 = 175$

$b = \sqrt{175}$

$= \sqrt{25}\sqrt{7}$

$= 5\sqrt{7}$

The missing length is $5\sqrt{7}$ meters.

49. $d = \sqrt{(x_2 - x_1)^2 + (y_2 - y_1)^2}$

$= \sqrt{(4-3)^2 + (1-5)^2}$

$= \sqrt{1^2 + (-4)^2}$

$= \sqrt{1+16} = \sqrt{17} \approx 4.12$

The distance between the two points is $\sqrt{17} \approx 4.12$ units.

51. $d = \sqrt{(x_2 - x_1)^2 + (y_2 - y_1)^2}$

$= \sqrt{(4-(-4))^2 + (17-2)^2}$

$= \sqrt{8^2 + 15^2} = \sqrt{64 + 225}$

$= \sqrt{289} = 17$

The distance between the two points is 17 units.

53. $d = \sqrt{(x_2 - x_1)^2 + (y_2 - y_1)^2}$

$= \sqrt{(9-6)^2 + (5-(-1))^2}$

$= \sqrt{3^2 + 6^2} = \sqrt{9 + 36}$

$= \sqrt{45} = 3\sqrt{5} \approx 6.71$

The distance between the two points is $3\sqrt{5} \approx 6.71$ units.

55. $d = \sqrt{(x_2 - x_1)^2 + (y_2 - y_1)^2}$

$= \sqrt{(-2-(-7))^2 + (-1-(-5))^2}$

$= \sqrt{5^2 + 4^2} = \sqrt{25 + 16}$

$= \sqrt{41} \approx 6.40$

The distance between the two points is $\sqrt{41} \approx 6.40$ units.

57. $d = \sqrt{(x_2 - x_1)^2 + (y_2 - y_1)^2}$

$= \sqrt{\left(4\sqrt{7} - (-2\sqrt{7})\right)^2 + (8-10)^2}$

$= \sqrt{(6\sqrt{7})^2 + (-2)^2} = \sqrt{252 + 4}$

$= \sqrt{256} = 16$

The distance between the two points is 16 units.

59. Let $x =$ the number.

$(x-3)^2 = 25$

$x - 3 = \sqrt{25}$ or $x - 3 = -\sqrt{25}$

$x = 3 + 5 \qquad\qquad x = 3 - 5$

$x = 8 \qquad\qquad\quad x = -2$

The number is -2 or 8.

61. Let $x =$ the number.

$(3x+2)^2 = 49$

$3x + 2 = \sqrt{49}$ or $3x + 2 = -\sqrt{49}$

$3x + 2 = 7 \qquad\qquad 3x + 2 = -7$

$3x = 5 \qquad\qquad\quad 3x = -9$

$x = \dfrac{5}{3} \qquad\qquad\quad x = -3$

The number is -3 or $\dfrac{5}{3}$.

63. $A = \pi r^2$

$\dfrac{A}{\pi} = r^2$

$r = \sqrt{\dfrac{A}{\pi}}$

$= \dfrac{\sqrt{A}}{\sqrt{\pi}} \cdot \dfrac{\sqrt{\pi}}{\sqrt{\pi}}$

$= \dfrac{\sqrt{A\pi}}{\pi}$

65. $I = \dfrac{k}{d^2}$

$d^2 I = k$

$d^2 = \dfrac{k}{I}$

$d = \sqrt{\dfrac{k}{I}} = \dfrac{\sqrt{k}}{\sqrt{I}}$

$ = \dfrac{\sqrt{k}}{\sqrt{I}} \cdot \dfrac{\sqrt{I}}{\sqrt{I}}$

$ = \dfrac{\sqrt{kI}}{I}$

67. Let x = the length of the ladder.

$8^2 + 10^2 = x^2$

$64 + 100 = x^2$

$164 = x^2$

$x = \sqrt{164} = \sqrt{4}\sqrt{41} = 2\sqrt{41}$

The length of the ladder is $2\sqrt{41}$ feet.

69. $90^2 + 90^2 = x^2$

$8100 + 8100 = x^2$

$16,200 = x^2$

$\sqrt{16,2000} = \sqrt{8100}\sqrt{2} = 90\sqrt{2}$

The distance from home plate to second base is $90\sqrt{2}$ feet.

71. $A = \pi r^2; A = 36\pi$

$36\pi = \pi r^2$

$36 = r^2$

$r = \sqrt{36} = 6$

The radius is 6 inches.

73. $W = 3t^2; W = 108$

$108 = 3t^2$

$36 = t^2$

$t = \sqrt{36} = 6$

The fetus weighs 108 grams after 6 weeks.

75. $d = 16t^2; d = 400$

The rock must fall 400 feet to hit the water.

$400 = 16t^2$

$25 = t^2$

$t = \sqrt{25} = 5$

It will take 5 seconds for the rock to hit the water.

77. The length of each side of the original garden is x meters. Therefore, the length of each larger side is $x + 4$ meters since 2 meters were added to each side.

Larger square: $A = (x+4)^2; A = 144$

$144 = (x+4)^2$

$x + 4 = \sqrt{144}$ or $x + 4 = -\sqrt{144}$

$x + 4 = 12$ or $x + 4 = -12$

$x = 8$ $x = -16$

Reject -16 because a length cannot be negative. The length of each side of the original square is 8 meters.

79. The volume of a rectangular solid is given by $V = lwh$. From the problem statement, we have $V = 200, l = x, w = x, h = 2$

$200 = x \cdot x \cdot 2$

$200 = 2x^2$

$100 = x^2$

$x = \sqrt{100} = 10$

The length and width of the open box are both 10 inches.

81. – 83. Answers will vary.

85. makes sense

87. makes sense

89. false; Changes to make the statement true will vary. A sample change is: They are not equivalent. The first equation has two solutions, $-5 \pm 2\sqrt{2}$, while the second equation is only satisfied by $-5 + 2\sqrt{2}$.

91. true

93.
$$d = \sqrt{(x_2 - x_1)^2 + (y_2 - y_1)^2}$$
$$5 = \sqrt{(x - (-3))^2 + (-5 - (-2))^2}$$
$$5 = \sqrt{(x + 3)^2 + (-3)^2}$$
$$5^2 = (x + 3)^2 + (-3)^2$$
$$25 = x^2 + 6x + 9 + 9$$
$$0 = x^2 + 6x - 7$$
$$0 = (x + 7)(x - 1)$$
$$x + 7 = 0 \quad \text{or} \quad x - 1 = 0$$
$$x = -7 \qquad\qquad x = 1$$
The possible values for x are -7 and 1.

95. $(x - 1)^2 - 9 = 0$

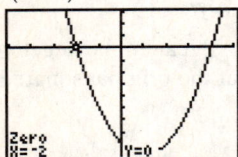 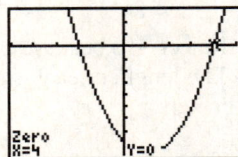

Check -2:
$$(-2 - 1)^2 - 9 = 0$$
$$(-3)^2 - 9 = 0$$
$$9 - 9 = 0$$
$$0 = 0, \text{ true}$$

Check 4:
$$(4 - 1)^2 - 9 = 0$$
$$3^2 - 9 = 0$$
$$9 - 9 = 0$$
$$0 = 0 \text{ true}$$

The solutions are $\{-2, 4\}$.

96. $12x^2 + 14x - 6 = 2\left(6x^2 + 7x - 3\right)$
$$= 2(2x + 3)(3x - 1)$$

97.
$$\frac{x^2 - x - 6}{3x - 3} \div \frac{x^2 - 4}{x - 1}$$
$$= \frac{x^2 - x - 6}{3x - 3} \cdot \frac{x - 1}{x^2 - 4}$$
$$= \frac{(x + 2)(x - 3)}{3(x - 1)} \cdot \frac{(x - 1)}{(x + 2)(x - 2)}$$
$$= \frac{x - 3}{3(x - 2)}$$

98.
$$4(x - 5) = 22 + 2(6x + 3)$$
$$4x - 20 = 22 + 12x + 6$$
$$4x - 20 = 28 + 12x$$
$$-8x - 20 = 28$$
$$-8x = 48$$
$$x = -6$$
The solution is set $\{-6\}$.

99. $x^2 + 8x + 16 = x^2 + 2 \cdot 4x + 4^2 = (x + 4)^2$

100. $x^2 - 14x + 49 = x^2 - 2 \cdot 7x + 7^2 = (x - 7)^2$

101. $x^2 + 5x + \dfrac{25}{4} = x^2 + 2 \cdot \dfrac{5}{2}x + \left(\dfrac{5}{2}\right)^2 = \left(x + \dfrac{5}{2}\right)^2$

9.2 Check Points

1. a. $x^2 + 10x$

The coefficient of the x-term is 10. Half of 10 is 5, and $5^2 = 25$. Add 25.
$$x^2 + 10x + 25 = (x + 5)^2$$

b. $x^2 - 6x$

The coefficient of the x-term is -6. Half of -6 is -3, and $(-3)^2 = 9$. Add 9.
$$x^2 - 6x + 9 = (x - 3)^2$$

c. $x^2 + 3x$

The coefficient of the x-term is 3. Half of 3 is $\dfrac{3}{2}$, and $\left(\dfrac{3}{2}\right)^2 = \dfrac{9}{4}$. Add $\dfrac{9}{4}$.
$$x^2 + 3x + \frac{9}{4} = \left(x + \frac{3}{2}\right)^2$$

2. a. The coefficient of the x-term is 6. Half of 6 is 3, and $3^2 = 9$. Add 9.

$$x^2 + 6x = 7$$
$$x^2 + 6x + 9 = 7 + 9$$
$$(x + 3)^2 = 16$$
$$x + 3 = \sqrt{16} \quad \text{or} \quad x + 3 = -\sqrt{16}$$
$$x + 3 = 4 \qquad\qquad x + 3 = -4$$
$$x = 1 \qquad\qquad\quad x = -7$$

The solution set is $\{-7, 1\}$.

b. $x^2 - 10x + 18 = 0$

First subtract 18 from both sides to isolate the binomial $x^2 - 10x$.

$$x^2 - 10x = -18$$

Next, complete the square.

$$x^2 - 10x + 25 = -18 + 25$$
$$x^2 - 10x + 25 = 7$$
$$(x - 5)^2 = 7$$
$$x - 5 = \sqrt{7} \quad \text{or} \quad x - 5 = -\sqrt{7}$$
$$x = 5 + \sqrt{7} \qquad\qquad x = 5 - \sqrt{7}$$

The solution set is $\left\{ 5 \pm \sqrt{7} \right\}$.

3. $2x^2 - 10x - 1 = 0$

First, divide both sides of the equation by 2 so that the coefficient of the x^2 term will be 1.

$$x^2 - 5x - \frac{1}{2} = 0$$

Next, add $\dfrac{1}{2}$ to both sides to isolate the binomial.

$$x^2 - 5x = \frac{1}{2}$$

Complete the square: The coefficient of the x-term is -5, and $\dfrac{1}{2}(-5) = -\dfrac{5}{2}$, so add $\left(-\dfrac{5}{2} \right)^2 = \dfrac{25}{4}$ to both sides.

$$x^2 - 5x + \frac{25}{4} = \frac{1}{2} + \frac{25}{4}$$
$$x^2 - 5x + \frac{25}{4} = \frac{2}{4} + \frac{25}{4}$$
$$\left(x - \frac{5}{2} \right)^2 = \frac{27}{4}$$

$$x - \frac{5}{2} = \sqrt{\frac{27}{4}} \quad \text{or} \quad x - \frac{5}{2} = -\sqrt{\frac{27}{4}}$$
$$x = \frac{5}{2} + \frac{\sqrt{27}}{2} \qquad\qquad x = \frac{5}{2} - \frac{\sqrt{27}}{2}$$
$$x = \frac{5}{2} + \frac{3\sqrt{3}}{2} \qquad\qquad x = \frac{5}{2} - \frac{3\sqrt{3}}{2}$$
$$x = \frac{5 + 3\sqrt{3}}{2} \qquad\qquad x = \frac{5 - 3\sqrt{3}}{2}$$

The solution set is $\left\{ \dfrac{5 \pm 3\sqrt{3}}{2} \right\}$.

9.2 Concept and Vocabulary Check

1. 81

2. 9

3. $\dfrac{49}{4}$

4. $\dfrac{25}{16}$

5. $\left(\dfrac{b}{2} \right)^2$ or $\dfrac{b^2}{4}$

6. 100

7. 1

8. $\dfrac{1}{4}$

9.2 Exercise Set

1. $x^2 + 10x$

The coefficient of the x-term is 10. Half of 10 is 5, and $5^2 = 25$. Add 25.

$$x^2 + 10x + 25 = (x + 5)^2$$

3. $x^2 - 2x$

The coefficient of the x-term is -2. Half of -2 is -1, and $(-1)^2 = 1$. Add 1.

$$x^2 - 2x + 1 = (x - 1)^2$$

5. $x^2 + 5x$

The coefficient of the x-term is 5. Half of 5 is $\dfrac{5}{2}$,

and $\left(\dfrac{5}{2}\right)^2 = \dfrac{25}{4}$. Add $\dfrac{25}{4}$.

$x^2 + 5x + \dfrac{25}{4} = \left(x + \dfrac{5}{2}\right)^2$

7. $x^2 - 7x$

The coefficient of the x-term is -7. Half of -7 is

$\dfrac{-7}{2}$, and $\left(\dfrac{-7}{2}\right)^2 = \dfrac{49}{4}$. Add $\dfrac{49}{4}$.

$x^2 - 7x + \dfrac{49}{4} = \left(x - \dfrac{7}{2}\right)^2$

9. $x^2 + \dfrac{1}{2}x$

The coefficient of the x-term is $\dfrac{1}{2}$. Half of

$\dfrac{1}{2}$ is $\dfrac{1}{2}\left(\dfrac{1}{2}\right) = \dfrac{1}{4}$, and $\left(\dfrac{1}{4}\right)^2 = \dfrac{1}{16}$. Add $\dfrac{1}{16}$.

$x^2 + \dfrac{1}{2}x + \dfrac{1}{16} = \left(x + \dfrac{1}{4}\right)^2$

11. $x^2 - \dfrac{4}{3}x$

The coefficient of the x-term is $-\dfrac{4}{3}$. Half of

$-\dfrac{4}{3}$ is $\dfrac{1}{2}\left(-\dfrac{4}{3}\right) = -\dfrac{2}{3}$, and $\left(-\dfrac{2}{3}\right)^2 = \dfrac{4}{9}$. Add $\dfrac{4}{9}$.

$x^2 - \dfrac{4}{3}x + \dfrac{4}{9} = \left(x - \dfrac{2}{3}\right)^2$

13. $x^2 + 4x = 5$

$x^2 + 4x + 4 = 5 + 4$

$(x + 2)^2 = 9$

$x + 2 = \sqrt{9}$ or $x + 2 = -\sqrt{9}$

$x + 2 = 3$ $x + 2 = -3$

$x = 1$ $x = -5$

The solution set is $\{-5, 1\}$.

15. $x^2 - 10x = -24$

$x^2 - 10x + 25 = -24 + 25$

$(x - 5)^2 = 1$

$x - 5 = \sqrt{1}$ or $x - 5 = -\sqrt{1}$

$x - 5 = 1$ $x - 5 = -1$

$x = 6$ $x = 4$

The solution set is $\{4, 6\}$.

17. $x^2 - 2x = 5$

$x^2 - 2x + 1 = 5 + 1$

$(x - 1)^2 = 6$

$x - 1 = \sqrt{6}$ or $x - 1 = -\sqrt{6}$

$x = 1 + \sqrt{6}$ or $x = 1 - \sqrt{6}$

The solution set is $\{1 \pm \sqrt{6}\}$.

19. $x^2 + 4x + 1 = 0$

First subtract 1 from both sides to isolate the binomial $x^2 + 4x$.

$x^2 + 4x = -1$

Next, complete the square.

$x^2 + 4x + 4 = -1 + 4$

$(x + 2)^2 = 3$

$x + 2 = \sqrt{3}$ or $x + 2 = -\sqrt{3}$

$x = -2 + \sqrt{3}$ or $x = -2 - \sqrt{3}$

The solution set is $\{-2 \pm \sqrt{3}\}$.

21. $x^2 - 3x = 28$

$x^2 - 3x + \dfrac{9}{4} = 28 + \dfrac{9}{4}$

$\left(x - \dfrac{3}{2}\right)^2 = \dfrac{121}{4}$

$x - \dfrac{3}{2} = \sqrt{\dfrac{121}{4}}$ or $x - \dfrac{3}{2} = -\sqrt{\dfrac{121}{4}}$

$x - \dfrac{3}{2} = \dfrac{11}{2}$ or $x - \dfrac{3}{2} = -\dfrac{11}{2}$

$x = \dfrac{14}{2} = 7$ $x = -\dfrac{8}{2} = -4$

The solution set is $\{-4, 7\}$.

23. $x^2 + 3x - 1 = 0$

$$x^2 + 3x = 1$$

$$x^2 + 3x + \frac{9}{4} = 1 + \frac{9}{4}$$

$$\left(x + \frac{3}{2}\right) = \frac{13}{4}$$

$$x + \frac{3}{2} = \sqrt{\frac{13}{4}} \qquad \text{or } x + \frac{3}{2} = -\sqrt{\frac{13}{4}}$$

$$x + \frac{3}{2} = \frac{\sqrt{13}}{2} \qquad \text{or } x + \frac{3}{2} = -\sqrt{\frac{13}{2}}$$

$$x = -\frac{3}{2} + \frac{\sqrt{13}}{2} \qquad \text{or } x = -\frac{3}{2} - \frac{\sqrt{13}}{2}$$

$$x = \frac{-3 + \sqrt{13}}{2} \qquad \text{or } x = \frac{-3 - \sqrt{13}}{2}$$

The solution set is $\left\{ \dfrac{-3 \pm \sqrt{13}}{2} \right\}$.

25. $x^2 = 7x - 3$

$$x^2 - 7x = -3$$

$$x^2 - 7x + \frac{49}{4} = -3 + \frac{49}{4}$$

$$\left(x - \frac{7}{2}\right)^2 = \frac{-12}{4} + \frac{49}{4}$$

$$\left(x - \frac{7}{2}\right)^2 = \frac{37}{4}$$

$$x - \frac{7}{2} = \sqrt{\frac{37}{4}} \qquad \text{or } x - \frac{7}{2} = -\sqrt{\frac{37}{4}}$$

$$x - \frac{7}{2} = \frac{\sqrt{37}}{2} \qquad \text{or } x - \frac{7}{2} = -\frac{\sqrt{37}}{2}$$

$$x = \frac{7}{2} + \frac{\sqrt{37}}{2} \qquad \text{or } x = \frac{7}{2} - \frac{\sqrt{37}}{2}$$

$$x = \frac{7 + \sqrt{37}}{2} \qquad \text{or } x = \frac{7 - \sqrt{37}}{2}$$

The solution set is $\left\{ \dfrac{7 \pm \sqrt{37}}{2} \right\}$.

27. $2x^2 - 2x - 6 = 0$

First, divide both sides of the equation by 2 so that the coefficient of the x^2 term will be 1.

$$x^2 - x - 3 = 0$$

Next, add 3 to both sides to isolate the binomial.

$$x^2 - x = 3$$

Complete the square: The coefficient of the x-term is -1, and $\frac{1}{2}(-1) = -\frac{1}{2}$, so add $\left(-\frac{1}{2}\right)^2 = \frac{1}{4}$ to both sides.

$$x^2 - x + \frac{1}{4} = 3 + \frac{1}{4}$$

$$\left(x - \frac{1}{2}\right)^2 = \frac{13}{4}$$

$$x - \frac{1}{2} = \sqrt{\frac{13}{4}} \qquad \text{or } x - \frac{1}{2} = -\sqrt{\frac{13}{4}}$$

$$x - \frac{1}{2} = \frac{\sqrt{13}}{2} \qquad \text{or } x - \frac{1}{2} = -\frac{\sqrt{13}}{2}$$

$$x = \frac{1 + \sqrt{13}}{2} \qquad \text{or } x = \frac{1 - \sqrt{13}}{2}$$

The solution set is $\left\{ \dfrac{1 \pm \sqrt{13}}{2} \right\}$.

29. $2x^2 - 3x + 1 = 0$

$$x^2 - \frac{3}{2}x + \frac{1}{2} = 0$$

$$x^2 - \frac{3}{2}x = -\frac{1}{2}$$

$$x^2 - \frac{3}{2}x + \frac{9}{16} = -\frac{1}{2} + \frac{9}{16}$$

$$\left(x - \frac{3}{4}\right)^2 = -\frac{8}{16} + \frac{9}{16}$$

$$\left(x - \frac{3}{4}\right)^2 = \frac{1}{16}$$

$$x - \frac{3}{4} = \sqrt{\frac{1}{16}} \qquad \text{or } x - \frac{3}{4} = -\sqrt{\frac{1}{16}}$$

$$x - \frac{3}{4} = \frac{1}{4} \qquad \text{or } x - \frac{3}{4} = -\frac{1}{4}$$

$$x = 1 \qquad\qquad x = \frac{2}{4} = \frac{1}{2}$$

The solution set is $\left\{ \dfrac{1}{2}, 1 \right\}$.

31. $2x^2 + 10x + 11 = 0$

$$x^2 + 5x + \frac{11}{2} = 0$$

$$x^2 + 5x = -\frac{11}{2}$$

$$x^2 + 5x + \frac{25}{4} = -\frac{11}{2} + \frac{25}{4}$$

$$\left(x + \frac{5}{2}\right)^2 = -\frac{22}{4} + \frac{25}{4}$$

$$\left(x + \frac{5}{2}\right)^2 = \frac{3}{4}$$

$$x + \frac{5}{2} = \sqrt{\frac{3}{4}} \qquad \text{or} \quad x + \frac{5}{2} = -\sqrt{\frac{3}{4}}$$

$$x + \frac{5}{2} = \frac{\sqrt{3}}{2} \qquad \text{or} \quad x + \frac{5}{2} = -\frac{\sqrt{3}}{2}$$

$$x = -\frac{5}{2} + \frac{\sqrt{3}}{2} \qquad \text{or} \quad x = -\frac{5}{2} - \frac{\sqrt{3}}{2}$$

$$x = \frac{-5 + \sqrt{3}}{2} \qquad \text{or} \quad x = \frac{-5 - \sqrt{3}}{2}$$

The solution set is $\left\{ \dfrac{-5 \pm \sqrt{3}}{2} \right\}$.

33. $4x^2 - 2x - 3 = 0$

$$x^2 - \frac{1}{2}x - \frac{3}{4} = 0$$

$$x^2 - \frac{1}{2}x = \frac{3}{4}$$

$$x^2 - \frac{1}{2}x + \frac{1}{16} = \frac{3}{4} + \frac{1}{16}$$

$$\left(x - \frac{1}{4}\right)^2 = \frac{12}{16} + \frac{1}{16}$$

$$\left(x - \frac{1}{4}\right)^2 = \frac{13}{16}$$

$$x - \frac{1}{4} = \sqrt{\frac{13}{16}} \qquad \text{or} \quad x - \frac{1}{4} = -\sqrt{\frac{13}{16}}$$

$$x - \frac{1}{4} = \frac{\sqrt{13}}{4} \qquad \text{or} \quad x - \frac{1}{4} = -\frac{\sqrt{13}}{4}$$

$$x = \frac{1}{4} + \frac{\sqrt{13}}{4} \qquad \text{or} \quad x = \frac{1}{4} - \frac{\sqrt{13}}{4}$$

$$x = \frac{1 + \sqrt{13}}{4} \qquad \text{or} \quad x = \frac{1 - \sqrt{13}}{4}$$

The solution set is $\left\{ \dfrac{1 \pm \sqrt{13}}{4} \right\}$.

35. $\dfrac{x^2}{6} - \dfrac{x}{3} - 1 = 0$

Multiply both sides of the equation by 6 to obtain

$$x^2 - 2x - 6 = 0$$

$$x^2 - 2x = 6$$

The coefficient on x is -2. Divide this by 2 and square the result.

$$\left(\frac{-2}{2}\right)^2 = (-1)^2 = 1$$

Add 1 to both sides of the equation:

$$x^2 - 2x + 1 = 6 + 1$$

$$(x - 1)^2 = 7$$

$$x - 1 = \pm\sqrt{7}$$

$$x = 1 \pm \sqrt{7}$$

The solution set is $\left\{ 1 \pm \sqrt{7} \right\}$.

37. $(x + 2)(x - 3) = 1$

$$x^2 - x - 6 = 1$$

$$x^2 - x = 7$$

The coefficient of x is -1. Divide this by 2 and square the result.

$$\left(\frac{-1}{2}\right)^2 = \frac{1}{4}$$

Add $\dfrac{1}{4}$ to both sides of the equation.

$$x^2 - x + \frac{1}{4} = 7 + \frac{1}{4}$$

$$\left(x - \frac{1}{2}\right)^2 = \frac{29}{4}$$

$$x - \frac{1}{2} = \pm\sqrt{\frac{29}{4}}$$

$$x = \frac{1}{2} \pm \frac{\sqrt{29}}{2} = \frac{1 \pm \sqrt{29}}{2}$$

The solution set is $\left\{ \dfrac{1 \pm \sqrt{29}}{2} \right\}$.

39. $x^2 + 4bx = 5b^2$

The coefficient on x is $4b$. Divide this by 2 and square the result.

$$\left(\frac{4b}{2}\right)^2 = (2b)^2 = 4b^2$$

Add $4b^2$ to both sides of the equation:

$$x^2 + 4bx + 4b^2 = 5b^2 + 4b^2$$

$$(x + 2b)^2 = 9b^2$$

$$x + 2b = \pm\sqrt{9b^2}$$

$$x = -2b \pm 3b$$

$x = -2b - 3b$ or $x = -2b + 3b$

$= -5b$ $\qquad = b$

The solution set is $\{-5b, b\}$.

41. Answers will vary.

43. makes sense

45. makes sense

47. false; Changes to make the statement true will vary. A sample change is: Completing the square is a method used for changing a binomial into a perfect square trinomial.

49. false; Changes to make the statement true will vary. A sample change is: Although not every quadratic equation can be solved by factoring, they can all be solved by completing the square.

51. Half of -20 is -10, and $(-10)^2 = 100$.

Therefore, the perfect square trinomial with middle term $-20x$ is $x^2 - 20x + 100$.

53. $x^2 + bx + c = 0$

Subtract c from both sides to isolate the binomial $x^2 + bx$.

$$x^2 + bx = -c$$

The coefficient of the x-term is b.

$\frac{1}{2}(b) = \frac{b}{2}$, and $\left(\frac{b}{2}\right)^2 = \frac{b^2}{4}$, so add $\frac{b^2}{4}$ to both sides.

$$x^2 + bx + \frac{b^2}{4} = -c + \frac{b^2}{4}$$

$$\left(x + \frac{b}{2}\right)^2 = \frac{-4c}{4} - \frac{b^2}{4}$$

$$\left(x + \frac{b}{2}\right)^2 = \frac{b^2 - 4c}{4}$$

$$\frac{x+b}{2} = \sqrt{\frac{b^2 - 4c}{4}}$$

$$x + \frac{b}{2} = \frac{\sqrt{b^2 - 4c}}{2}$$

$$x = -\frac{b}{2} + \frac{\sqrt{b^2 - 4c}}{2} = \frac{-b + \sqrt{b^2 - 4c}}{2}$$

or

$$\frac{x+b}{2} = -\sqrt{\frac{b^2 - 4c}{4}}$$

$$x + \frac{b}{2} = -\frac{\sqrt{b^2 - 4c}}{2}$$

$$x = -\frac{b}{2} - \frac{\sqrt{b^2 - 4c}}{2} = \frac{-b - \sqrt{b^2 - 4c}}{2}$$

The solution set is $\left\{\dfrac{-b \pm \sqrt{b^2 - 4c}}{2}\right\}$.

Note: If terms are not combined on the right and the radical is not simplified, the solutions will be written in a different, but equivalent, form:

$$\left\{-\frac{b}{2} \pm \sqrt{\frac{b^2}{4} - c}\right\}.$$

55.

$$\frac{2x+3}{x^2 - 7x + 12} - \frac{2}{x-3}$$

$$= \frac{2x+3}{(x-3)(x-4)} - \frac{2}{x-3}$$

$$= \frac{2x+3}{(x-3)(x-4)} - \frac{2(x-4)}{(x-3)(x-4)}$$

$$= \frac{(2x-3) - 2(x-4)}{(x-3)(x-4)}$$

$$= \frac{2x+3 - 2x + 8}{(x-3)(x-4)}$$

$$= \frac{11}{(x-3)(x-4)}$$

56.

$$\cfrac{x-\dfrac{1}{3}}{3-\dfrac{1}{x}}$$

LCD = $3x$

$$\cfrac{x-\dfrac{1}{3}}{3-\dfrac{1}{x}} = \frac{3x}{3x} \cdot \cfrac{\left(x-\dfrac{1}{3}\right)}{\left(3-\dfrac{1}{x}\right)}$$

$$= \cfrac{3x \cdot x - 3x \cdot \dfrac{1}{3}}{3x \cdot 3 - 3x \cdot \dfrac{1}{x}}$$

$$= \frac{3x^2 - x}{9x - 3}$$

$$= \frac{x(3x-1)}{3(3x-1)}$$

$$= \frac{x}{3}$$

57. $\sqrt{2x+3} = 2x-3$

Square both sides and solve for x:

$$\left(\sqrt{2x+3}\right)^2 = (2x-3)^2$$

$$2x+3 = (2x)^2 - 2 \cdot 2x \cdot 3 + 3^2$$

$$2x+3 = 4x^2 - 12x + 9$$

$$0 = 4x^2 - 14x + 6$$

$$0 = 2\left(2x^2 - 7x + 3\right)$$

$$0 = 2(2x-1)(x-3)$$

$$2x-1 = 0 \quad \text{or} \quad x-3 = 0$$

$$x = \frac{1}{2} \qquad\quad x = 3$$

Each of the proposed solutions must be checked in the original equation.

Check $\dfrac{1}{2}$: $\quad \sqrt{2x+3} = 2x-3$

$$\sqrt{2\left(\frac{1}{2}\right)+3} = 2\left(\frac{1}{2}\right)-3$$

$$\sqrt{1+3} = 1-3$$

$$= \sqrt{4} = 1-3$$

$$2 = -2, \text{ false}$$

Check 3: $\quad \sqrt{2x+3} = 2x-3$

$$\sqrt{2 \cdot 3 + 3} = 2 \cdot 3 - 3$$

$$\sqrt{6+3} = 6-3$$

$$\sqrt{9} = 3$$

$$3 = 3, \text{ true}$$

Thus, $\dfrac{1}{2}$ is an extraneous solution. The solution set is $\{3\}$.

58. $\dfrac{-b \pm \sqrt{b^2 - 4ac}}{2a} = \dfrac{-(9) \pm \sqrt{(9)^2 - 4(2)(-5)}}{2(2)}$

$$= \frac{-9 \pm \sqrt{81 + 40}}{4}$$

$$= \frac{-9 \pm \sqrt{121}}{4}$$

$$= \frac{-9 \pm 11}{4}$$

$$= \frac{-9+11}{4} \quad \text{or} \quad \frac{-9-11}{4}$$

$$= \frac{2}{4} \quad \text{or} \quad \frac{-20}{4}$$

$$= \frac{1}{2} \quad \text{or} \quad -5$$

59. $\dfrac{-b \pm \sqrt{b^2 - 4ac}}{2a} = \dfrac{-(-12) \pm \sqrt{(-12)^2 - 4(9)(4)}}{2(9)}$

$$= \frac{12 \pm \sqrt{144 - 144}}{18}$$

$$= \frac{12 \pm \sqrt{0}}{18}$$

$$= \frac{12 \pm 0}{18}$$

$$= \frac{12}{18}$$

$$= \frac{2}{3}$$

60. $\dfrac{-b \pm \sqrt{b^2 - 4ac}}{2a} = \dfrac{-(-2) \pm \sqrt{(-2)^2 - 4(1)(-6)}}{2(1)}$

$\phantom{\dfrac{-b \pm \sqrt{b^2 - 4ac}}{2a}} = \dfrac{2 \pm \sqrt{4 + 24}}{2(1)}$

$\phantom{\dfrac{-b \pm \sqrt{b^2 - 4ac}}{2a}} = \dfrac{2 \pm \sqrt{28}}{2}$

$\phantom{\dfrac{-b \pm \sqrt{b^2 - 4ac}}{2a}} = \dfrac{2 \pm 2\sqrt{7}}{2}$

$\phantom{\dfrac{-b \pm \sqrt{b^2 - 4ac}}{2a}} = \dfrac{2(1 \pm \sqrt{7})}{2}$

$\phantom{\dfrac{-b \pm \sqrt{b^2 - 4ac}}{2a}} = 1 \pm \sqrt{7}$

9.3 Check Points

1. $8x^2 + 2x - 1 = 0$

Identify the values of $a, b,$ and c:

$a = 8, b = 2,$ and $c = -1$.

Substitute these values into the quadratic formula and simplify to get the equation's solutions.

$x = \dfrac{-b \pm \sqrt{b^2 - 4ac}}{2a}$

$x = \dfrac{-2 \pm \sqrt{2^2 - 4(8)(-1)}}{2(8)}$

$ = \dfrac{-2 \pm \sqrt{4 + 32}}{16} = \dfrac{-2 \pm \sqrt{36}}{16} = \dfrac{-2 \pm 6}{16}$

$x = \dfrac{-2 + 6}{16}$ or $x = \dfrac{-2 - 6}{16}$

$ = \dfrac{4}{16} = \dfrac{1}{4} \qquad\qquad = \dfrac{-8}{16} = -\dfrac{1}{2}$

The solution set is $\left\{ -\dfrac{1}{2}, \dfrac{1}{4} \right\}$.

2. $x^2 = 6x - 4$

$x^2 - 6x + 4 = 0$

Identify the values of $a, b,$ and c:

$a = 1, b = -6,$ and $c = 4$.

Substitute these values into the quadratic formula and simplify to get the equation's solutions.

$x = \dfrac{-b \pm \sqrt{b^2 - 4ac}}{2a}$

$x = \dfrac{-(-6) \pm \sqrt{(-6)^2 - 4(1)(4)}}{2(1)}$

$ = \dfrac{6 \pm \sqrt{36 - 16}}{2}$

$ = \dfrac{6 \pm \sqrt{20}}{2}$

$ = \dfrac{6 \pm 2\sqrt{5}}{2}$

$ = 3 \pm \sqrt{5}$

The solution set is $\left\{ 3 \pm \sqrt{5} \right\}$.

3. $p = 0.004x^2 - 0.37x + 14.1$

$25 = 0.004x^2 - 0.37x + 14.1$

$0 = 0.004x^2 - 0.37x - 10.9$

Identify the values of $a, b,$ and c:

$a = 0.004, b = -0.37,$ and $c = -10.9$.

Substitute these values into the quadratic formula and simplify to get the equation's solutions.

$x = \dfrac{-b \pm \sqrt{b^2 - 4ac}}{2a}$

$x = \dfrac{-(-0.37) \pm \sqrt{(-0.37)^2 - 4(0.004)(-10.9)}}{2(0.004)}$

$ \approx -23$ or 116

Reject the negative solution to the quadratic equation because the time cannot be negative. 25% of the United States population will be foreign-born 116 years after 1920, or 2036.

9.3 Concept and Vocabulary Check

1. $\dfrac{-b \pm \sqrt{b^2 - 4ac}}{2a}$

2. $3; \ 10; \ -8$

3. $1; \ 8; \ -29$

4. 6; −3; −4

5. $-\dfrac{1}{3}$; 2

6. $-4 \pm 3\sqrt{5}$

7. the square root property

8. factoring and the zero-product principle

9. the quadratic formula

9.3 Exercise Set

1. $x^2 + 5x + 6 = 0$

Identify the values of $a, b,$ and c:

$a = 1, b = 5,$ and $c = 6.$

Substitute these values into the quadratic formula and simplify to get the equation's solutions.

$$x = \frac{-b \pm \sqrt{b^2 - 4ac}}{2a}$$

$$x = \frac{5 \pm \sqrt{5^2 - 4(1)(6)}}{2(1)}$$

$$= \frac{5 \pm \sqrt{25 - 24}}{2} = \frac{5 \pm \sqrt{1}}{2} = \frac{-5 \pm 1}{2}$$

$$x = \frac{-5 + 1}{2} \qquad \text{or} \quad x = \frac{-5 - 1}{2}$$

$$= \frac{-4}{2} = -2 \qquad\qquad = \frac{-6}{2} = -3$$

The solution set is $\{-3, -2\}$.

3. $x^2 + 5x + 3 = 0$

$a = 1, b = 5, c = 3$

$$x = \frac{-b \pm \sqrt{b^2 - 4ac}}{2a}$$

$$x = \frac{-5 \pm \sqrt{5^2 - 4(1)(3)}}{2 \cdot 1}$$

$$x = \frac{-5 \pm \sqrt{25 - 12}}{2}$$

$$= \frac{-5 \pm \sqrt{13}}{2}$$

The solution set is $\left\{\dfrac{-5 \pm \sqrt{13}}{2}\right\}$.

5. $x^2 + 4x - 6 = 0$

$a = 1, b = 4, c = -6$

$$x = \frac{-b \pm \sqrt{b^2 - 4ac}}{2a}$$

$$x = \frac{-4 \pm \sqrt{4^2 - 4(1)(-6)}}{2 \cdot 1}$$

$$= \frac{-4 \pm \sqrt{16 + 24}}{2}$$

$$= \frac{-4 \pm \sqrt{40}}{2}$$

$$= \frac{-4 \pm 2\sqrt{10}}{2}$$

$$= \frac{2\left(-2 \pm \sqrt{10}\right)}{2}$$

$$= -2 \pm \sqrt{10}$$

The solution set is $\left\{-2 \pm \sqrt{10}\right\}$.

7. $x^2 + 4x - 7 = 0$

$a = 1, b = 4, c = -7$

$$x = \frac{-b \pm \sqrt{b^2 - 4ac}}{2a}$$

$$= \frac{-4 \pm \sqrt{4^2 - 4(1)(-7)}}{2 \cdot 1}$$

$$= \frac{-4 \pm \sqrt{16 + 28}}{2}$$

$$= \frac{-4 \pm \sqrt{44}}{2}$$

$$= \frac{-4 \pm 2\sqrt{11}}{2}$$

$$= \frac{2\left(-2 \pm \sqrt{11}\right)}{2}$$

$$= -2 \pm \sqrt{11}$$

The solution set is $\left\{-2 \pm \sqrt{11}\right\}$.

9. $x^2 - 3x - 18 = 0$

$a = 1, b = -3, c = -18$

$x = \dfrac{-b \pm \sqrt{b^2 - 4ac}}{2a}$

$= \dfrac{-(-3) \pm \sqrt{(-3)^2 - 4(1)(-18)}}{2 \cdot 1}$

$= \dfrac{3 \pm \sqrt{9 + 72}}{2}$

$= \dfrac{3 \pm \sqrt{81}}{2}$

$= \dfrac{3 \pm 9}{2}$

$x = \dfrac{3 + 9}{2}$ or $x = \dfrac{3 - 9}{2}$

$x = \dfrac{12}{2} = 6$ or $x = \dfrac{-6}{2} = -3$

The solution set is $\{-3, 6\}$.

11. $6x^2 - 5x - 6 = 0$

$a = 6, b = -5, c = -6$

$x = \dfrac{-b \pm \sqrt{b^2 - 4ac}}{2a}$

$= \dfrac{-(-5) \pm \sqrt{(-5)^2 - 4(6)(-6)}}{2 \cdot 6}$

$= \dfrac{5 \pm \sqrt{25 + 144}}{12}$

$= \dfrac{5 \pm \sqrt{169}}{12} = \dfrac{5 \pm 13}{12}$

$x = \dfrac{5 + 13}{12}$ or $x = \dfrac{5 - 13}{12}$

$x = \dfrac{18}{12} = \dfrac{3}{2}$ or $x = \dfrac{-8}{12} = -\dfrac{2}{3}$

The solution set is $\left\{-\dfrac{2}{3}, \dfrac{3}{2}\right\}$.

13. $x^2 - 2x - 10 = 0$

$a = 1, b = -2, c = -10$

$x = \dfrac{-b \pm \sqrt{b^2 - 4ac}}{2a}$

$= \dfrac{-(-2) \pm \sqrt{(-2)^2 - 4(1)(-10)}}{2 \cdot 1}$

$= \dfrac{2 \pm \sqrt{4 + 40}}{2} = \dfrac{2 \pm \sqrt{44}}{2}$

$= \dfrac{2 \pm 2\sqrt{11}}{2}$

$= \dfrac{2(1 \pm \sqrt{11})}{2}$

$= 1 \pm \sqrt{11}$

The solution set is $\{1 \pm \sqrt{11}\}$.

15. $x^2 - x = 14$

Rewrite the equation in standard form.

Identify a, b, and c.

$a = 1, b = -1, c = -14$

Substitute these values into the quadratic formula.

$x = \dfrac{-b \pm \sqrt{b^2 - 4ac}}{2a}$

$= \dfrac{-(-1) \pm \sqrt{(-1)^2 - 4(1)(-14)}}{2 \cdot 1}$

$= \dfrac{1 \pm \sqrt{1 + 56}}{2} = \dfrac{1 \pm \sqrt{57}}{2}$

The radical cannot be simplified.

The solution set is $\left\{\dfrac{1 \pm \sqrt{57}}{2}\right\}$.

17. $6x^2 + 6x + 1 = 0$

$a = 6, b = 6, c = 1$

$x = \dfrac{-b \pm \sqrt{b^2 - 4ac}}{2a}$

$= \dfrac{-6 \pm \sqrt{6^2 - 4(6)(1)}}{2 \cdot 6}$

$= \dfrac{-6 \pm \sqrt{36 - 24}}{12} = \dfrac{-6 \pm \sqrt{12}}{12}$

$= \dfrac{-6 \pm 2\sqrt{3}}{12} = \dfrac{2\left(-3 \pm \sqrt{3}\right)}{12}$

$= \dfrac{-3 \pm \sqrt{3}}{6}$

The solution set is $\left\{ \dfrac{-3 \pm \sqrt{3}}{6} \right\}$.

19. $9x^2 - 12x + 4 = 0$

$a = 9, b = -12, c = 4$

$x = \dfrac{-b \pm \sqrt{b^2 - 4ac}}{2a}$

$= \dfrac{-(-12) \pm \sqrt{(-12)^2 - 4(9)(4)}}{2 \cdot 9}$

$= \dfrac{12 \pm \sqrt{144 - 144}}{18} = \dfrac{12 \pm \sqrt{0}}{18}$

$= \dfrac{12 \pm 0}{18} = \dfrac{12}{18}$

$= \dfrac{2}{3}$

The solution set is $\left\{ \dfrac{2}{3} \right\}$.

21. $4x^2 = 2x + 7$

Rewrite the equation in standard form.

$4x^2 - 2x - 7 = 0$

$a = 4, b = -2, c = -7$

$x = \dfrac{-b \pm \sqrt{b^2 - 4ac}}{2a}$

$= \dfrac{-(-2) \pm \sqrt{(-2)^2 - 4(4)(-7)}}{2 \cdot 4}$

$= \dfrac{2 \pm \sqrt{4 + 112}}{8} = \dfrac{2 \pm \sqrt{116}}{8}$

$= \dfrac{2 \pm 2\sqrt{29}}{8} = \dfrac{2\left(1 \pm \sqrt{29}\right)}{8}$

$= \dfrac{1 \pm \sqrt{29}}{4}$

The solution set is $\left\{ \dfrac{1 \pm \sqrt{29}}{4} \right\}$.

23. $2x^2 - x = 1$

Write the equation in standard form.

$2x^2 - x - 1 = 0$

Factor the trinomial.

$(2x + 1)(x - 1) = 0$

Use the zero-product property.

$2x + 1 = 0 \quad$ or $\quad x - 1 = 0$

$\quad 2x = -1 \qquad\qquad x = 1$

$\qquad x = -\dfrac{1}{2}$

The solution set is $\left\{ -\dfrac{1}{2}, 1 \right\}$.

25. $5x^2 + 2 = 11x$

Write the equation in standard form.

$5x^2 - 11x + 2 = 0$

Factor the trinomial.

$(5x - 1)(x - 2) = 0$

Use the zero-product property.

$5x - 1 = 0 \quad$ or $\quad x - 2 = 0$

$\quad 5x = 1 \qquad\qquad x = 2$

$\qquad x = \dfrac{1}{5}$

The solution set is $\left\{ \dfrac{1}{5}, 2 \right\}$.

27. $3x^2 = 60$

$x^2 = 20$

Use the square root property.

$x = \sqrt{20}$ or $x = -\sqrt{20}$

$x = 2\sqrt{5}$ $x = -2\sqrt{5}$

The solution set is $\left\{\pm 2\sqrt{5}\right\}$.

29. $x^2 - 2x = 1$

Write the equation in standard form.

$x^2 - 2x - 1 = 0$

The trinomial is prime, so we cannot factor and use the zero-product property. Instead, use the quadratic formula with $a = 1, b = -2, c = -1$

$x = \dfrac{-b \pm \sqrt{b^2 - 4ac}}{2a}$

$= \dfrac{-(-2) \pm \sqrt{(-2)^2 - 4(1)(-1)}}{2 \cdot 1}$

$= \dfrac{2 \pm \sqrt{4+4}}{2} = \dfrac{2 \pm \sqrt{8}}{2}$

$= \dfrac{2 \pm 2\sqrt{2}}{2} = \dfrac{2(1 \pm \sqrt{2})}{2}$

$= 1 \pm \sqrt{2}$

The solution set is $\left\{1 \pm \sqrt{2}\right\}$.

31. $(2x+3)(x+4) = 1$

Write the equation in standard form.

$2x^2 + 11x + 12 = 1$

$2x^2 + 11x + 11 = 0$

The trinomial cannot be factored, so use the quadratic formula with $a = 2, b = 11, c = 11$.

$x = \dfrac{-b \pm \sqrt{b^2 - 4ac}}{2a}$

$= \dfrac{-11 \pm \sqrt{11^2 - 4(2)(11)}}{2 \cdot 2}$

$= \dfrac{-11 \pm \sqrt{121 - 88}}{4}$

$= \dfrac{-11 \pm \sqrt{33}}{4}$

The solution set is $\left\{\dfrac{-11 \pm \sqrt{33}}{4}\right\}$.

33. $(3x-4)^2 = 16$

Use the square root property.

$3x - 4 = \sqrt{16}$ or $3x - 4 = -\sqrt{16}$

$3x - 4 = 4$ $3x - 4 = -4$

$3x = 8$ $3x = 0$

$x = \dfrac{8}{3}$ $x = 0$

The solution set is $\left\{0, \dfrac{8}{3}\right\}$.

35. $3x^2 - 12x + 12 = 0$

$3\left(x^2 - 4x + 4\right) = 0$

$3(x-2)^2 = 0$

$(x-2)^2 = 0$

$x - 2 = 0$

$x = 2$

The solution set is $\{2\}$.

37. $4x^2 - 16 = 0$

$4\left(x^2 - 4\right) = 0$

$4(x+2)(x-2) = 0$

$x + 2 = 0$ or $x - 2 = 0$

$x = -2$ $x = 2$

The solution set is $\{-2, 2\}$.

39. $x^2 + 9x = 0$

$x(x+9) = 0$

$x = 0$ or $x + 9 = 0$

$x = -9$

The solution set is $\{-9, 0\}$.

41. $\dfrac{3}{4}x^2 - \dfrac{5}{2}x - 2 = 0$

To clear fractions, multiply both sides by the LCD, 4.

$4\left(\dfrac{3}{4}x^2 - \dfrac{5}{2}x - 2\right) = 4 \cdot 0$

$3x^2 - 10x - 8 = 0$

$(3x+2)(x-4) = 0$

$3x+2=0$ or $x-4=0$

$3x=-2$ $x=4$

$x=-\dfrac{2}{3}$

The solution set is $\left\{-\dfrac{2}{3},4\right\}$.

43. $(3x-2)^2=10$

Use the square root property.

$3x-2=\sqrt{10}$ $3x-2=-\sqrt{10}$

$3x=2+\sqrt{10}$ or $3x=2-\sqrt{10}$

$x=\dfrac{2+\sqrt{10}}{3}$ $x=\dfrac{2-\sqrt{10}}{3}$

The solution set is $\left\{\dfrac{2\pm\sqrt{10}}{3}\right\}$.

45. $\dfrac{x^2}{x+7}-\dfrac{3}{x+7}=0$

Multiply both sides by the LCD, $x+7$.

$x^2-3=0$

$x^2=3$

$x=\pm\sqrt{3}$

The solution set is $\left\{\pm\sqrt{3}\right\}$.

47. $(x+2)^2+x(x+1)=4$

$x^2+4x+4+x^2+x=4$

$2x^2+5x=0$

$x(2x+5)=0$

$x=0$ or $2x+5=0$

$2x=-5$

$x=-\dfrac{5}{2}$

The solution set is $\left\{-\dfrac{5}{2},0\right\}$.

49. $2x^2-9x-3=9-9x$

$2x^2-12=0$

$x^2-6=0$

$x^2=6$

$x=\pm\sqrt{6}$

The solution set is $\left\{\pm\sqrt{6}\right\}$.

51. $\dfrac{1}{x}+\dfrac{1}{x+3}=\dfrac{1}{4}$

Multiply both sides by the LCD, $4x(x+3)$.

$$4x(x+3)\left(\dfrac{1}{x}+\dfrac{1}{x+3}\right)=4x(x+3)\dfrac{1}{4}$$

$4(x+3)+4x=x(x+3)$

$4x+12+4x=x^2+3x$

$0=x^2-5x-12$

Use the quadratic formula with $a=1$, $b=-5$, and $c=-12$.

$x=\dfrac{-(-5)\pm\sqrt{(-5)^2-4(1)(-12)}}{2(1)}$

$=\dfrac{5\pm\sqrt{25+48}}{2}$

$=\dfrac{5\pm\sqrt{73}}{2}$

The solution set is $\left\{\dfrac{5\pm\sqrt{73}}{2}\right\}$.

53. $h=-16t^2+60t+4$

$0=-16t^2+60t+4$

$16t^2-60t-4=0$

$4\left(4t^2-15t-1\right)=0$

$4t^2-15t-1=0$

Note: Dividing both sides by 4 is not necessary, but it results in smaller numbers to be substituted in the quadratic formula.

$a=4, b=-15, c=-1$

$t=\dfrac{-b\pm\sqrt{b^2-4ac}}{2a}$

$=\dfrac{-(-15)\pm\sqrt{(-15)^2-4(4)(-1)}}{2(4)}$

$=\dfrac{15\pm\sqrt{225+16}}{8}=\dfrac{15\pm\sqrt{241}}{8}$

$t=\dfrac{15+\sqrt{241}}{8}$ or $t=\dfrac{15-\sqrt{241}}{8}$

$t\approx 3.8$ or $t\approx-0.1$

Reject the negative solution to the quadratic equation because the time cannot be negative. It will take about 3.8 seconds for the football to hit the ground.

55.
$$h = -0.05x^2 + 27$$
$$22 = -0.05x^2 + 27$$
$$-5 = -0.05x^2$$
$$\frac{-5}{-0.05} = \frac{-0.05x^2}{-0.05}$$
$$100 = x^2$$
$$\sqrt{100} = x$$
$$10 = x$$

The arch is 22 feet high at a point that is 10 feet to the right of the center.

57. a. 2000 is 60 years after 1940. Substitute 60 for x.
$$N = -0.015x^2 + 1.15x + 41$$
$$N = -0.015(60)^2 + 1.15(60) + 41 = 56$$

According to the model, the average daily U.S. newspaper circulation was 56 million in 2000. The model's value matches the actual number displayed in the bar graph.

b.
$$N = -0.015x^2 + 1.15x + 41$$
$$37 = -0.015x^2 + 1.15x + 41$$
$$0 = -0.015x^2 + 1.15x + 4$$
$$a = -0.015, \ b = 1.15, \ c = 4$$
$$x = \frac{-b \pm \sqrt{b^2 - 4ac}}{2a}$$
$$x = \frac{-(1.15) \pm \sqrt{(1.15)^2 - 4(-0.015)(4)}}{2(-0.015)}$$
$$x \approx -3 \ \text{ or } \ x = 80$$

Reject the negative solution to the quadratic equation because the time cannot be negative. If trends continue, the average daily U.S. newspaper circulation will be 37 million 80 years after 1940, in 2020.

59. Let x = the width of the rectangle.
Then $x + 3 =$ the length.
$$A = l2$$
$$36 = (x + 3) \cdot x$$
$$36 = x^2 + 3x$$
$$0 = x^2 + 3x - 36$$
$$a = 1, b = 3, c = -36$$
$$x = \frac{-b \pm \sqrt{b^2 - 4ac}}{2a}$$
$$= \frac{-3 \pm \sqrt{(-3)^2 - 4(1)(-36)}}{2 \cdot 1}$$
$$= \frac{-3 \pm \sqrt{9 + 144}}{2} = \frac{-3 \pm \sqrt{153}}{2}$$
$$x = \frac{-3 + \sqrt{153}}{2} \ \text{ or } \ x = \frac{-3 - \sqrt{153}}{2}$$
$$x \approx 4.7 \qquad\qquad x \approx -7.7$$

Reject the negative solution of the quadratic equation because the width cannot be negative. If x = 4.7, then $x + 3$ = 7.7. Thus, to the nearest tenth of a meter, the width is 4.7 meters and the length is 7.7 meters.

61. Let x = the length of the shorter leg.
Then $x + 1$ = the length of the longer leg.
Use the Pythagorean Theorem.
$$x^2 + (x + 1)^2 = 4^2$$
$$x^2 + x^2 + 2x + 1 = 16$$
$$2x^2 + 2x - 15 = 0$$

Solve using the quadratic formula.
$$a = 2, b = 2, c = -15$$
$$x = \frac{-2 \pm \sqrt{2^2 - 4(2)(-15)}}{2 \cdot 2}$$
$$= \frac{2 \pm \sqrt{4 + 120}}{4}$$
$$= \frac{2 \pm \sqrt{124}}{4}$$
$$x = \frac{-2 + \sqrt{124}}{4} \ \text{ or } \ x = \frac{-2 - \sqrt{124}}{4}$$
$$x \approx 2.3 \qquad\qquad x \approx -3.3$$

The length of a leg cannot be negative, so reject −3.3. If x = 2.3, then $x + 1$ = 3.3. Thus, to the nearest tenth of a foot, the lengths of the legs are 2.3 feet and 3.3 feet.

63. a. $\dfrac{1}{\Phi-1}$

b.
$$\frac{\Phi}{1}=\frac{1}{\Phi-1}$$
$$(\Phi-1)\frac{\Phi}{1}=(\Phi-1)\frac{1}{\Phi-1}$$
$$\Phi^2-\Phi=1$$
$$\Phi^2-\Phi-1=0$$
$$\Phi=\frac{-b\pm\sqrt{b^2-4ac}}{2a}$$
$$\Phi=\frac{-(-1)\pm\sqrt{(-1)^2-4(1)(-1)}}{2(1)}$$
$$\Phi=\frac{1\pm\sqrt{1+4}}{2}$$
$$\Phi=\frac{1\pm\sqrt{5}}{2},\ \text{reject negative}$$
$$\Phi=\frac{1+\sqrt{5}}{2}$$

c. The golden ratio is $\dfrac{1+\sqrt{5}}{2}$ to 1.

65. – 67. Answers will vary.

69. does not make sense; Explanations will vary. Sample explanation: A quicker way to solve this equation is to use the square root property.
$$25x^2-49=0$$
$$25x^2=49$$
$$x^2=\frac{49}{25}$$
$$x=\pm\frac{7}{5}$$

71. does not make sense; Explanations will vary. Sample explanation: A linear model would be a less accurate model.

73. false; Changes to make the statement true will vary. A sample change is: The quadratic formula can be expressed as $x=\dfrac{-b}{2a}\pm\dfrac{\sqrt{b^2-4ac}}{2a}$.

75. true

77. If $b^2-4ac>0$, there are two real number solutions. If b^2-4ac is a perfect square, the solutions are rational; if it is not a perfect square, they are irrational.

If $b^2-4ac=0$, there is only one solution, which is rational.

If $b^2-4ac<0$, the solutions are not real numbers.

Therefore, by evaluating b^2-4ac, you can determine the kinds of solutions for any quadratic equation without actually solving it.

79. Let $x=$ the width of the border.

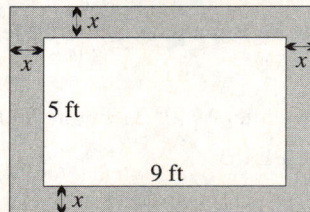

Area of large rectangle (garden plus border)
$$=(9+2x)(5+2x)$$
Area of small rectangle (garden) $=9\cdot5=45$
Area of border = Area of large rectangle - Area of small rectangle $=(9+2x)(5+2x)-45$
The area of the border is 40 square feet, so
$$(9+2x)(5+2x)-45=40.$$
Solve this equation.
$$(9+2x)(5+2x)-45=40.$$
$$45+28x+4x^2-45=40$$
$$4x^2+28x-40=0$$
$$4(x^2+7x-10)=0$$
$$x^2+7x-10=0$$
$$a=1,b=7,c=-10$$
$$x=\frac{-b\pm\sqrt{b^2-4ac}}{2a}$$
$$=\frac{-7\pm\sqrt{7^2-4(1)(-10)}}{2(1)}$$
$$=\frac{-7\pm\sqrt{49+40}}{2}$$
$$=\frac{-7\pm\sqrt{89}}{2}$$
$$x=\frac{-7+\sqrt{89}}{2}\quad\text{or}\quad x=\frac{-7-\sqrt{89}}{2}$$
$$x\approx1.2\qquad\qquad x\approx-8.2$$

Reject −8.2 because the width of the border must be positive. The width of the border is about 1.2 feet.

81. $y = -0.015x^2 + 1.15x + 41$

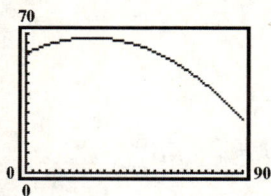

The graph verifies that the positive solution of the quadratic equation is approximately 80.

82. $125^{-\frac{2}{3}} = \dfrac{1}{125^{\frac{2}{3}}} = \dfrac{1}{\left(\sqrt[3]{125}\right)^2}$

$= \dfrac{1}{5^2} = \dfrac{1}{25}$

83. To rationalize the denominator, multiply numerator and denominator by the conjugate of the denominator.

$\dfrac{12}{3+\sqrt{5}} = \dfrac{12}{3+\sqrt{5}} \cdot \dfrac{3-\sqrt{5}}{3-\sqrt{5}}$

$= \dfrac{12\left(3-\sqrt{5}\right)}{3^2 - \left(\sqrt{5}\right)^2}$

$= \dfrac{12\left(3-\sqrt{5}\right)}{9-5}$

$= \dfrac{12\left(3-\sqrt{5}\right)}{4}$

$= 3\left(3-\sqrt{5}\right)$

$= 9 - 3\sqrt{5}$

84. $(x-y)\left(x^2 + xy + y^2\right)$

$= x\left(x^2 + xy + y^2\right) - y\left(x^2 + xy + y^2\right)$

$= x^3 + x^2 y + xy^2 - x^2 y - xy^2 - y^3$

$= x^3 - y^3$

85. No, squaring a real number cannot result in a negative number. Neither $\sqrt{-1}$ nor $\sqrt{-4}$ are real numbers.

86. a. $-\sqrt{4} = -2$, $\sqrt{1} = 1$

 b. $-\sqrt{5}$, $\sqrt{5}$

 c. $-\sqrt{5}$, $-\sqrt{4}$, $\sqrt{1}$, $\sqrt{5}$

 d. $\sqrt{-1}$, $\sqrt{-4}$

87. a. $-\sqrt{9} = -3$, $\sqrt{0} = 0$, $\sqrt{9} = 3$

 b. $-\sqrt{7}$, $\sqrt{7}$

 c. $-\sqrt{9}$, $-\sqrt{7}$, $\sqrt{0}$, $\sqrt{7}$, $\sqrt{9}$

 d. $\sqrt{-9}$, $\sqrt{-7}$

Mid-Chapter 9 Check Points

1. $(3x - 2)^2 = 100$

$3x - 2 = \pm\sqrt{100}$

$3x - 2 = \pm 10$

$3x = 2 \pm 10$

$x = \dfrac{2 \pm 10}{3}$

$x = \dfrac{2+10}{3} = 4$ or $x = \dfrac{2-10}{3} = -\dfrac{8}{3}$

The solution set is $\left\{-\dfrac{8}{3}, 4\right\}$.

2. $15x^2 = 5x$

$15x^2 - 5x = 0$

$5x(3x - 1) = 0$

$5x = 0$ or $3x - 1 = 0$

$x = 0$ $3x = 1$

$x = \dfrac{1}{3}$

The solution set is $\left\{0, \dfrac{1}{3}\right\}$.

3. $x^2 - 2x - 10 = 0$

$a = 1, b = -2, c = -10$

$x = \dfrac{-(-2) \pm \sqrt{(-2)^2 - 4(1)(-10)}}{2(1)}$

$= \dfrac{2 \pm \sqrt{4+40}}{2} = \dfrac{2 \pm \sqrt{44}}{2}$

$= \dfrac{2 \pm 2\sqrt{11}}{2} = 1 \pm \sqrt{11}$

The solution set is $\left\{1 \pm \sqrt{11}\right\}$.

4. $x^2 - 8x + 16 = 7$

$x^2 - 8x + 9 = 0$

$a = 1, b = -8, c = 9$

$x = \dfrac{-(-8) \pm \sqrt{(-8)^2 - 4(1)(9)}}{2(1)}$

$= \dfrac{8 \pm \sqrt{64 - 36}}{2} = \dfrac{8 \pm \sqrt{28}}{2}$

$= \dfrac{8 \pm 2\sqrt{7}}{2} = 4 \pm \sqrt{7}$

The solution set is $\left\{ 4 \pm \sqrt{7} \right\}$.

5. $3x^2 - x - 2 = 0$

$(3x + 2)(x - 1) = 0$

$3x + 2 = 0 \quad$ or $\quad x - 1 = 0$

$3x = -2 \qquad\qquad x = 1$

$x = -\dfrac{2}{3}$

The solution set is $\left\{ -\dfrac{2}{3}, 1 \right\}$.

6. $6x^2 = 10x - 3$

$6x^2 - 10x + 3 = 0$

$a = 6, b = -10, c = 3$

$x = \dfrac{-(-10) \pm \sqrt{(-10)^2 - 4(6)(3)}}{2(6)}$

$= \dfrac{10 \pm \sqrt{100 - 72}}{12} = \dfrac{10 \pm \sqrt{28}}{12}$

$= \dfrac{10 \pm 2\sqrt{7}}{12} = \dfrac{5 \pm \sqrt{7}}{6}$

The solution set is $\left\{ \dfrac{5 \pm \sqrt{7}}{6} \right\}$.

7. $x^2 + (x + 1)^2 = 25$

$x^2 + x^2 + 2x + 1 = 25$

$2x^2 + 2x - 24 = 0$

$x^2 + x - 12 = 0$

$(x + 4)(x - 3) = 0$

$x + 4 = 0 \quad$ or $\quad x - 3 = 0$

$x = -4 \qquad\qquad x = 3$

The solution set is $\{-4, 3\}$.

8. $(x + 5)^2 = 40$

$x + 5 = \pm\sqrt{40}$

$x + 5 = \pm 2\sqrt{10}$

$x = -5 \pm 2\sqrt{10}$

The solution set is $\left\{ -5 \pm 2\sqrt{10} \right\}$.

9. $2(x^2 - 8) = 11 - x^2$

$2x^2 - 16 = 11 - x^2$

$3x^2 = 27$

$x^2 = 9$

$x = \pm\sqrt{9}$

$x = \pm 3$

The solution set is $\{\pm 3\}$.

10. $2x^2 + 5x + 1 = 0$

$a = 2 \, , \; b = 5 \, , \; c = 1$

$x = \dfrac{-5 \pm \sqrt{5^2 - 4(2)(1)}}{2(2)}$

$= \dfrac{-5 \pm \sqrt{25 - 8}}{4} = \dfrac{-5 \pm \sqrt{17}}{4}$

The solution set is $\left\{ \dfrac{-5 \pm \sqrt{17}}{4} \right\}$.

11. $(x - 8)(2x - 3) = 34$

$2x^2 - 16x - 3x + 24 = 34$

$2x^2 - 19x - 10 = 0$

$(2x + 1)(x - 10) = 0$

$2x + 1 = 0 \quad$ or $\quad x - 10 = 0$

$2x = -1 \qquad\qquad x = 10$

$x = -\dfrac{1}{2}$

The solution set is $\left\{ -\dfrac{1}{2}, 10 \right\}$.

12. $x + \dfrac{16}{x} = 8$

Multiply both sides of the equation by x.

$$x\left(x + \dfrac{16}{x}\right) = x(8)$$

$$x^2 + 16 = 8x$$

$$x^2 - 8x + 16 = 0$$

$$(x-4)^2 = 0$$

$$x = 4$$

The solution set is $\{4\}$.

13. $x^2 + 14x - 32 = 0$

$$x^2 + 14x = 32$$

The coefficient on x is 14. Divide this by 2 and square the result.

$$\left(\dfrac{14}{2}\right)^2 = 7^2 = 49$$

Add 49 to both sides.

$$x^2 + 14x + 49 = 32 + 49$$

$$(x+7)^2 = 81$$

$$x + 7 = \pm\sqrt{81}$$

$$x + 7 = \pm 9$$

$$x = -7 \pm 9$$

$$x = -7 + 9 = 2 \quad \text{or} \quad x = -7 - 9 = -16$$

The solution set is $\{-16, 2\}$.

14. Let a = the length of the missing side.

$$a^2 + 6^2 = 8^2$$

$$a^2 + 36 = 64$$

$$a^2 = 28$$

$$a = \sqrt{28} = 2\sqrt{7}$$

The missing length is $2\sqrt{7}$ cm.

15. $d = \sqrt{(x_2 - x_1)^2 + (y_2 - y_1)^2}$

$$= \sqrt{(9 - (-3))^2 + (-3 - 2)^2}$$

$$= \sqrt{(12)^2 + (-5)^2} = \sqrt{144 + 25}$$

$$= \sqrt{169} = 13$$

The distance is 13 units.

16. $(4x)^2 + (3x)^2 = 20^2$

$$16x^2 + 9x^2 = 400$$

$$25x^2 = 400$$

$$x^2 = 16$$

$$x = \sqrt{16}$$

$$x = 4$$

$$3x = 12$$

$$4x = 16$$

The lengths of the legs are 12 inches and 16 inches.

9.4 Check Points

1. a. $\sqrt{-16} = \sqrt{16(-1)} = \sqrt{16}\sqrt{-1} = 4i$

b. $\sqrt{-5} = \sqrt{5(-1)} = \sqrt{5}\sqrt{-1} = i\sqrt{5}$

c. $\sqrt{-50} = \sqrt{50(-1)} = \sqrt{25 \cdot 2}\sqrt{-1} = 5i\sqrt{2}$

2. $(x+2)^2 = -25$

$$x + 2 = \sqrt{-25} \quad \text{or} \quad x + 2 = -\sqrt{-25}$$

$$x + 2 = 5i \qquad\qquad x + 2 = -5i$$

$$x = -2 + 5i \qquad\qquad x = -2 - 5i$$

The solution set is $\{-2 \pm 5i\}$.

3. $x^2 + 6x + 13 = 0$

$a = 1, \ b = 6, \ c = 13$

$$x = \dfrac{-b \pm \sqrt{b^2 - 4ac}}{2a}$$

$$= \dfrac{-6 \pm \sqrt{6^2 - 4(1)(13)}}{2 \cdot 1}$$

$$= \dfrac{-6 \pm \sqrt{36 - 52}}{2} = \dfrac{-6 \pm \sqrt{-16}}{2}$$

$$= \dfrac{-6 \pm 4i}{2} = \dfrac{2(-3 \pm 2i)}{2}$$

$$= -3 \pm 2i$$

The solution set is $\{-3 \pm 2i\}$.

9.4 Concept and Vocabulary Check

1. $\sqrt{-1}$; -1

2. $5i$

3. complex; imaginary; real

4. $2\pm i$

9.4 Exercise Set

1. $\sqrt{-36}=\sqrt{36(-1)}=\sqrt{36}\sqrt{-1}=6i$

3. $\sqrt{-13}=\sqrt{13(-1)}=\sqrt{13}\sqrt{-1}=i\sqrt{13}$

5. $\sqrt{-50}=\sqrt{50(-1)}=\sqrt{25\cdot2}\sqrt{-1}=5i\sqrt{2}$

7. $\sqrt{-20}=\sqrt{20(-1)}=\sqrt{4\cdot5}\sqrt{-1}=2i\sqrt{5}$

9. $-\sqrt{-28}=-\sqrt{28(-1)}$
   $=-\sqrt{4\cdot7}\sqrt{-1}$
   $=-2i\sqrt{7}$

11. $7+\sqrt{-16}=7+\sqrt{16}\sqrt{-1}=7+4i$

13. $10+\sqrt{-3}=10+\sqrt{3}\sqrt{-1}=10+i\sqrt{3}$

15. $6-\sqrt{-98}=6-\sqrt{98}\sqrt{-1}$
    $=6-\sqrt{49\cdot2}\sqrt{-1}$
    $=6-7i\sqrt{2}$

17. $(x-3)^2=-9$
    $x-3=\sqrt{-9}$ or $x-3=-\sqrt{-9}$
    $x-3=3i$ $\qquad x-3=-3i$
    $x=3+3i$ $\qquad x=3-3i$
    The solution set is $\{3\pm3i\}$.

19. $(x+7)^2=-64$
    $x+7=\sqrt{-64}$ or $x+7=-\sqrt{-64}$
    $x+7=8i$ $\qquad x+7=-8i$
    $x=-7+8i$ $\qquad x=-7-8i$
    The solution set is $\{-7\pm8i\}$.

21. $(x-2)^2=-7$
    $x-2=\sqrt{-7}$ or $x-2=-\sqrt{-7}$
    $x-2=i\sqrt{7}$ $\qquad x-2=-i\sqrt{7}$
    $x=2+i\sqrt{7}$ $\qquad x=2-i\sqrt{7}$
    The solution set is $\{2\pm i\sqrt{7}\}$.

23. $(y+3)^2=-18$
    $x+3=\sqrt{-18}$ or $x+3=-\sqrt{-18}$
    $x+3=3i\sqrt{2}$ $\qquad x+3=-3i\sqrt{2}$
    $x=-3+3i\sqrt{2}$ $\qquad x=-3-3i\sqrt{2}$
    The solution set is $\{-3\pm3i\sqrt{2}\}$.

25. $x^2+4x+5=0$
    $a=1,b=4,c=5$
    $x=\dfrac{-b\pm\sqrt{b^2-4ac}}{2a}$
    $=\dfrac{-4\pm\sqrt{4^2-4(1)(5)}}{2\cdot1}$
    $=\dfrac{-4\pm\sqrt{16-20}}{2}=\dfrac{-4\pm\sqrt{-4}}{2}$
    $=\dfrac{-4\pm2i}{2}=\dfrac{2(-2\pm i)}{2}$
    $=-2\pm i$
    The solution set is $\{-2\pm i\}$.

27. $x^2-6x+13=0$
    $a=1,b=-6,c=13$
    $x=\dfrac{-b\pm\sqrt{b^2-4ac}}{2a}$
    $=\dfrac{-(-6)\pm\sqrt{(-6)^2-4(1)(13)}}{2\cdot1}$
    $=\dfrac{6\pm\sqrt{36-52}}{2}=\dfrac{6\pm\sqrt{-16}}{2}$
    $=\dfrac{6\pm4i}{2}=\dfrac{2(3\pm2i)}{2}$
    $=3\pm2i$
    The solution set is $\{3\pm2i\}$.

29. $x^2 - 12x + 40 = 0$

$a = 1, b = -12, c = 40$

$x = \dfrac{-b \pm \sqrt{b^2 - 4ac}}{2a}$

$= \dfrac{-(-12) \pm \sqrt{(-12)^2 - 4(1)(40)}}{2 \cdot 1}$

$= \dfrac{12 \pm \sqrt{144 - 160}}{2} = \dfrac{12 \pm \sqrt{-16}}{2}$

$= \dfrac{12 \pm 4i}{2} = \dfrac{2(6 \pm 2i)}{2}$

$= 6 \pm 2i$

The solution set is $\{6 \pm 2i\}$.

31. $x^2 = 10x - 27$

Write the equation in standard form; then identify a, b, and c.

$x^2 - 10x + 27 = 0$

$a = 1, b = -10, c = 27$

$x = \dfrac{-b \pm \sqrt{b^2 - 4ac}}{2a}$

$= \dfrac{-(-10) \pm \sqrt{(-10)^2 - 4(1)(27)}}{2 \cdot 1}$

$= \dfrac{10 \pm \sqrt{100 - 108}}{2} = \dfrac{10 \pm \sqrt{-8}}{2}$

$= \dfrac{10 \pm 2i\sqrt{2}}{2} = \dfrac{2(5 \pm i\sqrt{2})}{2}$

$= 5 \pm i\sqrt{2}$

The solution set is $\left\{5 \pm i\sqrt{2}\right\}$.

33. $5x^2 = 2x - 3$

Write the equation in standard form.

$5x^2 - 2x + 3 = 0$

$a = 5, b = -2, c = 3$

$x = \dfrac{-b \pm \sqrt{b^2 - 4ac}}{2a}$

$= \dfrac{-(-2) \pm \sqrt{(-2)^2 - 4(5)(3)}}{2 \cdot 5}$

$= \dfrac{2 \pm \sqrt{4 - 60}}{10} = \dfrac{2 \pm \sqrt{-56}}{10}$

$= \dfrac{2 \pm 2i\sqrt{14}}{10} = \dfrac{2(1 \pm i\sqrt{14})}{10}$

$= \dfrac{1 \pm i\sqrt{14}}{5}$

The solution set is $\left\{\dfrac{1 \pm i\sqrt{14}}{5}\right\}$.

35. $2y^2 = 4y - 5$

Write the equation in standard form.

$2y^2 - 4y + 5 = 0$

$a = 2, b = -4, c = 5$

$y = \dfrac{-b \pm \sqrt{b^2 - 4ac}}{2a}$

$= \dfrac{-(-4) \pm \sqrt{(-4)^2 - 4(2)(5)}}{2 \cdot 2}$

$= \dfrac{4 \pm \sqrt{16 - 40}}{4} = \dfrac{4 \pm \sqrt{-24}}{4}$

$= \dfrac{4 \pm 2i\sqrt{6}}{4} = \dfrac{2(2 \pm i\sqrt{6})}{2 \cdot 2}$

$= \dfrac{2 \pm i\sqrt{6}}{2}$

The solution set is $\left\{\dfrac{2 \pm i\sqrt{6}}{2}\right\}$.

37. $12x^2 + 35 = 8x^2 + 15$

$4x^2 + 35 = 15$

$4x^2 = -20$

$x^2 = -5$

$x = \pm\sqrt{-5} = \pm i\sqrt{5}$

The solution set is $\left\{\pm i\sqrt{5}\right\}$.

39.

$$\frac{x+3}{5} = \frac{x-2}{x}$$

$$5x\left(\frac{x+3}{5}\right) = 5x\left(\frac{x-2}{x}\right)$$

$$x(x+3) = 5(x-2)$$

$$x^2 + 3x = 5x - 10$$

$$x^2 - 2x + 10 = 0$$
$$a = 1, b = -2, c = 10$$

$$x = \frac{-(-2) \pm \sqrt{(-2)^2 - 4(1)(10)}}{2(1)}$$

$$= \frac{2 \pm \sqrt{4-40}}{2} = \frac{2 \pm \sqrt{-36}}{2}$$

$$= \frac{2 \pm 6i}{2} = 1 \pm 3i$$

The solution set is $\{1 \pm 3i\}$.

41. $\dfrac{1}{x+1} - \dfrac{1}{2} = \dfrac{1}{x}$

Multiply both sides by the LCD, $2x(x+1)$.

$$2x(x+1)\left(\frac{1}{x+1} - \frac{1}{2}\right) = 2x(x+1)\left(\frac{1}{x}\right)$$

$$2x - x(x+1) = 2(x+1)$$

$$2x - x^2 - x = 2x + 2$$

$$-x^2 - x - 2 = 0$$

$$x^2 + x + 2 = 0$$
$$a = 1, b = 1, c = 2$$

$$x = \frac{-1 \pm \sqrt{(1)^2 - 4(1)(2)}}{2(1)}$$

$$= \frac{-1 \pm \sqrt{1-8}}{2} = \frac{-1 \pm \sqrt{-7}}{2}$$

$$= \frac{-1 \pm i\sqrt{7}}{2}$$

The solution set is $\left\{\dfrac{-1 \pm i\sqrt{7}}{2}\right\}$.

43.

$$R = -2x^2 + 36x$$

$$200 = -2x^2 + 36x$$

$$2x^2 + 36x + 200 = 0$$

$$x^2 + 18x + 100 = 0$$
$$a = 1, b = 18, c = 100$$

$$x = \frac{-b \pm \sqrt{b^2 - 4ac}}{2a}$$

$$= \frac{-18 \pm \sqrt{18^2 - 4(1)(100)}}{2 \cdot 1}$$

$$= \frac{-18 \pm \sqrt{324 - 400}}{2}$$

$$= \frac{-18 \pm \sqrt{-76}}{2}$$

$$= \frac{-18 \pm 2i\sqrt{19}}{2}$$

$$= \frac{2(-9 \pm i\sqrt{19})}{2}$$

$$= 9 \pm i\sqrt{19}$$

The solutions of the equation are $9 \pm i\sqrt{19}$, which are not real numbers. Because the equation has no real number solution, it is not possible to generate $200,000 in weekly revenue. The job applicant has guaranteed to do something that is impossible, so this person will not be hired.

45. – 49. Answers will vary.

51. does not make sense; Explanations will vary. Sample explanation: Imaginary numbers are not *undefined*.

53. does not make sense; Explanations will vary. Sample explanation: It is not a variable, $i = \sqrt{-1}$.

55. false; Changes to make the statement true will vary. A sample change is: $-\sqrt{-9} = -(3i) = -3i$.

57. false; Changes to make the statement true will vary. A sample change is: $2 + \sqrt{-4} = 2 + 2i$.

59. $x^2 - 2x + 2 = 0$
To show that $1+i$ is a solution of this equation substitute $1+i$ for x.

$$x^2 - 2x + 2 = 0$$

$$(1+i)^2 - 2(1+i) + 2 = 0$$

$$1 + 2i + i^2 - 2 - 2i + 2 = 0$$

$$1 + 2i + (-1) - 2 - 2i + 2 = 0$$

$$(1 - 1 - 2 + 2) + (2i - 2i) = 0$$

$$0 + 0 = 0$$

$$0 = 0, \text{ true}$$

To show that $1-i$ is the other solution, substitute $1-i$ for x.

$$x^2 - 2x + 2 = 0$$
$$(1-i)^2 - 2(1-i) + 2 = 0$$
$$1 - 2i + i^2 - 2 + 2i + 2 = 0$$
$$1 - 2i + (-1) - 2 + 2i + 2 = 0$$
$$(1 - 1 - 2 + 2) + (-2i + 2i) = 0$$
$$0 + 0 = 0$$
$$0 = 0, \text{ true}$$

61. The graph shows that the parabola and the horizontal line do not intersect. The vertex, which is the maximum point on the graph, has a y-coordinate that is less than 80. Therefore, the ball will never reach a height of 80 feet.

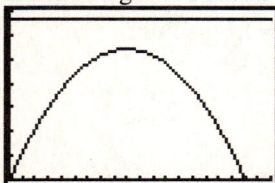

63. $y = \dfrac{1}{3}x - 2$

slope $= \dfrac{1}{3}$; y-intercept $= -2$

Plot $(0, -2)$. From this point, go 1 unit *up* and 3 units to the *right* to reach the point $(3, -1)$

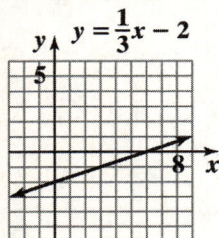

64. $2x - 3y = 6$

x-intercept: 3
y-intercept: -2
checkpoint: $(-3, -4)$
Draw a line through $(3,0)$, $(0, -2)$, and $(-3, -4)$.

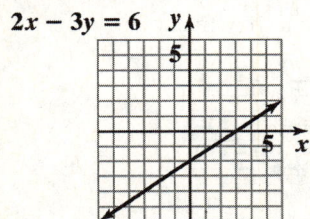

65. $x = -2$
Draw a vertical line with x-intercept -2.

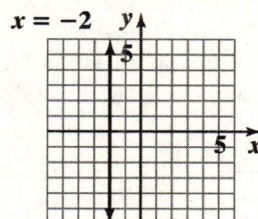

66. $y = x^2 + 4x + 3$

x	$x^2 + 4x + 3$	(x, y)
-5	$(-5)^2 + 4(-5) + 3 = 8$	$(-5, 8)$
-4	$(-4)^2 + 4(-4) + 3 = 3$	$(-4, 3)$
-3	$(-3)^2 + 4(-3) + 3 = 0$	$(-3, 0)$
-2	$(-2)^2 + 4(-2) + 3 = -1$	$(-2, -1)$
-1	$(-1)^2 + 4(-1) + 3 = 0$	$(-1, 0)$
0	$(0)^2 + 4(0) + 3 = 3$	$(0, 3)$
1	$(1)^2 + 4(1) + 3 = 8$	$(1, 8)$

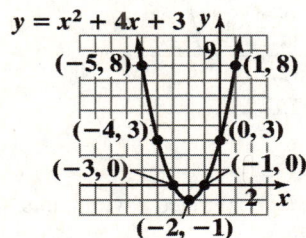

67. $y = x^2 - 2x - 3$

$$0 = x^2 - 2x - 3$$
$$0 = (x + 1)(x - 3)$$
$$x + 1 = 0 \quad \text{or} \quad x - 3 = 0$$
$$x = -1 \qquad\qquad x = 3$$

The x-intercepts are -1 and 3.

68. $y = (0)^2 - 2(0) - 3 = -3$
The y-intercept is -3.

9.5 Check Points

1. a. $y = x^2 - 6x + 8$

Because a, the coefficient of x^2, is 1, which is greater than 0, the parabola opens upward.

b. Make a table.

x	$y = x^2 - 6x + 8$	(x, y)
0	$y = (0)^2 - 6(0) + 8 = 8$	$(0, 8)$
1	$y = (1)^2 - 6(1) + 8 = 3$	$(1, 3)$
2	$y = (2)^2 - 6(2) + 8 = 0$	$(2, 0)$
3	$y = (3)^2 - 6(3) + 8 = -1$	$(3, -1)$
4	$y = (4)^2 - 6(4) + 8 = 0$	$(4, 0)$
5	$y = (5)^2 - 6(5) + 8 = 3$	$(5, 3)$
6	$y = (6)^2 - 6(6) + 8 = 8$	$(6, 8)$

Plot the points and connect them with a smooth curve.

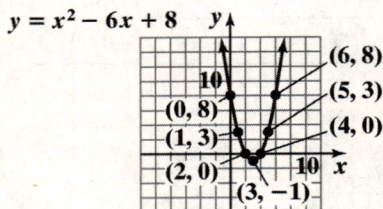

2. $y = x^2 - 6x + 8$

To find the x-intercepts, replace y with 0 and solve the resulting equation.

$0 = x^2 - 6x + 8$

$0 = (x - 4)(x - 2)$

$x - 4 = 0 \quad \text{or} \quad x - 2 = 0$

$\qquad x = 4 \qquad\qquad x = 2$

The x-intercepts are 2 and 4.

3. $y = x^2 - 6x + 8$

To find the y-intercept, replace x with 0.

$y = 0^2 - 6(0) + 8 = 8$

The y-intercept is 8.

4. $y = x^2 - 6x + 8$

$a = 1, b = -6, c = 8$

x-coordinate of vertex

$x = \dfrac{-b}{2a} = \dfrac{-(-6)}{2(1)} = \dfrac{6}{2} = 3$

y-coordinate of vertex

$y = 3^2 - 6(3) + 8 = 9 - 18 + 8 = -1$

The vertex is $(3, -1)$.

5. $y = x^2 + 6x + 5$

Step 1 Determine how the parabola opens.

Here a, the coefficient of x^2, is 1.

Because $a > 0$, the parabola opens upward.

Step 2 Find the vertex.

For this equation, $a = 1, b = 6$, and $c = 5$.

x-coordinate of vertex

$= \dfrac{-b}{2a} = \dfrac{-6}{2(1)} = \dfrac{-6}{2} = -3$

y-coordinate of vertex

$= (-3)^2 + 6(-3) + 5 = 9 - 18 + 5 = -4$

The vertex is $(-3, -4)$.

Step 3 Find the x-intercepts.

Replace y with 0 in $y = x^2 + 6x + 5$ and solve for x.

$x^2 + 6x + 5 = 0$

$(x + 5)(x + 1) = 0$

$x + 5 = 0 \quad \text{or} \quad x + 1 = 0$

$\quad x = -5 \qquad\qquad x = -1$

The x-intercepts are -5 and -1.

Step 4 Find the y-intercept.

Replace x with 0 and solve for y.

$y = 0^2 + 6(0) + 5 = 5$

The y-intercept is 5.

Steps 5 and 6 Plot the intercepts and the vertex.

Plot $(-3, -4), (-5, 0), (-1, 0)$, and $(0, 5)$, and connect them with a smooth curve.

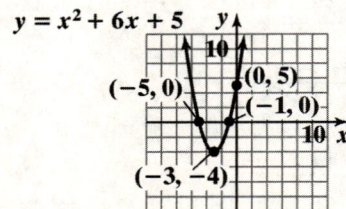

6. $y = -x^2 - 2x + 5$

Step 1 Determine how the parabola opens.
$a = -1 < 0$, so the parabola opens downward.

Step 2 Find the vertex.
$a = -1$, $b = -2$, $c = 5$
x-coordinate of vertex
$$x = \frac{-b}{2a} = \frac{-(-2)}{2(-1)} = \frac{2}{-2} = -1$$
y-coordinate of vertex
$$y = -(-1)^2 - 2(-1) + 5 = -1 + 2 + 5 = 6$$
The vertex is $(-1, 6)$.

Step 3 Find the x-intercepts.
$$y = -x^2 - 2x + 5$$
$$0 = -x^2 - 2x + 5$$
$a = -1$, $b = -2$, $c = 5$
$$x = \frac{-b \pm \sqrt{b^2 - 4ac}}{2a}$$
$$x = \frac{-(-2) \pm \sqrt{(-2)^2 - 4(-1)(5)}}{2(-1)}$$
$$x = \frac{2 \pm \sqrt{4 + 20}}{-2}$$
$$x = \frac{2 \pm \sqrt{24}}{-2}$$
$$x = \frac{2 \pm 2\sqrt{6}}{-2}$$
$$x = -1 \pm \sqrt{6}$$
$$x \approx -3.4 \ \text{ or } \ x \approx 1.4$$
The x-intercepts are approximately -3.4 and 1.4.
Step 4 Find the y-intercept.
$$y = -x^2 - 2x + 5$$
$$y = -0^2 - 2(0) + 5 = 5$$
The y-intercept is 5.
Steps 5 and 6
Plot $(-1, 6)$, $(-3.4, 0)$, $(1.4, 0)$, and $(0, 5)$, and connect them with a smooth curve.

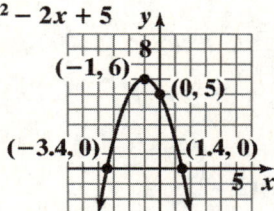

7. $y = -0.005x^2 + 2x + 5$

a. The information needed is found at the vertex.
x-coordinate of vertex
$$x = \frac{-b}{2a} = \frac{-2}{2(-0.005)} = 200$$
y-coordinate of vertex
$$y = -0.005(200)^2 + 2(200) + 5 = 205$$
The vertex is $(200, 205)$.
The maximum height of the arrow is 205 feet.
This occurs 200 feet from its release.

b. The arrow will hit the ground when the height reaches 0.
$$y = -0.005x^2 + 2x + 5$$
$$0 = -0.005x^2 + 2x + 5$$
$$x = \frac{-b \pm \sqrt{b^2 - 4ac}}{2a}$$
$$x = \frac{-2 \pm \sqrt{2^2 - 4(-0.005)(5)}}{2(-0.005)}$$
$$x \approx -2 \ \text{ or } \ x \approx 402$$
The arrow travels 402 feet before hitting the ground.

c. The starting point occurs when $x = 0$. Find the corresponding y-coordinate.
$$y = -0.005(0)^2 + 2(0) + 5 = 5$$
Plot $(0, 5)$, $(402, 0)$, and $(200, 205)$, and connect them with a smooth curve.

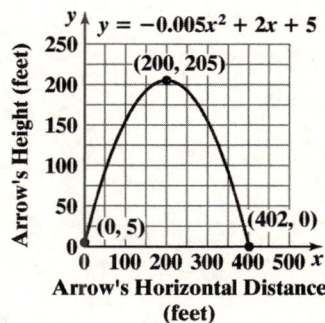

9.5 Concept and Vocabulary Check

1. parabola

2. $a > 0$; $a < 0$

3. $ax^2 + bx + c = 0$

4. 0

5. $\dfrac{-b}{2a}$; $\dfrac{-b}{2a}$ or the x-coordinate

9.5 Exercise Set

1. $y = x^2 - 4x + 3$

Because a, the coefficient of x^2, is 1, which is greater than 0, the parabola opens upward.

3. $y = -2x^2 + x + 6$

Because $a = -2$, which is less than 0, the parabola opens downward.

5. $y = x^2 - 4x + 3$

To find the x-intercepts, replace y with 0 and solve the resulting equation.

$x^2 - 4x + 3 = 0$

$(x-1)(x-3) = 0$

$x - 1 = 0$ or $x - 3 = 0$

$x = 1$ $x = 3$

The x-intercepts are 1 and 3.

7. $y = -x^2 + 8x - 12$

To find the x-intercepts, replace y with 0 and solve the resulting equation.

$0 = -x^2 + 8x - 12$

$x^2 - 8x + 12 = 0$

$(x-2)(x-6) = 0$

$x - 2 = 0$ or $x - 6 = 0$

$x = 2$ $x = 6$

The x-intercepts are 2 and 6.

9. $y = x^2 + 2x - 4$

To find the x-intercepts, replace y with 0 and solve the resulting equation.

$0 = x^2 + 2x - 4$

This equation cannot be solved by factoring, so use the quadratic formula with $a = 1, b = 2,$ and $c = -4$.

$x = \dfrac{-b \pm \sqrt{b^2 - 4ac}}{2a}$

$= \dfrac{-2 \pm \sqrt{2^2 - 4(1)(-4)}}{2(1)}$

$= \dfrac{-2 \pm \sqrt{4+16}}{2} = \dfrac{-2 \pm \sqrt{20}}{2}$

$= \dfrac{-2 \pm 2\sqrt{5}}{2} = -1 \pm \sqrt{5}$

$x = -1\sqrt{5} \approx -3.2$ or $x = -1 - \sqrt{5} \approx 1.2$

The x-intercepts are approximately -3.2 and 1.2.

11. $y = x^2 - 4x + 3$

To find the y-intercept, replace x with 0.

$y = 0^2 - 4 \cdot 0 + 3 = 0 - 0 + 3 = 3$

The y-intercept is 3.

13. $y = -x^2 + 8x - 12$

$y = -0^2 + 8 \cdot 0 - 12 = 0 + 0 - 12 = -12$

The y-intercept is -12.

15. $y = x^2 + 2x - 4$

$y = 0^2 + 2 \cdot 0 - 4 = -4$

The y-intercept is -4.

17. $y = x^2 + 6x$

$y = 0^2 + 6 \cdot 0 = 0$

The y-intercept is 0.

19. $y = x^2 - 4x + 3$

$a = 1, b = -4, c = 3$

x-coordinate of vertex

$= \dfrac{-b}{2a} = \dfrac{-(-4)}{2(1)} = \dfrac{4}{2} = 2$

y-coordinate of vertex

$= 2^2 - 4(2) + 3 = 4 - 8 + 3 = -1$

The vertex is $(2, -1)$.

21. $y = 2x^2 + 4x - 6$

$a = 2, b = 4, c = -6$

x-coordinate of vertex

$= \dfrac{-b}{2a} = \dfrac{-4}{2(2)} = \dfrac{-4}{4} = -1$

y-coordinate of vertex

$= 2(-1)^2 + 4(-1) - 6 = 2 - 4 - 6 = -8$

The vertex is $(-1, -8)$.

23. $y = x^2 + 6x$

$a = 1, b = 6, c = 0$

x-coordinate of vertex

$= \dfrac{-b}{2a} = \dfrac{-6}{2(1)} = \dfrac{-6}{2} = -3$

y-coordinate of vertex

$= (-3)^2 + 6(-3) = 9 - 18 = -9$

The vertex is $(-3, -9)$.

25. $y = x^2 + 8x + 7$

Step 1 Determine how the parabola opens.

Here a, the coefficient of x^2, is 1.
Because $a > 0$, the parabola opens upward.

Step 2 Find the vertex.

For this equation, $a = 1, b = 8$, and $c = 7$.

x-coordinate of vertex

$= \dfrac{-b}{2a} = \dfrac{-8}{2(1)} = \dfrac{-8}{2} = -4$

y-coordinate of vertex

$= (-4)^2 + 8(-4) + 7 = 16 - 32 + 7 = -9$

The vertex is $(-4, -9)$.

Step 3 Find the *x*-intercepts.

Replace y with 0 in $y = x^2 + 8x + 7$ and solve for x.

$x^2 + 8x + 7 = 0$

$(x + 7)(x + 1) = 0$

$x + 7 = 0$ or $x + 1 = 0$

$\quad x = -7 \qquad\quad x = -1$

The *x*-intercepts are −7 and −1.

Step 4 Find the *y*-intercept.

Replace x with 0 and solve for y.

$y = 0^2 + 8(0) + 7 = 0 + 0 + 7 = 7$

The *y*-intercept is 7.

Steps 5 and 6 Plot the intercepts and the vertex.
Connect these points with a smooth curve.

Plot $(-4, -9), (-7, 0), (-1, 0)$, and $(0, 7)$, and connect them with a smooth curve.

$y = x^2 + 8x + 7$

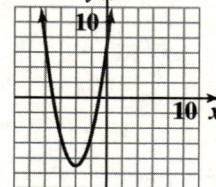

27. $y = x^2 - 2x - 8$

Step 1 $a = 1 > 0$, so the parabola opens upward.

Step 2 $a = 1, b = -2, c = -8$

x-coordinate of vertex

$= \dfrac{-b}{2a} = \dfrac{-2(-2)}{2(1)} = \dfrac{2}{2} = 1$

y-coordinate of vertex

$= 1^2 - 2(1) - 8 = 1 - 2 - 8 = -9$

The vertex is $(1, -9)$

Step 3

$x^2 - 2x - 8 = 0$

$(x + 2)(x - 4) = 0$

$x + 2 = 0$ or $x - 4 = 0$

$\quad x = -2 \qquad\quad x = 4$

The *x*-intercepts are −2 and 4.

Step 4 $y = 0^2 - 2(0) - 8 = 8$

The *y*-intercept is −8.

Steps 5 and 6 Plot

$(1, -9), (-2, 0), (4, 0)$, and $(0, -8)$, and connect them with a smooth curve.

$y = x^2 - 2x - 8$

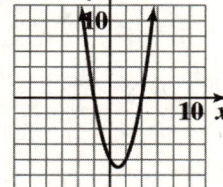

29. $y = -x^2 + 4x - 3$

Step 1 $a = -1 < 0$, so the parabola opens downward.

Step 2 $a = -1, b = 4, c = -3$

x-coordinate of vertex

$$= \frac{-b}{2a} = \frac{-4}{2(-1)} = \frac{-4}{-2} = 2$$

y-coordinate of vertex

$$= -2^2 + 4 \cdot 2 - -4 + 8 - 3 = 1$$

The vertex is $(2,1)$.

Step 3

$$-x^2 + 4x - 3 = 0$$
$$0 = x^2 - 4x + 3$$
$$0 = (x-1)(x-3)$$
$$x - 1 = 0 \quad \text{or} \quad x - 3 = 0$$
$$x = 1 \qquad\qquad x = 3$$

The x-intercepts are 1 and 3.

Step 4 $y = -0^2 + 4(0) - 3 = -3$

The y-intercept is -3.

Steps 5 and 6 Plot $(2,1), (1,0), (3,0)$, and $(0,-3)$, and connect them with a smooth curve.

$y = -x^2 + 4x - 3$

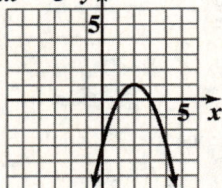

31. $y = x^2 - 1$

Step 1 $a = 1 > 0$, so the parabola opens upward.

Step 2 $a = 1, b = 0, c = -1$

x-coordinate of vertex

$$= \frac{-b}{2a} = \frac{-0}{2(1)} = \frac{0}{2} = 0$$

y-coordinate of vertex

$$= 0^2 - 1 = -1$$

The vertex is $(0,1)$.

Step 3

$$x^2 - 1 = 0$$
$$(x+1)(x-1) = 0$$
$$x + 1 = 0 \quad \text{or} \quad x - 1 = 0$$
$$x = -1 \qquad\qquad x = 1$$

The x-intercepts are -1 and 1.

Step 4 $y = 0^2 - 1 = -1$

The y-intercept is -1. Notice that this gives the same point as the vertex, $(0,-1)$.

Steps 5 and 6 Plot $(0,-1), (-1,0)$, and $(1,0)$, and connect them with a smooth curve.

$y = x^2 - 1$

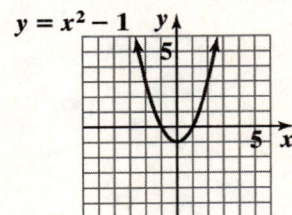

33. $y = x^2 + 2x + 1$

Step 1 $a - 1 > 0$, so the parabola opens upward.

Step 2 $a = 1, b = 2, c = 1$

x-coordinate of vertex

$$= \frac{-b}{2a} = \frac{-2}{2(1)} = \frac{-2}{2} = -1$$

y-coordinate of vertex

$$= (-1)^2 + 2(-1) + 1 = 1 - 2 + 1 = 0$$

The vertex is $(-1,0)$.

Step 3

$$x^2 + 2x + 1 = 0$$
$$(x+1)^2 = 0$$
$$x + 1 = 0$$
$$x = -1$$

There is only one x-intercept, -1. Notice that this gives the same point as the vertex, $(-1,0)$.

Step 4 $y = 0^2 + 2 \cdot 0 + 1 = 1$

The y-intercept is 1.

The work in Steps 1-4 has produced only two points, $(-1,0)$ and $(0,1)$. At least one additional point is needed. In order to have at least one point on each side of the vertex, choose an x-value less than -1 and find the corresponding y-value.

If $x = -2$, $y = (-2)^2 + 2(-2) + 1 = 4 - 4 + 1 = 1$.

Steps 5 and 6 Plot $(-2,1), (-1,0)$, and $(0,1)$, and connect them with a smooth curve.

$y = x^2 + 2x + 1$

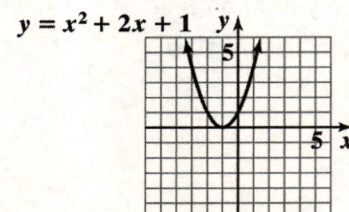

35. $y = -2x^2 + 4x + 5$

Step 1 $a = -2 < 0$, so the parabola opens downward.

Step 2 $a = -2, b = 4, c = 5$

x-coordinate of vertex

$= \dfrac{-b}{2a} = \dfrac{-4}{2(-2)} = \dfrac{-4}{-4} = 1$

y-coordinate of vertex

$= -2(1)^2 + 4(1) + 5 = -2 + 4 + 5 = 7$

The vertex is $(1, 7)$

Step 3

$-2x^2 + 4x + 5 = 0$

$\qquad 0 = 2x^2 - 4x - 5$

The trinomial cannot be factored, so use the quadratic formula with $a = 2, b = -4$, and $c = -5$.

$x = \dfrac{-b \pm \sqrt{b^2 - 4ac}}{2a}$

$= \dfrac{-(-4) \pm \sqrt{(-4)^2 - 4(2)(-5)}}{2 \cdot 2}$

$= \dfrac{4 \pm \sqrt{16 + 40}}{4} = \dfrac{4 \pm \sqrt{56}}{4}$

$x = \dfrac{4 + \sqrt{56}}{4} \approx 2.9$ or $x = \dfrac{4 - \sqrt{56}}{4} \approx -0.9$

The x-intercepts are approximately 2.9 and -0.9.

Step 4 $y = -2 \cdot 0^2 + 4.0 + 5 = 5$

The y-intercept is 5.

Steps 5 and 6 Plot

$(1, 7), (2.9, 0), (-0.9, 0)$, and $(0, 5)$, and connect them with a smooth curve.

$y = -2x^2 + 4x + 5$

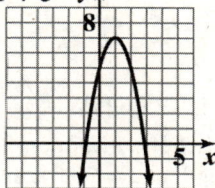

37. $y = (x - 3)^2 + 2$

$y = (x^2 - 6x + 9) + 2$

$y = x^2 - 6x + 11$

$a = 1, b = -6, c = 11$

$x = -\dfrac{b}{2a} = -\dfrac{(-6)}{2(1)} = \dfrac{6}{2} = 3$

$y = (3)^2 - 6(3) + 11 = 9 - 18 + 11 = 2$

The vertex is $(3, 2)$.

39. $y = (x + 5)^2 - 4$

$y = (x^2 + 10x + 25) - 4$

$y = x^2 + 10x + 21$

$a = 1, b = 10, c = 21$

$x = -\dfrac{b}{2a} = -\dfrac{10}{2(1)} = -5$

$y = (-5)^2 + 10(-5) + 21$

$\quad = 25 - 50 + 21 = -4$

The vertex is $(-5, -4)$.

41. $y = 2(x - 1)^2 - 3$

$y = 2(x^2 - 2x + 1) - 3$

$y = 2x^2 - 4x + 2 - 3$

$y = 2x^2 - 4x - 1$

$a = 2, b = -4, c = -1$

$x = -\dfrac{b}{2a} = -\dfrac{(-4)}{2(2)} = \dfrac{4}{4} = 1$

$y = 2(1)^2 - 4(1) - 1 = 2 - 4 - 1 = -3$

The vertex is $(1, -3)$.

43. $y = -3(x + 2)^2 + 5$

$y = -3(x^2 + 4x + 4) + 5$

$y = -3x^2 - 12x - 12 + 5$

$y = -3x^2 - 12x - 7$

$a = -3, b = -12, c = -7$

$x = -\dfrac{b}{2a} = -\dfrac{(-12)}{2(-3)} = -\dfrac{12}{6} = -2$

$y = -3(-2)^2 - 12(-2) - 7$

$\quad = -12 + 24 - 7 = 5$

The vertex is $(-2, 5)$.

45. For a parabola whose equation is $y = a(x-h)^2 + k$, the vertex is the point (h, k).

47. a. $y = -0.01x^2 + 0.7x + 6.1$

$a = -0.01, \ b = 0.7, \ c = 6.1$

x-coordinate of vertex

$= \dfrac{-b}{2a} = \dfrac{-0.7}{2(-0.01)} = 35$

y-coordinate of vertex

$y = -0.01x^2 + 0.7x + 6.1$

$y = -0.01(35)^2 + 0.7(35) + 6.1 = 18.35$

The maximum height of the shot is about 18.35 feet. This occurs 35 feet from its point of release.

b. The shot will reach the maximum horizontal distance when its height returns to 0.

$y = -0.01x^2 + 0.7x + 6.1$

$0 = -0.01x^2 + 0.7x + 6.1$

$a = -0.01, \ b = 0.7, \ c = 6.1$

$x = \dfrac{-b \pm \sqrt{b^2 - 4ac}}{2a}$

$x = \dfrac{-0.7 \pm \sqrt{0.7^2 - 4(-0.01)(6.1)}}{2(-0.01)}$

$x \approx 77.8 \ \text{ or } \ x \approx -7.8$

The maximum horizontal distance is 77.8 feet.

c. The initial height can be found at $x = 0$.

$y = -0.01x^2 + 0.7x + 6.1$

$y = -0.01(0)^2 + 0.7(0) + 6.1 = 6.1$

The shot was released at a height of 6.1 feet.

49. a. $y = -0.8x^2 + 3.2x + 6$

$a = -0.8, \ b = 3.2, \ c = 6$

x-coordinate of vertex

$= \dfrac{-b}{2a} = \dfrac{-3.2}{2(-0.8)} = 2$

y-coordinate of vertex

$y = -0.8x^2 + 3.2x + 6$

$y = -0.8(2)^2 + 3.2(2) + 6 = 9.2$

The maximum height of the ball is 9.2 feet. This occurs 2 feet from its point of release.

b. The ball hits the ground when $y = 0$.

$y = -0.8x^2 + 3.2x + 6$

$0 = -0.8x^2 + 3.2x + 6$

$a = -0.8, \ b = 3.2, \ c = 6$

$x = \dfrac{-b \pm \sqrt{b^2 - 4ac}}{2a}$

$x = \dfrac{-3.2 \pm \sqrt{3.2^2 - 4(-0.8)(6)}}{2(-0.8)}$

$x \approx 5.4 \ \text{ or } \ x \approx -1.4$

Reject the negative distance. The ball hits the ground 5.4 feet from where it is thrown.

c. $y = -0.8x^2 + 3.2x + 6$

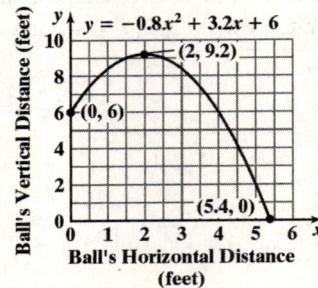

51. $A = l \cdot w$

$= x(50 - x)$

$= -x^2 + 50x$

The leading coefficient is negative, so the graph of the area equation opens down and the vertex is at the highest point.

$x = -\dfrac{b}{2a} = -\dfrac{50}{2(-1)} = 25$

$y = -(25)^2 + 50(25)$

$= 625$

The vertex is $(25, 625)$.

The area reaches a maximum value of 625 square yards if length is $x = 25$ yards and the width is $50 - 25 = 25$ yards.

53. – 57. Answers will vary.

59. makes sense

61. does not make sense; Explanations will vary. Sample explanation: Over the short portion of the path shown, it may appear linear. However the path is best described by a quadratic model.

63. true

65. false; Changes to make the statement true will vary. A sample change is: Since $a \geq 0$, the parabola opens upward and its vertex will be the lowest point on the graph.

67. Intersection points: $(-2,0)$ and $(2,0)$.

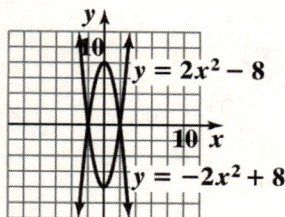

69. Answers will vary.

71. $y = -0.25x^2 + 40x$

$a = -0.25$, $b = 40$, $c = 0$

$x = -\dfrac{b}{2a} = -\dfrac{40}{2(-0.25)} = \dfrac{40}{0.5} = 80$

$y = -0.25(80)^2 + 40(80) = 1600$

The vertex is $(80,1600)$.

The y-intercept is 0 and the graph opens down.
Window settings will vary. One example would be:

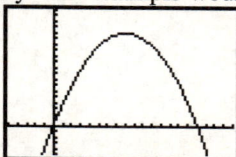

73. $y = 5x^2 + 40x + 600$

$a = 5$, $b = 40$, $c = 600$

$x = -\dfrac{b}{2a} = -\dfrac{40}{2(5)} = -\dfrac{40}{10} = -4$

$y = 5(-4)^2 + 40(-4) + 600 = 520$

The vertex is $(-4,520)$.

The y-intercept is 600 and the graph opens up.
Window settings will vary. One example would be:

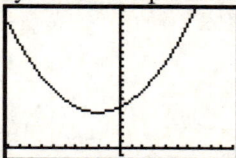

75. $7(x-2) = 10 - 2(x+3)$

$7x - 14 = 10 - 2x - 6$

$7x - 14 = 4 - 2x$

$9x - 14 = 4$

$9x = 18$

$x = 2$

The solution set is $\{2\}$.

76. $\dfrac{7}{x+2} + \dfrac{2}{x+3} = \dfrac{1}{x^2 + 5x + 6}$

$\dfrac{7}{x+2} + \dfrac{2}{x+3} = \dfrac{1}{(x+2)(x+3)}$

Restrictions: $x \neq -2, x \neq -3$

LCD $= (x+2)(x+3)$

$(x+2)(x+3)\left[\dfrac{7}{x+2} + \dfrac{2}{x+3}\right] = (x+2)(x+3)\left[\dfrac{1}{(x+2)(x+3)}\right]$

$7(x+3) + 2(x+2) = 1$

$7x + 21 + 2x + 4 = 1$

$9x + 25 = 1$

$9x = -24$

$x = \dfrac{-24}{9} = -\dfrac{8}{3}$

The solution set is $\left\{-\dfrac{8}{3}\right\}$.

77. $5x - 3y = -13$

$x = 2 - 4y$

To solve this system by the substitution method, substitute $2 - 4y$ for x in the first equation.

$5x - 3y = -13$

$5(2 - 4y) - 3y = -13$

$10 - 20y - 3y = -13$

$10 - 23y = -13$

$-23y = -23$

$y = 1$

Back-substitute 1 for y in the second equation of the given system.

$x = 2 - 4y = 2 - 4(1) = -2$

The solution set is $\{(-2, 1)\}$.

78. In set 1 each x-coordinate is paired with one and only one y-coordinate.

79. Graph (a) has each x-coordinate paired with one and only one y-coordinate.

80.　$x^2 + 3x + 5 = (-3)^2 + 3(-3) + 5$

$$= 9 - 9 + 5$$
$$= 5$$

9.6 Check Points

1. Domain: {golf, lawn mowing, water skiing, hiking, bicycling}
   Range: {250, 325, 430, 720}

2. **a.** This relation is not a function because the two ordered pairs (5,6) and (5,8) have the same first component but different second components.

   b. This relation is a function because no two ordered pairs have the same first component and different second components.

3. **a.**　$f(x) = 4x + 3$

   $f(5) = 4(5) + 3 = 20 + 3 = 23$

   b.　$f(x) = 4x + 3$

   $f(-2) = 4(-2) + 3 = -8 + 3 = -5$

   c.　$f(x) = 4x + 3$

   $f(0) = 4(0) + 3 = 0 + 3 = 3$

4. **a.**　$g(x) = x^2 + 4x + 3$

   $g(5) = (5)^2 + 4(5) + 3$
   $$= 25 + 20 + 3$$
   $$= 48$$

   b.　$g(x) = x^2 + 4x + 3$

   $g(-4) = (-4)^2 + 4(-4) + 3$
   $$= 16 - 16 + 3$$
   $$= 3$$

   c.　$g(x) = x^2 + 4x + 3$

   $g(0) = 0^2 + 4(0) + 3$
   $$= 0 + 0 + 3$$
   $$= 3$$

5. The first two graphs pass the vertical line and thus y is a function of x. The last graph fails the vertical line and thus y is not a function of x.

6. **a.**　$f(x) = -2.9x + 286$

   $f(90) = -2.9(90) + 286 = 25$

   According to the linear function, a 60-year-old has a 25% chance to survive to age 90.

   b.　$g(x) = 0.01x^2 - 4.9x + 370$

   $g(90) = 0.01(90)^2 - 4.9(90) + 370 = 10$

   According to the quadratic function, a 60-year-old has a 10% chance to survive to age 90.

   c. According to the bar graph, a 60-year-old has a 24% chance to survive to age 90. Thus the linear function serves as a better description of this data.

9.6 Concept and Vocabulary Check

1. relation; domain; range

2. function

3. f; x

4. linear

5. quadratic

6. more than once; function

9.6 Exercise Set

1. $\{(1,2),(3,4),(5,5)\}$ This relation is a function because no two ordered pairs have the same first component and different second components. The domain is the set of all first components: {1,3,5}. The range is the set of second components: {2,4,5}.

3. $\{(3,4),(3,5),(4,4),(4,5)\}$ This relation is not a function because the first two ordered pairs have the same first component but different second components. (The same applies to the third and fourth ordered pairs.)
   Domain: {3,4}
   Range: {4,5}

5. $\{(-3,-3),(-2,-2),(-1,-1),(0,0)\}$
   This relation is a function.
   Domain: {-3, -2, -1,0}
   Range: {-3, -2, -1,0}

7. $\{(1,4),(1,5),(1,6)\}$

This relation is not a function because all three ordered pairs have the same first component.
Domain: $\{1\}$
Range: $\{4,5,6\}$

9. $f(x) = x+5$

  a. $f(7) = 7+5 = 12$

  b. $f(-6) = -6+5 = -1$

  c. $f(0) = 0+5 = 5$

11. $f(x) = 7x$

  a $f(10) = 7 \cdot 10 = 70$

  b $f(-4) = 7(-4) = -28$

  c. $f(0) = 7 \cdot 0 = 0$

13. $f(x) = 8x-3$

  a. $f(12) = 8(12)-3 = 96-3 = 93$

  b. $f\left(-\dfrac{1}{2}\right) = 8\left(-\dfrac{1}{2}\right)-3 = -4-3 = -7$

  c. $f(0) = 8(0)-3 = 0-3 = -3$

15. $g(x) = x^2 + 3x$

  a. $g(2) = 2^2 + 3(2) = 4+6 = 10$

  b. $g(-2) = (-2)^2 + 3(-2) = 4-6 = -2$

  c. $g(0) = 0^2 + 3(0) = 0+0 = 0$

17. $h(x) = x^2 - 2x+3$

  a. $h(4) = 4^2 - 2 \cdot 4+3 = 16-8+3 = 11$

  b. $h(-4) = (-4)^2 - 2(-4)+3$
  $= 16+8+3 = 27$

  c. $h(0) = 0^2 - 2 \cdot 0+3 = 3$

19. $f(x) = 5$

The value of this function is 5 for every value of x.

  a. $f(9) = 5$

  b. $f(-9) = 5$

  c. $f(0) = 5$

21. $f(r) = \sqrt{r+6}+3$

  a. $f(-6) = \sqrt{-6+6}+3 = \sqrt{0}+3$
  $= 0+3 = 3$

  b. $f(10) = \sqrt{10+6}+3 = \sqrt{16}+3$
  $= 4+3 = 7$

23. $f(x) = \dfrac{x}{|x|}$

  a. $f(6) = \dfrac{6}{|6|} = \dfrac{6}{6} = 1$

  b. $f(-6) = \dfrac{-6}{|-6|} = \dfrac{-6}{6} = -1$

25. No vertical line will intersect this graph in more than one point so y is a function of x.

27. No vertical line will intersect this graph in more than one point, so y is a function of x.

29. Many vertical lines will intersect this graph in two points. One such line is the y-axis. Therefore, y is not a function of x.

31. No vertical line will intersect this graph in more than one point, so y is a function of x.

33. $f(x) = 2x+3$; $\{-1,0,1\}$
$f(-1) = 2(-1)+3 = -2+3 = 1$
$f(0) = 2(0)+3 = 0+3 = 3$
$f(1) = 2(1)+3 = 2+3 = 5$
$\{(-1,1),(0,3),(1,5)\}$

35. $g(x) = x - x^2$; $\{-2, -1, 0, 1, 2\}$

$g(-2) = (-2) - (-2)^2$
$= -2 - 4 = -6$

$g(-1) = (-1) - (-1)^2$
$= -1 - 1 = -2$

$g(0) = 0 - 0^2 = 0$

$g(1) = 1 - 1^2 = 1 - 1 = 0$

$g(2) = 2 - 2^2 = 2 - 4 = -2$

$\{(-2, -6), (-1, -2), (0, 0), (1, 0), (2, -2)\}$

37. $\dfrac{f(x) - f(h)}{x - h} = \dfrac{[6x+7] - [6h+7]}{x-h}$

$= \dfrac{6x + 7 - 6h - 7}{x - h}$

$= \dfrac{6x - 6h}{x - h} = \dfrac{6(x - h)}{(x - h)}$

$= 6$

where $x \neq h$.

39. $\dfrac{f(x) - f(h)}{x - h} = \dfrac{\left[x^2 - 1\right] - \left[h^2 - 1\right]}{x - h}$

$= \dfrac{x^2 - 1 - h^2 + 1}{x - h}$

$= \dfrac{x^2 - h^2}{x - h}$

$= \dfrac{(x - h)(x + h)}{x - h}$

$= x + h$

where $x \neq h$.

41. a. $\{(\text{Philippines, } 12), (\text{Spain, } 13),$

$(\text{Italy, } 14), (\text{Germany, } 14),$

$(\text{Russia, } 16)\}$

b. Yes, the relation is a function. Each country (element in the domain) corresponds to only one age (element in the range).

c. $\{(12, \text{ Philippines}), (13, \text{ Spain}),$

$(14, \text{ Italy}), (14, \text{ Germany}),$

$(16, \text{ Russia})\}$

b. No, the relation is not a function. 14 in the domain corresponds to two members in the range, Italy and Germany.

43. $f(x) = 0.76x + 171.4$

$f(20) = 0.76(20) + 171.4 = 186.6$

This means that at age 20, the average cholesterol level for an American man is 186.6, or approximately 187.

45. a. $C(x) = 0.28x^2 - 5.2x + 29$

$C(9) = 0.28(9)^2 - 5.2(9) + 29 \approx 5$

In 2009, cellphone use cost approximately 5¢ per minute.

b. This overestimates the value in the bar graph by 1¢.

c. $C(x) = 0.28x^2 - 5.2x + 29$

$C(0) = 0.28(0)^2 - 5.2(0) + 29 = 29$

In 2000, cellphone use cost approximately 29¢ per minute.
This is represented on the graph by the point $(0, 29)$.

47. – 51. Answers will vary.

53. makes sense

55. does not make sense; Explanations will vary. Sample explanation: The domain is the set of various ages.

57. false; Changes to make the statement true will vary. A sample change is: Two ordered pairs of a function can have the same second component and different first components.

59. true

61. Given $f(x) = ax^2 + bx + c$ and

$r = \dfrac{-b + \sqrt{b^2 - 4ac}}{2a}$, we can recognize that r is one

of the solutions of the quadratic equation $ax^2 + bx + c = 0$. Therefore, $f(r) = 0$.

63. 0.00397
To write this number in scientific notation, move the decimal point 3 places to the right. Because the number is between 0 and 1, the exponent will be positive.

$0.00397 = 3.97 \times 10^{-3}$

64.

$$x - 2 \overline{\smash{\big)}\, x^3 + 7x^2 - 2x + 3} \quad \overset{x^2 + 9x + 16}{}$$

$$\underline{x^3 - 2x^2}$$
$$9x^2 - 2x$$
$$\underline{9x^2 - 18x}$$
$$16x + 3$$
$$\underline{16x - 32}$$
$$35$$

$$\frac{x^3 + 7x^2 - 2x + 3}{x - 2} = x^2 + 9x + 16 + \frac{35}{x - 2}$$

65. $3x + 2y = 6$
$8x - 3y = 1$

To solve this system by the addition method, multiply the first equation by 3 and then second equation by 2; then add the equations.

$$9x + 6y = 18$$
$$\underline{16x - 6y = 2}$$
$$25x = 20$$

$$x = \frac{20}{25} = \frac{4}{5}$$

Instead of substituting $\frac{4}{5}$ for x for working with

fractions, go back to the original system and eliminate x.

To do this, multiply the first equation by 8 and the second equation by -3; then add.

$$24x + 16y = 48$$
$$\underline{-24x + 9y = -3}$$
$$25y = 45$$

$$y = \frac{45}{25} = \frac{9}{5}$$

The solution is $\left(\frac{4}{5}, \frac{9}{5} \right)$.

Chapter 9 Review Exercises

1. $x^2 = 64$
$x = \sqrt{64}$ or $x = -\sqrt{64}$
$x = 8 \qquad x = -8$
The solution set is $\{-8, 8\}$.

2. $x^2 = 17$
$x = \sqrt{17}$ or $x = -\sqrt{17}$
The solution set is $\left\{ \pm\sqrt{17} \right\}$.

3. $2x^2 = 150$
$x^2 = 75$
$x = \sqrt{75}$ or $x = -\sqrt{75}$
Simplify $\sqrt{75}$:
$\sqrt{75} = \sqrt{25}\sqrt{3} = 5\sqrt{3}$
$x = 5\sqrt{3}$ or $x = -5\sqrt{3}$
The solution set is $\left\{ \pm 5\sqrt{3} \right\}$.

4. $(x - 3)^2 = 9$
$x - 3 = \sqrt{9}$ or $x - 3 = -\sqrt{9}$
$x - 3 = 3 \qquad x - 3 = -3$
$x = 6 \qquad\qquad x = 0$
The solution set is $\{0, 6\}$.

5. $(y + 4)^2 = 5$
$y + 4 = \sqrt{5} \qquad$ or $y + 4 = -\sqrt{5}$
$y = -4 + \sqrt{5} \qquad y = -4 - \sqrt{5}$
The solution set is $\left\{ -4 \pm \sqrt{5} \right\}$.

6. $3y^2 - 5 = 0$
$3y^2 = 5$
$y^2 = \frac{5}{3}$

$$y = \sqrt{\frac{5}{3}} \text{ or } y = -\sqrt{\frac{5}{3}}$$

Rationalize the denominators.
$$\sqrt{\frac{5}{3}} = \frac{\sqrt{5}}{\sqrt{3}} \cdot \frac{\sqrt{3}}{\sqrt{3}} = \frac{\sqrt{15}}{3}$$
$$y = \frac{\sqrt{15}}{3} \text{ or } y = -\frac{\sqrt{15}}{3}$$

The solution set is $\left\{ \pm \dfrac{\sqrt{15}}{3} \right\}$.

7. $(2x-7)^2 = 25$

$2x-7 = \sqrt{25}$ or $2x-7 = -\sqrt{25}$

$2x-7 = 5$ $2x-7 = -5$

$2x = 12$ $2x = 2$

$x = 6$ $x = 1$

The solution set is $\{1,6\}$.

8. $(x+5)^2 = 12$

$x+5 = \sqrt{12}$ or $x+5 = -\sqrt{12}$

$x+5 = 2\sqrt{3}$ $x+5 = -2\sqrt{3}$

$x = -5 + 2\sqrt{3}$ $x = -5 - 2\sqrt{3}$

The solution set is $\left\{-5 \pm 2\sqrt{3}\right\}$.

9. Let c = the length of the hypotenuse.

$6^2 + 8^2 = c^2$

$36 + 64 = c^2$

$100 = c^2$

$c = \sqrt{100} = 10$

The missing length is 10 feet.

10. Let c = the length of the hypotenuse.

$4^2 + 6^2 = c^2$

$16 + 36 = c^2$

$c = \sqrt{52} = \sqrt{4}\sqrt{13} = 2\sqrt{13}$

The missing length is $2\sqrt{13}$ inches.

11. Let b = the missing length (length of one of the legs).

$11^2 + b^2 = 15^2$

$121 + b^2 = 225$

$b^2 = 104$

$b = \sqrt{104} = \sqrt{4}\sqrt{26} = 2\sqrt{26}$

The missing length is $2\sqrt{26}$ centimeters.

12. Let x = the distance between the base of the building and the bottom of the ladder.

$x^2 + 20^2 = 25^2$

$x^2 + 400 = 625$

$x = 225$

$x = \sqrt{225} = 15$

The bottom of the ladder is 15 feet away from the building.

13. Let h = the distance up the pole that the wires should be attached.

In the figure, the pole and one of the wires are shown.

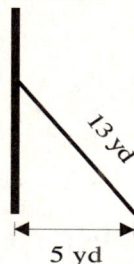

$5^2 + h^2 = 13^2$

$25 + h^2 = 169$

$h^2 = 144$

$h = \sqrt{144} = 12$

The wires will be attached 12 yards up the pole.

14. $W = 3t^2; W = 1200$

$1200 = 3t^2$

$400 = t^2$

$t = \sqrt{400} = 20$

The fetus weighs 1200 grams after 20 weeks.

15. $d = 16t^2; d = 100$

$100 = 16t^2$

$6.25 = t^2$

$t = \sqrt{6.25} = 2.5$

It will take the object 2.5 seconds to hit the water.

16. $d = \sqrt{(x_2 - x_1)^2 + (y_2 - y_1)^2}$

$= \sqrt{[1 - (-3)]^2 + [-5 - (-2)]^2}$

$= \sqrt{4^2 + (-3)^2}$

$= \sqrt{16 + 9}$

$= \sqrt{25}$

$= 5$

17. $d = \sqrt{(x_2 - x_1)^2 + (y_2 - y_1)^2}$

$= \sqrt{(5-3)^2 + (4-8)^2}$

$= \sqrt{2^2 + (-4)^2}$

$= \sqrt{4+16}$

$= \sqrt{20} = \sqrt{4 \cdot 5} = 2\sqrt{5} \approx 4.47$

18. $x^2 + 16x$

The coefficient of the *x*-term is 16. Half of 16 is 8, and $8^2 = 64$. Add 64.

$x^2 + 16x + 64 = (x+8)^2$

19. $x^2 - 6x$

The coefficient of the *x*-term is –6. Half of –6 is –3, and $(-3)^2 = 9$. Add 9.

$x^2 - 6x + 9 = (x-3)^2$

20. $x^2 + 3x$

The coefficient of the *x*-term is 3. Half of 3 is $\dfrac{3}{2}$

and $\left(\dfrac{3}{2}\right)^2 = \dfrac{9}{4}$. Add $\dfrac{9}{4}$.

$x^2 + 3x + \dfrac{9}{4} = \left(x + \dfrac{3}{2}\right)^2$

21. $x^2 - 5x$

The coefficient of the *x*-term is –5. Half of –5 is

$\dfrac{-5}{2}$ and $\left(\dfrac{-5}{2}\right)^2 = \dfrac{25}{4}$. Add $\dfrac{25}{4}$.

$x^2 - 5x + \dfrac{25}{4} = \left(x - \dfrac{5}{2}\right)^2$

22. $x^2 - 12x + 27 = 0$

First, subtract 27 from both sides to isolate the binomial $x^2 - 12x$.

$x^2 - 12x = -27$

Next, complete the square.
The coefficient of the *x*-term is –12. Half of –12 is –6 and $(-6)^2 = 36$. Add 36.

$x^2 - 12x + 36 = -27 + 36$

$(x-6)^2 = 9$

$x - 6 = \sqrt{9}$ or $x - 6 = -\sqrt{9}$

$x - 6 = 3$ $\qquad$ $x - 6 = -3$

$x = 9$ $\qquad\qquad$ $x = 3$

The solution set is $\{3, 9\}$.

23. $x^2 - 6x + 4 = 0$

$x^2 - 6x = -4$

$x^2 - 6x + 9 = -4 + 9$

$(x-3)^2 = 5$

$x - 3 = \sqrt{5}$ $\qquad$ or $x - 3 = -\sqrt{5}$

$x = 3 + \sqrt{5}$ $\qquad$ $x = 3 - \sqrt{5}$

The solution set is $\left\{3 \pm \sqrt{5}\right\}$.

24. $3x^2 - 12x + 11 = 0$

First, divide both sides of the equation by 3 so that the coefficient of the x^2-term will be 1.

$x^2 - 4x + \dfrac{11}{3} = 0$

$x^2 - 4x = -\dfrac{11}{3}$

$x^2 - 4x + 4 = -\dfrac{11}{3} + 4$

$(x-2)^2 = -\dfrac{11}{3} + \dfrac{12}{3}$

$(x-2)^2 = \dfrac{1}{3}$

$x - 2 = \sqrt{\dfrac{1}{3}}$ or $x - 2 = -\sqrt{\dfrac{1}{3}}$

Simplify the radicals.

$\sqrt{\dfrac{1}{3}} = \dfrac{\sqrt{1}}{\sqrt{3}} = \dfrac{1}{\sqrt{3}} \cdot \dfrac{\sqrt{3}}{\sqrt{3}} = \dfrac{\sqrt{3}}{3}$

$x - 2 = \dfrac{\sqrt{3}}{3}$ $\qquad$ or $x - 2 = -\dfrac{\sqrt{3}}{3}$

$x = 2 + \dfrac{\sqrt{3}}{3}$ $\qquad$ $x = 2 - \dfrac{\sqrt{3}}{3}$

$x = \dfrac{6}{3} + \dfrac{\sqrt{3}}{3}$ $\qquad$ $x = \dfrac{6}{3} - \dfrac{\sqrt{3}}{3}$

$x = \dfrac{6 + \sqrt{3}}{3}$ $\qquad$ $x = \dfrac{6 - \sqrt{3}}{3}$

The solution set is $\left\{\dfrac{6 \pm \sqrt{3}}{3}\right\}$.

25. $2x^2 + 5x - 3 = 0$

$a = 2, b = 5, c = -3$

Substitute these values into the quadratic formula.

$$x = \frac{-b \pm \sqrt{b^2 - 4ac}}{2a}$$

$$x = \frac{-5 \pm \sqrt{5^2 - 4(2)(-3)}}{2(2)}$$

$$= \frac{-5 \pm \sqrt{25 + 24}}{4}$$

$$= \frac{-5 \pm \sqrt{49}}{4} = \frac{-5 \pm 7}{4}$$

$$x = \frac{-5 + 7}{4} \text{ or } x = \frac{-5 - 7}{4}$$

$$= \frac{2}{4} = \frac{1}{2} \qquad = \frac{-12}{4} = -3$$

The solution set is $\left\{ -3, \dfrac{1}{2} \right\}$.

26. $x^2 = 2x + 4$

Rewrite the equation in standard form.

$x^2 - 2x - 4 = 0$

Identify $a, b,$ and c.

$a = 1, b = -2, c = -4$

Substitute these values into the quadratic formula.

$$x = \frac{-b \pm \sqrt{b^2 - 4ac}}{2a}$$

$$= \frac{-(-2) \pm \sqrt{(-2)^2 - 4(1)(-4)}}{2(1)}$$

$$= \frac{2 \pm \sqrt{4 + 16}}{2} = \frac{2 \pm \sqrt{20}}{2}$$

$$= \frac{2 \pm 2\sqrt{5}}{2} = \frac{2 \pm \sqrt{20}}{2}$$

$$= 1 \pm \sqrt{5}$$

The solution set is $\left\{ 1 \pm \sqrt{5} \right\}$.

27. $3x^2 + 5 = 9x$

Rewrite the equation in standard form.

$3x^2 - 9x + 5 = 0$

$a = 3, b = -9, c = 5$

$$x = \frac{-b \pm \sqrt{b^2 - 4ac}}{2a}$$

$$= \frac{-(-9) \pm \sqrt{(-9)^2 - 4(3)(5)}}{2 \cdot 3}$$

$$= \frac{9 \pm \sqrt{81 - 60}}{6} = \frac{9 \pm \sqrt{21}}{6}$$

The solution set is $\left\{ \dfrac{9 \pm \sqrt{21}}{6} \right\}$.

28. $2x^2 - 11x + 5 = 0$

This equation can be solved by the factoring method. Factor the trinomial.

$(2x - 1)(x - 5) = 0$

Use the zero-product principle.

$2x - 1 = 0 \text{ or } x - 5 = 0$

$\quad 2x = 1 \qquad\qquad x = 5$

$\quad x = \dfrac{1}{2}$

The solution set is $\left\{ \dfrac{1}{2}, 5 \right\}$.

29. $(3x + 5)(x - 3) = 5$

Write the equation in standard form.

$3x^2 - 4x - 15 = 5$

$3x^2 - 4x - 20 = 0$

This equation can be solved by the factoring method.

$(3x - 10)(x + 2) = 0$

$3x - 10 = 0 \quad \text{or } x + 2 = 0$

$\quad 3x = 10 \qquad\qquad x = -2$

$\quad x = \dfrac{10}{3}$

The solution set is $\left\{ -2, \dfrac{10}{3} \right\}$.

30. $3x^2 - 7x + 1 = 0$

The trinomial cannot be factored, so use the quadratic formula with $a = 3, b = -7,$ and $c = 1$.

$$x = \frac{-b \pm \sqrt{b^2 - 4ac}}{2a}$$

$$= \frac{-(-7) \pm \sqrt{(-7)^2 - 4(3)(1)}}{2(3)}$$

$$= \frac{7 \pm \sqrt{49 - 12}}{6}$$

$$= \frac{7 \pm \sqrt{37}}{6}$$

The solution set is $\left\{ \dfrac{7 \pm \sqrt{37}}{6} \right\}$.

31. $x^2 - 9 = 0$

Solve by the factoring method.

$(x + 3)(x - 3) = 0$

$x + 3 = 0 \quad$ or $\quad x - 3 = 0$

$\quad x = -3 \qquad\qquad x = 3$

The solution set is $\{-3, 3\}$.

32. $(2x - 3)^2 = 5$

Use the square root property.

$2x - 3 = \sqrt{5} \qquad$ or $\quad 2x - 3 = -\sqrt{5}$

$\quad 2x = 3 + \sqrt{5} \qquad\qquad 2x = 3 - \sqrt{5}$

$\qquad x = \dfrac{3 + \sqrt{5}}{2} \qquad\qquad x = \dfrac{3 - \sqrt{5}}{2}$

The solution set is $\left\{ \dfrac{3 \pm \sqrt{5}}{2} \right\}$.

33. a. 2007 is 4 years after 2003.

$T = 0.6x^2 + 5.4x + 28$

$T = 0.6(4)^2 + 5.4(4) + 28 = 59.2$

$59.2 billion was spent in 2007.

b. Let $T = 160$ and solve for x.

$T = 0.6x^2 + 5.4x + 28$

$160 = 0.6x^2 + 5.4x + 28$

$0 = 0.6x^2 + 5.4x - 132$

$$x = \frac{-b \pm \sqrt{b^2 - 4ac}}{2a}$$

$$x = \frac{-5.4 \pm \sqrt{5.4^2 - 4(0.6)(-132)}}{2(0.6)}$$

$$x = \frac{-5.4 \pm \sqrt{345.96}}{1.2}$$

$$x = \frac{-5.4 \pm 18.6}{1.2}$$

$x = -20 \quad$ or $\quad x = 11$

The amount spent to fight terrorism is expected to reach $160 billion 11 years after 2003, or 2014.

34. $\sqrt{-81} = \sqrt{81(-1)} = \sqrt{81}\sqrt{-1} = 9i$

35. $\sqrt{-23} = \sqrt{-23(-1)} = \sqrt{23}\sqrt{-1} = i\sqrt{23}$

36. $\sqrt{-48} = \sqrt{48}\sqrt{-1} = \sqrt{16 \cdot 3}\sqrt{-1} = 4i\sqrt{3}$

37. $3 + \sqrt{-49} = 3 + \sqrt{49}\sqrt{-1} = 3 + 7i$

38. $x^2 = -100$

Use the square root property.

$x = \sqrt{-100} \quad$ or $\quad x = -\sqrt{-100}$

$x = 10i \qquad\qquad x = -10i$

The solutions are $\pm 10i$.

39. $5x^2 = -125$

$x^2 = -25$

$x = \sqrt{-25} \quad$ or $\quad x = -\sqrt{-25}$

$x = 5i \qquad\qquad x = -5i$

The solution set is $\{\pm 5i\}$.

40. $(2x + 1)^2 = -8$

$2x + 1 = \pm\sqrt{-8}$

$2x + 1 = \pm 2i\sqrt{2}$

$\quad 2x = -1 \pm 2i\sqrt{2}$

$\qquad x = \dfrac{-1 \pm 2i\sqrt{2}}{2}$

The solution set is $\left\{ \dfrac{-1 \pm 2i\sqrt{2}}{2} \right\}$.

41. $x^2 - 4x + 13 = 0$

Use the quadratic formula with
$a = 1, b = -4,$ and $c = 13.$

$$x = \frac{-b \pm \sqrt{b^2 - 4ac}}{2a}$$

$$= \frac{-(-4) \pm \sqrt{(-4)^2 - 4(1)(13)}}{2 \cdot 1}$$

$$= \frac{4 \pm \sqrt{16 - 52}}{2} = \frac{4 \pm \sqrt{-36}}{2}$$

$$= \frac{4 \pm 6i}{2} = \frac{2(2 \pm 3i)}{2}$$

$$= 2 \pm 3i$$

The solution set is $\{2 \pm 3i\}.$

42. $3x^2 - x + 2 = 0$

$a = 3, b = -1, c = 2$

$$x = \frac{-b \pm \sqrt{b^2 - 4ac}}{2a}$$

$$= \frac{-(-1) \pm \sqrt{(-1)^2 - 4(3)(2)}}{2 \cdot 3}$$

$$= \frac{1 \pm \sqrt{1 - 24}}{6} = \frac{1 \pm \sqrt{-23}}{6}$$

$$= \frac{1 \pm i\sqrt{23}}{6}$$

The solution set is $\left\{ \dfrac{1 \pm i\sqrt{23}}{6} \right\}.$

43. $y = x^2 - 6x - 7$

a. Because a, the coefficient of x^2, is 1, which is greater than 0, the parabola opens upward.

b. $x^2 - 6x - 7 = 0$

$(x+1)(x-7) = 0$

$x + 1 = 0$ or $x - 7 = 0$

     $x = -1$ $x = 7$

The x-intercepts are -1 and 7.

c. $y = 0^2 - 6 \cdot 0 - 7 = -7$

The y-intercept is -7.

d. $a = 1, b = -6$

x-coordinate of vertex

$$= \frac{-b}{2a} = \frac{-(-6)}{2(1)} = \frac{6}{2} = 3$$

y-coordinate of vertex

$$= 3^2 - 6 \cdot 3 - 7$$

$$= 9 - 18 - 7$$

$$= -16$$

The vertex is $(3, -16)$

e. Plot the points $(-1, 0), (7, 0), (0, -7),$
and $(3, -16)$, and connect these points with a smooth curve.

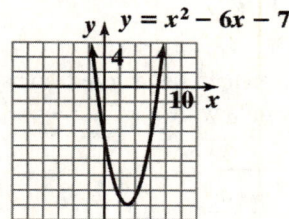

44. $y = -x^2 - 2x + 3$

a. $a = -1 < 0$, so the parabola opens downward.

b. $-x^2 - 2x + 3 = 0$

$$0 = x^2 + 2x - 3$$

$$0 = (x+3)(x-1)$$

$x + 3 = 0$ or $x - 1 = 0$

     $x = -3$ $x = 1$

The x-intercepts are -3 and 1.

c. $y = -0^2 - 2 \cdot 0 + 3 = 3$

The y-intercept is 3.

d. $a = -1, b = -2$

x-intercept of vertex

$$= \frac{-b}{2a} = \frac{-(-2)}{2(-1)} = \frac{2}{-2} = -1$$

y-coordinate of vertex

$$= -(-1)^2 - 2(-1) + 3$$

$$= -1 + 2 + 3$$

$$= 4$$

The vertex is $(-1, 4)$.

e. Plot the points $(-3,0),(1,0),(0,3)$, and $(-1,4)$, and connect them with a smooth curve.

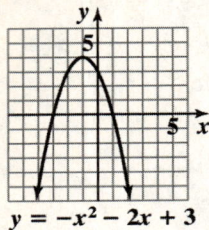

$y = -x^2 - 2x + 3$

45. $y = -3x^2 + 6x + 1$

a. $a = -3 < 0$, so the parabola opens downward.

b. $-3x^2 + 6x + 1 = 0$

$$0 = 3x^2 - 6x - 1$$

The trinomial cannot be factored, so use the quadratic formula with $a = 3, b = -6,$ and $c = -1$.

$$x = \frac{-b \pm \sqrt{b^2 - 4ac}}{2a}$$

$$= \frac{-(-6) \pm \sqrt{(-6)^2 - 4(3)(-1)}}{2 \cdot 3}$$

$$= \frac{6 \pm \sqrt{36 + 12}}{6} = \frac{6 \pm \sqrt{48}}{6}$$

$$= \frac{6 \pm 4\sqrt{3}}{6} = \frac{2(3 \pm 2\sqrt{3})}{6}$$

$$= \frac{3 \pm 2\sqrt{3}}{3}$$

$$x = \frac{3 + 2\sqrt{3}}{3} \quad \text{or} \quad x = \frac{3 - 2\sqrt{3}}{3}$$

$$x \approx 2.2 \qquad\qquad x \approx -0.2$$

The x-intercepts are approximately 2.2 and -0.2.

c. $y = -3 \cdot 0^2 + 6 \cdot 0 + 1 = 1$

The y-intercept is 1.

d. $a = -3, b = 6$

$$x = \frac{-b}{2a} = \frac{-6}{2(-3)} = \frac{-6}{-6} = 1$$

$$y = -3(1)^2 + 6(1) + 1 = -3 + 6 + 1 = 4$$

The vertex is $(1,4)$.

e. Plot the points $(2.2,0),(-0.2,0),(0,1)$, and $(1,4)$, and connect them with a smooth curve.

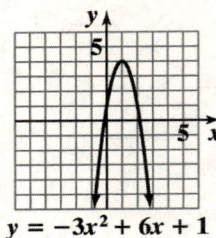

$y = -3x^2 + 6x + 1$

46. $y = x^2 - 4x$

a. $a = 1 > 0$, so the parabola opens upward.

b. $x^2 - 4x = 0$

$$x(x-4) = 0$$

$$x = 0 \quad \text{or} \quad x - 4 = 0$$

$$x = 4$$

The x-intercepts are 0 and 4.

c. $y = 0^2 - 4 \cdot 0 = 0$

The y-intercept is 0. This does not provide an additional point because part (b) already showed that the parabola passes through the origin.

d. $a = 1, b = -4$

$$x = \frac{-b}{2a} = \frac{-(-4)}{2(1)} = \frac{4}{2} = 2$$

$$y = 2^2 - 4 \cdot 2 = 4 - 8 = -4$$

The vertex is $(2,-4)$.

e. Plot the points $(0,0),(4,0)$, and $(2,-4)$, and draw a smooth curve through them.

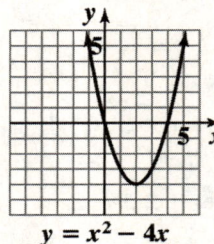

$y = x^2 - 4x$

47. The information needed occurs at the vertex.

x-coordinate.

$$x = \frac{-b}{2a} = \frac{-200}{2(-16)} = 6.25$$

y-coordinate.

$$y = -16x^2 + 200x + 4$$

$$y = -16(6.25)^2 + 200(6.25) + 4 = 629$$

The fireworks reach their greatest height 6.25 seconds after they are launched. The height reached is 629 feet.

48. a. The information needed occurs at the vertex.

x-coordinate.

$$x = \frac{-b}{2a} = \frac{-1}{2(-0.025)} = 20$$

y-coordinate.

$$y = -0.025x^2 + x + 6$$

$$y = -0.025(20)^2 + 20 + 6 = 16$$

The football will reach its greatest height 20 yards after it is tossed. The height reached is 16 feet.

b. $y = -0.025x^2 + x + 6$

$$y = -0.025(38)^2 + 38 + 6 = 7.9$$

The defender must reach 7.9 feet.

c. The football will hit the ground when $y = 0$.

$$y = -0.025x^2 + x + 6$$

$$0 = -0.025x^2 + x + 6$$

$$x = \frac{-b \pm \sqrt{b^2 - 4ac}}{2a}$$

$$x = \frac{-1 \pm \sqrt{1^2 - 4(-0.025)(6)}}{2(-0.025)}$$

$$x \approx -5.3 \text{ or } x \approx 45.3$$

The football will go 45.3 yards before it hits the ground.

d. Find the initial height.

$$y = -0.025x^2 + x + 6$$

$$y = -0.025(0)^2 + 0 + 6 = 6$$

Plot (20,16), (45.3,0), and (0,6). Connect the points with a smooth curve.

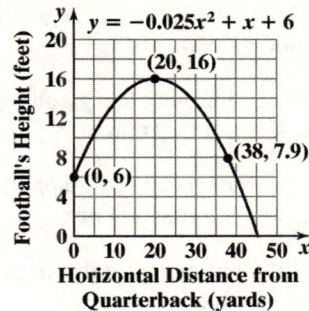

49. $\{(2,7),(3,7),(5,7)\}$

This relation is a function because no two ordered pairs have the same first component and different second components.

The domain is the set of first components: {2, 3, 5}. The range is the set of second components: {7}

50. $\{(1,10),(2,500),(3,\pi)\}$

This relation is a function.

Domain: $\{1,2,3\}$

Range: $\{10,500,\pi\}$

51. $\{(12,13),(14,15),(12,19)\}$

This relation is not a function because two of the ordered pairs have the same first component, 12, but different second components.

Domain: $\{12,14\}$

Range: $\{13,15,19\}$

52. $f(x) = 3x - 4$

a. $f(-5) = 3(-5) - 4 = -15 - 4 = -19$

b. $f(6) = 3 \cdot 6 - 4 = 18 - 4 = 14$

c. $f(0) = 3 \cdot 0 - 4 = 0 - 4 = -4$

53. $g(x) = x^2 - 5x + 2$

a. $g(-4) = (-4)^2 - 5(-4) + 2$

$$= 16 + 20 + 2 = 38$$

b. $g(3) = 3^2 - 5 \cdot 3 + 2$
$= 9 - 15 + 2 = -4$

c. $g(0) = 0^2 - 5 \cdot 0 + 2 = -0 + 2 = 2$

54. Many vertical lines will intersect the graph in two points, so y is not a function of x.

55. No vertical line will intersect this graph in more than one point, so y is a function of x.

56. No vertical line will intersect this graph in more than one point, so y is a function of x.

57. Many vertical lines will intersect the graph in two points, so y is not a function of x.

58. a. $D(x) = 0.8x^2 - 17x + 109$

$D(1) = 0.8(1)^2 - 17(1) + 109 = 92.8$

Skin damage begins for burn-prone people after 92.8 minutes when the sun's UV index is 1. This is represented by the point (1, 92.8).

b. $D(x) = 0.8x^2 - 17x + 109$

$D(10) = 0.8(10)^2 - 17(10) + 109 = 19$

Skin damage begins for burn-prone people after 19 minutes when the sun's UV index is 10. This is represented by the point (10, 19).

Chapter 9 Test

1. $3x^2 = 48$

$x^2 = 16$

$x = \sqrt{16}$ or $x = -\sqrt{16}$

$x = 4$ or $x = -4$

The solution set is $\{-4, 4\}$.

2. $(x-3)^2 = 5$

$x - 3 = \sqrt{5}$ or $x - 3 = -\sqrt{5}$

$x = 3 + \sqrt{5}$ $x = 3 - \sqrt{5}$

The solution set is $\{3 \pm \sqrt{5}\}$.

3. Let b = distance PQ across the lake. Use the Pythagorean Theorem.

$8^2 + b^2 = 12^2$

$64 + b^2 = 144$

$b^2 = 80$

$b = \sqrt{80} = \sqrt{16}\sqrt{5} = 4\sqrt{5}$

The distance across the lake is $4\sqrt{5}$ yards.

4. $d = \sqrt{(x_2 - x_1)^2 + (y_2 - y_1)^2}$

$= \sqrt{(-4-3)^2 + [1-(-2)]^2}$

$= \sqrt{(-7)^2 + 3^2}$

$= \sqrt{49 + 9} = \sqrt{58} \approx 7.62$

5. $x^2 + 4x - 3 = 0$

$x^2 + 4x = 3$ ← complete the square

$x^2 + 4x + 4 = 3 + 4$

$(x+2)^2 = 7$

$x + 2 = \sqrt{7}$ or $x + 2 = -\sqrt{7}$

$x = -2 + \sqrt{7}$ $x = -2 - \sqrt{7}$

The solution set is $\{-2 \pm \sqrt{7}\}$.

6. $3x^2 + 5x + 1 = 0$

The trinomial cannot be factored, so use the quadratic formula with $a = 3, b = 5$, and $c = 1$.

$x = \dfrac{-b \pm \sqrt{b^2 - 4ac}}{2a}$

$= \dfrac{-5 \pm \sqrt{5^2 - 4(3)(1)}}{2 \cdot 3}$

$= \dfrac{-5 \pm \sqrt{25 - 12}}{6}$

$= \dfrac{-5 \pm \sqrt{13}}{6}$

The solution set is $\left\{\dfrac{-5 \pm \sqrt{13}}{6}\right\}$.

7. $(3x-5)(x+2)=-6$

Write the equation in standard form.

$3x^2+x-10=-6$

$3x^2+x-4=0$

Factor the trinomial and use the zero-product principle.

$(3x+4)(x-1)=0$

$3x+4=0 \quad$ or $\quad x-1=0$

$3x=-4 \qquad\qquad x=1$

$x=-\dfrac{4}{3}$

The solution set is $\left\{-\dfrac{4}{3},1\right\}$.

8. $(2x+1)^2=36$

$2x+1=\sqrt{36}\quad$ or $\quad 2x+1=-\sqrt{36}$

$2x+1=6 \qquad\qquad 2x+1=-6$

$2x=5 \qquad\qquad\quad 2x=-7$

$x=\dfrac{5}{2} \qquad\qquad\quad x=-\dfrac{7}{2}$

The solution set is $\left\{-\dfrac{7}{2},\dfrac{5}{2}\right\}$.

9. $2x^2=6x-1$

Write the equation in standard form.

$2x^2-6x+1=0$

The trinomial cannot be factored so use the quadratic formula with $a=2, b=-6,$ and $c=1$.

$x=\dfrac{-b\pm\sqrt{b^2-4ac}}{2a}$

$=\dfrac{-(-6)\pm\sqrt{(-6)^2-4(2)(1)}}{2\cdot 2}$

$=\dfrac{6\pm\sqrt{36-8}}{4}=\dfrac{6\pm\sqrt{28}}{4}$

$=\dfrac{6\pm 2\sqrt{7}}{4}=\dfrac{2\left(3\pm\sqrt{7}\right)}{4}$

$=\dfrac{3\pm\sqrt{7}}{2}$

The solution set is $\left\{\dfrac{3\pm\sqrt{7}}{2}\right\}$.

10. $2x^2+9x=5$

Write the equation in standard form.

$2x^2+9x-5=0$

Solve by the factoring method.

$(2x-1)(x+5)=0$

$2x-1=0 \quad$ or $\quad x+5=0$

$2x=2 \qquad\qquad x=-5$

$x=\dfrac{1}{2}$

The solution set is $\left\{-5,\dfrac{1}{2}\right\}$.

11. $\sqrt{-121}=\sqrt{121(-1)}=\sqrt{121}\sqrt{-1}=11i$

12. $\sqrt{-75}=\sqrt{75(-1)}=\sqrt{25\cdot 3}\sqrt{-1}$

$\qquad\quad=5i\sqrt{3}$

13. $x^2+36=0$

$x^2=-36$

$x=\sqrt{-36}\quad$ or $\quad x=-\sqrt{-36}$

$x=6i \qquad\qquad x=-6i$

The solution set is $\{\pm 6i\}$.

14. $(x-5)^2=25$

$x-5=\sqrt{-25}\quad$ or $\quad x-5=-\sqrt{-25}$

$x-5=5i \qquad\quad$ or $\quad x-5=-5i$

$x=5+5i \qquad\qquad x=5-5i$

The solution set is $\{5\pm 5i\}$.

15. $x^2-2x+5=0$

The trinomial cannot be factored, so use the quadratic formula with $a=1, b=-2$, and $c=5$.

$x=\dfrac{-b\pm\sqrt{b^2-4ac}}{2a}$

$=\dfrac{-(-2)\pm\sqrt{(-2)^2-4(1)(5)}}{2\cdot 1}$

$=\dfrac{2\pm\sqrt{4-20}}{2}=\dfrac{2\pm\sqrt{-16}}{2}$

$=\dfrac{2\pm 4i}{2}=\dfrac{2(1\pm 2i)}{2}$

$=1\pm 2i$

The solution set is $\{1\pm 2i\}$.

16. $y = x^2 + 2x - 8$

Step 1 $a = 1 > 0$, so the parabola opens upward.

Step 2 $a = 1, b = 2, c = -8$

x-coordinate of vertex

$= \dfrac{-b}{2a} = \dfrac{-2}{2(1)} = \dfrac{-2}{2} = -1$

y-coordinate of vertex

$= (-1)^2 + 2(-1) - 8 = 1 - 2 - 8 = -9$

The vertex is $(-1, -9)$.

Step 3 $x^2 + 2x - 8 = 0$

$(x + 4)(x - 2) = 0$

$x + 4 = 0$ or $x - 2 = 0$

$x = -4 \qquad x = 2$

The x-intercepts are -4 and 2.

Step 4 $y = 0^2 + 2 \cdot 0 - 8 = -8$

The y-intercept is -8.

Steps 5 and 6 Plot $(-1, -9), (-4, 0), (2, 0)$, and

$(0, -8)$, then connect them with a smooth curve.

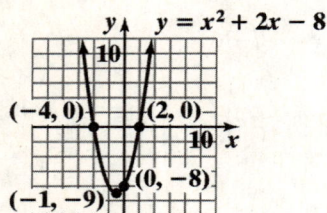

17. $y = -2x^2 + 16x - 24$

Step 1 $a = -2 < 0$, so the parabola opens downward.

Step 2 $a = -2, b = 16, c = -24$

$x = \dfrac{-b}{2a} = \dfrac{-16}{2(-2)} = \dfrac{-16}{-4} = 4$

$y = -2 \cdot 4^2 + 16 \cdot 4 - 24$

$= -32 + 64 - 24 = 8$

The vertex is $(4, 8)$.

Step 3

$-2x^2 + 16x - 24 = 0$

$0 = 2x^2 - 16x + 24$

$0 = 2(x^2 - 8x + 12)$

$0 = 2(x - 2)(x - 6)$

$x - 2 = 0$ or $x - 6 = 0$

$x = 2 \qquad x = 6$

The x-intercepts are 2 and 6.

Step 4

$y = -2 \cdot 0^2 + 16 \cdot 0 - 24 = -24$

The y-intercept is -24.

Steps 5 and 6 Plot $(4, 8), (2, 0), (6, 0)$, and

$(0, -24)$, then connect them with a smooth curve.

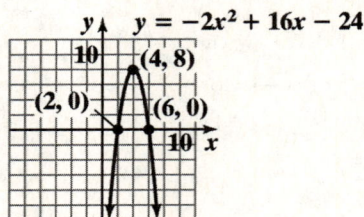

18. $y = -16x^2 + 64x + 5$

The graph of this equation is a parabola opening downward, so its vertex is a maximum point. Find the coordinates of the vertex by using

$a = -16, b = 64$.

$x = \dfrac{-b}{2a} = \dfrac{-64}{2(-16)} = \dfrac{-64}{-32} = 2$

$y = -16 \cdot 2^2 + 64 \cdot 2 + 5$

$= -64 + 128 + 5 = 69$

The baseball reaches its maximum height 2 seconds after it is hit. The maximum height is 69 feet.

19. The baseball will hit the ground when $y = 0$, so

solve the quadratic equation $0 = -16x^2 + 64x + 5$ or

$16x^2 - 64x - 5 = 0$.

The trinomial cannot be factored, so use the quadratic formula with $a = 16$,

$b = -64$, and $c = -5$.

$x = \dfrac{-b \pm \sqrt{b^2 - 4ac}}{2a}$

$= \dfrac{-(-64) \pm \sqrt{(-64)^2 - 4(16)(-5)}}{2 \cdot 16}$

$= \dfrac{64 \pm \sqrt{4096 + 320}}{32} = \dfrac{64 \pm \sqrt{4416}}{32}$

$x = \dfrac{64 + \sqrt{4416}}{32}$ or $x = \dfrac{64 - \sqrt{4416}}{32}$

$x \approx 4.1 \qquad\qquad x \approx -0.08$

Reject -0.08 because it represents a time before the baseball was hit. The baseball hit the ground 4.1 seconds after it was hit.

20. $\{(1,2),(3,4),(5,6),(6,6)\}$

This relation is a function because no two ordered pairs have the same first component and different second components.
Domain: $\{1,3,5,6\}$
Range: $\{2,4,6\}$

21. $\{(2,1),(4,3),(6,5),(6,6)\}$

This relation is not a function because two ordered pairs have the same first component, 6, but different second components.
Domain: $\{2,4,6\}$
Range: $\{1,3,5,6\}$

22. $f(x) = 7x - 3$
$f(10) = 7 \cdot 10 - 3 = 70 - 3 = 67$

23. $g(x) = x^2 - 3x + 7$
$$g(-2) = (-2)^2 - 3(-2) + 7$$
$$= 4 + 6 + 7 = 17$$

24. No vertical line will intersect this graph in more than one point, so y is a function of x.

25. Many vertical lines will intersect this graph in more than one point. One such line is the y-axis. Therefore, y is not a function of x.

26. $f(x) = \dfrac{x^2 - x}{2}$

$$f(9) = \frac{9^2 - 9}{2} = \frac{81 - 9}{2} = \frac{72}{2} = 36$$

This means that in a tournament with 9 chess players, 36 games must be played.

Cumulative Review Exercises (Chapters 1-9)

1. $2 - 4(x+2) = 5 - 3(2x+1)$
$$2 - 4x - 8 = 5 - 6x - 3$$
$$-4x - 6 = 2 - 6x$$
$$2x - 6 = 2$$
$$2x = 8$$
$$x = 4$$

The solution set is $\{4\}$.

2. $\dfrac{x}{2} - 3 = \dfrac{x}{5}$

Multiply both sides by the LCD, 10.

$$10\left(\frac{x}{2} - 3\right) = 10\left(\frac{x}{5}\right)$$
$$5x - 30 = 2x$$
$$3x - 30 = 0$$
$$3x = 30$$
$$x = 10$$

The solution set is $\{10\}$.

3. $3x + 9 \geq 5(x-1)$
$$3x + 9 \geq 5x - 5$$
$$-2x + 9 \geq -5$$
$$-2x \geq -14$$
$$\frac{-2x}{-2} \leq \frac{-14}{-2}$$
$$x \leq 7$$
$$(-\infty, 7]$$

4. $2x + 3y = 6$
   $x + 2y = 5$

To solve this system by the addition method, multiply the second equation by -2; then add the equations.

$$\begin{aligned} 2x + 3y &= 6 \\ \underline{-2x - 4y} &= \underline{-10} \\ -y &= -4 \\ y &= 4 \end{aligned}$$

Back-substitute 4 for y into the second original equation and solve for x.

$$x + 2y = 5$$
$$x + 2 \cdot 4 = 5$$
$$x + 8 = 5$$
$$x = -3$$

The solution set is $\{(-3, 4)\}$.

5. $3x - 2y = 1$

$y = 10 - 2x$

To solve this system by the substitution method, substitute $10 - 2x$ for y into the first equation.

$$3x - 2y = 1$$
$$3x - 2(10 - 2x) = 1$$
$$3x - 20 + 4x = 1$$
$$7x - 20 = 1$$
$$7x = 21$$
$$x = 3$$

Back-substitute 3 for x into the second equation of the given system to find in value of y.

$$y = 10 - 2(3) = 10 - 6 = 4$$

The ordered pair solution is $(3, 4)$ and the solution

set is $\{(3, 4)\}$.

6. $\dfrac{3}{x+5} - 1 = \dfrac{4-x}{2x+10}$

$\dfrac{3}{x+5} - 1 = \dfrac{4-x}{2(x+5)}$

Restriction: $x \neq -5$

LCD = $2(x+5)$

$$2(x+5)\left[\dfrac{3}{x+5} - 1\right] = 2(x+5)\left[\dfrac{4-x}{2(x+5)}\right]$$
$$6 - 2(x+5) = 4 - x$$
$$6 - 2x - 10 = 4 - x$$
$$-4 - 2x = 4 - x$$
$$-4 - x = 4$$
$$-x = 8$$
$$x = -8$$

The solution set is $\{-8\}$.

7. $x + \dfrac{6}{x} = -5$

Restriction: $x \neq 0$

LCD = x

$$x\left(x + \dfrac{6}{x}\right) = x(-5)$$
$$x^2 + 6 = -5x$$
$$x^2 + 5x + 6 = 0$$
$$(x+3)(x+2) = 0$$
$$x + 3 = 0 \quad \text{or} \quad x + 2 = 0$$
$$x = -3 \qquad\quad x = -2$$

The solution set is $\{-3, -2\}$.

8. $$x - 5 = \sqrt{x+7}$$
$$(x-5)^2 = \left(\sqrt{x+7}\right)^2$$
$$x^2 - 10x + 25 = x + 7$$
$$x^2 - 11x + 18 = 0$$
$$(x-9)(x-2) = 0$$
$$x - 9 = 0 \quad \text{or} \quad x - 2 = 0$$
$$x = 9 \qquad\qquad x = 2$$

Check each proposed solution in the original equation.

Check 9:   Check 2:

$x - 5 = \sqrt{x+7}$ $x - 5 = \sqrt{x+7}$

$9 - 5 = \sqrt{9+7}$ $2 - 5 = \sqrt{2+7}$

$\quad 4 = \sqrt{16}$ $-3 = \sqrt{9}$

$\quad 4 = 4$, true $-3 = 3$, false

The checks show that 9 satisfies the equation, while 2 is an extraneous solution. The solution set is $\{9\}$.

9. $(x-2)^2 = 20$

To solve this quadratic equation, use the square root property.

$x - 2 = \sqrt{20}$ or $x - 2 = -\sqrt{20}$

$x - 2 = 2\sqrt{5}$ $x - 2 = -2\sqrt{5}$

$\quad x = 2 + 2\sqrt{5}$ $x = 2 - 2\sqrt{5}$

The solution set is $\{2 \pm 2\sqrt{5}.\}$.

10. $3x^2 - 6x + 2 = 0$

Use the quadratic formula with $a = 3$, $b = -6$, and $c = 2$.

$$x = \dfrac{-b \pm \sqrt{b^2 - 4ac}}{2a}$$

$$= \dfrac{-(-6) \pm \sqrt{(-6)^2 - 4(3)(2)}}{2 \cdot 3}$$

$$= \dfrac{6 \pm \sqrt{36 - 24}}{6} = \dfrac{6 \pm \sqrt{12}}{6}$$

$$= \dfrac{6 \pm 2\sqrt{3}}{6} = \dfrac{2(3 \pm \sqrt{3})}{6}$$

$$= \dfrac{3 \pm \sqrt{3}}{3}$$

The solution set is $\left\{\dfrac{3 \pm \sqrt{3}}{3}\right\}$.

11. $A = \dfrac{5r+2}{t}$ for t

$tA = t\left(\dfrac{5r+2}{t}\right)$

$tA = 5r+2$

$\dfrac{tA}{A} = \dfrac{5r+2}{A}$

$t = \dfrac{5r+2}{A}$

12. $\dfrac{12x^3}{3x^{12}} = \dfrac{12}{3} \cdot \dfrac{x^3}{x^{12}}$

$= 4x^{3-12}$

$= 4x^{-9}$

$= \dfrac{4}{x^9}$

13. $4 \cdot 6 \div 2 \cdot 3 + (-5)$

$= 24 \div 2 \cdot 3 + (-5)$

$= 12 \cdot 3 + (-5)$

$= 36 + (-5) = 31$

14. $\left(6x^2 - 8x + 3\right) - \left(-4x^2 + x - 1\right)$

$= \left(6x^2 - 8x + 3\right) + \left(4x^2 - x + 1\right)$

$= \left(6x^2 + 4x^2\right) + (-8x - x) + (3 + 1)$

$= 10x^2 - 9x + 4$

15. $(7x + 4)(3x - 5)$

$= 7x(3x) + 7x(-5) + 4(3x) + 4(-5)$

$= 21x^2 - 35x + 12x - 20$

$= 21x^2 - 23x - 20$

16. $(5x - 2)^2 = (5x)^2 - 2 \cdot 5x \cdot 2 + 2^2$

$= 25x^2 - 20x + 4$

17. $(x + y)\left(x^2 - xy + y^2\right)$

$= x\left(x^2 - xy + y^2\right) + y\left(x^2 - xy + y^2\right)$

$= x^3 - x^2 y + xy^2 + x^2 y - xy^2 + y^3$

$= x^3 + y^3$

18. $\dfrac{x^2 + 6x + 8}{x^2} \div \left(3x^2 + 6x\right)$

$= \dfrac{x^2 + 6x + 8}{x^2} \cdot \dfrac{1}{3x^2 + 6x}$

$= \dfrac{(x + 2)(x + 4)}{x^2} \cdot \dfrac{1}{3x(x + 2)}$

$= \dfrac{x + 4}{3x^3}$

19. $\dfrac{x}{x^2 + 2x - 3} - \dfrac{x}{x^2 - 5x + 4}$

To find the LCD, factor the denominators.

$x^2 + 2x - 3 = (x + 3)(x - 1)$

$x^2 - 5x + 4 = (x - 1)(x - 4)$

LCD $= (x + 3)(x - 1)(x - 4)$

$\dfrac{x}{(x + 3)(x - 1)} - \dfrac{x}{(x - 1)(x - 4)}$

$= \dfrac{x(x - 4)}{(x + 3)(x - 1)(x - 4)}$

$- \dfrac{x(x + 3)}{(x + 3)(x - 1)(x - 4)}$

$= \dfrac{x(x - 4) - x(x + 3)}{(x + 3)(x - 1)(x - 4)}$

$= \dfrac{x^2 - 4x - x^2 - 3x}{(x + 3)(x - 1)(x - 4)}$

$= \dfrac{-7x}{(x + 3)(x - 1)(x - 4)}$

20. $\dfrac{x-\dfrac{1}{5}}{5-\dfrac{1}{x}}$

LCD = $5x$

$\dfrac{x-\dfrac{1}{5}}{5-\dfrac{1}{x}} = \dfrac{5x}{5x} \cdot \dfrac{\left(x-\dfrac{1}{5}\right)}{\left(5-\dfrac{1}{x}\right)}$

$= \dfrac{5x \cdot x - 5x \cdot \dfrac{1}{5}}{5x \cdot 5 - 5x \cdot \dfrac{1}{x}}$

$= \dfrac{5x^2 - x}{25x - 5}$

$= \dfrac{x(5x-1)}{5(5x-1)} = \dfrac{x}{5}$

21. $3\sqrt{20} + 2\sqrt{45} = 3\sqrt{4}\sqrt{5} + 2\sqrt{9}\sqrt{5}$

$= 3 \cdot 2\sqrt{5} + 2 \cdot 3\sqrt{5}$

$= 6\sqrt{5} + 6\sqrt{5} = 12\sqrt{5}$

22. $\sqrt{3x} \cdot \sqrt{6x} = \sqrt{18x^2} = \sqrt{9x^2}\sqrt{2}$

$= 3x\sqrt{2}$

23. $\dfrac{2}{\sqrt{3}} = \dfrac{2}{\sqrt{3}} \cdot \dfrac{\sqrt{3}}{\sqrt{3}} = \dfrac{2\sqrt{3}}{3}$

24. $\dfrac{8}{3-\sqrt{5}} = \dfrac{8}{3-\sqrt{5}} \cdot \dfrac{3+\sqrt{5}}{3+\sqrt{5}}$

$= \dfrac{8\left(3+\sqrt{5}\right)}{3^2 - \left(\sqrt{5}\right)^2}$

$= \dfrac{8\left(3+\sqrt{5}\right)}{9-5}$

$= \dfrac{8\left(3+\sqrt{5}\right)}{4}$

$= 2\left(3+\sqrt{5}\right) = 6+2\sqrt{5}$

25. $4x^2 - 49 = (2x)^2 - 7^2$

$= (2x+7)(2x-7)$

26. Factor by grouping.

$x^3 + 3x^2 - x - 3 = \left(x^3 + 3x^2\right) + (-x-3)$

$= x^2(x+3) - 1(x+3)$

$= (x+3)\left(x^2 - 1\right)$

$= (x+3)(x+1)(x-1)$

27. $2x^2 + 8x - 42 = 2\left(x^2 + 4x - 21\right)$

$= 2(x-3)(x+7)$

28. $x^5 - 16x = x\left(x^4 - 16\right)$

$= \left[\left(x^2\right)^2 - 4^2\right]$

$= x\left(x^2 + 4\right)\left(x^2 - 4\right)$

$= x\left(x^2 + 4\right)(x+2)(x-2)$

29. $x^3 - 10x^2 + 25x = x\left(x^2 - 10x + 25\right)$

$= x(x-5)^2$

30. $x^3 - 8$

Use the formula for factoring the difference of two cubes.

$x^3 - 8 = x^3 - 2^3$

$= (x-2)\left(x^2 + 2 \cdot x + 2^2\right)$

$= (x-2)\left(x^2 + 2x + 4\right)$

31. $8^{-\frac{2}{3}} = \dfrac{1}{8^{\frac{2}{3}}} = \dfrac{1}{\left(\sqrt[3]{8}\right)^2} = \dfrac{1}{2^2} = \dfrac{1}{4}$

32. $y = \dfrac{1}{3}x - 1$

slope $= \dfrac{1}{3}$; y-intercept $= -1$

Plot $(0, -1)$. From the point, move 1 unit *up* and 3 units to the *right* to reach the point $(3, 0)$. Draw a line through $(0, -1)$ and $(3, 0)$.

$y = \dfrac{1}{3}x - 1$

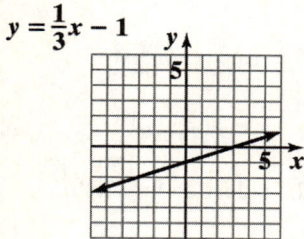

33. $3x + 2y = -6$

x-intercept: -2
y-intercept: -3
checkpoint: $(-4, 3)$
Draw a line through $(-2, 0), (0, -3)$, and $(-4, 3)$.

$3x + 2y = -6$

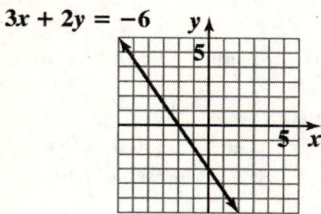

34. $y = -2$

The graph is a horizontal line with y-intercept -2.

$y = -2$

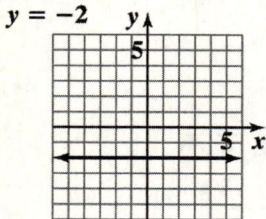

35. $3x - 4y \le 12$

Graph $3x - 4y = 12$ as a solid line using its x-intercept, 4, and its y-intercept -3. Use $(0, 0)$ as a test point. Because $3 \cdot 0 - 4 \cdot 0 \le 12$ is true, shade the half-plane containing $(0, 0)$.

$3x - 4y \le 12$

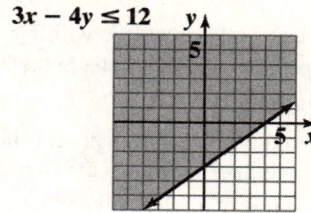

36. $y = x^2 - 2x - 3$

The graph is a parabola opening upward.
Find the vertex: $a = 1, b = -2$

$$x = \frac{-b}{2a} = \frac{-(-2)}{2(1)} = \frac{2}{2} = 1$$

$$y = 1^2 - 2 \cdot 1 - 3 = 1 - 2 - 3 = -4$$

The vertex is $(1, -4)$.
Find the x-intercepts:

$$x^2 - 2x - 3 = 0$$
$$(x + 1)(x - 3) = 0$$
$$x + 1 = 0 \quad \text{or} \quad x - 3 = 0$$
$$x = -1 \qquad\qquad x = 3$$

The x-intercepts are -1 and 3.
Find the y-intercept:

$$y = 0^2 - 2 \cdot 0 - 3 = 3$$

The y-intercept is -3.
Plot the points $(1, -4), (-1, 0), (3, 0)$, and $(0, -3)$, then connect them with a smooth curve.

$y = x^2 - 2x - 3$

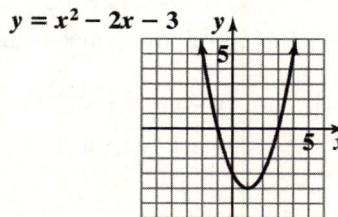

37. $2x + y < 4$

$\qquad x > 2$

Graph $2x + y = 4$ as a dashed line using its x-intercept, 2, and y-intercept, 4.

Because $2 \cdot 0 + 0 < 4$ is true, shade the half-plane containing $(0,0)$.

Graph $x = 2$ as a dashed vertical line with x-intercept 2. Because $0 > 2$ is false, shade the half-plane *not* containing $(0,0)$.

This is the region to the right of the vertical line. The solution set is the intersection of the two shaded regions. The open circle at $(2,0)$ shows that is not included in the graph.

$2x + y < 4$
$x > 2$

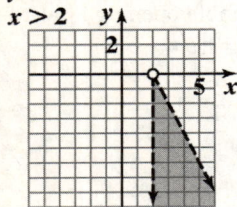

38. $(-1, 3)$ and $(2, -3)$

$$m = \frac{y_2 - y_1}{x_2 - x_1} = \frac{-3 - 3}{2 - (-1)} = \frac{-6}{3} = -2$$

39. Line passing through $(1, 2)$ and $(3, 6)$

First find the slope.

$$m = \frac{6 - 2}{3 - 1} = \frac{4}{2} = 2$$

Use the slope and either point in the point-slope form.

$y - y_1 = m(x - x_1)$ or $y - y_1 = m(x - x_1)$

$\quad y - 2 = 2(x - 1) \qquad\quad y - 6 = 2(x - 3)$

Rewrite either of these equations in slope-intercept form.

$y - 2 = 2(x - 1)$ or $y - 6 = 2(x - 3)$

$y - 2 = 2x - 2 \qquad\quad y - 6 = 2x - 6$

$\quad y = 2x \qquad\qquad\quad\; y = 2x$

Notice that the results are the same. The slope-intercept form is $y = 2x$.

40. Let x = the unknown number.

$5x - 7 = 208$

$5x = 215$

$\quad x = 43$

The number is 43.

41. Let x = the price of the camera before the reduction.

$x - 0.20x = 256$

$1x - 0.20x = 256$

$0.80x = 256$

$$\frac{0.80x}{0.80} = \frac{256}{0.80}$$

$x = 320$

The price before the reduction was $320.

42. Let x = the width of the rectangle.

Then $3x$ = the length.

$2x + 2(3x) = 400$

$2x + 6x = 400$

$\qquad 8x = 400$

$\qquad\; x = 50$

$\quad 3x = 150$

The length is 150 yards and the width is 50 yards.

43. Let x = amount invested at 7%

Then $20,000 - x$ = amount invested at 9%.

The total interest earned in one year is 1550, so the equation is

$0.07x + 0.09(20,000 - x) = 1550$.

Solve this equation.

$0.07x + 1800 - 0.09x = 1550$

$\qquad -0.02x + 1800 = 1550$

$\qquad\qquad -0.02x = -250$

$$\frac{-0.02x}{-0.02} = \frac{-250}{-0.02}$$

$\qquad\qquad\qquad x = 12,000$

$12,500 was invested at 7% and $20,000 - $12,500 = $7500 was invested at 9%.

44. Let x = the number of liters of 40% acid solution. Then $12 - x$ = the number of liters of 70% acid solution.

	Number of liters	$\times$ Percent of Acid =	Amount of Acid
40% Acid solution	x	0.40	$0.4x$
70% Acid solution	$12 - x$	0.70	$0.7(12 - x)$
50% Acid solution	12	0.50	$0.5(12)$

$$0.40x + 0.70(12 - x) = 0.50(12)$$
$$0.40x + 8.4 - 0.70x = 6$$
$$-0.30x + 8.4 = 6$$
$$-0.30x = -2.4$$
$$\frac{-0.30x}{-0.30} = \frac{-2.4}{-0.30}$$
$$x = 8$$

8 liters of 40% acid solution and $12 - 8 = 4$ liters of 70% acid solution should be used.

45. Let x = the time it will take working together.

In 1 hour, you can paint $\dfrac{1}{6}$ of the room.

In x hours, you can paint $\dfrac{x}{6}$ of the room.

In 1 hour, your friend can paint $\dfrac{1}{12}$ of the room.

In x hours, your friend can paint $\dfrac{x}{12}$ of the room.

Since together you and your friend must paint 1 room, then that is given by the equation $\dfrac{x}{6} + \dfrac{x}{12} = 1$.

$$\frac{x}{6} + \frac{x}{12} = 1$$
$$12\left(\frac{x}{6} + \frac{x}{12}\right) = 12(1)$$
$$2x + x = 12$$
$$3x = 12$$
$$x = 4$$

Together you and your friend can paint the room in 4 hours.

46. Let x = the number of students to be enrolled.

$$\frac{x \text{ students}}{176 \text{ faculty members}} = \frac{23}{2}$$
$$\frac{x}{176} = \frac{23}{2}$$
$$2x = 176 \cdot 23$$
$$2x = 4048$$
$$x = 2024$$

The university should enroll 2024 students.

47. Let x = the height of the sail.
Use the formula for the area of a triangle.

$$A = \frac{1}{2}bh$$
$$120 = \frac{1}{2} \cdot 15 \cdot x$$
$$2 \cdot 120 = 2\left(\frac{1}{2} \cdot 15 \cdot x\right)$$
$$240 = 15x$$
$$16 = x$$

The height of the sail is 16 feet.

48. Let x = the measure of the second angle.
Then $x + 10$ = the measure of the first angle.
$4x + 20$ = the measure of the third angle.
$$x + (x + 10) + (4x + 20) = 180$$
$$6x + 30 = 180$$
$$6x = 150$$
$$x = 25$$

Measure of the first angle = $x + 10 = 35°$
Measure of the second angle = $x = 25°$
Measure of the third angle = $4x + 20 = 120°$

49. Let x = the price of a TV.
Then y = the price of a stereo.

$3x + 4y = 2530$

$4x + 3y = 2510$

To solve this system by the addition method, multiply the first equation by 4 and the second equation by -3; then add the resulting equations.

$12x + 16y = 10,120$

$\underline{-12x - 9y = -7530}$

$7y = 2590$

$y = 370$

Back-substitute 370 for y in the first equation of the original system.

$3x + 4y = 2530$

$3x + 4(370) = 2530$

$3x + 1480 = 2530$

$3x = 1050$

$x = 350$

The price of a TV is \$350 and the price of a stereo is \$370.

50. Let x = the width of the rectangle.
Then $x + 6$ = the length of the rectangle.
The area of the rectangle is 55 square meters, so the equation is $(x + 6)(x) = 55$.

Solve this equation.

$x^2 + 6x = 55$

$x^2 + 6x - 55 = 0$

$(x + 11)(x - 5) = 0$

$x + 11 = 0 \quad$ or $\quad x - 5 = 0$

$x = -11 \qquad\qquad x = 5$

Reject 11 because the width cannot be negative. The width is 5 meters and the length is 11 meters.

Appendix
Mean, Median, Mode

Check Points Appendix

1. **a.** $\dfrac{10+20+30+40+50}{5} = \dfrac{150}{5} = 30$

   b. $\dfrac{3+10+10+10+117}{5} = \dfrac{150}{5} = 30$

2. **a.** First arrange the data items from smallest to largest: 25, 28, 35, 40, 42.
      The number of data items is odd, so the median is the middle number. The median is 35.

   b. First arrange the data items from smallest to largest: 61, 72, 79, 85, 87, 93.
      The number of data items is even, so the median is the mean of the two middle data items.
      The median is $\dfrac{79+85}{2} = \dfrac{164}{2} = 82$.

3. **a.** Mean $= \dfrac{0+19.6+21.0+23.9+24.7+25.1}{6} = \dfrac{114.3}{6} = \$19.05$ million

   b. Position of mean: $\dfrac{n+1}{2} = \dfrac{6+1}{2} = 3.5$ position
      The median is the mean of the data items in positions 3 and 4.
      Thus, the median is $\dfrac{21.0+23.9}{2} = \$22.45$ million.

   c. The median is greater than the mean because one data item ($0) is much less than the others.

4. **a.** 3, 8, 5, 8, 9, 10
      8 occurs most often. The mode is 8.

   b. 3, 8, 5, 8, 9, 3
      Both 3 and 8 occur most often. The modes are 3 and 8.

   c. 3, 8, 5, 6, 9, 10
      Each data item occurs the same number of times. There is no mode.

5. *Mean:* Mean $= \dfrac{107+136+138+138+172+173+190+191}{8} = \dfrac{1245}{8} \approx 155.6$

   Median: The data items are arranged in order: 107, 136, 138, 138, 172, 173, 190, 191
   The number of data items is even, so the median is the mean of the two middle data items.
   The median is $\dfrac{138+172}{2} = \dfrac{310}{2} = 155$.

   Mode: The number 138 occurs more often than any other. The mode is 138.

Exercise Set Appendix

1. $\dfrac{7+4+3+2+8+5+1+3}{8} = \dfrac{33}{8} = 4.125$

3. $\dfrac{91+95+99+97+93+95}{6} = \dfrac{570}{6} = 95$

5. $\dfrac{100+40+70+40+60}{5} = \dfrac{310}{5} = 62$

7. $\dfrac{1.6+3.8+5.0+2.7+4.2+4.2+3.2+4.7+3.6+2.5+2.5}{11} = \dfrac{38}{11} \approx 3.45$

9. First arrange the data items from smallest to largest: 1, 2, 3, 3, 4, 5, 7, 8
   The number of data items is even, so the median is the mean of the two middle data items. The median is 3.5.

11. First arrange the data items from smallest to largest: 91, 93, 95, 95, 97, 99
    The number of data items is even, so the median is the mean of the two middle data items.
    Median $= \dfrac{95+95}{2} = 95$

13. First arrange the data items from smallest to largest: 40, 40, 60, 70, 100
    The number of data items is odd, so the median is the middle number. The median is 60.

15. First arrange the data items from smallest to largest: 1.6, 2.5, 2.5, 2.7, 3.2, 3.6, 3.8, 4.2, 4.2, 4.7, 5.0
    The number of data items is odd, so the median is the middle number. The median is 3.6.

17. The mode is 3.

19. The mode is 95.

21. The mode is 40.

23. The modes are 2.5 and 4.2.

25. The data items are 42, 43, 46, 46, 47, 47.
    Mean: Mean $= \dfrac{42+43+46+46+47+47}{6} = \dfrac{271}{6} \approx 45.2$
    Median: The data items are arranged in order. The number of data items is even, so the median is the mean of the two middle data items. Median $= \dfrac{46+46}{2} = 46$
    Mode: The numbers 46 and 47 occur more often than any other. The modes are 46 and 47.

27. The data items are $17,500, $19,000, $22,000, $27,500, $98,500
    Mean: Mean $= \dfrac{\$17,500+\$19,000+\$22,000+\$27,500+\$98,500}{5} = \dfrac{\$184,500}{5} \approx \$36,900$
    Median: The data items are arranged in order. The number of data items is odd, so the median is the middle number. The median is $22,000.
    The median is more representative. Explanations will vary.